TRENDS IN COLLISION AND GROUNDING OF SHIPS AND OFFSHORE STRUCTURES

This compilation of papers describes the latest research results and innovations presented at the 10th International Conference on Collision and Grounding of Ships and Offshore Structures (ICCGS 2025, Shanghai, China, 16–19 September 2025). The contributions cover a wide range of topics, including:

- behaviour of vessels in collision and grounding;
- collision and grounding experiments;
- behaviour of structures and materials under impact loadings;
- ultimate strength of ship structures and components;
- new designs for structural improvement;
- risk assessment and innovative navigation systems;
- collision between ships and offshore structures.

This publication is an important tool for academics, engineers and professionals involved in developing new trends in collision and grounding of ships and offshore structures.

The Proceedings in Marine Technology and Ocean Engineering series is devoted to the publication of proceedings of peer-reviewed international conferences dealing with various aspects of marine technology and ocean engineering. The series includes the proceedings of the following conferences: the Marine Structures (MARSTRUCT) Conferences, the Renewable Energies Offshore (RENEW) Conferences, the Maritime Technology (MARTECH) Conferences, the Collision and Grounding of Ships and Offshore Structures (ICCGS) Conferences, and the International Maritime Association of the Mediterranean (IMAM) Conferences.

Proceedings in Marine Technology and Ocean Engineering

BOOK SERIES EDITOR

EDITORIAL BOARD MEMBERS

ABOUT THE SERIES

The 'Proceedings in Marine Technology and Ocean Engineering' series is devoted to the publication of proceedings of peer-reviewed international conferences dealing with various aspects of 'Marine Technology and Ocean Engineering'. The Series includes the proceedings of the following conferences: the Marine Structures (MARSTRUCT) Conferences, the Renewable Energies Offshore (RENEW) Conferences, the Maritime Technology (MARTECH) Conferences, the Collision and Grounding of Ships and Offshore Structures (ICCGS) Conferences, and the International Maritime Association of the Mediterranean (IMAM) Conferences. The 'Marine Technology and Ocean Engineering' series is also open to new conferences that cover topics on the sustainable exploration and exploitation of marine resources in various fields, such as maritime transport and ports, usage of the ocean including coastal areas, nautical activities, the exploration and exploitation of mineral resources, the protection of the marine environment and its resources, and risk analysis, safety and reliability. The aim of the series is to stimulate advanced education and training through the wide dissemination of the results of scientific research.

BOOKS IN THE SERIES

Volume 1: Advances in Renewable Energies Offshore, 2019, C. Guedes Soares (Ed.)
Volume 2: Trends in the Analysis and Design of Marine Structures, 2019, J. Parunov and C. Guedes Soares (Eds)
Volume 3: Sustainable Development and Innovations in Marine Technologies, 2020, P. Georgiev and C. Guedes Soares (Eds)
Volume 4: Developments in the Collision and Grounding of Ships and Offshore Structures, 2020, C. Guedes Soares (Ed.)
Volume 5: Developments in Renewable Energies Offshore, 2021, C. Guedes Soares (Ed.)
Volume 6: Developments in Maritime Technology and Engineering, 2021, C. Guedes Soares and T.A. Santos (Eds)
Volume 7: Developments in the Analysis and Design of Marine Structures, 2021, J. Amdahl and C. Guedes Soares (Eds)
Volume 8: Trends in Maritime Technology and Engineering, 2022, C. Guedes Soares and T.A. Santos (Eds)
Volume 9: Sustainable Development and Innovations in Marine Technologies, 2023, Selma Ergin and C. Guedes Soares (Eds)
Volume 10: Trends in Renewable Energies Offshore, 2023, C. Guedes Soares (Ed.)
Volume 11: Advances in the Analysis and Design of Marine Structures, 2023, J.W. Ringsberg and C. Guedes Soares (Eds)
Volume 12: Advances in the Collision and Grounding of Ships and Offshore Structures, 2024, H. Le Sourne and C. Guedes Soares (Eds)
Volume 13: Advances in Maritime Technology and Engineering, 2024, C. Guedes Soares and T.A. Santos (Eds.)
Volume 14: Innovations in Renewable Energies Offshore, 2025, C. Guedes Soares and S. Wang (Eds)
Volume 15: Innovations in the Analysis and Design of Marine Structures, 2025, Y. Garbatov & C. Guedes Soares (Eds)
Volume 16: Trends in Collision and Grounding of Ships and Offshore Structures, 2026, Ling Zhu and C. Guedes Soares (Eds)

Proceedings in Marine Technology and Ocean Engineering (Print): ISSN: 2638-647X
Proceedings in Marine Technology and Ocean Engineering (Online): eISSN: 2638-6461

PROCEEDINGS OF THE 10TH INTERNATIONAL CONFERENCE ON COLLISION AND GROUNDING OF SHIPS AND OFFSHORE STRUCTURES (ICCGS 2025), SHANGHAI, CHINA, 16–19 SEPTEMBER 2025

Trends in Collision and Grounding of Ships and Offshore Structures

Edited by

Ling Zhu

Wuhan University of Technology, China

C. Guedes Soares

Técnico Lisboa, Lisbon, Portugal

CRC Press
Taylor & Francis Group
Boca Raton London New York

CRC Press is an imprint of the
Taylor & Francis Group, an **informa** business

A BALKEMA BOOK

Designed front cover image: Collision between a roll-on/roll-off ship and container ship. Image provided by the French Navy, reference from Offshore Energy.
Designed back cover image: Wuhan University of Technology, Wuhan, China. Used with permission of WUT.

First published 2026
by CRC Press/Balkema
4 Park Square, Milton Park, Abingdon, Oxon, OX14 4RN

and by CRC Press/Balkema
2385 NW Executive Center Drive, Suite 320, Boca Raton FL 33431

CRC Press/Balkema is an imprint of the Taylor & Francis Group, an informa business

British Library Cataloguing-in-Publication Data
A catalogue record for this book is available from the British Library

ISBN: 978-1-041-16361-9 (hbk)
ISBN: 978-1-041-16400-5 (pbk)
ISBN: 978-1-003-68440-4 (ebk)

DOI: 10.1201/9781003684404

Typeset in Times New Roman
by Integra Software Services Pvt. Ltd.

Trends in Collision and Grounding of Ships and Offshore Structures – Zhu & Guedes Soares (Eds)
© 2026 The Editor(s), ISBN 978-1-041-16361-9

Table of contents

Collision and grounding experiments

Collision and grounding in arctic conditions

Collision between ships and offshore structures

Ultimate strength of ship structures and components

New designs for structural improvement

Risk assessment and innovative navigation systems

Trends in Collision and Grounding of Ships and Offshore Structures – Zhu & Guedes Soares (Eds)
© 2026 The Editor(s), ISBN 978-1-041-16361-9

Preface

This book contains the papers presented at the 10th International Conference on Collision and Grounding of Ships and Offshore Structures, ICCGS 2025, held on 16 to 19 September 2025 in Shanghai, China, and organised by Wuhan University of Technology, Wuhan, China and co-organized by Shanghai Jiao Tong University, Shanghai, China and the China Ship Scientific Research Center, Wuxi, China.

This is the tenth event in the ICCGS series of Conferences and follows on from previous events held every three years in San Francisco, USA in 1996; Copenhagen, Denmark in 2001; Tokyo, Japan in 2004; Hamburg, Germany in 2007; Helsinki, Finland in 2010; Trondheim, Norway in 2013; Ulsan, South Korea in 2016; Lisbon, Portugal in 2019; and, Nantes, France in 2023.

The aim of ICCGS conferences is to present state-of-the-art methods for analysis and design against collision and grounding of ships, collisions between ships and icebergs, offshore structures, floating wind turbines, bridges, submerged tunnels and waterfront structures.

The main objective of the ICCGS Conferences is to provide an international forum for naval architects, marine engineers and scientists, from academia and industry, to present state-of-the-art methods for analysis and design against collision and grounding of ships, collisions between ships and icebergs, offshore structures, floating wind turbines, bridges, submerged tunnels and waterfront structures. This promotes and encourages the discussion of the latest innovations and achievements, to share current trends, emerging issues and experiences in the naval architects and ocean engineering communities, and promote developments and information exchange in this field.

The 2025 Conference reflects the innovation in the work conducted and the current trends in the areas of collision and grounding of ships and offshore structures, including methods, modelling procedures and experimental results. The book contains 56 papers that are categorized into the following themes and areas of research:

- Behaviour of vessels in collision and grounding
- Collision and grounding experiments
- Collision and grounding in arctic conditions
- Behaviour of structures and materials under impact loadings
- Ultimate strength of ship structures and components
- New designs for structural improvement
- Risk assessment and innovative navigation systems
- Collision between ships and offshore structures

We hope that the overview coming from experts of different sectors will be an important contribution to the audience.

The articles in this book were accepted after a peer-review process based on the full text of the papers. We thank the Steering and Scientific Committees, who undertook most of the responsibility for reviewing the papers. We are also grateful to the additional anonymous reviewers who helped the authors deliver better papers by providing them with constructive comments. We hope this process contributed to a consistently good level of the papers included in the book.

Ling Zhu
Carlos Guedes Soares

Trends in Collision and Grounding of Ships and Offshore Structures – Zhu & Guedes Soares (Eds)
© 2026 The Editor(s), ISBN 978-1-041-16361-9

Organisation

Conference Chairman

Ling Zhu, *Wuhan University of Technology, Wuhan, P.R. China*

Steering Committee

Carlos Guedes Soares, *Portugal (Chairman)*
Sören Ehlers, *Germany*
Zhiqiang Hu, *UK*
Pentti Kujala, *Finland*
Kun Liu, *P.R. China*
Jakub Montewka, *Poland*
Jonas Ringsberg, *Sweden*
Manolis Samuelides, *Greece*
Sung Chul Shin, *Korea*
Hervé Le Sourne, *France*
Martjin Hoogeland, *The Netherland*
Kristjan Tabri, *Estonia*
Deyu Wang, *P.R. China*
Yasuhira Yamada, *Japan*
Zhaolong Yu, *Norway*
Shengming Zhang, *UK*
Ling Zhu, *P.R. China*

Scientific Committee

Ling Zhu, *P.R. China (Chairman)*
Tore Borvik, *Norway*
Loïc Buldgen, *Belgium*
Gang Chen, *P.R. China*
Nianzhong Chen, *P.R. China*
Yuansheng Cheng, *P.R. China*
Sang-Rai Cho, *South Korea*
Joonmo Choung, *South Korea*
Floris Goerlandt, *Canada*
Lucjan Gucma, *Poland*
Spyros Hirdaris, *Finland*
Kazuhiro Iijima, *Japan*

Hiroko Itoh, *Japan*
Yvonne Koldenhof, *The Netherlands*
Mihkel Kõrgesaar, *Estonia*
Lucas Marquez, *Belgium*
Wengang Mao, *Sweden*
Stephane Paboeuf, *France*
Philippe Rigo, *Belgium*
Siljko Rudan, *Croatia*
Ângelo Palos Teixeira, *Portugal*
Carey Walters, *The Netherlands*
Bing Wu, *P.R. China*
Zhihu Zhan, *P.R. China*
Yanjie Zhao, *P.R. China*

Organising Committee

Ling Zhu, *Wuhan University of Technology, P.R. China (Chairman)*
Deyu Wang, *Shanghai Jiao Tong University, P.R. China*
Yanjie Zhao, *China Ship Scientific Research Center, P.R. China*
Qiyu Liang, *Wuhan University of Technology, P.R. China*
Guijie Shi, *Shanghai Jiao Tong University, P.R. China*
Jiangyan Shao, *Wuhan University of Technology, P.R. China*
Kailing Guo, *Wuhan University of Technology, P.R. China*

Technical Programme & Conference Secretariat

Maria de Fátima Pina, *Técnico Lisboa, Universidade de Lisboa, Portugal*
Jiangyan Shao, *Wuhan University of Technology, P.R. China*
Kailing Guo, *Wuhan University of Technology, P.R. China*
Guijie Shi, *Shanghai Jiao Tong University, P.R. China*

Behaviour of vessels in collision and grounding

Trends in Collision and Grounding of Ships and Offshore Structures – Zhu & Guedes Soares (Eds)

A coupled ANSYS LS-DYNA® MCOL and ANSYS AQWA™ solution for ship collision and grounding analysis

Y. Bi & S.X. Du
ANSYS UK LIMITED, Horsham, West Sussex, UK

Z. Cui & Y. Huang
ANSYS, Inc., Livermore, California, USA

ABSTRACT: Ship collisions and groundings remain critical challenges in the shipbuilding industry, necessitating advanced analysis tools to enhance structural design and safety. This study presents an integrated numerical framework that couples Ansys LS-DYNA® MCOL and Ansys Aqwa™ to simulate and analyze ship collision and grounding scenarios. The proposed approach captures both internal and external collision dynamics by leveraging state-of-the-art solvers specialized in finite element (FE) analysis for highly nonlinear structural responses and hydrodynamic response modeling. This coupling ensures high accuracy and computational efficiency in predicting ship behavior under extreme loading conditions. A case study involving an 80-meter tanker is conducted to evaluate the effects of hydrodynamic coupling and FE model truncation on simulation accuracy and computational cost. Results demonstrate that an appropriately truncated FE model, accounting for Saint-Venant's Principle, improves computational efficiency without compromising accuracy. Additionally, the influence of hydrodynamic database with forward speed on coupled FE analysis is examined.

1 INTRODUCTION

Maritime accidents, particularly ship collisions and groundings, continue to pose significant challenges to the shipbuilding industry, leading to substantial environmental, economic, and human losses. A recent incident underscoring these risks occurred in the North Sea, where the cargo ship *Solong* collided with the oil tanker *Stena Immaculate*. This collision resulted in an oil spill, raising environmental concerns, and tragically led to the presumed death of a crew member (The Guardian, 2025). Similarly, the 1989 Exxon Valdez disaster spilled approximately 37,000 metric tons of crude oil into Prince William Sound, Alaska, causing extensive ecological and economic damage (Wikipedia, 2024). These incidents highlight the imperative for accurate analyses to enhance the crashworthiness of ship structures and mitigate the consequences of such accidents.

Ship collision and grounding assessments have evolved through various methodologies, including empirical, experimental, numerical, analytical, and semi-analytical approaches. Classification societies have developed guidelines to standardize these assessments. For instance, in 2016, Lloyd's Register issued the Guidance Notes for Collision Assessment for the Location of Low-flashpoint Fuel Tanks, aiming to enhance the safety of alternative fuel-powered vessels (Lloyd's Register, 2016). Similarly, DNV GL has provided specific guidelines and general requirements for determining structural capacity using nonlinear finite element analysis (NLFEA) software, emphasizing the importance of advanced simulation tools in structural assessments (DNV GL, 2016).

The analysis of ship collisions and grounding encompasses both internal mechanics and external dynamics. Internal mechanics involve structural responses such as material yielding, buckling, large plastic deformations, and progressive failure mechanisms. External dynamics pertain to the rigid body motions influenced by transient contact forces and hydrodynamic loads. These aspects can be analyzed independently (decoupled method) or concurrently (coupled method). Studies by Tabri (2012) and Liu et al. (2017) have demonstrated that coupled methods, which account for the interaction between rigid body motions and structural deformations, provide more accurate predictions for oblique collisions than decoupled methods. Among coupled methods, Shiomitus et al., (2023) demonstrated that the LS-DYNA MCOL method yields sway velocity, contact force, and absorbed energy results in close agreement with those from the S-ALE method, while requiring slightly less computation time. Since their study, recent optimizations to the MCOL solver have significantly improved its performance—achieving

DOI: 10.1201/9781003684404-1

a 15-fold increase in solving speed in our grounding test case.

Scenario-based accidental analyses have frequently utilized LS-DYNA due to its robust coupled nonlinear finite element analysis capabilities. However, LS-DYNA lacks intrinsic hydrodynamic parameter computation necessary for external dynamics analysis (MCOL 2015). Integrating a specialized hydrodynamic solver can enhance the accuracy and feasibility of these simulations. This paper proposes a workflow that couples LS-DYNA's Multi-Physics Collision (MCOL) module with Aqwa, a state-of-the -art hydrodynamic response solver, to simulate ship collision and grounding scenarios. The study introduces the core methodologies of these solvers and examines modelling considerations, such as the effects of forward speed on hydrodynamic responses and internal dynamics, finite element model truncation selection. The objective is to provide recommendations for modelling strategies that balance accuracy and computational efficiency.

2 METHODOLOGY

2.1 *Ls-Dyna/Mcol approach*

The ship-ship collision mechanics can be subdivided into inner collision mechanics and outer collision dynamics (Pedersen 1995, Pedersen 1982). The inner collision mechanics is a crash problem governed by buckling, material yielding and assemblies. It is investigated using the explicit finite element code LS-DYNA. Collision forces are computed at this step. Ignoring the influence of the structural local deformation on the global structural mass distribution, the outer collision dynamics, i.e. global motions of the two ships considered as rigid bodies under collision forces and hydrodynamic pressure forces, are computed with the program MCOL, coupled with LS-DYNA code. Hydrodynamic loads in MCOL are calculated with the hydrostatic restoring matrix, the added mass matrix at infinite frequency and the frequency dependent added damping matrices computed for each ship by Aqwa. The inner collision analysis step and the outer rigid body hydrodynamic analysis step are bidirectional coupled, in which the outer rigid body response provides the boundary conditions (position and velocity) at the FE model truncated sectional end for the inner collision analysis step, on the other hand the inner collision step gives the equivalent collision forces for the outer step of the whole ship rigid body hydrodynamic motion simulation. Note that the inner collision analysis step does not directly include the hydrostatic and hydrodynamic forces acting on the truncated FE model (deformable partition), these forces' effects are implicitly applied by the truncated FE model boundary conditions provided by the outer rigid body hydrodynamic response step. This approach will introduce numerical errors if the truncated deformable partition is unnecessarily large, which is discussed in Section 3.4.1.

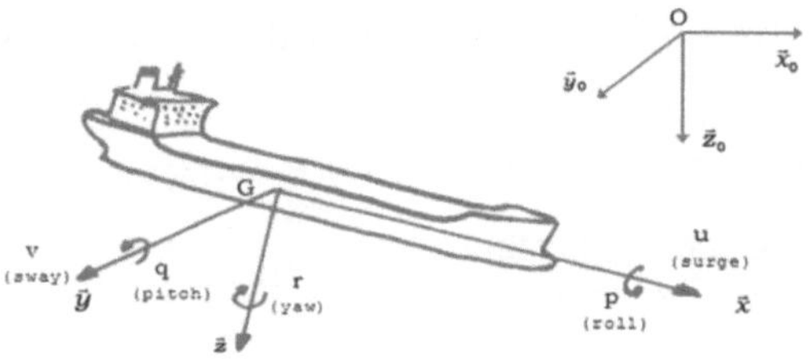

Figure1. Body-fixed and earth-fixed reference frame (MCOL 2015).

For the MCOL method, each ship uses two reference frames. The first one is a body-fixed frame with its origin in the center of mass of the ship and with an x-axis along the forward axis of the ship, a starboard y-axis and a downward z-axis. The second frame is an earth-fixed frame defined as the initial position of the body-fixed frame (Figure 1). The motion of a ship is defined by its roll, pitch and yaw, and by the translation of tis center of mass from its initial position. Hence the general motion is described with the following equations.

$$
\begin{aligned}
x &= (x_{OG}, y_{OG}, z_{OG}, \phi, \theta, \psi)^T \\
y &= (u, v, \omega, p, q, r)^T = (v^T, \omega^T)^T \\
f &= (X, Y, Z, K, M, N)^T
\end{aligned}
\quad (1)
$$

where x denotes the earth-fixed position of the center of mass and the Eulerian angles, y the body-fixed components of the absolute velocity v of the center of mass and of the angular velocity vector w, f the body-fixed components of the forces and of the moments, relative to the center of mass, acting on the body.

The free rotation of a body in space involves the description of the rigid body, then the rotation matrices are required for the transformation from body-fixed to earth-fixed frame components (MCOL, 2015). Once this is achieved, the Newton's law is used to define the rigid body motion concerning the body fixed rotation.

The basis for MCOL is a convolution integral approach for simulating the equations of motion. A mass and inertia tensor are required as input for each ship. The masses are then augmented to include the effects of the added mass of the surrounding water. Aqwa determines the various terms of the damping/buoyancy force formulas which are also input to MCOL. The coupling is accomplished in a simple manner: at each timestep LS-DYNA computes the resultant forces and moments on the MCOL rigid bodies and passes them to MCOL. MCOL then updates the positions of the ships and returns the new rigid body locations to LS-DYNA.

The resulting equation of ship motion can be written in the body fixed reference frame with the following form (MCOL, 2015):

$$[M+M_A][\dot{y}]+[G(y)][y]=[F_w]+[F_H]+[F_v]+[F_c] \quad (2)$$

where M and M_A are the structural mass matrix and added mass matrix, G is skew-symmetrical the gyroscopic matrix, F_w, F_H, Fv, and Fc are the wave damping force, restoring force, viscous force and contact force respectively. The wave damping force is calculated by the following equation:

$$\begin{aligned} H(\tau) &= \tfrac{2}{\pi}\int_0^{\infty}[C(\omega)]\cos(\omega)\,d\omega \\ F_w &= -{}_0^t[H(\tau)][y(t-\tau)-y(0)]d\tau \end{aligned} \quad (3)$$

where $C(\omega)$represents the wave damping coefficient matrix.

A model of the collision analysis is shown on Figure 2. Only the collision areas of ships and ice impinged on are meshed, the remaining structure being modeled by a rigid body described by its inertia matrix and the position of its center of mass.

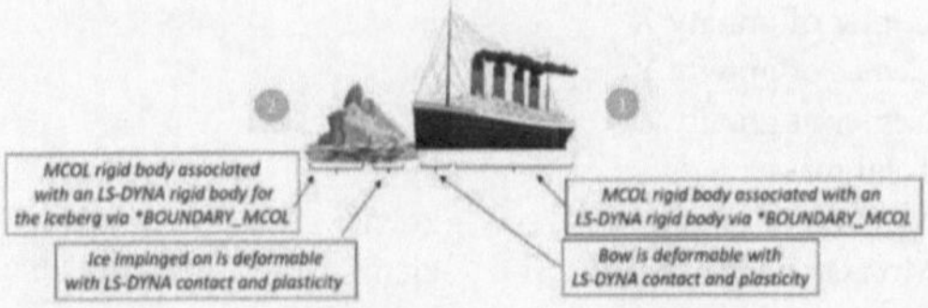

Figure 2. Ls-Dyna/Mcol collision simulation approach (Cui 2024).

Card 1	1	2	3	4	5	6	7	8
Variable	NMCOL	MXSTEP	ETMCOL	TSUBC	PRTMCOL			
Type	I	I	F	F	F			
Default	2	none	0.0	0.0	none			
Remarks			2					

Figure 3. *BOUNDARY_MCOL (LS-DYNA 2023).

The MCOL solver is a rigid body mechanics program for modeling the dynamics of ships and is activated by the keyword ***BOUNDARY_MCOL** (shown in Figure 3) in LS-DYNA which is used to define parameters for MCOL coupling.

2.2 *Hydrodynamic properties calculation*

Aqwa application is an industry recognized software, which provides a toolset for investigating the effects of environmental loads on floating and fixed offshore and marine structures in both frequency domain and time domain simulations.

The three-dimensional panel method is employed to estimate the hydrostatic and hydrodynamic behaviour of large-volume structures in waves.

The hydrostatic stiffness matrix is expressed in terms of motions about the centre of gravity, the matrix will take the form

$$K_{hys} = \begin{bmatrix} 0 & 0 & 0 & 0 & 0 & 0 \\ 0 & 0 & 0 & 0 & 0' & 0 \\ 0 & 0 & K_{33} & K_{34} & K_{35} & 0 \\ 0 & 0 & K_{43} & K_{44} & K_{45} & K_{46} \\ 0 & 0 & K_{53} & K_{54} & K_{55} & K_{56} \\ 0 & 0 & 0 & 0 & 0 & 0 \end{bmatrix} \quad (4)$$

where the various terms in the stiffness matrix are

$$\begin{aligned} K_{33} &= -\rho g \textstyle\int_{S_0} n_3 dS = \rho g A \\ K_{34} = K_{43} &= -\rho g \textstyle\int_{S_0} (Y-Y_g) n_3 dS \\ K_{35} = K_{53} &= \rho g \textstyle\int_{S_0} (X-X_g) n_3 dS \\ K_{44} &= -\rho g \textstyle\int_{S_0} (Y-Y_g)^2 n_3 dS + \rho g (Z_B - Z_g)\nabla \\ K_{45} = K_{54} &= -\rho g \textstyle\int_{S_0} (X-X_g)(Y-Y_g) n_3 dS \\ K_{55} &= -\rho g \textstyle\int_{S_0} (X-X_g)^2 n_3 dS + \rho g (Z_B - Z_g)\nabla \\ K_{46} &= -\rho g (X_B - X_g)\nabla \\ K_{56} &= -\rho g (Y_B - Y_g)\nabla \end{aligned}$$

in which A is the cut water-plane area; ∇ is the displacement; S_0 is the wetted surface of the body in still water; $n = (n_1, n_2, n_3)$ is the unit normal vector of the body surface pointing outwards; where $X = X_g$ is the position vector of a point on the hull surface with respect to the centre of gravity; $X_B = (X_B, Y_B, Z_B)$ is the entre of buoyancy.

The frequency domain hydrodynamic analysis is based on the linear fluid potential theory with the source strength distribution integration approach over the mean wetted structure surface presented by a series of diffraction panels.

Once the radiation wave potential, φ_{rk}, due to the kth unit basic translational or rotational rigid body motion (surge, sway, heave, roll, pitch and yaw with respect to the structural COG) are calculated, the radiation wave force can be expressed in the real and imaginary parts, corresponding to the contributions of the added mass and wave damping coefficients

$$\begin{aligned} F_{rjk} &= \omega_e^2 A_{jk} + i\omega_e B_{jk} \\ &= -\rho \textstyle\int_{S_0} \left\{ \left(i\omega_e + \vec{U}\cdot\nabla \right) \varphi_{rk}(\vec{x}) \right\} n_j dS \\ &= -i\omega_e \rho \textstyle\int_{S_0} \left(n_j + \frac{i}{\omega_e} \vec{U}\cdot\nabla n_j \right) \varphi_{rk}(\vec{x}) dS \end{aligned} \quad (5)$$

in which ω_e is the encounter wave frequency; U is the forward speed of the structure; $(n_4, n_5, n_6) = (X - X_g) \times n$ is the extended unit normal vector corresponding to three basic rigid body rotational motions.

More detailed description is available in Aqwa Theory Manual (Ansys Aqwa 2024).

2.3 *Solution process*

In the proposed solution, Aqwa, LS-DYNA MCOL, and the LS-DYNA finite element (FE) solver work together to simulate ship collision and grounding scenarios. Aqwa first generates the hydrodynamic database, which includes key parameters such as added mass, radiation damping, and hydrostatic restoring stiffness. This database serves as a critical input for MCOL, enabling it to compute hydrodynamic loads and rigid body motions. While hydrodynamic coefficients can vary due to changes in a ship's position and orientation during collision or grounding events, multiple hydrodynamic databases generated by Aqwa—spanning a range of heave, roll, and pitch conditions—can be utilized by MCOL. Each motion condition is associated with a predefined range that is segmented into intervals, and a separate database is generated for each combination of these segmented conditions. During simulation, MCOL determines the ship's real-time motion and selects the appropriate hydrodynamic dataset based on which interval each motion condition falls into. This approach allows MCOL to compute hydrodynamic loads that reflect the ship's instantaneous position and orientation, while maintaining computational efficiency.

During the accident scenario analysis, MCOL and the FE solver operate in a coupled manner. MCOL applies external hydrodynamic forces based on the ship's motion state, while the FE solver handles internal structural deformation computation. Data is exchanged between MCOL and the FE solver throughout the simulation, ensuring a more accurate representation of the interaction between rigid body motion and structural response. The data transfer interactions between the three solvers are illustrated in Figure 4.

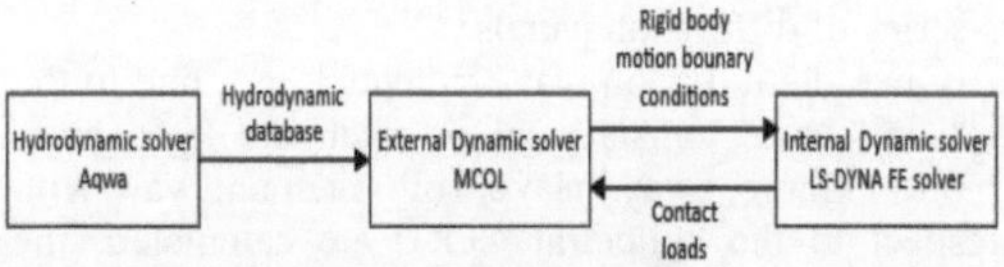

Figure 4. Data transfer between the solvers.

3 CASE STUDIES

To validate the functionalities of this solution and to find the best modelling approach from both the finite element (FE) model and hydrodynamic database perspectives, a case study was conducted on an 80-meter tanker. This ship was designed by a professional ship structural designer from Seanaptic Ltd. Although it has not been manufactured, its design serves as a good reference simulation model (Cui 2024). Dimension details are provided in Table 1.

Table 1. Model ship dimensions.

	Value
Dimension	m
Length	80
Width	16
Draft	5.12

3.1 *Ship model for hydrodynamic property calculations*

To support FE model analysis, a ship hydrodynamic database at the mean water surface was prepared. The primary values determining the hydrodynamic database calculations include wetted surface geometry and the center of gravity position. The total mass properties, derived from the full structural geometry, are listed in Table 2.

Table 2. Mass properties of the model ship.

Mass properties	Units	Values
Center of gravity X	m	38.75
Center of gravity Y	m	0
Center of gravity Z	m	-0.32
Total mass	kg	5213764
Moment of inertia Ixx	kg.m^2	199944116
Moment of inertia Iyy	kg.m^2	1694037489
Moment of inertia Izz	kg.m^2	1673054216

The hydrodynamic database, calculated using the Aqwa solver and later used to support FE coupled calculations, includes:

- Mass matrix of the ship
- Hydrostatic restoring matrix
- Buoyancy reference parameters
- Added mass matrix at infinite frequency
- Wave damping matrices across different frequencies

These datasets enable the MCOL solver to compute rigid body motion by providing necessary inputs for gravity force, hydrostatic restoring force and moments, buoyancy force, and radiation forces and moments. Incident wave loads are excluded; however, viscous forces and moments can be incorporated by defining viscous damping and lifting surface parameters, which are not provided by the potential flow solver, Aqwa.

3.2 *Collision/grounding scenario definition*

A standard scenario of collision or grounding at the bulbous bow was designed to analyze the solution. The bulbous bow has a radius of around 2 meters.

The distance from the bow tip to the collision bulkhead is approximately 6.4 meters. The ship travels at a service speed of 11 knots before the collision or grounding event occurs.

Figure 5. Tanker bow bulb structure.

In the collision scenario, the bulbous bow strikes the center of the sidewall surface of a cylindrical tube, which is fixed at both ends for simplification. The dimensions and material properties of the cylindrical tube are provided in Table 3.

Table 3.

Properties	Units	Values
Length	m	12.25
Diameter	m	2.78
Young's Modulus	MPa	2.1E5
Yield Stress	MPa	350
Tangent Modulus	MPa	782

Figure 6. Collision scenario setup.

In the grounding scenario, the rock is modeled as a rigid body fixed to the seabed. The rock has a conical shape, with detailed dimensions shown in Table 4. Its centerline is aligned with the ship's centerline. The grounded depth, defined as the vertical distance between the rock tip and the ship's keel, is 1.35 meters.

Table 4.

	Value
Dimesion	m
Rock height	5.1
Bottom diameter	4.42

Figure 7. Grounding scenario setup.

3.3 *Material properties*

The tanker is composed of NV A steel plates, with mechanical properties summarized in Table 5. In the simulation, plastic strain failure is permitted, and elements are deleted when plastic strain reaches 0.1. Under high-speed loading, strain rate effects are modeled using the Cowper-Symonds equation, with a strain rate parameter (c) of 40.5 /s and a strain hardening exponent (p) of 5. To simplify material modeling, the material is assumed to be perfectly plastic, meaning that after reaching the yield point, stress remains constant instead of increasing with strain.

Table 5. Mechanical properties of the material.

Property	Units	Value
Young's modulus	MPa	2E5
Poisson's ratio	–	0.3
Yield stress	MPa	235
Tensile ultimate Strength	MPa	235

3.4 *Nonlinear transient MCOL-FE coupled analysis*

The target structures in the collision/grounding scenario were analyzed numerically using a finite element model. A trade-off exists between fidelity and computational cost when selecting the mesh size. In this study, the adopted mesh size for the tanker was 0.2 meters, with approximately 362k mesh panels covering the entire ship. However, a proper FE model should only include the structural sections directly influenced by the collision and grounding events in the coupled solution. The following section elaborates on the influence of different FE model truncations.

3.4.1 *Influence of FE model truncation length*

To assess the impact of truncation length on the FE model, four modeling approaches with varying truncation distances from the bow bulb were compared for both collision and grounding scenarios. All truncation distances exceeded the collision bulkhead location. Additionally, a pure FE analysis without coupling with the hydrodynamic database was conducted to examine the influence of the MCOL solver on the FE results. Figure 8 illustrates the different truncation cases and simulation setups.

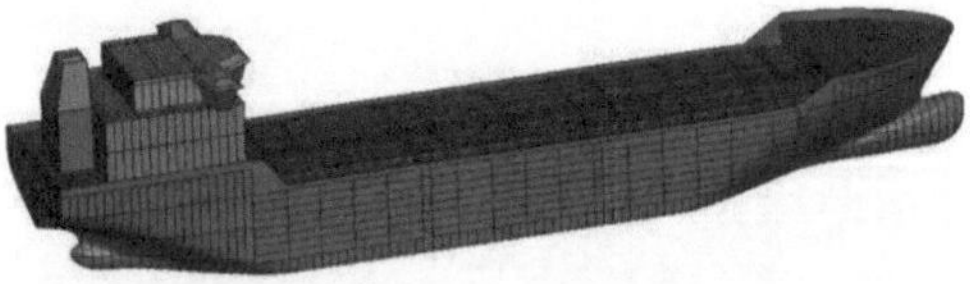

Case 1: a purely FEM model of the ship

Case 2: the MCOL deformable partition is 16% of the ship

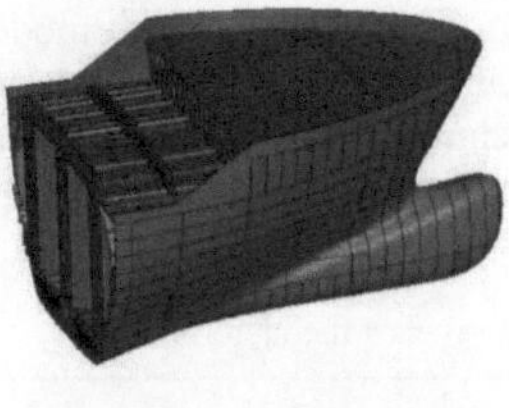

Case 3: the MCOL deformable partition is 24% of the ship

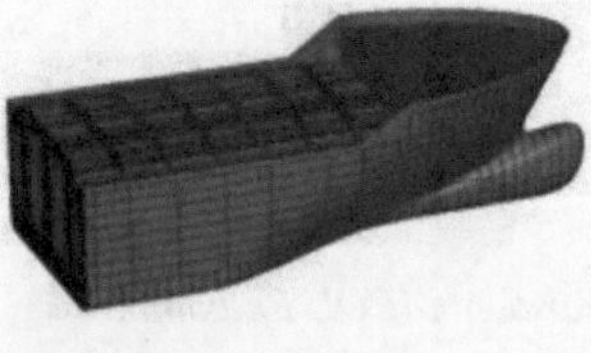

Case 4: the MCOL deformable partition is 42% of the ship

Case 5: the MCOL deformable partition is 100% of the ship

Figure 8. Different truncation cases and simulation setups of the FE model.

The collision and grounding events occur within a short duration. The total simulation time was set to 1 second for the collision scenario and 2 seconds for the grounding scenario to capture structural deformation. The simulated von Mises stress contours for the collision and grounding cases are displayed in Figures 9 and 10, respectively.

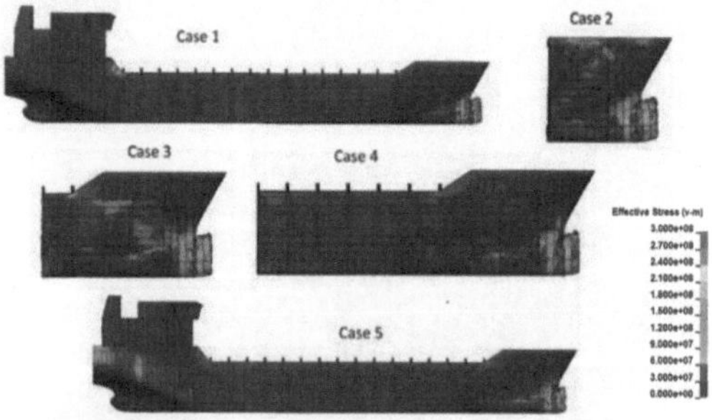

a). Side view

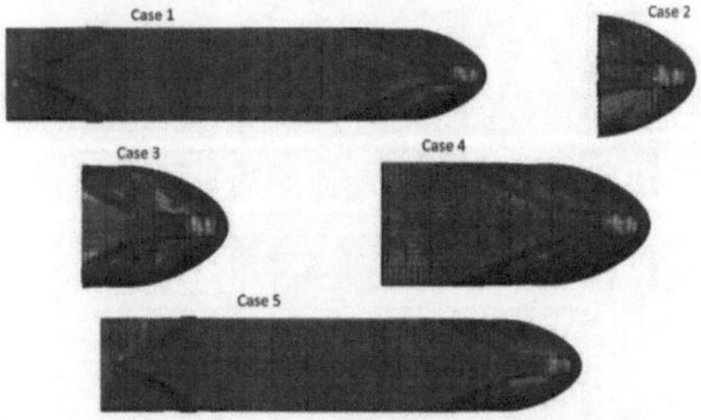

b). Bottom view

Figure 9. Von-mises stress contour of collision scenario at end of the simulation.

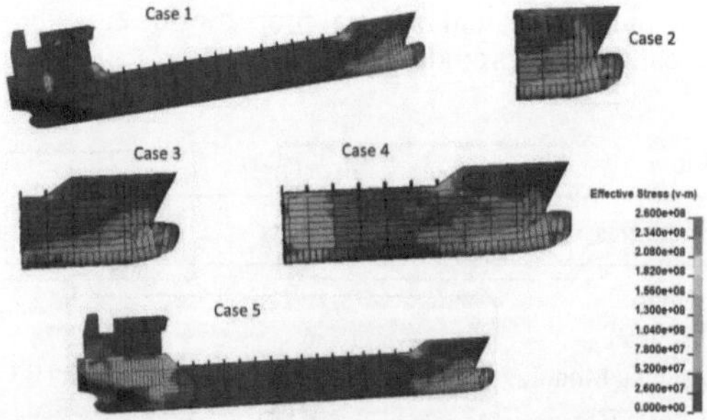

a). Side view

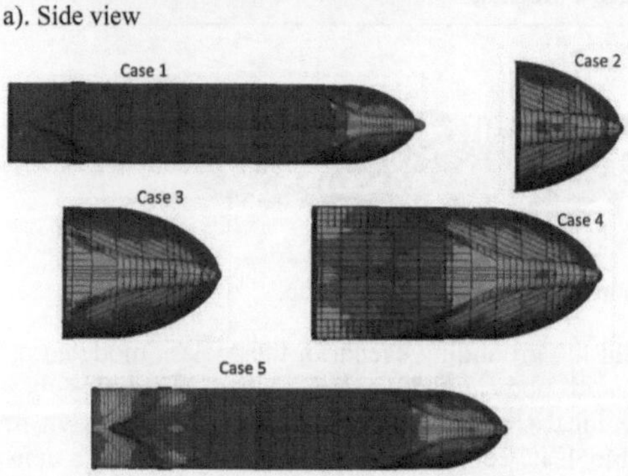

b). Bottom view

Figure 10. Von-mises stress contour of grounding scenario at end of the simulation.

In Cases 4 and 5, where longer FE model sections were included, artificial stress concentrations emerged near the truncation area, deviating from expected physical behavior. These effects were particularly noticeable in the grounding scenario, where the ship's rigid body velocity had decreased but had not yet reached zero. Because hydrodynamic effects were not incorporated into the FE model during internal deformation calculations, this likely led to force imbalances. Consequently, incorrect boundary conditions at the coupling interface between external and internal dynamics led to these artificial stress concentrations at the FE model's truncation boundary. Conversely, in Cases 2 and 3, such artificial stress concentrations were not observed, which indicates that such truncation configurations provide more accurate results.

The energy evolution over time for different cases is plotted in Figures 11 and 12 for the grounding and collision scenarios, respectively. Note the internal energy calculation excludes the failed elements that are eroded. The energy results for Cases 2 and 3 closely align. Similarly, Figures 13 and 14 depict the plastic strain evolution of a selected element near the bow's high-stress region, where the plastic strain curves of Cases 2 and 3 exhibit significant similarity but differ notably from the other cases.

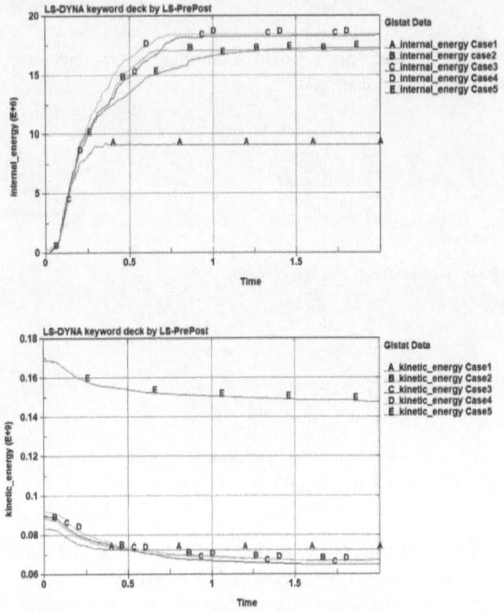

Figure 11. Internal and kinetic energy evolution in grounding scenarios.

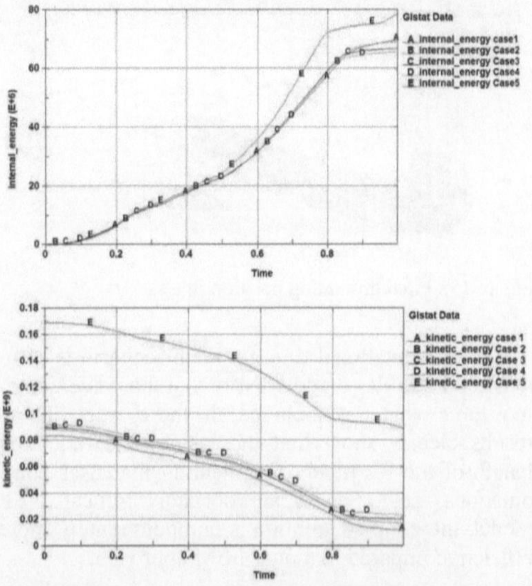

Figure 12. Internal and kinetic energy evolution in collision scenarios.

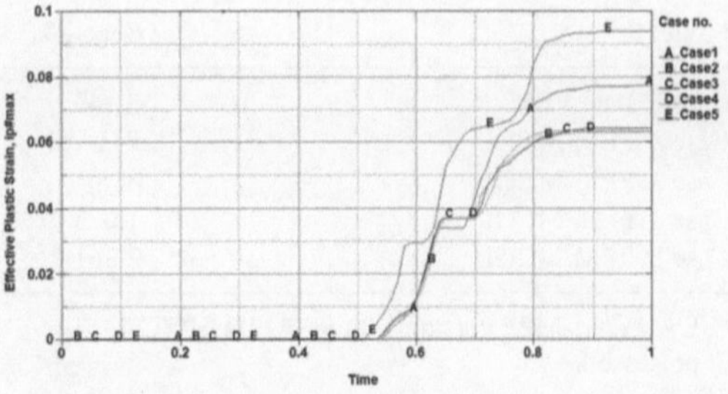

Figure 13. Effective plastic strain of element 257579 in collision scenarios.

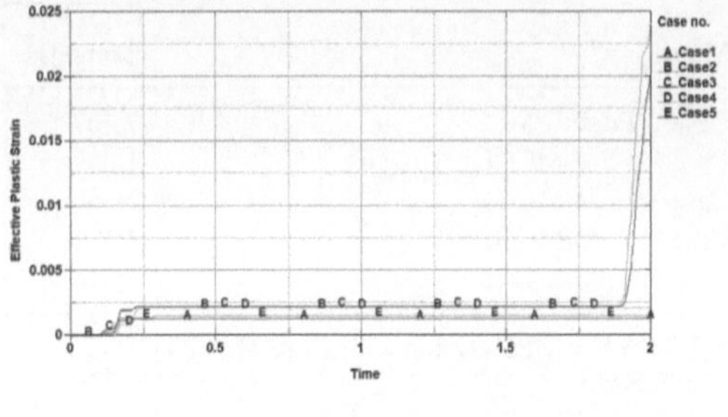

Figure 14. Effective plastic strain of element 251949 in grounding scenarios.

When comparing the pure FE model with the MCOL-coupled system, Case 1 demonstrates significantly lower energy levels due to the exclusion of hydrodynamic effects. This discrepancy is even more pronounced in the grounding scenario, where the absence of hydrostatic restoring moments results in exaggerated and unrealistic vertical motion. Figure 15 illustrates the final position of Case 1 at the end of the analysis.

Figure 15. Final grounding position in Case 1.

The computational time for each scenario is summarized in Tables 6 and 7. Note that the wave damping force was not included in these tests. These results clearly show that increasing the truncation length of the FE model significantly increases computational cost. For an appropriately truncated FE model, the coupled solution is computationally more efficient compared to a pure FEM approach.

Table 6. CPU time comparison for collision scenarios.

Model	CPU time* Seconds
Case 1: Pure FEM	2,149
Case 2: FEM/MCOL	741
Case 3: FEM/MCOL	818
Case 4: FEM/MCOL	1,078
Case 5: FEM/MCOL	2,361

* LS-DYNA MPP version with 96 AMD EPYC 7451 processors.

Table 7. CPU time comparison for grounding scenarios.

Model	CPU time* Seconds
Case 1: Pure FEM	7,507
Case 2: FEM/MCOL	1,639
Case 3: FEM/MCOL	2,369
Case 4: FEM/MCOL	3,181
Case 5: FEM/MCOL	7,614

* LS-DYNA MPP version with 96 AMD EPYC 7451 processors.

This comparison highlights that, for both result accuracy and computational efficiency, the FE model should include only the structural sections directly affected by the collision and grounding events. This truncation configuration is consistent with Saint-Venant's Principle, which states that localized effects dissipate over a short distance, making it unnecessary to extend the FE model beyond the region of primary structural response.

3.4.2 *Influences of the hydrodynamic database with forward speed effect*

To further examine the impact of the hydrodynamic database, this section investigates how hydrodynamic databases with different forward speeds influence the coupled FE solution results. The grounding scenario was used for this analysis. Based on the ship's service speed and its final stable speed after grounding, three different forward speeds were considered to generate the hydrodynamic database: zero forward speed, 5 m/s, and 5.659 m/s (11 knots). In the primary pitch motion degree of freedom, the radiation damping coefficients exhibit slight but noticeable variations across lower encounter wave frequencies, as shown in Figure 16.

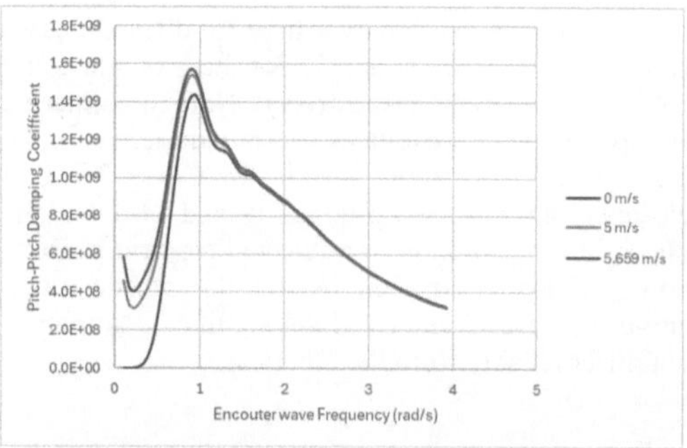

Figure 16. Pitch-pitch radiation damping coefficient under different forward speeds.

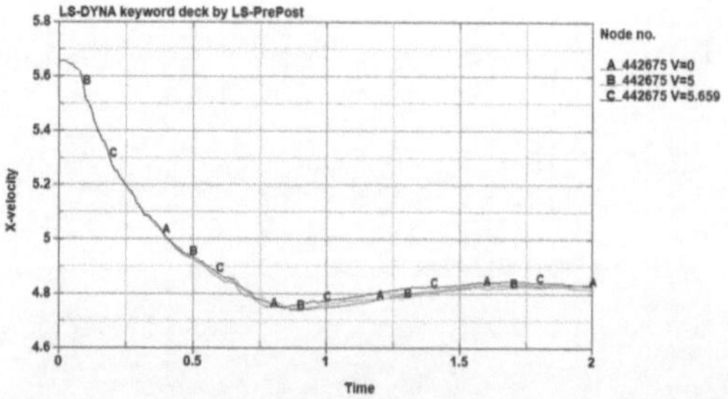

Figure 17. Velocity at maneuvering direction of element 442675 at truncation boundary.

These differences in the hydrodynamic database are reflected in the coupled FE analysis results. Figure 17 illustrates the velocity in the surge direction for an element at the truncation boundary of the FE model, indicating that the rigid body motion in the ship longitudinal direction is only slightly affected by the choice of forward-speed database. In terms of internal structural response, the plastic strain contours are generally similar across cases, though models incorporating a non-zero forward-speed hydrodynamic database exhibit slightly more severe plastic deformation at the grounding contact area (Figure 18). The energy (total, internal, kinetic) differences between the various hydrodynamic database cases, shown in Figure 19, are relatively small, though the zero forward speed database case deviates more noticeably from the other two cases. The decrease in total energy shown in Figure 19 is due to energy dissipation into the surrounding water.

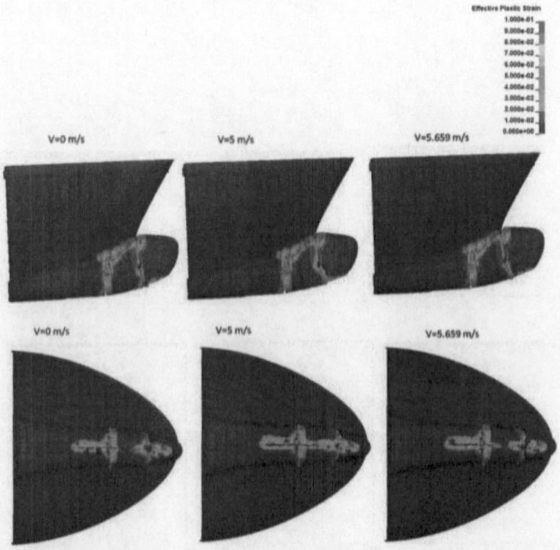

Figure 18. Plastic strain contour solved with different forward speed databases.

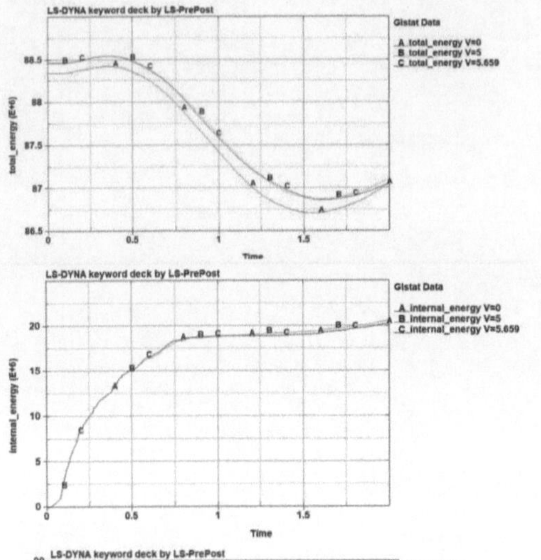

Figure 19. Energy evolution solved with different forward speed databases.

Overall, the influence of forward speed on the hydrodynamic database and its subsequent effect on the coupled FE analysis results is consistent. When forward speed has only a minor effect on the hydrodynamic database, its subsequent effect on the FE collision/grounding simulation results is similarly limited.

4 CONCLUSIONS

This study presents an integrated simulation framework combining Aqwa, LS-DYNA MCOL, and the LS-DYNA FE solver to enable a smooth and user-friendly workflow for ship collision and grounding analyses. By incorporating Aqwa, a state-of-the-art hydrodynamic solver, this workflow ensures a robust estimation of hydrodynamic influences throughout the analysis.

The case studies on an 80-meter tanker highlight the importance of FE model truncation and the effect of forward-speed-dependent hydrodynamic databases on internal structural dynamics. Results indicate that an appropriately truncated FE model, following Saint-Venant's Principle, significantly improves computational efficiency and accuracy. Additionally, the impact of hydrodynamic databases with forward speed effect on internal structural response remains insignificant in our testing scenarios.

Future work could focus on further investigations in hydrodynamic modeling, including generating and utilizing multiple databases to account for large positional variations in collision and grounding scenarios, as well as incorporating wave-induced motion in the initial conditions of collision and grounding.

REFERENCES

Ansys Aqwa. 2024. Ansys Aqwa Theory Manual, 2024R2. Canonsburg: Ansys Inc.

Cui, Z., Bi, Y., Davies, A., Du, S. & Huang, Y. 2024. Introducing Ansys LS-DYNA MCOL solver coupling with Ansys Aqwa software for shipbuilding analysis. In *2024 International LS-DYNA Conference, Metro Detroit, Michigan, USA*.

DNV GL. 2016. Recommended Practice, DNVGL-RP-C208: Determination of Structural Capacity Using Non-linear Finite Element Analysis (NLFEA). Høvik: DNV GL.

Liu, B., Villavicencio, R., Zhang, S. & Guedes Soares, C. 2017. Assessment of external dynamics and internal mechanics in ship collisions. *Ocean Engineering* 141: 326–36.

Lloyd's Register. 2016. Guidance Notes for Collision Assessment for the Location of Low-flashpoint Fuel Tanks. London: Lloyd's Register.

LS-DYNA®. 2023. R15 Keyword Manual, Vol I. Livermore: Livermore Software Technology Corporation.

MCOL. 2015. Theoretical Manual. La Ciotat: Principia Marine.

Pedersen, M.J. 1982. Dynamics of ship collisions. *Ocean Engineering* 9(2): 295–329.

Pedersen, P.T. 1995. Collision and grounding mechanics. In *Proceedings of WEMT'95*: 125–157. Copenhagen: WEMT.
Shiomitsu, D., Kokudai, M. & Fukui, T. 2023. Research and development on ship collision – Consideration of the equivalent added mass coefficient of a struck ship. *ClassNK Technical Journal* 8(II): 65–76.
Tabri, K. 2012. Influence of coupling in the prediction of ship collision damage. *Ships and Offshore Structures* 7(1): 47–54.
The Guardian. 2025. North Sea collision: Ship's master arrested as fears grow for marine life. March 11. [Online]. Available: https://www.theguardian.com/uk-news/2025/mar/11/north-sea-collision-man-arrested-on-suspicion-of-gross-negligence-manslaughter.
Wikipedia. 2024. Exxon Valdez oil spill. February 26. [Online]. Available: https://en.wikipedia.org/wiki/Exxon_Valdez.

Trends in Collision and Grounding of Ships and Offshore Structures – Zhu & Guedes Soares (Eds)
© 2026 The Author(s), ISBN 978-1-041-16361-9

An analytical method to predict the crashworthiness of the VLGC ship collided by the ship bow

Z. Chen
School of Mechanical Engineering, Wuhan Polytechnic University, Wuhan, P.R. China

M. Zhang
School of Mechanical Engineering, Wuhan Polytechnic University, Wuhan, P.R. China
Shanghai Key Laboratory of Ship Engineering, Shanghai, P.R. China

L. Lin & H. Wang
Shanghai Rules & Research Institute, China Classification Society, Shanghai, P.R, China

ABSTRACT: This paper presents a simplified analytical methodology, with the objective of forecasting the critical dissipated energy of the VLGC ship's side structure during a head-on collision with the bow of another vessel, wherein the colliding bow incorporates both a bulbous bow and a raked bow. The critical penetration depth and the corresponding resistance force for a sphere colliding with a plate are employed to determine the energy dissipation attributed to the bulbous bow. In parallel, the material damage volume concept is utilized to calculate the energy dissipation incurred by the ship's side structure due to the impact of the raked bow. The analytical prediction method is verified by the numerical simulation. A typical collision scenario is defined and the results obtained by the proposed analytical method and that of the numerical simulation are compared. The results correspond well, which affirms the efficacy of the proposed analytical method.

1 INTRODUCTION

VLGC (Very Large Gas Carrier) is a specialized large-scale liquefied gas tanker designed for transporting liquefied petroleum gas (LPG) or ammonia, typically with a cargo capacity ranging between 70,000 to 85,000 cubic meters. VLGC ships can pose significant risks in collision scenarios, due to their massive size and cargo of liquefied petroleum gas or ammonia.

The simplified analytical method serves as a valuable tool for preliminary ship collision-resistant design, enabling rapid estimation of the ultimate energy absorption capacity of hull structures (Zhang, 1999; Liu et al., 2018). However, the diversity in ship types leads to variations in hull structures and bow configurations, posing challenges for analytical predictions of the collision resistance. Typically, hull structures employ stiffened plate constructions, while bow forms mainly include bulbous bows and raked bows. Consequently, current research primarily focuses on predicting the energy absorption capacity of hull plating under impacts from these two bow types. Until now, many researchers have proposed analytical formulas for the crashworthiness prediction of stiffened plates under bulbous bow impact. For examples, Simonsen & Lauridsen (2000) proposed analytical expressions to predict the plate failure and the energy absorption up to the plate failure. (Liu et al., 2014) proposed an analytical criterion to evaluate the critical deflection, force and absorbed energy of the steel or aluminium materials for thin plates subjected to impact loads. Zhang et al. (2019) developed analytical expressions to calculate the whole collapse resistance for the initial fracture prediction and crack propagation resistance of the side plate induced by the lateral impact loading from a full-shape bulbous bow indenter. In addition, some researchers have proposed corresponding analytical formulas for the scenario of stiffened plates subjected to raked bow impact. Sun et al. (2015) obtained the resistance and energy dissipation of the individual components of the ship side structure to assess the complete crashworthiness of the side structure of the struck ship impacted by a raked bow. Zhang et al. (2024) proposed a new analytical method to rapidly evaluate side structural responses impacted by a rigid raked bow in right-angle ship collisions with large indentation. The scenario investigated in this study involves a VLGC ship struck by another vessel equipped with both the bulbous bow and the raked bow. Specifically, the VLGC experiences bulbous bow impacts below the waterline and raked bow impacts above the waterline. The collision scenario is complex, which can lead to

DOI: 10.1201/9781003684404-2

computationally intensive predictions for engineering analyses. Therefore, this study aims to develop a simplified analytical formula to estimate the ultimate energy absorption capacity of struck ships in complex collision scenarios, ultimately facilitating practical application.

The accuracy of analytical predictions for real-world ship collision resistance is typically validated through numerical simulations where the definition of failure criteria is the key factor. The failure criteria determine the moment of hull plate rupture, and the energy absorbed by the hull structure before this moment is generally regarded as the ultimate energy absorption capacity of the ship (Chen et al., 2025). Therefore, selecting appropriate failure criteria is crucial. To date, researchers have developed numerous failure criteria to simulate the failure of structural elements during ship collisions, such as the BWH criterion (Alsos et al., 2008), RTCL criterion (Tørnqvist, 2003), Hosford-Coulomb criterion (Cerik et al., 2019) and MSRT criterion (Lu et al., 2022), etc. The application of failure criteria typically requires the implementation of user subroutines to determine element failure, which is not convenient for engineers. Moreover, the failure criteria are generally sensitive to mesh size. Hull structures are usually modeled using coarse-mesh shell elements due to their large size. Currently, the relationship between failure parameters and mesh size is primarily determined using Barba's law (Kõrgesaar & Romanoff, 2014). Besides, the rupture of the hull shell plates caused by the bulbous bow is attributed to biaxial tensile failure. Therefore, this study specifically employs the relationship between the element failure strain under biaxial tension and the ratio of l_e/t (characteristic element length to plate thickness) to evaluate the failure of outer hull plates

This study aims to develop an analytical method for predicting the collision resistance of the VLGC ship. The research includes: 1) deriving energy absorption-penetration depth formulas for hull plating under impact from the bulbous bows and the raked bows; 2) Establishing a finite element model of ship collisions and validating the analytical formulas through three working conditions. The simplified analytical method proposed in this paper can serve as a practical tool for the preliminary collision-resistant design of the VLGC ship.

2 METHODOLOGY

2.1 *Collision scenario*

The collision simulation involves the struck ship, which is a VLGCship, and the striking ship, which is a 5618 TEU container vessel. The displacement, draft, and length of both ships are shown in Table 1. The 5618 TEU container vessel strikes the VLGC ship at a speed of 11 knots with a 90-degree impact angle, and the collision location is at the cargo tank area, as illustrated in Figure 2.

Table 1. Parameters of the Struck Ship and Striking Ship.

ship	Displacement (tons)	Draft(m)	Length verall(m)
VLGC	70827.3	11.7	215
5618TEU	74193.5	11.27	265.8

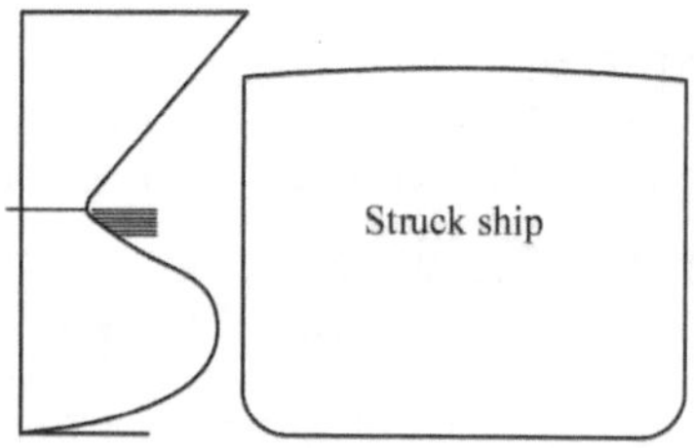

Figure 1. The ship bow side collision scenario.

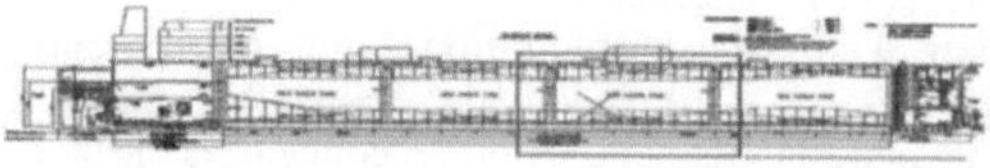

Figure 2. The impact location of the VLGC ship.

2.2 *Equivalent treatment method for stiffened plates*

To simplify the analysis process, the stiffening members on the shell plate are considered as an increase in the shell thickness. The equivalent shell plate thickness t_{eq} is calculated as follows:

$$t_{eq} = t_o + k\frac{A_s}{2b_0} \tag{1}$$

where t_0 is the original thickness of the shell plate; As is the cross-sectional area of the stiffening member; k is a coefficient taken as 1.05; $2b_0$ is the spacing between stiffening members.

2.3 *Energy dissipation by the bulbous bow*

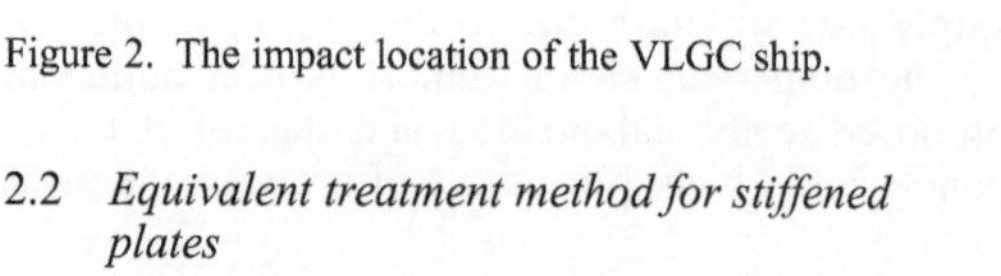

Figure 3. Deformation mode of the plate punched by a sphere.

The bulbous bow is simplified as a sphere with a radius of r_0. The sphere acts on a rectangular plate with dimensions of $2a \times 2b$ (where $2a$ is the long side and $2b$ is the short side). According to Zhang et al. (2019), the impact force F pressed by the sphere on the rectangular plate is given by:

$$F = 2\pi ktr_0(1/\cos\varphi - 1)^n \cos\varphi \sin^2\varphi \quad (2)$$

where t is the equivalent shell plate thickness; φ is the rotation angle of the shell plate at point C in Figure 3; k is the material strengthening coefficient; n is the material hardening exponent. According to Reference Villavicencio & Guedes Soares (2012), k and n can be calculated as:

$$n = 1.5\ln[1 + 1/(0.24 + 0.01395\sigma_b)] \quad (3)$$

$$k = 0.85\sigma_t(e/n)^n \quad (4)$$

where σ_b is the tensile strength; e is the natural constant。

Zhang et al. (2019) also provides the critical rotation angle φ_{cf} at point C and the penetration depth w_p at shell fracture, they are expressed respectively as:

$$\cos\varphi_{cf} = 0.58 - 0.48n \quad (5)$$

$$w_p = 0.5\sqrt{(7n + 0.7)br_0} \quad (6)$$

A simplified treatment is applied to the collision force of Eq. (1). Since the material hardening exponent n ranges between 0.1 and 0.3, the value of $2(1/\cos\varphi - 1)^n \cos\varphi \sin^2\varphi$ falls within a narrow interval of 0.753–0.763. Therefore, the impact force F_p at the moment of rectangular plate fracture can be expressed as:

$$F_p = 0.76\pi tr_0 k \quad (7)$$

The energy value during the plate deformation-to-fracture process is expressed as:

$$E_p = 0.5F_p w_p \quad (8)$$

2.4 *Energy dissipation by the raked bow*

For collisions where a striking ship with a raked bow impacts the side structure of a struck ship, the calculation of energy absorption by the damaged area of the struck ship depends on the relative vertical position between the struck and striking ships. This is because different relative positions result in varying damage zones on the struck ship, as illustrated in Figure 4. When the raked bow is lower than the deck of the struck ship (denoted as collision scenario 1), the damage zone of the struck ship is concentrated on the side shell. Conversely, when the raked bow is higher than the struck ship (collision scenario 2), both the side and deck structures of the struck ship sustain collision damage. Thus, the energy absorption calculation must account for damaged members in both the side and deck shells.

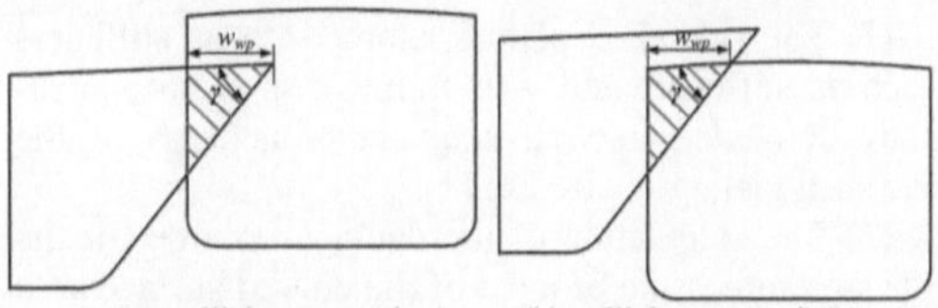

(a) collision scenario 1 (b) collision scenario2

Figure 4. Two raked bow collision scenarios.

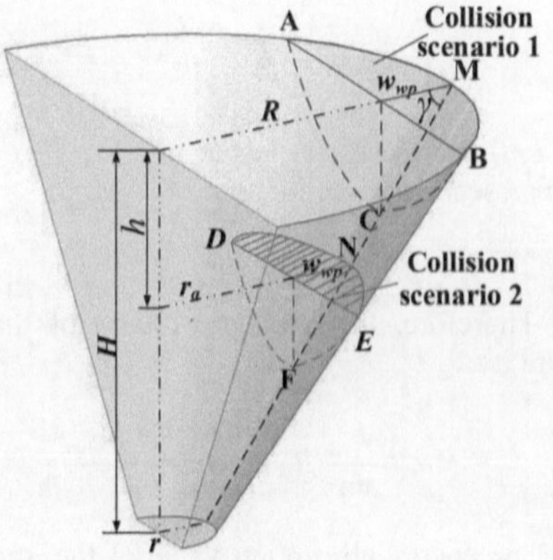

Figure 5. Damage zone of the struck ship collided by the raked bow.

In Figure 4, w_{wp} represents the penetration depth of the bow, γ denotes the rake angle of the raked bow. To simplify calculations, the raked bow section is idealized as a conical surface with an upper-end radius R and lower-end radius r, as illustrated in Figure 5, where H indicates the height of the raked bow section.

According to the theory proposed by Minorsky (1959), the energy absorption of ship structures during collisions is related to the volume of damaged material, expressed as:

$$E_{wp} = \delta\sigma_0 V \quad (9)$$

where δ is a coefficient; V represents the volume of damaged material; σ_0 denotes the flow stress of the material, calculated as the average of the material's yield strength σ_y and tensile strength σ_u.

The calculation of the damaged material volume V depends on the collision scenario, as illustrated by the areas enclosed by the blue (collision scenario 1) and red (collision scenario 2) lines in Figure 5.

2.4.1 *Energy dissipation in collision scenario 1*

For collision scenario 1, the volume of damaged material (V) corresponds to the structural material volume in the side-shell and internal compartment regions that are in contact with the striking ship's bow (AMBC). A step-by-step calculation procedure for estimating the damaged volume (V) and structural energy absorption is expressed as below:

(1) The side shell plating, along with its stiffeners (such as stiffeners and web frames), shall be equivalently converted into an equivalent thickness of the side shell plating t_{eqs} by Eq. (1).

(2) The calculation of the damage volume for the side structure is the product of the curved surface area enclosed by AMBC and the equivalent thickness of the outer plate. The curved surface area enclosed by AMBC is simplified to twice the area of triangle S_{AMC}. The calculation of S_{AMC} is as follows:

$$S_{AMC} = \frac{1}{2} L_{CM} \times L_{\overset{\frown}{AM}} \tag{10}$$

where L_{CM} represents the length of CM, $L_{CM} = \frac{w_{wp}}{\sin\gamma}$; $L_{\overset{\frown}{AM}}$ represents the arc length of AM. $L_{\overset{\frown}{AM}} = R \times \arccos\frac{R - w_{wp}}{R}$。

(3) Therefore, the damage volume of the side structure is:

$$V = t_{eqs} \frac{w_{wp}}{\sin\gamma} R \cdot \arccos\frac{R - w_{wp}}{R} \tag{11}$$

(4) The energy absorption value of the impacted ship's side structure in collision scenario 1 is:

$$E_{wp1} = \delta_1 \sigma_0 t_{eqs} \frac{w_{wp}}{\sin\gamma} R \cdot \arccos\frac{R - w_{wp}}{R} \tag{12}$$

where δ_1 is the energy absorption coefficient for collision scenario 1, taken as 1; t_{eqs} is the equivalent thickness of the impacted ship's side shell plate.

2.4.2 *Energy dissipation in collision scenario 2*

For collision scenario 2, the damaged material volume corresponds to the material volume of the damaged structures in the side, the deck, and the internal compartments that are in contact with the surface DNEF in Figure 5. A calculation procedure for estimating the damaged volume (V) and structural energy absorption is as below.

(1) The side shell and deck shell, along with their stiffeners are equivalently converted into an equivalent thickness of t_{eqs} and t_{eqd}, respectively.

(2) The damaged volume of the side shell structure is calculated as the product of the equivalent thickness of the side shell plating (t_{eqs}) and the surface area enclosed by the DNEF. The DNEF-enclosed surface area can be simplified as twice the area of triangle S_{DNF}. The calculation of S_{DNF} is as follows:

$$S_{DNF} = \frac{1}{2} L_{NF} \times L_{\overset{\frown}{DN}} \tag{13}$$

where L_{NF} is the Length of NF, $L_{NF} = \frac{w_{wp}}{\sin\gamma}$; $L_{\overset{\frown}{DN}}$ is the arc length of DN, $L_{\overset{\frown}{DN}} = r_a \times \arccos\frac{r_a - w_{wp}}{r_a}$, r_a represents the bow radius at the impact location in collision scenario 2,and it can be expressed as:

$$r_a = \frac{HR - Rh + rh}{H} \tag{14}$$

where h is the vertical height from the impact point on the ship's bow flare section to its topmost point; R and r are the upper and lower base radius of the conical section, respectively; H is the total height of the conical section.

(3) The damaged volume of the side shell structure is then calculated as:

$$V_s = t_{eqs} \times L_{NF} \times L_{\overset{\frown}{DN}}$$

$$= t_{eq} \times \frac{w_{wp}}{\sin\gamma} \times r_a \times \arccos\frac{r_a - w_{wp}}{r_a} \tag{15}$$

(4) The damaged volume of the deck structure shall be determined as the product of the area of the DNE plane and the equivalent thickness of the deck plating (t_{eqd}). The area of the DNE plane can be simplified as twice the area of triangle SDNF. The calculation of SDNF is as follows:

$$S_{DNF} = \frac{r_a \times L_{\overset{\frown}{DN}}}{2} - (r_a - w_{wp})\sqrt{2r_a w_{wp} - w_{wp}^2}$$

$$= \frac{{r_a}^2 \times \arccos\frac{r_a - w_{wp}}{r_a}}{2} - (r_a - w_{wp})\sqrt{2r_a w_{wp} - w_{wp}^2} \tag{16}$$

(5) The damaged volume of the deck-side shell structure is given by:

$$V_d = t_{eqd} S_{DNF} = t_{eqd} \times \frac{{r_a}^2 \times \arccos\frac{r_a - w_{wp}}{r_a}}{2}$$

$$- (r_a - w_{wp})\sqrt{2r_a w_{wp} - w_{wp}^2} \tag{17}$$

Thus, the structural energy absorption value for collision scenario 2 can be estimated as:

$$E_{wp2} = \delta_2 \sigma_0 (V_s + V_d) = \delta_2 \sigma_0 \Big(\frac{t_{eq} w_{wp} r}{\sin\gamma} \arccos\frac{r_a - w_{wp}}{r_a}$$

$$+ \frac{t_{eqd} {r_a}^2}{2} \arccos\frac{r_a - w_{wp}}{r_a} - (r_a - w_{wp})\sqrt{2r_a w_{wp} - w_{wp}^2}\Big) \tag{18}$$

where δ_2 denotes the energy absorption coefficient, taken as 0.1。

In summary, when impacted by the raked bow, the energy absorption value E_{wp} of the struck vessel can be expressed as:

$$E_{wp} = \begin{cases} E_{wp1} & \text{collision scenario1} \\ E_{wp2} & \text{collision scenario2} \end{cases} \quad (19)$$

2.5 *Total energy dissipation by the struck ship*

When the bulbous bow causes the ship structure to rupture, the total energy absorption value is the energy absorbed by the struck vessel's side shell structure under the combined action of the bulbous bow and the raked bow. This value represents the critical deformation energy (E_c) of the struck vessel and is expressed as:

$$E_c = E_p + E_{wp} \quad (20)$$

3 VERIFIED BY THE NUMERICAL SIMULATION

3.1 *Finite element model*

Figure 6 shows the finite element models of the two ships, with an added mass coefficient of 0.04 for the striking ship and 0.56 for the struck ship (Zhu & Wang, 2025). The masses of the striking ship and struck ship are 77,161 tons and 110,490.6 tons, respectively. The mesh in the contact area between the two ships is refined to 100 mm, while the mesh size in other areas is 800 mm. The hull structure plates are made of H32 high-strength steel, and the power-law hardening model is used to account for the plastic hardening effect of the steel. The material curve is shown in Figure 7. It should be noted that the strain rate effect on material strengthening is not considered because its influence on the load is limited (Gruben et al., 2017), and secondly to simplify the calculation process.

(a) VLGC ship

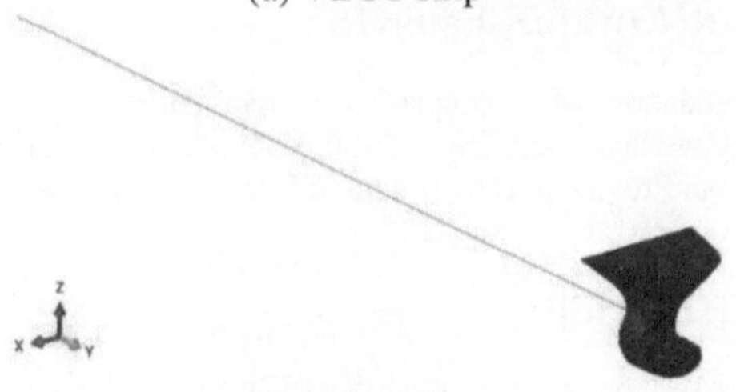

(b) Striking ship

Figure 6. Finite element model of the VLGC ship and the striking ship.

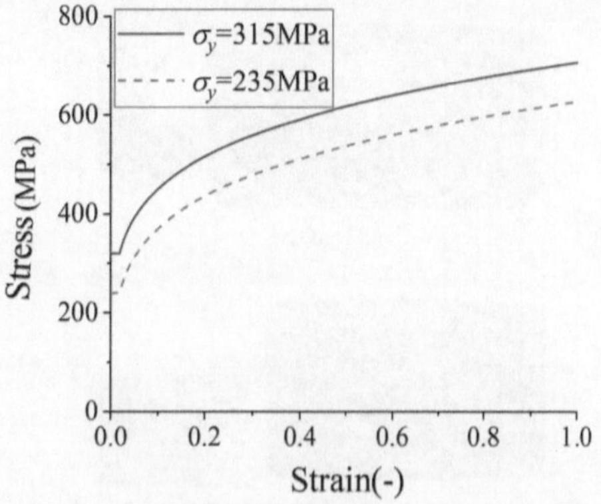

Figure 7. True stress-strain curves of the H32 high strength steel and the Q235 steel.

Considering biaxial tensile failure, the defined failure strain ε_f for the struck ship is:

$$\varepsilon_f(l_e, t) = 2n + (\varepsilon_u - 2n)\frac{t}{l_e} \quad (21)$$

where l_e/t represents the ratio of element size to thickness; ε_u is the failure strain when l_e/t=1, taken as 0.58.

The element-size-to-thickness ratio for the damaged area of the side structure is 6. Based on Equation (21), the calculated failure strain is approximately 0.43. The element damage time is about 1 s, resulting in an average strain rate of 0.4 s^{-1}. According to Paik et al. (2017), the ratio of dynamic failure strain to static failure strain is=0.92. Therefore, the dynamic failure strain is calculated as: ε_{fd}=0.43×0.92=0.4.

Self-contact was defined with a friction coefficient of 0.3. This contact setting can simulate both the collision interaction between the two ships and the contact behavior among structural components of the hull. For the struck ship, rotational degrees of freedom in the depth direction were applied at both ends of the hull girder, while for the striking ship, the degrees of freedom in the depth direction were constrained, and an initial velocity of 11 knots was applied. An explicit dynamic analysis was performed in ABAQUS with a total computation time of 5 seconds.

3.2 *Collision cases*

Numerical simulations were conducted for three cases. In case 1, the draught of the striking ship and the struck ship was the same, while cases 2 and 3 involved the raked bow impacting the outer shell plate and the deck, in order to verify the accuracy of the analytical solution under the impact of a raked bow.

3.3 *Comparison of results*

Figure 9 shows the damage shapes of the struck ship and the comparison curves of energy versus

(a) case1

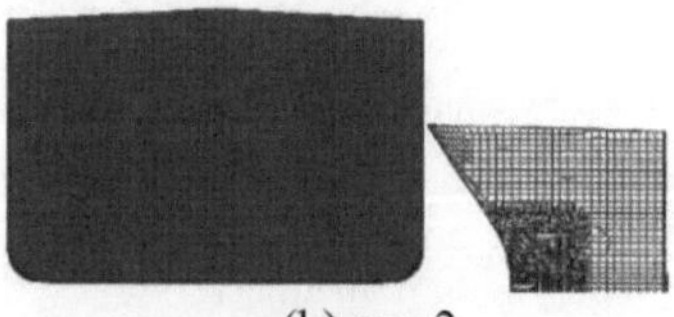

(b) case2

(c) case3

Figure 8. Collision cases.

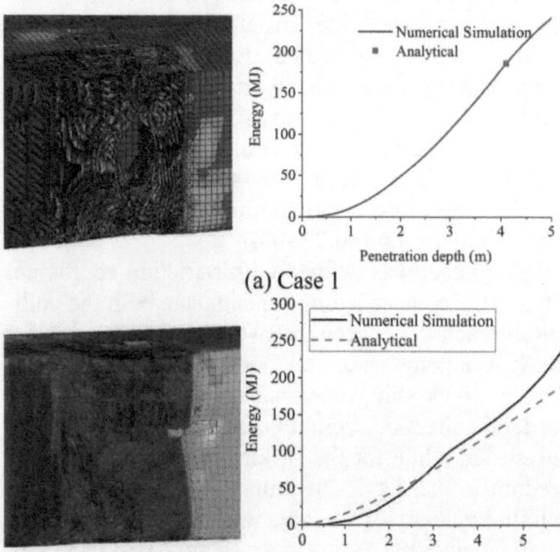

(a) Case 1

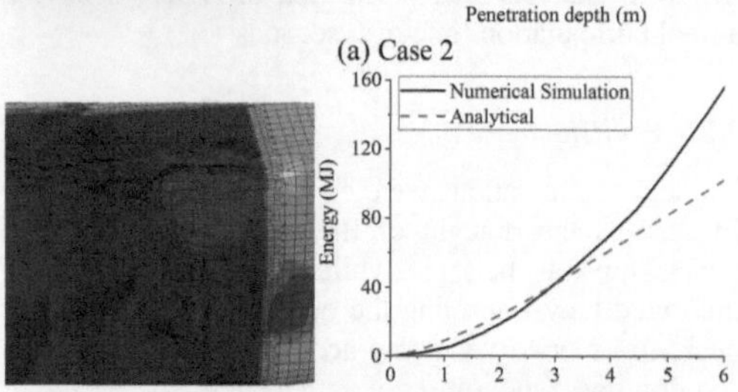

(a) Case 2

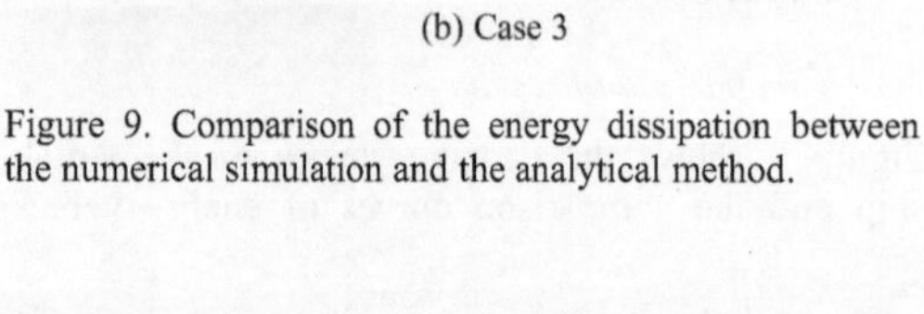

(b) Case 3

Figure 9. Comparison of the energy dissipation between the numerical simulation and the analytical method.

penetration depth between the analytical solution and the simulation. The curves are close when the penetration depth is small, but as the penetration depth increases, the analytical solution yields lower values compared to the simulation. Compared with the numerical simulation results, the simplified analytical method incorporates simplifications for both the striking ship's bow and the struck ship's structure, such as: 1) the impacted region is considered as the area constrained by transverse bulkheads, and the stiffened plate structure is simplified as a flat plate with clamped boundary conditions on all sides. 2) The bulbous bow of the striking ship is simplified as a spherical shape, and the raked bow is simplified as a conical shape. These simplifications result in lower energy absorption values for the side structure in the simplified analytical method compared to the numerical simulation, meaning the simplified method provides conservative (safer) results.

4 SUMMARY

This paper proposes a simplified analytical method for VLGC ship collisions. The simplified analytical method provides formulas for calculating the energy absorption values of the ship's side shell structure under impacts from both the bulbous bow and the raked bow. For bulbous bow impacts, the bow is simplified as a rigid spherical head, and formulas are given for calculating the maximum impact force on the shell plate, critical penetration depth, and critical deformation energy. For raked bow impacts, the bow is simplified as a conical body, which interacts with the side shell structure in two forms. The energy absorption value of the shell plate is proportional to the damaged volume. In the numerical simulation method, finite element models are established, taking into account the material model, failure criterion, element size, boundary conditions, and added mass coefficient. Explicit dynamic analysis is then performed to obtain the energy absorption values of the side shell. Under the three impact scenarios, the numerical simulation results closely match those of the simplified analytical method, with the simplified analytical method yielding more conservative estimates.

ACKNOWLEDGEMENTS.

Foundation of Shanghai Key Laboratory of Ship Engineering (NO. SE 202302). Guiding Project of Hubei Provincial Department of Education (B2023049.

REFERENCES

Alsos, H. S., Hopperstad, O. S., Törnqvist, R., Amdahl, J. (2008). Analytical and numerical analysis of sheet metal instability using a stress based criterion. *International Journal of Solids and Structures*, 45(7-8), 2042–2055.

Cerik, B. C., Lee, K., Park, S. J., Choung, J. (2019). Simulation of ship collision and grounding damage using Hosford-Coulomb fracture model for shell elements. *Ocean Engineering*, 173, 415–432.

Chen, W., Chen, C, Zhang, Y. H., Huang, H., Wu, X. X., Li, X. B. (2025). Research on the damage and energy absorption characteristics of spiral gradient composite structures subjected to impact loading. *Engineering Fracture Mechanics*, 320, 111077.

Gruben, G., Sølvernes, S., Berstad, T., Morin, D. (2017). Low-velocity impact behaviour and failure of stiffened steel plates. Marine Structures, 54, 73–91.

Kõrgesaar, M., Romanoff, J. (2014). Influence of mesh size, stress triaxiality and damage induced softening on ductile fracture of large-scale shell structures. *Marine structures*, 38, 1–17.

Liu, B., Pedersen, P. T., Zhu, L., Zhang, S. M. (2018). Review of experiments and calculation procedures for ship collision and grounding damage. *Marine Structures*, 59,105–121.

Liu, B., Villavicencio, R., Guedes Soares, C. (2014). On the failure criterion of aluminum and steel plates subjected to low-velocity impact by a spherical indenter. *International Journal of Mechanical Sciences*, 80, 1–15.

Lu, Y., Liu, K., Wang, Z. L., Tang, W. Y., Amdahl, J. (2022). Development of ductile fracture modelling approach in ship impact simulations. *Ocean engineering*, 252, 111173.

Minorsky, V. (1959). An analysis of ship collisions with reference to nuclear power plants. *Journal of Ship Research*, 3(2),1–4.

Paik, K. J.,Kim, J. K., Lee, H. J., BG Jung, B. G., SJ Kim, S. J. (2017). Test database of the mechanical properties of mild, high-tensile and stainless steel and aluminium alloy associated with cold temperatures and strain rates. *Ships and Offshore Structures*,12, sup1.

Simonsen, B. C., Lauridsen, L. P., (2000). Energy absorption and ductile failure in metalsheets under lateral indentation by a sphere. *International Journal of Impact Engineering*. 24 (10), 1017–1039.

Sun B, Hu, Z. Q., Wang, G. (2015). An analytical method for predicting the ship side structure response in raked bow collisions. *Marine Structures*, 41, 288–311.

Tørnqvist, R. (2003). Design of crashworthy ship structures. *PhD thesis*. Technical University of Denmark.

Villavicencio, R., Guedes Soares, C. (2012). Numerical modelling of laterally impacted plates reinforced by free and end connected stiffeners. *Engineering Structures*, 44, 46–62.

Zhang, M, Sun, Q. B., Liu, J. X., Hu, Z. Q., Zhang, S. M. (2019). A study of the rupture behavior of a ship side plate laterally punched by a full-shape bulbous bow indenter. *Ocean Engineering*, 182, 48–60.

Zhang, Y, Yuan, Y.C., Zhou, J.Y., Tang, W. Y. (2024). A new analytical model to evaluate ship side collisions with large indentation considering interaction effects between structural components. *Marine Structures*, 95, 103596.

Zhang, S.M. (1999). The Mechanics of Ship Collisions. *PhD Thesis*. Technical University of Denmark.

Zhu, L, Wang, X. B. (2025). Ship collision model tests in a water tank. *Marine Structures*, 103, 103791.

Trends in Collision and Grounding of Ships and Offshore Structures – Zhu & Guedes Soares (Eds)
© 2026 The Author(s), ISBN 978-1-041-16361-9

Machine learning-based prediction of structural damage in ship intentional collision scenarios

W.J. Qiu, S. Zong & K. Liu
School of Naval Architecture and Ocean Engineering, Jiangsu University of Science and Technology, Zhenjiang, China

B.Q. Chen & C. Guedes Soares
Centre for Marine Technology and Ocean Engineering (CENTEC), Instituto Superior Técnico, Universidade de Lisboa, Lisbon, Portugal

ABSTRACT: This study investigates typical intentional collision scenarios involving maritime law enforcement vessels. Latin Hypercube Sampling is utilized to systematically generate multiple collision scenarios, and finite element simulations are conducted to build a comprehensive collision simulation database. Based on this database, a Back-Propagation neural network model is applied to predict and evaluate structural dam-age characteristics under different collision scenarios. Moreover, Kendall's coefficient analysis is employed to quantify the influence of various collision parameters on structural damage, identifying key factors. The results indicate excellent agreement between predictions by the neural network model and finite element simulation outcomes, validating the effectiveness of machine learning approaches in collision prediction. Sensitivity analysis further highlights impact velocity as a critical parameter requiring particular attention when formulating intentional collision scenarios strategies.

1 INTRODUCTION

With the escalating international competition for maritime rights and interests, maritime law enforcement agencies are increasingly adopting intentional collision scenarios tactics in response to non-military maritime conflicts. Intentional collision scenarios, involving deliberate pushing or direct impacts, are more effective than conventional interception methods in forcing a targeted vessel to alter its course or lose maneuverability. Nevertheless, precise control of impact velocity and collision angles is crucial to effectively disrupt target vessels while minimizing risks to personnel safety. Hence, accurately predicting and assessing structural damage resulting from collisions is of significant practical importance for enhancing maritime law enforcement efficiency, optimizing collision strategies, and formulating effective protective measures.

In recent years, research in the field of ship collisions has primarily focused on collision damage mechanisms, finite element numerical simulations, and collision risk assessments. When ships collide it is necessary to consider both the external dynamics of the ships as rigid bodies, but also the internal dynamics that are dominated by the energy absorption in structural deformation (Liu et al, 2017).

Numerous studies have employed the Finite Element Method (FEM) to finely model the ship collision process, evaluating hull deformation, structural failure, and energy absorption characteristics under various collision conditions (Ringsberg et al. 2018). These investigations not only elucidate the stress distribution and local damage mechanisms at critical areas during collisions but also provide theoretical foundations and technical support for structural design optimization and enhanced collision safety performance.

Liu et al. (2022) studied structural responses of ship stiffened panels subjected to inclined collisions, performing impact tests at a 75° angle and analyzing damage characteristics of stiffened plates using validated FEMs, resulting in ultimate bearing capacity ratio curves. Cheng et al. (2022) conducted numerical simulations of wedge-shaped impactors colliding with stiffened sandwich panels, determining the energy absorption ratios of each structural component and investigating dynamic responses under varying impact velocities and locations. Chen et al. (2022) have made experiments and FEM calculation on a on a double hull structure subject to collision, which are a representative case (Liu et al., 2021).

Liu & Guedes Soares (2019) developed a model to represent the effect of strain rate on dynamic responses of laterally impacted steel plates. Liu et al. (2024) introduced a modified Mohr–Coulomb (MMC) failure criterion specifically for metallic materials, validated through finite element simulations to enhance the accuracy of collision damage predictions.

Wang et al. (2023) proposed a simplified analytical method to predict structural responses of side structures impacted by bulbous bows under

DOI: 10.1201/9781003684404-3

oblique collision scenarios, achieving results consistent with numerical simulations. Xiong et al. (2024) used numerical simulations to investigate structural responses and energy transformation characteristics during collisions involving single-hulled and double-hulled ships, discovering that adding connections between inner and outer hulls significantly improves collision resistance. An overview of approaches adopted in ship collision studies are provided by Liu & Guedes Soares (2023).

Additionally, to mitigate the computational cost associated with finite element analyses, researchers have introduced machine learning methods to enhance computational efficiency and rapidly predict collision outcomes. Machine learning effectively processes large simulation datasets, capturing complex nonlinear behaviors inherent in collision processes, thereby constructing efficient and accurate predictive models and significantly reducing dependence on high-performance computational resources (Feng et al., 2025). Specifically, models combining Back-Propagation (BP) neural networks with optimization algorithms have been successfully applied to collision damage prediction and structural optimization. Qiu et al. (2023) integrated orthogonal experimental design, BP neural networks (BPNN), and Sparrow Search Algorithm (SSA) to optimize the crashworthiness of ship structures under multiple operating conditions, with predictions showing high consistency with finite element simulations. Sahib & Kovács (2024) developed a multi-objective optimization model using Genetic Algorithms (GA) and Artificial Neural Networks (ANN) to optimize the weight and cost of composite sandwich panel structures.

Despite significant advances in ship collision research, most studies have focused on accidental collisions or traditional maritime incidents, leaving the specific scenario of intentional collision scenarios by maritime law enforcement vessels underexplored. To ensure targeted vessels avoid severe structural damage during intentional collision scenarios in maritime enforcement operations, it is crucial to develop scientifically sound collision strategies.

Addressing this research gap, this paper focuses on typical intentional collision scenarios involving maritime law enforcement vessels. Latin Hypercube Sampling (LHS) is employed to quantitatively generate various collision scenarios, and finite element simulations are performed to establish a comprehensive simulation database. Based on this database, a Back-Propagation (BP) neural network model is adopted to predict and evaluate structural damage characteristics under different collision conditions. Furthermore, Kendall's coefficient analysis is utilized to determine the influence of collision parameters on structural damage and to identify key factors.

2 SHIP COLLISION SAMPLE DATABASE

2.1 *Impact parameters*

The intentional collision scenarios studied in this paper represent typical ship-to-ship collision events, assuming both the impacting and impacted vessels have a displacement of 3,000 tons. The primary impact parameters considered during the intentional collision scenarios simulations include impact velocity, impact angle, and impact location, as illustrated in Figure 1.

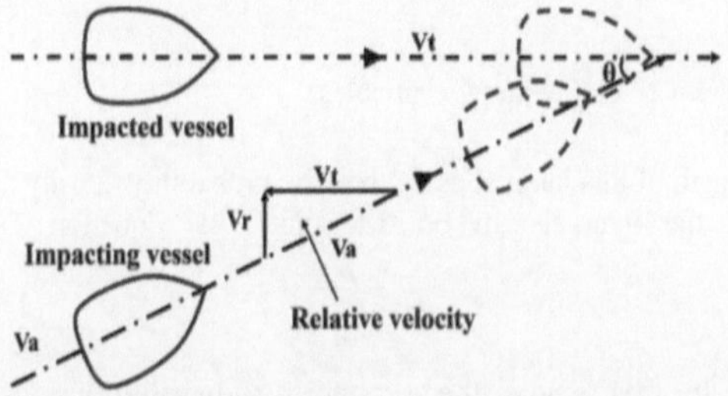

Figure 1. Illustration of impact parameters.

(1) Impact Velocity: This study focuses on simulating collision scenarios in which both vessels are in motion, with vessel speeds ranging from 0 to 26 knots. In the finite element simulations, the impact velocity is defined as the velocity of the impacting vessel relative to the impacted vessel within the global coordinate system.
(2) Impact Angle: The impact angle is defined as the angle formed between the longitudinal axes of the impacting and impacted vessels. Due to the symmetry of ship collisions with respect to impact angles, this study only analyzes collision scenarios within an impact angle range of 0° to 90°.
(3) Impact Location: The primary impact locations considered include the bow, midship, and stern regions of the impacted vessel. Additionally, this study further examines specific contact points between the vessels, including the main deck, Bulkheads, and Side shells.

2.2 *Latin hypercube sampling*

This study employs the LHS method to generate multiple combinations of ship collision parameters. The core principle of the LHS approach can be summarized as follows: each input dimension's distribution interval is divided into several sub-intervals of equal probability, from which precisely one sample point is randomly selected per sub-interval. These samples are then matched to form parameter combinations, ensuring comprehensive coverage of the parameter space (Lu et al., 2024).

Figure 2 illustrates the principle of LHS with two variables.

The LSH method requires prior knowledge of the probability distribution for each parameter. Suppose the probability of a particular interval is denoted as P_i and the

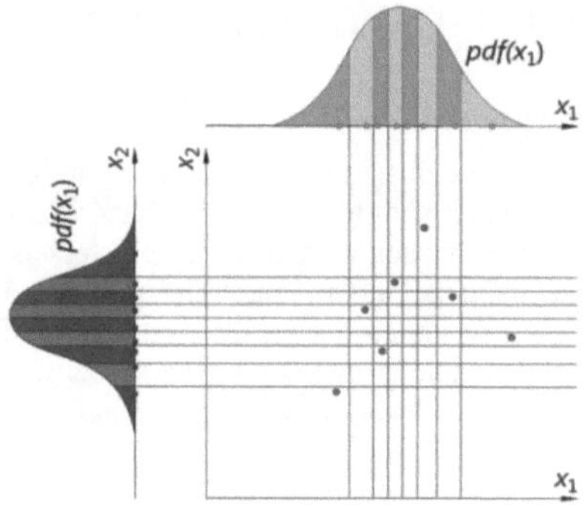

Figure 2. LHS with two variables.

length of this interval as L_i, then the probability density f_i of the interval can be determined by Equation 1:

$$f_i = \frac{P_i}{L_i} \quad (1)$$

In this study, the probability distributions of impact parameters during active ship collision scenarios are derived by analyzing and organizing 90 sets of collision data, as illustrated in Figures 3 to 5.

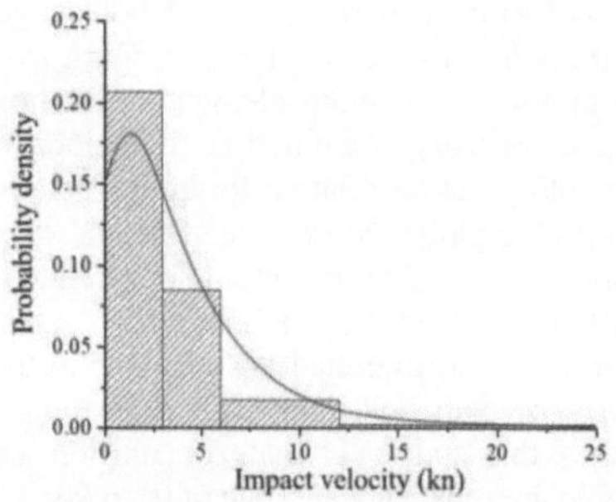

Figure 3. Probability distribution of impact velocity.

Based on these probability distribution diagrams, the LHS method is employed to generate 18 collision samples each for the bow, midship, and stern impact locations. The impact velocities range from 1.5 kn to 12 kn, while impact angles range from 15° to 90°. Each location includes 18 samples, selected to balance sufficient coverage of the parameter space with the high computational cost of finite element simulations. Preliminary tests showed that this sample size was adequate to capture typical damage patterns and support effective neural network training. Some of the sampled collision scenarios are presented in Table 1.

2.3 *Collision simulation calculations*

In this study, a specific 3000-ton maritime public-service vessel is selected to establish intentional collision simulation scenarios. Some hull parameters are presented in Table 2.

Figure 6 presents the ship finite element model established using ABAQUS finite element software.

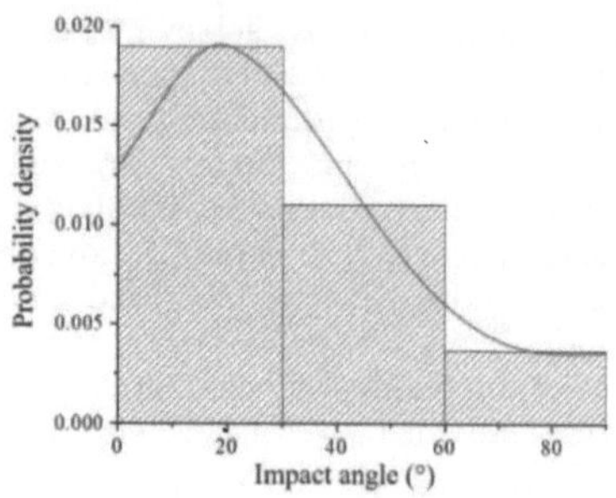

Figure 4. Probability distribution of impact angle.

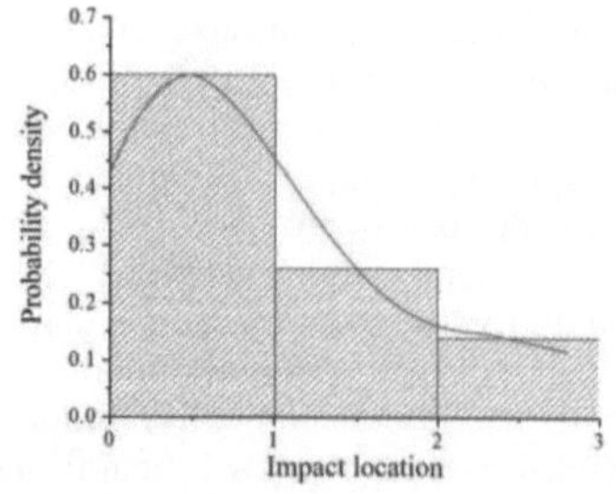

Figure 5. Probability distribution of impact location.

Table 1. Some of the sampled collision scenarios.

Condition	Location	Angle (°)	Velocity (kn)	Contact
LC1	Bow	15	1.5	Main deck
LC2	Bow	15	6	Main deck
LC3	Bow	30	3	Side shell
LC4	Bow	45	3	Side shell
LC5	Bow	45	6	Side shell
LC6	Bow	60	3	Main deck
LC7	Bow	60	6	Main deck
LC8	Bow	90	6	Side shell
LC9	Bow	90	12	Side shell
LC10	Midship	15	3	Side shell
LC11	Midship	15	6	Side shell
LC12	Midship	30	1.5	Bulkhead
LC13	Midship	30	3	Bulkhead
LC14	Midship	30	6	Bulkhead
LC15	Midship	45	6	Side shell
LC16	Midship	45	12	Side shell
LC17	Midship	60	1.5	Main deck
LC18	Midship	90	6	Side shell
LC19	Midship	90	12	Side shell
LC20	Stern	15	1.5	Side shell
LC21	Stern	15	3	Side shell
LC22	Stern	30	6	Side shell
LC23	Stern	45	1.5	Bulkhead
LC24	Stern	45	3	Bulkhead
LC25	Stern	60	6	Main deck
LC26	Stern	60	12	Main deck
LC27	Stern	90	1.5	Side shell
LC28	Stern	90	6	Side shell

Table 2. Partial hull parameters.

Main Dimensions	Value
Total Length	132.8 m
Breadth	16.0 m
Depth	8.3 m
Displacement	3000 t
Speed	0~26 kn

Given that this study primarily focuses on the structural response of the collision region of the impacted vessel, the impacting vessel is modeled as a rigid body, whereas the impacted vessel is set as a deformable body with its collision region meshed. To optimize simulation computational time, a finer mesh with a size of 0.05 m is applied to the collision region of the impacted vessel, while a coarser mesh with a size of 0.2 m is used for the remaining areas of both ships. Both ships are modeled using the four-node shell element S4R from the ABAQUS element library.

Figure 6. Ship finite element model.

Considering the strain rate sensitivity of the material, the Cowper-Symonds model is adopted as the constitutive model for the hull material (Cowper & Symonds, 1957):

$$\frac{\sigma_d}{\sigma_v} = 1 + (\frac{\varepsilon}{D})^{\frac{1}{q}} \quad (2)$$

where σ_d is dynamic yield stress, σ_y is static yield stress, ε is equivalent strain rate, D and q are constants set to 40.4 and 5, respectively.

To simulate ship collisions in water, the effect of water on the collision is incorporated by considering the added water mass. In this study, the added mass is set as $0.6m$ for the impacted vessel and $0.03m$ for the impacting vessel, where m denotes the hull mass.

Numerical simulations are conducted using ABAQUS finite element software. The collision duration is set to 1.5 s, and the damage area and breach area of the struck ship under various collision conditions are recorded. Some of the computed results for the collision scenarios are presented in Table 3.

Table 3. Some of the computed results.

Condition	Damage Area (m^2)	Breach Area (m^2)
LC1	0.000	N/A
LC3	0.150	N/A
LC5	0.290	N/A
LC8	0.332	N/A
LC9	0.980	N/A
LC12	0.590	N/A
LC13	0.740	0.050
LC17	1.420	1.200
LC18	8.770	6.580
LC20	0.119	N/A
LC21	0.178	N/A
LC22	0.214	N/A
LC23	0.455	N/A
LC24	5.150	0.530
LC27	3.900	3.150
LC28	23.380	18.260
LC29	0.126	N/A
LC35	5.020	4.100
LC36	17.780	15.990
LC37	0.160	N/A
LC38	0.240	N/A
LC42	0.700	0.240
LC43	0.290	N/A
LC44	1.620	0.077
LC49	0.700	0.240
LC50	0.290	N/A
LC51	1.620	0.077
LC53	4.230	2.760

3 PREDICTION OF STRUCTURAL DAMAGE IN PROACTIVE SHIP COLLISIONS

3.1 *Machine learning*

Based on the sample database established from the two-ship collision scenarios, this study employs a BPNN from machine learning to effectively predict the collision damage characteristics of the target ship.

The BPNN is a typical multilayer feedforward network without feedback, characterized by its highly flexible nonlinear mapping capability, which can effectively describe the complex nonlinear relationships between input and output parameters. A typical BP network structure consists of an input layer, a hidden layer, and an output layer. Generally, increasing the number of hidden layers can enhance the model's fitting accuracy; however, it also increases the network's complexity and computational cost. To balance model performance and computational efficiency, this study prioritizes a three-layer BP neural network with a relatively simple structure and high training efficiency, and improves the prediction accuracy by adjusting the number of nodes in the hidden layer, as shown in Figure 7.

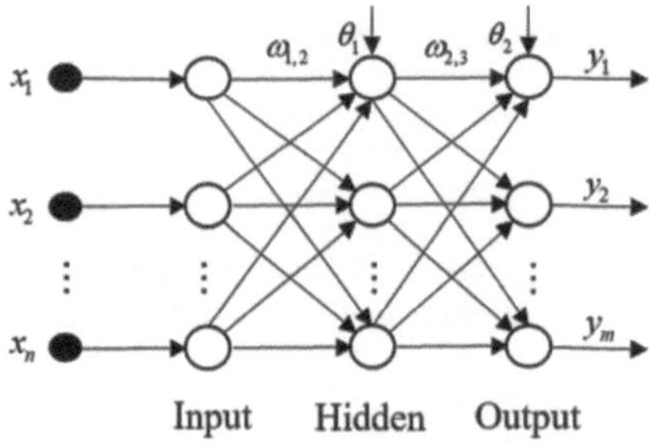

Figure 7. Three layers BP neural network.

The BP neural network employs the error backpropagation algorithm as its training method. During the training process, the output error is successively propagated from the output layer back to the input layer, while the weights are gradually updated, enabling the output to converge toward the expected value.

Given that the sample database constructed in this study comprises both continuous data (e.g., impact angle, impact velocity) and discrete data (e.g., impact location, contact location), this paper proposes a BP neural network model within a multitask learning framework. Specifically, the output layer explicitly distinguishes between a continuous prediction task and a discrete classification task to address the prediction requirements of different data types. To effectively handle discrete input parameters, categorical encoding is applied, while continuous input parameters are directly fed into the network for numerical prediction. Moreover, considering that the "breach area" output parameter is unique—indicating either a non-breach state ("N/A") or a specific numerical value in the event of a breach —a two-stage prediction strategy is designed: first, the "breach area" is treated as a classification problem to determine whether a breach occurs, and if a breach is predicted, a subsequent numerical prediction is conducted to determine the exact breach area.

3.2 *Model training and validation*

In the construction of the neural network model, the sigmoid activation function is employed to enhance the nonlinear representational capability of the network. Prior to model training, all data underwent standardization to accelerate convergence and improve training outcomes. During the training phase, mean squared error (MSE) is utilized as the loss function for continuous outputs, while binary cross-entropy loss (BCE) is applied to discrete outputs. The overall network loss function is then formulated by a weighted combination of these two losses.

The Adam optimization algorithm (Kingma & Ba, 2014), known for its excellent optimization performance, is selected to optimize network parameters. By incorporating momentum terms and adaptive learning rates, the Adam algorithm significantly improves the convergence speed and prediction stability, thus enhancing both training efficiency and generalization capability. During training, the maximum number of iterations is set to 20,000, with an initial learning rate of 0.001. The training process demonstrated stable convergence behavior, with the validation error steadily decreasing and eventually stabilizing, indicating robust learning and good generalization.

Upon completion of training, the predictive performance of the machine learning model is validated using simulation calculations under two collision conditions distinct from those in the training dataset. The collision scenarios and corresponding simulation results are detailed in Tables 4 and Table 5, respectively.

Table 4. Collision scenarios for validation.

Condition	Location	Angle (°)	Velocity (kn)	Contact
LC55	Bow	30	4.5	Side shell
LC56	Midship	45	9	Side shell

Table 5. Comparative results.

Condition		LC55	LC56
Damage Area (m^2)	Prediction	0.27	9.01
	Simulation	0.3	9.54
	Error (%)	10	5.6
Breach Area (m^2)	Prediction	N/A	7.9
	Simulation	N/A	8
	Error (%)	N/A	1.25

From the comparative results presented in Table 5, it can be observed that the neural network model demonstrates high predictive accuracy and reliability under both working conditions. Overall, the prediction errors for all considered scenarios remain within reasonable limits. Specifically, when the damage area is relatively small, the prediction error is approximately 10%. Conversely, under the LC56 condition with more severe damage, the prediction error decreases further, highlighting the model's superior generalization capability in scenarios involving complex or severe damage.

3.3 *Prediction of structural breach conditions*

The neural network model established in this study is utilized to predict structural damage outcomes under various collision scenarios. The occurrence of structural breach is adopted as the criterion to determine hazardous conditions. By separately fitting two distinct collision scenarios—one involving structural breach and the other characterized only by failure strain without breach—the critical boundaries distinguishing hazardous conditions are obtained for collision impacts occurring at the bow, midship, and stern sections of the impacted vessel. These boundaries are illustrated in Figures 8~10, where the yellow points represent sample predictions generated by the neural network model.

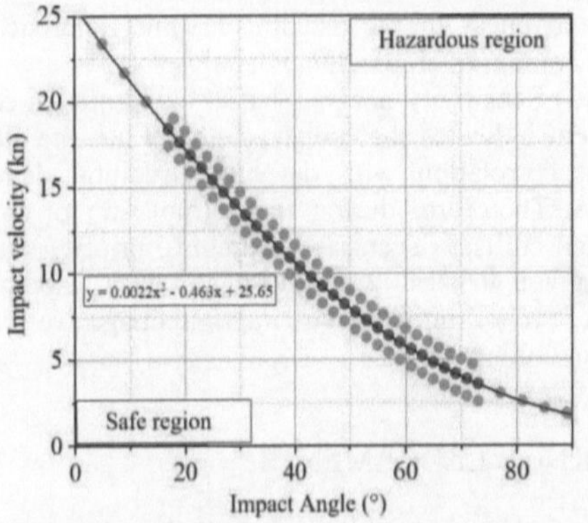

Figure 8. Impact location at the bow of the impacted vessel.

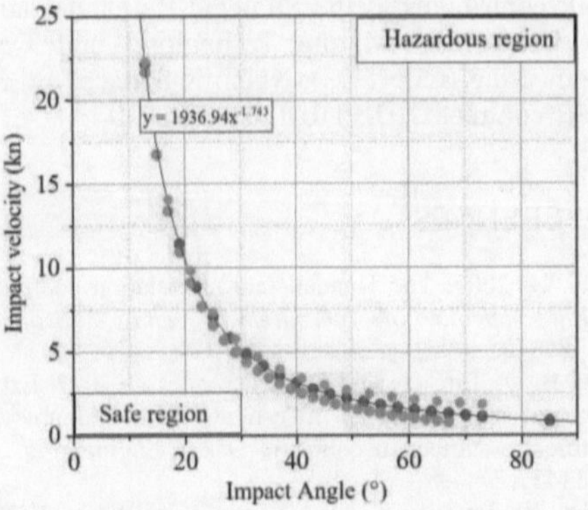

Figure 9. Impact location at the midship of the impacted vessel.

Based on the hazard condition boundary diagram, it is possible to roughly determine whether the combination of different impact velocity and impact angle falls under hazardous conditions: impact parameter combinations above the boundary may lead to structural rupture, whereas those below the boundary will not result in structural rupture following a collision. When the actual impact parameters (e.g., impact velocity and impact angle) exceed the delineated boundary, timely measures must be implemented to reduce the risk of structural rupture in the target vessel.

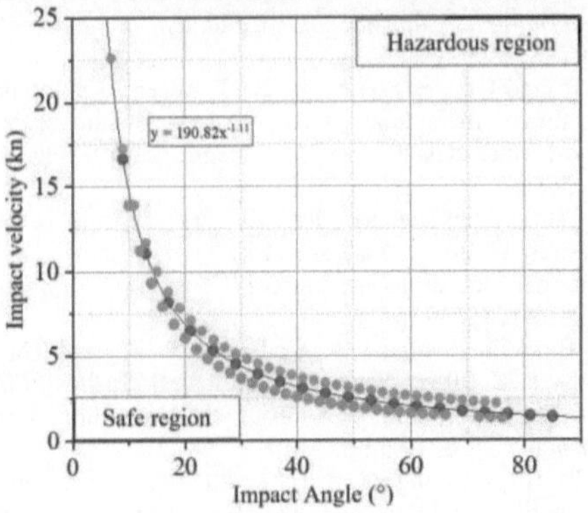

Figure 10. Impact location at the stern of the impacted vessel.

4 SENSITIVITY ANALYSIS OF VARIOUS IMPACT PARAMETERS

4.1 *Kendall coefficient*

In order to investigate the correlation between the damage sustained by the impacted vessel and impact parameters such as impact velocity and impact angle in compressive collision scenarios, this study introduces the Kendall coefficient for analysis. The Kendall coefficient quantitatively evaluates the degree of correlation between two variables by counting the number of concordant and discordant pairs in the sample data (Abdi, 2007). The specific calculation equation is as follows:

$$\tau = \frac{n_c - n_d}{\frac{1}{2}n(n-1)} \tag{3}$$

where n_c denotes the number of concordant pairs, n_d is the number of discordant pairs, and n represents the total number of samples. For n samples, a total of $n(n\text{-}1)/2$ pairs of samples are constructed. Given any two samples, (x_i, y_i) and (x_j, y_j), if $x_i > x_j$ and $y_i > y_j$, the data pair is considered concordant. Conversely, if $x_i > x_j$ but $y_i < y_j$, the data pair is considered discordant.

A Kendall coefficient approaching 1.0 indicates a strong positive correlation between the two variables, whereas a coefficient approaching -1.0 indicates a strong negative correlation. The closer the coefficient is to 0, the weaker the correlation between the variables.

4.2 *Sensitivity analysis*

To facilitate statistical analysis, this paper selects the damage area as an indicator to evaluate the extent of hull damage. Based on Equation 3, Kendall coefficients between the collision parameters and the hull damage area are calculated separately for impact locations at the bow, midship, and stern of the vessel. The calculated results are illustrated in Figures 11 to 13.

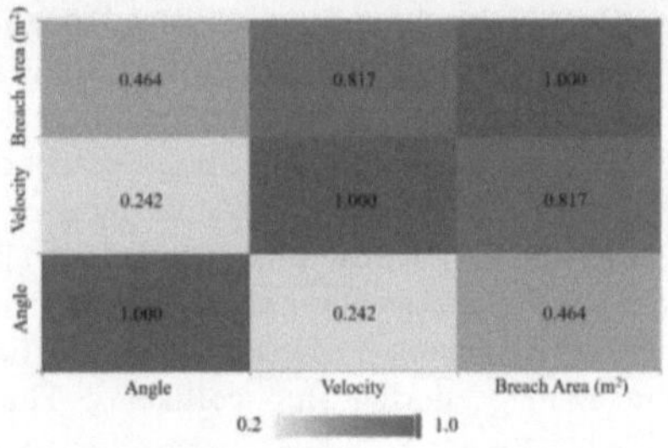

Figure 11. Kendall coefficient at the bow.

According to the calculated results shown in the figures, the impact velocity exhibits the strongest correlation with the extent of hull damage, regardless of the collision location along the vessel, indicating

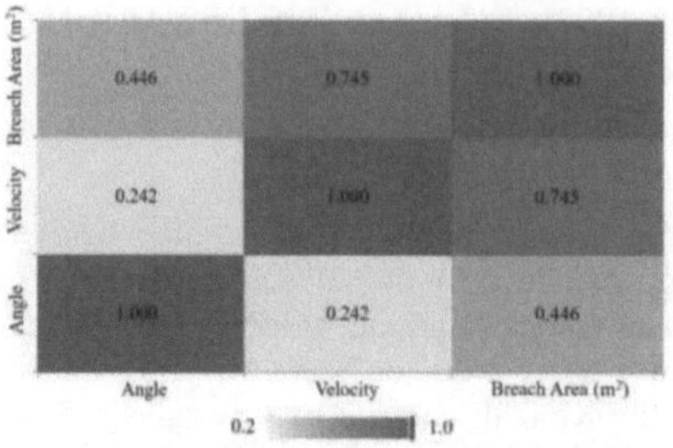

Figure 12. Kendall coefficient at midship.

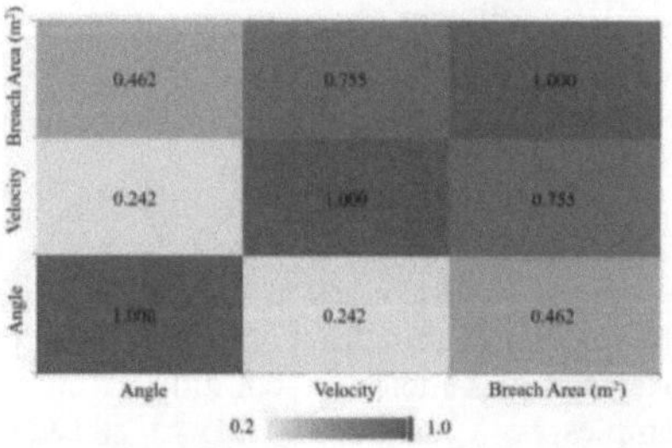

Figure 13. Kendall coefficient at the stern.

a significant concordance. Meanwhile, the correlation between the impact angle and the extent of hull damage remains at a moderate level, though it still notably influences whether damage occurs. Additionally, by comparing the results among the bow, midship, and stern positions, it is evident that impact velocity has a more pronounced impact on hull damage at the bow region. Therefore, when formulating practical plans for proactive compressive collisions, particular attention should be given to selecting an appropriate impact velocity.

5 CONCLUSIONS

This paper focuses on the scenario of intentional collision scenarios involving maritime service vessels and proposes a machine learning-based prediction method for ship collision damage. Using Latin Hypercube Sampling, various collision scenarios are generated and simulated through finite element simulations, creating a comprehensive database of proactive ship collision cases. Based on this database, a three-layer backpropagation neural network model is developed, achieving prediction and evaluation of structural damage during ship collisions. The primary conclusions are as follows:

(1) The BP neural network model demonstrates high accuracy in predicting ship collision damage. Compared to finite element simulation results, the prediction error for the damage area ranges from 5.6% to 10%, and the error for breach area prediction is only 1.25%, confirming the reliability and effectiveness of the machine learning approach in the domain of ship collision analysis.

(2) Sensitivity analysis based on Kendall's coefficient indicates that impact velocity has the strongest correlation with the hull structural damage area. Therefore, during the formulation of intentional collision scenarios strategies, priority should be given to strictly controlling impact velocity to effectively mitigate structural damage resulting from collisions.

ACKNOWLEDGEMENTS

The authors gratefully acknowledge the financial support of the National Natural Science Foundation of China (Grant No. 52171311, 52271279). This work contributes to the Strategic Research Plan of the Centre for Marine Technology and Ocean Engineering (CENTEC), which is financed by FCT under contract UIDB/UIDP/00134/2020.

REFERENCES

Abdi, H. 2007. The Kendall rank correlation coefficient. *Encyclopedia of measurement and statistics* 2: 508–510.

Chen, B. Q.; Liu, B., and Guedes Soares, C. 2022. Experimental and numerical investigation on a double hull structure subject to collision. *Ocean Engineering*. 256; 111437.

Cheng, Y., Liu, K., Li, Y., Wang, Z. & Wang, J. 2022. Experimental and numerical simulation of dynamic response of U-type corrugated sandwich panels under low-velocity impact. *Ocean Engineering* 245: 110492.

Cowper G, and Symonds P. 1957. Strain hardening and strain rate effects in the loading of cantilever beams. *Report, No.* 28, Division of Applied Mathematics, Brown University.

Feng, Y., Wang, H., Xia, G., Cao, W., Li, T., Wang, X. & Liu, Z. 2025. A machine learning-based data-driven method for risk analysis of marine accidents. *Journal of Marine Engineering & Technology, 24*(2): 147–158.

Kingma, D. P., & Ba, J. (2014). Adam: A method for stochastic optimization. arXiv preprint arXiv:1412.6980.

Liu, B. and Guedes Soares, C. 2019. Effect of strain rate on dynamic responses of laterally impacted steel plates. *International Journal of Mechanical Sciences*. 160; 307–317.

Liu, B. and Guedes Soares, C. 2023. Recent developments in ship collision analysis and challenges to an accidental limit state design method. *Ocean Engineering*. 270; 113636.

Liu, B.; Villavicencio, R.; Zhang, S., and Guedes Soares, C. 2017. Assessment of external dynamics and internal mechanics in ship collisions. *Ocean Engineering*. 141; 326–336.

Liu, B.; Villavicencio, R.; Pedersen, P. T., and Guedes Soares, C. 2021. Analysis of structural crashworthiness of double-hull ships in collision and grounding. *Marine Structures*. 76: 102898.

Liu, F., Hu, Y., Feng, G. & Li, C. 2022. Experimental and numerical study on the penetration of the inclined stiffened plate. *Ocean Engineering*, 258: 111792.

Liu, S., Liu, K., Liu, H., Zong, S., Lu, Y. & Liu, C. 2024. Calibration of the Modified Mohr–Coulomb Failure Criterion and Its Application in the Study of Collision Response of Ship Hull Plate Frame Structures. *Journal of Marine Science and Engineering* 12(5): 805.

Lu, L., Li, G., Xing, P., Gao, H., Song, Y., & Zhang, H. 2024. Numerical calculation and experimental investigation of the dynamic alignment of ship propulsion shafting based on Latin hypercube stochastic finite element. *Ocean Engineering* 296: 116935.

Qiu, W., Liu, K., Liu, H., Zong, S., Wang, J. & Gao, Z. 2023. Crashworthiness Optimization Method of Ship Structure under Multi-Working Conditions. *Journal of Marine Science and Engineering* 11(7): 1335.

Ringsberg, J. W.; Amdahl, J.; Chen, B. Q.; Cho, S. R.; Ehlers, S.; Hu, ZQ.; Kubiczek, J. M.; Kõrgesaar, M.; Liu, B.; Marinatos, J. N.; Nicklas, K.; Parunov, J.; Quinton, B. W.; Rudan, S.; Samuelides, M.; Guedes Soares, C.; Tabri, K.; Villavicencio, R.; Yamada, Y.; Yu, Z., and Zhang, S. 2018. MARSTRUCT benchmark study on nonlinear FE simulation of an experiment of an indenter impact with a ship side-shell structure. *Marine Structures*. 59; 142–157.

Sahib, M. M. & Kovács, G. 2024. Multi-objective optimization of composite sandwich structures using Artificial Neural Networks and Genetic Algorithm. *Results in Engineering* 21: 101937.

Xiong, Z., Ma, M., Guo, Y. & Li, R. 2024. Study of energy absorption characteristics of coastal ship side structures based on collision accidents. *Ocean Engineering* 307: 118079.

Wang, Z., Guo, C., Wang, C., Chen, G., Xu, Y. & Li, Q. 2024. An analytical method for predicting the structural response of ship side structures by bulbous bow in oblique collision scenarios. *Ships and Offshore Structures* 19(10): 1490–1503.

Trends in Collision and Grounding of Ships and Offshore Structures – Zhu & Guedes Soares (Eds)
© 2026 The Author(s), ISBN 978-1-041-16361-9

Study on the crashworthiness of nuclear-powered vessels with internal reinforcement in the double hull structure during maritime collision accidents

D-H. Seo
Bureau Veritas, Busan, South Korea

C. Lim, S-J. Oh & S-C. Shin
Pusan National University, Busan, South Korea

ABSTRACT: This study evaluates the crashworthiness of a hypothetical nuclear-powered merchant ship under ship-to-ship and aircraft collision scenarios. A nonlinear finite element model based on a 300K VLCC was developed, with and without reinforced concrete in the double-hull side structure. Material strain-rate effects and contact interactions were considered. Validation was conducted using four benchmark experiments on steel and concrete structures. Results show that reinforced concrete significantly improves impact resistance, particularly under high-velocity impact, suggesting its potential for enhancing the safety of nuclear-powered vessels.

1 INTRODUCTION

In response to increasingly stringent environmental regulations, nuclear-powered vessels are being actively explored as a sustainable alternative to reduce carbon emissions in the maritime industry. These vessels offer numerous advantages, including low operational emissions and extended cruising range without the need for refueling. As interest in their practical deployment grows, ensuring the structural safety of nuclear-powered vessels has become a key area of research.

Unlike conventional ships, nuclear-powered vessels require a significantly higher level of structural integrity, particularly in regions surrounding the reactor silo. In the event of a collision or deliberate sabotage, the consequences can be severe ranging from human casualties and environmental damage to major economic losses. Therefore, it is essential to assess and enhance the crashworthiness of these vessels under worst-case impact scenarios.

This study aims to evaluate whether the incorporation of concrete structures can serve as an effective solution to enhance the crashworthiness of nuclear-powered vessels. Focusing on double-hull structure, the analysis investigates the protective role of internal concrete elements particularly in way of the reactor silo under worst-case impact scenarios. The study considers high-energy collisions involving large vessels and aircraft, with variations in impactor mass, velocity, and strain rate. The study focuses on high-energy collision scenarios involving large vessels and aircraft. The findings are expected to offer practical insights into the structural design of next-generation nuclear-powered ships with improved impact.

While reinforced concrete has long been used in land-based infrastructure to provide impact resistance and structural robustness, its application in marine structures particularly within hull structures has remained largely unexplored. In the context of commercial ship design, structural reinforcements have traditionally relied on steel-based systems. This study introduces an alternative approach by investigating the feasibility of embedding reinforced concrete between the inner and outer hulls of nuclear-powered vessels to improve crashworthiness against high-energy impacts.

2 LITERATURE REVIEW

Numerous experimental and numerical studies have been conducted to understand the structural response of ship structures and composite systems under axial and lateral impact loads. These studies have focused on critical aspects such as material modeling, failure criteria, strain rate effects, and structural crashworthiness in both conventional and nuclear-powered ship designs.

Choung & Cho (2008) emphasized the need for true stress correction in tensile tests to improve material accuracy in finite element analysis. Ehlers et al. (2008) benchmarked failure criteria for ship side structures, showing their influence on crashworthiness predictions.

Hogström et al. (2009) and Hogström & Ringsberg (2012) studied strain-rate sensitivity and fracture behavior under complex stress states, highlighting the role of strain state and size effects in thin-walled steel structures.

DOI: 10.1201/9781003684404-4

Huang et al. (2024) investigated the axial performance of square concrete-filled double-skin tubular columns, offering insights into hybrid concrete–steel behavior. Kim et al. (2017) conducted projectile impact tests on SC walls, identifying failure modes such as penetration, debonding, and spalling under high-speed loading.

Kim et al. (2024) proposed a collision-resistant design for nuclear-powered containership using small modular reactors. Paik (2007a, 2007b) presented key modeling and verification methods for ship collision simulations, forming a basis for current nonlinear FE techniques.

The MARSTRUCT benchmark (Ringsberg et al., 2018) validated impact simulations across multiple researchers using a standardized indenter test. Lastly, NRC (2009) guidelines stress the need for assessing aircraft impacts in nuclear designs, which is directly relevant to the present study.

3 SCENARIO FOR CASE STUDIES ON STRUCTURAL CRASHWORTHINESS ANALYSIS

A nuclear-powered vessel can cause catastrophic consequences in the event of a collision or terrorist attack, including radioactive leakage, severe structural damage, and large-scale environmental disasters. Considering these risks, the present study assumes a worst-case scenario to evaluate the structural crashworthiness of such a vessel.

The struck structure is a hypothetical nuclear-powered merchant ship with a double-hull. The model was developed based on the mid cargo hold of a Very Large Crude Oil Carrier (VLCC) and was extended longitudinally to include three cargo holds in total, ensuring sufficient simulation domain for the impact analysis. Additionally, to enhance the impact resistance of the struck structure, a scenario incorporating concrete reinforcement within the double-hull side structure was also considered.

Two types of striking structures were assumed in this study. The first striking structure is a VLCC-class double-hull oil tanker, with its bow structure colliding against the side structure of the struck vessel. This scenario represents the most severe ship-to-ship collision case, reflecting maximum mass and kinetic energy transfer, and aims to replicate worst-case structural damage.

The second striking structure is an aircraft colliding with the struck vessel. This scenario is inspired by real-world incidents such as the 9/11 terrorist attacks. While the probability of such an event is low, its potential consequences are catastrophic, involving high-speed impact and severe localized structural failure. This scenario aims to assess the crashworthiness of the nuclear-powered vessels under extreme but plausible threat conditions.

In the present study, four cases are investigated. In Case 1, the space in the double-hull side structure is void of any ballasting material, and the striking structure is a VLCC. In Case 2, the space is filled with concrete, and the striking structure is the same VLCC.

In Case 3, the space is void, but the striking structure is an aircraft. In Case 4, the space is filled with concrete, and the striking structure is an aircraft. The details of each case are summarized in Table 1.

Table 1. Summary of impact scenarios considered in the present study.

Case	Striking structure	Ballasting material
1	VLCC	Empty (no ballasting material)
2	VLCC	Reinforced concrete
3	Aircraft	Empty (no ballasting material)
4	Aircraft	Reinforced concrete

As the internal stiffeners are embedded in the concrete and interact structurally during impact, the system can be regarded as a composite structure equivalent to reinforced concrete.

4 VALIDATION OF NUMERICAL METHOD

To ensure the reliability of the numerical approach, key aspects requiring validation include strain rate-dependent material behavior, failure mechanisms, and the interaction between different materials under impact loading. In the present study, the numerical method was validated through four representative experiments covering a range of loading rates and structural configurations:

1. Concrete-filled Steel Tubular Column (Quasi-static strain rates)
2. Projectile Impact Test on Steel-plate Concrete Wall (High strain rates)

These cases were selected to cover the critical deformation and failure modes of both steel and reinforced concrete structures across different strain rate regimes, providing a robust foundation for validating the proposed numerical methodology.

4.1 *Concrete-filled steel tubular column (Quasi-static strain rates)*

To validate the numerical method under quasi-static conditions, a compression test on a concrete-filled steel tubular (CFST) column was simulated and compared against the experimental results reported by Huang et al. (2024). The same loading and boundary conditions were applied in the present study to ensure a direct validation of the numerical simulation technique under axial compression. The interaction between the concrete core and the steel tube was modeled using surface contact, which was deemed sufficient to capture the confinement effect and load transfer mechanism observed in the experiment.

The numerical model for the CFST column, including the validation against experimental results under quasi-static loading, was previously established and verified in our earlier study Kim et al. (2024). In the present work, a summary is provided for continuity.

4.1.1 *Experimental description*

The experimental setup investigates the axial compressive behavior of short concrete-filled steel tubular (CFST) columns under quasi-static loading conditions. The specimens consist of circular hollow steel tubes filled with normal-strength concrete. Axial compression is applied using a hydraulic actuator at a slow loading rate of approximately 0.2 mm/min to simulate quasi-static strain rates. Load–displacement curves, failure modes, and strain distributions are recorded during the test. Typical cross-sections of the stiffened CFST columns are shown in Figure 1, and the detailed specifications of the test specimens are summarized in Table 2, both reproduced from Huang et al. (2024).

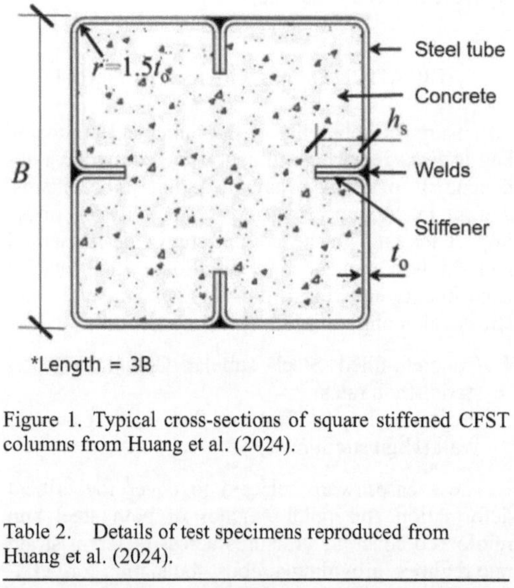

Figure 1. Typical cross-sections of square stiffened CFST columns from Huang et al. (2024).

Table 2. Details of test specimens reproduced from Huang et al. (2024).

	B (mm)	L (mm)	t_o (mm)	h_s (mm)
SC10	160	480	2	30

4.1.2 *Numerical modeling*

A finite element model is developed using LS-DYNA to simulate the compressive behavior of CFST columns. The steel tube is modeled using shell elements with an elastoplastic material model that includes strain hardening. The concrete core is modeled with solid elements using *MAT159 (CSCM Concrete)* to account for crushing and confinement under multiaxial stress states.

The material properties of the steel and concrete used in the tests are presented in Tables 3 and 4, respectively.

The steel-concrete interaction is defined with LS-DYNA's *Automatic Surface to Surface* contact. The bottom end of the column is fixed, while displacement is applied on the top end to replicate experimental boundary conditions.

Table 3. Material properties of steel.

Position	σ_Y	σ_U (MPa)	E_s (GPa)	δ (%)
Flat plate	280	421	207	51.5
Corner	380	436	205	26.3

Table 4. Concrete mix proportions (kg/m^3).

Cement	Water	Medium sand	Coarse aggregate	Water reducer	f_{cu} (MPa)
410	190	720	1080	0.41	41.2

4.1.3 *Comparison between experiment and simulation*

The ultimate strength of numerical results shows good agreement with the experimental load-displacement curves.

As shown in Figure 2, snap-through behavior was observed in the post-buckling response of the NLFEA, which is a known numerical artifact commonly associated with explicit finite element solvers. Since the focus of this study is on high-energy impact scenarios rather than detailed post-buckling behavior, this phenomenon does not significantly affect the overall conclusions.

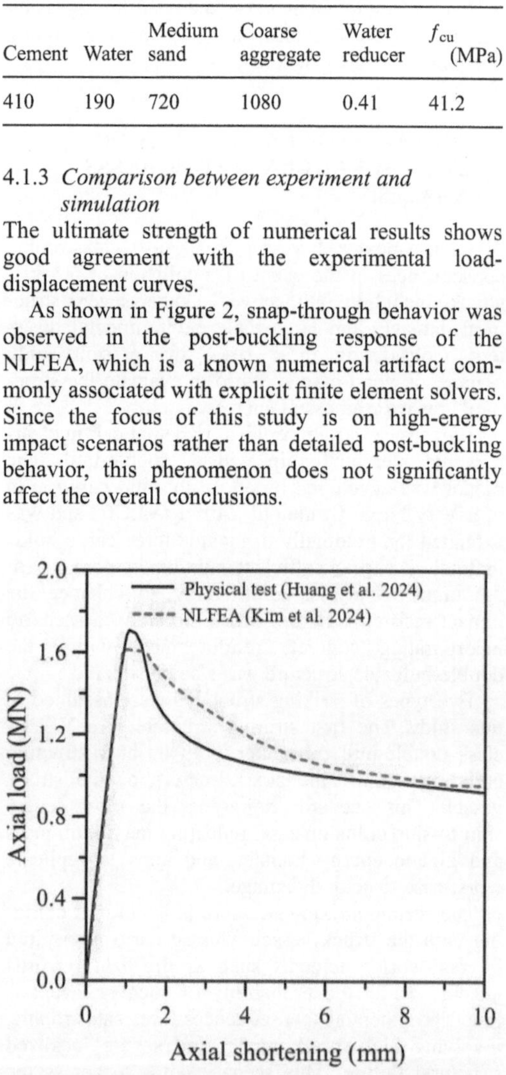

Figure 2. Comparison of the deformed shapes and the ultimate compressive strength behavior obtained from the physical test performed by Huang et al. (2024) and the NLFEA conducted by Kim et al. (2024).

4.1.4 *Discussion*

The CFST demonstrates strong confinement characteristics under quasi-static loading. The steel tube not only enhances the axial strength but also significantly improves ductility by preventing early concrete failure. The numerical simulation using the CSCM concrete model effectively replicates the experimentally observed global behavior and local failure mechanisms.

This validation case supports the applicability of the modeling technique for simulating confined concrete behavior, and it forms a foundational benchmark for subsequent dynamic impact studies.

4.2 *Projectile impact test on steel-plate concrete well (high strain rates)*

To simulate an aircraft impact scenario on a nuclear-powered merchant ship, it is essential to accurately capture material behavior under high strain rate conditions. Aircraft collisions involve extremely high impact velocities, resulting in dynamic structural responses that differ significantly from quasi-static conditions. Therefore, validation of the numerical model in the high strain rate regime is crucial to ensure reliable prediction of structural performance.

Of particular importance is the interaction between the concrete core and the stud bolts, which play a critical role in ensuring composite action between the steel plates and the concrete. Accurately modeling this interaction is essential to replicate the local failure mechanisms and overall structural behavior under impact loading.

For validation, the residual velocity of the projectile after perforation was used as the key metric, allowing direct comparison between the experimental results and the numerical simulation. The present validation uses this specific configuration as the benchmark for confirming the numerical model's capability to accurately simulate high-velocity impact scenarios. Through this validation process, the reliability of the high strain rate analysis methodology is confirmed, supporting the assessment of the crashworthiness of the nuclear-powered merchant ship under extreme impact conditions.

4.2.1 *Experimental description*

In this study, validation is conducted based on the projectile impact test of Steel-Plate Concrete (SC) walls reported by Kim et al. (2017). The referenced study performed full-scale experiments and numerical analyses on SC walls subjected to high-speed projectile impacts, focusing on a configuration with a 175 mm concrete thickness and a 4 mm steel plate. The experiments provided detailed data including initial projectile velocity (approximately 150 m/s), penetration and perforation behavior, plate damage patterns, rear plate deformation, and concrete failure modes.

Figure 3. illustrates the configuration of the steel-plate concrete (SC) well specimen used in the experiment, as reported by Kim et al. (2017). This experiment investigates the high-strain-rate behavior of an SC composite wall subjected to projectile impact. The wall consists of front and back steel plates sandwiching a reinforced concrete core.

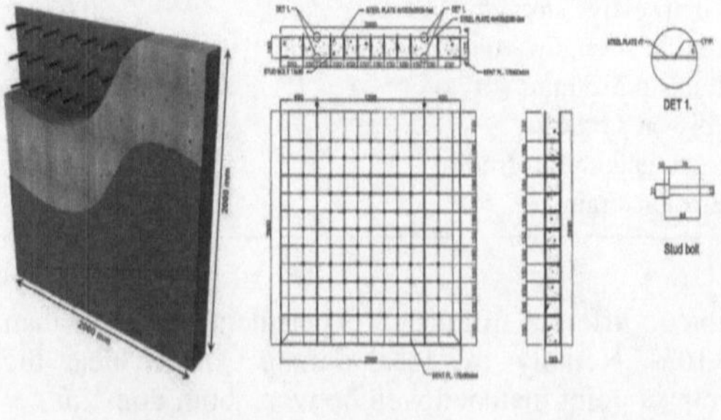

Figure 3. Configuration of the SC Panel Specimen from Kim et al. (2017).

4.2.2 *Numerical modeling*

Numerical analysis was conducted using LS-DYNA with explicit dynamic simulation settings. The SC wall was modeled with shell elements for the steel plates, solid elements for the concrete core, and truss elements for internal reinforcements. Material models included:

- **Steel plates and shear stud**: *MAT24* Cowper–Symonds model incorporating strain-rate effects.
- **Concrete**: *MAT159 CSCM Concrete* with rate-dependent softening and damage evolution.
- **Contact interactions**: The steel-concrete interface was modeled using *CONSTRAINED_LAGRANGE_IN_SOLID* or *TIED_SURFACE_TO_SURFACE* depending on the interface behavior.

The projectile was defined with solid elements and given an initial velocity corresponding to the experimental conditions.

Table 5. Material properties of steel.

Material property		Plate	Shear stud	Missile
Density ρ (ton/m^3)		7.80	7.80	7.80
Elastic modulus (MPa)		210,000	210,000	210,000
Poisson's ratio ν		0.3	0.3	0.3
Yield stress σ_Y (MPa)		326	345	250
Tangent modulus (MPa)		387	504	502
Cowper-Symonds parameter	C	40	40	40
	q	5	5	5

4.2.3 *Comparison between experiment and simulation*

The numerical results were compared with the experimental data in terms of:

The simulations reproduced the onset of penetration, localization of stress near the impact site, and progressive failure, showing good agreement with the experimental results. The predicted peak impact

Table 6. Material properties of concrete.

Material property	Concrete
Density ρ (ton/m^3)	2.37
Compressive strength (MPa)	50.075
Tensile strength (MPa)	5.075
Tangent Modulus (MPa)	31000
Poisson's ratio ν	0.19
Aggregate radius (mm)	10
Erosion strain	0.3

force differed from the experiment by less than ±10%. Notably, the local damage shapes near the impact point matched well between both domains.

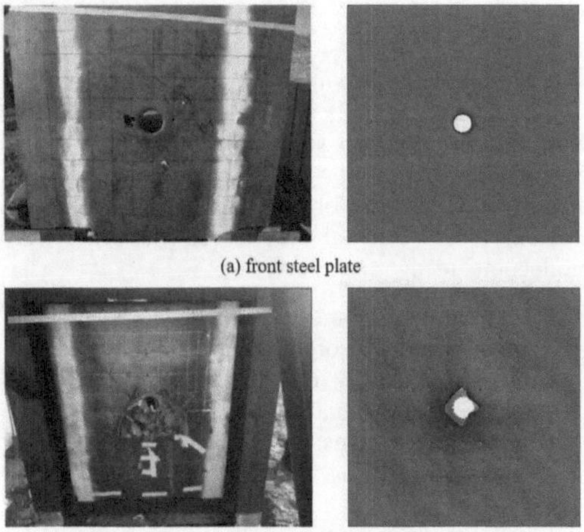

Figure 4. Local impact behavior of SC Panel Specimen.

Table 7. Summary of Test and Simulation Results.

	Initial velocity	Residual velocity
Experiment	147.5	85.0
Kim et al. (2017)		58.3
Present study		76.8

* Unit: m/s

4.2.4 *Discussion*

The validated numerical model successfully captured the dynamic response of the SC wall under high-speed impact conditions. The inclusion of strain-rate sensitivity and nonlinear failure criteria for concrete and steel was critical for accurate predictions. Sensitivity analysis showed that variations in rear plate thickness and concrete strength significantly influenced energy absorption and back-face damage. These insights support the applicability of such modeling approaches for design and optimization of impact-resistant SC structural components.

5 NUMERICAL MODELING FOR CRASHWORTHINESS ANALYSIS

5.1 *Striking vessel - hypothetical 300K VLCC*

The striking vessel considered in this study is a hypothetical 300K VLCC, designed in compliance with the Common Structural Rules (IACS, 2020) using the principal dimensions and hull form of the KVLCC2. The KVLCC2 model was originally developed and published by the Korea Research Institute of Ships and Ocean Engineering (KRISO), and its geometric data are openly available.

The structural scantlings were determined based on global longitudinal strength, local strength, and midship cargo hold FE analysis, conducted using Bureau Veritas software tools MARS and VeriSTAR Hull. It should be noted that fatigue strength assessment was not included in the current scope.

5.2 *Striking aircraft - Hypothetical Boeing 777*

The aircraft model used as the striking structure is a hypothetical Boeing 777, directly adopted from the model developed and proposed by Paik (2022). The geometric configuration and principal specifications of the aircraft are illustrated in the following Figure 5 and summarized in Table 8.

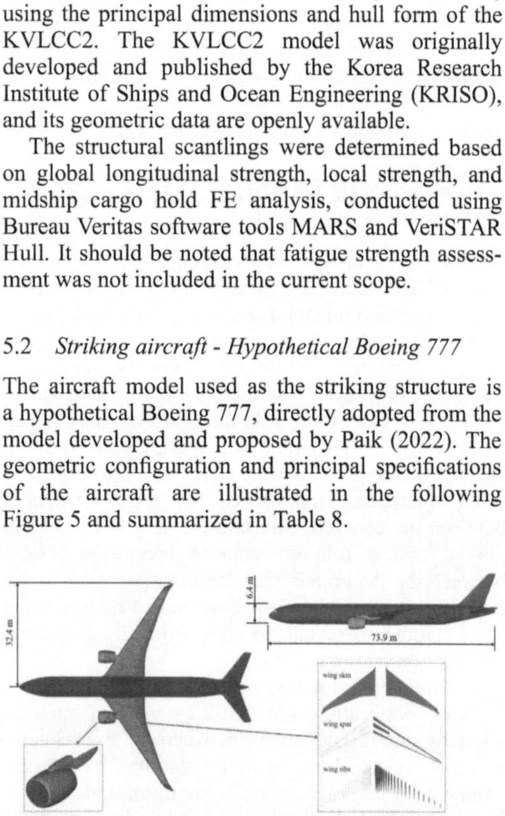

Figure 5. Principal dimensions of a hypothetical Boeing 777, adapted from Paik (2022).

Table 8. Structural scantlings of the fuselage and wings of a hypothetical Boeing 777, adapted from Paik (2022).

Member	Thickness	Member	Thickness
Outer skin	6	Deck	10
Inner skin	6	Bulkhead	20
Frame	10	Rudder	6
Stringer	6	Stabilizer	6

* Unit: mm

5.3 *Struck vessel - hypothetical nuclear-powered merchant ship*

The struck vessel in this study is a hypothetical nuclear-powered merchant ship, modeled by longitudinally extending the mid cargo hold of the striking

VLCC to include three cargo holds (Figure 6). While actual design nuclear-powered merchant ships have not been opened to the public, and detailed scantling information such as the location of the reactor silo and surrounding structural arrangements is unavailable, the proposed model is considered sufficient to examine the overall trend in crashworthiness. A schematic view of the modeled struck structure is shown in Figure 7. This model enables a comparative evaluation of the structural response with and without reinforced concrete in the double-hull side structure.

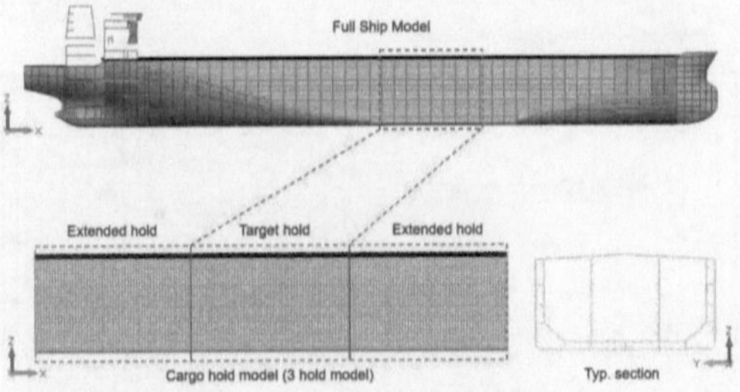

Figure 6. Extension of the mid cargo hold from the VLCC to construct a three-hold model for the struck nuclear-powered merchant ship.

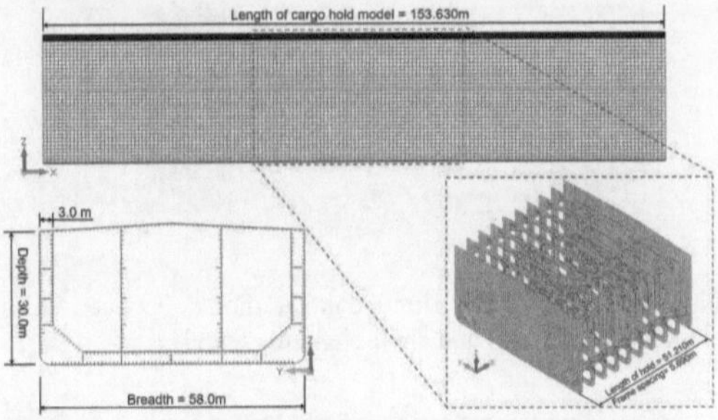

Figure 7. Main dimensions of the hypothetical nuclear-powered merchant ship used as the struck structure.

In the present study, the midship region of the struck vessel is assumed to house the reactor silo, as illustrated in Figure 8. The reinforced concrete is placed between the shell and the inner hull

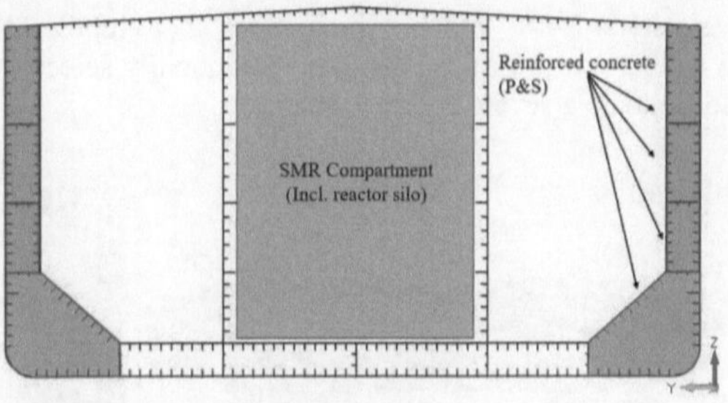

Figure 8. SMR compartment and reinforced concrete region of the hypothetical nuclear-powered merchant ship used as the struck structure.

5.4 *Material models*

The material properties used in the numerical models are summarized in Tables 6, 9, and 10. The steel structures of both the striking and struck ships were modeled using an elastic–plastic material with strain rate sensitivity, as detailed in Table 9. This includes properties such as yield strength, ultimate tensile strength, Young's modulus, and strain-rate parameters.

The reinforced concrete used in the double-hull side structure was modeled with a strain rate sensitive concrete material model that accounts for compressive and tensile behavior, which is same model as *4.2 Projectile Impact Test on Steel-plate Concrete well.* Material properties are listed in Table 6

For the aircraft model, aluminum alloy and composite materials were assumed for the fuselage and wing structures, respectively. The material properties, including density, modulus, are listed in Table 10.

Table 9. Material properties of the striking and struck ship structures.

Material property		Mild steel	High tensile steel	
			HT32	HT36
Density ρ (ton/m^3)		7.85	7.85	7.85
Elastic modulus (MPa)		210,000	210,000	210,000
Poisson's ratio ν		0.3	0.3	0.3
Yield stress σ_Y (MPa)		235	315	250
Tangent modulus (MPa)		387	504	502
Cowper-Symonds	C	40	3200	3200
parameter	q	5	5	5

Table 10. Material properties of aircraft.

Material property		Al2024-T3	Mild	HT36
Density ρ (ton/m^3)		2.85	7.85	7.85
Elastic modulus (MPa)		72,000	210,000	210,000
Poisson's ratio ν		0.33	0.3	0.3
Yield stress σ_Y (MPa)		400	235	355
Tangent modulus (MPa)		600	387	504
Cowper-Symonds	C	610.4	3200	3200
parameter	q	3.6	5	5

5.5 *Boundary and initial condition*

The boundary and initial conditions applied to the collision simulation are illustrated in Figure 9. To constrain global motion, all six degrees of freedom were fixed at both longitudinal ends of the struck vessel. The striking VLCC was assigned an initial velocity of 18 knots, which does not represent a typical cruising speed but was intentionally selected to simulate a severe collision scenario.

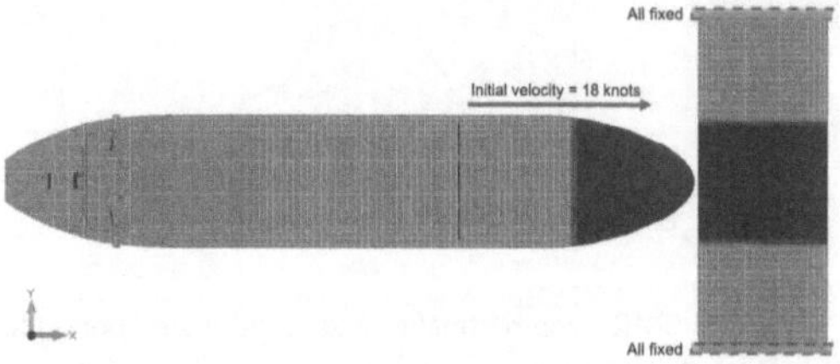

Figure 9. Boundary and initial condition of striking vessel and struck vessel.

The boundary and initial conditions for the aircraft impact simulation are shown in Figure 10. As in the ship collision scenario, all six degrees of freedom were constrained at both longitudinal ends of the struck vessel to prevent global motion. The striking aircraft was assigned an initial velocity of 200 m/s, representing a high-speed impact condition relevant to external threats such as missile or aircraft debris.

Striking aircraft is positioned above 24 m from baseline considering draft of struck vessel, diameter of aircraft and spacing between upper deck and no.1 stringer.

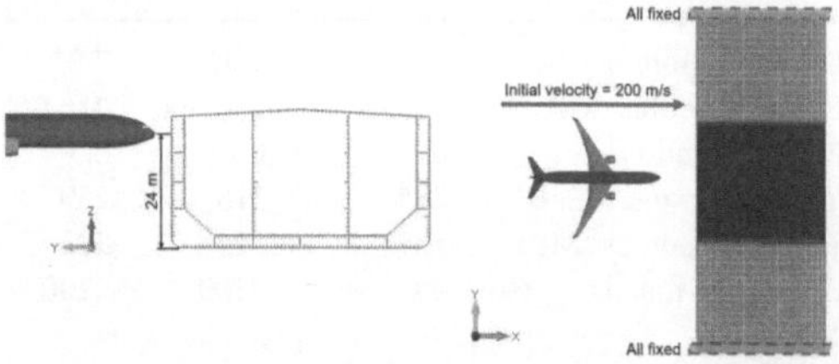

Figure 10. Boundary and initial condition of striking aircraft and struck vessel.

5.6 *Contact model*

Contact interactions were defined to simulate the physical interaction between the striking and struck structures during impact. As shown in Figure 11, contact was modeled between the hull of the striking vessel and the side structure of the struck vessel. For the aircraft impact scenario, a similar contact definition was applied between the aircraft body and the ship structure, as illustrated in Figure 12.

To accurately represent the interaction between dissimilar materials, constrained-based contact models were used between solid elements representing reinforced concrete and shell elements representing steel plates. This coupling approach, depicted in Figure 13, allows for effective force transmission and relative motion constraints at the interface between concrete and steel components.

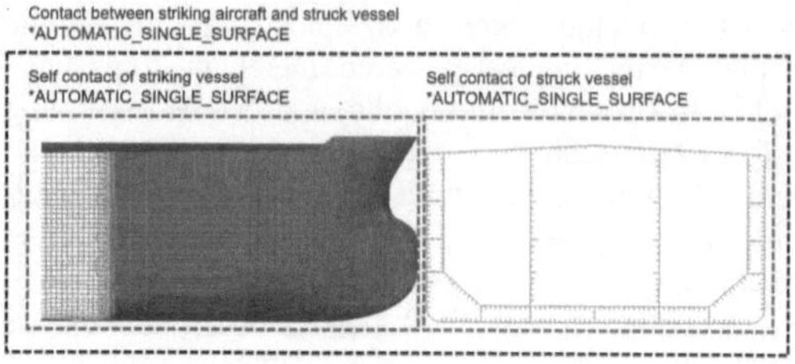

Figure 11. Contact model between striking vessel and struck vessel.

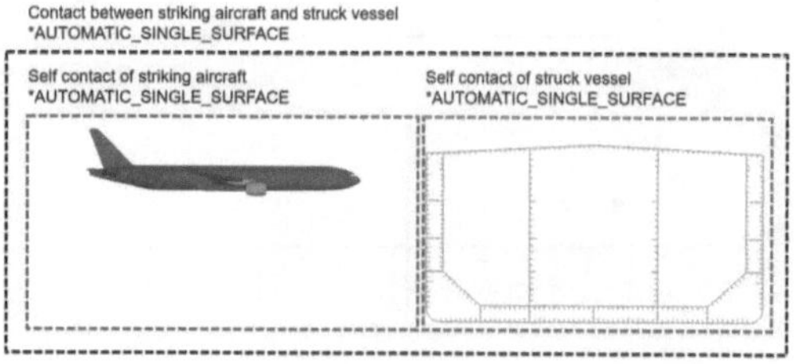

Figure 12. Contact model between striking aircraft and struck vessel.

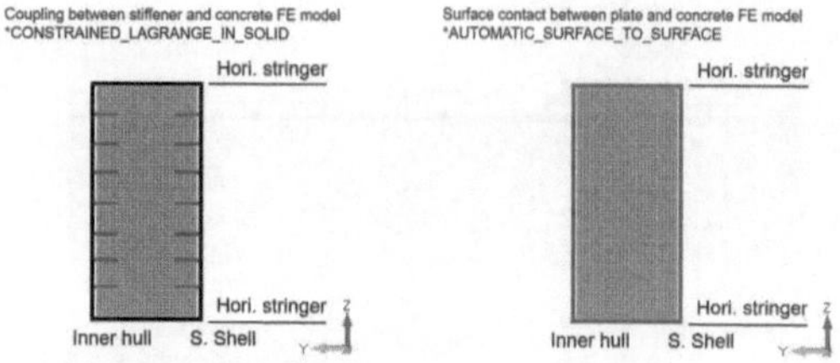

Figure 13. Constrained and contact model between solid elements (concrete) and shell elements (steel).

6 CASE STUDIES ON STRUCTURAL CRASHWORTHINESS ANALYSIS

Figure 14 illustrates the relationship between absorbed energy and penetration depth in a ship-to-ship collision scenario, where a VLCC strikes the side of a nuclear-powered merchant ship. Case 1 assumes no ballasting material in the side structure, while Case 2 includes reinforced concrete. The results show that concrete ballast enhances the structure's energy absorption capacity, although it also permits greater deformation. This highlights the role of internal fill in improving crashworthiness under lateral ship impact.

Figure 15 presents the absorbed energy versus penetration response for an aircraft impact scenario, where a hypothetical Boeing 777 collides with a nuclear-powered merchant vessel. Case 3 represents the unreinforced double-hull configuration, and Case 4 incorporates reinforced concrete. The reinforced configuration (Case 4) demonstrates significantly greater energy absorption capacity, particularly under high-speed impact conditions. This emphasizes the structural advantage of concrete reinforcement in mitigating damage from extreme external threats such as aircraft collision.

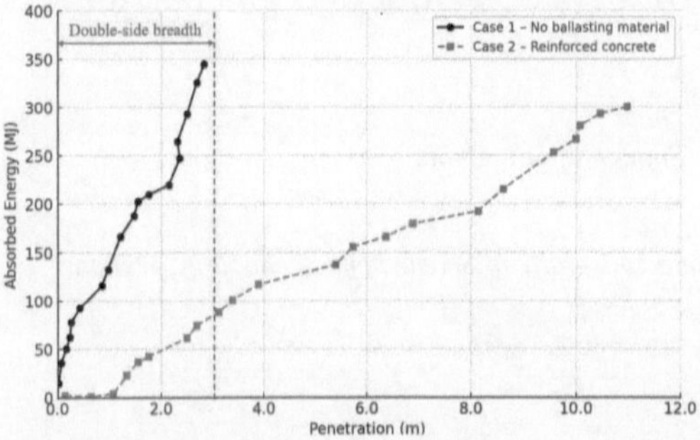

Figure 14. Comparison of absorbed energy versus penetration for Case 1 (no ballasting) and Case 2 (reinforced concrete).

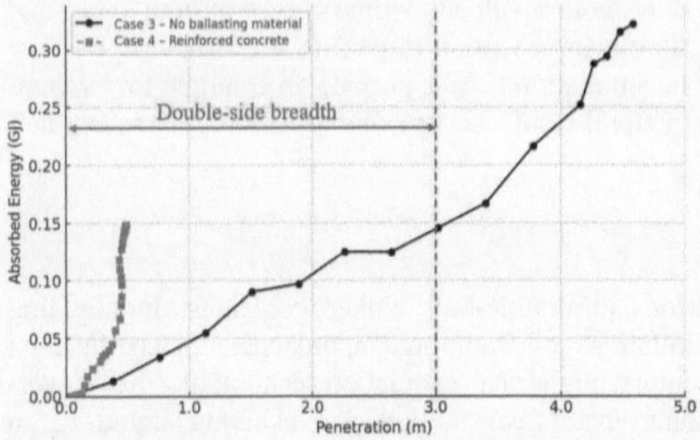

Figure 15. Comparison of absorbed energy versus penetration for Case 3 (no ballasting) and Case 4 (reinforced concrete).

7 CONCLUSIONS

This study evaluated the crashworthiness of double-hull nuclear-powered vessels by incorporating reinforced concrete structures between the inner and outer hulls. Finite element simulations of high-energy collisions, including ship and aircraft impacts, demonstrated that the presence of internal concrete significantly enhanced the structural resistance to penetration.

Notably, cases with concrete-filled compartments did not exhibit perforation of the inner hull, whereas models without concrete reinforcement failed under similar impact conditions. These findings suggest that reinforced concrete can act as an effective protective layer in worst-case collision scenarios involving nuclear-powered ships.

While the analysis provides valuable insights, the modeling still involves assumptions regarding fracture criteria, strain-rate effects, and material behavior under extreme loading.

This study is focused on crashworthiness, therefore the additional weight due to concrete was not calculated. Future study will aim to investigate the hull girder strength due to the weight of concrete and the additional strength provided by the concrete girder in terms of bending, shear, and buckling.

REFERENCES

Choung, J.M. and Cho, S.R. (2008). Study on true stress correction from tensile tests. Journal of Mechanical Science and Technology, 22, pp.1039–1051.

Ehlers, S., Broekhuijsen, J., Alsos, H.S., Biehl, F. and Tabri, K. (2008). Simulating the collision response of ship side structures: a failure criteria benchmark study. International Shipbuilding Progress, 55(1), pp.127–144.

Hogström, P., Ringsberg, J.W. and Johnson, E. (2009). An experimental and numerical study of the effects of length scale and strain state on the necking and fracture behaviours in sheet metals. International Journal of Impact Engineering, 36(1), pp.1194–1203.

Hogström, P. and Ringsberg, J.W. (2012). An extensive study of a ship's survivability after collision: a parameter study of material characteristics, nonlinear FEA and damage stability analyses. Marine Structures, 27(1), pp.1–28.

Huang, W.F., Shao, Y.B., Hassanein, M.F., Hadzima-Nyarko, M., Radu, D. and Cashell, K.A. (2024). Experimental and numerical investigation of square concrete-filled double-skin steel stiffened tubular stub columns with CHS inner tubes under axial compression. Thin-Walled Structures, 2024, 111792.

Kim, H.J., Seo, D.H., Mujeeb-Ahmed, M.P., He, K., Dong, Y., Wang, F. and Paik, J.K. (2024). Collision-accidental safety design and engineering of a 25,800 TEU containership powered by small modular reactors. In: Proceedings of the International Conference on Ships and Offshore Structures (ICSOS 2024), 8–12 September 2024, Chennai, India.

Kim, K.S., Moon, I.H., Choi, H.J. and Nam, D.W. (2016). A preliminary study on the local impact behavior of steel-plate concrete walls. Nuclear Engineering and Design, 310, pp.336–351.

NRC (2009). Consideration of aircraft impacts for new nuclear power reactors. Nuclear Regulatory Commission, RIN 3150 –AI19, Washington, DC.

Paik, J.K. (2007a). Practical techniques for finite element modeling to simulate structural crashworthiness in ship collisions and grounding (Part I: Theory). Ships and Offshore Structures, 2(1), pp.69–80.

Paik, J.K. (2007b). Practical techniques for finite element modeling to simulate structural crashworthiness in ship collisions and grounding (Part II: Verification). Ships and Offshore Structures, 2(1), pp.81–86.

Paik, J.K. (2022). *Ship-Shaped Offshore Installations: Design, Construction, Operation, Healthcare and Decommissioning*. 2nd ed. Cambridge: Cambridge University Press.

Ringsberg, J.W., Amdahl, J., Chen, B.Q., Cho, S.R., Ehlers, S., Hu, Z., Kubiczek, J.M., Kõrgesaar, M., Liu, B., Marinatos, J.N., Niklas, K., Parunov, J., Rudan, S., Samuelides, M., Guedes Soares, C., Tabri, K., Villavicencio, R., Yamada, Y., Yu, Z. and Zhang, S. (2018). MARSTRUCT benchmark study on nonlinear FE simulation of an experiment of an indenter impact with a ship side-shell structure. Marine Structures, 59, pp.142–157.

Trends in Collision and Grounding of Ships and Offshore Structures – Zhu & Guedes Soares (Eds)
© 2026 The Author(s), ISBN 978-1-041-16361-9

Numerical simulation study on collision of a type A liquefied gas carrier

H. Wang, L. Lin & G. Wang
Shanghai Rules and Research Institute CCS, Shanghai, P.R. China
Ships and Offshore Installations Structural Safety Laboratory, Shanghai, P.R. China

Y.J. Zhao & L. Zhu
School of Naval Architecture and Ocean Engineering, Huazhong University of Science and Technology, Wuhan, P.R. China

ABSTRACT: This paper takes a single side Type A VLGC (Very Large Gas Carrier) as the target vessel and establishes a ship collision simulation analysis model. Numerical simulations were conducted on the collision process where the VLGC is struck respectively by two container ships with a bulbous bow or straight bow structure. The study investigates the motion and structural response of the struck ship under different bows and varying striking speeds. Key results include the impact force - time history curves, damage characteristics of the struck ship's side structures, roll angle - time history curves of the struck ship, motion velocity - time curves of both ships, and energy absorption curves of specific structural components. Particular attention is paid to the structural damage locations, severity, and breach shapes of the struck VLGC. The simulation results provide an example for evaluating the anti-collision capacity of the VLGC and the breach structural results serve as input for the subsequent natural leakage of cryogenic cargo and flooding on VLGC.

1 INTRODUCTION

In recent years, with the development of economic globalization and the maritime shipping industry, the global market has imposed higher demands on maritime transportation volume and efficiency. However, this has also brought about multifaceted safety risks that cannot be ignored. According to data from the IMO Global Integrated Shipping Information System (GISIS) database, ship collisions account for approximately 20% of all maritime accidents, making them the most frequent type of accident. These collisions not only result in casualties and economic losses but also cause severe damage to marine ecosystems due to onboard cargo and fossil fuel leakage. If a ship sinks, it poses significant threats to the navigation safety of subsequent vessels. Therefore, research on ship collision risk analysis is crucial for ensuring the safety of liquefied gas carrier transportation and enhancing the economic efficiency of maritime operations.

The ship collision research framework originated from Minorsky (1959) deformation energy theory, which decomposes collision energy into external hydrodynamic forces and structural dissipation components. During the 1970s, the seven-dimensional parameter space theory (considering displacement, relative velocity, bow type, etc.) and Woisin (1979) empirical formula established a classical analytical framework. However, these methods still exhibited 35% prediction errors in double-hull tanker collisions, highlighting limitations of traditional approaches. Since the 21st century, numerical simulation technologies have overcome theoretical bottlenecks, ushering liquid cargo vessel collision research into a high-precision computational era. The core of structural collision simulation calculations lies in fluid-structure interaction (FSI) modeling and nonlinear dynamic algorithms. Wang et al. (2001) established a full-scale refined ship model using explicit transient algorithms, confirming that the kinetic energy loss of the striking vessel is predominantly converted into plastic deformation energy of the struck ship's side structure, with minimal energy absorbed by rigid-body motion of the struck ship or fluid media. In terms of algorithmic innovations: Song et al. (2017) applied ALE FSI analysis technique using LS-DYNA code to numerically simulate collisions between two identical ships, explicitly modeling surrounding fluid flow and investigating the effects of the struck ship's forward speed, mass, and collision angle on collision responses; Feng et al. (2023) simulated the collision process between large floating ice and marine structures using a coupled SPH-FEM model, with numerical results for impact forces, wave disturbances, and structural deformations aligning closely with experimental data. According to Qu et al. (2022), quasi-static compression ratio tests showed strong consistency with LS-DYNA numerical simulations, accurately reflecting low-speed collision characteristics and

DOI: 10.1201/9781003684404-5

structural deformation modes, thereby validating the numerical method. Using of local structural models achieved sufficient accuracy while significantly reducing modeling and computational effort. Additionally, related research focuses on data-driven intelligent simulation: Zhang et al. (2025) developed deep learning-based surrogate models using AIS data to identify collision scenarios, simulated damage extents via the super-element method, and established a database linking collision scenarios with damage assessments. The trained model completed predictions in under 0.1 seconds, a drastic improvement over traditional 10-minute simulations.

This paper takes a single side Type A VLGC (Very Large Gas Carrier) as the target vessel and establishes a ship collision simulation analysis model. Numerical simulations were conducted on the collision process where the VLGC is struck by two types of the bow of a container ship with a bulbous bow or straight bow structure. The study investigates the motion and structural response of the struck ship under different bows and varying striking speeds. Particular attention is paid to the structural damage locations, severity, and breach shapes of the struck VLGC. The simulation results provide an example for evaluating the anti-collision capacity of the VLGC and the breach structural results serve as input for the subsequent natural leakage of cryogenic cargo and flooding on VLGC.

2 SHIP COLLISION SIMULATION METHODS

Ship collision is an extremely complex nonlinear transient response process, with the entire event lasting a short duration. During this period, structural components rapidly transition from the elastic stage to the plastic stage, undergoing significant plastic deformation, damage, and eventual rupture. It also represents a classic fluid-structure interaction (FSI) problem, where the influence of the surrounding fluid field on the collision cannot be ignored. The most precise numerical simulation method for ship collisions is the fluid-structure coupling approach. In such models, the ship structure is simulated using a Lagrangian mesh to track material deformation, while the surrounding water domain is modeled with an Eulerian mesh or Smoothed Particle Hydrodynamics (SPH). The interaction between these two domains is established through coupling surfaces to transfer loads. However, due to the time-consuming nature of fluid-structure coupling modeling and computations, simplified methods are often adopted for batch scenario analyses. Specifically, the undamaged hull structure is simplified using the equivalent hull girder method, where intact sections are abstracted as beam elements and integrated into the finite element model of the collision zone. Meanwhile, hydrodynamic effects are approximated through the added mass method, which accounts for water influence by assigning virtual mass to the hull structure, eliminating the need for explicit fluid domain modeling. This simplification significantly reduces computational resources and time while maintaining engineering accuracy. In this study, the equivalent hull girder method and added mass method will be employed to simulate and analyze ship collisions.

3 SIMULATION SETUP

3.1 *Ship information*

Select a medium-sized VLGC (Very Large Gas Carrier) ship as the struck vessel, and conduct collision simulation analysis to study the structural damage characteristics and motion response of its No.2 cargo tank area after collision. The ship adopts the IMO Type A cargo containment system for transporting liquefied petroleum gas (LPG), with the general arrangement shown in Figure 1. Its characteristic is having 4 pairs of cargo tanks arranged from stern to bow. Since the damage characteristics and breach shapes of the struck tank area are closely related to the bow shape of the striking ship, there are significant differences in damage patterns under impacts from different bow types: bulbous bows may cause large breaches on the struck ship's side structure, while straight bows may induce extensive plastic deformation. Based on domestic and international research, the most severe structural damage occurs when the striking and struck ships have comparable displacements. Therefore, two container ships with similar displacement to the Type A VLGC are selected as striking ships: 5618 TEU (Striking Ship 1 with bulbous bow) and 4600 TEU (Striking Ship 2 with straight bow). The displacements, drafts, and lengths of all three ships are listed in Table 1.

Table 1. Principal dimensions of the struck and striking ships.

Ships	Disp./ton	Draft/m	Length/m
Struck Ship	70800	11.6	220
Striking Ship 1	74200	11.3	260
Striking Ship 2	103200	13.5	225

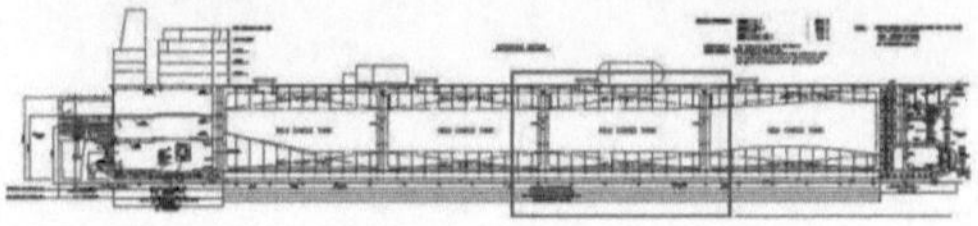

Figure 1. General arrangement plan of the struck Type A VLGC.

3.2 *Modelling of collision models*

A finite element model was established for the collided VLGC, simulating hydrodynamic effects using the added mass method. To simplify calculations, the hull was modeled differently before and after amidship: the forward section incorporated all hull structural components, with both the shell plates and their stiffeners modeled using S4 shell elements, while the aft section retained only the outer hull structure, with adjusted material density properties to ensure accurate mass distribution. MPC coupling was applied to connect the outer hull nodes with structural unit nodes. To capture large deformations and fractures in the impact zone, a 100mm element size was adopted for the longitudinal range from FR135 to FR163, whereas the rest of the hull used an 800mm element size aligned with longitudinal stiffener spacing. The completed VLGC finite element model is shown in Figure 2.

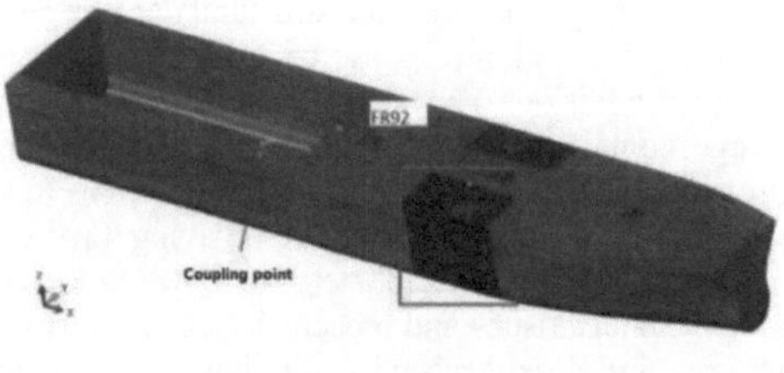

Figure 2. FE model of the struck ship.

A finite element model was established for two types of striking ships, where the hydrodynamic effects were simulated using the added mass method. To enhance computational efficiency, the finite element model comprised two distinct sections: the bow structure was modeled with shell elements, while the section aft of the bow employed an equivalent hull girder method. The beam elements representing the aft section were positioned at the center of gravity height of the striking ship, with nodes near the bow coupled to those of the bow-end face. Mesh refinement was implemented in the forepeak tank area with an element size of 100 mm, while the remaining bow shell elements adopted a coarser 800 mm mesh size. Striking Ship 1 is modeled, as shown in Figure 3.

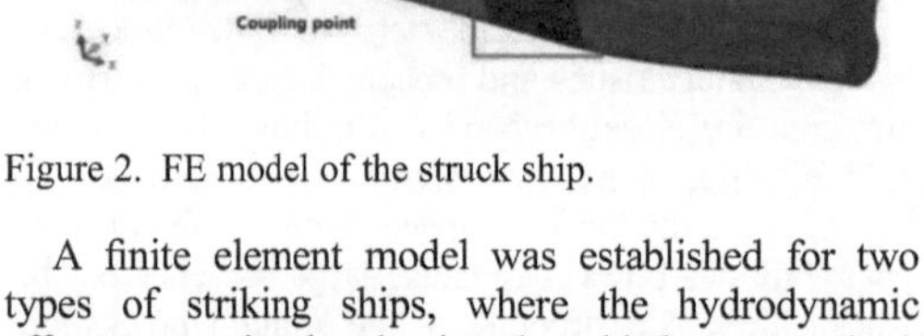

Figure 3. FE Model of the striking ships.

3.3 *Selection of key parameters*

(1) Material Constitutive Models and Failure Criteria

Based on the related studies by Park (2017) and Liu et al. (2013), the bow of the striking ship and the struck cabin model in the present simulation were assigned power hardening material properties. The material reinforcement formula is as follows:

$$\sigma_{eq} = \begin{cases} R_{eH} & \varepsilon_{eq} \leq \varepsilon_{plat} \\ k\varepsilon_{eq}^{n} & \varepsilon_{eq} > \varepsilon_{plat} \end{cases} \quad (1)$$

Where: σ_{eq} and ε_{eq} are the equivalent stress and equivalent strain of the material, respectively, and ε_{plat} is the strain value at the end of the yield stage; R_{eH} is the material's yielding strength, n is the hardening index and k is the reinforced strength.

For the element failure strain in numerical simulations, the following formula is adopted:

$$\varepsilon_F(l_e, t) = 2n + (\varepsilon_u - 2\text{n})\frac{t}{l_e} \quad (2)$$

Where: l_e and t represent the size and thickness of the element, respectively, and ε_u denotes the failure strain value when $l_e/t = 1$, taken as 0.58.

(2) Collision angle and impact location

According to relevant collision studies, when a striking ship impacts the midship area of a struck ship at a perpendicular angle, the majority of the striking ship's lost kinetic energy will be absorbed by the struck ship, resulting in the most severe structural damage to the struck vessel. This paper selects a 90° angle between the striking and struck ships. The longitudinal impact position on the struck ship is set at the midpoint between the transverse bulkhead of No.2 liquid cargo tank and the swash bulkhead, which represents the weakest point in the lateral structure of the entire cargo hold area.

(3) Added mass coefficient

For the added mass coefficients of striking and struck ships in ship collision simulations, Minorsky (1959) proposed recommended ranges for horizontal plane motions: 0.4–1.3 for sway motion and 0.02–0.07 for surge motion. Based on comparative analysis of experimental results, this study adopts an added mass coefficient of 1.0 for the struck ship and 0.04 for the striking ship, both within Minorsky's recommended ranges.

(4) Boundary conditions and load application

No constraints are applied to the struck ship, while the striking ship has its vertical and rotational degrees of freedom constrained. The load is applied by assigning an initial velocity to the striking ship, with the impact simulation time set to 10 seconds. The contact between the two ships is defined as self-contact.

(5)Set of coordinate system

The coordinate system for ship collision adopted in this study is illustrated in Figure 4. The origin is located at the baseline of the mid-longitudinal section at the stern of the struck ship, and a right-handed coordinate system is adopted. The y-axis corresponds to the transverse direction of the struck ship, oriented from the starboard to the port side, representing the impact direction of the striking ship in collision analysis.

4 COLLISION SIMULATION AND ANALYSIS

4.1 *Simulation cases*

The simulation cases for the computational analysis in this paper are listed in Table 2. For Striking Ship 1 with a bulbous bow, there are 4 striking speed cases; for Striking Ship 2 with a straight bow, more high speed 20kn and 25kn are considered to capture the

Figure 4. Coordinate system of collision.

characteristics of structural rupture. In all cases, the striking ship collides with the struck ship at a 90 degree angle, and the impact position along the moulded depth direction is determined by the drafts of both ships.

Table 2. Simulation cases lists of collision simulation.

Striking ship	Simulation case	Striking speed/kn
5618TEU (Bulbous bow)	Case 1-1	5
	Case 1-2	8
	Case 1-3	10
	Case 1-4	11

4.2 *Bulbous bow collision simulation and analysis*

Using the nonlinear finite element analysis method, simulation calculations were conducted for the process of Striking Ship 1 with a bulbous bow structure vertically impacting the struck ship at different speeds. The structural damage to the hull shell plate and cargo tank side plating of the struck ship is shown in Table 3 and Figure 5. From the damage diagrams, it can be observed that at an striking speed of 5 knots, both the hull shell and cargo tank side plating of the struck ship exhibited significant deformation but no rupture. At striking speeds of 8 knots and 10 knots, the hull shell developed a large breach while the cargo tank side plating showed significant deformation without rupture. When the striking speed reached 11 knots, both the hull shell and cargo tank side plating displayed distinct tearing characteristics, the shape and size of two breaches shown in Figure 6.

Table 3. Structural Damage to the VLGC.

Striking speed/kn	Outer-shell plating	Cargo tank side plating
5	Plastic deformation No breakage	Plastic deformation No breakage
8	Breach: 8.45m*5.28m	Plastic deformation No breakage
10	Breach: 11.0m*5.52m	Plastic deformation No breakage
11	Breach: 11.92m*5.67m	Breach: 11.2m*6.25m

(a)Damage to the VLGC outer-shell plating and cargo tank side plating at 5 knot striking speed.

(b)Damage to the VLGC outer-shell plating and cargo tank shell plating at 8 knot striking speed.

(c)Damage to the VLGC outer-shell plating and cargo tank side plating at 10 knot striking speed.

(d)Damage to the VLGC outer-shell plating and cargo tank side plating at 11 knot striking speed.

Figure 5. Structural damage at various striking speed.

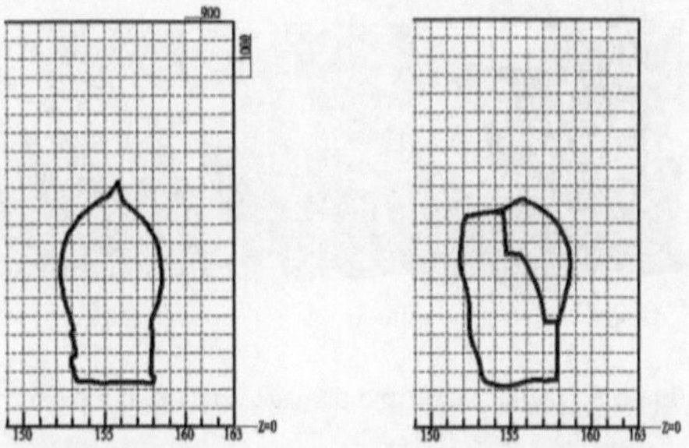

(a)Outer-shell plating (b)Cargo tank side plating

Figure 6. Breach size of VLGC at 11 knot striking speed.

Taking Case 1-4 with a striking speed of 11 kn as an example, analyze the collision process, hull motion, and structural damage.

(1) Process of the collision

In Case 1-4, the relative positions of the striking ship and the struck ship at different moments are shown in Figure 7. The figure reveals that due to the relatively high collision speed, the striking ship's bulbous bow penetrates deeply into the struck ship's side shell. Consequently, the struck ship exhibits significant translational and rotational motion. Moreover, as the bulbous bow deeply intrudes into the struck ship's internal structure, the two ships become difficult to separate.

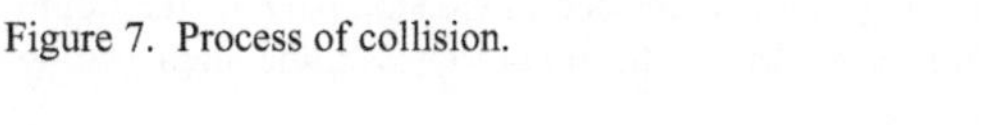

(a)t=0s (b)t=2s (c)t=4s

(d)t=6s (e)t=8s (f)t=10s

Figure 7. Process of collision.

In Cases 1-4, cross-sectional views of the deformation and damage process of the liquid cargo tank at different time steps are shown in Figure 8. As illustrated, after the bulbous bow of the striking ship impacts the hull side and cargo tank side of the struck ship, both the structures of hull side and cargo tank begin to exhibit indentation deformations. At t=2s, the upper inclined bow of the striking ship collides with the deck structure and tank top plate of the struck ship, causing extensive deformation across the entire area. By t=4s, the bulbous bow penetrates deeply into the struck ship, making separation between the two vessels extremely challenging.

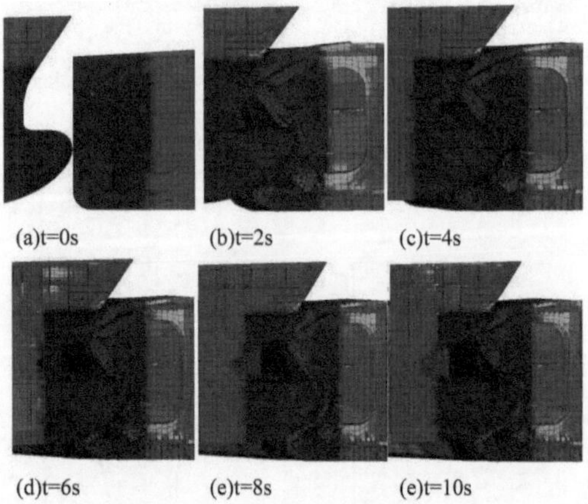

Figure 8. Deformation and damage process of the cargo tank.

(2) Impact force - time history curves

In Cases 1-4, the impact force-time history curves of the two ships in the x-, y-, and z-directions are shown in Figure 9, where the impact force refers to the interaction force acting on the bow hull of the striking ship. It can also be observed from the figure that the impact forces exhibit a fluctuating decrease over a prolonged period after the collision. This is primarily because the high speed of the striking ship results in deep penetration of its bulbous bow into the struck ship, making separation difficult. The rolling motion of the struck ship under the impact force entrains the striking ship, leading to minor variations in Fx and Fz. The larger Fy is attributed to the intense interaction between the flare of the striking ship's bow and the deck and hull side of the struck ship.

(3) Damage of the structures

In Cases 1-4, the damage stress contour plots of the shell plating of the struck ship, the side plate of No.2 liquid cargo tank, and the bow of the striking ship are shown in Figure 10. It can be observed that the shell plating of the struck ship forms a large opening extending 11.92 m along the moulded depth direction and 5.67 m along the ship length direction under the impact of the bulbous bow. Simultaneously, larger openings are also formed in the side plate of the liquid cargo tank, with both the breaches in the shell plating and the liquid cargo tank side exhibiting distinct tearing characteristics.

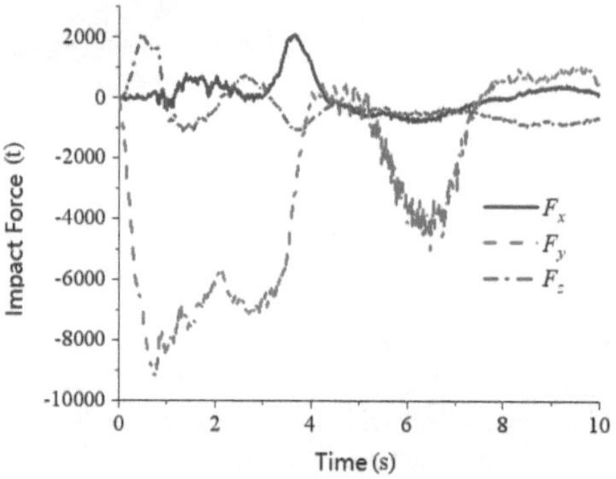

Figure 9. Impact force - time history curves.

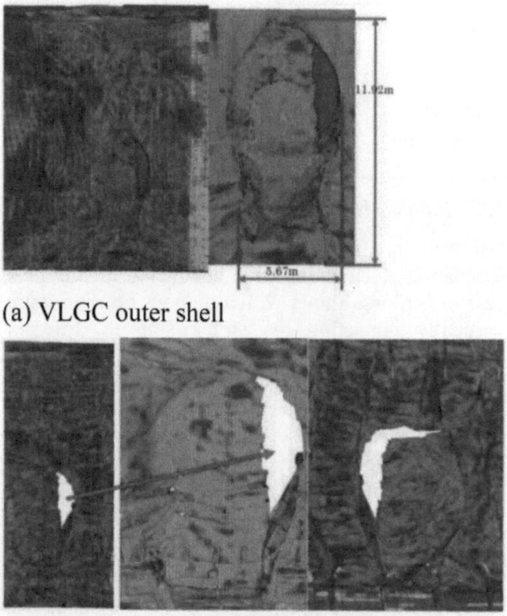

Figure 10. Damage of the structures.

(4) Roll angle/Motion velocity - time history curves

In Case 1-4, the roll angle-time history curve of the struck ship is shown in Figure 11. From the figure, it can be seen that during the collision process, the roll angle of the struck ship continuously increases until reaching a maximum of 0.8° at 3.3 second, then begins to decrease to -0.24° at 7.4 second, and subsequently rises again. It should be noted that during the collision process, the penetration of the striking ship through the struck ship will restrict the rolling motion of the struck ship, resulting in relatively small rolling movements.

The velocity-time history curves of both ships during collision are shown in Figure 12, where Node 1 represents the velocity-time history curve of the striking ship, while Nodes 2 and 3 represent the velocity-time history curves at the upper and lower positions of the non-impacted side shell plate of the struck ship. From the figure, it can be observed that after collision occurrence, the velocity of the striking ship decreases from 5.6584 m/s to 2.3 m/s, while the translation velocity of the struck ship increases from stationary to 2.7 m/s. The velocity at Node 2 remains lower than that at Node 3, indicating greater damage to the lower hull of the struck ship and significant influence of the striking ship on the entire vessel's posture. After 3.8 second post-collision, the motion velocity of the struck ship exceeds that of the striking ship, indicating the separation between the two vessels.

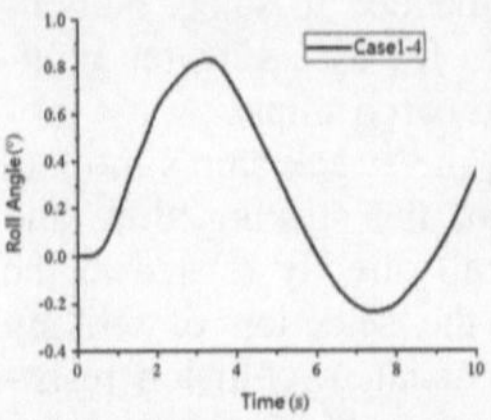

Figure 11. Roll angle - time history curves.

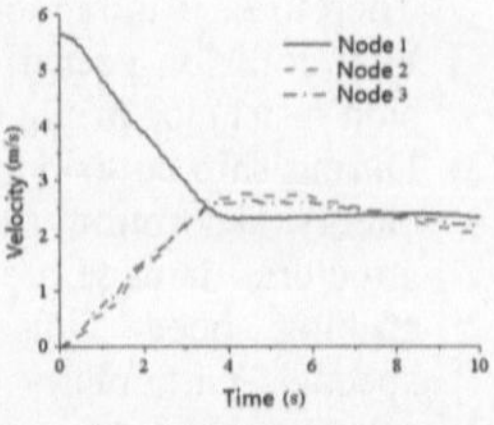

Figure 12. Motion velocity - time history curves.

(5) Energy absorption - time history curves

In Case 1-4, the energy absorption-time curves of certain structures are shown in Figure 13. From the figure, it can be observed that compared to the striking bow, the struck ship absorbs the majority of the energy, primarily through the side shell plate, bottom plate, and No.2 liquid cargo tank side plate, which aligns with the observed damage patterns.

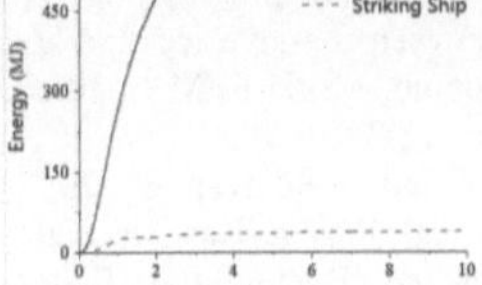

(a)The striking and struck ship

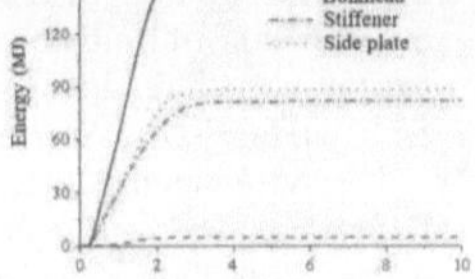

(b)The Cargo tank 2 of VLGC

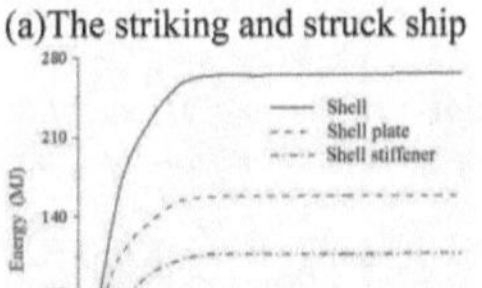

(c)The shell of VLGC

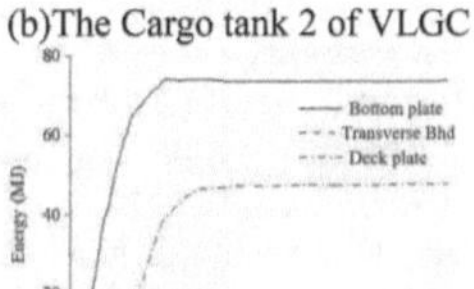

(e)Other structures of VLGC

Figure 13. Energy absorption - time history curves.

4.3 *Straight bow collision simulation and analysis*

The simulation analysis of a Type A VLGC vessel being impacted by a 4600TEU container ship at different speeds and collision positions followed the same computational process as described in Section 3.2. Analysis results indicate that the straight bow of the impacting ship increased the collision contact area. Even when the 4600TEU container ship collided at an extremely high speed of 25kn, only localized structural crushing occurred on the side structure of the impacted VLGC vessel without any rupture formation, as shown in Figure 14. The straight bow collision analysis serves solely as a comparative and auxiliary study to bulbous bow impacts and will not be elaborated further here.

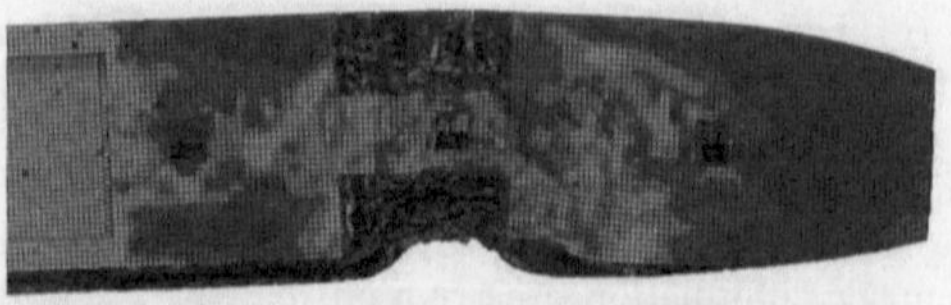

(a)Top view of the stuck ship

(b)Outer shell of the stuck ship

(c)Cargo tank of the stuck ship

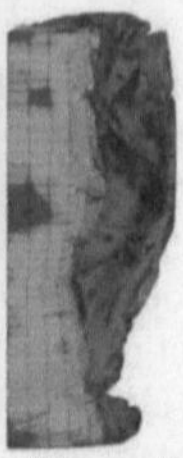

(d)Bow of the striking ship

Figure 14. Structural damage after impact by straight bow at 25 kn striking speed.

5 SUMMARY OF COLLISION SIMULATION

According to the collision simulation analysis of the Type A VLGC in this paper, the relevant summaries are as follows:

(1) Through simulation analysis of container ships with bulbous bow structures colliding at different speed at FR156 (midpoint between the aft transverse bulkhead of No.2 cargo tank and the swash bulkhead) of the Type A VLGC, it can be concluded that:

(a) The structural damage form of the struck ship is related to the striking speed.When the striking speed is 5 kn, the VLGC side structure only exhibits significant plastic deformation damage, with the largest deformation occurring in the bulbous bow contact area, forming a large dent but no tearing rupture. When the striking speed reaches 8 kn or higher, the VLGC side structure develops breach damage on the outer shell firstly in the bulbous bow contact area, and the breach size increases with speed, particularly expanding more significantly in the molded depth direction. At a collision speed of 11 kn, breach damage also appears on the side plating of cargo tank on the VLGC.

(b) The impact force is related to the striking speed and breach size.When the striking speed is 5 kn, significant plastic deformation occurs in the VLGC side structure, with an impact force of 7,771 tons. When the striking speed reaches 8 kn or higher, the impact force increases substantially compared to the 5 kn scenario but stabilizes around 9,300 tons, indicating that this value represents the maximum impact force the VLGC side structure can withstand. Further increases in striking speed will only deepen the penetration without amplifying the impact force.

(c) The energy absorption by the struck ship's structure during collision is related to the striking speed.Higher striking speeds result in greater kinetic energy absorption by both the struck and striking ships, primarily due to the expansion of the damage deformation area, which increases energy absorption. The majority of the kinetic energy is absorbed by the deformation of the struck ship, with the striking ship absorbing approximately 5% of that energy.

(d) The velocity of the struck ship is related to the striking speed.Higher striking speed lead to greater velocities of the struck ship and an earlier onset of joint motion between the two vessels.

(e) The roll angle of the struck VLGC ship is related to the striking speed.Higher striking speed reduce the roll angle of the struck VLGC. This occurs because the increased speed of the striking ship causes deeper penetration of its bulbous bow into the VLGC, promoting synchronized motion between the ships and thereby reducing the roll angle.

(2) When a striking ship with a straight bow structure impacts the VLGC at a high speed, the side structure of the struck ship only exhibits local structural crushing without breach formation, primarily due to the large contact area by the straight bow dispersing the localized forces.

6 CONCLUSIONS

Based on the theory of ship collision simulation, this paper conducted simulation analysis on the impact process of two types of container ships with different bow structures colliding with the cargo tank region of a Type A large liquefied gas carrier. The main conclusions are as follows:

(1) The equivalent beam method and added mass method employed in this paper for simulation effectively balance computational accuracy and efficiency, successfully modeling the ship collision process, reflecting the essence of collision mechanisms, and yielding reasonable simulation results. These achievements have been incorporated into the Rules for Steel Sea-Going Ships of China Classification Society (2025). The methodology and computational procedures can serve as reference cases for assessing the collision resistance of liquid cargo ships, such as liquefied gas carrier through simulation.

(2) Research example shows that when a striking ship with a bulbous bow form collides vertically at a high striking speed against the side shell structure of the struck liquid cargo ship's cargo hold, the side structure may suffer severe breach damage, which should be given sufficient attention. When the striking ship has a straight bow form, the resulting damage to the struck ship is less severe compared to bulbous bow collisions. Therefore, it is recommended to select bulbous bow collision scenarios for assessing the collision resistance of liquid cargo ships.

(3) During ship collisions, the struck ship's motion, energy absorption from the striking ship, and structural damage are all directly related to the striking speed. Thus, the selection of striking speed is particularly crucial in collision resistance assessments.

(4) For Type A independent tank liquefied gas carriers, the full damage scenario of the side shell structure can be selected to evaluate post-collision residual strength. The calculated residual hull girder strength of the VLGC in this example under side collisions meets regulatory criteria requirements.

REFERENCES

China Classification Society. 2025. *Rules for Classification of Sea-going Steel Ships*. Beijing: China Communications Press.

Feng, Y.; Li, H.; Ong, M.C.; Wang, W.W. 2023. Numerical investigation of collision between massive ice floe and marine structure using coupled SPH-FEM method. *Ships and Offshore Structures* 18:380–390.

Liu, B.; Villavicencio, R.; Guedes Soares, C. 2013. Experimen -tal and Numerical Plastic Response and Failure of Laterally Impacted Rectangular Plates. *Journal of Offshore Mechanics and Arctic Engineering* 135(4).

Minorsky, V. U. 1959. An analysis of ship collisions with reference to nuclear power plants. *Journal of Ship Research* 3(1):1–10.

Paik, K. J.; Kim, J. K.; Lee, H. J.; Jung, B.G.; Kim, S.J. 2017. Test database of the mechanical properties of mild, high-tensile and stainless steel and aluminium alloy associated with cold temperatures and strain rates. *Ships and Offshore Structures* 12(Suppl).

Qu, Y.; Chen, G. Y.; Liu, J.; Zhang, P.; Cheng, Y.S. 2022. Experiment and numerical analysis of crashworthiness of the double-skin side structure on a nuclear power platform [in Chinese]. *Chinese Journal of Ship Research* 17 (Suppl):59-65.

Song, M.; Ma, J.; Huang, Y. 2017. Fluid-structure interaction analysis of ship-ship collisions. *Marine Structures* 55: 121–136.

Wang, Z. L.; Gu, Y. N. 2001. Numerical Simulation Research on the Dynamic Process of Ship Collision [in Chinese]. *Explosion and Shock Waves* 21(1):29–34.

Youssef, S. A. M.; Noh, S. H.; Paik, J. K. 2017. A new method for assessing the safety of ships damaged by collisions. *Ships and Offshore Structures* 12(5-6):862–872.

Zhang, M. Y.; Wang, H. D.; Conti, F.; Manderbacka, T.; Remes, H.; Hirdaris, S. 2025. A hybrid deep learning method for the real-time prediction of collision damage consequences in operational conditions. *EngineeringApplications of Artificial Intelligence* 145: 110158.

Trends in Collision and Grounding of Ships and Offshore Structures – Zhu & Guedes Soares (Eds)
© 2026 The Author(s), ISBN 978-1-041-16361-9

Development and verification of a numerical model for ships collision

Y. Wang, M.S. Chen, Y.C. Tang & L. Zhu
Wuhan University of Technology, Wuhan, P.R. China

Y.J. Zhao
China Ship Scientific Research Center, Wuxi, P.R. China

ABSTRACT: This article develops a coupled dynamic numerical model for ship collision problems. The collision problem is divided into internal and external dynamics problems for analysis, followed by fluid-structure coupling motion response analysis based on the Cummins equation. In the coupled external hydrodynamic solution, a state-space model simplifies the process of solving the convolution term in the Cummins equation. The state-space model's feasibility and accuracy are also validated. Nonlinear springs model simulates internal structural responses, with stiffness coefficients determined from experimental force-displacement curves. Deviations in motion response between the model and AQWA results are compared for various speeds and stiffness coefficients. Energy dissipation changes before and after collisions are also analyzed in different collision speeds conditions.

1 INTRODUCTION

1.1 *Motivation of research*

Collision problem has always been a key research direction in the field of ship and ocean engineering. Ship collision is a complex nonlinear dynamic response process of ship structure under huge collision load in a very short time. There are a lot of nonlinear problems, and it involves the cross integration of structural impact dynamics, ship kinematics, hydrodynamics, plastic mechanics, fracture mechanics and other disciplines. Many scholars have carried out lots of researches and proposed many theories and methods. Recently, with the rapid development of ship intelligence and unmanned ships, the problem of ship collision avoidance has once again become a research hotspot. This requires rapid assessment of collision damage, then the corresponding collision avoidance strategy is given. Therefore, it is urgent to give a feasible numerical model for rapid prediction of ship collision.

1.2 *Literature review*

Zhang (2022) summarized the relevant research on collision problems in his doctoral thesis. In the past few decades, a common theoretical approach to studying ship collision problems has been to decouple the collision problem into two parts: external dynamic and internal dynamic (Minorsky, 1958). External dynamic analysis mainly focuses on the global six degree of freedom motion and energy dissipation during the collision process, while internal dynamic analysis mainly solves plastic deformation theoretical problems such as structural damage and structural energy absorption during the collision process.

Many scholars have conducted relevant research on external dynamic analysis. Popov et al., (1969) established an analytical collision model for solving external dynamic problems in ship ice collisions. The model only considers the normal collision response at the collision point and simplifies the collision process into a one-dimensional problem. Pedersen & Zhang (1998) developed a two-dimensional collision model, based on which further study of the external dynamics and energy dissipation of sticking and sliding scenarios after the collision has been done. However, all the above researches simplify the difficulty of solving the collision problem by reducing the dimensionality. For 3D collision problem, Liang (1986) and Liang & Chen (1992) gave the basic equations of three-dimensional collision dynamic analysis and the solving process of the equations, and then studied the change process of collision force and energy during ship collision. Liu & Amdahl (2010) extended the general analytical collision model proposed by Stronge (2004) to study ship collision. The paper further compared the results with the research conducted by Petersen & Zhang (1998), and found that capturing the characteristics of vertical eccentric collision can improve the calculation accuracy of energy dissipation in the collision process. Other recent studies on external motivation analysis include (Liu & Amdahl 2019; Wang et al., 2021; Pineau et al., 2022; Ragulskis et al., 2022). However, it is important to note that, in the context of external dynamics analysis using decoupled methods, hydrodynamic effects

DOI: 10.1201/9781003684404-6

are typically simplified. Consequently, the accurate determination of the constant added mass factor is crucial for obtaining reliable energy dissipation results. Motora et al., (1971) introduced the concept of equivalent added mass to more accurately estimate hydrodynamic forces. The equations for three distinct concepts of equivalent added mass were proposed. Other relevant researches encompass (Jia & Moan 2010; Zhao, et al. 2020).

The internal dynamics primarily pertain to local structural characteristics, including structural deformation and impact resistance, etc. Liu et al., (2018). provided a comprehensive summary of the various theoretical approaches employed in internal dynamics analysis, and critically evaluated the strengths and limitations of each method. The empirical formula method integrates the ship's energy dissipation with post-collision factors such as collapse volume and deformation mode to derive an approximate expression. Minorsky, (1958) proposed the initial empirical formula by considering the relationship between the damaged volume and energy dissipation. Pedersen & Zhang (2000) further refined the empirical formula, establishing relationships between energy dissipation and various deformation modes, damaged volumes, and structural material properties. However, the empirical formula can only account for energy dissipation during the collision process unilaterally and does not provide a detailed representation of the specific structural damage incurred. According to the experiment, Calle et al., (2017) conducted studies on straight ship bow structures, and the experimental results were utilized to validate the simplified analytical method. Chen et al., (2022) investigated the collision damage of double-layer structures, specifically the main side structures of modern ships, through experimental studies. In addition to the aforementioned experimental studies, numerous investigations were conducted to examine specific structural components within the bow and hull structures: for example, shell plating or stiffened panels (Kim et al., 2016; Gruben et al., 2017; Zeng et al., 2020), and web gieders (Liu & Guedes Soares, 2016; Chen et al., 2019; Liu et al., 2019). Nevertheless, with the advancement of computational power, the numerical simulation method has progressively evolved into the predominant approach for internal dynamics analysis. Abu & Dow (2013) utilized the finite element method and finite element analysis software to investigate the penetration damage in ships. Prabowo et al. (2019) utilized the finite element analysis method to simulate numerical parameters associated with the response and deformation of the hull structure under impact loading. They further investigated the influence of the element length-to-thickness ratio (ELT) on the efficiency of numerical analysis. Their results indicate that when ELT is set to 10, the analysis time can be reduced by approximately 20%. Sha & Amdahl (2019) established a high-fidelity finite element model of the bow of a container ship and the prestressed concrete pontoon wall, and investigated the structural deformation and energy absorption characteristics of both the impacted ship and the impacted pontoon.

This paper is structured into four sections. Section 1 provides an overview of the current research status on ship collisions. Section 2 introduces the theoretical framework of the ship collision model presented in this study, establishing a numerical model for two-ship collisions based on external dynamic analysis theory through decoupling methods. Section 3 validates the numerical model by comparing it with small-scale wet collision test data from full-scale ships. Section 4 summarizes the key findings and contributions of this research and give a conclusion.

2 MATHEMATICAL THEORY

2.1 *Coordinates system of ships collision*

Figure 1 presents a schematic diagram of the coordinate system utilized in the numerical model. In this figure, OXYZ represents the global coordinate system of the model, with its origin is positioned at a fixed point in the space, which is at the center of gravity of the struck ship in this paper. The local coordinate system $G_Ax_1y_1z_1$ for striking ship's (i.e. Ship A) motion response has its origin positioned at the center of gravity. When in static water equilibrium, the XOY plane of the global coordinate system is parallel to the local coordinate plane $x_1G_Ay_1$. Similarly, the local coordinate system $G_Bx_2y_2z_2$ for struck ship's (i.e. Ship B) motion response originates from the center of gravity, and in the state of static water equilibrium, the XOY plane of the global coordinate system also remains parallel to the local coordinate plane $x_2G_By_2$.

2.2 *Frequency domain analysis*

Potential flow theory is commonly employed to analyze the interaction between wave structures in external dynamics problems(Tabri, Määttänen et al. 2008). In the theory of linear potential flow, several fundamental assumptions are made: 1) the fluid is assumed to be irrotational, inviscid, and incompressible; 2) the amplitude of regular incident waves is significantly smaller than the wavelength, implying a low wave steepness; 3) all structural motions are considered to be harmonic. Under these assumptions, the velocity potential function of the flow field must satisfy the Laplace equation, as presented in Equation 1.

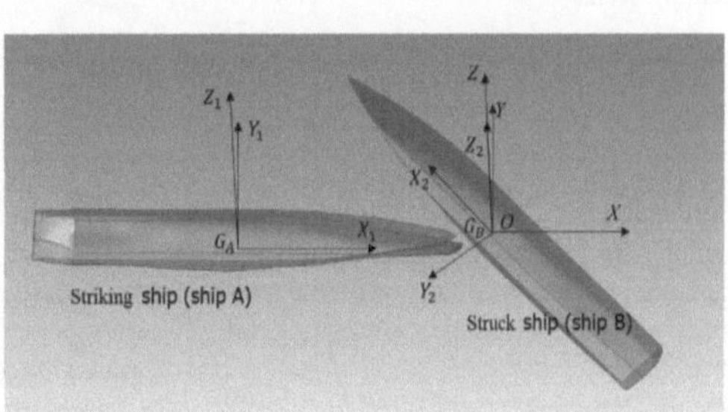

Figure 1. Schematic diagram of collision model coordinate system.

$$\nabla\mathbf{u}(x,y,z,t) = \nabla^2\phi(x,y,z,t) = \frac{\partial^2\phi}{\partial x^2}\vec{i} + \frac{\partial^2\phi}{\partial y^2}\vec{j} + \frac{\partial^2\phi}{\partial z^2}\vec{k} = 0\frac{\partial^2\Omega}{\partial u^2} \quad (1)$$

Where u = the velocity vector; ϕ = the velocity potential function; (x, y, z) = the position coordinate of a particle of the fluid; t = the time; $(\vec{i},\vec{j},\vec{k})$ = the unit vector of the three coordinate axes of the coordinate system.

For a single regular wave incident, the first-order velocity potential can be expressed in complex form as given by Equation 2.

$$\phi(x,y,z,t) = \mathrm{Re}\left[\varphi(x,y,z)e^{i\omega t}\right] \quad (2)$$

Where ω = the wave excitation frequency; φ = the complex form of velocity potential.

The complex form of the velocity potential function also satisfies the Laplace equation. In numerical simulations, this complex velocity potential can be decomposed into three components: incident potential, diffracted potential, and radiated potential. When solving potential flow problems using the boundary element method, Green Function can be employed to derive solutions. By addressing the boundary conditions, the corresponding velocity potential can be determined, which subsequently allows for the calculation of frequency-domain outputs such as hydrodynamic coefficients, excitation forces, and response amplitude operators (i.e. RAO).

The additional mass and damping coefficients of the floating body are derived from the radiation velocity potential, as shown in Equation 3.

$$\begin{cases} \mathrm{A}_{ij}(\omega) = -\frac{1}{\omega}\mathrm{Im}\left[i\rho\omega \iint_{S_{MB}} n_i\varphi_j ds\right] \\ \mathrm{B}_{ij}(\omega) = \mathrm{Re}\left[i\rho\omega \iint_{S_{MB}} n_i\varphi_j ds\right] \end{cases} \quad (3)$$

Where A_{ij} = the added mass produced in the i-direction due to motion in the j-direction; B_{ij} = the added damping in the i-direction caused by motion in the j-direction; *Im* = the imaginary part of a complex number; Re = the real part of a complex number.

The wave excitation force is derived from the incident and diffraction potentials, as presented in Equation 4.

$$\widehat{\mathrm{f}}^{\mathrm{exc}}(\mathrm{i}\omega) = -\mathrm{i}\omega\rho \iint_{\mathrm{S_{MB}}} (\varphi_{\mathrm{I}} + \varphi_{\mathrm{D}}) \cdot \mathrm{n}dS \quad (4)$$

Where $\widehat{\mathrm{f}}^{\mathrm{exc}}$ = the wave excitation force; ρ = the liquid density (typically taken as 1025 kg/m³ for seawater); S_{MB} = the average wetted surface area of the floating body; n = the normalized normal vector matrix for the six degrees of freedom.

Based on the excitation force, added mass, and added damping, the Response Amplitude Operator (RAO) can be determined, as presented in Equation 5.

$$\widehat{\mathrm{X}}(i\omega) = \widehat{\mathrm{f}}^{\mathrm{exc}}(i\omega)/\{-\omega^2[\mathrm{M} + \mathrm{A}(\omega)] + i\omega\mathrm{B}(\omega) + \mathrm{C}\} \quad (5)$$

Where $\widehat{\mathrm{X}}(i\omega)$ = the amplitude and phase of the floating body's response to wave action; M = the mass matrix of the floating body; C = the hydrostatic stiffness matrix of the floating body.

2.3 *Time domain analysis*

After completing the frequency domain calculations, the hydrodynamic coefficients of the two-ship coupling system, including the added mass and added damping coefficients, are substituted into the Cummins equation of motion to formulate the time-domain motion equation. In the case of zero speed, the added damping coefficient B converges to 0 as frequency increases, i.e. $B(\infty) = 0$. Consequently, when the frequency range is small, the impulse response function $K(t)$ can be determined using the added damping coefficient. Then added mass $A(\infty)$ can be derived from $K(t)$. Therefore, the terms related to added damping are omitted in the Cummins equation. However, in the Cummins equation involving speed, $A(\infty)$, $B(\infty)$, and $K(t)$ require modification. Since $B(\infty)$ does not converge to zero but rather to a specific value, $K(t)$ cannot be directly solved using the added damping coefficient B. Instead, polynomial fitting must be applied to the added damping coefficient to obtain $B(\infty)$, followed by solving the impulse response function $K(t)$, and ultimately determining $A(\infty)$. Additionally, the relevant terms of added damping should be considered in the Cummins equation.

The calculation formula for the additional damping term is presented in the following Equation 6.

$$B(\infty) = \lim_{\omega\to\infty}\left(\frac{a}{\omega^2} + \frac{b}{\omega^4} + c\right) \quad (6)$$

Where a, b, c = the constant coefficients related to the additional damping coefficient.

The impulse response function term is calculated as follows Equation 7.

$$\begin{gathered} K(t) = \frac{2}{\pi}\int_0^\infty [B(\omega) - B(\infty)]\cos(\omega_e t)d\omega \\ \omega_e = \omega - kU\cos\beta \end{gathered} \quad (7)$$

Where k = the wave number; U = the incident wave speed; β = the incident wave direction angle; ω_e = the encounter frequency.

The added mass item is determined by the following Equation 8.

$$A(\infty) = A(\omega) + \frac{1}{\omega}\int_0^\infty K(\tau)\sin(\omega t)d\tau \quad (8)$$

The Cummins equation involving speed is presented in Equation 9.

$$\begin{bmatrix} M_1 + A_{1,1}(\infty) & 0 \\ 0 & M_2 + A_{2,2}(\infty) \end{bmatrix} \begin{bmatrix} \ddot{x}_1 \\ \ddot{x}_2 \end{bmatrix} + \begin{bmatrix} \int_0^t K_{1,1}(t-\tau) \cdot \dot{x}_1 d\tau & 0 \\ 0 & \int_0^t K_{2,2}(t-\tau) \cdot \dot{x}_2 d\tau \end{bmatrix} + \begin{bmatrix} B_{1,1}(\infty) & 0 \\ 0 & B_{2,2}(\infty) \end{bmatrix} \begin{bmatrix} \dot{x}_1 \\ \dot{x}_2 \end{bmatrix} + \begin{bmatrix} C_{1,1} + c_k & 0 \\ 0 & C_{2,2} \end{bmatrix} \begin{bmatrix} x_1 \\ x_2 \end{bmatrix} = \begin{bmatrix} F_1^{exc} \\ F_2^{exc} \end{bmatrix} + \begin{bmatrix} -F_1^{impact} \\ F_2^{impact} \end{bmatrix} \quad (9)$$

Where M_1 and M_2 = the mass of the two ships, respectively; A_{11} and A_{22} = the added mass coefficient of the two ships, respectively; K_{11} and K_{22} = the impulse response function of the two ships, respectively; B_{11} and B_{22} = the added damping coefficient of the two ships, respectively; C_{11} and C_{22} = the restoration coefficient in still water of the two ships, respectively; F_1^{exc} and F_2^{exc} = the excitation force of the two ships; F_1^{impact} and F_2^{impact} = the impact force of the two ships.

2.4 *Convolution term solution*

Convolution is extensively utilized to describe linear time-invariant systems in the context of time-domain linear system theory (Kristiansen, Hjulstad et al. 2005). Direct computation of the convolution term in the Cummins equation is both complex and computationally intensive. Given that the convolution term represents a linear interconnected system, it can be effectively replaced by an equivalent constant-coefficient state-space model. This substitution ensures accuracy while significantly enhancing computational efficiency (Chen, Taylor et al. 2014). For a single-degree-of-freedom linear time-invariant system, the convolution takes the following Equation 10.

$$y(t) = \int_0^t h(t-\tau)u(\tau)d\tau \quad (10)$$

Where $h(t)$ is the impulse response function, $u(t)$ is the input term of a linear system, representing the velocity of the incident wave, also referred to as the ship speed, $y(t)$ is the output term of the linear system, representing the radiation damping force.

For single-input single-output (SISO) linear systems, the state-space model typically assumes the following form (Friedland 2005).

$$\begin{aligned} \dot{z}(t) &= {}_c z(t) + B_c u(t) \\ y(t) &= C_c z(t) + D_c u(t) \end{aligned} \quad (11)$$

Where $z(t)$ = the state vector; A_c = the state matrix; B_c = the input matrix; C_c = the output matrix; D_c = the feedthrough matrix; n = the order of the SISO state-space model; c = a time-continuous state-space model.

The response of this system to any input can be uniquely determined by the input value and the initial value of the state vector. Under the initial condition of zero, Equation 10. and Equation 11. have consistent representations for the system input and output, so the one model can be transformed into the other. The solution to Equation 11. can be obtained by the following Equation 12.

$$\begin{aligned} y(t) = C_c \exp(A_c t) z(0) + \\ C_c \int_0^t \exp(A_c(t-\tau)) B_c u(\tau) d\tau + D_c u(t) \end{aligned} \quad (12)$$

This approach provides a feasible method for solving ordinary differential equations. According to the Equation 12, treating the input term as an impulse with an initial condition of zero yields the impulse response function $h(t)$.

$$h(t) = C_c \exp(A_c t) B_c + D_c \delta(t) \quad (13)$$

Where *exp* = the mathematical symbol for the exponential function of e; $\delta(t)$ = the Dirac function.

For multi-degree-of-freedom (MDOF) problems, the parameter matrix is composed of the parameter matrices of the SISO state-space model, as shown in Equation 14.

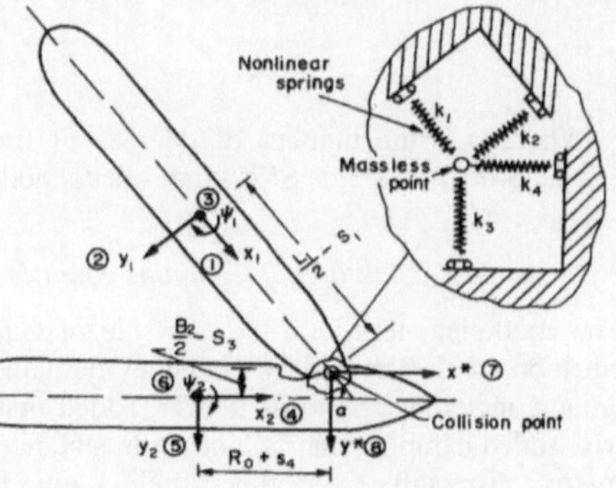

Figure 2. Collision model in Petersen's paper (Petersen 1982).

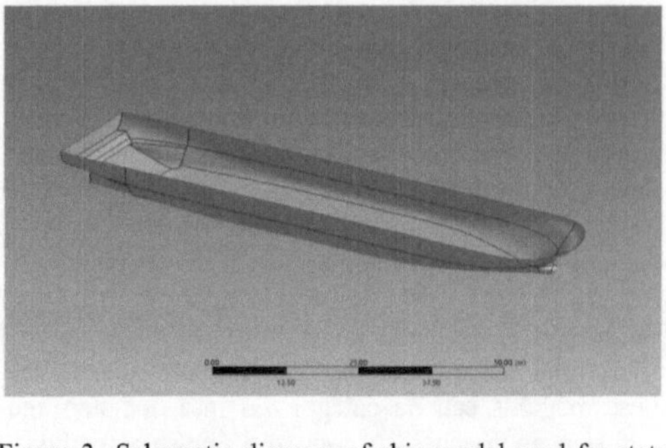

Figure 3. Schematic diagram of ship model used for state space model validation.

$$\mathrm{A}_M = \begin{bmatrix} \mathrm{A}_{i1} & 0 & \cdots & 0 \\ 0 & \mathrm{A}_{i2} & \cdots & 0 \\ 0 & 0 & \ddots & 0 \\ 0 & 0 & \cdots & \mathrm{A}_{iq} \end{bmatrix},$$
$$\mathrm{A}_M \in \mathrm{R}^{(q\times n)\times(q\times n)} \quad (i = 1, \ldots, q)$$
$$\mathrm{B}_M = \begin{bmatrix} \mathrm{B}_{11} & 0 & \cdots & 0 \\ 0 & \mathrm{B}_{12} & \ddots & \vdots \\ \vdots & \ddots & \ddots & 0 \\ 0 & 0 & \cdots & \mathrm{B}_{1q} \\ \vdots & \vdots & \vdots & \vdots \\ \mathrm{B}_{q1} & 0 & \cdots & 0 \\ 0 & \mathrm{B}_{q2} & \ddots & \vdots \\ \vdots & \ddots & \ddots & 0 \\ 0 & 0 & \cdots & \mathrm{B}_{qq} \end{bmatrix},$$
$$\mathrm{B}_M \in \mathrm{R}^{(q\times 11)}$$
$$\mathrm{C}_M = \begin{bmatrix} \mathrm{C}_{11} & \cdots & \mathrm{C}_{1q} & 0 & \cdots & \cdots & 0 \\ \vdots & \vdots & \vdots & \ddots & \vdots & \vdots & \vdots \\ 0 & \cdots & 0 & \cdots & \mathrm{C}_{q1} & \cdots & \mathrm{C}_{qq} \end{bmatrix},$$
$$\mathrm{C}_M \in \mathrm{R}^{(q)\times(q\times n)}$$
$$\mathrm{D}_M = \begin{bmatrix} \mathrm{D}_{11} & \cdots & \mathrm{D}_{1q} \\ \vdots & \ddots & \vdots \\ \mathrm{D}_{q1} & \cdots & \mathrm{D}_{qq} \end{bmatrix},$$
$$\mathrm{D}_M \in \mathrm{R}^{q\times q} \tag{14}$$

Where q = the number of degrees of freedom; n = the order of a single SISO state-space model.

2.5 *Numerical solution of Cummins equation*

The coefficient terms on the left side of Cummins equation are determined by frequency-domain calculations, including the mass matrix, added mass matrix, added damping matrix, and hydrostatic stiffness matrix. To solve for the time-domain motion response of a floating body under complex nonlinear conditions, it is necessary to define only the external force term on the right side. Since ship collisions are characterized by rapid and violent dynamics phenomenon, the predominant external force is the collision force. In this paper, the collision force during the collision process is calculated using the nonlinear spring force model proposed by Petersen, as shown in Figure 2. For the specific derivation, refer to the relevant literature.

When solving nonlinear differential equations, obtaining an analytical solution is often prohibitively difficult. Therefore, in engineering applications, iterative numerical methods are commonly employed to solve initial value problems of differential equations. These methods can be categorized into one-step and multi-step approaches. Among the one-step methods, notable examples include the Euler method, the improved Euler method, and the Runge-Kutta method. This study employs the classical fourth-order Runge-Kutta method, which is briefly introduced below. The core idea of the Runge-Kutta method is as follows: let the step size be h, and artificially introduce L intermediate nodes within the interval $[x_n, x_{n+1}]$. The value of y_{n+1} is obtained by y_n adding the weighted average of the approximate derivative values at corresponding nodes, with multiplied by the step size, The derivative at each node K_{n+1} is computed based on the derivative K_n.

$$\begin{cases} y_{n+1} = y_n + h\sum_{i=1}^{L}\lambda_i K_i \\ K_1 = f(x_n, y_n) \\ K_2 = f(x_n + c_2 h, y_n + c_2 h K_1) \\ K_3 = f(x_n + c_3 h, y_n + c_3 h(a_{31}K_1 + a_{32}K_2)) \\ \cdots\cdots\cdots\cdots\cdots\cdots \\ K_i = f\left(x_n + c_i h, y_n + c_i h\sum_{j=1}^{i-1} a_{ij}K_j\right) \quad i = 2, 3, \cdots, L \end{cases} \tag{15}$$

Where h = the computational step size; L = the series; K_i = approximate values of derivatives at various series; λ_i, c_i, a_{ij} = indeterminate coefficients and they have the following relationship expression $\sum_{i=1}^{L}\lambda_i = 1, c_i \leq 1, \sum_{j=1}^{i-1} a_{ij} = 1$.

This process leads to the general form of the Runge-Kutta method, as shown in Equation 16. And each undetermined coefficient can be determined through Taylor expansion and truncation error analysis. For the explicit fourth-order Runge-Kutta method, specifically the classical fourth-order Runge-Kutta method, it can be expressed in the form of Equation 17.

$$\begin{cases} y_{n+1} = y_n + \frac{h}{6}(K_1 + 2K_2 + 2K_3 + K_4) \\ K_1 = f(x_n, y_n) \\ K_2 = f\left(x_n + \frac{h}{2}, y_n + \frac{h}{2}K_1\right) \\ K_3 = f\left(x_n + \frac{h}{2}, y_n + \frac{h}{2}K_2\right) \\ K_4 = f(x_n + h, y_n + hK_3) \end{cases} \tag{16}$$

After obtaining the time-domain results of the motion parameters for both ships during the collision process, it is necessary to analyze the energy dissipation before and after the collision. In this context, the energy dissipated throughout the collision is quantified by calculating the difference in kinetic energy prior to and following the impact. The expression for energy dissipation is presented in the following Equation 18.

$$\begin{gathered} E_{K0} = \tfrac{1}{2}(M_a + m_{ax})V_a^2 + \tfrac{1}{2}(M_b + m_{by})V_b^2 \\ E_{Kt} = \tfrac{1}{2}\left[(M_a + m_{ax}) + (M_b + m_{by})\right]V_c^2 \\ + \tfrac{1}{2}(I_{yy} + J_{yy})\omega_{\mathrm{roll}}^2 + \tfrac{1}{2}(I_{zz} + J_{zz})\omega_{\mathrm{yaw}}^2 \\ \Delta E = E_{K0} - E_{Kt} \end{gathered} \tag{17}$$

Where a b c = the symbol of striking ship, the struck ship and the two ships after the collision; E_{K0} and E_{Kt} = the kinetic energy of the two ships before and after collision; m_{ax} = attaching water mass to the striking ship; m_{by} = the quality of the water attached to the struck ship; I_{yy}, I_{zz} = the roll and yaw moments of inertia of the struck ship; J_{yy}, J_{zz} = associated with products of water mass concerning roll and yaw dynamics for that vessel; ΔE is the energy dissipated value.

3 NUMERICAL SIMULATION RESULTS

3.1 *Verification of state-space model*

As input parameters for the numerical model, the frequency-domain output terms can be solved using software AQWA. However, when solving for frequency-domain parameters in AQWA, the coupling model does not support setting different incident wave velocities simultaneously. To obtain the frequency domain output terms corresponding to the two ships at different speeds, the frequency-domain coupling terms must be solved separately twice.

After determining the required hydrodynamic coefficients at the specified location, a constant-coefficient time-domain calculation model is constructed by employing a state-space model to substitute for the convolution term in the Cummins equation of motion. This substitution significantly enhances the computational efficiency of the model. To ensure the accuracy of the state space model, after obtaining the hydrodynamic coefficients in the frequency domain, it is essential to apply the Fourier transform to these coefficients to derive the delay function required for solving the Cummins equation. When fitting the delay function using the state space the coefficient of determination (R^2) can be directly assessed by comparing the fitted delay function with the original hydrodynamic coefficients. To maintain the accuracy of the state space model, the coefficient of determination R^2 should exceed 0.97. The R^2 value can be calculated using the following Equation 19.

$$R^2 = 1 - \frac{\sum_i (K_A(t) - K_C(t))^2}{\sum_i (K_A(t) - \bar{K}_A)^2} \quad (18)$$

Where $K_A(t)$ is the delay function fitted by the state space model, $K_c(t)$ is the delay function calculated directly from the hydrodynamic coefficient, $\overline{K_A}$ is the average value of $K_c(t)$.

Table 1. Main parameters of the ship model.

Terms	Values
Length L(m)	96.97
Width B(m)	20.78
Depth D(m)	5.58
Design draft d(m)	3.5
LCG(m)	0.024
TCG(m)	0
VCG(m)	3.3315
Displacement(ton)	5598.263

Subsequently, the reliability of the time delay function derived from the state space model is validated by comparing the coefficient of determination. The condition with weak nonlinearity at 8 kn is selected for further analysis. By incorporating the delay function, a more accurate fourth-order state space model can be established. The delay functions and coefficients of determination for the heave-heave heave-pitch, and pitch-pitch modes are illustrated in Figures 4-6.

To verify the accuracy of the time-domain model in the software AQWA and the numerical program CPTDM, the frequency-domain and time-domain solutions from the software AQWA and the time-domain solution from the program CPTDM for the ship's heave and pitch motions are compared after removing the nonlinear influence (specifically, the collision force). As shown in Figures 7-8, the three curves align closely, indicating that the results from.

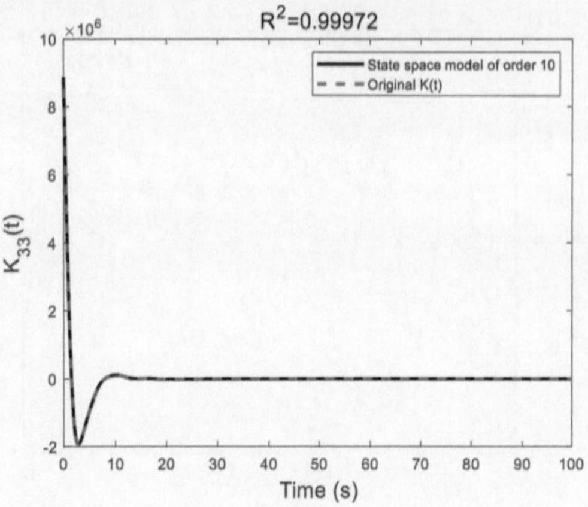

Figure 4. Comparison of Time Delay Function Curves and Determination Coefficient (R^2) Values of heave-heave freedom.

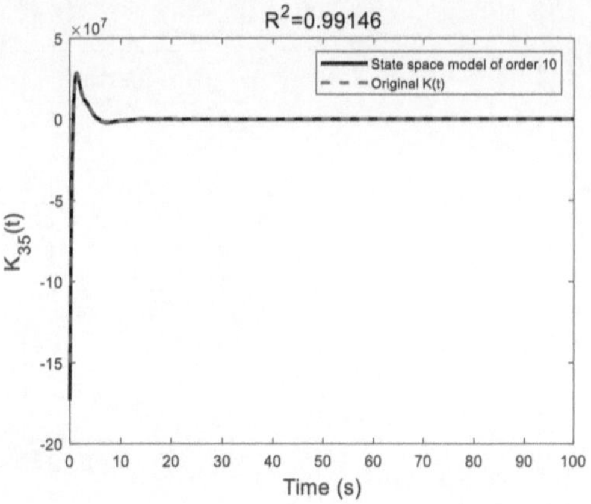

Figure 5. Comparison of Time Delay Function Curves and Determination Coefficient (R^2) Values of heave-pitch freedom.

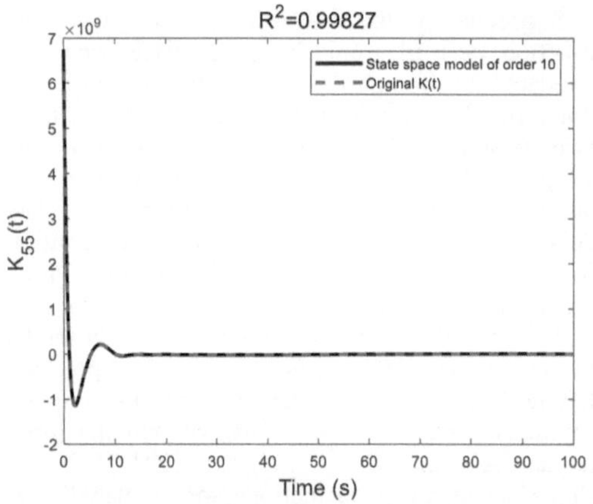

Figure 6. Comparison of Time Delay Function Curves and Determination Coefficient (R^2) Values of pitch-pitch freedom.

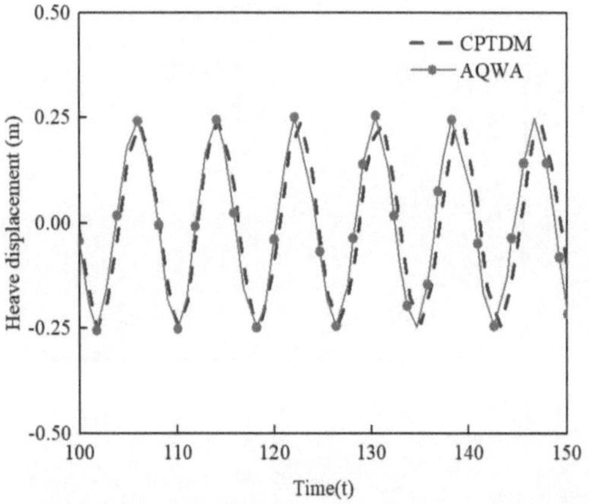

Figure 7. Comparison of CPTDM Frequency-to-Time Domain Results with AQWA Time-Domain Results of heave motion.

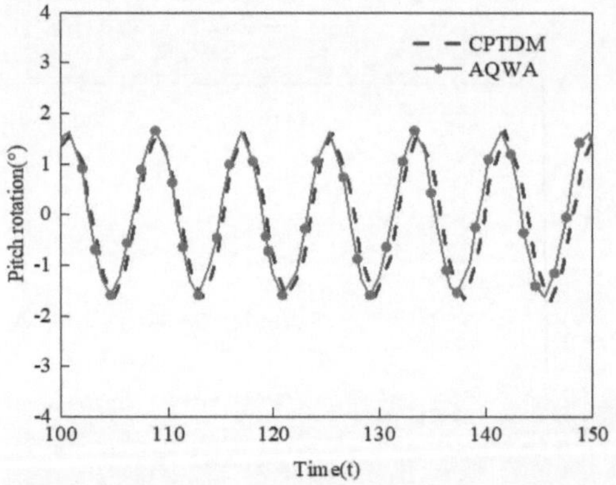

Figure 8. Comparison of CPTDM Frequency-to-Time Domain Results with AQWA Time-Domain Results of pitch motion.

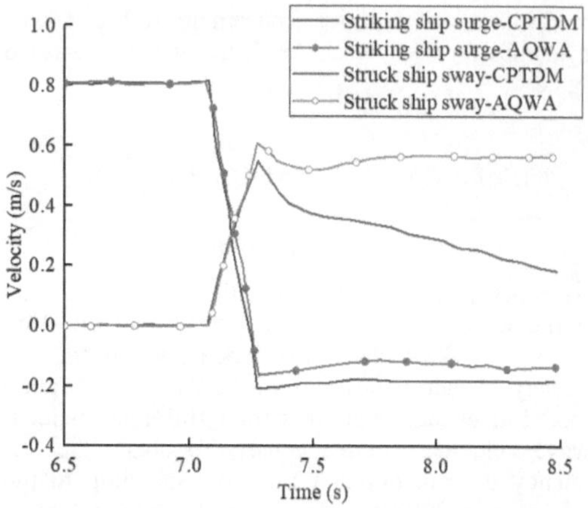

Figure 9. Curve of ships speed of 0.8m/s collision condition.

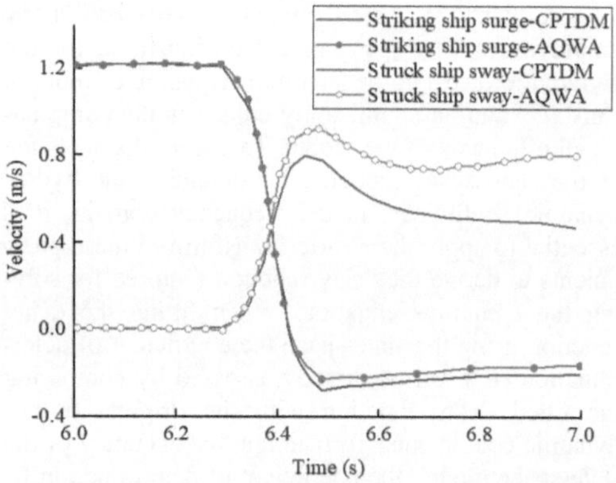

Figure 10. Curve of ships speed of 1.22m/s collision condition.

the time-domain model of the program CPTDM are consistent with those from the software AQWA. This confirms the good computational accuracy of the program CPTDM numerical program.

3.2 *Impact motion response analysis*

Based on the results of the full-scale wet collision test of the ship model, this study compares the motion response characteristics during double ship collisions. Figure 11 illustrates the frequency-domain coupling model for double ship collisions developed in the Aqwa software. Table 2 presents the primary parameter data for the two ships.

In order to verify the feasibility of the numerical model CPTDM, this paper compares the time domain motion response results of small-scale twin-ship collision scenarios Figures 9-10 and Figures 12-13 present the comparison curves of collision time-domain parameters. Two different collision speed conditions are selected for analysis, corresponding to the collision speeds of 10 kn and 15 kn for real ships. Based on the comparison results, it is demonstrated that the time-domain results of collision parameters obtained using the numerical model CPTDM align well with those calculated by the software Aqwa. This indicates that the numerical model CPTDM exhibits a certain level of reliability in addressing double ships collision problem.

Table 2. Main parameters of two ship models.

Terms	Values of striking ship	Values of struck ship
Length L(m)	3.154	3.232
Width B(m)	0.4	0.405
Depth D(m)	0.198	0.3
Draft d(m)	0.145	0.112
Displacement(kg)	60	65

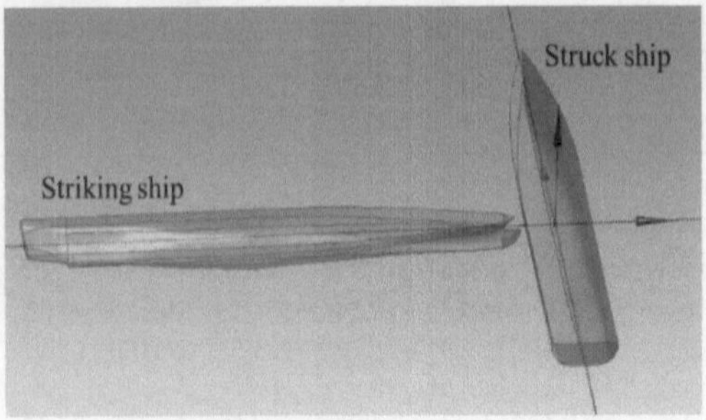

Figure 11. Schematic representation of the Aqwa frequency domain coupling model.

The focus of ships collision research lies in the energy dissipation during the collision process. In this paper, the time-domain parameters of twin-ship collisions under various speed conditions are analyzed followed by a further examination of the energy dissipation results under these conditions. Based on velocities of 0.8m/s and 1.22m/s, combined with the screening of reliable results from ship model tests, five different collision velocities were selected for comparative analysis. Figures 14-17 present the curves of time-domain results for parameters during the ship-ship collision process under five different collision speed conditions.

The comparison results demonstrate that as the collision speed increases, the curvature of the change in the surge speed of the striking ship and the sway speed of the struck ship becomes more pronounced. Specifically, higher speeds lead to more intense variations in velocity during the collision be ween the two vessels. Additionally, it is evident that an increase in collision speed corresponds to a greater kinetic energy of the striking ship, which in turn results in larger collision forces and penetration depths post-collision—a finding consistent with real-world observations. By analyzing the collision force time-history curve presented in Figure 16, it can be observed that the maximum collision force exhibits a sudden change when the ship model's speed reaches 1.22 m/s or higher. Further investigation reveals that the collision depth exceeds the thickness of the ship model's outer plating, causing the outer plating to rupture. This rupture leads to a sudden increase in contact force at the point of fracture, thereby resulting in a sharp rise in the collision force.

Table 3, in the below, presents the energy dissipation of two ships before and after collision at various speeds. The analysis results indicate that as the collision velocity increase, the energy dissipation during the collision also escalates. Furthermore, the ratio of dissipated energy across five different velocity conditions as presented in the Table 3, remains relatively consistent with respect to the pre-collision energy. This percentage stabilizes around 58%, with a calculated average of 57.9%.

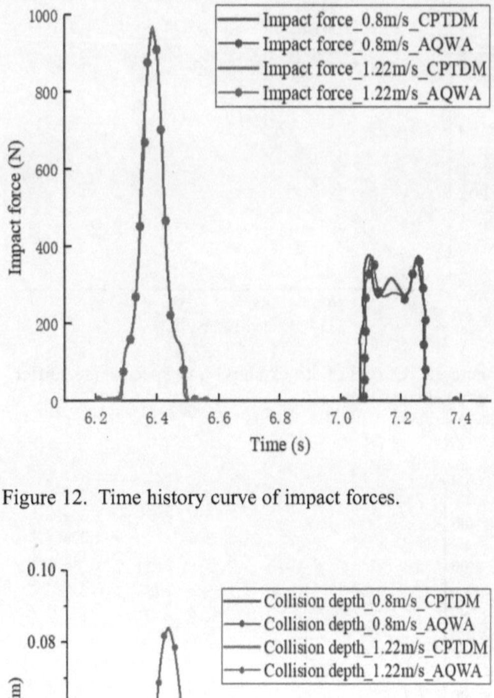

Figure 12. Time history curve of impact forces.

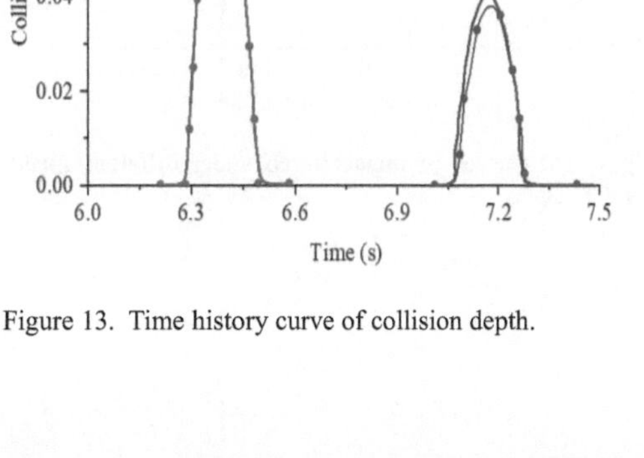

Figure 13. Time history curve of collision depth.

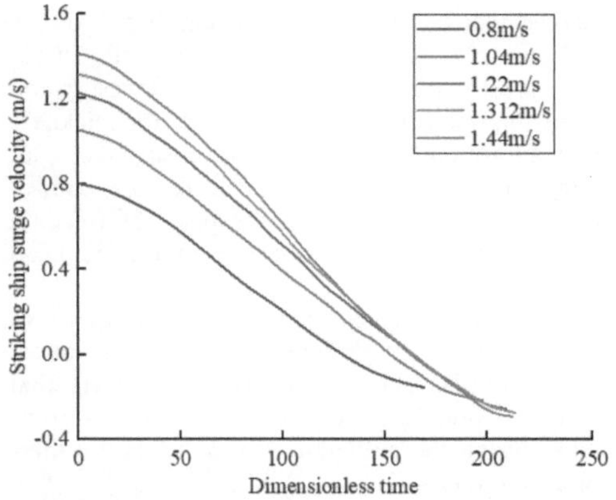

Figure 14. Curves of striking ship surge speed under different speed conditions.

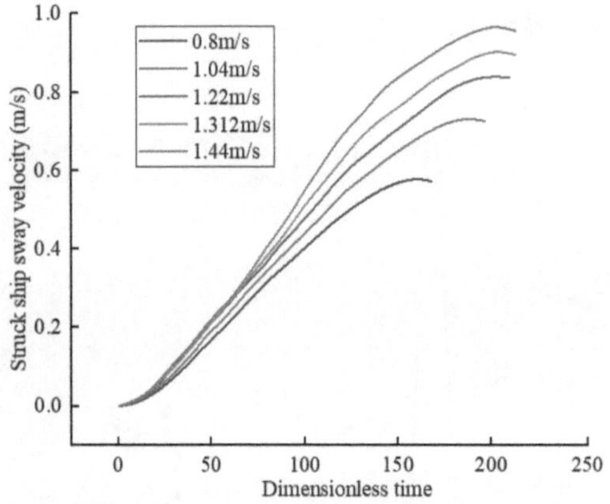

Figure 15. Curves of struck ship sway speed under different speed conditions.

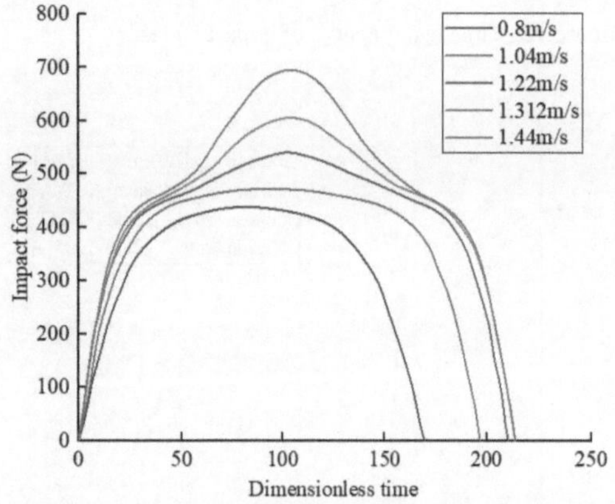

Figure 16. Curves of impact force under different speed conditions.

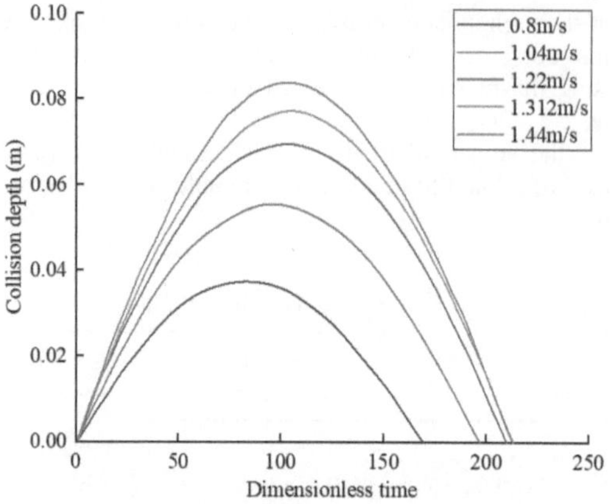

Figure 17. Curves of collision depth under different speed conditions.

Table 3. Energy dissipation results under different speed conditions.

Term A*	Term B*	Term C*	Term D*	Term E*	Term F*	Term G*
m/s	J	J	J	J	J	J
0.8	19.714	8.246	11.288	57.3%	11.086	0.202
1.04	33.317	14.171	19.146	57.5%	19.176	-0.03
1.22	45.848	19.09	26.789	58.4%	26.188	0.57
1.312	53.024	22.105	30.919	58.3%	30.338	0.58
1.44	60.375	25.43	34.946	57.9%	34.735	0.211

* Term A is the Speed conditions,
* Term B is the Energy before collision,
* Term C is the Energy after collision,
* Term D is the Dissipated energy,
* Term E is the Ratio of dissipated energy to Energy before collision,
* Term F is the Collision force work energy,
* Term G is the Difference between dissipated energy and work energy of collision force.

These findings suggest that when the structural parameters of the ship model are held constant during collisions, the energy dissipated is primarily determined by this consistency and exhibits minimal dependence on collision velocity. The present paper quantifies the work energy of the collision force and further compares the difference between the dissipated energy and the work energy of the collision force. It is observed that the difference is negligible, with the calculated average difference being 0.306 J. This suggests that the work energy of the collision force obtained through the numerical model CPTDM can be considered as the energy dissipated during the collision process. These findings further validate the reliability of the collision simulation results produced by the numerical model.

4 CONCLUSIONS

Based on the nonlinear spring model and Cummins equation, this paper presents a numerical model for simulating the ship-ship collision process. To validate the feasibility of the proposed model, time-domain response results from two small-sized ship models were compared between AQWA software and the CPTDM numerical model. The curves representing collision velocity, collision force, and collision depth for both ships exhibited strong consistency, indicating that the CPTDM numerical model is reliable for simulating ship collisions.

Utilizing the CPTDM numerical model, further analysis was conducted on time-domain results concerning collision parameters as well as energy dissipation during collisions. The findings reveal that higher velocities lead to more pronounced changes in collision dynamics and increased energy dissipation values. However, additional analysis indicates that while energy dissipation increases with rising collision velocities under varying conditions, its proportion relative to pre-collision energy remains relatively stable across different scenarios. The calculated average proportion stands at 57.9%, suggesting that variations in energy dissipation proportions during collisions are independent of impact velocity magnitudes. Furthermore, it was observed that the difference between dissipated energy and impact force is minimal, with an average discrepancy of 0.306 J; this further substantiates the reliability of the CPTDM numerical model in simulating ship collisions.

REFERENCES

AbuBakar, A. and R. S. Dow 2013. Simulation of ship grounding damage using the finite element method. *International Journal of Solids Structures* **50**(5): 623–636.

Calle, M. A. G., R. E. Oshiro and M. Alves 2017. Ship collision and grounding: Scaled experiments and numerical analysis. *International journal of impact engineering* **103**: 195–210.

Chen, B. Q., B. Liu and C. Guedes Soares 2019. Experimental and numerical investigation on the nfluence of stiffeners on the crushing resistance of web girders in ship grounding. *Marine Structures* **63**: 351–363.

Chen, B. Q., B. Liu and C. Guedes Soares 2022. Experimental and numerical investigation on a double hull structure subject to collision. *Ocean Engineering* **256**: 111437.

Chen, M. S., R. E. Taylor and Y. S. Choo 2014. Time domain modeling of a dynamic impact oscillator under wave excitations. *Ocean engineering* **76**: 40–51.

Friedland, B. 2005. An introduction to state-space methods. *Control System Design*: 3–16.

Gruben, G., S. Sølvernes, T. Berstad, D. Morin, O. S. Hopperstad and M. Langseth 2017. Low-velocity impact behaviour and failure of stiffened steel plates. *Marine Structures* **54**: 73–91.

Jia, H. R. and T. Moan 2010. Global responses of struck ships in collision with emphasis on hydrodynamic effects. *International Conference on Offshore Mechanics and Arctic Engineering.*

Kim, K. J., J. H. Lee, D. K. Park, B. G. Jung, X. Han and J. K. Paik 2016. An experimental and numerical study on nonlinear impact responses of steel-plated structures in an Arctic environment. *International Journal of Impact Engineering* **93**: 99–115.

Kristiansen, E., Å. Hjulstad and O. Egeland 2005. State-space representation of radiation forces in time-domain vessel models. *Ocean Engineering* **32**, 17-18: 2195–2216.

Liang, W. J. 1986. Three-dimensional analysis of ship/ship collision. *Journal of Shanghai Ship Shipping Research Institute* **9**, 1: 80–93.

Liang, W. J. and G. Z. Chen 1992. Ship collision force and energy absorbability. *Journal of Shanghai Ship and Shipping Resaerch Institute* **15**, **2**: 23–35.

Liu, B., P. T. Pedersen, L. Zhu and S. M. Zhang 2018. Review of experiments and calculation procedures for ship collision and grounding damage. *Marine Structures* **59**: 105–121.

Liu, B. and C. Guedes Soares 2016. Experimental and numerical analysis of the crushing behaviour of stiffened web girders. *International Journal of Impact Engineering* **88**: 22–38.

Liu, K., Y. Lu, Z. L. Wang and G. Wang 2019. An experimental, numerical and analytical study on deformation mechanisms of web girders in a collision or grounding incident. *Ships Offshore Structures* **14**, 8: 839–852.

Liu, Z. H. and J. Amdahl 2010. A new formulation of the impact mechanics of ship collisions and its application to a ship–iceberg collision. *Marine Structures* **23**, 3: 360–384.

iu, Z. H. and J. Amdahl 2019. On multi-planar impact mechanics in ship collisions. *Marine Structures* **63**: 364–383.

Minorsky, V. U. 1958. An analysis of ship collisions with reference to protection of nuclear power plants, *Sharp (George G.) Inc*., New York.

Motora, S., M. Fujino, M. Sugiura and M. Sugita 1971. "Equivalent added mass of ships in collisions. *Selected Papers, J of Soc of Naval Arch of Japan* **7**: 138–148.

Pedersen, P. T. and S. M. Zhang 1998. On impact mechanics in ship collisions. *Marine structures* **11**, 10: 429–449.

Pedersen, P. T. and S. M. Zhang 2000. Absorbed energy in ship collisions and grounding—revising Minorsky's empirical method. *Journal of Ship Research* **44**, 02: 140–154.

Petersen, M. J. 1982. Dynamics of ship collisions. *Ocean Engineering* **9**, 4: 295–329.

Pineau, J. P., F. Conti, H. Le Sourne and T. Looten 2022. A fast simulation tool for ship grounding damage analysis. *Ocean Engineering* **262**: 112248.

Popov, Y. N., O. V. Faddeev, D. E. Kheisin and A. A. Yakovlev 1969. *Strength of ships sailing in ice. Leningrad.*

Prabowo, A. R., T. Putranto and J. M. Sohn 2019. Simulation of the behavior of a ship hull under grounding: Effect of applied element size on structural crashworthiness. *Journal of Marine Science Engineering* **7**, 8: 270.

Ragulskis, K., A. Pauliukas, A. Bubulis, P. Paškevičius, R. Maskeliūnas and L. M. Ragulskis 2022. Investigation of dynamics of a system with two sided impacts. *Advances in mechanical engineering and transport* **1**, 18: 24–30.

Sha, Y. and J. Amdahl 2019. Numerical investigations of a prestressed pontoon wall subjected to ship collision loads. *Ocean Engineering* **172**: 234–244.

Stronge, W. J. 2004. Impact mechanics, *Cambridge university press*.

Tabri, K., J. Määttänen and J. Ranta 2008. Model-scale experiments of symmetric ship collisions. *Journal of marine science technology* **13**: 71–84.

Wang, X. B., C. B. Li and L. Zhu 2021. Numerical simulation of ship collision and validations with experimental results. *ASME 2021 40th International Conference on Ocean*, Offshore and Arctic Engineering.

Zeng, Y., H. Chen, R. Yu, Z. Y. Shen, Z. X. Yu and J. X. Liu 2020. Experimental research on dynamic behavior of circular mild steel plates with surface cracks subjected to repeated impacts in low temperature. *Shock Vibration* **2020**, 1: 3713709.

Zhang, Y. C. 2022. Dynamic responses analysis for accidental collision scenarios of floating offshore wind turbine. PH. D., Newcastle University.

Zhao, Z. J., S. H. Yang, Y. F. Suo and Z. Y. Zhang 2020. Impact analysis of the dynamic added mass coefficients of ship collision based on the external analytical method. *2020 Chinese Automation Congress (CAC)*, IEEE.

Trends in Collision and Grounding of Ships and Offshore Structures – Zhu & Guedes Soares (Eds)
© 2026 The Author(s), ISBN 978-1-041-16361-9

Structural denting and tearing damage under basalt reef grounding

Y.W. Wang
Key Laboratory of High Performance Ship Technology (Wuhan University of Technology), Ministry of Education, Wuhan, P.R. China

L. Zhu
School of Naval Architecture, Ocean and Energy Power Engineering, Wuhan University of Technology, Wuhan, P. R. China

Z.H. Zhou
Key Laboratory of High Performance Ship Technology (Wuhan University of Technology), Ministry of Education, Wuhan, P.R. China

ABSTRACT: Ship grounding has always been one of the most frequent and severe marine accidents in ship navigation. Grounding not only results in damage of hull structure and equipment, but also causes total damage or capsize of the ship, and even accompanied by casualties and environmental pollution. The obstructions causing a ship to run aground are the top of some isolated rocks, submerged bedrock hill or ridge protruding from the seabed. These rocks could be basalt, granite or coral reef. In this paper, cone-shaped rock with a semi-apex angle of 30 ° is made of basalt. The bottom plating specimen embedded into the ship model is made of galvanized carbon steel sheet. The ship grounding model test is carried out in an indoor water tank. The results of ship global motion, damage modes of ship bottom plating and basalt rock, and horizontal grounding resistance forces under basalt rock are obtained in the model test. The experimental study shows the damage modes of bottom plating and basalt rock. In this paper the Abaqus and the CEL FEM (Finite Element Method) is used to study the structural dynamic response of a ship grounding on basalt reef under the same conditions as the experiment. Then the results are compared with the experimental results followed by some concluding remarks.

1 INTRODUCTION

Ship grounding is an hazardous maritime accident. The ship's hull structure is prone to damage in such incident, such as cracks and perforations. The inflow of seawater poses a threat to the lives of those on board. If the ship is carrying oil or other hazardous substances, it may also lead to leaks, severely damaging the marine ecological environment. Meanwhile, high costs will be incurred for ship repair, salvage operations, and environmental compensation, seriously endangering lives and property. In - depth exploration of the mechanisms and consequences of ship grounding is of great significance for enhancing maritime safety, reducing environmental risks, and minimizing economic losses. On March 18, 1967, the supertanker SS Torrey Canyon ran aground on the Seven Stones Reef off Cornwall, UK. Due to deviation from the main shipping lane, its hull struck a basalt reef at high speed, tearing the keel and rupturing 8 oil tanks. Approximately 80,000 metric tons of crude oil spilled, contaminating the British and French coastlines, marking one of the most severe marine ecological disasters in global history.

Ship grounding has always been a matter with great concern. With the support of the Japan Shipbuilding Industry Structural Improvement Association (ASIS, 1994), two large - scale grounding tests were carried out to simulate grounding accidents with varying degrees of damage. Rodd (1996) reported that the center had conducted four dry grounding tests on double - bottom structures with different stiffener arrangements. Muscat-Fenech & Atkins (1998) conducted quasi - static cutting tests on thin plates using sharp and blunt cone rocks respectively, obtaining the major and minor strains of the thin Zhu et al. (2002) analyzed the distribution and extent of damage in ship grounding accidents based on the statistical data of Lloyd's Register.

DOI: 10.1201/9781003684404-7

Figure 1. The SS Torrey Canyon Grounding, 1967.

They also used empirical formulas and theoretical models to evaluate the grounding force of the bottom structure when the ship ran aground on multiple rocks. Zhang (2002) idealized ship grounding as a rigid wedge tearing through a plate, assuming the plate curls into two curved surfaces behind the wedge tip while material ahead of the wedge is stretched and fractured by direct pushing. A semi-empirical formula was proposed to determine grounding forces during high-energy ship-rock collisions. Considering the influence of the surrounding water, Zhu et al. (2018) conducted ship model grounding tests with different initial relative grounding heights in a water tank and obtained the longitudinal grounding force. To investigate the influence of friction on structural damage during ship grounding, Zhou et al. (2020) conducted three quasi-static cutting tests under distinct friction conditions, revealing that reduced friction coefficients significantly decrease grounding forces and energy absorption during the stable cutting phase. Zhou et al. (2024) conducted numerical simulation of ship model grounding tests in a water tank using the Coupled Eulerian- Lagrangian method, demonstrating close agreement with experimental data.

Ocean reefs not only have diverse shapes, but their compositions and material properties also vary depending on the geographical location of the sea area. Reefs can be composed of biological reef bodies, or formed by the extension of volcanic rock masses or continental rock masses underwater. In the Arctic region, they may even be ice reefs. Since the ship grounding process is a highly nonlinear problem, and the interaction mechanism among the reef, the ship, and the water area is rather complex. In previous studies, reefs have always been simplified as rigid bodies. In this paper, a ship grounding model test is conducted in a water tank using crushable basalt to simulate reef rocks. Hu et al. (2019) proposed an elastoplastic constitutive model for rocks based on elastoplastic mechanics theory. This model adopts the unified strength criterion as the yield criterion, incorporating strain hardening/softening behaviors and strain rate effects. Building on this framework, Xu (2022) conducted laboratory-scale hydro-mechanical triaxial compression tests on basalt specimens integrated with acoustic emission monitoring. The study revealed the mechanical properties and failure mechanisms of the specimens under hydro- mechanical coupling conditions. Further extending the research scope, Zhu et al. (2020) combined experimental and numerical methods to investigate the nonlinear elastoplastic responses of plates subjected to wedge-shaped ice impact. The work explored the plastic deformation and energy absorption characteristics of the plates, as well as the energy dissipation mechanisms induced by ice damage.

In numerical simulations of ship grounding scenarios, the failure criteria for hull and reef structures require special attention. For the hull structure, the strain-based Forming Limit Diagram (FLD) failure criterion proposed by Zhu & Atkins (1998) was implemented. Structural elements are eliminated when the failure criteria were met. The failure characterization of basalt reefs draws upon ice material failure mechanisms. The numerical ice model developed by Cai et al. (2020) provides useful insights, which synergistically integrates the Derradji-Aouat multi-yield surface constitutive model with empirical failure criteria. This model demonstrates strong consistency with the pressure- area (P-A) relationship specified in the ISO 19906 standard, exhibiting a deviation of less than 5%.

2 EXPERIMENTAL DETAILS

2.1 *Test setup*

The schematic diagram of the ship model grounding experimental device is shown in Figure 2, mainly including the ship model, basalt reef model, Motion Capture System-Qualysis,, and a motor. The motor and strings are used to accelerate the ship model. In this experiment, the ship model was towed at a speed of 2.05 m/s by motorand then moved across the basalt reef freely. The Motion Capture System is used to measure and record ship models' displacement, velocity, and acceleration at a frequency of 1000Hz.

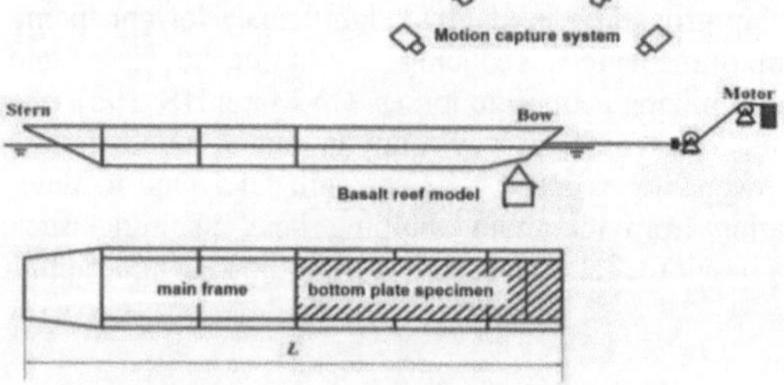

Figure 2. The schematic diagram of ship grounding test.

2.2 *Ship model description*

The main dimensions of the ship model are designed at a scale of 1:65 relative to a VLCC (Very Large Crude Carrier), with a simplified structure consisting of a box-shaped midsection and inclined plates forming the bow, as shown in Figure 3.

Figure 3. The 3D modeling illustration of the ship model.

The model has a block coefficient of approximately 0.88, with a length of 3.14 m, a width of 0.734 m, a height of 0.320 m, and a total weight of 421 kg. The test model comprises two structural components: the main hull frame made of 5.8 mm-thick low-carbon steel plates and the bottom plating specimen fabricated from 0.5 mm-thick galvanized carbon steel sheets.

2.3 *Bottom plate specimen*

In the test, the bottom plating specimen—made of a 0.5 mm-thick galvanized carbon steel sheet— consists of three sections (Figure 4): the first section is a 45° inclined plate (311.1 mm long), the second a 15° inclined plate (373.2 mm long), and the third a horizontal bottom plate (800 mm effective length), where the latter two sections form the primary raking damage zones. Gridlines (50 mm × 50 mm) are marked on the specimens, and the transversely stiffened plate incorporates seven stiffeners with rectangular cross-sections (32 mm × 0.5 mm) spaced at 100 mm intervals.

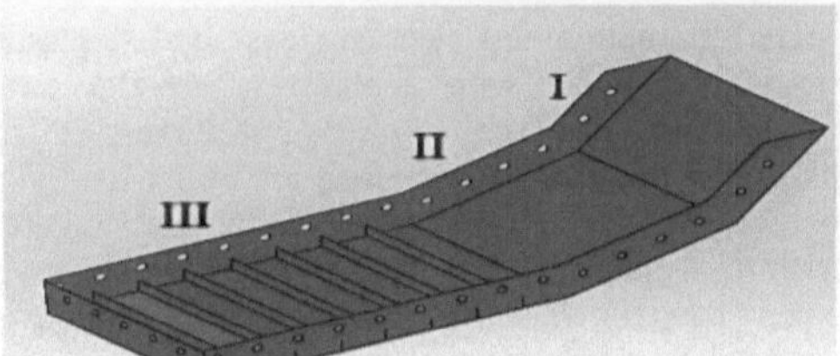

Figure 4. The schematic diagram of stiffened plate.

2.4 *Basalt reef model*

The cone-shaped rock model, with a semi-apex angle of 30 ° and made of basalt, has a height of 181 mm, and its tip is chamfered to a radius of 2 mm. To secure the basalt reef to the base, the fixture for the reef model is constructed from 5-mm-thick high-strength stainless steel, comprising two clamps and four M20 bolts. Each clamp consists of two semi-conical shells joined by a connecting screw at the bottom. During testing, the reef model is positioned between the clamps, tightly fitted together, and secured with the M20 bolts, while an M30 screw at the lower end of the clamp anchors it to the reef base at the tank floor, preventing movement or rotation during ship grounding tests.

2.5 *Test scenarios*

The ship model has a draft of 190 mm, with its bottom baseline positioned 82 mm below the top of the basalt reef installed on the grounding base. During the experiment, the model was connected via a traction cable routed through a fixed pulley to a motor, which provided forward propulsion. Once the ship model reached the target initial speed of 2.05 m/s, the cable was severed, allowing the model to coast at a constant velocity before grounding onto the basalt reef at the predetermined speed.

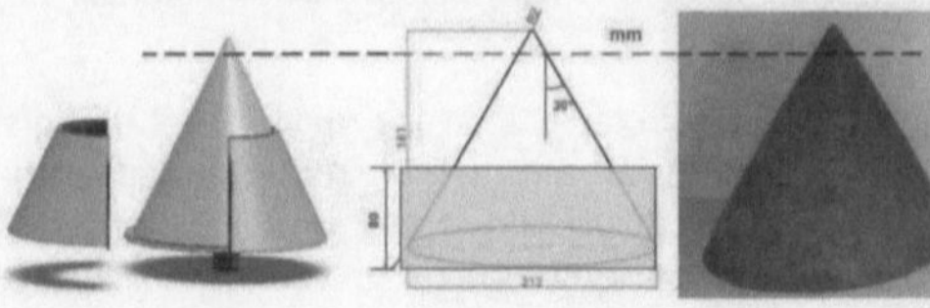

Figure 5. The schematic diagram of basalt reef model and fixture.

3 EXPERIMENTAL RESULTS

The results of the grounding test is shown in Figure 6, including the damage of the bottom plating specimen, longitudinal grounding force, energy integrated by longitudinal grounding force, and the motion of the ship model.

3.1 *Damage mode*

The damage and deformation of the transversely stiffened bottom plating specimen are shown in Figure 6. The specimen exhibits a continuous denting/tearing mode, with wavy, upward-curling edges along both sides of the torn area. The first transverse stiffener fractured and bent sternward, while the total grounding damage spans 488.5 mm in length and 56.8 mm in maximum width. The entire grounding event lasted 0.505 seconds.

3.2 *Longitudinal grounding resistance*

The motion capture system monitors changes in the longitudinal grounding speed of the ship model, enabling calculation of acceleration and longitudinal grounding force variations. Initially, the ship model

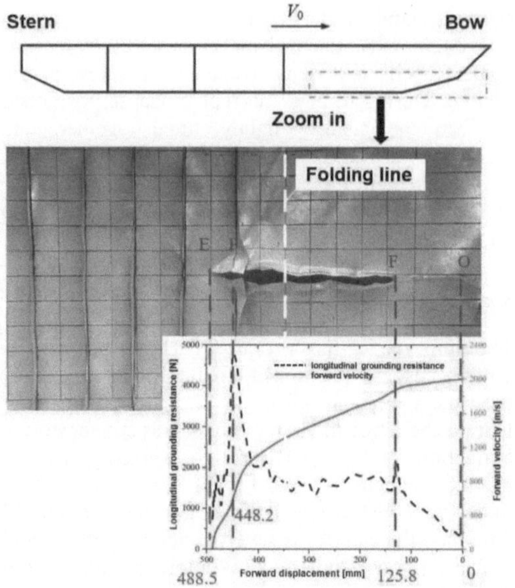

Figure 6. The damage longitudinal resistance and the history of the forward velocity of stiffened ship bottom.

contacted the reef at a forward speed of 2.05 m/s (marked as point O), after which continuous denting damage occurred on the bottom plate alongside a gradual rise in longitudinal grounding force. At 0.064 s (grounding displacement: 125.8 mm, point F), the forward speed decreased to 1.87 m/s, the bottom plate began fracturing, and the force reached a minor peak (1998.6 N). Subsequent tearing damage progressed until grounding reached the III section of the stiffened plate, where speed reduction accelerated and force surged sharply. By 0.314 s (displacement: 449.6 mm, point P), the force peaked at 4798.3 N, the first transverse stiffener sheared off, and fracture edges bent sternward as the speed dropped to 0.62 m/s. Finally, the bottom plate tore an additional 50 mm, the ship's speed reached zero, and the force rapidly diminished to 0.

3.3 *Basalt reef damage*

During the grounding, the top of the basalt reef fractured. The fracture surface was inclined and uneven, as shown in Figure 7. The original height of the reef was 181 mm, and the height of the reef after the test was 158 mm.

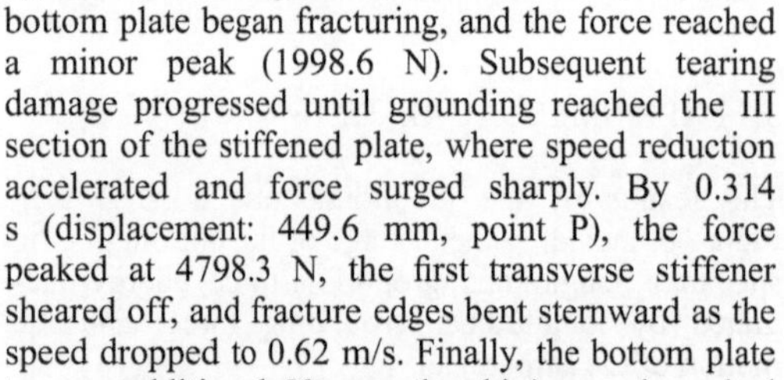

Figure 7. The damage of basalt reef.

4 SIMULATION

The finite element software Abaqus is used to conduct a simulation of the basalt reef grounding under the same experimental conditions.

4.1 *Constitutive model of material*

In this paper, basalt is modeled as an elastoplastic material with a linearly elastic component and a plastic component represented by the linear Drucker-Prager model in Abaqus/Explicit, as illustrated in Figure 8.

The failure of basalt reef is characterized by the maximum normal stress criterion and shear failure criterion, where the former postulates that failure occurs when any of the three principal stresses reaches the basalt reef's compressive or tensile strength, and the latter defines failure based on shear failure strain. The failure of basalt reef is defined based on shear failure strain, which is set to 0.02 in this study. The relevant parameters are provided in Table 1.

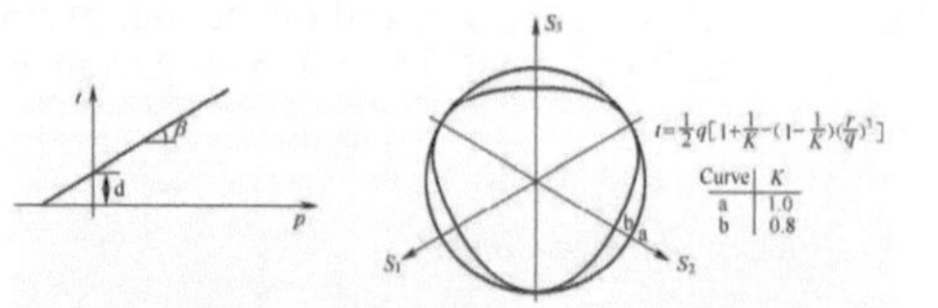

Figure 8. Yield surface of the linear Drucker-Prager model.

Table 1. The parameters of the basalt reef.

Density/kg/m^3	2900
Young'smodulus/Gpa	48.75
Poisson'sratio	0.26
compressivestrength/Mpa	122.6
tensilestrength/Mpa	10

The ship bottom test plate employs an elastoplastic constitutive model with a thickness of 0.5 mm, Young's modulus of 199,000 MPa, yield stress of 315 MPa, and ultimate tensile stress of 365 MPa. The Johnson-Cook (JC) constitutive model accounts for strain-rate effects, while ductile fracture failure is implemented in Abaqus, where fracture initiation is governed by a predefined failure strain. once reached, the material fractures.

4.2 *Simulation settings*

The ship-basalt reef grounding model test is simulated using the CEL (Coupled Eulerian- Lagrangian) method in Abaqus/Explicit, where the Eulerian mesh represents the water area and the Lagrangian mesh models the ship and basalt reef. The relative positions of the ship model and reef within the water domain are illustrated in Figure 9.

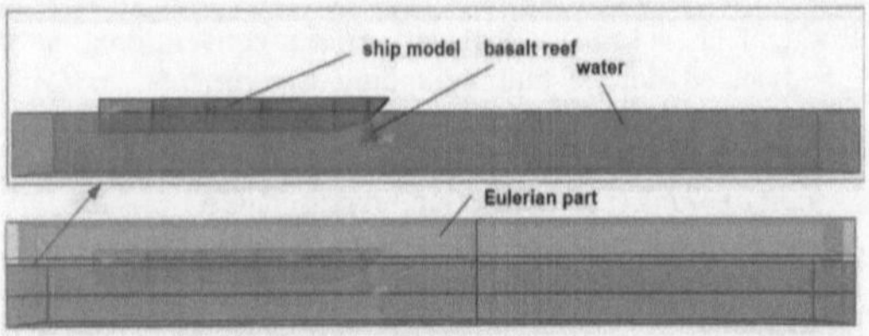

Figure 9. The position of the ship model and the basalt reef Simulation results.

4.3 *Simulation results*

The damage modes of the ship's bottom plate in both the grounding experiment and simulation are consistent, exhibiting continuous denting/tearing patterns. The experimental and simulated damage lengths are 488.5 mm and 459.6 mm, respectively, with the shorter simulated length attributed to earlier fracture initiation, higher initial grounding forces, and premature energy dissipation in the simulation.

Comparison of the basalt reef damage post-grounding shows fractured reef tips and more severe front-side damage in both experimental and simulation results, as illustrated in Figure 11.

Figure 10. The comparison between experiment and simulation.

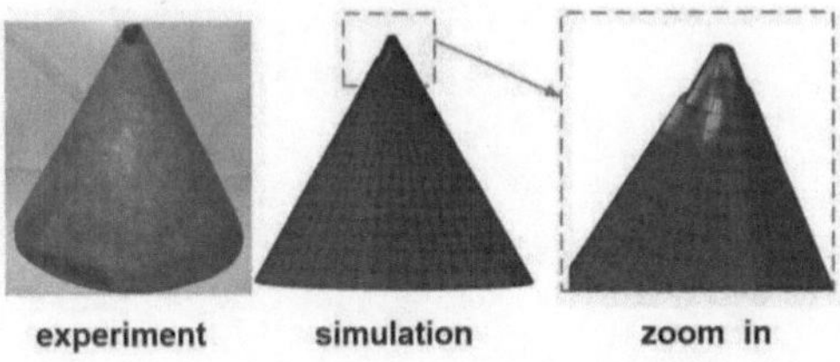

Figure 11. The comparison between experiment and simulation.

5 GROUNDING ENERGY DISSIPATION ANALYSIS

During ship grounding, part of the ship's initial kinetic energy is dissipated through water resistance, friction between the basalt reef and the bottom plate specimen, energy absorption by the reef's damage/deformation, and the plate's own deformation/fracture. The remaining energy splits into potential energy of the ship model and residual kinetic energy, as summarized in the formula below.

$$E_{k0} = E_f + E_D + E_R + E_{\text{Pot}} + E_{\text{hydro}} + E_k$$

In this context, E_{k0} represents the ship model's initial kinetic energy, E_f the frictional energy dissipation, E_D the energy consumed by the bottom plate's deformation/fracture, E_R the energy absorbed by the basalt reef's damage/deformation, E_{Pot} the ship model's potential energy increment, E_{hydro} the fluid-induced energy dissipation, and E_k the ship's residual kinetic energy. The grounding process concludes when the ship model's velocity reaches zero (kinetic energy $E_k = 0$).

Through the simulation calculation of the ship grounding on the basalt reef, the energy dissipation situation of the ship grounding under the test conditions is obtained, as shown in Figure 12.

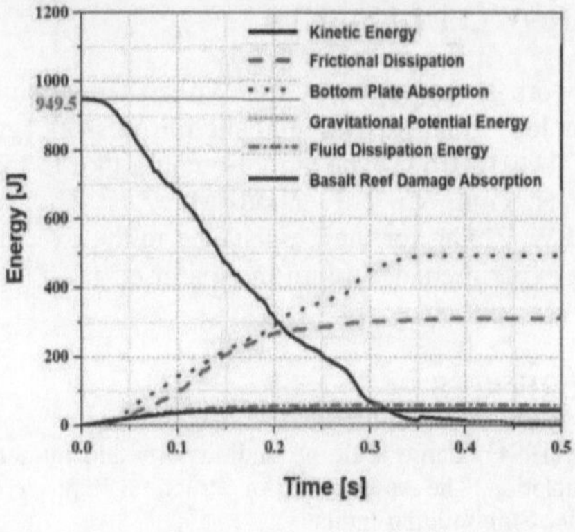

Figure 12. The energy conversion during ship grounding.

During the grounding simulation, the energy absorbed through hull structural deformation and fracture accounts for 52.0% (494.0 J), frictional energy dissipation accounts for 33.1% (314.0 J), the energy dissipated by fluid interaction accounts for 6.2% (58.7 J) of the initial kinetic energy, the energy consumed by the deformation and damage of the crushable basalt reef represents 4.8% (45.8 J), and the gravitational potential energy gained by the vessel comprises 3.9% (37.05 J).

6 CONCLUDING REMARKS

In this study, a ship grounding experiment was conducted using a conical basalt reef, followed by numerical simulations via the Coupled Eulerian-Lagrangian (CEL) method in Abaqus under identical experimental conditions to validate and expand the findings.

(1) During the test, the stiffened ship bottom plate exhibited continuous denting/tearing damage with wavy edges along the torn area, while the first transverse stiffener fractured and the basalt reef's top developed an inclined, uneven fracture surface.
(2) Both experimental and simulated results showed consistent continuous denting/tearing damage modes in the ship bottom plate, though the simulated damage length (459.6 mm) was shorter than the experimental value (488.5 mm) due to earlier fracture initiation and higher initial forces in the simulation. Post-grounding comparisons of the basalt reef revealed fractured tips and more severe front-side damage in both cases.
(3) Throughout the grounding process, the ship's initial kinetic energy dissipated through bottom plate deformation/fracture (52.0%), friction (33.1%), fluid resistance (6.2%), reef damage/deformation (4.8%), and potential energy conversion (3.9%), with residual kinetic energy persisting until the ship's velocity reached zero.

ACKNOWLEDGEMENTS

This work is supported by the Wuhan University of Technology fund for Distinguished Professors (grant no. 471-40120163). The authors would like to thank Mr. Z.H. Guo, Mr. H.F. Zhou, Mr. Z.H. Zhou and Mr. H.W. Zhang for their assistance in the execution of the experimental work of the manuscript.

REFERENCES

ASIS. (1994). Large scale grounding tests and numerical simulation. The Association for Structural Improvement of the Shipbuilding Industry.

Rodd, J. L. (1996). Observations on conventional and advanced double hull grounding experiments. *In Proceedings of the International Conference on Designs and Methodologies for Collision and Grounding Protection of Ships* (pp. 1–10). San Francisco, CA.

Muscat-Fenech, C. & Atkins, A. G. (1998). Denting and fracture of sheet steel by blunt and sharp obstacles in glancing collisions. *International Journal of Impact Engineering*, 21(7), 499–519.

Zhu, L & Atkins, A. G. (1998). Failure criteria for ship collision and grounding. *In Proceedings of the International Conference on Practical Design of Ships and Mobile Units* (pp. 1–12). The Hague, Netherlands.

Wang, G., Arita, K. & Liu, D. (2000). Behavior of a double hull in a variety of stranding or collision scenarios. *Marine Structures*, 13(3), 147–187.

Zhu, L, James, P & Zhang, S. (2002). Statistics and damage assessment of ship grounding. *Marine Structures*, 15(4), 515–530.

Zhang, S. (2002). Plate tearing and bottom damage in ship grounding. *Marine Structures*, 15(2), 101–117.

Zhu, L, Zhou, Q. & Chen, M. (2018). Grounding experiments of a ship model in a water tank. *In Proceedings of the ASME 2018 37th International Conference on Ocean, Offshore and Arctic Engineering*. Madrid, Spain.

Zhou, Z., Zhu, L & Guo, K. (2020). Plate tearing experiments over a cone-shaped rock considering friction in ship grounding. *In Proceedings of the 30th International Ocean and Polar Engineering Conference* (pp. 1–8). Shanghai, China.

Zhou Z, Zhu L & Li C. (2024). Ship Grounding Simulation with Different Rock Shapes and its Verification[J]. *Ships and Offshore Structures*,1,1–13.

Hu, X., Li, K. & Qu, S. (2019). Elastoplastic constitutive model of rock based on unified strength theory and its numerical implementation. *Explosion and Shock Waves*, 39(8), 130–138.

Xu, Q. (2022). Mechanical properties and numerical simulation of basalt under hydro-mechanical coupling [Doctoral dissertation, Shenyang University of Technology].

Zhu L, Cai W & Chen M. (2020). Experimental and numerical analyses of elastic-plastic responses of ship plates under ice floe impacts. *Ocean Engineering*, 218:108174.

Cai W, Zhu L & Yu T. (2020). Numerical simulation for plates under ice impact based on a concrete constitutive ice mode. *International Journal of Impact Engineering*, 143:103594.

Trends in Collision and Grounding of Ships and Offshore Structures – Zhu & Guedes Soares (Eds)

Study on multi-mode failure of ship side structures under wedge-shaped bow collision based on numerical simulation

K. Xu, G. Shi & D. Wang
State Key Laboratory of Ocean Engineering, Shanghai Jiao Tong University, Shanghai, P.R. China
Institute of Marine Equipment, Shanghai Jiao Tong University, Shanghai, P.R. China

ABSTRACT: Ship collision accidents occur frequently, and such collisions often result in damage to multiple structural components of the vessel. Different components exhibit unique characteristics and deformation modes. Additionally, the coupling effects between components and the evolution of damage leads to a highly complex failure of the ship's structure. This paper establishes multiple collision scenarios between the bow and side structures of the vessel based on varying collision angles, speeds, and positions. Considering that the actual ship's bow is not completely rigid, each scenario use elastic bow to better simulate real-world collision scenarios. Finite element software ABAQUS is utilized for numerical simulations of these conditions. Combining the results of the numerical simulations, this study investigates the multi-mode failure evolution process of the hull structure during the collision and analyzes the failure modes of different components.

1 INTRODUCTION

As the most common type of navigation accident, ship collisions can lead to severe consequences such as casualties, environmental pollution, and economic losses. The raked bow, a common type of ship bow, can effectively prevent waves from coming aboard. However, this type of bow structure is relatively sharper and often causes more severe consequences when colliding with the ship's side structure. To minimize the losses caused by collision accidents, it is necessary to study the crashworthiness of ship structures.

The main research methods are divided into three categories: experimental methods, simplified analytical methods, and numerical simulations (Hong, 2009). Experiments can provide reliable evidence for the other two methods, but they are costly and time-consuming. Most experiments involve the interaction between rigid bows and typical components of ship structures. Sahu et al. (2018) studied the failure of plates under different pressure heads through experiments. Liu et al. (2022) conducted penetration tests on reinforced plates at different angles. Qiu et al. (2024) performed small-scale quasi-static tests on L-shaped structures composed of decks and sides. Compared to experimental methods, simplified analytical methods have the advantages of low cost and high efficiency, and can provide some explanations for the physical principles behind structural failure. Sun et al. (2015), Wang et al. (2021), and Zhou et al. (2023) derived formulas for the forces on different structures of the ship's side under oblique collision conditions by combining finite element results. Numerical simulation methods, while also being low-cost and efficient, can flexibly construct different collision scenarios and visualize the collision process. Liu et al. (2021) investigated the structural damage patterns of double-hull ships under collision and grounding scenarios based on numerical simulation. Ringsberg et al. (2018) and Che et al. (2022) used numerical methods to study the failure sequence and modes of different components during the collision process between a spherical bow and a double-layer side structure.

This study mainly uses the nonlinear finite element software ABAQUS to investigate the collision process between a raked bow and side structures. Various collision scenarios are set up based on collision location, collision speed, and collision angle, and the effects of different influencing factors on various structural failure modes are studied. This provides a reference for the study of ship crashworthiness.

2 FINITE ELEMENT SETTINGS

2.1 *Bow and compartment segment finite element models*

The bow model's shape and dimensions are as shown in Figure 1. The bow's outer contour is based on the dimensions of a real ship's bow scaled down to 1/20th size, with the internal framework simplified to two deck structures and longitudinal bulkhead structures. The steel thickness of the bow is 5mm. To

DOI: 10.1201/9781003684404-8

better simulate the rotation of the bow, none of the six degrees of freedom of the bow were constrained.

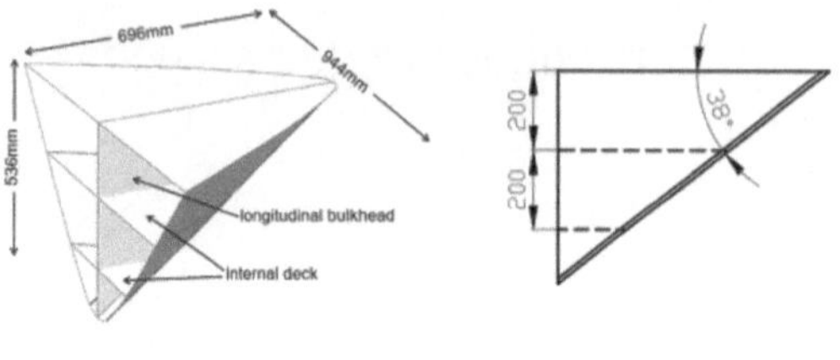

Figure 1. Shape and size of bow.

For the compartment structure, the parallel mid-body section of the ship was selected. The impact points were designed at the intersection of the transverse bulkhead and the longitudinal girder, as well as at the longitudinal girder between two transverse bulkheads. Considering the structure of the transverse bulkhead, which has high rigidity and strength, it can limit the lateral deformation and torsion at both ends of the compartment. Therefore, only the compartment between the transverse bulkheads at both ends of the impact points was modeled. The thickness of the model steel was set to 3mm, and the shape and dimensions are shown in Figure 2.

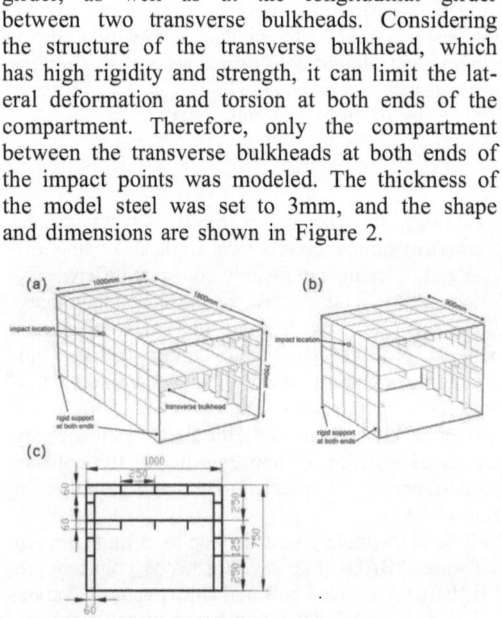

Figure 2. Compartment model shape and dimensions: (a) Model used when impacting the transverse bulkhead. (b) Model used when impacting the truss. (c) Section dimensions of the cabin.

2.2 *Working condition setting*

This paper mainly analyzes the impact location, impact velocity and angle. When analyzing the impact position effect, the fixed impact velocity was 5 m/s and the angle was 90 degrees. Two typical impact locations were selected for analysis:

1) Impact location 1: the intersection of the transverse bulkhead and the longitudinal girder.
2) Impact location 2: the longitudinal girder in the middle of the two transverse bulkheads.

When analyzing the impact of velocity, the initial collision location is fixed at the intersection of the transverse bulkhead and the longitudinal girder, with the collision angle set to 90 degrees. Three collision speeds are set: 3m/s, 5m/s, and 7m/s.

When analyzing the impact of angle, the initial collision location is fixed at the intersection of the transverse bulkhead and the longitudinal girder, with the collision speed set to 5m/s. Three collision angles are designed: 45 degrees, 60 degrees, and 75 degrees.All design conditions are shown in Table 1:

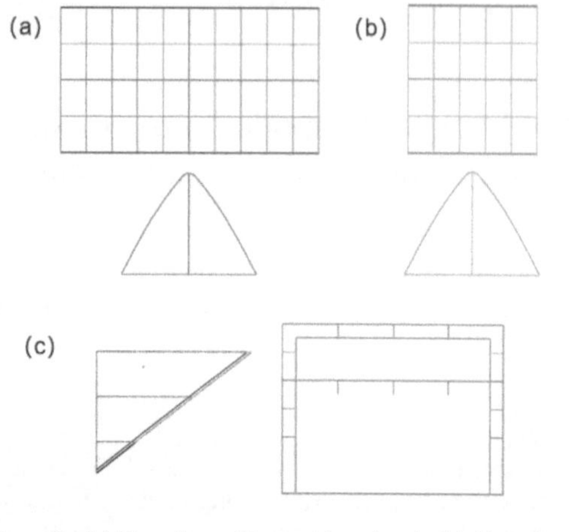

Figure 3. (a) Top view of impact location 1. (b) Top view of impact location 2. (c) Side view of two impact location.

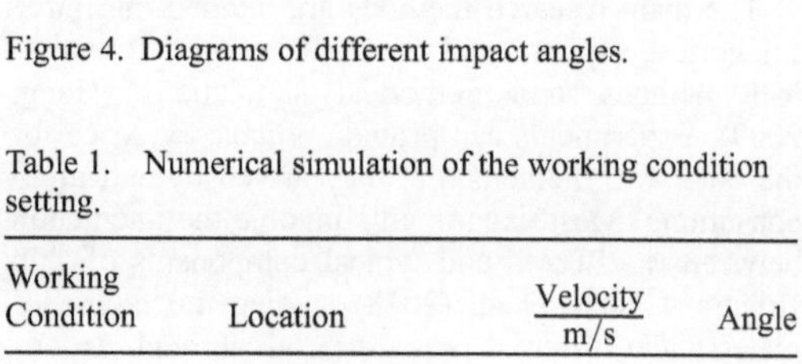

Figure 4. Diagrams of different impact angles.

Table 1. Numerical simulation of the working condition setting.

Working Condition	Location	Velocity m/s	Angle
a1	impact location 1	5	90°
a2	impact location 2	5	90°
b1	impact location 1	3	90°
b2	impact location 1	5	90°
b3	impact location 1	7	90°
c1	impact location 1	5	15°
c2	impact location 1	5	30°
c3	impact location 1	5	45°

2.3 *Numerical simulation material parameter setting*

The simulation employed a trilinear constitutive model considering element failure for the steel material. The material's constitutive relationship was derived from uniaxial tensile tests of steel (Figure 5). The study assumed the material to be isotropic, and the true stress-strain curve of the material was obtained based on the uniaxial tensile test results, with calculations as shown in the following equations:

$$\sigma_{ture} = \sigma_0(1 + \varepsilon_0) \tag{1}$$

$$\varepsilon_{ture} = \ln(1 + \varepsilon_0) \tag{2}$$

Among them, σ_0 and ε_0 represent the stress and strain in tensile tests, while σ_{ture} and ε_{ture} denote the true stress and true strain.

The true stress-strain curve was fitted using a trilinear curve, with the fitting formula as shown in the equation 3, and the results are presented in Figure 6.

$$\sigma = \begin{cases} E\varepsilon, & 0<\varepsilon<\varepsilon_0 \\ \sigma_S + K_1(\varepsilon - \varepsilon_0), & \varepsilon_0<\varepsilon<\varepsilon_{P1} \\ \sigma_{P1} + K_2(\varepsilon - \varepsilon_{P1}), & \varepsilon_{P1}<\varepsilon \end{cases} \tag{3}$$

Figure 5. Steel Tensile Test.

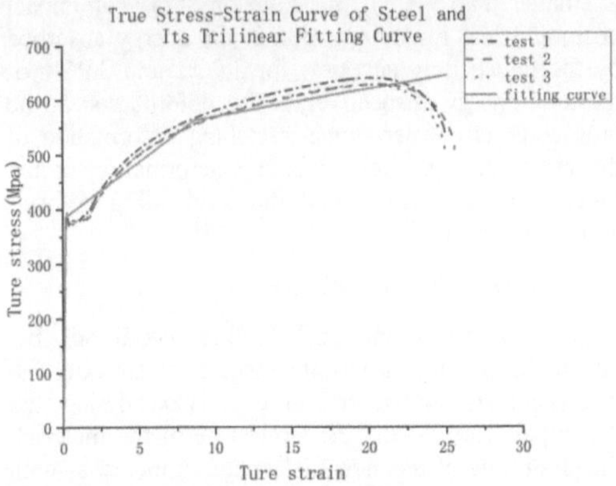

Figure 6. True stress-strain curve of steel and its trilinear fitting curve.

Since steel is highly sensitive to strain rate, an increase in strain rate leads to a rise in its yield strength. Under impact conditions, the strain rate significantly influences the dynamic response and damage of the structure. Therefore, the Cowper-Symonds model is employed to describe the strain rate hardening effect of the material. The model is expressed by the following equation:

$$\sigma_{yd} = \sigma_y\left(1 + \left(\frac{\dot{\varepsilon}}{D}\right)^{\frac{1}{q}}\right) \tag{4}$$

Among them, σ_y is the material yield strength, σ_{yd} is the material dynamic yield strength, $\dot{\varepsilon}$ is the material strain rate, and D and q are material constants. For this calculation, D=3200 and q=5 are adopted. The parameters of the metal material in the finite element calculation are shown in the table:

Table 2. Material parameter settings.

Physical quantity	Symbol	Unit
Density	Kg/m^3	7850
Young's modulus	EPa	2.05×10^{11}
Poisson ratio	v	0.3
Yield stress	Pa	3.26×10^{8}
Second yield	Pa	5.63×10^{8}
Tangent modulus	K$_1$Pa	2.07×10^{9}
Tangent modulus	K$_2$Pa	5.04×10^{8}
Hardening parameters	D	3200
Strain rate parameter	q	5
Equivalent plastic failure strain		0.25

3 INTERPRETATION OF RESULT

3.1 *Different impact locations*

Simulate the two conditions of a1 and a2, and analyze the resistance-ship bow displacement curve in conjunction with the collision process in numerical simulation.

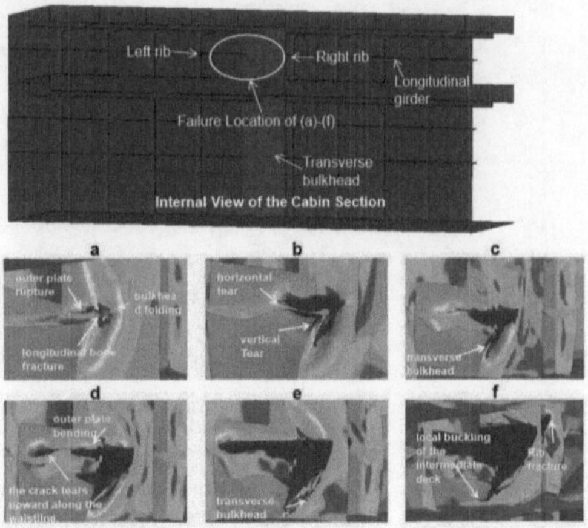

Figure 7. Failure sequence of various components at impact location 1.

Record the failure conditions of each component during the collision process at different impact positions, as shown in Figure 7-Figure 8, and mark the corresponding positions of the failure events on the resistance-ship bow displacement curve, as shown in Figure 9.

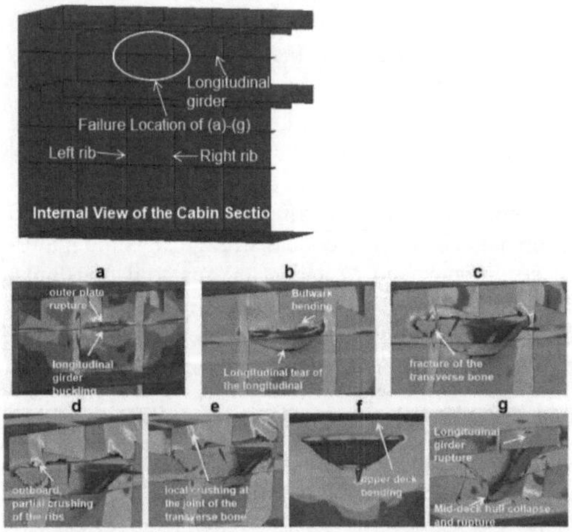

Figure 8. Failure sequence of various components at impact location 2.

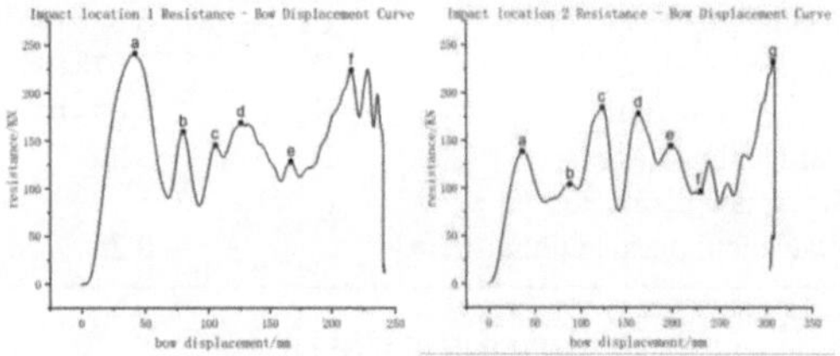

Figure 9. Resistance - Bow Position curve under different impact locations.

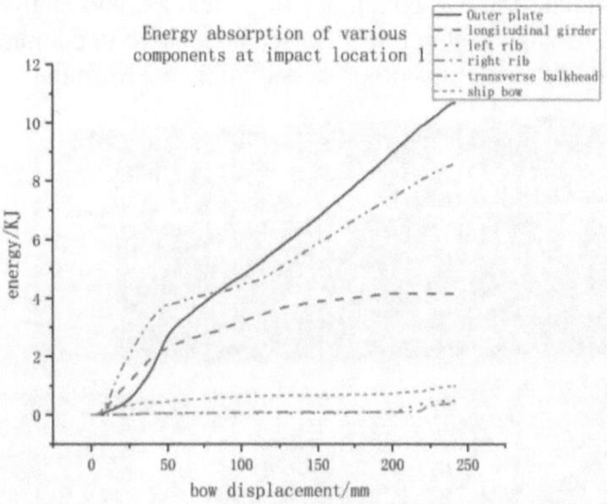

Figure 10. Energy absorption of various components at impact location 1.

Then calculate the changes of the deformation energy of different components with the increased bow displacement, as shown in Figure Figure 10-Figure 11. Combined with the deformation and energy curve of each component in the collision process, it can be seen that for the impact position 1, at the early stage of impact, the folding deformation of the transverse bulkhead and the flexion and bending of the longitudinal truss are mainly dissipated, followed by the membrane tensile deformation of the outer plate. With the further invasion of the bow, the membrane stretching range is further increased and the steel plate with a week of breaking bends all around, and the outer plate becomes the dominant energy dissipation.

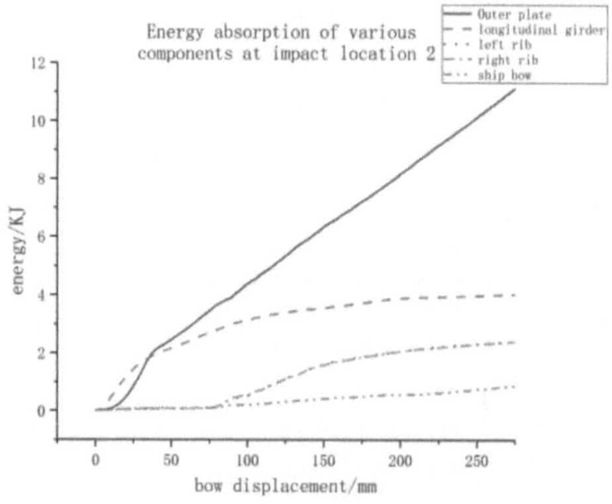

Figure 11. Energy absorption of various components at impact location 2.

At this time, the longitudinal truss is locally bending and deformed, and the transverse bulkhead is mainly torn, and the contribution to the energy dissipation is smaller. Subsequently, a large area of curved folding occurs in the transverse bulkhead, which increases the energy absorption once again. For the impact position 2, the flexion and bending of the longus are mainly for energy dissipation, followed by the membrane tensile deformation of the outer plate. Similarly, with the increase of the impact range, the outer plate replaces the truss as the main energy dissipation member. When the bow displacement increases further, the ribs on both sides of the impact position undergo flexion, bend and tear, and the dissipation of energy increases. However, the bow of the impact ship contributes less to the energy dissipation during the collision process, which is smaller than that of the collision ship (For impact position 1 and impact position 2, the energy absorbed by the punch only accounts for 3.96% and 3.95% of the total energy, respectively). The deformation at the bow consists of membrane stretching deformation of the outer plating, slight buckling deformation of the longitudinal bulkheads, and minor buckling deformation of the deck.

3.2 *Different impact velocity*

Simulations were conducted for three conditions, b1, b2, and b3, and the curves of resistance versus bow displacement are shown in Figure 12. Considering that the impact rate affects the strain rate of the material, the strain rate of the side shell material increases with the increase in impact velocity. From the three sets of resistance-ship bow curves, it can be seen that as the impact velocity increases, both the magnitude of the first peak resistance and the corresponding bow displacement increase. This indicates that a higher strain rate enhances the strength of the hull structure. Furthermore, the overlapping sections of the subsequent three curves exhibit a consistent trend, suggesting that the failure sequence and failure modes of the different components of the side structure are not significantly influenced by the material strain rate sensitivity. The final damage to the lower hull under different conditions is shown in Figure 13. When the collision speed is 3m/s, the deformation failure of the side structure is basically limited to the rectangular area formed by two decks and the ribs on both sides of the collision location. When the collision speed is 7m/s, as the ram penetrates deeper, the ribs on both sides of the impact point are torn, and the tip of the ram strikes the longitudinal girder of the deck, causing it to bend and tear.

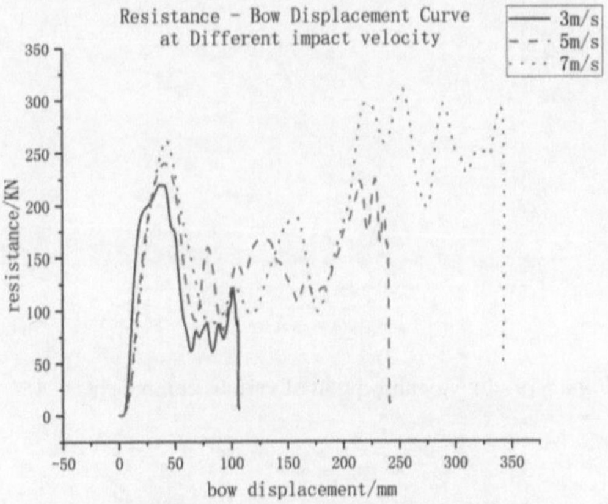

Figure 12. Resistance vs. bow position curve at different impact velocity.

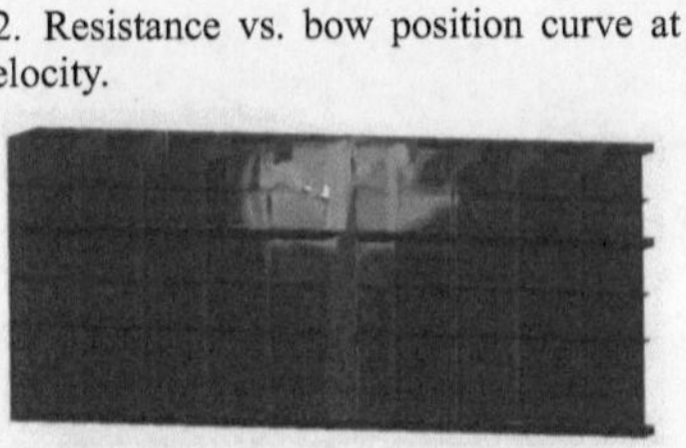

(a)3m/s

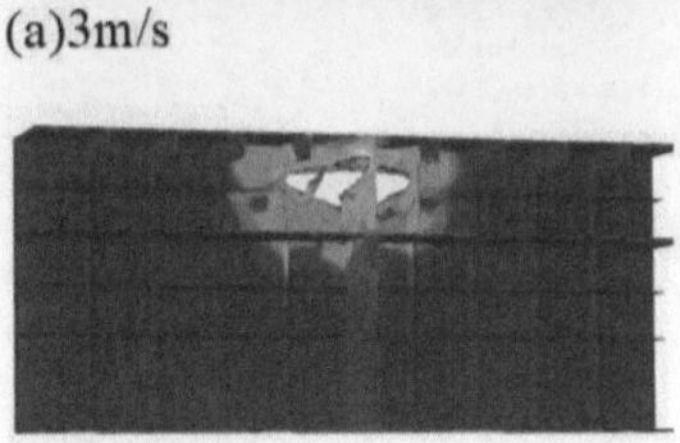

(b)5m/s

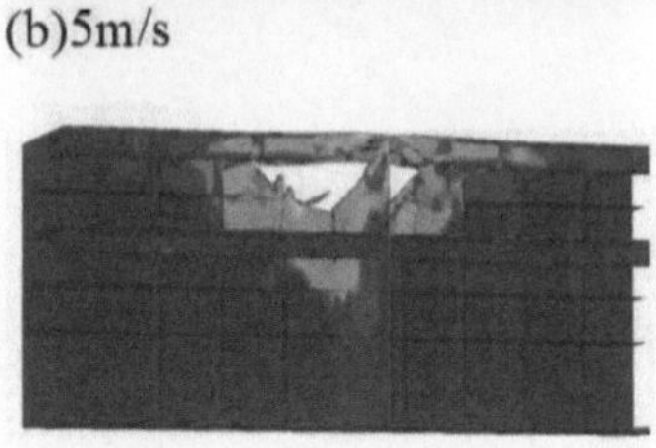

(c)7m/s

Figure 13. Final damage state of compartments at different impact velocity.

Observing the energy absorption conditions of the three working conditions as shown in Figure 14 to Figure 16. With the increase of impact velocity, the energy absorption of the outer plate and transverse bulkhead keeps increasing. However, the energy absorption of the longitudinal stringer does not show a significant increase when the impact velocity rises from 5 m/s to 7 m/s. This observation indicates that there is an upper limit to the energy absorption capacity of the longitudinal stiffener, and once it is completely fractured, its ability to absorb energy cannot be further enhanced.

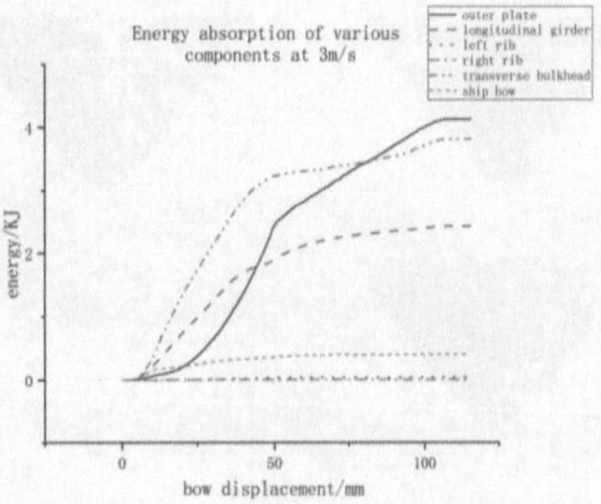

Figure 14. Energy absorption of various components at 3m/s.

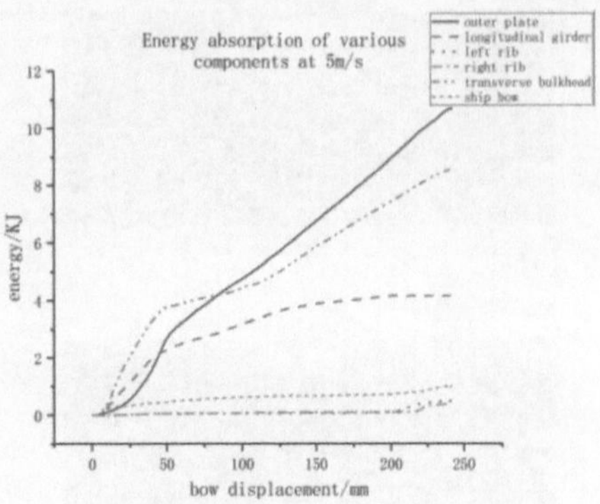

Figure 15. Final damage state of compartments at 5m/s.

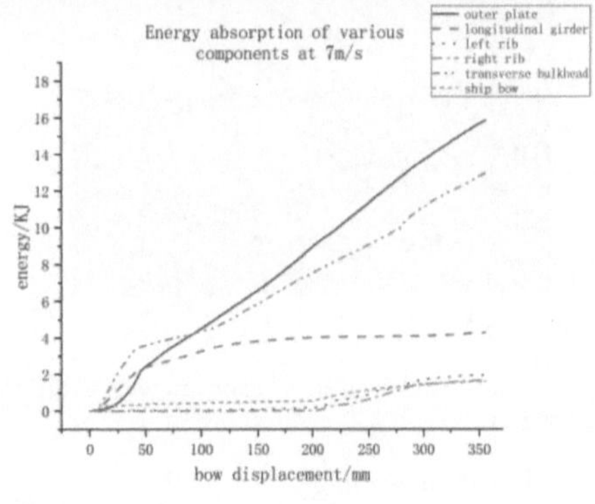

Figure 16. Final damage state of compartments at different impact velocity.

The damage to the bow of the ship, as shown in Figure 17, significantly increases with the increase in impact speed. At an impact speed of 3m/s, only a slight plastic deformation occurs at the tip of the bow. At an impact speed of 5m/s, the bow column and the outer plate of the side structure, as well as the connection positions between the transverse bulkhead and the internal deck, experience more pronounced plastic deformation upon contact. At an impact speed of 7m/s, the outer plates on both sides of the bow and the areas where the side frames and longitudinal girders form a cross structure are severely indented inward.

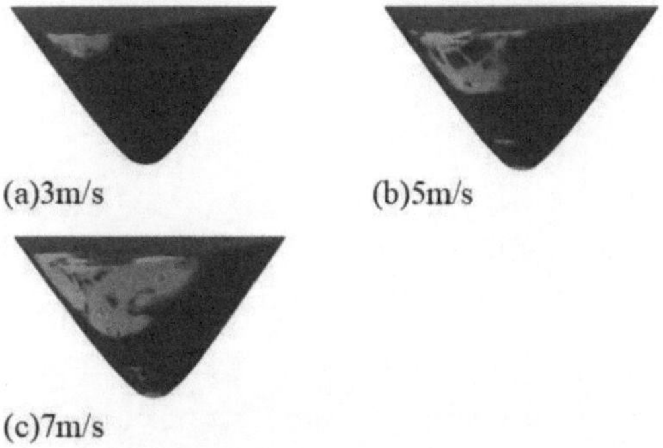

Figure 17. Final damage state of ship bow at different impact velocity.

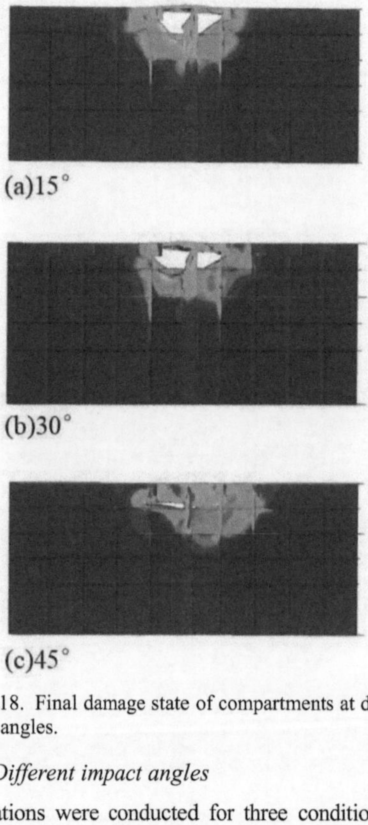

Figure 18. Final damage state of compartments at different impact angles.

3.3 *Different impact angles*

Simulations were conducted for three conditions, c1, c2, and c3. The final damage to the lower compartment under different conditions is shown in Figure18. In the oblique collision condition, the ram would deflect upwards, with the breach area located between the upper deck and the lower longitudinal girder, where the longitudinal girder primarily experienced buckling bending without folding. As the collision angle increased, the shape of the breach in the outer plate gradually became more elongated. When the collision angle was 45 degrees, the bow did not penetrate the hull but instead slid along the side. Compared to head-on collisions, the impact range of oblique collisions is greater at the same speed.

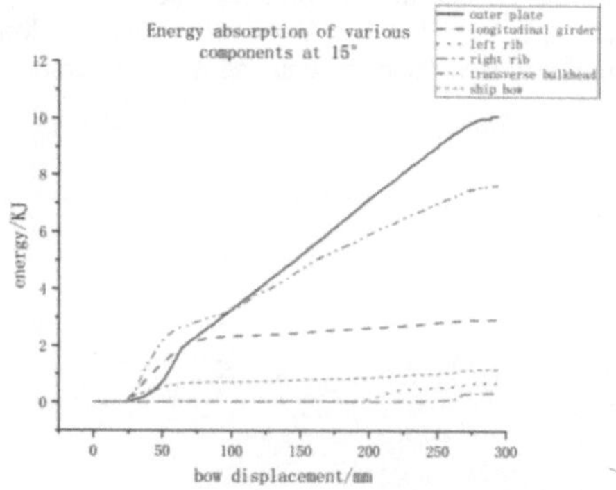

Figure 19. Energy absorption of various components at 15°.

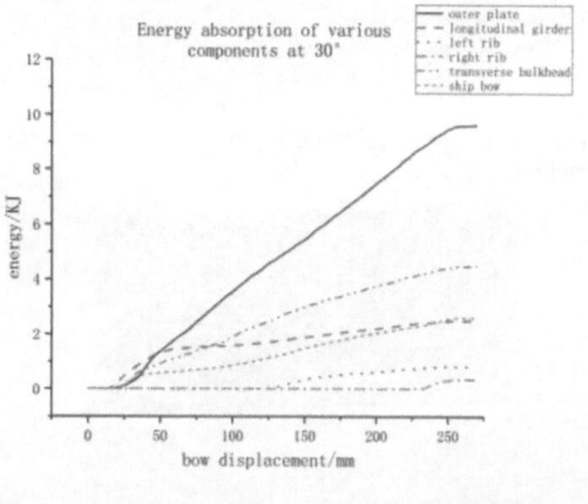

Figure 20. Energy absorption of various components at 30°.

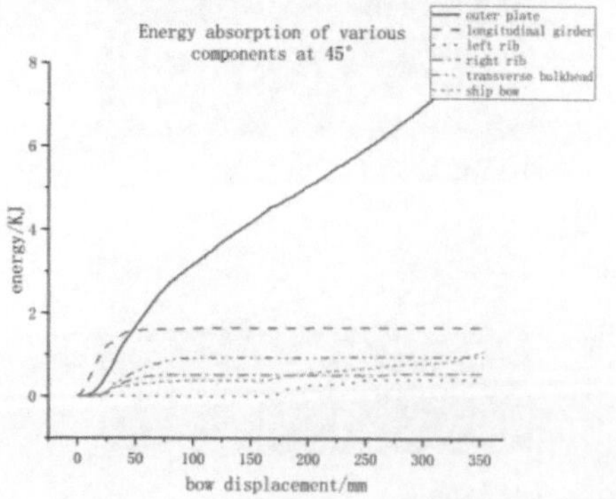

Figure 21. Energy absorption of various components at 45°.

Observing the energy absorption conditions of the three working conditions as shown in Figure 19 to Figure 21, the side shell plating remains the primary energy-absorbing component. With the increasing of initial collision angle, the deformation of the transverse bulkhead shifts from longitudinal folding and crushing to Lateral stretching, resulting in a significant reduction in energy absorption. At the initial impact angle of 45°, the bow deflects due to the collision, slides along the side shell, and separates from it. Consequently, the input energy is not fully dissipated during the collision process.

The damage to the bow of the ship, as shown in Figure 22. Compared to the head-on collision condition, the damage to the ship's bow is lighter in the oblique collision condition. There was a slight plastic deformation at the bow tip, and a tensile deformation inward occurred on the side of the bow outer plate that came into contact with the side of the hull.

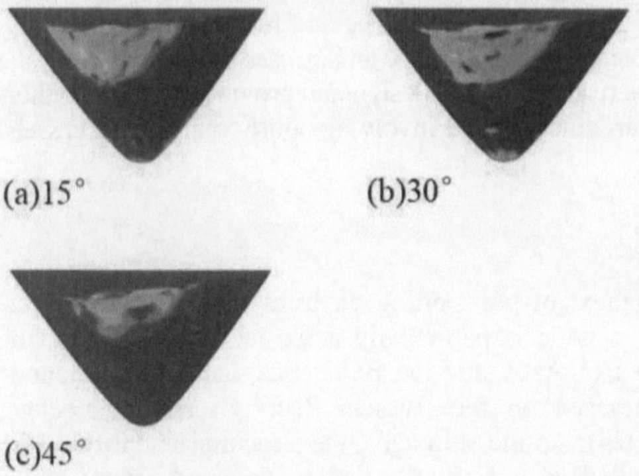

Figure 22. Final damage state of ship bow at different impact angles.

4 CONCLUSIONS

This article employs the finite element software ABAQUS to conduct numerical simulations of collisions between a wedge-shaped bow and the ship's side, generating virtual experimental data. Based on the numerical simulation results, the deformation and failure sequence of different components of the struck ship at various collision locations were organized, and the energy absorption of different components during the collision process was analyzed. Furthermore, the similarities and differences in failure modes under different collision angles and speeds were discussed. The deformation of the elastic bow during the collision process was also considered. The research findings indicate that:

1. During the collision process, the energy is primarily absorbed by the side structure of the struck vessel, while the contribution of the bow to energy absorption is relatively small.
2. Throughout the entire collision process, the outer shell and transverse bulkheads of the struck vessel serve as the main energy-absorbing components. Although the longitudinal frames contribute significantly to energy absorption in the initial stages of the collision, they have a maximum energy absorption capacity.
3. The strain rate influences the strength of the ship's side structure. As the impact velocity increases, the strength of the side structure also increases; however, this does not alter the failure modes or the sequence of failure of the different components of the side structure.
4. Compared to head-on collisions, oblique impacts at the same speed result in a larger damage extent on the side structure.

The research results have certain value for the study of ship collision resistance.

REFERENCES

Hong, L. 2009. Simplified analysis and design of ships subjected to collision and grounding. Doctoral thesis. Trondheim: Department of Marine Technology, Norwegian University of Science and Technology.

Chen, B.-Q., Liu, B. & Guedes Soares, C. 2022. Experimental and numerical investigation on a double hull structure subject to collision. *Ocean Engineering*, 256: 1–12.

Liu, F., Hu, Y., Feng, G. & Li, C. 2022. Experimental and numerical study on the penetration of the inclined stiffened plate. *Ocean Engineering*, 258: 1–12.

Liu, B., Villavicencio, R., Pedersen, P.T. & Guedes Soares, C. 2021. Analysis of structural crashworthiness of double-hull ships in collision and grounding. *Marine Structures*, 76: 1–12.

Qiu, X., Zhao, Y. & Wang, D. 2024. Mechanical analysis and interior stiffeners arrangement discussion of a specific ship side collision situation. *Ocean Engineering*, 309: 1–12.

Ringsberg, J.W., Amdahl, J., Chen, B.Q., Cho, S.-R., Ehlers, S., Hu, Z., Kubiczek, J.M., Kõrgesaar, M., Liu, B., Marinatos, J.N., Niklas, K., Parunov, J., Quinton, B.W.T., Rudan, S., Samuelides, M., Guedes Soares, C., Tabri, K., Villavicencio, R., Yamada, Y., Yu, Z., Zhang, S. 2018. MARSTRUCT benchmark study on nonlinear FE simulation of an experiment of an indenter impact with a ship side-shell structure. *Marine Structures*, 59: 142–157.

Sahu, S. & Ansari, M.Z. 2018. Effect of impactor shapes on low-velocity impact characteristics of Al plate. *Ships and Offshore Structures*, 14: 53–63.

Sun, B., Hu, Z. & Wang, G. 2015. An analytical method for predicting the ship side structure response in raked bow collisions. Marine Structures, 41: 288–311.

Wang, Z., Liu, K., Chen, G. & Hu, Z. 2020. An analytical method to assess the structural responses of ship side structures by raked bow under oblique collision scenarios. *Proceedings of the Institution of Mechanical Engineers, Part M: Journal of Engineering for the Maritime Environment*, 235: 773–791.

Zhou, J., Zhao, Y., Zhang, Y., Zhang, L., Yuan, Y. & Tang, W. 2023. An analytical method for evaluating the structural resistance of plate structures under in-plane impact load in ship-side collision scenarios. *Ocean Engineering*, 288: 1–12.

Trends in Collision and Grounding of Ships and Offshore Structures – Zhu & Guedes Soares (Eds)

An enhanced model of hull fracture in ship-ship collisions considering impact of corrosion on structural integrity

F. Zarzycki, K. Woloszyk & J. Montewka
Institute of Naval Architecture, Gdansk University of Technology, Gdansk, Poland

ABSTRACT: Ship-ship collisions can lead to severe structural failure, with hull fracture being the most critical due to its potential to cause environmental disaster and loss of life. Corrosion can significantly reduce a hull's structural integrity, making it more vulnerable to such events. The problem is more serious than ever with the rise of the "shadow fleet" of poorly maintained vessels. For this reason, this study develops a model to analyze hull damage behaviour in ship-ship collisions, considering an enhanced model of corrosion effects on structural integrity. The semi-analytical approach, considering external dynamics and internal structural mechanics during collisions, is extended by a novel approach accounting for corrosion degradation. A tanker case study illustrates the damage risk associated with corrosion. The approach supports risk-based ship design and provide maritime authorities with a tool to evaluate the consequences of collisions, particularly those involving poorly maintained vessels. This improves safety oversight and mitigation strategies.

1 INTRODUCTION

The hull of a ship is subjected to factors that make it vulnerable. These are physical stresses, such as waves or changes in the ship's loading conditions, which constantly affect the bending moments and shear forces of the hull (Liu & Guedes Soares 2020, Moreira & Guedes Soares, 2020). In addition, the salty seawater environment significantly influences the effects of corrosion (Zhang 2019). Proper maintenance and regular inspection of the condition of the hulls are therefore essential for the long-term and safe operation of ships. Ship design and construction must meet a number of standards for safe maintenance, including ultimate strength criteria. These are defined by the minimum class requirements set out in the rules of the classification societies (International Association of Classification Societies 2024).

However, a rising problem at the moment is the "*shadow fleet*", which consists of a growing number of ageing and poorly maintained vessels that do not meet minimum safety requirements. This poses significant environmental and maritime safety. The "*shadow fleet*" refers to a group of vessels that operate outside traditional regulatory, insurance, and tracking systems. These ships are often used to bypass international sanctions or restrictions, particularly in the context of transporting oil from sanctioned countries. The issue is particularly acute in the Black Sea, the Sea of Azov and the Baltic Sea due to the sanctions imposed on the Russian fleet (Caprile & Leclerc, 2024). As the ship ages, the structural integrity of the hull decreases. Lack of inspection and proper maintenance leading to corrosion increases the risk of fracture (Zhang et al. 2019, Feng et al. 2020). As a recent case shows, severe weather conditions combined with the aforementioned factors recently led to hull collapse of m/t Volgoneft 212 and m/t Volgoneft 239, resulting in an oil spill and massive pollution of marine wildlife (Harding 2024). The risk and consequences of such an event increase even more in the event of a ship collision, when greater forces arise (Goerlandt & Montewka, 2015). Some of the models for hull damage assessment consider configuration between striking and struck vessel (Ringsberg et al. 2023) or the structural configuration of the impacted hull (Hogstro¨m & Ringsberg, 2013). However, the effect of corrosion is not adequately considered in most damage risk models, which focus primarily on collision and grounding events (Prestileo; Luıs et al., 2009) and are rarely enhanced by corrosion effects, for example, by using advanced FE simulations (Kuznecovs et al., 2021). Furthermore, existing risk models typically ignore the subsequent hazards of collapse of the hull girder and only account for oil spills caused by perforations on the ship.

DOI: 10.1201/9781003684404-9

For this reason, in this paper, we introduce a model of hull damage of a tanker in a ship-ship collision, considering the impact of corrosion on structural integrity. This can be further implemented in risk-based ship design and can be useful to maritime authorities or class societies in improving maritime safety by identifying the actual risk level or possible consequences of ship-to-ship collisions, including the "*shadow fleet*". The paper is structured as follows: in Section 2, the method of updating Zhang (1999) model to include the loss of hull resistance due to corrosion is presented. Section 3 presents models, followed by a case study. Subsequently, Section 4 describes the results and discusses their application. Finally, Section 5 concludes the paper.

2 METHODS

Recent scientific research has shown that corrosion models that do not consider non-uniform thinning of components and reduction in mechanical properties can greatly overestimate the strength characteristics of the structure (Saad-Eldeen et al. 2013). For this reason, different methods have been proposed to consider corrosion as a more complex phenomenon; see for example (Woloszyk & Garbatov 2023, Woloszyk et al. 2024).

Similarly to the above-mentioned works, we pro- pose a collision model which accounts for the redution in hull thickness due to corrosion which in turn decreases the mechanical properties of the hull material. The external dynamics and internal mechanics of the structure during ship collisions are based on the method proposed by Zhang (1999). The energy absorbed by the struck ship is related to the volume of material destroyed, considering the thickness and material properties of the crushed plates and stiffeners, with a simple function. Knowing the strength characteristics of the hull, it is possible to determine the maximum velocity values for different damage modes. As the speed is raised to a power of two in the formula, its change has a significant effect on the penetration value.

2.1 *Damage size in collision*

The extent of damage in a collision is determined by the strength characteristics of the hull material and its structure, such as double hull or reinforced frames. In the case of a striking ship, the relevant factors are: dimensions of the bow, the impact area resulting from the shape of the ship's bow, the bow stem angle and bulbous bow, the impact height, or the angle of entry. In addition, the mass and speed of both vessels are decisive.

To determine the amount of hull damage in a collision between two ships, it is assumed that the energy to be dissipated, resulting from Equation 1, which takes into account the mass and speed of the ships, must be equal to the energy absorbed by the hull of the struck ship, resulting from Equations 2 and 3:

$$E_1 = \left(\frac{M}{M + 0.6M_0}\right) * 0.5 M_0 U_0^2 \qquad (1)$$

$$E_2 = 0.77 \epsilon_C \sigma_0 R_t \qquad (2)$$

$$E_3 = 3.50 \left(\frac{t}{d}\right)^{0.67} \sigma_0 R_t \qquad (3)$$

where M [t] and M_0 [t] are the masses of the struck and striking ship respectively; U_0 [m/s] is the speed of the striking ship; ϵ_c [%] is the critical strain; σ_0 [Pa] is the flow stress; t [m] is the average thickness of the plating d [m] is the average crushing width of the plates in the crushed sections and R_t [m^3] is a damage volume.

The coupling between external dynamics and internal mechanics is achieved through a sequential approach, in which the external collision parameters such as mass and speed of the vessel resulting in impact energy are first calculated and then applied as boundary conditions in the internal structural response model. This energy-based transfer ensures consistency between the dynamic loading phase and the structural deformation phase, enabling a semi-analytical estimation of damage progression.

Total absorbed energy is a combination of several modes of destruction such as plastic tension, crushing or folding modes. The Equation 2 refers to the energy which is absorbed by the plastic tension mode while Equation 3 by the crushing and folding. Each of them can be calculated separately and the integration of them in the calculation of the total absorbed energy depends on the ship's design and impact characteristics.

To compensate for the impact energy, the volume of damage (R_t) to individual components such as the outer shell, inner shell, transverse webs, longitudinal stiffeners or decks should be estimated taking into account their dimensions and thicknesses. By knowing the energy that must be absorbed by the hull of an impacted ship, it is possible to determine the volume that must be destroyed by the above modes. On this basis, the size of the hull rupture can be determined. In the case of an ageing ship, both thickness and mechanical properties, such as yield stress or critical strain, deteriorate.

2.2 *Material properties degradation*

To consider the impact of corrosion degradation, the mechanical properties (yield stress, ultimate tensile stress, and total elongation) were determined in each structural element separately, depending on the local degradation level. In this paper, the flow stress (σ_0) of material should be considered as an average of yield stress (R_{e0}) and ultimate tensile (R_{m0}).

Moreover, the critical strain value (ϵ_c) is subject to change in the ma- terial ductility expressed as δ, where δ_0 is the ductility of non-corroded material (Zhang, 1999):

$$\in_c = 0.10\left(\frac{\delta_0}{0.32}\right) \quad (4)$$

The change in mechanical properties is considered as a function of mechanical properties of non-corroded material and degree of degradation (*DoD*) and can be expressed by the following formulas (Woloszyk et al. 2022):

$$R_{e,\text{corr}} = R_{e0}(1 - 0.0056DoD) \quad (5)$$

$$R_{m,\text{corr}} = R_{m0}(1 - 0.0032DoD) \quad (6)$$

$$\delta_{\text{corr}} = \delta_0 e^{-0.0038DoD} \quad (7)$$

$$\in_{c,\text{corr}} = 0.10\left(\frac{\delta_{\text{corr}}}{0.32}\right) \quad (8)$$

where new critical strain value ($\epsilon_{c,corr}$) is adjusted by corroded material ductility (δ_{corr}) and *DoD* is expressed as the loss of plating/stiffener thickness.

3 CASE STUDY

This section presents a case study of the Ro-Pax ship as the striking and the double-hull oil product tanker as the struck vessel. Both are treated as critical to safety due to the cargo they carry and the potentially devastating consequences of their collision. Ro-Pax vessels operate with high frequency in areas of European waters, typically operating at high speed and on fixed routes that regularly cross major shipping lanes. On the other hand, the case of the "*shadow fleet*" mainly concerns tankers. It is, therefore, reasonable to consider such a case as a possible occurrence, but also as one with serious consequences. Additionally, we assume a worst-case scenario where the bow of the ship is rigid, and the impact is perpendicular in the central part of the hull.

In this section we consider three different cases of the same double hull tanker, but with different hull strength properties due to the effect of corrosion. These are:

1. newly designed double hull oil tanker with the thickness and strength of materials specified in the vessel's technical documentation;
2. 25-year-old double hull oil tanker resulting in 10 % reduction in thickness elements;
3. 25-year-old double hull oil tanker resulting in a 10 % reduction in thickness elements and degradation of material properties, expressed as a DoD equal to 10 %.

3.1 *Ships' dimensions and initial mechanical properties*

The main dimensions were taken from the ships' documentation and technical drawings, as shown in Table 1, along with other operational parameters. Therein #1 should be interpreted as the striking Ro-Ro and #2 as the struck oil product tanker.

Table 1. Main dimensions of the Ro-Ro (#1) and Tanker (#2) with corresponding mechanical properties of the materials considered in the presented case study.

Ships' characteristics			
Dimension	Symbol	Value	Unit
Lenght overall #1	LOA_1	199.8	m
Length overall #2	LOA_2	138.1	m
Breadth moulded #1	B_1	26.5	m
Breadth moulded #2	B_2	22.5	m
Draught #1	D_1	6.95	m
Draught #2	D_2	8.00	m
Displacement (Mass) #1	M_0	19690	t
Displacement (Mass) #2	M	17950	t
Speed #1	U_0	10.29	m/s
Initial mechanical properties (non-corroded)			
Yield stress	$Re0$	235	MPa
Ultimate tensile stress	$Rm0$	400	MPa
Flow stress	σ_0	318	MPa
Material ductility	δ_0	22	%
Critical strain	ϵ_c	6.9	%
Mechanical properties DoD = 10 %			
Yield stress	$Re,corr$	221.8	Mpa
Ultimate tensile stress	$Rm,corr$	387.2	Mpa
Flow stress	$\sigma corr$	304.5	MPa
Material ductility	$\delta corr$	21	%
Critical strain	$\epsilon c,corr$	6.6	%

3.2 *Dissipated energy and corresponding impact area*

The value of the energy to be dissipated in this case study was calculated taking into account the mass and speed of the concerned ships as explained in Section 2 and is equal to E_1 = 628.66 *MJ*. This energy should therefore be used as the basis for determining the energy of destruction to compensate for it.

3.2.1 *Damage height (vertical) and length (along the struck ship)*

The first step is to determine the possible surface area over which the energy can be absorbed. The height of the Ro-Pax bow above the baseline was assumed to be 16.95 *m*, the draught was 6.95 *m* and the draught of the tanker to be 8.00 *m*. The calculation also took into account the characteristics and positions of the bulbous bow. Therefore, in this case, it extends the vertical position of the strike which starts at the tanker's double bottom and extends up to 18.00 *m* above the tanker's baseline, covering the entire side surface, as shown in Figure 1.

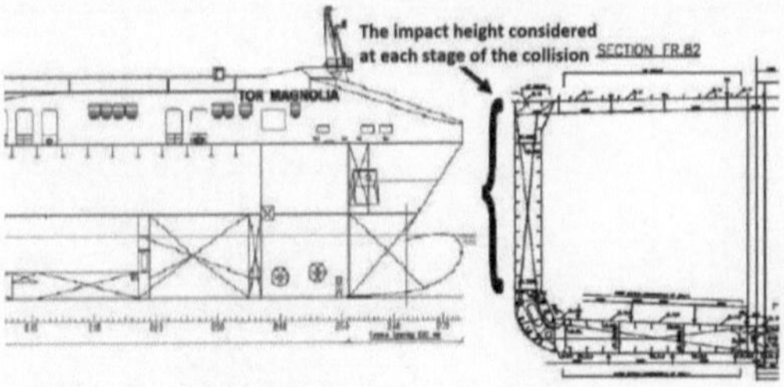

Figure 1. Collision situation - damage height.

Since the bow of a ship is not a cuboid, the length of the damage area (along the longitudinal axis of the struck ship) must be iteratively adjusted to match the width of the striking ship at a given penetration depth (Figure 2). It is, therefore, not a constant value for a particular ship, and it is incorporated into the calculation method.

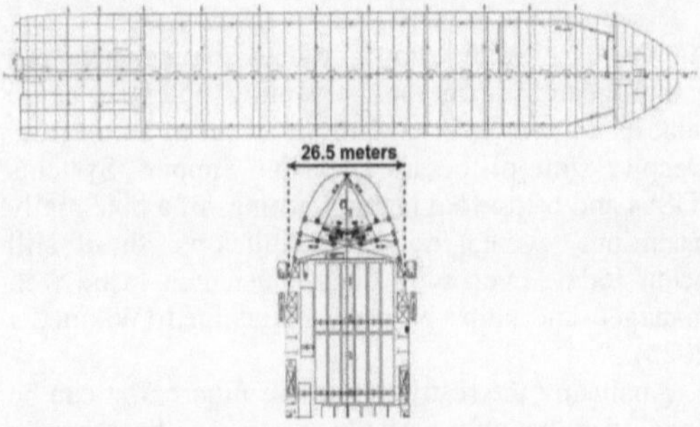

Figure 2. Collision situation - damage length.

3.3 *Penetration depth*

The depth of penetration and the volume of the damaged hull are related to the energy that needs to be dissipated. Different formulas are applied as the striking ship passes through different parts of the hull.

First, the outer shell plating is ruptured. To account for the presence of stiffeners, including their average spacing, the equivalent thickness method is used.

Next, the striking ship penetrates the space between the outer and inner shells, which consists of tweendecks, transverse webs, and longitudinal stringers. In this region, energy is primarily dissipated through crashing and folding, followed by tension damage mode. The number of transverse webs must be adjusted according to their spacing, taking into account the penetration depth and the width of the hull. It must be adjusted iteratively, similar to the length of the hull fracture, as described above. If the total energy has not been absorbed by this point, the inner hull plating will collapse. At this stage, a critical limit is reached beyond which the ruptured tank may begin to leak. The remaining barrier before reaching the other side of the hull is the corrugated bulkhead placed along the centreline of the tanker, and the main deck impacted by a crushing mode (across full breadth).

3.3.1 *Results*

Firstly, we have considered the case of a newly designed tanker. In this scenario, we assume that there is no change in thickness or material properties. In such a case, the calculated penetration depth at an impact speed of 20 *kn* was 11.97 *m*, which is slightly more than half the breadth of the tanker ship.

Secondly, the calculations took into account a 10% reduction in the average thickness of the plating and stiffeners of each element concerned. This value corresponds to the average loss after 25 years of ship operation (International Association of Classification Societies 2018). As a result of the loss in thickness of the hull elements, the penetration depth compensating for the impact energy increased by 3.85 *m*, reaching a total value of 15.82 *m*.

For the third case, not only the thickness reduction but also the degradation of material properties as described in section 2 was considered. The yield stress, ultimate tensile strength and critical strain were changed by DoD of 10 %. In this case the depth of penetration was the greatest, reaching 17.47 *m*.

In all three cases, the energy to be absorbed is the same, but they differ in their variability characteristics, as shown in Figure 3. In the first case the volume of destruction is 17.61 m^3 and in the second case 17.30 m^3. In the third case, however, the damage volume is more than 2% greater than in the second case because it must take into account the reduction in the strength properties of the materials (Figure 5).

Nevertheless, all three cases differ in the value of the penetration depth. The destroyed hull sections, according to the described scenarios and at a striking speed of 20 *kn*, are subsequently marked as 1, 2 and 3 in Figure 4. These relationships are shown in the Figure 5. As illustrated in the graph, the distinction between various damage modes can be determined based on the specific tanker hull elements that are encountered.

3.4 *Effect of impact speed on cargo tank integrity*

In this work, we also investigate the relationship between the penetration depth and the speed of the striking vessel. For instance, for the new ship analysed, the maximum impact speed that does not result in the rupture of the inner hull plating and the possible spillage of cargo from the tank is 13.2 *kn*. At this speed, the energy to be dissipated is 188.82 *MJ*, which only damages the outer hull and the folding of materials up to the inner hull plating.

However, the maximum speed of the ship, taking into account the effects of corrosion and changes in

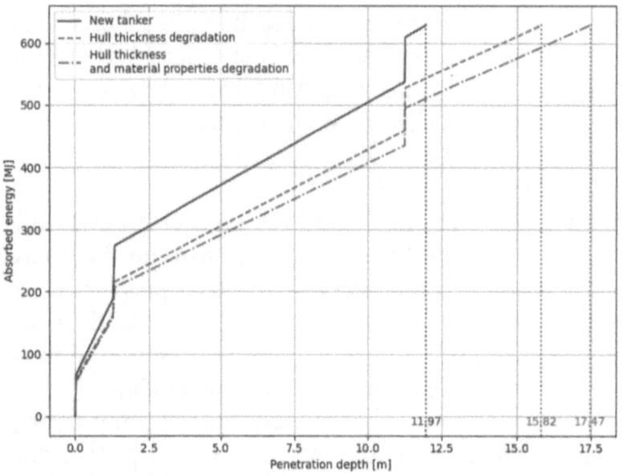

Figure 3. Absorbed energy for different scenarios at a speed of 20 *kn*.

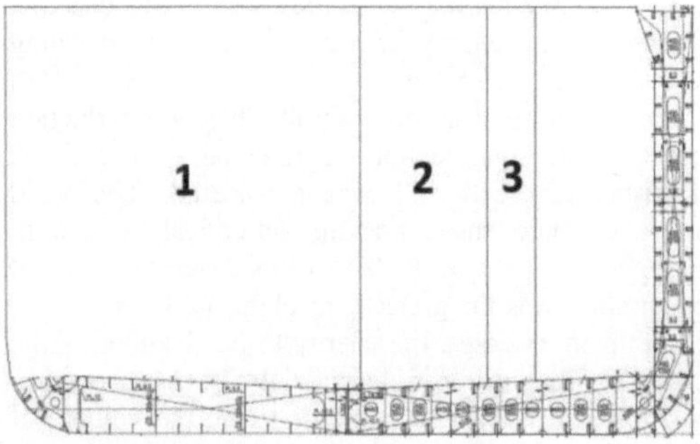

Figure 4. Destroyed hull volume for different scenarios at a speed of 20 *kn*.

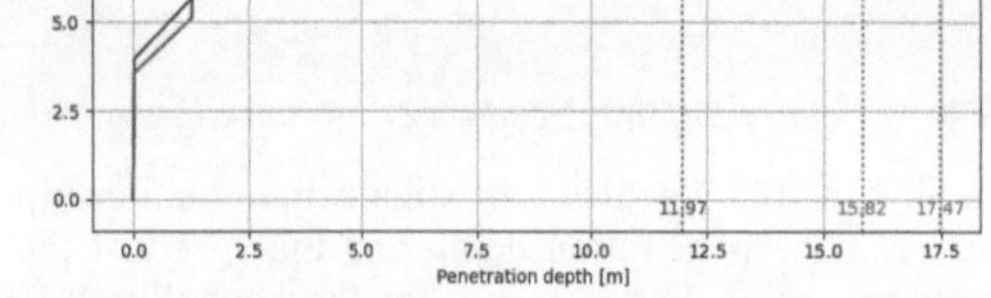

Figure 5. Penetration depth for different scenarios at a speed of 20 *kn*.

the strength properties of the materials, must be lower to prevent the cargo tank from leaking. The results are shown in Table 2, where speed is expressed in knots, for three different hull strength cases presented in the case study. A similar analysis can be carried out for a given impact speed to check the predicted penetration depth (Figure 6).

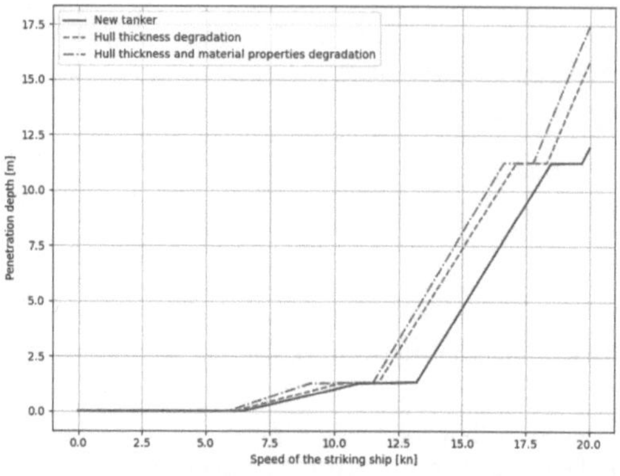

Figure 6. The relationship between the speed of the striking vessel and the depth of penetration.

4 DISCUSSIONS

As the case study shows, proper maintenance has a significant impact on preventing potential cargo tank damage and spillage in the event of a collision. Despite state-of-the-art Decision Support Systems (DSS) and bridge navigator warnings of a potentially hazardous event, including collisions, these still occur today, even with flags considered to be well managed and ships properly maintained (Wikipedia 2025).

Analysing the results shown in Figure 5, it can be concluded that a 25 year old ship is significantly less resistant to impact. The penetration depth for the new ship was 11.97 *m*, while for the second and third cases it was 15.82 *m* and 17.47 *m* respectively. At a constant impact speed of 20 *kn*, the difference in penetration depth between a new and a 25-year-old ship was almost 4 *m*. Taking into account the width of the tanker, which is 22.5 *m*, this is a change of almost 20%. However, it is not only the reduction in thickness that needs to be taken into account in ageing ships, but also the deterioration in material properties. This has also been included in the calculation and the results show a change of up to 1.65 *m* compared to a simple thickness reduction. Such a model is particularly useful given that corrosion levels can be even higher on poorly maintained ships, with an increasing proportion of the fleet being over 25 years old.

Moreover, in most cases ships do not collide at full sea speed. For example, the collision may be the result of a late reaction by the navigator to change course or reduce speed. For this reason, we have also investigated the relationship between different striking velocities and their corresponding penetration depths. This is essential for predicting the threshold value at which the inner hull shell remains intact, taking into account different material properties. For the new ship, the threshold speed was 13.2 *kn*, while for the third case analysed it was 11.7 *kn*. It can also be seen from Figure 6 that for impact velocities up to 5.9 *kn* in each case, the outer shell of the tanker only deforms without causing significant damage expressed in terms of penetration depth.

These and many other analyses can be performed for this particular case. Importantly, the method presented in this article not only provides useful predictive results, but is also scalable. By having information on the basic properties of ships potentially involved in collisions, the extent of damage can be determined. In addition, different hull strength properties can be explored by using the DoD index for calculations. In the case study section, the change due to corrosion expressed by the DoD value was based on existing literature. However, further validation using representative data from the ship in question and additional numerical work are needed. Similarly, the assumed mechanical properties of the corroded material should be verified through dedicated testing and analysis.

It is, therefore, possible to easily adjust parameters such as the age of the ship or the degree of corrosion in relation to, for example, the flag of the ship, which can also be taken into account when analysing the so-called "shadow fleet" and its potential accident consequences. Therefore, future research should focus on providing a comprehensive method that eliminates the need for manual data entry of ship specifications, such as possible thickness reduction, and instead incorporates a probabilistic model of impact scenarios that takes this into account. Additionally, the proposed model should be generalised to incorporate a wider range of ship types and localised corrosion patterns, such as pitting. Furthermore, the calculations should take into account that the hull of the striking ship is not rigid and that the impact is not necessarily perpendicular.

Additionally, the method can be integrated into DSS to provide navigators with real-time updates on the potential consequences of a collision. This could serve as a first step toward the Corporate Social Responsibility (CSR) paradigm, where decision-making on board should take into account not only the likelihood of an accident, but also its potential consequences. However, its practical application may be constrained by current limitations in real-time corrosion monitoring and the availability of high-resolution structural data. It is currently not feasible to obtain such measurements for every vessel. Therefore, numerical methods and predictive models could be used to estimate structural degradation and support the decision-making process. Additionally, a deterministic model may fail to capture the full range of possible outcomes, leading to either overly conservative or insufficiently robust results. By employing probabilistic methods, such as Monte Carlo simulations or Bayesian Network, a meta-model can provide a more comprehensive risk assessment, identifying both collision scenarios and possible outcomes and providing the full spectrum of risk-based collision analysis.

Table 2. Maximum speed in knots for the 3 different cases up to the damage considered modes.

Case \ Damage Mode	Outer Shell rapture	Transverse webs and longitudinal stringers crushing	Inner shell rupture (cargo tank leak)	Corrugated bulkhead crushing
New tanker	6.5	11.0	13.2	19.7
25 years old tanker with thickness degradation	6.2	10.2	11.7	18.3
25 years old tanker with thickness and material properties degradation	5.9	9.0	11.5	17.8

5 CONCLUSION

This paper presents a model of hull fracture in ship-to-ship collisions that takes into account the effect of corrosion on structural integrity. This was achieved through applying different properties of the damaged material in ship collision volume calculation. The most critical consequence is hull fracture, which can lead to catastrophic environmental damage in the event of an oil spill or loss of life if the ship sinks. The present study determines the increased vulnerability of ship hulls due to corrosion, reducing structural strength and increasing damage under impact conditions.

The developed semi-analytical model, which incorporates external collision dynamics and internal structural mechanics, has been extended with a novel approach that considers non-uniform corrosion degradation. The case study of a tanker collision demonstrates the impact of corrosion on hull integrity. For example, at the same impact speed, the penetration depth of the corroded hull is up to 5.5 m greater than that of a new ship. Furthermore, a new tanker with an intact hull can withstand an impact velocity of up to 13.2 kn before the cargo tank ruptures, whereas the corroded hull can only withstand 11.5 kn.

The results of this study emphasis the need to include corrosion effects in ship collision risk assessments. The proposed approach can be applied to risk-based ship design and serve as a valuable tool for maritime authorities to improve safety standards. In addition, this methodology can be used to assess the risks associated with the "shadow fleet" by identifying potential hazards associated with the operation of poorly maintained ships.

ACKNOWLEDGEMENTS

The funding to carry out this research was provided by Interreg Baltic Sea Region (co-funded by EU) via Open Risk II project "Tool for shared and dynamic maritime traffic risk picture of the Baltic Sea Region" under grant No #C041. Financial support of second author from Gdan´sk University of Technology by the DEC-3/2/2023/IDUB/II.1b/Am grant un- der the AMERICIUM - 'Excellence Initiative - Re- search University' program is gratefully acknowledged.

REFERENCES

Caprile A. & Leclerc G. 2024. Russia's 'shadow fleet': Bringing the threat to light. Briefing, European Parliament.

Feng, L., Li, D., Shi, H., Zhang, Q. & Wang, S. 2020. A study on the ultimate strength of ship plate with coupled corrosion and crack damage. *Ocean Engineering* 200: 106950.

Goerlandt, F. & Montewka, J. 2015. A framework for risk analysis of maritime transportation systems: A case study for oil spill from tankers in a ship–ship collision. *Safety Science* 76: 42–66.

Harding, L. 2024. Russian tanker sinks in Black Sea spilling 4,300 tonnes of oil. *The Guardian*.

Hogström, P. & Ringsberg, J. W. 2013. Assessment of the crashworthiness of a selection of innovative ship structures. *Ocean Engineering* 59: 58–72.

International Association of Classification Societies 2018. Corrosion Additions and Wastage Allowances. Technical Report Pt 1, Ch 3, Sec 3, London.

International Association of Classification Societies 2024. Common Structural Rules for Bulk Carriers and Oil Tankers Archives.

Kuznecovs, A., Schreuder, M. & Ringsberg, J. W. 2021. Methodology for the simulation of a ship's damage stability and ultimate strength conditions following a collision. *Marine Structures* 79: 103027.

Liu, B. & Guedes Soares, C. 2020. Ultimate strength assessment of ship hull structures subjected to cyclic bending moments. *Ocean Engineering* 215: 107685.

Luıs, R. M., Teixeira, A. P. & Guedes Soares, C. 2009. Longitudinal strength reliability of a tanker hull accidentally grounded. *Structural Safety* 31(3): 224–233.

Moreira, L. & Guedes Soares, C. 2020. Neural network model for estimation of hull bending moment and shear force of ships in waves. *Ocean Engineering* 206: 107347.

Prestileo, A., Rizzuto, E., Teixeira, A. P. & Guedes Soares, C. 2013. Bottom damage scenarios for the hull girder structural assessment. *Marine Structures* 33: 33–55.

Ringsberg, J.W., Kuznecovs, A. & Johnson, E. 2023. Analysis of how the conditions in a collision scenario affect the size of a struck vessel's damage opening and ultimate strength. In *Advances in the Analysis and Design of Marine Structures*. CRC Press. Num Pages: 9.

Saad-Eldeen, S., Garbatov, Y. & Guedes Soares, C. 2013. Ultimate strength assessment of corroded box girders. *Ocean Engineering* 58: 35–47.

Wikipedia, C. 2025. 2025 North Sea ship collision. Page Version ID: 1280519354.

Woloszyk, K. & Garbatov, Y. 2023. Advanced numerical modelling for predicting residual compressive strength of corroded stiffened plates. *Thin-Walled Structures* 183: 110380.

Woloszyk, K., Garbatov, Y., & Kłosowski, P. 2022. Stress–strain model of lower corroded steel plates of normal strength for fitness-for-purpose analyses. *Construction and Building Materials* 323: 126560.

Woloszyk, K., Goerlandt, F. & Montewka, J. 2024. A methodology for ultimate strength assessment of ship hull girder accounting for enhanced corrosion degradation modelling. *Marine Structures* 93: 103530.

Zhang, H. 2019. Study on the Influence of Galvanic Corrosion on Ship Structural Materials. *IOP Conference Series: Earth and Environmental Science* 252(2): 022015. Publisher: IOP Publishing.

Zhang, H., Xu, S., Nie, B. & Wen, Y. 2019. Effect of corrosion on the fracture properties of steel plates. *Construction and Building Materials* 225: 1202–1213.

Zhang, S. 1999. *The Mechanics of Ship Collisions*. Kgs. Lyngby, Denmark: Technical University of Denmark.

Trends in Collision and Grounding of Ships and Offshore Structures – Zhu & Guedes Soares (Eds)
© 2026 The Author(s), ISBN 978-1-041-16361-9

Critical deflection of shell rupture in ship collision and energy absorption capability

S.M. Zhang & H. Ocakli
Lloyd's Register, Global Technology Centre, Southampton, UK

R. Villavicencio
Lloyd's Register, Global Technical Support Office, Rotterdam, NL

ABSTRACT: One of the key elements in ship collision analyses is to estimate the critical deflection at the rupture of the shell plating, as this critical point will determine whether the ship hull is still intact, or its watertightness will be lost leading to the ingress of water and/or oil outflow. This paper presents a review of existing methods and introduces a new method to determine the critical deflection at the rupture of shell plating. The new method is validated with experimental test results and is tested and verified with ship collision accident data. The new method is applied to ship collision case studies to estimate the critical deflection, the collision energy and the critical collision speed that will result in the rupture of the ship hull. This analysis should provide relevant information to manage the risk and the potential significant consequences in the event of a ship collision accident.

1 INTRODUCTION

Ship structures are very complicated systems. Thus, the deformed, destroyed, and crushed modes of the structures due to collision are also very complex. A ship can be viewed as an assembly of welded plated structures where the shell plating is constructed from various plates stiffened by stiffeners.

Observations from full-scale ship accidents and model experiments reveal that the primary energy-absorbing mechanisms of ship structures in collision are: (1) membrane deformation of the shell plating and attached stiffeners, (2) folding and crushing of transverse web frames and longitudinal stringers, (3) folding, cutting, and crushing of horizontal decks, (4) crushing of transverse bulkheads.

The methods for the analysis of structural damage in ship collisions can be divided into four categories: (1) Empirical methods, (2) Finite element methods, (3) Experimental methods, (4) Simplified analytical methods. Here, the focus is on simplified analytical methods which are based on the upper-bound theorem and assumptions concerning folding patterns from observations of accidental damages, experimental studies, or numerical simulations. Usually, these methods can give good predictions through a fast simple analysis, and thus they are commonly used for analysis of ship collisions.

A major assumption in the simplified analytical methods is that the different structural members, such as the side shell, decks, and web frames, do not interact, except at their joint boundaries, but contribute independently to the total collision resistance.

In ship collisions, when the penetration or deflection is large enough, the shell plate will rupture. The critical deflection of shell at which point rupture starts is one of the most challenging issues in collision damage assessments. This limit is also important to assess whether liquids or oil outflow will occur and to quantify the amount of oil that will flow out.

The fracture and rupture of shell structures in ship collisions are influenced by many factors, such as the hull material, plate thicknesses, structural arrangements, welding, shape of the indenter and the impact speed. It is an extremely complex problem to predict the rupture of structures accurately. Jones & Wierzbicki (1993) discussed the rupture criteria of ductile metal beams subjected to large dynamic loads. Three major failure criteria of the metal beams were discussed. The first is the tensile tearing failure mode, which occurs when the maximum tensile strain equals the critical rupture strain of the material. The second failure model is the transverse shear failure mode which develops in a beam when large transverse shear deformations occur within a very short region of the plastic beam, then when the total transverse shear displacement in a particular location equals a critical value, the beam or plate ruptures. The third failure criterion is the energy density failure mode, where it is assumed that rupture occurs in a rigid-plastic structure when the

DOI: 10.1201/9781003684404-10

absorption of plastic work per unit volume reaches a critical value.

The simplified analytical methods (e.g. Zhang et al. 2019) are based on overall deformation mechanisms. In this case it is not possible to trace the strain history of material elements at a very detailed level. Therefore, many authors use the 'maximum strain failure criteria'. That is, when the maximum strain in a structure reaches a critical strain, the structure ruptures.

When the critical rupture strain (ε_f) is known, the critical deflection or penetration of the shell plating (δ_f) can be determined. For example, consider a point load acting in the middle of a plate strip with a span $2b$. The critical rupture deflection can be then determined from: $\delta_f = b\,(2\varepsilon_f)^{0.5}$.

A typical critical rupture strain of mild steels from standard uniaxial tensile tests is around 0.35. If this critical strain $\varepsilon_f = 0.35$ is used, then the critical deflection or penetration is $\delta_f = 0.84b$.

This value of critical strain is far too large in the assessment of full-scale collisions. The critical strain value suggested for side collisions is between 5% to 10% if the simple beam theory used (Amdahl, 1995). In the minor collision analyses performed by McDermott et al. (1974), the critical rupture strain for mild steel material for side collisions is evaluated from $\varepsilon_{cr} = 0.31\ \varepsilon_f$. The main reason of having to use such smaller value of critical rupture strain is that the theoretical model used to estimate a critical deflection is too simple. Thus, to get a good agreement with the experimental data, the critical rupture strain must be tuned and reduced.

Therefore, over the past 20 years new simplified methods have been proposed to determine the critical deflection of the shell plate at initial rupture. While those methods have been validated with detailed experiments and finite element simulations, they are based on plate models disregarding the effect of the secondary and primary structural components. In the present paper, some of those existing methods are reviewed and a new method to determine the critical deflection at the rupture of shell plating is introduced. The new method is validated with a large number of experimental test results of plates, stiffened panels and double hull structures and it is tested and verified with ship collision accident data as a practical application. The new method is also applied to ship collision case studies to estimate the critical deflection, the collision energy and the critical collision speed that will result in the rupture of the ship hull.

2 CRITICAL DEFLECTION OF SHELL RUPTURE

This section presents the review of the existing methods and introduces the new method to determine the critical deflection at the rupture of the shell plating. In addition, this section summarizes the comparison of the new method with various experimental results and with two ship collision accidents.

2.1 *Analysis methods*

Through theoretical analysis, finite element simulations and experimental studies, Simonsen & Lauridsen (2000), Lee et al. (2004), Liu et al. (2014), Zhang et al. (2019) and Zhang M et al. (2019) have investigated the critical defection of shell plates initial rupture. It has been found that the radius of the indenter and the plate size have a significant influence on the critical deflection. The expressions from their studies are given below:

(1) Simonsen et al (2000):

$$\delta_f = 1.41 n^{0.33} b^{0.48} {R_b}^{0.52} \tag{1}$$

(2) Lee et al (2004):

$$\delta_f = (2D_c)^{0.35} b^{0.62} {R_b}^{0.38} \tag{2}$$

(3) Liu et al (2014):

$$\delta_f = 1.15 n^{0.18} b^{0.5} {R_b}^{0.5} \tag{3}$$

(4) Zhang et al (2019):

$$\delta_f = {\varepsilon_f}^{0.4} b^{0.4} {R_b}^{0.6} \tag{4}$$

(5) Zhang M et al (2019):

$$\delta_f = 0.5(7n + 0.67)^{0.5} b^{0.5} {R_b}^{0.5} \tag{5}$$

where n is the material strain hardening exponent, b is the radius of a circular plate or half width of a square/rectangular plate short side if the impact is at the center or the plate width if the impact is at the edge (i.e. the shortest distance to the support edge), and R_b is the radius of the indenter, see Figure 1. D_c is the material critical damage value for ductile fracture, defined as the product of the failure strain and the average stress triaxiality which can be taken as 0.27 for punch indentations, and ε_f is the rupture strain.

These methods (Eq. 1 to Eq. 5) have been validated by some experimental data using four edge-clamped plates (or circular plates) indented by rigid spheres.

Following the similar principle, as the methods summarised in Eq. (1) to Eq. (5), a new empirical formula for determining the critical deflection of shell plate at initial rupture is proposed in Eq. (6), where t is the shell plate thickness.

$$\delta_f = t^{0.1} b^{0.45} R_b{}^{0.45} \tag{6}$$

The new formula (Eq. 6) disregards the fracture strain and the material strain hardening exponent. In general, Eq. (6) follows the same approach of Eq. (4), which is presented in Zhang et al (2019). It is verified with a wide range of experimental test results which include plate models and stiffened panel models and double hull models. The method aims at considering geometric parameters only and avoiding the hull material properties because 'accurate' material properties are often unknown in ship collision accidents or full-scale ship collision analysis. Still the new method manages to predict the ship collision response as demonstrated in the next sections.

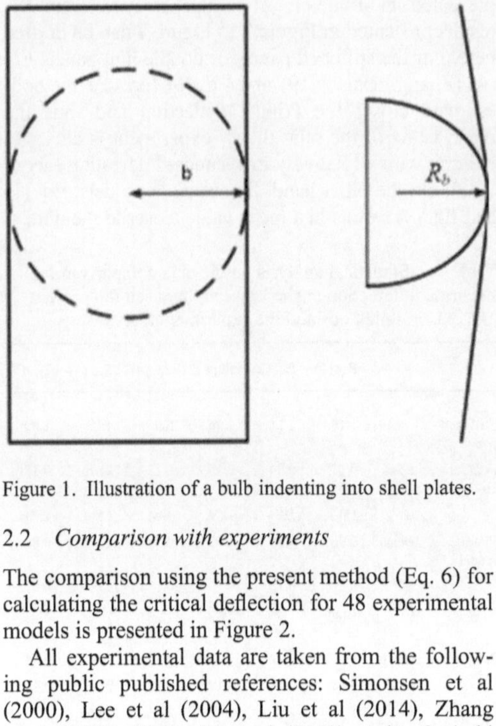

Figure 1. Illustration of a bulb indenting into shell plates.

2.2 *Comparison with experiments*

The comparison using the present method (Eq. 6) for calculating the critical deflection for 48 experimental models is presented in Figure 2.

All experimental data are taken from the following public published references: Simonsen et al (2000), Lee et al (2004), Liu et al (2014), Zhang et al (2019), Zhang M et al (2019), Arita & Aoki (1985), Wang et al (2000), Tautz et al (2013), Lehmann & Peschmann (2002), Paik et al (1999), Alsos & Amdahl (2009). Details of these experimental data can be also found in Zhang et al (2019).

The comparisons using the existing formulae (Eq. 1 to Eq. 5) for calculating the critical deflection for the experimental models are also carried out and presented in Figure 3 to Figure 7.

It can be seen from the comparisons that the methods proposed by Simonsen et al (2000), Lee et al. (2004) and Liu et al. (2014) give a similar level of agreement for the plate models (triangle symbols in Figure 3 to Figure 5) as these methods were originally validated using plate models. However, for stiffened panels models or double hull models (diamond symbols in Figure 3 to Figure 5), those existing methods overestimate the results largely.

The newly developed method (Eq. 6) generally underestimates the results for the plate models, but it gives reasonable predictions for the stiffened panel models and the double hull models.

It should be noted that for the plate models, the ideal boundary condition of fully clamped edges may be difficult to achieve in laboratory model tests, as the in-plane displacements occur, and consequently this increases the experimental results of the critical deflection. This could be one of the reasons that the existing methods, which were calibrated with plate models, give higher results for the double hull models.

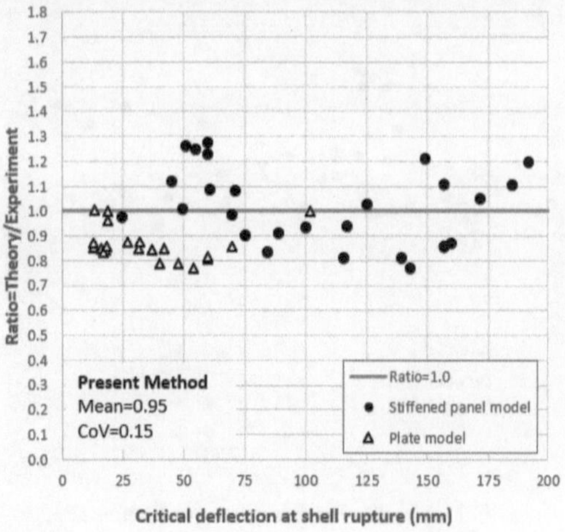

Figure 2. Comparison of critical deflection: present method (Eq. 6) and experiments.

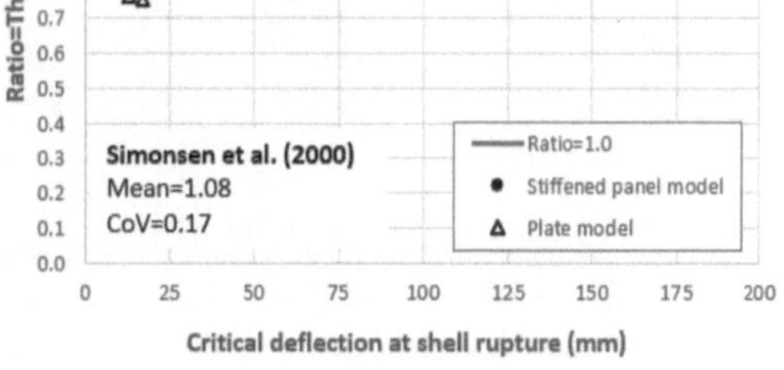

Figure 3. Comparison of critical deflection: Simonsen et al. (2000) (Eq. 1) and experiments.

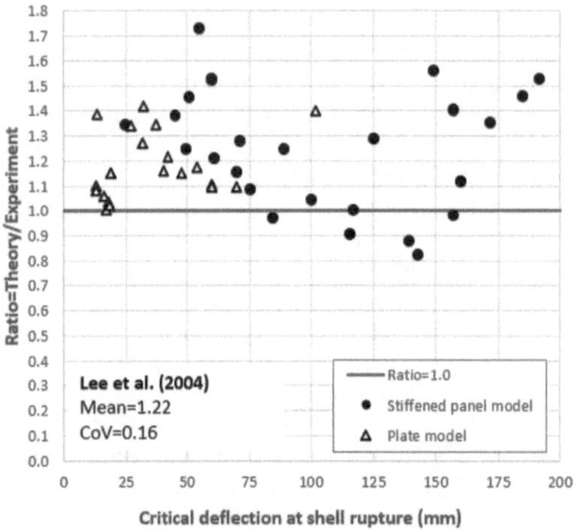

Figure 4. Comparison of critical deflection: Lee et al. (2004) (Eq, 2) and experiments.

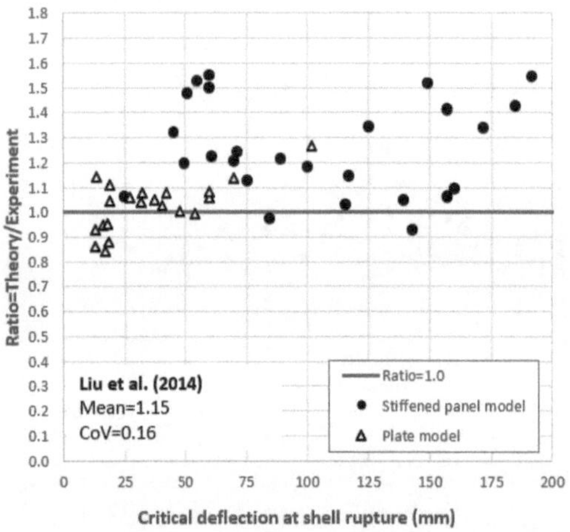

Figure 5. Comparison of critical deflection: Liu et al. (2014) (Eq. 3) and experiments.

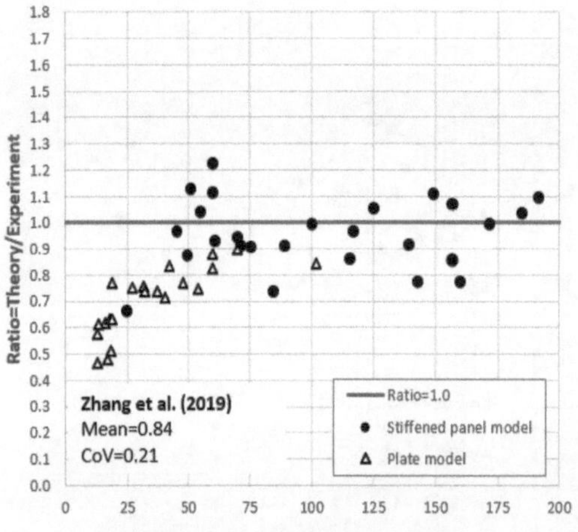

Figure 6. Comparison of critical deflection: Zhang et al. (2019) (Eq. 4) and experiments.

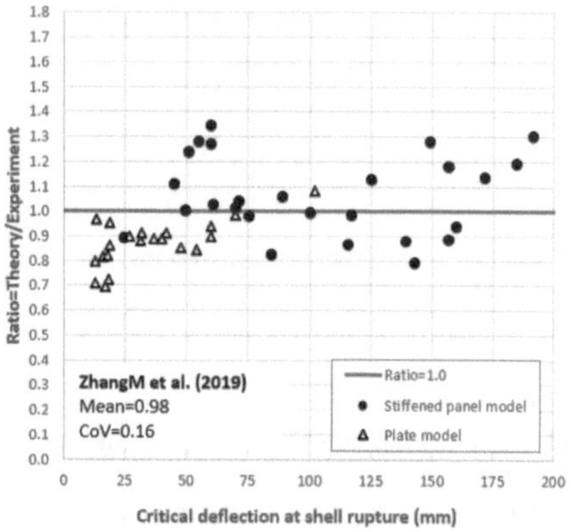

Figure 7. Comparison of critical deflection: Zhang M et al. (2019) (Eq. 5) and experiments.

A statistical analysis on the data comparison for the critical defection at shell rupture between the present and existing methods and the experimental results is presented in Table 1. A similar analysis from the results presented in Figure 2 to Figure 7 can be drafted here. For the stiffened panels or double hull panels Eq. (4) (Zhang et al. 2019) and Eq. (6) (present method) estimate closer the critical deflection and thus the mean value of the ratio theory-experiment is closer to the unit with a relatively low standard deviation (about 0.15). On the other hand, for the plate models, Eq. (1) and Eq. (3) results in a mean value closer to the unit.

Table 1. Statistical analysis on the data comparison for the critical defection at shell rupture between the present and existing methods and the experimental results.

		Eq. (1)	Eq. (2)	Eq. (3)	Eq. (4)	Eq. (5)	Eq. (6)
Stiffened panel	Mean	1.19	1.25	1.26	0.95	1.06	1.02
or Double Hull	STD	0.17	0.24	0.19	0.13	0.16	0.15
	CoV	15%	19%	15%	14%	15%	15%
Plate model	Mean	0.96	1.18	1.03	0.71	0.88	0.87
	STD	0.1	0.12	0.1	0.13	0.09	0.07
	CoV	10.90-%	11%	10%	18%	10%	8%
All model	Mean	1.08	1.22	1.15	0.84	0.98	0.95
	STD	0.18	0.2	0.19	0.18	0.16	0.14
	CoV	17%	16%	16%	21%	16%	15%

Note: Mean is the average of ratio of calculated to experimental results, STD is the standard deviation, CoV is the coefficient of variation.

2.3 *Verification with collision accident data*

The new and existing methods are also used to verify two collision accidents: (1) a ferry ship struck by a cargo ship. (2) a large LNG ship struck by an oil tanker. Details of the collision accidents can be found in Zhang et al (2019). Both the ferry and the LNG carrier resulted with a big hole on the side shell which was created by the bulbous bow of striking ships during the collision accidents.

The critical defection at the side shell rupture calculated using the new and existing methods are presented in Table 2. The methods of Simonsen et al. (2000) (Eq. 1), Lee et al. (2004) (Eq. 2) and Liu et al. (2014) (Eq. 3) overestimates the accident values by 30% to 50%. The present method (Eq. 6) and the method by Zhang et al (2019) (Eq. 4) give reasonable estimation (about 10% deviation).

It should be noted that predicting the collision response of actual accidental data is a rather difficult task. Not only theoretical methods result in large deviations with accidental data, but also advanced nonlinear finite element simulations may also result in differences in the range of 30% to 50% (see, for example, Zhang et al. 2019). Therefore, the difference of about 5% to 10% predicted by Eq. (4) and Eq. (6) can be considered as a reliable tool for ship collision assessments.

Table 2. Critical defection at the shell rupture calculated using the new and existing methods (unit: mm).

Case	In accident	Eq. (1)	Eq. (2)	Eq. (3)	Eq. (4)	Eq. (5)	Eq. (6)
Ferry	550	766	820	821	593	685	613
A LNGC	2450	3294	3319	3486	2580	2922	2402
		Ratio = calculated result to actual accident					
A ferry	1	1.39	1.49	1.49	1.08	1.24	1.11
A LNGC	1	1.34	1.35	1.42	1.05	1.19	0.98

3 CASE STUDIES

3.1 *LNG carrier stuck by a similar size container ship*

It is a concern in the industry that LNG carriers during the loading and/or unloading at ports could be struck by ships entering the ports. For a relatively sharp bow, the consequence of such collision would be significant with potential gas leakage into the port and surrounding areas. To avoid such situation, it is of interest to determine the maximum critical speed that a typical striking ship will have before the LNG cargo tank boundary is reached.

The collision scenario assumes that the LNG carrier is fully loaded, and it is hit by a fully loaded container ship while waiting for discharging cargo. It is also assumed that the LNG carrier is stationary and floating freely before the perpendicular collision occurs amidship (see Figure 8). The main dimensions of the ships are presented in Table 3.

The task here is to determine the critical rupture defection, absorbed energy and critical speed of the striking ship at three critical stages: side shell rupture, striking bow just reached the inner skin and inner skin ruptured.

The critical rupture defections are calculated using Eq. (6) and the absorbed energy and critical speed of the striking ship are calculated using the theory in Zhang et al (2019). The parameters used for the calculations are summarized in Table 4 and the results are presented in Table 5. It is seen from the results that the inner skin of the LNG carrier can be ruptured if collision speeds are over 5 kt.

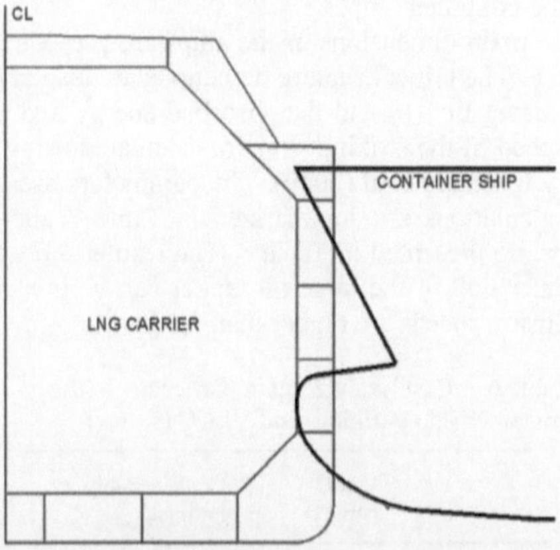

Figure 8. Illustration of the LNG ship and the container ship collision scenario.

Table 3. Case study 1: main particulars of the container ship (striking) and LNG carrier (struck).

	LNG (struck)	Container ship (striking)
L	275	276
B	44	40
D	26	24.3
T	12	13
Displacement (ton)	110000	100000

Table 4. Parameters for case study 1.

Item	Case study 1LNGC (struck ship)
b (mm)	1650
Rb (mm)	2235
t (mm)	15.5
Side shell rupture (m)	1.186
Double side width (m)	2.437
Inner skin rupture (m)	3.623

Table 5. Case study 1: analysis results of LNG carrier struck by a container ship.

	Penetration (m)	Absorbed Energy (MJ)	Critical striking speed (kt)
Side shell rupture	1.186	41.9	2.2
Double side width	2.437	106.6	3.4
Inner skin rupture	3.623	221.3	4.9

3.2 *VLCC oil tanker stuck by a similar size container ship*

The collision scenario of this case study is like the case study of the collision between the LNG carrier and the container ship.

The main dimensions of the ships are presented in Table 6. The critical rupture defections are also calculated using Eq. (6) and the absorbed energy and critical speed of the striking ship are calculated using the theory in Zhang et al (2019). The parameters used for the calculations are summarized in Table 7 and the results are presented in Table 8. The results show that the inner hull of the large oil tanker can be breached if collision speeds are higher than 3.7 kt.

Table 6. Case study 2: main particulars of the container ship (striking) and VLCC (struck).

	VLCC (struck)	Container ship (striking)
L	320	320
B	58	48.8
D	31	27.3
T	20.8	16
Displacement (ton)	314275	178000

Table 7. Parameters for case study 2.

Item	Case study 2 VLCC (struck ship)
b (mm)	2560
Rb (mm)	2490
t (mm)	20.8
Side shell rupture (m)	1.562
Double side width (m)	3.52
Inner skin rupture (m)	5.082

Table 8. Case study 2: analysis results of VLCC struck by
a container ship.

	Penetration (m)	Absorbed Energy (MJ)	Critical striking speed (kt)
Side shell rupture	1.562	39.4	1.5
Double side width	3.520	119.9	2.5
Inner skin rupture	5.082	253.9	3.7

4 CONCLUSIONS

Ship collision analysis requires to assess relatively complex ship structures, naturally considering stiffened panels and double hull configurations as those engineered for tankers and LNG carriers. The new method presented herein (Eq. 6) manages to predict with reasonable accuracy on the critical deflection of experiments of scaled ship structures and actual accidental data. The accuracy of the new method is comparable with one of the existing methods (Eq. 4). The fact that the new method can predict collision responses comparable with rather complex finite element simulations, and this indicates that Eq. (6) can be used as a good tool for ship collision assessments.

However, the use of Eq. (6), and all previous methods (Eq. 1 to Eq. 5), is mainly limited to ships where bulbous bow is present. In principle, this is not a limitation as most commercial vessels are designed with bulbous bow. Still, simplified calculation methods should be developed to consider other stem shapes, like ships with vertical bow, where the shape can be assessed with a wedge-shaped indenter.

REFERENCES

Alsos HS, Amdahl J. 2009. On the resistance to penetration on stiffened plates, Part I – Experiment. International Journal of Impact Engineering 36: 799–807.

Amdahl J. 1995. Side Collision. 22nd WEGEMT Graduate School, Technical University of Denmark.

Arita K, Aoki G. 1985. Strength of ship bottom grounding (1st report)-an investigation into the case of a ship stranded on a rock. (in Japanese). J. Soc. Nav. Archit. Jpn. 158: 359–367.

Jones N, Wierzbicki T. 1993. Structural Crashworthiness and Failure. Elsevier Applied Science.

Lee YW, Woertz JC, Wierzbicki T. 2004. Fracture prediction of thin plates under hemispherical punch with calibration and experimental verification. Int. J. Mech. Sci. 46: 751–781.

Lehmann E, Peschmann J. 2002. Energy absorption by the steel structure of ships in the event of collisions. Mar. Struct. 15: 429–441.

Liu B, Villavicencio R, Guedes Soares C. 2014. On the failure criterion of aluminium and steel plates subjected to low-velocity impact by a spherical indenter. Int. J. Mech. Sci. 80: 1–15.

McDermott J, Kline R, Jones E, Maniar N, Chiang W. 1974. Tanker structural analysis for minor collisions. In: SNAME Transactions.

Paik JK, Chung JY, Choe IH, Thayamballi AK, Pedersen PT, Wang G. 1999. On rational design of double hull tanker structures against collision. In: The Society of Naval Architects and Marine Engineers. Transactions SNAME.vol. 107. paper number 14.

Simonsen BC, Lauridsen LP. 2000. Energy absorption and ductile failure in metal sheets under lateral indentation by a sphere. Int. J. Impact Eng. 24: 1017–1039.

Tautz I, Schottelndreyer M, Lehmann E, Fricke W. 2013. Collision tests with rigid and deformable bulbous bows driven against double hull side structures. In: Amdahl, Ehlers, Leira (Eds.). Proceedings of the 6th International Conference on Collision and Grounding of Ships and Offshore Structures. Taylor & Francis Group, London. ISBN 978-1-138-00059-9, pp. 93–100.

Wang G, Arita K, Liu D. 2000. Behavior of a double hull in a variety of stranding or collision scenarios. Mar. Struct. 13: 147–187.

Zhang S, Pedersen PT, Villavicencio R. 2019. Probability and Mechanics of Ship Collision and Grounding. Elsevier Inc.

Zhang M, Sun Q, Liu J, Hu Z, Zhang S. 2019. A study of the rupture behaviour of a ship side plate laterally punched by a full-shape bulbous bow indenter. Ocean Engineering 182: 48–60.

Trends in Collision and Grounding of Ships and Offshore Structures – Zhu & Guedes Soares (Eds)

Numerical simulation of ship grounding on a soft slope

H.W. Zhang
Key Laboratory of High Performance Ship Technology (Wuhan University of Technology), Ministry of Education, Wuhan, Hubei, P.R. China

L. Zhu
School of Naval Architecture, Ocean and Energy Power Engineering, Wuhan University of Technology, Wuhan, Hubei, P.R. China

P.T. Pedersen
Department of Civil and Mechanical Engineering, Technical University of Denmark, Lyngby, Denmark

ABSTRACT: The terrain in coastal shallow waters is mainly flat sandy slopes, which are highly susceptible to ship soft grounding accidents. During this process, the ship often beaches on the sand and mud of the shallow bank at a certain speed, which may not only cause local structural damage, but also cause the ship hull to bend or even break, resulting in serious consequences. This study endeavors to carry out a series of numerical simulations on the movement response of a model ship soft grounding on a sand slope. In the finite element software ABAQUS, the ship is modeled as a rigid body and it is hypothesized that the water around the ship hull is a still water area. The test cases of the model ship grounding on a sandy slope at different initial speeds in a water tank are reproduced. The calculation results are compared with the experimental measurements to verify the feasibility of the simulation technology. Furthermore, based on the simulation results, the influence of different grounding speeds on grounding force and ship's movement response is discussed.

1 INTRODUCTION

Ship grounding, as a significant catastrophic event, has garnered considerable attention from researchers. Early studies on ship grounding primarily focused on accident statistics and theoretical analyses of grounding damage. For instance, Zhu et al. (2002) examined the distribution of grounding damage on Ro-Ro ships using statistical data from the Lloyd's Register's damage database. They also developed theoretical models and semi-empirical formulae to assess two real grounding incidents.

Experiments play a crucial role in understanding the mechanics of ship grounding, providing a reliable foundation for theoretical derivations and numerical simulations. Numerous studies have employed experimental methods to investigate ship grounding. Calle et al. (2017, 2019), for example, simplified the research by avoiding complex fluid interactions, conducting dry grounding experiments, and comparing the model results with finite element simulation data. Zhu et al. (2018) designed a ship grounding experiment in a water tank and compared the experimental results of different rock penetrations. Zhu et al. (2024) reported an experimental ship grounding study in a controlled water tank, where a model ship passed over a rigid reef, resulting in the tearing damage of the ship's bottom plate with the interaction of water and structure taken into account. Additionally, Zhou et al., (2024) carried out extensive experiments on ship grounding under multi-rock conditions, analyzing the ship's dynamic response and the plate tearing modes. Zhang et al. (2025) conducted a soft grounding experiment of a model ship in an experimental water tank and analyzed the external dynamic motion characteristics of the ship.

With the advancement of computer technology, numerical simulations of ship grounding have become increasingly prevalent. Alsos & Amdahl, (2007) employed finite element software to investigate the parameters of ship grounding damage induced by seabed topographical features. Abubakar & Dow (2013) utilized the finite element method to study the resistance of ship bottom stiffeners to penetration damage, comparing their findings with experimental grounding results. Lee et al., (2017) introduced simulation results for multiple marine accidents, including collisions and groundings, using FSI technology. Their work demonstrated the feasibility of using scientific methods to investigate the causes of marine accidents and systematically reproduce accident damage scenarios. Additionally, Zhou et al. (2024)

DOI: 10.1201/9781003684404-11

conducted numerical simulations on ship grounding under various rock shapes and explored how the initial state of rocks influences ship damage modes, based on experimental findings.

Ship grounding can be generally classified into two categories: hard grounding and soft grounding, depending on the grounding scenario. However, the majority of research has predominantly focused on hard grounding, which typically occurs when ships ground on rocks or reefs, leading to localized structural damage. In contrast, soft grounding, which involves ships beaching on flat sandy slopes, has received relatively little attention. When a ship runs aground in shallow waters, the bow makes immediate contact with the sand, pressing into the seabed. Following a brief impact phase, the ship gradually slides up the slope. During this process, the bow may undergo local concave deformation, and the hull may experience bending or fracture due to excessive bending moments.

Pedersen (1994) initially proposed a simplified computational program for calculating ship soft grounding, which can rapidly calculate the ship motion responses and load under grounding. Building on this foundational work, Pedersen et al. expanded their research into soil mechanics. In 1995, Pedersen & Simonsen (1995) integrated soil dynamics into their computational model, adopting a more precise calculation method that incorporated hydrodynamics and the elastic vibration of the ship hull girder. Sterndorff & Pedersen (1996) conducted a large-scale ship grounding experiment on an artificial island, comparing the experimental results with theoretical calculations to validate the accuracy of the proposed models. Despite these advancements, there remains relatively limited simulation-based research on soft grounding. Feddersen & Lehmann (2007) employed the finite element method to numerically simulate soft grounding scenarios, where the soil was modeled as an elastic body and the hull was represented as the thin outer plate. Similarly, Quémener et al. (2012) conducted finite element simulations of a loose cargo ship running aground on a smooth seabed, accounting for the buckling deformation of the bow during the grounding process.

This study aims to perform numerical simulations of ship soft grounding. The model ship is represented as a rigid body in the finite element software ABAQUS, while the surrounding water is assumed to be still water. Dynamic analysis steps are employed to reproduce experimental scenarios of the ship grounding on a sandy slope at various speeds. The simulation results are compared with the experimental measurements obtained from model tests, thereby validating the feasibility and accuracy of the simulation approach.

2 TEST CASE

A model ship soft grounding experiment was conducted in the experimental water tank of the Impact and Fluid Structure Interaction (IFSI) Lab at Wuhan University of Technology. Given the water tank's size and practical constraints, the experimental water depth was maintained at approximately 0.6m. The model ship was scaled from a Very Large Crude Carrier (VLCC) with a scaling ratio of 1:117. The main data of the model ship is provided in the table below.

Table 1. Main data of the model ship.

Property	Value
Length	2.75 m
Beam	0.5 m
Depth	0.25 m
Mass	201 kg
Density	7.85×10^3 kg/m^3
Block coefficient	0.84
Water line area	1.375m^2

Figure 1 illustrates the set-up of the experiments, before the experiment, the trim angle of the model ship was adjusted using weight blocks to ensure the ship floated at the design draft of 0.165 meters. The angle of the sandy slope was also carefully controlled. During the experiment, the model ship was pulled horizontally by a motor and traction rope until the bow approached the front edge of the sandy slope. At that point, the rope was released, allowing the model ship to rush up the sandy slope with an initial velocity.

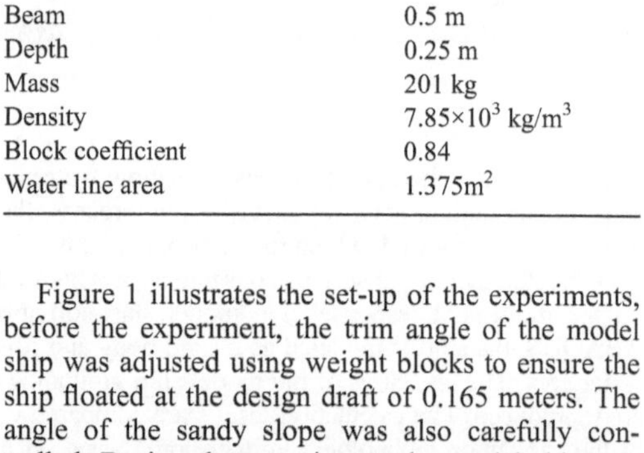

Figure 1. Experimental scene in the water tank.

3 NUMERICAL MODEL

3.1 *Material model*

In simulating ship soft grounding, the representation of the sandy slope is critical, and selecting appropriate soil materials plays a key role in influencing the entire grounding process. In this study, the Drucker-Prager (D-P) material model, as implemented in Abaqus, was employed for the elastoplastic analysis of geotechnical media to simulate sandy slope behavior.

The D-P model is derived from the generalized Mises yield criterion or D-P yield criterion, which accounts for hydrostatic pressure, and is thus often referred to as the generalized Mises model. In ABAQUS, this model has been enhanced to form a modified D-P model, which accurately predicts the stress state of soil at its ultimate strength. The geotechnical parameters used in engineering practice, typically expressed as cohesive force and friction angle, can be represented in linear form. The soil parameters of the sandy slope in this study are summarized in Table 2.

Table 2. Soil properties for the sandy slope.

Property	Value
Thickness d	0.1m(front)~0.3m(end)
Density ρ	1800kg/m^3
Friction angle β	36°
Flow stress ratio K	1
Dilation angle ψ	0°
Yield stress σ_c	500Pa
Young's modulus E	18MPa
Poisson's ratio ν	0.28

3.2 *F.E model*

In this study, a ship model was developed using the finite element software ABAQUS, and the ABAQUS/Explicit module was employed to simulate the soft grounding process of the ship. The model consists of two types of elements: the four-node simplified integral Lagrangian element (S4R) for modeling the ship's structure and the eight-node hexahedral element (C3D8R) for modeling the sandy slope. The grid size and number of elements in the sandy slope significantly influence both the computational efficiency and accuracy of the simulation. In this study, the sandy slope was modeled as an equal-section, right-angled ladder platform with dimensions of 1.6 m in length, 1.5 m in width, and a slope angle of 1:8. The finite element model of the model ship's soft grounding is illustrated in Figure 2.

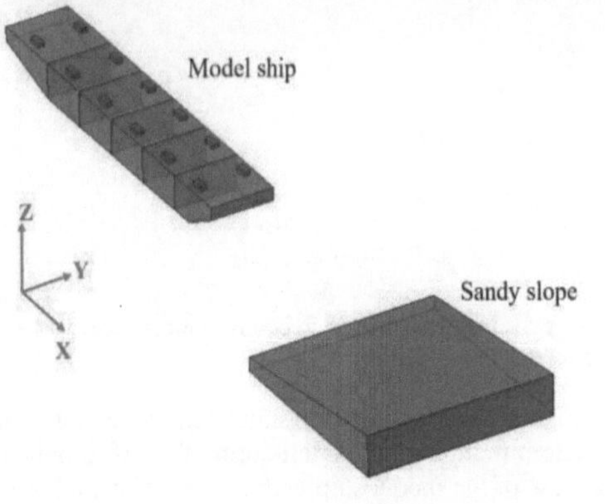

Figure 2. Finite element model of the soft grounding.

In the simulation, the model ship is treated as a rigid body, and the water surface is assumed to remain static throughout the entire soft grounding process. The user-defined subroutine in ABAQUS is employed to apply static water pressure to the ship's surface. A surface-to-surface contact interaction, with a friction coefficient of 0.6, is defined between the sandy slope and the model ship. The contact algorithm automatically tracks the interface between the two after the calculation begins. To ensure the simulation is stable and realistic, and to fully capture the motion and dynamic responses of the ship during soft grounding, the remaining five degrees of freedom of the model ship, excluding motion in the x-direction, are initially constrained. This allows the ship to travel horizontally at a constant speed for a certain distance. When the bow of the ship nears the sandy slope, the control of the ship's degrees of freedom is released, enabling the ship to soft ground on the slope at an initial speed. During this process, the displacement of the surrounding and bottom surfaces of the sandy slope is constrained, and set the additional mass of the model ship in the direction of surge to 0.1M.

4 RESULTS AND DISCUSSION

4.1 *Mesh convergence analysis*

In numerical simulation, the size and number of grids are crucial for balancing calculation efficiency and accuracy. To reliably assess simulation errors, a grid convergence analysis is essential for finite element models. Since the hull is modeled as a rigid body, shell elements with a size of approximately 0.15m are chosen for the model ship to reduce computational costs. The focus is primarily on the grid selection for the sandy slope. To assess the grid sensitivity of the sandy slope, three grid sizes were tested—coarse, medium, and fine: 10mm × 10mm × 10mm, 15mm × 15mm × 15mm, and 20mm × 20mm × 20mm. The calculation results are shown in Figure 3. Taking the displacement of the center of gravity as an example, it can be found that the overall difference in the calculation results under the three grids is small. The calculation results under the 20mm grid fluctuate greatly, while the curves under the smaller 10mm and 15mm grids are relatively stable and very similar. Consequently, to optimize both calculation time and efficiency, this study selects the medium grid size of 15mm × 15mm × 15mm for meshing the sandy slope.

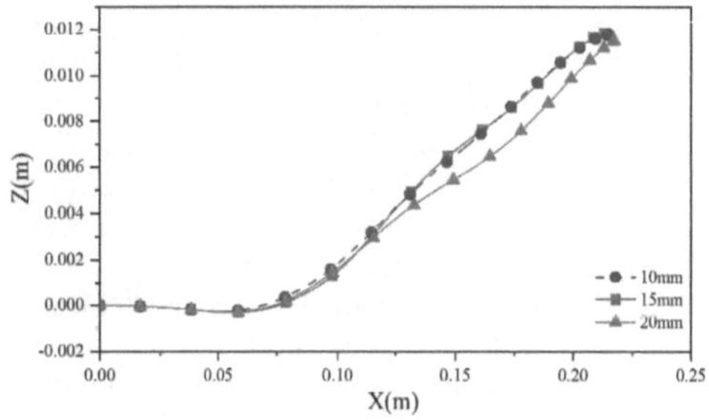

Figure 3. Grid convergence analysis results: displacement of the center of gravity.

4.2 *Typical grounding scenario*

A soft grounding test case of a model ship was simulated in ABAQUS, where the model ship ran aground on a sandy slope at a speed of 1.183 m/s. The three degree of freedom displacement and acceleration of the model ship during this process were calculated and compared with experimental results to validate the accuracy of the numerical simulations. Figure 4 illustrates the entire soft grounding process. It is evident from the figure that as the grounding progresses, compression occurs between the bow of the ship and the sandy slope. As the ship moves upward along the slope, the contact point pushes sand forward. The bow experiences a distinct movement pattern, first rising, then falling, and finally rising again. This behavior is driven by a combination of soil interaction and the ship's inertia.

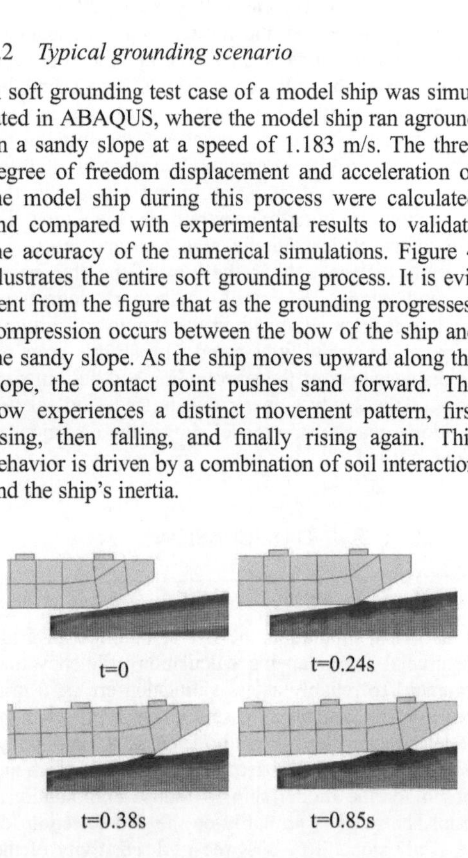

Figure 4. Model ship soft grounding process.

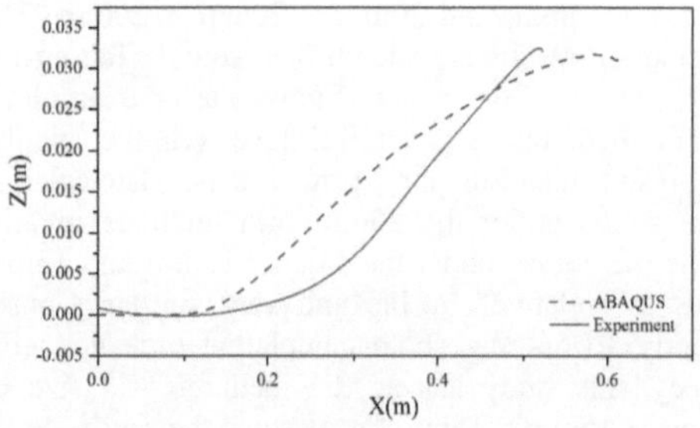

Figure 5. The motion trajectory of the center of gravity of the model ship.

Figure 5 shows the motion trajectory of the center of gravity of the model ship during the soft grounding process. The blue curve represents the simulation results, while the red curve corresponds to the experimental measurement results. From the figure, it is evident that the simulated ship moves a significantly greater distance in the x direction and the curve has a clear 'bulge' compared to the experimental results. This suggests that the ship experienced relatively higher grounding forces in the simulation. One possible explanation for this discrepancy is the challenge in accurately modeling the impact of additional mass during the simulation, which likely resulted in a smaller grounding force in the numerical calculations.

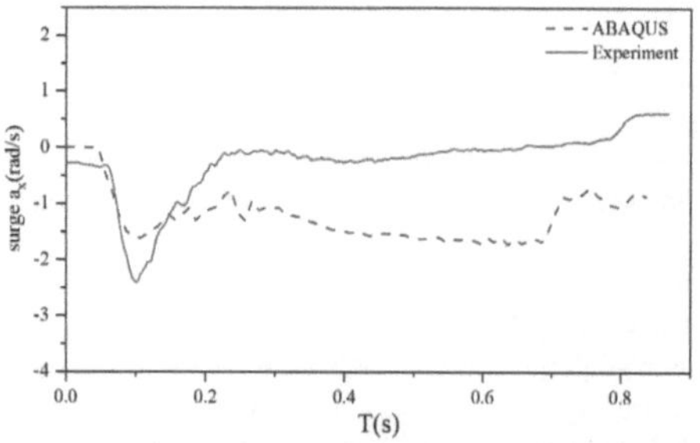

(a) surge: a_x

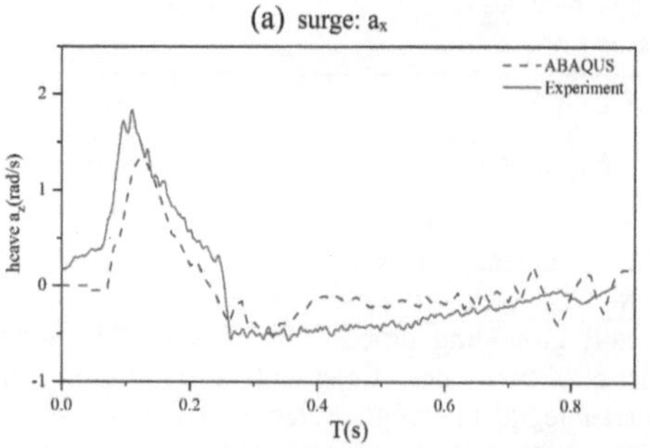

(b) heave: a_z

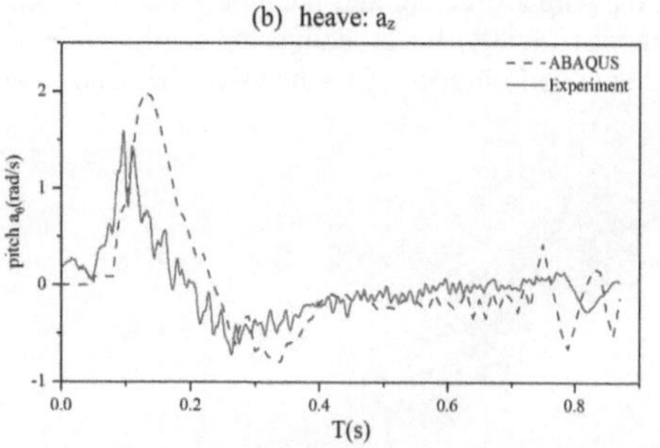

(c) pitch: a_θ

Figure 6. Three degree of freedom motion acceleration of the model ship.

Figure 6 compares the simulated three-degree-of-freedom motion acceleration during the soft grounding process of the model ship with the experimental measurement results. While the overall trend of the motion acceleration in the simulation closely aligns with the experimental data, there are noticeable differences in the numerical amplitudes. For the surge degree of freedom, the simulated acceleration is slightly smaller than the experimental data at the first peak, but diverges significantly after that. In the simulation, the ship experiences a second notable deceleration following the initial impact, whereas the experimental data shows a flattening of the curve. This difference in behavior is also reflected in the discrepancy in the trajectory of the ship's center of gravity. One possible explanation for this variation could be an error in the slope angle of the sandy slope in the experimental setup, which may have influenced the collision dynamics between the ship and the sand. Additionally, the simulation modeled the surrounding water as a hydrostatic pressure, which likely introduced some simplifications, affecting the accuracy of the calculation, especially when considering the added mass.

From the calculation results, it is evident that the simulation in this study successfully reproduces most of the experimental observations of model ship soft grounding, including the displacement and acceleration of the ship's motion. Based on these results, it can be concluded that the numerical simulation method for ship soft grounding proposed in this study is feasible and effective.

4.3 *Effect of initial velocity on dynamic response*

This section explores the influence of initial velocity on the soft grounding response of the model ship by considering four different initial velocity conditions. The initial velocity of the model ship during grounding is varied to assess its impact on the grounding behavior. The calculation cases are outlined in Table 3, with the typical grounding scenario referred to as Case 4.

Table 3. Different cases of soft grounding.

Case	Velocity(m/s)	Slope angle
1	0.421	
2	0.676	
3	0.964	1:8
4	1.183	

The first peak values of the three degree of freedom displacement and acceleration of the model ship during the grounding process are presented in Table 4. From the results, it can be observed that the displacement values at the center of gravity of the ship, obtained from both simulation and experiment, are very similar. However, for motion acceleration, a noticeable trend emerges: as the initial speed of the ship increases, the difference of the peak acceleration also increases. This suggests that hydrodynamic forces around the ship may have a more significant impact at higher speeds, and simulation calculations, which do not fully account for these hydrodynamic effects, tend to result in greater errors.

5 CONCLUDING REMARKS

The model ship was treated as a rigid body in the finite element software ABAQUS, and simulations were performed for experimental cases where the ship ran aground on a sandy slope at different initial velocities in a water tank. The simulated results were compared with the experimental measurements, and the findings indicate that finite element simulations can, to a certain extent, accurately reproduce the soft grounding behavior of the model ship.

While the motion trajectories of the ship's center of gravity in both the simulations and experiments show slight differences, the displacement values are generally quite close. As for the motion acceleration, although some numerical discrepancies remain, both the simulation and experimental results clearly capture the intense peak motion caused by the instantaneous dynamic impact during grounding. However, the simulation appears unable to fully represent the flattening phase following this brief impact. This limitation is likely due to the challenges in accurately accounting for hydrodynamic effects, such as the added mass, in finite element simulations. Additionally, while the slope angle of the sand was controlled during the experiments, the surface of the sandy slope may not have been perfectly uniform. Future research will focus on developing more accurate and efficient simulation methods using

Table 4. The first peak values of the motion displacement and acceleration of the model ship during the grounding process.

Case	X(m)		Z(m)		$\|a_x\|(m/s^2)$		$a_z(m/s^2)$		$a_\theta(rad/s)$	
	Exp	Aba	Exp	Aba	Exp	Aba	Exp	Aba	Exp	Aba
1	0.140	0.167	0.0095	0.012	0.608	0.616	0.319	0.321	0.252	0.458
2	0.288	0.297	0.017	0.019	0.788	0.881	0.589	0.558	0.901	0.494
3	0.430	0.404	0.027	0.026	1.962	1.256	1.67	0.97	0.932	1.437
4	0.522	0.614	0.032	0.031	2.591	1.612	1.824	1.323	1.574	1.990

* 'Exp' means the experimental measurements, 'Aba' means the simulation results.

fluid-structure interaction (FSI) techniques, to better account for hydrodynamic effects in the grounding process.

ACKNOWLEDGEMENTS

This work was supported by the Wuhan University of Technology start-up fund for Distinguished Professors (grant no. 471-40120163).

REFERENCES

Abubakar, A. & Dow, R.S. 2013. Simulation of ship grounding damage using the finite element method. *International Journal of Solids & Structures* 50(5): 623–636.

Alsos, H.S., Jø. & Amdahl, R. 2007. On the resistance of tanker bottom structures during stranding. *Marine Structures* 20(4): 218–237.

Calle, M.A.G., Oshiro, R.E. & Alves, M. 2017. Ship collision and grounding: scaled experiments and numerical analysis. *International Journal of Impact Engineering* 103: 195–210.

Calle, M.A.G., Oshiro, R.E., Korgesaar, M., Alves, M. & Kujala, P. 2019. Combined strain rate, mesh size and calibration test influence on structural failure: miniature ship grounding test. *Ocean Engineering* 173: 215–226.

Feddersen, G.J. & Lehmann, E.A. 2007. The Effects to a Container Ship due to Soft Grounding.

Lee, S.G., Lee, J.S., Lee, H.S., Park, J.H. & Jung, T.Y. 2017. Full-scale ship collision, grounding and sinking simulation using highly advanced M&S system of FSI analysis technique. Procedia Engineering 173(Complete): 1507–1514.

Pedersen, P.T. 1994. Ship grounding and hull-girder strength. *Marine Structures* 7(1): 1–29.

Pedersen, P.T. & Simonsen, B.C. 1995. Dynamics of ships running aground. *Journal of Marine Science and Technology* 1(1): 37–45.

Quéméner, Y., Huang, C.H. & Chen K.C. 2012. Critical Scenario Prediction of a Bulk Carrier Subjected to Soft Grounding. *Proceedings of 5th PAAMES and AMEC.*

Sterndorff, M.J. & Pedersen, P.T. 1996. Grounding experiments on soft bottoms. *Journal of Marine Science and Technology* 1(3): 174–181.

Zhang, H.W., Zhu, L. & Pedersen, P.T. 2025. An experimental study on soft grounding of a model ship in water tank. *Proceedings of 35th International Ocean and Polar Engineering Conference.*

Zhou, Z.H., Zhu, L. & Li, C.B. 2024. Ship grounding simulation with different rock shapes and its verification. *Ships and Offshore Structures*: 1–13.

Zhou, Z.H., Zhu, L. & Liang, Q.Y. 2024. Dynamic Responses and Damage of a Model Ship in Multi-Rock Grounding. *Journal of Marine Science and Engineering* 12(11): 1908.

Zhu, L., James, P. & Zhang, S.M. 2002. Statistics and damage assessment of ship grounding. *Marine structures* 15: 515–530.

Zhu, L, Zhou, Q.W, Chen, M.S. & Chen, X.Q. 2018. Grounding Experiments of a Ship Model in Water Tank. *Proceedings of the ASME 2018 37th International Conference on Ocean, Offshore and Arctic Engineering.* Volume 11A: 17–22.

Zhu, L., Zhou, Z.H. & Pedersen, P.T. 2024. Ship grounding model tests in a water tank: an experimental study. *Marine structures* 93(Jan): 1.1–1.24.

Trends in Collision and Grounding of Ships and Offshore Structures – Zhu & Guedes Soares (Eds)
© 2026 The Author(s), ISBN 978-1-041-16361-9

Research on the assessment method for ship collision damage levels

Y.J. Zhao, T. Xie & Z.J. Li
China Ship Scientific Research Center, Wuxi, P.R. China
National Key Laboratory of Ship Structural Safety, Wuxi, P.R. China
Taihu Lake Laboratory of Deep Sea Technology and Science, Wuxi, P.R. China

J. Lu
China Ship Scientific Research Center, Wuxi, P.R. China

Z.C. Hong
Jiangsu University of Science and Technology, Zhenjiang, P.R. China

ABSTRACT: Following a ship collision accident, in addition to structural damage, catastrophic events such as compartment flooding, capsizing, and loss of structural integrity leading to breakage may occur. Even if such disasters are avoided, the initial damage caused by the collision can degrade the ship's speed, maneuverability, seakeeping performance, and other operational capabilities compared to an intact ship. The shock environment or flooding resulting from the collision may also challenge the normal operation of onboard equipment, thereby affecting the ship's overall performance. This paper proposes a method for assessing ship collision damage levels by integrating damage tree analysis and quantitative vulnerability analysis. This approach comprehensively evaluates both structural damage and the resulting performance degradation. The method facilitates a holistic assessment of the impact of collisions on ship performance.

1 INTRODUCTION

Research on structural damage caused by ship collisions is well-established (Minorsky, 1958; Zhang, 2019; Kim, 2022; Zhang, 2024; Zhu, 2025), with most studies focusing on external and internal dynamics, which will not be reiterated here. However, post-collision performance evaluation remains insufficiently explored. In terms of damaged hull stability, the flooding process and capsizing behavior of a damaged hull in waves are highly complex. Bu (2022) proposed a viscous/potential unified model for predicting wave-induced stability, validating its accuracy for both intact and damaged ships. Lu (2024) combined an intact stability time-domain motion model with experimental results on damaged hull roll periods and damping to compute time-domain roll motions under various flooding scenarios. For residual strength assessment of damaged hulls, the Smith method (Maestro, 1989) and finite element analysis (Notaro, 2010) are commonly used, but evaluations of asymmetric loading conditions (bending, shear, and torsion) remain limited. Damaged hull stability and residual strength directly determine survivability, while degraded speed, maneuverability, and seakeeping performance affect operational functionality. Most existing studies focus on seakeeping (Korkut, 2004; Palazzo, 2004; Papanikolaou, 2001; Gao, 2001), with limited research on speed and maneuverability (Hoshino, 2002; Shoichi, 2002).

Regarding damage level classification, the 1989 Exxon Valdez oil spill prompted the U.S. to enact OPA-90 (Oil Pollution Act of 1990), mandating double-hull structures for oil tankers entering U.S. waters—an early example of collision resistance grading. Germanischer Lloyd uses the "COLL" notation to classify collision resistance, comparing it to a standard single-side hull. The American Bureau of Shipping employs the "RES" notation to grade residual strength post-collision or grounding. However, collision resistance grading should not rely solely on structural damage but also consider performance loss. Structural damage is a superficial outcome; its impact on functionality is critical. Thus, a performance-based damage level assessment method is essential. Current research in this area remains inadequate.

This paper introduces the concept of "vulnerability" from ship survivability assessment. A structural tree model and collision damage network are established, followed by damage level classification and criteria. By integrating damage tree analysis and quantitative vulnerability assessment, structural damage and functional performance loss are evaluated

DOI: 10.1201/9781003684404-12

holistically. This method provides a scientific basis for assessing the comprehensive impact of collisions on ships.

2 SHIP STRUCTURAL TREE MODEL AND DAMAGE NETWORK

To facilitate analysis, all compartments of the target ship are numbered (e.g., from stern to bow, top to bottom), and deck plates and bulkheads are labeled, as shown in Figure 1.

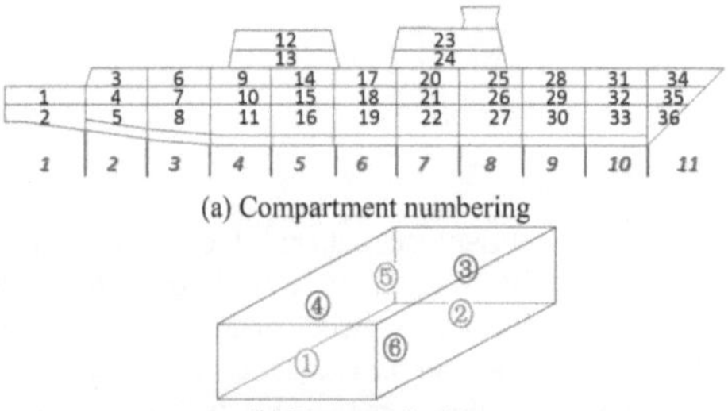

(a) Compartment numbering

(b) Plate numbering

Figure 1. Compartment and plate numbering for ship-ship collision analysis.

Based on the mission, establishing a damage relationship network for the ship that links mission capability to system functions and ultimately to specific structures/equipment. When mission attributes differ, the target ship has varying functional requirements for its systems, resulting in distinct damage trees. In this study, the ship's mission is assumed to be patrol. The primary first-level systems related to this mission include: structural system, propulsion system, electrical system and detection system.

- Structural system: Primarily affects the ship's hydrodynamic performance (including overall navigational performance: speed, maneuverability, seakeeping) as well as buoyancy, stability, and residual strength.
- Propulsion system: Mainly influences mobility parameters such as sailing speed and turning radius.
- Electrical system: Determines power supply and distribution capabilities, affecting all onboard electrical equipment.
- Detection system: A critical system determining whether patrol missions can be successfully accomplished.

As shown in Table 1, the damage relationship network for the patrol ship is established accordingly. Notably, since both the propulsion system and detection system require electrical power to operate normally, the electrical system (B(3)) is incorporated as a third-level subsystem within both systems.

Table 1. Damage relationship network of patrol ship.

mission	Level 1	Level 2	Level 3	Level 4
Patrol A(1)	Structural System B(1)	Overall Navigational Performance C(1)	Speed D(1)	/
			Maneuverability D(2)	/
			Seakeeping D(3)	/
		Buoyancy C(2)	/	/
		Stability C(3)	/	/
		Residual Strength C(4)	/	/
	Propulsion System B(2)	Propulsion Capacity C(5)	Port Propulsion D(4)	Propeller E(1)
				Shaft E(2)
				Gearbox E(3)
				Clutch E(4)
				Gas Turbine E(5)
				Diesel Engine E(6)
			Starboard Propulsion D(5)	Propeller E(7)
				Shaft E(8)
				Gearbox E(9)
				Clutch E(10)
				Gas Turbine E(11)
				Diesel Engine E(12)
			Electrical System B(3)	/
		Steering Capacity C(6)	Port Steering D(6)	Rudder E(13)
				Steering Gear E(14)
			Starboard Steering D(7)	Rudder E(15)
				Steering Gear E (16)
			Electrical System B(3)	/
	Electrical System B(3)	/	Power Generation D(8)	Generator 1 E(17)
				Generator 2 E(18)
				Generator 3 E(19)
			Power Distribution D(9)	Distribution unit E(20)
	Detection System B(4)	/	Helicopter Detection D(10)	Helicopter 1 E(21)
				Helicopter 2 E(22)
				Navigation Radar E(23)
				Search Radar E(24)
			Electrical Sys-tem B(3)	/

3 DAMAGE LEVELS AND CRITERIA

3.1 *Damage levels*

In order to classify and evaluate the damage during ship collision, it is necessary to put forward the damage level for all levels of systems and capabilities in the damage relationship network, and in this paper, the damage level is classified into three levels. For the ability to accomplish the patrol mission, the three levels of damage are defined as follows:

- Level 1: Structural system, power system or electric system level 1 damage to the patrol task can not be performed;
- Level 2: Grade 2 damage to structural system, power system or electric system to the point where the ability to accomplish the patrol mission is reduced;
- Level 3: structural system, power system or electric power system are not higher than level 2 damage, and basically have no effect on the ability to accomplish the patrol task.

Taking a level 1 structural system as an example, the following damage levels are given:

Table 2. Structural system damage levels.

Damage level	Definition
1	Buoyancy, stability, residual strength, or overall navigational performance at level 1 damage.
2	Buoyancy, stability, residual strength, or overall navigational performance at level 2 damage.
3	All of Buoyancy, stability, residual strength, or overall navigational performance at level 3 damage.

For propulsion capability, ship-handling capability, power generation and distribution capability, etc., 66% and 33% reductions are used as the damage thresholds for Levels 1, 2, and 3.

3.2 *Damage criteria*

3.2.1 *Buoyancy loss*

Assuming that the unsinkability of the target ship is specified as any three compartments are unsinkable after water ingress, the loss of displacement is m1 +m2+m3. According to the stipulated provision, in the event that any three compartments X1, X2, X3 all enter the water, the loss of displacement is equivalent to m1+m2+m3. The damage level is divided as follows, with the degree of loss of displacement serving as the limit value for the classification of buoyancy.

Table 3. Buoyancy loss levels.

Damage level	Definition
1	Buoyancy loss ≥(m1+m2+m3)
2	0.5*(m1+m2+m3)≤Buoyancy loss<(m1+m2 +m3)
3	Buoyancy loss<0.5*(m1+m2+m3)

3.2.2 *Stability loss*

The maximum transverse angle of θ2 and the maximum longitudinal angle of φ2 for any three-compartment inlet submerged ship are taken into consideration to define the stability loss level as follows:

Table 4. Stability loss levels.

Damage level	Definition
1	Transverse inclination $\theta \geq \theta 2$ or longitudinal inclination $\varphi \geq \varphi 2$
2	$\theta 1 \leq \theta < \theta 2$ or $\varphi 1 \leq \varphi < \varphi 2$ ($\theta 1 = \theta 2/2$, $\varphi 1 = \varphi 2/2$)
3	$\theta < \theta 1$ and $\varphi < \varphi 1$

3.2.3 *Residual Strength loss*

Taking the residual load-bearing moment of the ship after collision damage as an index, the residual strength loss level is defined as follows:

Table 5. Residual Strength loss levels.

Damage level	Definition
1	Over 66% reduction in ultimate load-bearing moment M0
2	Reduction of ultimate load-bearing moment M0 [33%, 66%]
3	Reduction of ultimate load-bearing moment M0 by 33% or less

3.2.4 *Overall navigational performance loss*

In this paper, speed, maneuverability, and wave resistance are collectively referred to as overall navigational performance. The reason for this is that the loss of buoyancy, stability and residual strength will directly lead to the loss of ship's vitality. The overall navigational performance does not directly determine the ship's vitality. There is no mature method for calculating the overall navigational performance of a damaged ship, and it mostly relies on CFD calculations. By weighting the variation rates of maximum speed, slewing period, longitudinal distance, roll angel, we can obtain the characterization parameter of the overall navigational performance S. The loss level of the overall navigational performance is defined as follows:

Table 6. Overall navigational performance loss levels.

Damage level	Definition
1	The rate of change in overall navigational performance S exceeds 66%
2	The rate of change in overall navigational perfor-mance S [33%, 66%]
3	The rate of change in overall navigational perfor-mance S less than 33%

3.2.5 *Equipment damage*

Given that the phenomenon of ship collision can be categorized as a low-speed impact, the calculation of equipment damage can be made with reference to the design requirements for impact resistance of equipment. The requirements for equipment impact resistance level vary among countries, though the overall requirements are analogous. These requirements are contingent upon the equipment installation location and installation mode, which consequently stipulate the relevant design load requirements. This safety threshold, designated as V1, signifies the commencement of the damage limit value. The damage threshold, designated as V2, is determined empirically. For elastic mounting equipment, V2=4V1, and for rigid mounting equipment, V2=3V1. The equipment flooding damage is determined by the damage judgment index, which categorizes the equipment as either damaged by water immersion or undamaged.

Table 7. Equipment damage levels.

Damage level	Definition
1	Equipment submerged or $V \geq V2$
2	Equipment not submerged and $V1 \leq V < V2$
3	Equipment not submerged and $V < V1$

The damage levels and criteria for each level in the damage relationship network have been defined. These definitions will provide inputs for the construction of subsequent damage tree models and the analysis of damage levels.

4 DAMAGE TREE MODEL CONSTRUCTION

The damage tree model of the target ship is established based on the damage relationship network. In order to facilitate the damage tree characterization, the elements in the damage relationship network are characterized by the following symbols.

- Patrol mission: A;
- Level 1 ~ Level 4: B, C, D, E;
- The numbers 1~n in parentheses after the letters are the sequential numbering of specific elements at each level;
- The upper right scale of the letters is the damage level, and 1~3 indicates the damage of level 1~3.

The following figure gives the damage tree model of the first, second and third level damage of the patrol ship respectively. Only the second level system is given in the figure, and the damage tree of the second level system to the fourth level system can be given in a similar logical relationship.

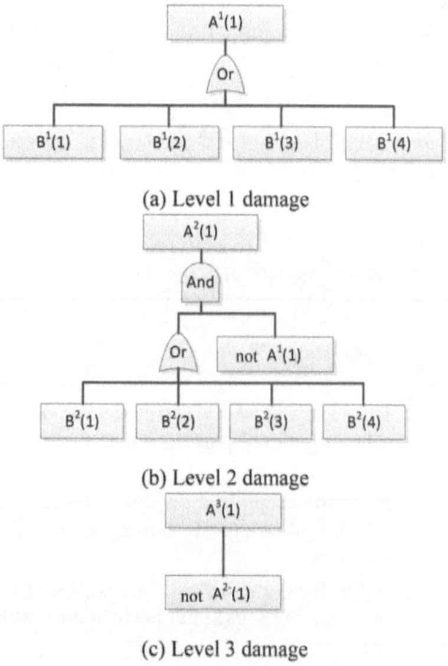

(a) Level 1 damage

(b) Level 2 damage

(c) Level 3 damage

Figure 2. Damage tree model for patrol mission.

5 VULNERABILITY QUANTIFICATION ANALYSIS

The construction of the target ship damage tree previously mentioned enables the qualitative definition and establishment of the damage level and damage relationship of the entire ship, each system, and each piece of equipment. However, if it is necessary to quantitatively determine the probability of a certain damage level of the entire ship and each system, then it is necessary to establish the corresponding probability calculation model. This paper uses the fuzzy stochastic theory to construct a quantitative analysis model from the bottom event probability level, level by level, to the top event probability, according to the analysis of the damage tree.

5.1 *Fuzzy random probability model*

5.1.1 *Randomness of damage*

Affected by the production process, the use of environmental conditions, etc., in the ship collision conditions, structure, equipment damage thresholds are objective uncertainty, random fluctuations within a certain range. Random variables with the above characteristics can generally be considered to obey the continuous normal distribution function:

$$\phi(\xi) = \frac{1}{\sqrt{2\pi}\sigma} e^{-\frac{(\xi-a)^2}{2\sigma^2}} \text{ (a>0, } \sigma\text{>0, const)} \quad (1)$$

Therefore, the response values ξ_{res} and damage thresholds ξ_0 are normally distributed. The probability density functions are respectively

$$\phi(\xi_{res}) = \frac{1}{\sqrt{2\pi}\sigma_1} e^{-\frac{(\xi_{res}-a_1)^2}{2\sigma_1{}^2}} \quad (2)$$

$$\phi(\xi_0) = \frac{1}{\sqrt{2\pi}\sigma_2} e^{-\frac{(\xi_0-a_2)^2}{2\sigma_2{}^2}} \quad (3)$$

Then the state function of the structure or equipment at the damage criticality is

$$Z = \xi_{res} - \xi_0 \quad (4)$$

$Z > 0$, means damaged; $Z<0$ means intact, as shown below.

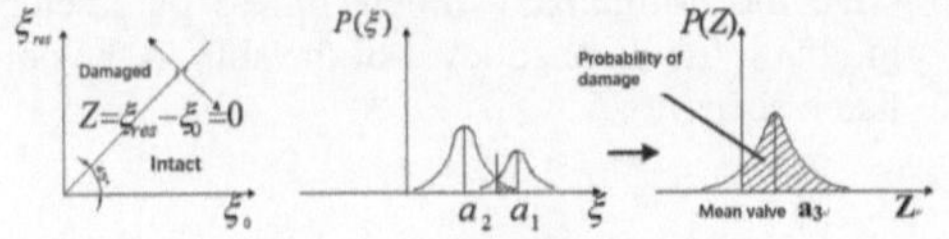

Figure 3. Models for calculating the criticality factor for damage and the probability of damage.

The probability of damage can thus be expressed as

$$P_d = P(Z>0) = P((\xi_{res} - \xi_o)>0) \quad (5)$$

As ξ_0 and ξ_{res} obey a normal distribution, the damage probability is obtained from the fact that a linear combination of a finite number of normally distributed random variables still has the properties of a normal distribution:

$$P_d = 1 - \int_{-\infty}^{0} \frac{1}{\sqrt{2\pi}\sigma_3} \exp(-\frac{(Z-a_3)^2}{2\sigma_3{}^2})dZ \quad (6)$$

In which a_3 and σ_3 are constant greater than 0, calculated by the addition theorem of the normal distribution.

5.1.2 *Ambiguity of damage*

In addition to randomness, another form of uncertainty exists: the equipment or structure damage threshold according to the relevant standards of artificially determined ambiguity and subjective ambiguity on the classification of equipment damage. The presence or absence of damage is determined by subjective ambiguity. For instance, the deformation of the joints in the equipment is regarded as damage, yet the plastic strain, which can reach up to 2% or 5%, is considered a certain level of damage. This is due to the lack of sufficient theoretical and experimental basis for such a determination, which is characterized by a certain degree of subjectivity. It is imperative to consider the human element in the interpretation of the ambiguity, given the inherently random and uncertain nature of the damage process. Employing a dual-faceted approach that incorporates both analytical and computational methodologies will ensure a more rational and comprehensive evaluation.

It can be assumed that in the intact domain there is still a potential for damage; on the contrary, the structure or equipment in the damage domain also has a certain degree of safety, the demarcation line between damage and no damage is in fact a non-explicit, mutually inclusive fuzzy gradient domain. The damage criticality coefficient Z_0 is equal to 0 indicating that it is in a critical state, and the affiliation degree of damage and integrity is 50%. The boundary of non-certain damage is actually a fuzzy region with intermediary transition. This requires the application of fuzzy mathematical theory of specific equipment or structure to study the degree of its affiliation with the safety or damage to the affiliation function to analyze and describe, as shown in the figure below. The size of the fuzzy region range is determined by the upper and lower boundaries of the value Z_a and Z_b. Z_a and Z_b can take the upper and lower limits of the damage level.

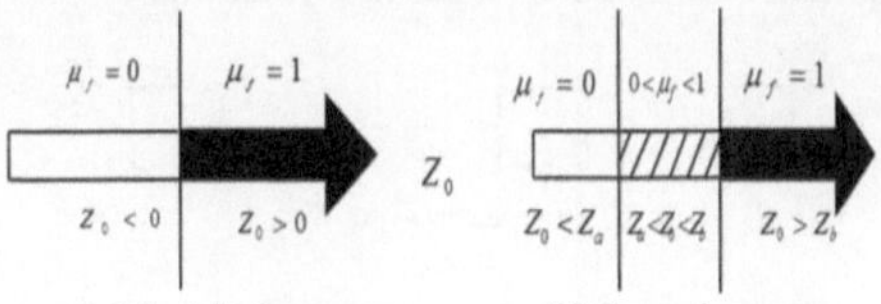

Figure 4. Relationship between certain damage and fuzzy damage and their affiliation functions.

5.1.3 *Analysis model*

Randomness and fuzziness reflect the two uncertain phenomena of things from different sides, and there are both differences and connections between them, which can be interpenetrated, while the latter is the promotion and development of the former. Since the fuzzy probability analysis considers the potential danger in the safety domain and the non-injury potential in the damage domain, the calculation results will be more comprehensive and reasonable.

Assuming that the affiliation function of the damage critical coefficient Z_0 obeys a symmetric ascending half-normal distribution about the point (0,0.5), the affiliation function of the damage indicator X_{iq} attached to the damage is:

$$\mu_f(Z_0) = \begin{cases} 0, Z_0 < Z_a \\ 1 - e^{-k(Z_0 - Z_a)^2}, Z_a \le Z_0 < 0, k > 0 \\ 0.5, Z_0 = 0 \\ e^{-k'(Z_0 - Z_b)^2}, 0 < Z_0 < Z_b, k' > 0 \\ 1.0, Z_0 \ge Z_b \end{cases} \tag{7}$$

The distribution of the affiliation function expressed in the above equation is shown below.

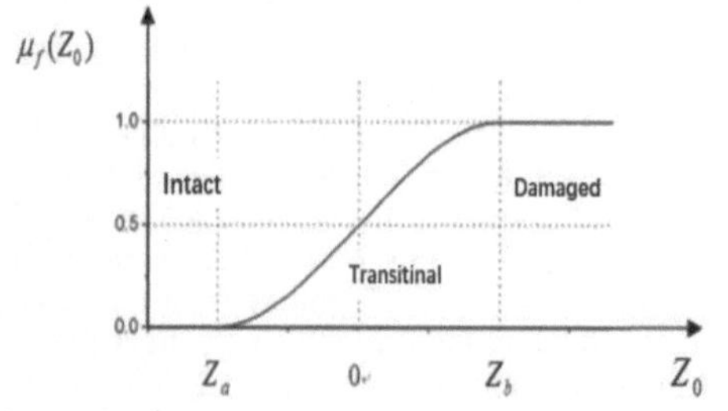

Figure 5. Distribution of damage coefficients Z_0.

5.2 *Probability of damage to patrol ships*

According to the damage tree, from the equipment to the system, from the system to the whole boat damage level, probability can be calculated. The concept of damage tree originates from system reliability theory, the top event probability is derived through a series of bottom events, through the combination of logical relationships with and/or gates, shown in the following figure. Consider non-redundant equipment as a series relationship while redundant equipment as a parallel relationship.

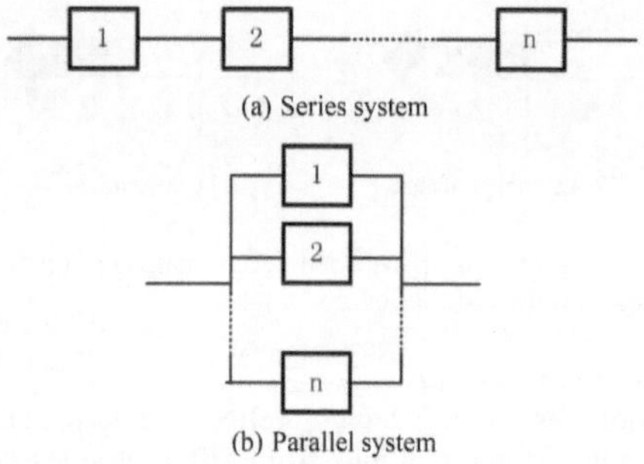

Figure 6. Series and parallel systems.

As the system and equipment composition of typical ships are very complex, the loss of function relationship between some systems and equipment sometimes cannot be calculated by simple methods, and thus the determination of the weighting factor is also more difficult. The expert scoring method combined with interval hierarchical analysis can be used to determine the weighting factors, which will not be discussed here due to space constraint.

5.3 *Quantitative assessment of damage levels*

Through the classification of the damage level and the calculation of the damage probability of each level, the probability that the whole ship and system on each damage level can be obtained, and then the normalized dimensionless parameter is used to represent each damage level. The specific steps of the damage level assessing method are as follows:

(1) Determine the collision scenario: the relative angle of collision, relative speed and relative position of the two identified ships;
(2) Determine the impact plate number and calculate the structural damage;
(3) Calculate the residual stability, residual strength and overall navigational performance under the damage state of the structure;
(4) Calculate the shock environment of the equipment in the collision scenario, and clarify whether the compartment where the equipment is located is waterlogged;
(5) Calculate the damage probability and damage level of the structure and equipment based on the fuzzy random probability model;
(6) Combine the weighting factors and calculate the damage probability of each level of the system based on the damage tree step by step upward;
(7) Give the normalized dimensionless parameters to obtain the damage level of the ship in the collision scenario.

6 CONCLUSIONS

The definition of ship collision damage levels cannot be determined by relying only on the degree of structural damage. Rather, it is a more scientific evaluation method to expand from structural damage rating to functional damage rating according to the actual mission requirements of ships. This paper presents a novel ship collision damage rating method. The method integrates structural damage analysis, equipment collision impact environment analysis, damage tree model, and quantitative vulnerability analysis method. It grades ship collision damage, thereby comprehensively considering structural damage and performance loss resulting from such damage. The method is instrumental in providing a more comprehensive assessment of the impact of ship collisions on ship performance.

7 DISCUSSIONS

The application of this method has certain limitations and potential challenges:

(1) To conduct ship collision damage level assessment based on this method, extensive ship-related information must be obtained. Specifically, when constructing the damage tree, comprehensive system and equipment data related to ship operational functions (beyond structural information alone) are required, which imposes higher demands on data collection and analysis.
(2) The establishment of damage criteria is subject to significant human intervention. Different scholars may propose varying criteria for the same event, leading to inconsistencies in evaluation results. Particularly for shock damage criteria of onboard equipment, reliable data support is often lacking.
(3) Weight factors are another critical element heavily reliant on human expertise. Their values vary substantially across ship types and mission scenarios. Currently, the most reasonable approach to determine weight factors remains the expert scoring method.
(4) The evaluation process requires substantial time, making it impractical for engineering applications. Ideally, a comprehensive database of computational or experimental samples should be established. Structural damage, residual stability, residual strength, overall navigational performance, and shock environments under various collision scenarios could then be rapidly extracted from this database to enable swift assessments. Since the process partially depends on human judgment, the absolute values of evaluation results are less critical. What matters most are the relative differences in outcomes across scenarios and the analysis of underlying causes for these differences.

REFERENCES

Bu S X, Gu M, Zhan J H, Zhang P J. Experimental and numerical investigations on a damaged passenger ship motion in regular beam waves. Proceeding of the 14th International Conference on Hydrodynamics, 2022.

Gao Q X. CFD simulation of water ingress into damaged ships[J]. Journal of Ship Mechanics, 2001, 5(6): 9–18.

Hoshino K, Hara S, Yamakawa K, Yukawa K. Wave drift force and drift motion of the bow part of broken ship[C]. Proceedings of the twelfth international offshore and polar engineering conference. Kitakyushu, Japan, 2002: 26–31.

Kim S J, Taimuri G, Kujala P, Conti F, Le Sourne H, Pineau, J-P, Looten T, Bae H S, Mujeeb-Ahmed M P, Vassalos D, Kaydihan L, Hirdaris S. Comparison of numerical approaches for structural response analysis of passenger ships in collisions and groundings. Marine Structures, 81 (2022) 103125.

Korkut E, Atla M, I Ncecik A. An experimental study of motion behavior with an intact and damaged Ro-Ro ship model[J]. Ocean Engineering, 2004(31): 483–512.

Lu, J, Zhao Y J, Shi, G, Yu T J, Gu, M. A numerical method on large roll motion in beam seas under intact and damaged conditions. Journal of Marine Science and Engineering. 2024, 12(11), 2043.

Maestro M. An assessment of the structural capability of damaged ships: the plastic approach in longitudinal unsymmetrical bending and the influence of buckling [J]. Int Shipbuilding Progrss, 1989, 36(8):733–737.

Minorsky V U. An analysis of ship collisions with reference to protection of nuclear power plants[R]. Sharp (George G) Inc., New York, 1958.

Notaro G, Kippenes J, Amlashi H, Russo, M Residual hull girder strength of ships with collision or grounding damages [C]. 11th Inte. Symp. On Practical Design of Ships and Other Floating Structures, Rio, Brazil, 2010.

Palazzo L, Kat J. Model experiments and simulations of a damaged ship with air-flow taken into account [J]. Marine technology, 2004, 41(1): 38–44.

Papanikolaou A. Benchmark study on the capsizing of a damaged Ro-Ro passenger ship in waves-final report [R]. Report of the Second ITTC Damage Stability Benchmark Study. [S. 1.]2001.

Shoichi H. Drift prevent of disabled ship in rough sea [R]. The Japanese research project on disabled ships. Fy. 1998-2002.

Shoichi H, Kenji Y, Hoshino, K., Yukawa, K. Towline tension and unstable motion of damaged and capsized geometric ship during towing[A]. Proceedings of the twelfth international offshore and polar engineering conference. Kitakyushu, Japan, 2002: 322–329.

Zhang S. M., Pedersen P.T., Villavicencio R. 2019. Probability and mechanics of ship collision and grounding. Butter-worth-Heinemann.

Zhu L, Wang X B. Ship collision model tests in a water tank. Marine Structures, 103 (2025) 103791.

Zhang Y C, Yuan Y C, Zhou, J Y, Tang, W Y. A new analytical model to evaluate ship side collisions with large indentation considering interaction effects between structural components. Marine Structures, 95 (2024) 103596.

Collision and grounding experiments

Trends in Collision and Grounding of Ships and Offshore Structures – Zhu & Guedes Soares (Eds)
© 2026 The Author(s), ISBN 978-1-041-16361-9

Full-field measurement of residual strains in ship steel during bending operations using digital image correlation

A.A. Elsafty, A.Y. Elruby & S. Nakhla
Memorial University, St John's, Newfoundland, Canada

ABSTRACT: The aim of this work is to analyze the residual strains of ship steel during cold forming processes. Residual stresses can lead to several undesirable effects such as plastic deformation affecting ship structures' load bearing capacity. The cold forming processes, such as air bending and press braking, consist of two primary stages. The first stage involves loading under pure bending where plastic deformation occurs. The second stage is an unloading reverse elastic process known as springback. Extensive testing is performed on two commonly used materials in ship building, namely, steel 44w and aluminum 6061. Three thick plates for each material were fabricated. Three-point bending was utilized to mimic cold forming process with a maximum displacement reaching double the thickness. Through the thickness residual strains are captured by noncontact 3D-digital image correlation (DIC) stereo system using ARAMIS Adjustable 12M setup. 3D-DIC is chosen for its ability to provide a continuous deformation history throughout the loading and unloading stages. In contrast, traditional residual strain measurements methods, such as deep hole drilling, contouring and x-ray diffraction only evaluate strains at the time of measurement. The initial results of the current study are presented and discussed including mechanical characterization as well as full-field strain measurements.

1 INTRODUCTION

During shipbuilding, forming processes whether cold forming or hot forming are utilized for ship structures. Hot forming has accuracy limitations when dealing with saddle shaped plates. As a result, cold forming is usually preferred. However, the effects of residual stresses on the strength of formed steel must be examined carefully especially since residual stresses and strains vary through the thickness unlike hot forming (Shen et al., 2018). Residual stresses are defined as those that remain in a material or body after processing in the absence of external forces (Lord & Fry 2002). According to (Denton, 1966). the typical distribution of residual strains in the axial direction is illustrated in Figure 1(a). The typical normal stress due to bending distribution is shown in Figure 1(b) becomes the final residual stress distribution in Figure 1(d). This is due to the springback which is assumed as reverse elastic behavior as shown in Figure 1(c). It can be demonstrated that the final residual distribution follows a compressive, tensile, compressive then tensile pattern. This is due to the unequal plastic deformation caused by removal of load resulting in misfit of elements (Denton, 1966). The balance between tension and compression in residual stress distribution

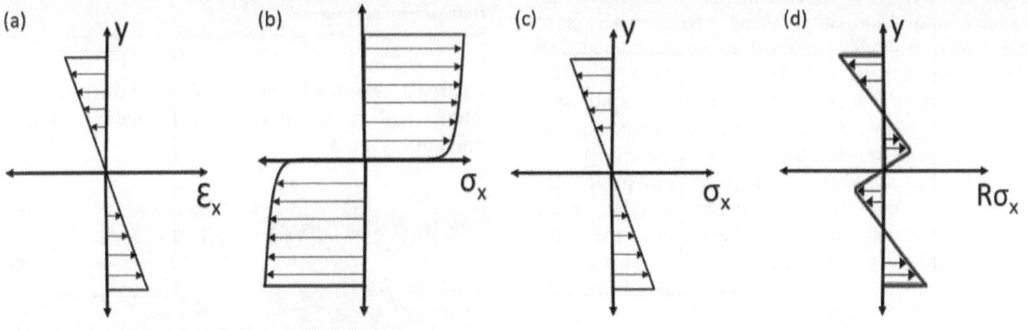

Figure 1. (a) Normal strain distribution for bent beam, (b) Elastic-plastic bending stress distribution, (c) Stress due to springback and (d) Final residual stress distribution.

DOI: 10.1201/9781003684404-13

emerge because in any free-standing body stress equilibrium must be maintained.Industrially, tensile residual stresses are undesirable since they are the prime cause of fatigue failure, quench cracking and stress corrosion cracking. On the other hand, compressive residual stresses are favored as they increase fatigue strength, improve resistance to stress corrosion cracking (Kobayashi et al., 1998).

Accordingly, the analysis of residual stresses has been the aim of several studies. The engineering theory for plastic bending of non-work hardening materials was deduced by Ludwick (1903). Assumptions include cross sections plane remains plane and the shear stress and its distribution through sheet thickness is neglected.

In addition, neutral plane shifting is neglected since neutral plane layer is assumed to coincide with the mid plane. The bending moment during unloading should have the same magnitude but an opposite direction to the applied elastic bending moment (Wang et al., 1993). Hill presented the first meticulous solution for plane strain bending for isotropic rigid perfectly plastic materials incorporating the yield criterion and its associated flow rule (Hill, 1950). Based on which, Quach et al. proposed an analytical method with the effect of material anisotropy considered employing plane strain assumption (Quach et al., 2009). Moreover, Zheng et al. (2019) proposed a simplified theoretical method on the residual stresses in press-braked stainless steels sections assuming isotropic behavior.

On the other hand, several studies have investigated the residual stresses numerically using finite element (FE) analysis. For instance, Mutafi et al. (2024) recommended using 3D-Shell. However, 3D-shell model does not predict the shift in neutral plane.

Numerous experimental methods were utilized to analyze residual stresses. These include destructive methods such as sectioning and contouring, semi-destructive methods such as hole drilling and ring core and non-destructive methods such as x-ray diffraction and neutron diffraction (Díaz et al., 2021). Alternatively, three-dimensional digital image correlation (3D-DIC) has been proposed as a method to determine the residual strains and can evaluate stresses when combined with methods mentioned above as conducted in (Harrington & Schajer, 2017). This is because unlike the traditional methods, 3D-DIC provides a continuous deformation history throughout the measurement rather than evaluating strains at the time of measurement.

(Gerbo et al., 2016) and (Gothivarekar et al., 2020) employed 3D-DIC to predict residual strains in axial direction based on post bending measurements of metal plates. However, to the best of the authors' knowledge, very limited studies have captured the full through-thickness evolution of residual strain during the entire loading and unloading history of shipbuilding plates during cold forming. Thus, the chief objective of this study is to utilize the stereo 3D-DIC system to provide continuous full-field deformation history of thick 44W steel and 6061 aluminum plates under three point bending for residual strain measurements. The approach enables the observation of neutral axis shifts and the springback behavior before and after unloading.

2 MATERIALS AND METHODS

First, material properties of both steel 44W and aluminum 6061 are characterized. Both materials are used in the ship building industry where steel 44W exhibits significant ductility when compared to aluminum 6061. This is done to validate the proposed methodology for different material behaviors. Flat standard tensile specimens were fabricated as per the ASTM E8/E8M – 24 adhering to its recommended dimensions for sheet type specimens where sufficient material having a width of 12.5 mm or less is available, as illustrated in the schematic diagram in Figure 2 and the tensile specimen dimensions summary in Table 1.

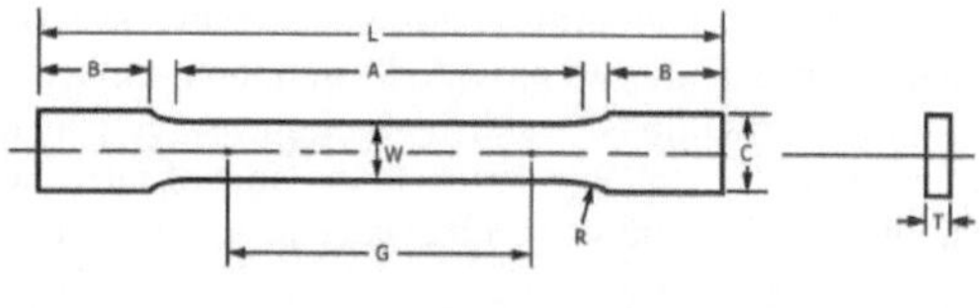

Figure 2. Flat (rectangular) specimen geometry (ASTM E8/E8M-24).

Table 1. Tensile Specimen Dimensions.

Dimension	Symbol	Value
		mm
Gauge length	G	50
Width	W	12.5
Thickness	T	6.0
Radius	R	12.5
Overall Length	L	200
Length of test section	A	57
Length of grip section	B	50
Width of grip section	C	20

Three 25mm thick plates were machined for each material and the specimens before bending are presented in Figure 3.

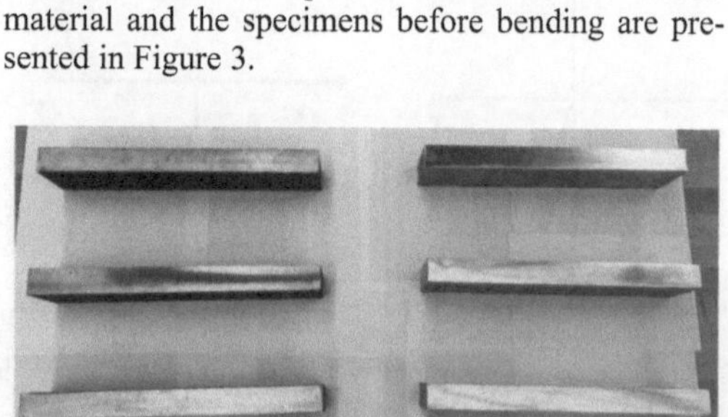

Figure 3. Thick Plates (44W Steel on the left and AL 6061 on the right).

The thick plates dimensions are summarized in Table2 where three measurements of the thickness and widths at three different locations: left end, mid-span and right end of the plates were measured. Noteworthy to mention that the thick plates were cut using a waterjet to minimize residual stresses that may arise from conventional machining processes. This was sought to minimize biasing the test results in the current work.

Table 2. Bending Specimens Dimensions.

		Width			Thickness		
Specimen	Length	W1	W2	W3	T1	T2	T3
				Mm			
S1	300.0	59.9	59.9	59.8	26.7	26.6	26.5
S2	300.0	59.3	59.9	59.0	26.1	26.0	26.1
S3	300.0	60.0	60.0	59.9	26.5	26.2	26.4
Al1	300.0	59.8	59.9	59.9	25.6	25.6	25.6
Al2	300.0	60.0	60.1	60.0	25.8	25.8	25.8
Al3	300.0	60.0	60.0	60.1	25.7	25.7	25.9

3 EXPERIMENTAL SETUP

3.1 *Tensile Test Setup*

The 5585H electromechanical Instron load frame was used for mechanical characterization of the tensile specimens. The machine has a maximum load capacity of 250 kN and the complete test setup is shown in Figure 4. The DIC system used for tensile test consists of two Mega Speed cameras, model MS130K and LED lights were used.

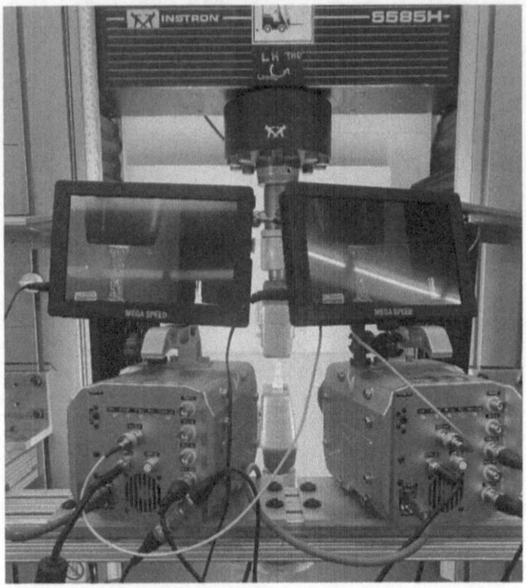

Figure 4. Tensile test setup with 3D-DIC cameras and high resolution screens for focusing.

The sensor pixel size of the identical cameras is 12 μm. The 40 mm lenses were used to provide the required focus on the specimens. The 3D DIC system is calibrated using GOM correlate standard coded panel MV 55x44. The 3D version of GOM correlate software was used for the analysis.

The tensile specimens, Aluminum in Figure 5(a) and Steel in Figure 5(b), showing the representative speckled patterns just before testing in the load frame.

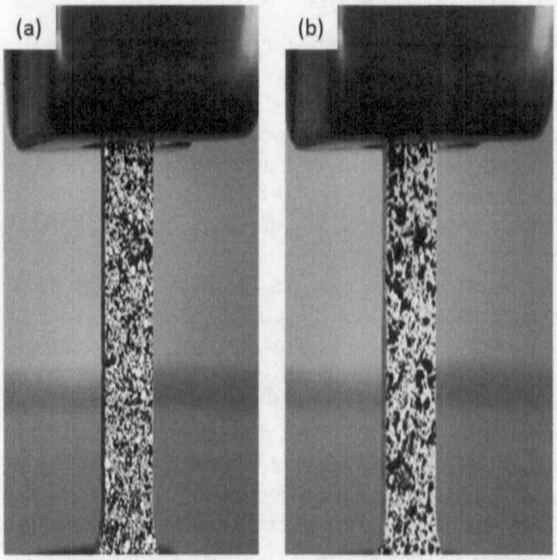

Figure 5. Tensile specimens (a) Al6061 (b)S44W.

3.2 *Three-Point Bending Setup*

Three-point bending was used to mimic the cold process of bending thick plates. The three-point bending was conducted using the Tinius Olsen Machine as shown in Figure 6. The Tinius Olsen machine is a servo-hydraulic machine with a maximum load capacity of 2 MN.

Figure 6. Three-point bending setup.

The three-point bending required a larger field of view (FOV). Accordingly, the ZEISS ARAMIS Adjustable 12M system was used to measure the strains of plates during bending. The system consists of two 12M manual focus cameras with a 4096 x 3000 pixels resolution and a pixel size of 3.45 μm placed at an angle of 25°. Additionally, the lighting is provided by two LED light sources. The system is calibrated using the CP40 standard calibration panel for a measuring volume 320 mm^3. The calibration deviation was 50 μm.

3.3 *3D-Digital Image Correlation Details*

DIC relies on the contrasting surface caused by the randomly speckled pattern of the digital image to track the subtle motion. However, achieving the optimal speckle size can be challenging. Having speckles that are smaller than the minimum of two-pixels causes the image to be aliased. The optimal speckle size can be calculated from the relationship between the camera resolution and the field of view (FOV). Previous works has recommended the ideal speckle size to be between 3-5 pixels (Szalai & Dogossy, 2021). Additionally, the measurements accuracy can be compromised by the frequency of grey regions caused by the blending of the black and white (Reu, 2015). Therefore, the tensile specimens and the plates were speckled using black and white spray paint by layering one colour on top of the other until the required pattern quality is achieved as validated by the software. The speckle pattern used followed the suggestions proposed by (Lionello & Cristofolini, 2014) where matt paint was used to prevent reflections caused by the metals' shiny surface. Flexible polymer paint was used to allow for deformations without sacrificing the accuracy. In the current work, the DIC calibration enabled displacement measurements precision of 10.7 μm.

The speckled plate side was aligned with the three-point bending fixture plane to avoid any shadow on the plate especially beneath the top roller.

The captured digital images are divided into pixelated regions called facets. Each facet consists of 19 by 16-pixel squares. The 3D-DIC system tracks throughout the testing based on the contrast in the speckle. The 3D-DIC provides full field non-contact strain measurements unlike strain gauges which only provide discrete point(s) measurements. Moreover, it eliminates the risk of losing the data by virtue of damaging the strain gauge during testing.

4 RESULT S

4.1 *Tensile Test*

For tensile testing, a rate of 5mm/min was used. The representative tensile behaviors of steel and aluminum are shown in Figure 7. The steel specimens showed a linear behavior followed by distinct yielding exhibiting a region of upper and lower yielding. Followed by strain hardening. The steel material was significantly ductile showing a failure strain exceeding 25% strain. As for the aluminum behavior, it didn't show a distinct yielding region. In addition, limited strain hardening was observed. Noteworthy to mention that the aluminum recorded failure strain significantly lower than that of steel. Particular details on the mechanical properties of both materials are documented in Table 3.

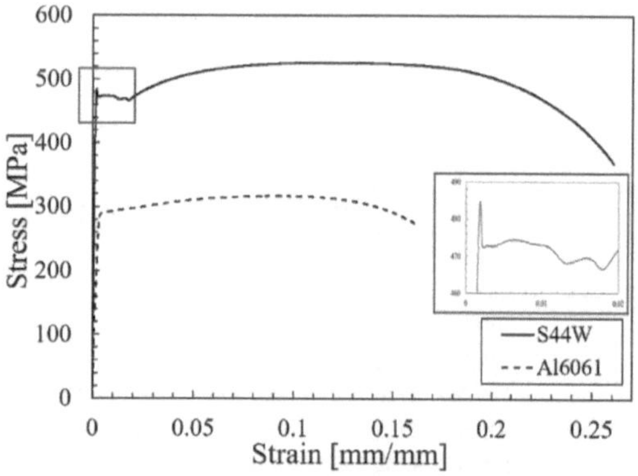

Figure 7. Tensile stress-strain curves.

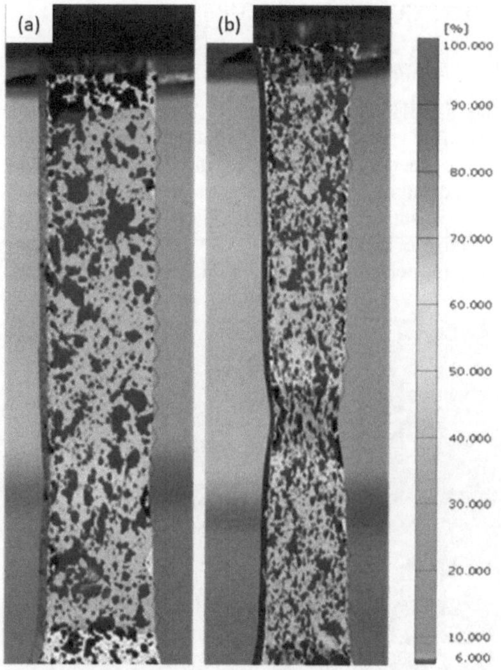

Figure 8. Tensile Specimen behavior (a) Al6061 (b) S44W.

Figures 8 (a) and (b) show the full field strain measurements along the longitudinal axis of each specimen for Al6061 and S44W, respectively. The same legend was used to provide comprehensive comparison. The figure demonstrates that S44W undergoes a stronger necking effect before failure than Al6061. In addition, steel has a more localized deformation specifically in the necking regions compared to Al 6061 which undergoes a more uniform strain distribution. Since the same legend was used, steel undergoes higher strain values than Al6061.

Table 3. Materials properties summary.

Material	Elastic Modulus	Yield Strength	Ultimate Strength	Strain to Failure
	GPa	MPa	MPa	mm/mm
Steel 44W	205	480	526	0.26
Aluminium 6061	70	270	316	0.16

4.2 *Three Point Bending*

For the three-point bending test, an initial rate of 0.25 mm/s until a displacement of 5 mm is reached in order to capture significant data within the elastic region. Afterwards, a rate of 0.5 mm/s was applied until a total displacement of twice the thickness was reached as illustrated in the load displacement curves in Figure 9.

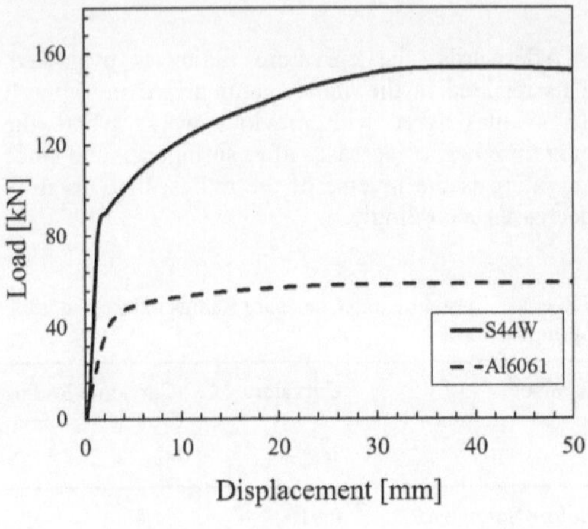

Figure 9. Three-point bending load displacement curves.

It can be demonstrated that S44W withstand higher load bearing capacity, reaching almost 160 kN, compared to approximately 50 kN for Al6061. Initially, both materials show a rapid increase in load indicating elastic deformation. Upon reaching the onset of plastic deformation the S44W exhibited significant strain hardening while the AL6061 showed almost no hardening effect. This can be attributed to the different natures of both metals. It is well known that steel alloys typically show upper and lower yielding ahead of strain hardening as can be observed from Figure 7. On the other hand, the aluminum alloy did not show this effect as can be observed by the stress plateau beyond yielding. The difference in the load displacement linear regions' slopes can be correlated to the different moduli of elasticity since both plates have identical dimensions.

The Zeiss correlate software is used for the data analysis as illustrated in Figure 10 (a) and (b) for Al6061 and S44W, respectively. A surface component covering the plate was created to capture real time measurements during deformation.

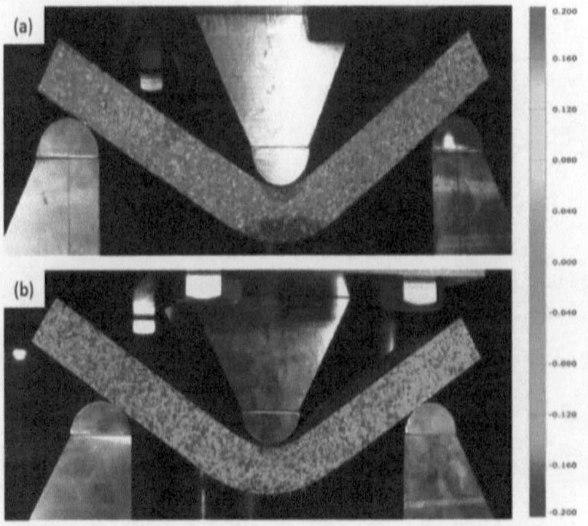

Figure 10. Full-field strain measurements using 3D-DIC (a) Al6061 (b) S44W.

It can be demonstrated that the 3D-DIC full-field strains illustrate the typical bending behavior where the upper portion of the plate is under compressive stresses and the lower portion is under tensile stresses. Typically, the stresses become higher towards the top and bottom surfaces and decrease towards the neutral plane. This is indicated in the contour plot of both plates where the S44W shows less intensity of normal strains when compared to Al6061. For both plates, normal strains due to bending were significantly present in the plates' midspan while away from vicinity of the top roller they only underwent little to no deformation. To capture the through the thickness strains, a section was created at the midspan. As a result, the normal strains in the axial direction i.e., along the longitudinal axis, obtained using the 3D-DIC under three-point bending for both plates are displayed in Figure 11.

Two trend lines were superimposed on the normal strains to better illustrate the difference in their behaviors. The red solid line shows the trend of the S44W plates behavior. The trend line is approximately linear. On the other hand, the Al6061 trend line is presented by a solid blue line on Figure 11. The trend line of Al6061 is significantly nonlinear when compared to S44W. It can be denoted that aluminum specimens endure more strains due to bending than steel. Moreover, because of plastic deformation the neutral plane is significantly shifted as can be observed from Figure 11. The shift in the neutral plane is evident for Al6061 plates while the shift for S44W is minimal.

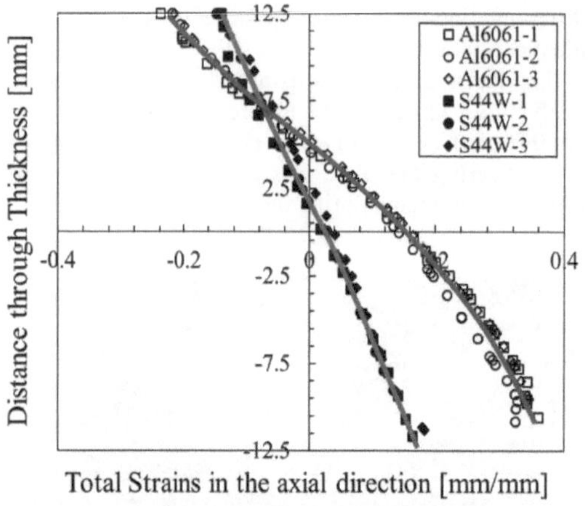

Figure 11. Through the thickness normal strains.

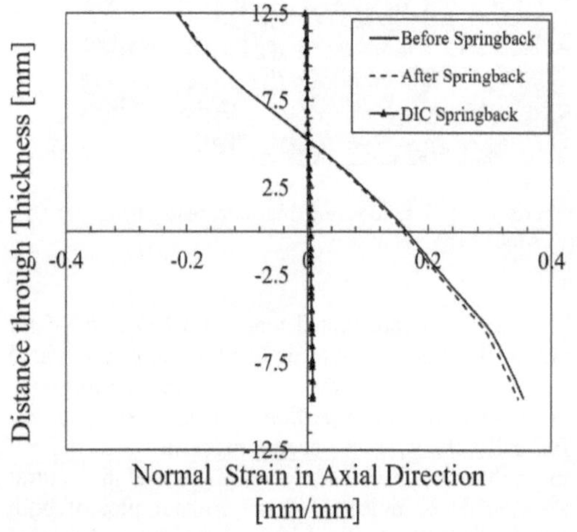

Figure 12. Strains due to springback by 3D-DIC.

Finally, the steel plates showed almost similar strains in both tension and compression unlike aluminum plates which undergo higher strains in compression than in tension. The difference in strains before and after unloading due to springback for plate Al6061-3 is presented in Figure 12, validating the assumption of the springback having a reverse elastic bending. The springback effect as presented is shown by the line almost aligned with the vertical axis on Figure 12. This can be attributed to the elastic strains being minimal when compared to plastic ones. Generally, curvature measurements are essential to determine the residual stresses after forming processes as most analytical models rely on them (Shen et al., 2018). Curvatures can be measured using sections along the x-axis. Thus, the curvature before and after springback shown in Figure 13 were extracted by obtaining the measurements once before and once after unloading. The curvature values are approximately deduced using equation (1):

$$K = 8d/\left(C^2 + 4d^2\right) \tag{1}$$

where d is the vertical distance between the minimum and maximum point (mm) and C is the horizontal distance between the two points at the far ends of the curve (mm).

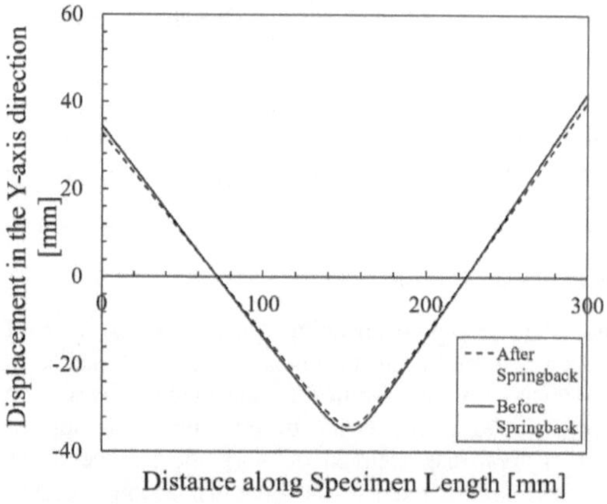

Figure 13. Curvature before and after springback.

Afterwards, the curvature radius is evalauted. This resulted in the values summarized in Table 4. The values agree with previous works where the curvature radius increases after springback and since curvature is the inverse of the radius, the curvature decreases accordingly.

Table 4. Curavture and Curvature Radius before and after Springback.

Parameter	Curvature	Curvature Radius
	1/mm	mm
Before Springback	0.01629	61.4
After Springback	0.01598	62.6

5 CONCLUSIONS

The cold forming processes, such as air bending and press braking, consist of two primary stages. The first stage involves loading under pure bending, where plastic deformation occurs. The second stage is an unloading reverse elastic process known as springback. The aim of this study is to analyze the residual strains of thick steel plates namely S44W and aluminum plates namely Al6061 plates. The wo selected metals for the proposed work were characterized and the two materials were utilized to investigate the plastic deformation due to bending employing 3D-DIC measurements through the history of the bending.

Accordingly, three-point bending was utilized to mimic the cold forming process. The results of mechanical characterization as well as the plastic deformation process using three-point bending were presented and discussed. In addition, full-field strain measurments were essential to better understand the mechanisms of the process pre and post plastic deformation. Consequently, springback process and neutral-axis migration were captured in real time. Finally, approximate the curvature and curvature radius values before and after springback were evaluated.

REFERENCES

Denton, A. A. 1966. "Determination of Residual Stresses." *Metallurgical Reviews* 11(1):1–23. doi:10.1179/MTLR.1966.11.1.1.

Díaz, A., Cuesta, I. I., Alegre, J. M., de Jesus, A. M. P., and Manso, J. M.. 2021. "Residual Stresses in Cold-Formed Steel Members: Review of Measurement Methods and Numerical Modelling." *Thin-Walled Structures* 159. doi:10.1016/J.TWS.2020.107335.

Gerbo, E. J., Thrall, A. P., Smith, B. J., and Zoli, T. P.. 2016. "Full-Field Measurement of Residual Strains in Cold Bent Steel Plates." *Journal of Constructional Steel Research* 127:187–203. doi:10.1016/J.JCSR.2016.07.026.

Gothivarekar, S., Coppieters, S., Van de Velde, A., and D. Debruyne. 2020. "Advanced FE Model Validation of Cold-Forming Process Using DIC: Air Bending of High Strength Steel." *International Journal of Material Forming* 13(3):409–21. doi:10.1007/S12289-020-01536-1/METRICS.

Harrington, J. S., and Schajer, G. S.. 2017. "Measurement of Structural Stresses by Hole-Drilling and DIC." *Experimental Mechanics* 57(4):559–67. doi:10.1007/S11340-016-0247-X/FIGURES/11.

Hill, R. 1950. "The Mathematical Theory of Plasticity." *The Mathematical Gazette* 35(313):208–10. doi:10.2307/3611266.

Kobayashi, M., Matsui, T., and Y. Murakami. 1998. "Mechanism of Creation of Compressive Residual Stress by Shot Peening." *International Journal of Fatigue* 20 (5):351–57. doi:10.1016/S0142-1123(98)00002-4.

Lord, Jerry David, and Fry, Antony Thomas. 2002. *A Review of Residual Stress Measurement Methods. Measurement of Residual Stress in Components*.

Ludwik P. 1903. "Technologische Studie Über Blechbiegung" *Technische Blatter*.

Mutafi, Ayad, Irwan, J. M., Yidris, Noorfaizal, Ghaleb, Atef M., Al-Alimi, Sami, Amran, Mugahed, Maged Qasem, Mousa Hasan, and Al-Fakih, Amin. 2024. "An In-Depth Comparative FEA on through-Thickness Residual Stresses in Press-Braked Cold-Formed Steel Section." *Results in Engineering* 22:102124. doi:10.1016/J.RINENG.2024.102124.

Quach, W. M., Teng, J. G., and Chung, K. F.. 2009. "Residual Stresses in Press-Braked Stainless Steel Sections, II: Press-Braking Operations." *Journal of Constructional Steel Research* 65(8–9):1816–26. doi:10.1016/J.JCSR.2009.04.011.

Shen, Wei, jun Yan, Ren, Lin, Yue, and qi Fu, He. 2018. "Residual Stress Analysis of Hull Plate in Multi-Point Forming." *Journal of Constructional Steel Research* 148:65–76. doi:10.1016/J.JCSR.2018.05.017.

Wang, Chuantao, Kinzel, Gary, and Altan, Taylan. 1993. "Mathematical Modeling of Plane-Strain Bending of Sheet and Plate." *Journal of Materials Processing Technology* 39(3-4):279–304. doi:10.1016/0924-0136(93)90164-2.

Trends in Collision and Grounding of Ships and Offshore Structures – Zhu & Guedes Soares (Eds)
© 2026 The Author(s), ISBN 978-1-041-16361-9

Study on the strain rate effect in the design of ship collision similarity tests

T.T. Guo, Y.J. Zhao, L.P. Zhang, H.Y. Zou, D.Y. Zhang & H.K. Wang
China Ship Scientific Research Center, Wuxi, Jiangsu, P.R. China
National Key Laboratory of Ship Structural Safety, Wuxi, Jiangsu, P.R. China
Taihu Laboratory of Deepsea Technological Science, Wuxi, Jiangsu, P.R. China

ABSTRACT: In order to establish a conversion method for predicting actual ship collision results from scaled model test outcomes, it is necessary to focus on the issue of strain rate effects not satisfying similarity in the scaling process of scaled model test results. The strain rate effect is analysed in the design of ship collision similarity tests based on numerical methods. A modification method based on the J-C constitutive model and the average strain rate of collision is proposed. By discussing the influencing factors of strain rate effects in ship collision test design in combination with the analysis and calculation results, the relationship between the strain rate strengthening coefficient and the static yield strength of commonly used metal materials in ships is provided. The findings of this study can offer important references for the design of scale test models for ship collisions.

1 INTRODUCTION

To establish a conversion method for deriving full-scale ship collision results from model test data, it is essential to address the issue of non-similar strain rate effects during the scaling and similarity transformation processes. The non-similar strain rate effects refers to the fact that during the scaling process of the ship collision test model, the influence of the strain rate on the structure does not satisfy the similarity relationship, that is, a geometric scaling ratio β . This paper analyzes the strain rate effect in the design of ship collision similarity tests and proposes a correction method based on the Johnson-cook constitutive model and the average strain rate during collision.

Since the 1960s, pioneers in the field of ship collisions, such as Minorsky (1958), have initiated research into ship collision problems through a series of experiments, first categorizing ship collision issues into internal and external dynamics. By now, various methods have been developed for studying ship collisions, including empirical formula methods, experimental approaches, numerical simulations, and theoretical analyses. Among these, the test method remains the most direct and effective way to study ship collisions and evaluate ship structural crashworthiness. However, due to limitations in test site availability and costs, conducting full-scale ship collision tests is extremely challenging. Therefore, scaled model tests are often used for substitutive research. Nevertheless, deriving prototype results from scaled model test data involves complex similarity issues, and currently, there is no widely accepted conversion method. Additionally, existing similarity transformation methods, such as Hopkins' similarity law, fail to satisfy linear similarity relationships in terms of material strain rate effects, gravitational effects, and fracture behavior (Jones, 1989). This phenomenon is generally referred to as "material distortion". Barenblatt (2003) pointed out that strain rate effects and material fracture do not conform to linear similarity relationships, making them two critical factors affecting the design of scaled ship collision tests. Booth (1983) conducted drop hammer experiments comparing structural deformation results between prototypes and 1/4-scaled models, demonstrating that due to strain rate strengthening effects, scaling model test results lead to smaller damage prediction for prototype results.

To correct the influence of strain rate effects caused by scale effects in scaled impact tests, Oshiro & Alves (2006, 2007) adopted a velocity correction factor based on material power-law constitutive equations to modify the strain rate effect. Besides increasing the impact velocity of scaled models, Jiang & Wei (2016, 2019) proposed an effective method by increasing the mass (or density) of the impact head to correct the strain rate effects. Furthermore, Wang et al. (2021), combining dimensional analysis, proposed a dimensionless number for strain rate strengthening effects, ensuring similarity in strain rate effects between scaled models and prototypes through material substitution under the condition of maintaining equal dimensionless numbers. Among these methods, increasing the velocity or mass of scaled models is applicable to simple impact problems, requiring theoretical derivation to establish the relationship between impact velocity and structural average strain rate. The method of altering model

DOI: 10.1201/9781003684404-14

materials, while capable of achieving similar strain rate effects, necessitates precise analysis of strain rate strengthening capabilities for various materials to select appropriate substitutes. However, differences in material properties between prototypes and models may also lead to discrepancies in damage forms and other issues, making it unsuitable for general ship collision problems.

In this paper, first, a similarity test design method for ship collisions is presented based on dimensional analysis. Then, damage transformation relationships between prototypes and scaled models are provided using the stiffness characteristics of simple side shell plating frames. Subsequently, numerical calculation methods are employed to compare and validate similarity test results, analyzing the influence of strain rate effects during prototype-scaled model result conversion. Furthermore, a strain rate effect correction method based on the Johnson-Cook constitutive model and collision velocity is proposed using numerical computation tools, enabling the transformation of scaled model results to prototype results. Finally, factors influencing strain rate effects in ship collision test design are discussed based on analysis and calculation results, and relationships between strain rate strengthening coefficients and static yield strengths for commonly used metallic materials in ships are provided. For the ship collision situation where the rigid bow collides with a single side plate without damage, the research results of this paper can provide a reference for the similar test design considering the strain rate effect.

2 SHIP COLLISION SIMILARITY TEST DESIGN

2.1 *Dimensional analysis*

Define the ratio of damage deformation energy to total initial kinetic energy during ship collision as the energy absorption rate η. In Equation 1, E_i represents the energy absorbed by structural deformation and E_t represents the total initial kinetic energy. To ensure similarity in damage, the full-scale ship collision model and scaled collision model must maintain similar energy absorption rates, that is :

$$\eta = \frac{E_i}{E_t} \times 100\% \tag{1}$$

Define the geometric scaling coefficient between the scaled model and prototype as β. Using Buckingham π theorem for dimensional analysis of ship collisions, input parameters include structural mass M (kg), initial impact velocity v_0 (m/s), material dynamic yield strength σ_d (Pa), plate thickness t(m), ship length L (m), ship height H (m), and material density ρ (kg/m³).

For ship collision problems, take the energy absorption rate η as the overall output quantity. Conduct dimensional analysis with length l, time s, and mass m as base quantities, as shown in Table 1.

Table 1. Dimensional analysis of ship collision.

	M	v_0	σ_d	t	L	H	ρ
m	1	0	1	0	0	0	1
l	0	1	-1	1	1	1	-3
s	0	-1	-2	0	0	0	0

Using the Gibbings transformation:

$$\eta = f(\frac{\rho v_0{}^2}{\sigma_d}, \frac{L}{t}, \frac{H}{\text{t}}) \tag{2}$$

$$\frac{\rho v_0{}^2}{\sigma_d} = \frac{LHt\rho v_0{}^2}{\sigma_d LHt} = \frac{M v_0{}^2}{\sigma_d LHt} \tag{3}$$

Define $\frac{\rho v_0{}^2}{\sigma_d}$ and $\frac{M v_0{}^2}{\sigma_d LHt}$ as damage similarity numbers Π_1 and Π_2, respectively.

Table 2. Similarity of physical quantities in ship collision scaled model tests.

NO.	Physical Quantity	Scaling Coefficient
1	Volume/mass/energy	β^3
2	Force	β^2
3	Length	β
4	Time	β
5	Strain/Stress	1
6	Velocity	1
7	Acceleration	$1/\beta$
8	Strain rate	$1/\beta$
9	Added mass coefficient	1

To ensure equal energy absorption rates in scaled models, i.e., similar structural damage, the damage similarity numbers must be equal. Based on the damage similarity numbers, scaling relationships for various physical quantities can be derived. The similarity relationships for physical quantities in ship collision scaled model tests are shown in Table 2.

In subsequent dimensional analyses, prototype quantities are denoted with subscript p, and scaled model quantities with subscript s.

2.2 *Stiffness characteristics of the side plating*

Taking the single-layer outer plate of the struck ship's side plating as the research object, assuming boundary conditions with fixed support, and considering the loading form under wedge-shaped bow collision as a point load, this section employs the Upper-bound theorem to analyze the stiffness characteristics of the struck ship's side plating. With the impact point as the origin, a coordinate system is established as shown in Figure 1-a.

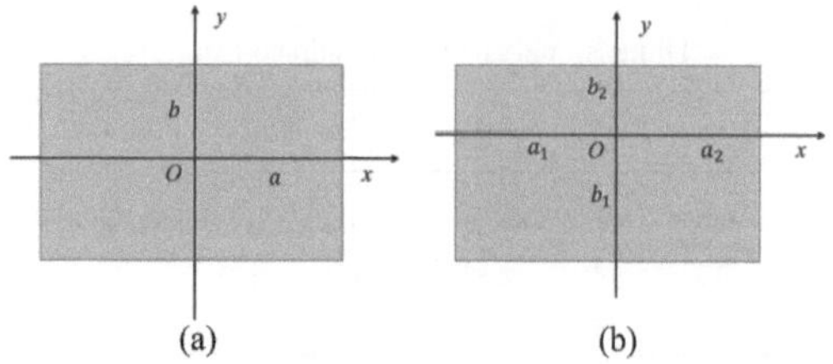

Figure 1. Coordinate system schematic.

The deformation W of the fixed plate must satisfy:

$$W=\delta_{\max}\cdot f(x,y) \tag{4}$$

In Equation (4), $\delta_{\max}$ represents the maximum deflection at the impact point. The deformation mode $W(x)$ is selected as a quadratic polynomial to satisfy:

$$f(x,y)=\frac{(a^2-x^2)(b^2-y^2)}{a^2b^2} \tag{5}$$

Here, a and b are the distances from the impact point to the edges of the side plating. After a series of formulas, it is easy to get the energy absorbed by structural deformation E_i:

$$E_i = C(\frac{a}{b}+\frac{b}{a})t\delta^2 \tag{6}$$

In Equation (6):

$$C=\frac{16\sqrt{3}}{45}\sigma_0 \tag{7}$$

If the impact point is not at the center of the plate (Figure 1-b), a further relationship can be derived:

$$E_i = \frac{C}{4}(\frac{S}{b_1 b_2}+\frac{S}{a_1 a_2})t\delta^2 \tag{8}$$

Here, S represents the plate area and a_1、a_2、b_1、b_2 represent distances from the impact point to four edges of the side plating.

For side plating with stiffeners, the plate thickness in Equations (6) and (8) can be replaced by an equivalent plate thickness t_{eq}, satisfying:

$$t_{eq}=\frac{t_{fb}h_{fb}+t_{mb}h_{mb}}{D_j} \tag{9}$$

In Equation (9): h_{fb} and h_{mb} are the widths of the stiffener flange and panel, respectively, t_{fb} and t_{mb} are the thicknesses of the stiffener flange and panel, respectively, D_j is the spacing between stiffeners.

2.3 *Damage transformation relationships in similarity tests*

Given the geometric scaling ratio between the scaled model and prototype as β, that is:

$$\beta=\frac{L_s}{L_p}=\frac{t_p}{t_s} \tag{10}$$

then, based on Equation (6), the structural deformation energy ratio under the same collision velocity satisfies:

$$\frac{E_p}{E_s}=\frac{t_p}{t_s}(\frac{\delta_p}{\delta_s})^2 \tag{11}$$

Letting the total input energy be E_t, and considering equal velocities and energy absorption rates between the prototype and scaled model, we have:

$$\frac{E_p}{E_s}=\frac{\eta E_{tp}}{\eta E_{ts}}=\frac{\eta m_p {v_p}^2}{\eta m_s {v_s}^2}=\frac{m_p}{m_s}=\beta^3 \tag{12}$$

By combining Equations (10) and (12), it follows that:

$$\frac{\delta_p}{\delta_s}=\beta \tag{13}$$

Clearly, for single-layer plates or plate-frame structures in ship collisions, if the structure undergoes large deformation without failure, tensile deformation energy is the primary energy absorption form. In similarity tests satisfying the scaling ratios as shown in Table 2, the deflection of the prototype and scaled model should conform to geometric similarity relationships.

3 NUMERICAL CALCULATION MODEL OF SHIP COLLISION SIMILARITY TEST

A pair of numerical ship collision similarity tests are designed with a geometric scaling ratio of 1:3. Numerical calculation methods are employed to analyze the influence of strain rate effects in these similarity tests.

3.1 *Model design*

The collision prototype consists of a pair of steel ship models made of Q235 steel, with the impact bow in the form of a wedge-shaped bow (treated as a rigid body). The main dimensions and plate thicknesses of the impacting and struck ships are listed in Table 3. The collision velocity is set to 3 m/s with an incidence angle of 90°, and the impact point is at the midpoint of the ship's length, located 0.4 m below the upper deck. After applying a direct scaling ratio of 1:3, the main dimensions and plate thicknesses of the scaled model are shown in Table 4. According to the similarity relationships in Table 2, to ensure equal energy absorption rates and similar damage, the collision velocity for the scaled model should be set to 3 m/s without considering the strain effect.

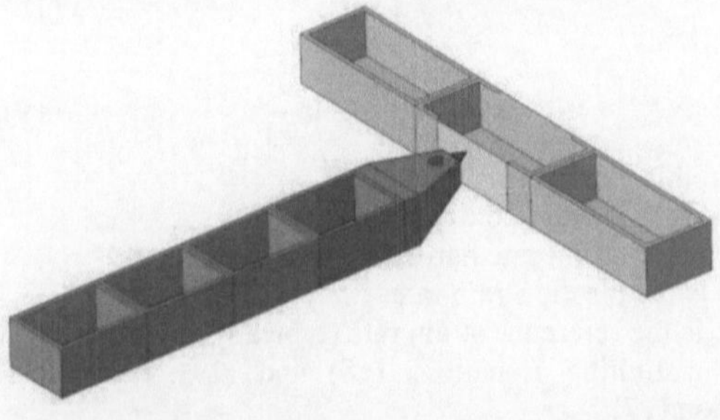

Figure 2. Schematic diagram of collision ship models.

Table 3. Main dimensions and plate thicknesses of prototype ships.

Main Dimensions (m)			Thickness (m)	
Length	Breadth	Height	Struck ship plate	Other Plate
3	0.45	0.3	0.003	0.006

Table 4. Main dimensions and plate thicknesses of scaled ships.

Main Dimensions (m)			Thickness (m)	
Length	Breadth	Height	Struck ship plate	Other Plate
1	0.15	0.1	0.001	0.002

Taking the ship collision prototype as an example, the simulation calculation process is introduced below; the scaled model simulation follows a similar procedure.

3.2 *Model establishment and meshing*

In ABAQUS, the ship collision model is established as shown in Figure 3. The model is meshed with a grid size of 10 mm, using S4R element types, resulting in a total of 79,946 elements.

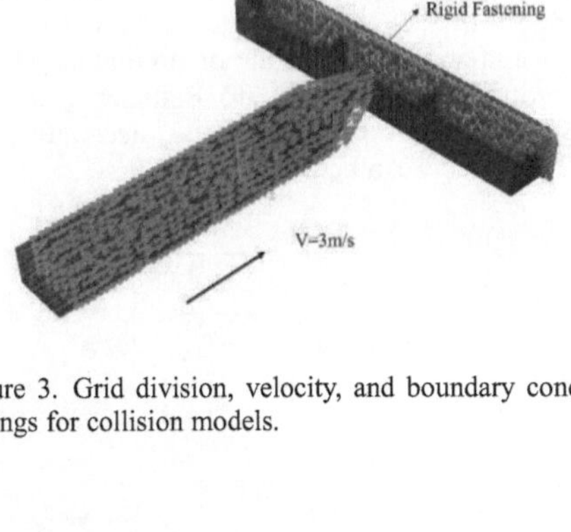

Figure 3. Grid division, velocity, and boundary condition settings for collision models.

3.3 *Material and boundary settings*

The impacting ship's bow is set as a rigid body, while the remaining parts are assigned material properties of Q235 steel following the Johnson-Cook (J-C) constitutive and failure models. The constitutive and failure model parameters are listed in Table 5.

The J-C constitutive model is expressed as:

$$\bar{\sigma} = \left[A + B\left(\bar{\varepsilon}^{pl}\right)^n\right]\left[1 + C\ln\left(\frac{\dot{\bar{\varepsilon}}^{pl}}{\dot{\varepsilon}_0}\right)\right]\left(1 - \left(\frac{T - T_m}{T_r - T_m}\right)^m\right) \tag{14}$$

The J-C failure model considers the effects of stress triaxiality, temperature, and strain rate on fracture strain:

$$\bar{\varepsilon}_f^{pl} = [D_1 + D_2\exp(D_3\eta)]\left[1 + D_4\ln\left(\frac{\dot{\bar{\varepsilon}}^{pl}}{\dot{\varepsilon}_0}\right)\right]\left(1 + D_5\frac{T - T_m}{T_r - T_m}\right) \tag{15}$$

Table 5. J-C Constitutive and Failure Parameters for Steel.

Constitutive parameter	A/ MPa	B/ MPa	c	n	m
Value	292	889	0.058	0.746	0.94
Failure parameter	D_1	D_2	D_3	D_4	D_5
Value	0.38	1.47	2.58	0.014	8.07

To facilitate calculations, the boundaries of both prototype and scaled models for the struck ship are set as rigidly fixed, as shown in Figure 3. The initial velocity of the impacting ship is set to 3 m/s, with a collision calculation duration of 0.8 seconds. Surface-to-surface automatic contact algorithms are employed, and results such as deflection, strain, stress, etc., are outputted.

3.4 *Comparative analysis of calculation results*

Using identical computational settings, numerical calculations are performed for the scaled ship collision model, with deflection, equivalent plastic strain, collision force, and energy conversion results compared between prototypes and scaled models. Figures 4 and 5 show the deflection cloud diagrams of the prototype and scaled model of struck ships, respectively. Under the same collision velocity, their maximum deflections are 55.9 mm and 17.5 mm respectively, which do not satisfy the geometric similarity relationship defined by Equation (13). The analysis reveals that due to strain rate hardening effects, the scaled model's maximum deflection of 17.5 mm is lower than the theoretical value (55.9/3 = 18.6 mm), indicating that under constant velocity conditions, deriving prototype results from scaled model tests would be dangerous.

Figure 4. Deflection cloud diagram of struck ship prototype.

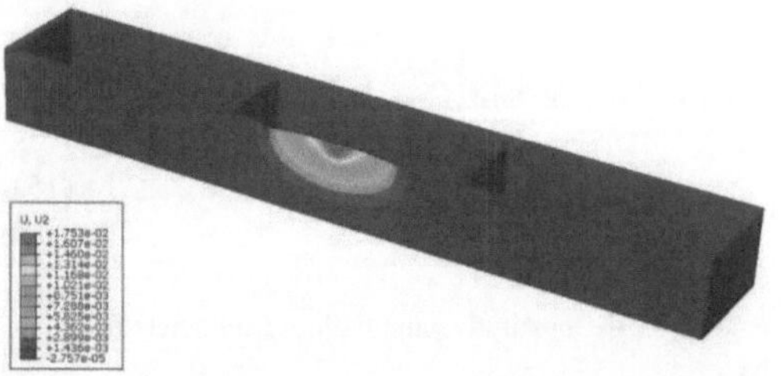

Figure 5. Deflection cloud diagram of scaled model struck ship.

4 STRAIN RATE EFFECT CORRECTION

4.1 *Equivalent strain rate correction method*

In the design of scaled model tests, if it is feasible to fabricate models using prototype materials, identical materials should be selected. In such cases, stress wave propagation velocities in models and prototypes are the same, satisfying the strain similarity relationships as shown in Table 2. However, strain rate effects on yield strength do not conform to these relationships. Therefore, to ensure equivalent dynamic responses, corrections to the collision velocity specified in Table 2 are necessary.

Scaled models adhering to Table 2 without velocity correction are referred to hereafter with subscript s, while corrected models are denoted with subscript a (Amended model). The similarity numbers for strain rate effects must be equal between velocity corrected models and prototypes:

$$\frac{\Pi_{1p}}{\Pi_{1a}} = \frac{\frac{\rho_p {v_{0p}}^2}{\sigma_{dp}}}{\frac{\rho_a {v_{0a}}^2}{\sigma_{da}}} = 1 \tag{16}$$

With identical materials, we have:

$$\frac{{v_{0p}}^2}{{v_{0a}}^2} = \frac{\sigma_{dp}}{\sigma_{da}} \tag{17}$$

From Equation (14), the strain rate effect in the Johnson-Cook constitutive model is represented by the hardening term θ

$$\sigma_d = \theta \sigma_s \tag{18}$$

$$\theta = 1 + c \ln \frac{\dot{\varepsilon}}{\varepsilon_0} \tag{19}$$

In this expression:
σ_s is the static yield strength (MPa),
c is the strain rate hardening coefficient,
$\dot{\varepsilon}$ is the plastic strain rate,
$\varepsilon_0$is the reference strain rate (typically 1/s).
Substituting Equations (18) and (19) into Equation (17):

$$\frac{{v_{0a}}^2}{{v_{0p}}^2} = \frac{\sigma_{da}}{\sigma_{dp}} = \frac{1 + c \ln \frac{\dot{\varepsilon}_a}{\varepsilon_0} \sigma_{sa}}{1 + c \ln \frac{\dot{\varepsilon}_p}{\varepsilon_0} \sigma_{sp}} \tag{20}$$

Here, c is a known material constant. To determine the corrected collision velocity v_{0a} for scaled models, relationships between strain rate$\dot{\varepsilon}_a$,$\dot{\varepsilon}_p$and impact velocity v_0 must be established:

$$\dot{\varepsilon}_a, \dot{\varepsilon}_p = f(v_0) \tag{21}$$

Assuming constant strain rates during impact and proportionality between deformation and impact velocity, with Δt as a proportionality factor. Based on the strain rate definition:

$$\dot{\varepsilon} = \frac{\Delta \varepsilon}{\Delta t} = \frac{\Delta l / L_f}{\Delta t} \tag{22}$$

$$\Delta l \propto v_0 \cdot \Delta t \tag{23}$$

Considering instantaneous collision conditions:

$$\Delta t \approx \frac{L_f}{v_0} \tag{24}$$

Substituting into Equation (22):

$$\dot{\varepsilon} = \frac{\Delta \varepsilon}{\Delta t} = \frac{\Delta l / L_f}{\Delta t} = \frac{v_0 \Delta t / L_f}{\Delta t} = \frac{v_0}{L_f} \tag{25}$$

It should be noted that L_f is the model's feature size (not a specific structural dimension) and is independent of impact velocity. It can be determined through numerical simulations or preliminary tests. Substituting Equation (25) into Equation (20), the corrected initial collision velocity v_{0a} accounting for strain rate effects can be derived:

$$\frac{v_{0a}}{v_{0p}} = \sqrt{\frac{1 + c \ln \frac{v_{0a}}{L_{fa}}}{1 + c \ln \frac{v_{0p}}{L_{fp}}}} \tag{26}$$

4.2 *Numerical verification of strain rate correction*

Through simulation calculations, the time history curves of equivalent plastic strain at the collision point for prototype and scaled models are obtained. During the collision duration, the average strain rate at the impact point is computed. Based on Equation (25), characteristic scales for prototype and scaled models are derived, as shown in Table 6.

Table 6. Calculation results for prototype and scaled model characteristic Scales.

Parameter	Scaled Model	Prototype	Ratio (Scaled/ Prototype)
Average Strain Rate $\dot{\varepsilon}$	6.1	16.4	0.37
Feature Size L_f/m	0.49	0.18	2.72

Using Equation (26), the corrected collision velocity for scaled models satisfies:

$$\frac{{v_{0a}}^2}{3^2} = \frac{1 + 0.058 \ln \frac{v_{0a}}{0.18}}{1 + 0.058 \ln \frac{3}{0.49}} \tag{27}$$

Solving this transcendental equation numerically yields:

$$v_{0a} = 3.08\text{m/s} \tag{28}$$

Adjusting the collision velocity of the scaled model, the deflection cloud diagram for the struck ship after correction is shown in Figure 6. Comparing results between uncorrected and corrected models with prototype data reveals: The maximum deflection for the corrected model is 18.4 mm, while the theoretical value derived from the prototype (55.9/3=18.6 mm 55.9/3=18.6mm) has an error of only 1.1%.

This demonstrates that strain rate corrected methodbased velocity significantly improves the similarity between scaled model and prototype results.

Figure 6. Deflection cloud diagram of scaled model struck Ship after velocity correction.

Table 7. Comparison between corrected and uncorrected models.

	Maximum Deflection (mm)	Theoretical Maximum Deflection (mm)	Relative Error
Uncorrected Model	17.5	55.9/3=18.6	5.9%
Corrected Model	18.4		1.1%

4.3 *Similarity test design method considering strain rate effects*

Above all, a method for designing similarity tests considering strain rate effects can be established, following the steps below:

1. Determine Uncorrected Velocity: Based on test conditions and requirements, establish the geometric scaling ratio β. Using the similarity relationships in Table 2, an uncorrected velocity v_m of scaled model test can be designed (equal to prototype velocity).

2. Feature Size Calibration: Conduct numerical simulations for the prototype and uncorrected scaled model to output the average strain rate at the impact point. Based on Equation (25), calibrate the characteristic scales L_f for the prototype and scaled model. (Pre-test experiments can also be used, where strain gauges measure the average strain rate at the impact point to determine the feature size.)

3. Determine Corrected Velocity: Combining collision velocity, material strain rate strengthening coefficient c, and model feature size, calculate the corrected scaled model collision velocity using Equation (26). After correction, the damage deflection of the scaled model can be transformed to prototype results.

4.4 *Further discussion*

From prior calculations and analyses, it is evident that strain rate effects in ship collision test design are influenced by collision velocity and model scale β. Generally, higher collision velocities and smaller geometric scaling ratios β result in greater strain rate effects. For low-speed collisions or large-scaled model tests, strain rate effects can be neglected.

Additionally, material strain rate strengthening capability is a critical factor affecting strain rate effects. In this study, the Johnson-Cook constitutive model's strain rate strengthening coefficient c is used to characterize material behavior. Table 8 lists static yield strengths and strain rate strengthening coefficients for several commonly used metallic materials of ships. Clearly, for metallic materials, higher static yield strength corresponds to weaker strain rate strengthening capability. Figure 7 and Equation (29) provide an empirical relationship between yield strength and strain rate strengthening coefficient, which can be interpolated for practical applications.

Therefore, in the design of high-speed ship collision similarity tests with materials exhibiting good plasticity and low static yield strength, as well as small geometric scaling ratios, it is essential to account for strain rate effects. The correction method provided in the paper can be employed to determine corrected initial velocities.

Table 8. Static yield strengths and strain rate strengthening coefficients for common ship metallic materials.

	Q235	Q355B	Q690	S7	TC4
σ_s (MPa)	235	355	690	724	893
c	0.058	0.045	0.024	0.012	0.007

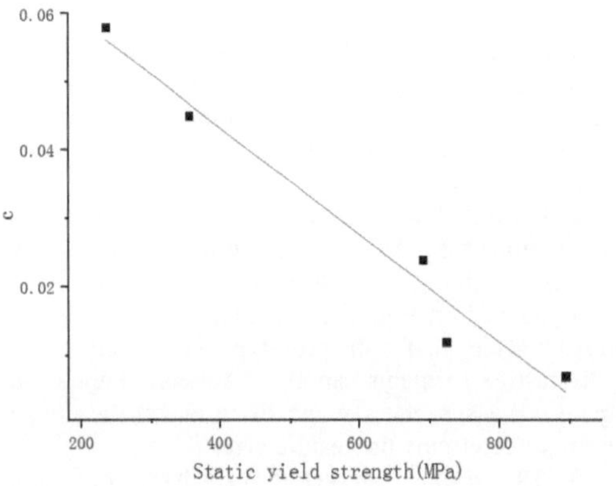

Figure 7. Relationship between strain rate strengthening coefficient and static yield strength.

$$c = -7.82\sigma_s + 0.07(235 \leq \sigma_s \leq 893) \quad (29)$$

5 CONCLUSIONS

This report presents a similarity test design method for ship collisions based on dimensional analysis. Numerical simulations are employed to analyze strain rate effects in ship collision tests, and a velocity correction method is proposed using the Johnson-Cook constitutive model and average strain rate at the impact point. For the ship collision situation where the rigid bow collides with a single side plate without damage, the research results of this paper can provide a reference for the similar test design considering the strain rate effect. Main conclusions include:

1. To ensure equal energy absorption rates and damage similarity, the dimensionless damage numbers $\frac{\rho v_0{}^2}{\sigma_d}$ and $\frac{M v_0{}^2}{\sigma_d LHt}$ must be equal.

2.For single-layer plates or plate-frame structures in ship collisions, if large deformations occur without failure, tensile deformation energy is the primary energy absorption mechanism. In such cases, prototype and scaled model deflections satisfy geometric similarity relationships.

3. Numerical results demonstrate that neglecting strain rate effects leads to smaller damage prediction results when deriving prototype results from scaled model tests. The proposed correction method, based on the Johnson-Cook constitutive model and average impact point strain rate, enables accurate transformation of scaled model deflection and energy results to prototype values.

4. Strain rate effects in ship collision test design are influenced by collision velocity and geometric scaling ratios β. Higher velocities and smaller scaling ratios result in greater strain rate effects, which can be neglected for low-speed collisions or large-scaled models. Additionally, materials with lower static yield strength exhibit stronger strain rate effects.

REFERENCES

Alves, M., Oshiro, R.E., 2006. Scaling the impact of a mass on a structure. Int. J.Impact Eng. 32, 1158–1173.

Barenblatt, 2003. Scaling. Cambridge: Cambridge University Press.,

Booth E, Collier D, Miles J.,1983, Impact scalability of plated steel structures. Structural crashworthiness. London: Butterworks,136-74.

Jiang, Z.R., Zhong, Y.K., Shi K.R., Luo, B. 2016. Gravity-based impact comparability rule of single-layer reticulated shells and its numerical verification. South China Univ. Techno. 44: 43–48.

Jones, N., 1989. Structural impact. Cambridge University Press, Cambridge.

Minorsky V U, 1958. An analysis of ship collisions with reference to protection of nuclear power plants. Sharp (George G.) Inc., New York.

Oshiro, R.E., Alves, M., 2007. Scaling of cylindrical shells under axial impact. Int.J. Impact Eng. 34, 89–103.

Wang, S., Xu, F., Zhang, X., Yang,L., Liu, X., 2021 Material Similarity of Scaled Models. International Journal of Impact Engineering.

Wei, D., Hu, C., 2019. Scaling of an impacted reticulated dome using partial similitude method. Lat. Am. J. Solids Struct. 16.

Trends in Collision and Grounding of Ships and Offshore Structures – Zhu & Guedes Soares (Eds)
© 2026 The Author(s), ISBN 978-1-041-16361-9

Overcoming scale effects in collision analysis

S.-H. Park
UlsanLab Inc, Ulsan, Republic of Korea

S.-R. Cho
UlsanLab Inc. and University of Ulsan, Ulsan, Republic of Korea

ABSTRACT: Collision analysis of ship structures has become commonplace in design offices thanks to the availability of high-speed personal computers. However, a significant challenge remains unresolved: the scale effect on the damage of steel-plated structures subjected to lateral collisions. Under static loading, small-scale model structures often exhibit lower ultimate strengths compared to full-scale structures due to comparatively greater initial shape imperfections. Conversely, under impact loading, the damage to small-scale models may be less severe than that observed in full-scale structures. In other words, full-scale structures may demonstrate weaker performance under impact loading, potentially leading to overly optimistic designs. Many researchers validate their numerical methods using small-scale collision test data. Therefore, the validated numerical methods possibly provide underestimations of the extents of damage of full-scale structures. This paper proposes a practical approach to mitigate scale effects in ship structure analysis. The methodology accounts for strain-rate effects on both the yield strength and tensile strength of materials, enabling more accurate and reliable predictions of plastic damage. Additionally, a new fracture criterion is introduced, enhancing the prediction of fracture damage in ship structures subjected to collisions. Some calculation examples are also provided in the paper to evaluate the accuracy and reliability of the proposed method.

1 INTRODUCTION

Collision analysis of ship structures has become commonplace in design offices due to the availability of high-speed personal computers. However, a significant challenge remains unresolved: the scale effect on the damage of steel-plated structures subjected to lateral collisions. Small-scale model structures, often used for experimental validation due to the high cost and time-consuming nature of full-scale impact tests, may exhibit different behaviors compared to their full-scale counterparts. Specifically, under impact loading, the damage to small-scale models may be less severe than that observed in full-scale structures. This can lead to an overestimation of the impact resistance of actual structures if small-scale results are directly applied without considering the scale effect, potentially leading to overly optimistic designs. Therefore, validated numerical methods based on small-scale data possibly provide underestimations of the extents of damage of full-scale structures.

Many researchers have experimentally investigated the scale effect of dynamic loads such as explosion or impact to address this challenge. Duffey et al. (1984) evaluated the size effect on the response of circular plates to punch impact loading, observing more plastic deformation and lower energy absorption in larger scale ratio models. Drazetic et al. (1994) conducted impact tests on hollow steel beams, confirming a decrease in bending angle with a decrease in scale ratio. Wen & Jones (1993) investigated impact on circular plates, noting denting and fracture damage, and the strain-rate sensitivity. Schleyer et al. (2004) explored transverse impact pressure on square models, reporting that the scale ratio effect varied with boundary conditions. Jiang et al. (2006) observed a decrease in normalized displacement as the scale ratio decreased in solid cylindrical models. While some studies, like Jones & Birch (2008), suggested that geometrically similar scaling could reliably predict full-scale behavior for specific perforation scenarios, others, such as Wang et al. (2012) on reinforced concrete slabs, highlighted the need for adjustments when scaling from model to prototype due to more severe damage in larger slabs. More recently, Fu et al. (2021) showed significant influence of strain-rate effect on the scaling of armor steel under explosive loads, and Yao et al. (2022) proposed an equivalent scaling method for steel box structures under internal blast, demonstrating its effectiveness for similar damage modes. Despite these valuable contributions, a comprehensive and practical approach to consistently mitigate scale effects across various damage modes

DOI: 10.1201/9781003684404-15

(denting and fracture) and material behaviors in ship structure analysis remains a critical need.

This paper proposes a practical approach to mitigate scale effects in ship structure analysis, specifically focusing on overcoming the underestimation of full-scale damage from small-scale model experiments. The methodology accounts for strain-rate effects on both the yield strength and tensile strength of materials, enabling more accurate and reliable predictions of plastic damage. Additionally, a new fracture criterion is introduced, enhancing the prediction of fracture damage in ship structures subjected to collisions. Some calculation examples are also provided in the paper to evaluate the accuracy and reliability of the proposed method. In this study, we first investigated the scale effect on the structural behavior of unstiffened steel plates subjected to lateral impact loading. Numerical analyses were conducted on 27 unstiffened square plate test models, as reported by Park et al. (2023), including models with only denting damage and those with combined denting and fracture damage. The adopted dynamic constitutive relationships were validated using the dented models, and the combined damage models were utilized to develop the shear fracture criterion. Finally, a practical analysis guide was developed and proposed for predicting combined denting and fracture damage in laterally impacted plated steel structures. The proposed analysis guidance was applied to three double hull models, showing good correlation with actual damage.

2 SUMMARY OF SCALE EFFECT TESTS ON UNSTIFFENED PLATES

Park et al. (2023) conducted impact tests on 27 unstiffened steel plates. This section presents a brief overview of the test models, material properties, and results.

2.1 *Test models*

Three nominally identical models were fabricated for each size. The dimensions of the test sections were determined based on the nominal scale ratio of each model. The middle part of the unstiffened plate served as the test area, while the surrounding region was secured to the foundation using bolts covered by angle bars. The dimensions of the test sections and the overall sizes of each series are provided in Table 1.

Table 1. Dimensions of test models.

Model	t_p* (mm)	Test area (mm)	Overall area (mm)
SE-3	3.0	600 x 600	800 x 800
SE-4	4.0	800 x 800	1,000 x 1,000
SE-5	5.0	1,000 x 1,000	1,150 x 1,150

* Nominal thickness of plate

2.2 *Material properties*

To determine the material properties of the test models, tensile tests were conducted using a universal testing machine in accordance with KS B 0801 (2007) and KS B 0802 (2018). Five tensile test coupons were extracted from each parent plate. The average values of the five tensile test results for each material, including their actual thicknesses, are presented in Table 2.

Table 2. Actual thicknesses and material properties of test models.

Model	Actual thickness (mm)	Young's modulus (MPa)	Yield strength (MPa)	Ultimate strength (MPa)
SE-3	2.85	201,700	327.7	498.9
SE-4	3.85	207,900	331.5	485.7
SE-5	4.71	211,000	359.3	489.9

2.3 *Impact tests*

The impact tests were performed using a free-drop testing machine. To examine the scale effects of different deformation shapes, three types of striker headers were utilized: knife-edge, hemispherical, and pyramid, as shown in Figure 1. The knife-edge and hemispherical strikers were used to induce local denting damage in the model, whereas the pyramid striker was designed to cause fracture damage.

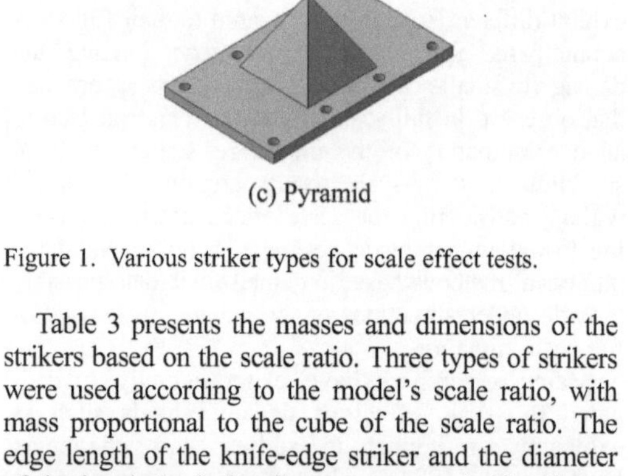

(a) Knife-edge (b) Hemisphere

(c) Pyramid

Figure 1. Various striker types for scale effect tests.

Table 3 presents the masses and dimensions of the strikers based on the scale ratio. Three types of strikers were used according to the model's scale ratio, with mass proportional to the cube of the scale ratio. The edge length of the knife-edge striker and the diameter of the hemispherical striker were set to half and one-tenth of the test section size, respectively.

To examine the scaling effect at different velocities, three drop heights were selected: 500, 1,000, and 1,500 mm, corresponding to impact velocities of 3.13, 4.43, and 5.43 m/s, respectively.

Table 3. Dimensions of the various strikers.

		Striker header type		
		Knife-edge	Hemisphere	Pyramid
Scale factor	Mass (kg)	Edge length (mm)	Diameter (mm)	Side length (mm)
1.00	86	300	60	140
1.33	206	400	80	170
1.67	400	500	100	200

2.4 *Naming of test models*

The names of the test models consist of four components: SE – (scale) – (striker header type) – (drop height).

- The first component, SE, stands for Scale Effect.
- The second component represents the scale factor, where 3 corresponds to the base scale (1.0), 4 corresponds to 1.33, and 5 corresponds to 1.67.
- The third component denotes the striker header type, categorized as knife-edge (K), hemisphere (H), and pyramid (P).
- The final component indicates the drop height in meters (0.5, 1.0, and 1.5), which corresponds to impact velocities of 3.13 m/s, 4.43 m/s, and 5.43 m/s, respectively.

For instance, if the model is named SE-4-H-1.0, it indicates a scale factor of 1.33, a hemisphere-type striker header, and a drop height of 1.0 m.

2.5 *Test results*

Figure 2 illustrates the deformed shapes of the models following the impact tests. Figure 2(a) depicts the denting damage caused by the knife-edge impact in model SE-5-K-1.5. Points P1, P2, and P3 in the figure indicate the locations where permanent deflection was measured after impact. P1 and P3 correspond to the contact points at both ends of the striker, while P2 represents the central part of the striker edge. Figure 2(b) shows the deformed shape of model SE-5-H-1.5, with point P1 marking the location where permanent deflection was measured. Figure 2(c) presents the fracture damage caused by the pyramid-shaped striker, with D1 and D2 denoting the diagonal lengths measured after the experiment.

The measurement results for denting and fracture damage are shown in Figure 3. In this figure, δ is a non-dimensional damage parameter, and it is defined as follows:

$$\delta = \frac{d}{t_p}\sqrt{\frac{\sigma_Y}{E}} \tag{1}$$

where d is the permanent deflection of denting damage or diagonal length of fracture-damage, t_p is the plate thickness, σ_Y is the yield strength and E is the Young's modulus. Furthermore, δ_E and are the δ values at the end and central parts, respectively. As shown in these figures, the nondimensional denting and fracture damage increased as the scale of the model increased. The impact tests results confirmed the scale effect of unstiffened steel plates subjected to a lateral impact load.

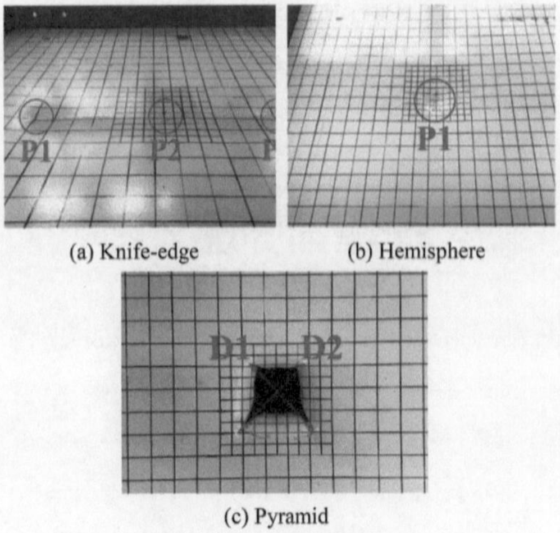

Figure 2. Deformed shape of test models.

3 VALIDATION OF DYNAMIC CONSTITUTIVE RELATION

3.1 *Finite element modelling*

Impact simulations were performed using the commercial software package ABAQUS (2022). The striker was modelled as a simple rigid body. The mass and impact velocity were provided as representative points. Additionally, modelling was performed considering the actual impact location and approximate tilted striker angle. Boundary conditions constrained both the displacement and rotation of the bolts that fixed the model in the experiment, as shown in Figure 4. The models comprise a four-node shell element (S4R) with an hourglass regulation and reduced integration. Five integration points were used throughout the thickness range.

3.2 *Dynamic constitutive equations*

The strain and strain-rate-hardening sensitivities of the materials were considered in this study. The results of static tensile tests were used to define the strain-hardening properties of the material. The formula reflecting a dynamic hardening mechanism

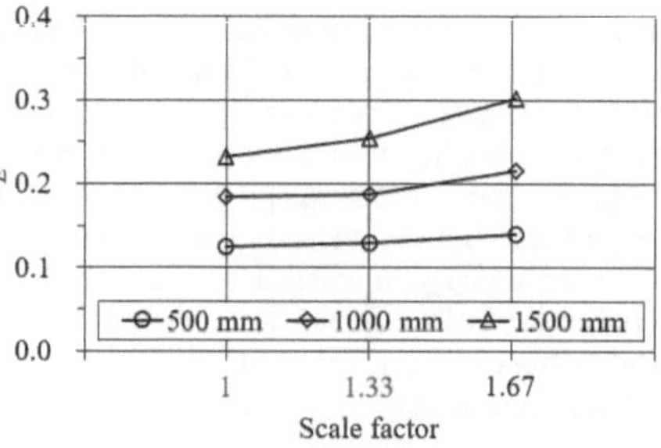

(a) Permanent deflections at ends of striker (P1 & P3): knife-edge header

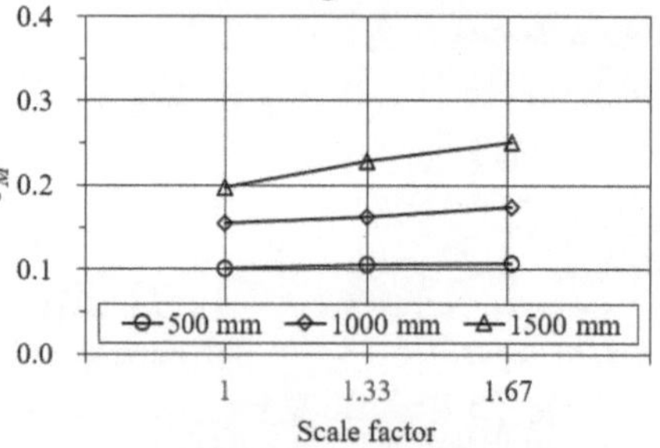

(b) Permanent deflection at central part of striker edge (P2): knife-edge header

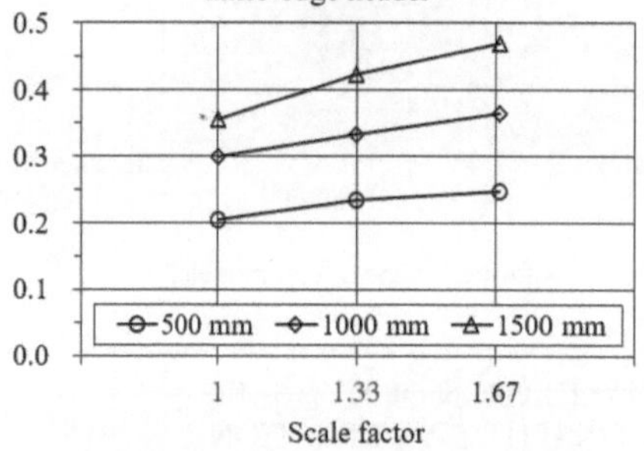

(c) Permanent deflections: hemisphere header

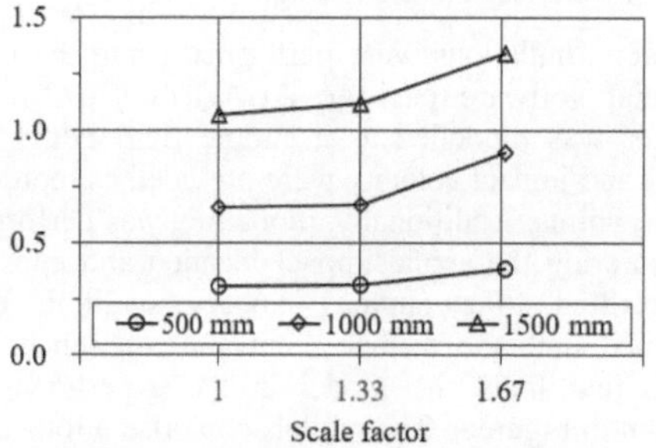

(d) Fractured diagonal length: Pyramid header

Figure 3. Plots of non-dimensional damage against scale factors, (a) Permanent deflections at ends of striker (P1 & P3): knife-edge header, (b) Permanent deflection at central part of striker edge (P2): knife-edge header, (c) Permanent deflections: hemisphere header, (d) Fractured diagonal length: Pyramid header.

proposed by Cho et al. (2015) was obtained using Equations. (2) – (5). The dynamic values of yield strength, ultimate tensile strength, hardening start

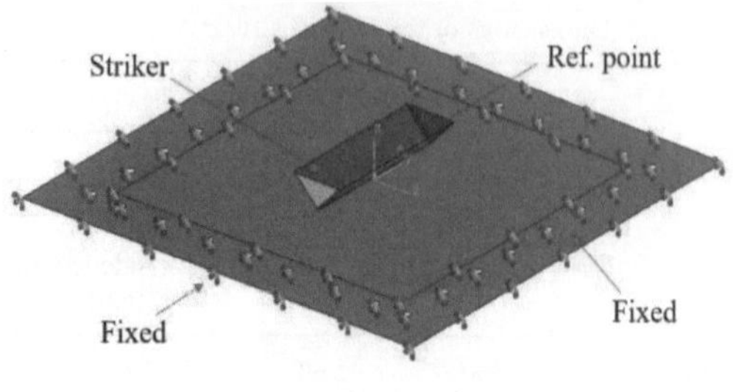

Figure 4. Impact simulation conditions.

strain, and ultimate tensile strain were expressed in terms of the strain-rate ($\dot{\varepsilon}$). The true stress–strain dynamic material curves were, then, obtained with the strain-rate (s^{-1}) set at 0 (static tensile), 0.01, 0.1, 1, 30, 80, 100, and 300 (dynamic tensile). Figure 5 shows the stress-strain curve considering the strain-rate.

Jones & Jones (2002) indicated that the strain-rate sensitivity of the ultimate tensile strength can be less than that of the yield strength. Therefore, the dynamic tensile strength needs to be determined differently. Using Equation (3), the strain-rate sensitivity difference can be considered.

$$\frac{\sigma_{YD}}{\sigma_Y} = 1 + \left\{0.3\left(\frac{E}{1000\sigma_Y}\right)^{0.5}\right\}(\dot{\varepsilon})^{0.25} \quad (2)$$

$$\frac{\sigma_{TD}}{\sigma_{YD}} = 1 + \left\{0.16\left(\frac{\sigma_T}{\sigma_{YD}}\right)^{3.325}\right\}(\dot{\varepsilon})^{0.023} \quad (3)$$

$$\frac{\varepsilon_{HSD}}{\varepsilon_{HS}} = 1 + \left\{0.1\left(\frac{E}{1000\sigma_Y}\right)^{1.73}\right\}(\dot{\varepsilon})^{0.333} \quad (4)$$

$$\frac{\varepsilon_{TD}}{\varepsilon_T} = 1 - \left\{0.117\left(\frac{E}{1000\sigma_Y}\right)^{2.352}\left(\frac{\sigma_T}{\sigma_Y}\right)^{0.588}\right\}(\dot{\varepsilon})^{0.2} \quad (5)$$

where σ_{YD} = dynamic yield strength; σ_{TD} = dynamic tensile strength; ε_{HS} = hardening start strain; ε_{HSD} = dynamic hardening start strain; ε_{TD} = dynamic ultimate tensile strain.

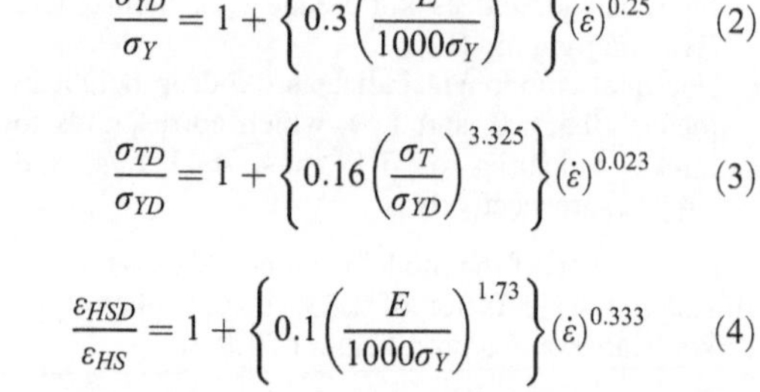

Figure 5. Dynamic constitutive relationships for simulation.

3.3 *Mesh convergence tests*

Mesh convergence tests were performed to obtain reasonable numerical results and reduce the analysis time. The tests were conducted on the model SE-H-5-0.5, which was not fractured. Figure 6 shows the convergence test results. The selected element size was approximately twice the shell thickness.

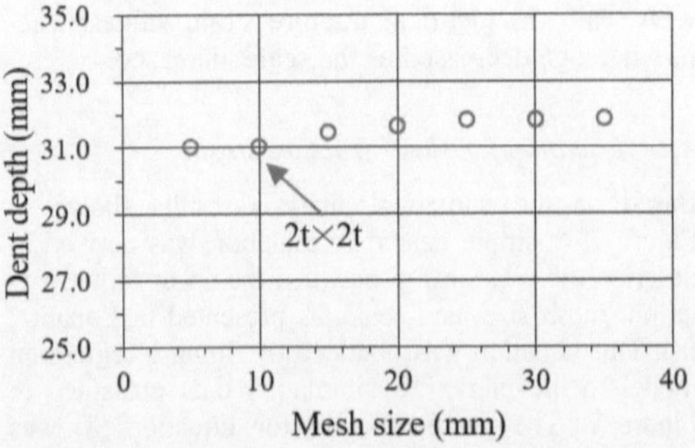

Figure 6. Results of the mesh convergence tests.

3.4 *Numerical simulation of denting damage*

As mentioned earlier, the dent depths of nine models impacted by knife-edge header and another nine models impacted by hemisphere header having only denting damage were numerically predicted adopting the dynamic constitutive equations. Table 4 presents the comparison results for the knife-edge header. These results allow for a comparison of the mean X_m and COV for end and central damages. For the end damage, the mean and COV were 1.08 and 2.1%, respectively. For the central damage, the mean and COV were 1.11 and 1.8%, respectively. Table 5 presents the comparison results for the hemispherical header. The mean and COV were 1.08 and 2.3%, respectively.

Table 4. Comparison of the predicted dent depth from numerical analysis with that of experiment: Knife-edge header.

	δ_E			δ_M		
Model	Exp.	Num.	X_m	Exp.	Num.	X_m
SE-3-K-0.5	0.125	0.132	1.06	0.101	0.113	1.12
SE-4-K-0.5	0.129	0.138	1.07	0.105	0.117	1.08
SE-5-K-0.5	0.140	0.154	1.10	0.108	0.119	1.14
SE-3-K-1.0	0.184	0.195	1.06	0.155	0.175	1.13
SE-4-K-1.0	0.187	0.203	1.08	0.162	0.184	1.13
SE-5-K-1.0	0.216	0.231	1.07	0.174	0.192	1.10
SE-3-K-1.5	0.231	0.258	1.11	0.198	0.215	1.09
SE-4-K-1.5	0.253	0.280	1.11	0.228	0.255	1.12
SE-5-K-1.5	0.301	0.318	1.06	0.251	0.279	1.11
	Mean		1.08	Mean		1.11
	COV		2.1%	COV		1.8%

For both cases, the COVs are quite acceptable, though there is a slight conservatism in the denting damage predictions. However, this conservatism is within a desirable range for design. Based on the comparison results shown in Tables 4 and 5, it can be concluded that the adopted dynamic constitutive equations appropriately consider strain and strain-rate hardening in numerical impact analyses. This implies that the scale effect of strain-rate on denting damage can be effectively addressed by using dynamic constitutive equations that appropriately consider the strain-rate effects on both yield strength and tensile ultimate strength.

Table 5. Comparison of the predicted dent depth from numerical analysis with that of experiment: Hemisphere header.

	δ		
Model	Exp.	Num.	X_m
SE-3-H-0.5	0.204	0.226	1.11
SE-4-H-0.5	0.235	0.252	1.07
SE-5-H-0.5	0.248	0.272	1.10
SE-3-H-1.0	0.300	0.327	1.09
SE-4-H-1.0	0.333	0.361	1.08
SE-5-H-1.0	0.365	0.396	1.08
SE-3-H-1.5	0.355	0.393	1.11
SE-4-H-1.5	0.423	0.436	1.03
SE-5-H-1.5	0.470	0.496	1.05
	Mean		1.08
	COV		2.3%

4 DEVELOPMENT OF FRACTURE CRITERION

The selection of an appropriate fracture criterion and the determination of its corresponding critical value are crucial for the fracture analysis of impacted structures. In this study, the shear fracture criterion was adopted to monitor fracture occurrence (Ringsberg et al., 2018; Do et al., 2019; Park et al., 2023). This criterion provides a phenomenological representation of damage initiation due to shear band localization. It assumes that fracture occurs when the accumulated equivalent plastic strain reaches the material's failure shear strain. This approach is one of the most practical and user-friendly methods for conducting fracture analysis.

The Norwegian Technical Standard Institution (NTSI) (2004) has proposed fixed values of 0.2 and 0.15 for mild and high-tensile steels, respectively, in the context of the shear fracture criterion. However, it is well known that these fracture constant values are significantly influenced by element size in numerical analysis (Samuelides et al., 2007).

Therefore, applying the values provided by NTSI to different element sizes is not appropriate. To address this element size dependency, several researchers have investigated the relationship between element size and the fracture criterion. Samuelides et al. (2007) expressed the shear fracture strain as shown in Equation (6). Choi (2014) derived the equation in Equation (7) for an unstiffened plate model by varying the element size from 1.0 to 5.0 times the plate thickness to determine the fracture constant value.

$$\varepsilon_r = 0.056 + 0.54\frac{t_p}{l} \quad (6)$$

$$\varepsilon_r = 0.1 + 0.26\left(\frac{t_p}{l}\right) \quad (7)$$

where ε_r is the shear fracture strain and l is the mesh size.

In this study, based on the results of convergence tests, an element size of twice the plate thickness was selected, and the appropriate shear failure criterion was derived.

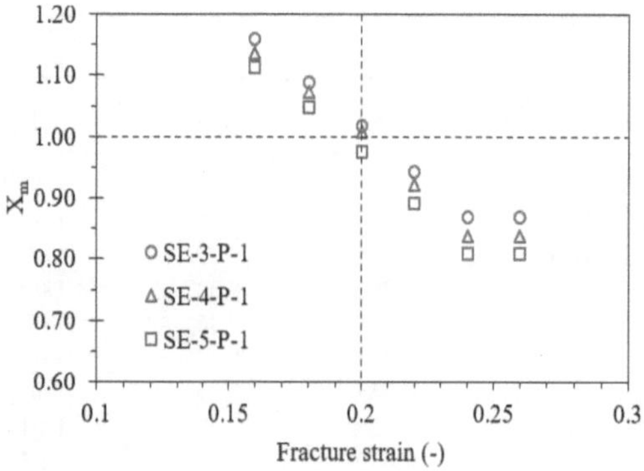

(a) Velocity: 3.13 m/s

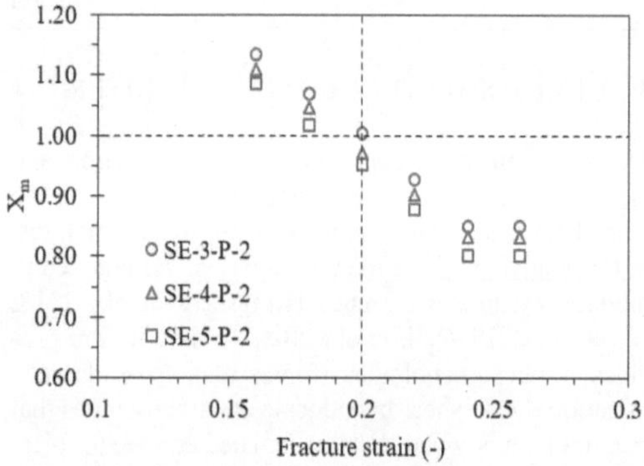

(b) Velocity: 4.43 m/s

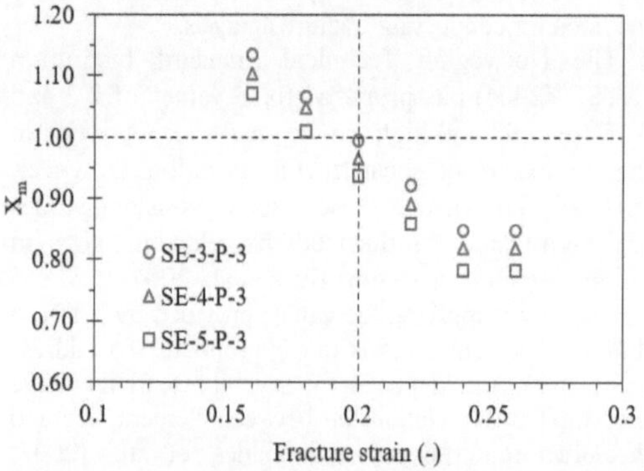

(c) Velocity: 5.43 m/s

Figure 7. Damage ratio according to changes in fracture strain, (a) Velocity: 3.13 m/s, (b) Velocity: 4.43 m/s, (c) Velocity: 5.43 m/s.

4.1 *Numerical analysis*

An appropriate shear fracture criterion for each experimental model was determined through fracture analysis. By varying the fracture strain from 0.16 to 0.26, the value that best matched the diagonal length of the experimental results was identified. Figure 7 presents the results of the parametric study, which confirmed that the shear fracture criterion varies with scale. At identical fracture strain values, fracture damage decreased as the scale increased.

4.2 *Equation for shear fracture strain*

Based on the numerical analysis results shown in Figure 7, a simple design formulation was derived to describe the relationship between the shear failure criterion, mesh size, and scale, as presented in Equation (8). This equation was obtained by fitting a regression model to the numerical simulation data presented in Figure 7. The goodness-of-fit for Equation (8) was quantitatively assessed by the coefficient of determination (R^2), which was found to be 0.9725. This high R^2 value indicates that the proposed equation accurately captures the trends observed in the numerical analysis results, providing a reliable predictive tool for the shear fracture strain.

Once the mesh size for the numerical model is determined, obtaining the shear fracture criterion for numerical studies becomes straightforward. Equation (8) contains two terms within the curly brackets on the right-hand side. The first term represents the effect of element thinness, while the second term accounts for the scale effect.

$$\varepsilon_r = 0.3 \times exp\left\{\left(-0.1\frac{l}{t_p}\right) + \left(-60.0\frac{l}{l_C}\right)\right\} \quad (8)$$

$$l_C = C \times 1\ (ms) \quad (9)$$

$$C = \sqrt{E/\rho} \quad (10)$$

where l_c is the travelling distance of elastic shock wave during 1 ms, C is the elastic wave speed and ρ is the density of material. For steel, assuming the elastic modulus of 207,000 MPa and density of 7,850 kg/mm^3, C becomes 5,135 m/s.

5 APPLICATION EXAMPLES

5.1 *Application example models*

To evaluate the applicability of the proposed analysis guidelines for predicting denting and fracture damage in plated steel structures, numerical simulations were conducted using reported test results. Cho et al. (2017) performed lateral collision tests on steel double-hull side structure models using a rigid striker with a hemispherical nose.

Figure 8. Overall view of car crash testing facilities together with a model.

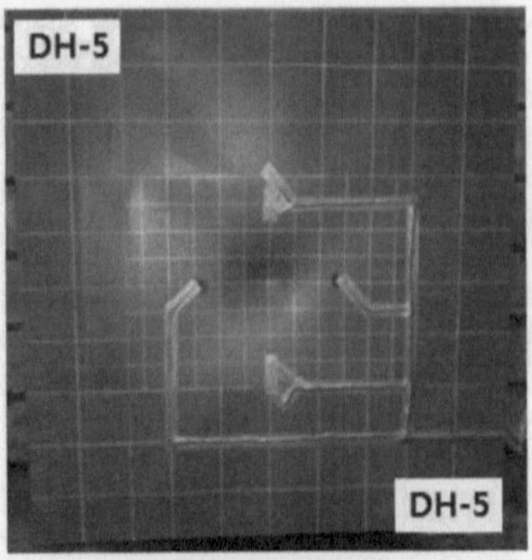

(a) Model DH-5

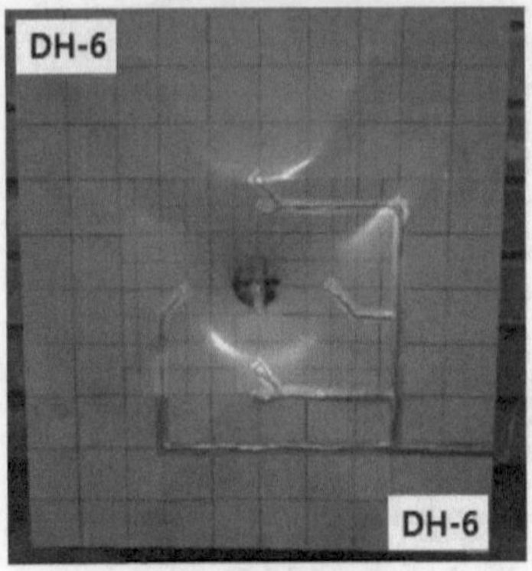

(b) Model DH-6

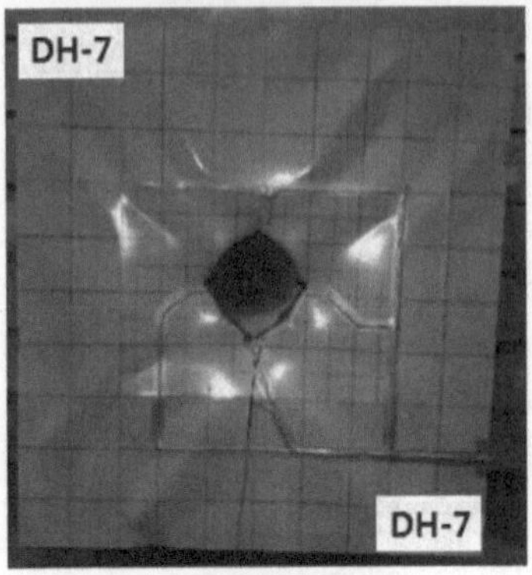

(c) Model DH-7

Figure 9. Damaged models after collision tests.

Three double-hull models were constructed at approximately one-seventh the scale of a Very Large Crude Oil Carrier (VLCC). The models were made of mild steel, with a nominal plate thickness of 3.0 mm. The overall height and width of the models were 1,950 mm and 1,830 mm, respectively.

The three models were tested using the car crash testing facilities at the Automotive Parts Institute Center of Ulsan Techno Park. A solid hemispherical header with a diameter of 300 mm was attached to the front of the striker, which had a mass of 1,694 kg. The collision velocities were set to 3 m/s, 5 m/s, and 9 m/s for models DH-5, DH-6, and DH-7, respectively. Figure 8 provides an overview of the car crash testing facilities along with one of the test models. Figure 9 presents the damaged models after the collision tests.

5.2 *Numerical simulation*

Numerical simulations were conducted following the analysis guidelines proposed in this study, incorporating the dynamic constitutive equations (Equations 2–5) and utilizing Equation (8) to determine the shear fracture criterion. The FE models used for the impact simulations are shown in Figure 10.

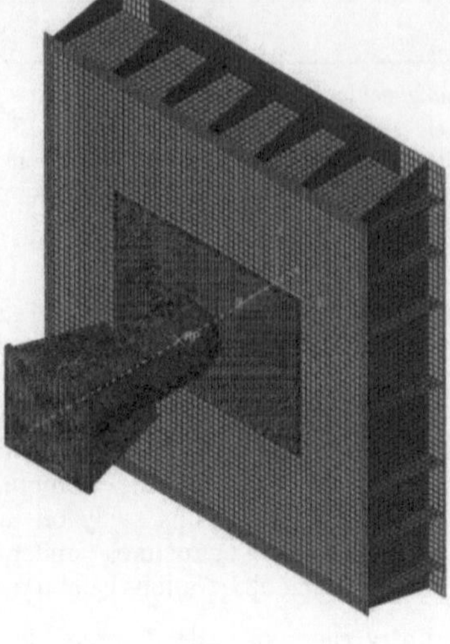

Figure 10. FE model of double hull model.

A four-node shell element was used for the FE model, with the striker assumed to be rigid. Following the proposed analysis guidance, the element size in the impact region was set to approximately twice the shell thickness (mesh size: 6×6 mm), while a coarser mesh with an element size of 15×15 mm was applied farther from the impact region. In this case, a shear failure criterion of 0.23 was used, as determined by Equation (8).

5.3 *Simulation results*

The results of the numerical analysis are summarized in Tables 6 and 7 for models DH-5 and DH-6, and

DH-7, respectively. When comparing the predicted dent depth of the front plate with the actual dent depth, the differences are only 1.5% for model DH-5 and 2.0% for model DH-6. These results suggest that the assumptions adopted in the dynamic constitutive equations for the analysis were largely valid.

Table 6. Numerical analysis results of models DH-5 and DH-6.

Model	Dent depth		
	Exp.	Num.	X_m
DH-5	68 mm	69.0 mm	1.015
DH-6	112 mm	114.2 mm	1.020

The analysis results for model DH-7 are presented in Table 7, where the fracture diagonal length of the front plate and the dent depth of the rear plate are compared with the experimental results. The difference in fracture diagonal length is 1.7%, while the difference in dent depth is 6.9%. Overall, the agreement between the experimental results and numerical predictions is quite good.

Table 7. Numerical analysis results of model DH-7.

Damage	Exp.	Num.	X_m
Fracture diagonal length of front plate	353 mm	346.2 mm	0.983
Dent depth of rear plate	32 mm	34.2 mm	1.069

6 CONCLUSIONS

This study examined the scale effect on the structural behavior of unstiffened steel plates subjected to lateral impact loading. Rigorous numerical simulations were conducted using unstiffened plate models to quantitatively assess the scale effect on denting and fracture damage. Based on the comprehensive numerical investigations in this study on the scale effects of plated steel structures under lateral impacts, the following observations can be made:

- The primary cause of scale distortion in denting damage is the strain-rate sensitivity of the material. By properly accounting for strain-rate effects on both yield strength and ultimate tensile strength using appropriate dynamic constitutive equations, scale distortion in denting damage can be mitigated.
- The scale effects on fracture damage are more significant than those on plastic damage.

Based on the results and findings of this study, a practical analysis guideline is proposed for simulating large-scale plated steel structures under lateral impact, enabling the prediction of plastic and fracture damage. This guideline consists of the following:

- Model the impacted regions with an element size of twice the plate thickness.
- Employ the dynamic constitutive equations, Equations (2) – (5), to account for the strain and strain-rate hardenings of the material. Other dynamic constitutive equations can be used, but they must consider the strain-rate effects on both yield strength and ultimate tensile strength.
- Monitor the occurrence of fracture using the shear fracture strain predicted by Equation (8).

The proposed analysis guidance was applied to three double hull models with dimensions of 1,830 mm in width and 1,950 mm in height. The results from these applications were reasonably close to the experimental data, demonstrating that the proposed guidance can be confidently used to perform numerical simulations with a commercial package to predict denting and fracture damage in larger-scale plated steel structures under lateral impacts.

This study primarily focused on the application of the proposed methodology to mild steel plates. While the results demonstrate good applicability for this material, the universality of the method across various material types, such as high-tensile steels or aluminum alloys commonly used in ship structures, was not extensively investigated. Therefore, a limitation of the current study is the lack of comprehensive validation concerning the material universality of the proposed dynamic constitutive equations and fracture criterion. Future research will aim to address this limitation by conducting experimental and numerical investigations on a wider range of materials. This will involve verifying the applicability and extending the proposed methodology to different steel grades and other metallic materials. Such efforts will enhance the robustness and practical utility of the analysis guideline, allowing for more reliable collision damage predictions for a broader spectrum of shipbuilding materials.

ACKNOWLEDGEMENTS

This research was supported by Korea Institute of Marine Science & Technology Promotion (KIMST) funded by the Ministry of Oceans and Fisheries (2520000248).

REFERENCES

ABAQUS, 2022. User manual version 6.22. 2022.

Choi SI., 2014. Experimental and numerical investigations on the scaling effect on permanent deflections and fractures of plates. MSc. Thesis University of Ulsan.

Cho S-R, Choi, SI, Son SK., 2015. Dynamic material properties of marine steels under impact loadings. In ASEM15 Incheon Republic of Korea 25–29.

Cho S-R, Song S-U, Park S-H, Shin HK., 2017. Plastic and fracture damages of double hull structures under lateral collisions. In: Marstruct Glasgow UK.

Do QT, Le DNC, Seo BS, Shin HK, Cho S-R., 2019. Fracture response of tubular T-joints under dynamic mass impact. In: ICCGS 2019 Lisbon Portugal:75–84.

Do QT, Muttaqie T, Park S-H, Shin HK, Cho S-R., 2018. Predicting the collision damage of steel ring-stiffened cylinders and their residual strength under hydrostatic pressure. Ocean Eng; 169:326–43.

Do QT, Muttaqie T, Shin HK, Cho S-R., 2018. Dynamic lateral mass impact on steel stringer-stiffened cylinders. Int J Impact Eng;116:105–126.

Drazetic P, Ravalard Y, Dacheux F, Marguet B., 1994. Applying non-direct similitude technique to the dynamic bending collapse of rectangular section tubes. Int J Impact Eng; 15(6):797–814.

Duffey TA, Cheresh MC, Sutherland SH., 1984. Experimental Verification of Scaling Laws for Punch-Impact-Loaded Structures. Int J Impact Eng;2(1):103–117.

Fu T, Zhang M, Zheng Q, Zhou D, Sun X, Wang X., 2021. Scaling the response of armor steel subjected to blast loading. Int J Impact Eng; 153:103863.

Jiang P, Tian CJ, Xie RZ, Meng DS., 2006. Experimental investigation into scaling laws for conical shells struck by projectiles. Int J Impact Eng;32(8):1284–1298.

Jones N, Birch RS., 2008. On the scaling of low-velocity perforation of mild steel plates. J. Pressure Vessel Technol; 130(3):031207.

Jones N, Jones C., 2002. Inelastic failure of fully clamped beams and circular plates under impact loading. Proc Instn Mech Engrs, Part C: J Mech Engrs Sci; 216:133–149.

KS. 2007. B 0801. Test pieces for tensile test for metallic materials. Korean Agency for Technology and Standards.

KS, 2018. B 0802. Method of tensile test for metallic materials. Korean Agency for Technology and Standards.

Samuelides M, Georgios V, Michael T, Jorgen A, Robert D., 2007. Simulation of the behavior of double bottoms subjected to grounding actions. In: ICCGS 2007 Hamburg Germany: 93–102.

NTSI, 2004. 2004. Design of steel structures. NORSOK Standard N-004, Norwegian Technology Standard Institution, Oslo Norway.

Paik JK, Lee TK., 1996. Scaling and dynamic effects on the plate cutting response. J Soc Naval Arc Korea in Korean; 33(3):48–55.

Park S-H, Kim DU, Jeong D, Cho S-R., 2023. Experimental investigations on the scale effect of unstiffened steel plates under lateral mass impact. In ICCGS 2023 Nantes France.

Park S-H, Yoon S-H, Muttaqie T, Do QT, Cho S-R, 2023. Effects of Local Denting and Fracture Damage on the Residual Longitudinal Strength of Box Girders. J Mar Sci Eng; 11(1).

Ringsberg JW, Amdahl J, Chen BQ, Cho S-R, Ehlers S, Hu Z, Kubiczek JM, Kõrgesaar M, Liu Q, Marinatos JM., 2018. MARSTRUCT benchmark study on nonlinear FE simulation of an experiment of an indenter impact with a ship side-shell structure. Mar Struct; 59:142–157.

Schleyer GK, Hsu SS, White MD., 2004. Scaling of pulse loaded mild steel plates with different edge restraint. Int J Mechanical Sci; 46:1267–1287.

Shen WQ, Fung KW, Triantafyllos P, Wajid NM, Nordin N., 1998. An experimental study on the scaling of plate cutting. Int J Impact Eng; 21(8):645–662.

Wang W, Zhang D, Lu F, Wang S-C, Tang F., 2012. Experimental study on scaling the explosion resistance of a one-way square reinforced concrete slab under a close-in blast loading. Int J Impact Eng; 49:158–164.

Wen HM, Jones N., 1993. Experimental investigation of the scaling laws for metal plates struck by large masses. Int J Impact Eng; 13(3):485–505.

Yao S, Ma Y, Zhao N, Wang Z, Zhang D, Lu F., 2022. Equivalent scaling method on the dynamic response of box-shaped structures under internal blast. Int J Impact Eng.; 160:104074.

Trends in Collision and Grounding of Ships and Offshore Structures – Zhu & Guedes Soares (Eds)
© 2026 The Author(s), ISBN 978-1-041-16361-9

Model tests of ship collisions for the effects of different impact location in a water tank

X.B. Wang, H.F. Zhou & L. Zhu
Key Laboratory of High Performance Ship Technology (Wuhan University of Technology), Ministry of Education, Wuhan, P.R. China
School of Naval Architecture, Ocean and Energy Power Engineering, Wuhan University of Technology, Wuhan, P.R. China

Y.J. Zhao & L.P. Zhang
China Ship Scientific Research Center, Wuxi, P.R. China
Taihu Lake Laboratory of Deep Sea Technology and Science, Wuxi, P.R. China

ABSTRACT: In recent years, the frequency of ship collisions has increased, underscoring the need for further investigation into the mechanics of such incidents. While collisions can occur at various locations on a vessel, statistical data indicate that very serious marine casualties are more frequently associated with damage in the fore part and amidships region of the ship. This paper presents perpendicular collision model tests conducted in a water tank, focusing on two typical collision scenarios, one is that a striking ship impacting the amidship area of the struck ship's side structure, and another is that colliding with the fore section. This study couples' external dynamics with internal mechanics to analyze the collision dynamic responses, including structural damage modes, six-degree-of-freedom (6DOF) motion characteristics, and energy dissipation mechanisms across different impact locations. The findings reveal that the energy absorbed through deformation and damage of the side structure varies significantly between these two collision scenarios. This study provides a technical foundation for improving ship crashworthiness under severe collision loads, contributing to the enhancement of impact resistance in ship design.

1 INTRODUCTION

Because of the increasing demand for transportation, maritime traffic in many regions of the world has been growing annually and is expected to increase significantly in the coming decades. Numerous existing research reports indicate that ship collisions account for a large proportion of major accidents in fairways (EMSA, 2023; Zhu et al., 2002). Collision risk can be mitigated by minimizing the probability of collisions or reducing their consequences; however, reducing the consequences of collisions requires a good understanding of the physical phenomena involved.

In order to assess the physical phenomena of actual collision phenomena of ships in service, empirical formulas, simplified analytical methods, finite element simulations, and experiments have been employed. Empirical formulas and simplified methods based on some important assumptions have been used as useful tools for quickly evaluating ship performance in such incidents (Minorsky, 1958; Liu & Amdahl, 2010). With advancements in computer hardware and software, finite element simulations of full-scale ship collisions have become feasible (Ehlers et al., 2012; Yu et al., 2015; Song et al., 2017; Ringsberg et al., 2018; Wang et al., 2021; Kim et al., 2022). The advantage of empirical methods is their simplicity and convenience. However, due to limited accident data, they can only be applied to specific ship types in collision studies. The simplified analytical methods are highly efficient but are generally based on idealized assumptions, which significantly affect their accuracy. The numerical simulation methods are powerful, but their accuracy is highly dependent on input parameters, require skilled technical personnel, and involve long computation times in complex multi-parameter scenarios. More importantly, both analytical and numerical methods require validation through experimental approaches.

Experimental methods are crucial for revealing new phenomena and mechanisms. However, experimental data for studying these phenomena and validating calculation models have been consistently scarce. Large-scale model tests and full-scale ship collision experiments involve numerous scenarios and complex phenomena, making their design and

DOI: 10.1201/9781003684404-16

execution processes both time-consuming and labor-intensive (Wevers & Vredeveldt, 1999). Additionally, these tests are difficult to control and operate. As a result, laboratory model tests remain the most practical method for investigating side collision phenomena, which including "dry tests" and tank tests. In the "dry tests," the quasi-static loading is often applied, where the load is carefully controlled and apply on the structure using different shapes of indenters (Paik et al., 1999; Wang et al., 2000; Zhang et al., 2018). It should be noted that quasi-static tests cannot accurately capture the dynamic response of structures involving inertial effects (Jones, 2006).

Some of the dynamic model tests have been reported by Zhu et al. (1994), Liu et al. (2015) and Calle et al. (2017). In these experiments, the struck structure is often fixed, without considering the influence of surrounding fluid around the ship. To make collision experiments more representative of real-world scenarios, other researchers conducted dynamic collision tests in water (Tabri et al., 2008; Tabri et al., 2010; Zhu et al., 2024; Zhu & Wang, 2025; Zhou et al., 2025). These tests aimed to obtain precise measurements not only during contact but also for a period after the impact. Since perpendicular collisions at the midsection are generally considered the most hazardous in ship collisions, studies on collisions involving the fore part of a ship are extremely rare.

This study primarily investigates perpendicular collisions through scaled model experiments, focusing on the effects of different impact locations on collision characteristics. In the scaled model experiments, honeycomb sandwich panels were used to replace part of the side structure of the struck ship to simulate the damage characteristics of a real ship. Through these experiments, data were obtained on the motions of the ship models, impact forces, deformation and damage modes and the energy absorption of the side structure under collision loads at different locations. Furthermore, some damage mechanisms, as well as energy conversion and dissipation mechanisms under fully coupled conditions in ship collisions, were summarized and discussed.

2 EXPERIMENT MODEL

Ship collision model experiments were conducted in the Impact and Fluid-Structure Interaction Laboratory at Wuhan University of Technology. The six-degree-of-freedom (6DOF) motions of the ships and other dynamic response during collision were recorded.

The experimental water tank has a length of 35.1 m, a width of 3.8 m, and a depth of 1.5 m, and it is equipped with an in-tank ship collision testing system (ISCTS). In this study, the water depth was set to 0.8 m. The layout of the experimental system is shown in Figure 1. The striking ship model was accelerated by traction, with the traction force direction always remaining in the same horizontal plane as the center of gravity of the model. Once the striking ship model reached the desired velocity, the power supply was cut off, and the constraints at both ends of the struck ship were simultaneously released, allowing the two ships to collide freely.

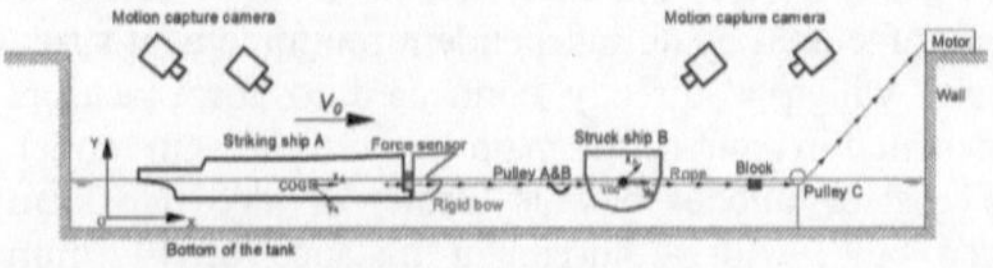

Figure 1. Layout of the experimental system.

2.1 *Ship models*

2.1.1 *Model of the striking ship*

The model of the striking ship is shown in Figure 2, and its main particulars are listed in Table 1. The overall length of the model is 3.24 m, and the displacement is 60 kg. Specifically, the added mass coefficients used in this study were calculated using the ANSYS-AQWA software based on three-dimensional potential flow theory. The values correspond to the added mass at infinite frequency, which is commonly used in impact analyses where the collision duration is extremely short. This approach has been widely adopted in previous studies, such as Yu et al. (2019) and Ye et al. (2023).

Figure 2. Model of the striking ship (Ship A).

Table 1. Main particulars of the model of the striking ship.

Particular	Ship A	Units
Overall length, $\boldsymbol{L}$	3.24	m
Breadth, $\boldsymbol{B}$	0.41	m
Depth, $\boldsymbol{D}$	0.23	m
Design draught, $\boldsymbol{T}$	0.14	m
Displacement, Δ	60.00	kg
Moment of inertia in the x-direction around its own coordinate system, I_{xx}	1.18	kg·m^2
Moment of inertia in the y and z directions around its own coordinate system, I_{yy}, I_{zz}	38.76	kg·m^2
Added mass coefficient for surge, m_{ax}	0.023	-
Added mass coefficient for sway, m_{ay}	0.570	-
Added mass coefficient for heave, m_{az}	1.784	-
Added mass coefficient for pitch, J_{ax}	0.873	-
Added mass coefficient for roll, J_{ay}	0.003	-
Added mass coefficient for yaw, J_{az}	0.335	-

The model is capable of measuring the impact force acting on the bulbous bow of the ship during collision. Figure 3a illustrates the shape of the bow and the arrangement of dynamic compression and tension force sensor. The striking ship model (Ship A) consists of two main components: a wooden main hull and a steel bulbous bow. The bow section is manufactured by an independent reinforcement structure, which is directly connected to force sensors mounted in front of the main body of the ship model. The steel bulbous bow is further divided into stem and bulb, with a horizontal distance of 147 mm between their front ends. In the experiments, only the upper stem made contact with the side shell of the struck ship. The schematic diagram of the contact region is shown in Figure 3b, illustrating the interaction between the two ships.

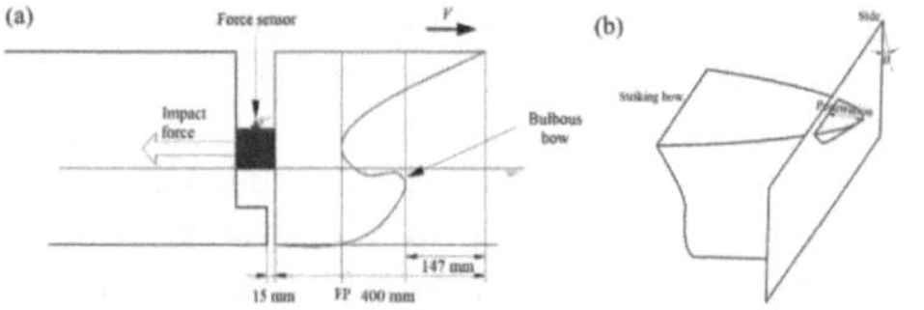

Figure 3. (a) Illustration of the connection of the force sensor at the bow of the strike ship model. (b) Contact zone between striking and struck ship.

A Kistler 9327C force sensor was mounted between the bulbous bow and the main body of the striking ship model to enable real-time monitoring of collision forces. The sensor has an axial measurement range of ±8 kN and a measurement range of ±4 kN in the other two directions. During the experiments, the data sampling frequency was set to 3.2 kHz.

2.1.2 *Models of the struck ships*

2.1.2.1 *Main dimensions*

The models of the two struck ships are shown in Figure 4, and their main particulars are listed in Table 2. To account for the influence of fluid flow variations around the ship caused by changes in the shape of different impact locations on the same model, and to ensure that the scaled model maintains sufficient structural strength after further cutting and installation of the struck structure, two separate struck ship models were constructed. The struck ship models are also divided into two components: the wooden main body and the struck structure on the side.

2.1.2.2 *Side structure*

To simulate the stiffness and energy absorption characteristics of the side structure of a full-scale ship, while also considering cost reduction and improved experimental efficiency, the side structure of the struck ship was designed as a replaceable honeycomb sandwich structure. The primary design concept of the replaceable structure is to ensure that the force-displacement curve of the full-scale ship structure under actual collision velocity is similar to that of the scaled model structure under scaled collision velocity.

(a) Experimental model of the struck ship with the impact location amidships (Ship B_1)

(b) Experimental model of the struck ship with the impact location at the fore part (Ship B_2)

Figure 4. Models of the struck ships.

Table 2. Main particulars of the models of the struck ships.

Particular	Ship B_1	Ship B_2	Units
Overall length, $\boldsymbol{L}$	3.18	3.18	m
Breadth, $\boldsymbol{B}$	0.42	0.42	m
Depth, $\boldsymbol{D}$	0.32	0.32	m
Design draught, $\boldsymbol{T}$	0.11	0.11	m
Displacement, $\boldsymbol{\Delta}$	65	62.8	kg
Moment of inertia in the x-direction around its own coordinate system, $\boldsymbol{I_{xx}}$	1.34	1.29	kg·m^2
Moment of inertia in the *y* and *z* directions around its own coordinate system, $\boldsymbol{I_{yy}}$, $\boldsymbol{I_{zz}}$	44.29	42.79	kg·m^2
Added mass coefficient for surge, $\boldsymbol{m_{bx}}$	0.008	0.008	-
Added mass coefficient for sway, $\boldsymbol{m_{by}}$	0.348	0.348	-
Added mass coefficient for heave, $\boldsymbol{m_{bz}}$	1.649	1.649	-
Added mass coefficient for pitch, $\boldsymbol{J_{bx}}$	0.893	0.893	-
Added mass coefficient for roll, $\boldsymbol{J_{by}}$	0.002	0.002	-
Added mass coefficient for yaw, $\boldsymbol{J_{bz}}$	0.181	0.181	-

The replaceable amidships structure designed for this experiment is shown in Figure 5. The overall side structure consists of the hull, watertight base, a structural support, and a honeycomb sandwich panel, arranged from amidships to port. Among these, the hull, watertight base, and structural support were designed as rigid components in this experiment. Honeycomb sandwich panels consist of two metallic facesheets and an aluminum honeycomb core. The material properties and dimensions of the equivalent structures are listed in Table 3.

Table 3. Dimensions and material properties of the side structure of the struck ship.

	Size		
Components	Amidships	Fore-part	Material
Honeycomb core	680×270×150 mm (effective: 620×230×150 mm) Cell edge length: 17 mm Foil thickness: 0.04 mm	370×250×100 mm (effective: 310×210×100 mm) Cell edge length: 17 mm Foil thickness: 0.04 mm	3003 Al (**E**: 70 GPa, σ_y: 145 MPa)
Front-face sheet	680×470×0.1 mm (effective: 620×420×0.1 mm)	370×400×0.1 mm (effective: 310×350×0.1 mm)	1080 Al (**E**: 68 GPa, σ_y: 50 MPa)
Back-face sheet	680×270×1 mm (effective: 620×230×1 mm)	680×270×1 mm (effective: 620×230×1 mm)	Q235 steel (**E**: 210 GPa, σ_y: 240 MPa)

where **E** is the Young'smodulus, σ_y is the Yield strength.

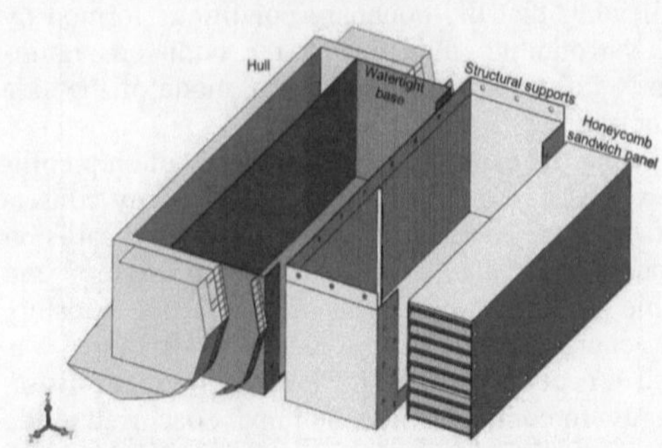

Figure 5. Struck structure in the middle region of struck ship B_1.

2.2 *Collision scenarios in experiment*

The test cases are shown in Table 4, The ship coordinate system is shown in Figure 6. Two impact locations were considered: the fore region and amidships. Under both test cases, the mass of the two ships remained constant, while the collision angle and impact velocity were kept similar.

Table 4. Test cases considering different impact locations.

Case	Impact location	Impact velocity	Impact point	Impact angle
M1	Amidships	1.19 m/s	Middle of the side structure	90 °
M2	Fore part	1.29 m/s		

3 RESULTS AND DISCUSSION

3.1 *The six DOF motions of ship*

The six DOF motion time-history results of the three ship models before and after the collision, measured using the 3D motion capture system Qualisys, are shown in Figsure 7 and 8. The collision occurs at 0.2 seconds. The velocity variations during the collision can be obtained separately by differentiating the motion time curves.

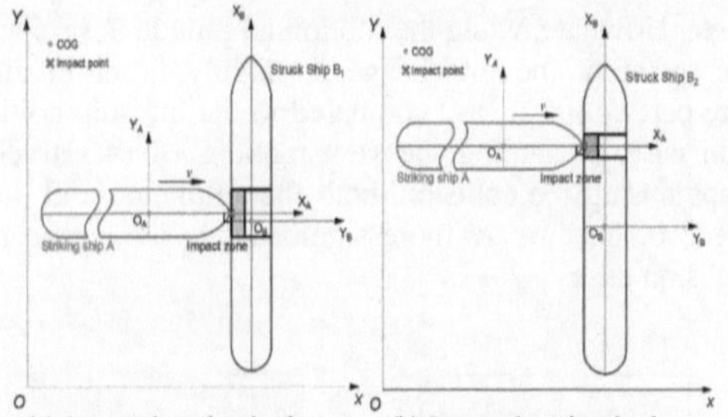

(a) Impact location in the middle of the struck ship model (b) Impact location in the fore part of the struck ship model

Figure 6. Illustration of the ship coordinate systems.

It can be observed that in both test cases, the primary translational motion of the striking ship during the collision is surge. Additionally, compared to the midship collision case M1, the striking ship in the fore-part collision case M2 exhibits a more obvious rotational motion for yaw. For the struck ships, the most significant translational motion is sway, while the most obvious rotational motions are roll and yaw. Due to the impact point being significantly offset from the center of gravity of the struck ship model in the vertical direction, noticeable roll motion occurs in both

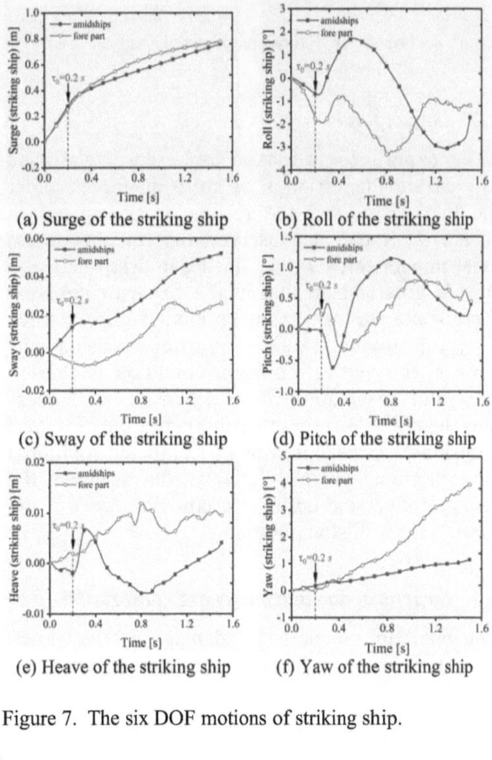

(a) Surge of the striking ship (b) Roll of the striking ship (c) Sway of the striking ship (d) Pitch of the striking ship (e) Heave of the striking ship (f) Yaw of the striking ship

Figure 7. The six DOF motions of striking ship.

cases. However, while the motion amplitude is similar, the speed of motion change is slightly faster in the fore-part collision case compared to the midship collision case. Regarding the yaw motions of the struck ships during the collision, both the amplitude and the speed of change are more significant in the fore-part collision case.

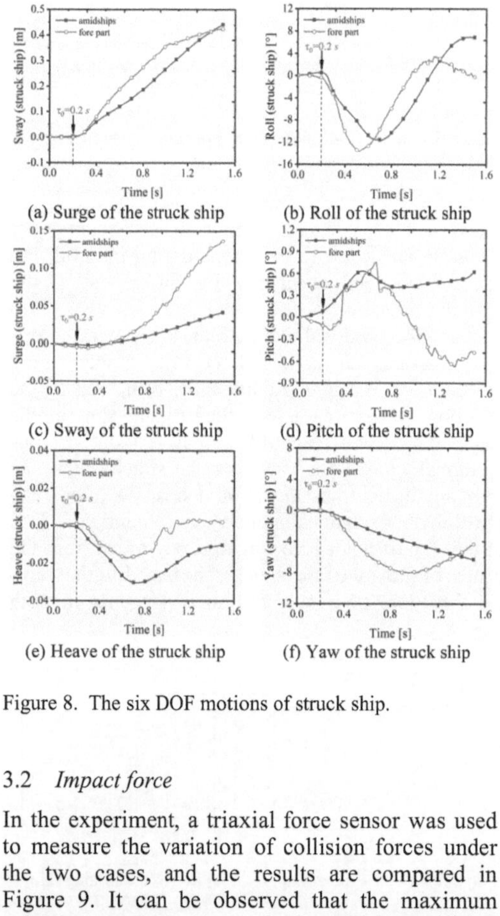

(a) Surge of the struck ship (b) Roll of the struck ship

(c) Sway of the struck ship (d) Pitch of the struck ship

(e) Heave of the struck ship (f) Yaw of the struck ship

Figure 8. The six DOF motions of struck ship.

3.2 *Impact force*

In the experiment, a triaxial force sensor was used to measure the variation of collision forces under the two cases, and the results are compared in Figure 9. It can be observed that the maximum axial impact force $\boldsymbol{F_{x\_max}}$ in the midship collision case is greater than that in the fore-part collision case, while the maximum horizontal impact force $\boldsymbol{F_{y\_max}}$ is smaller. Additionally, the collision duration in the vertical collision condition is shorter. A careful observation of the experimental process may help explain these phenomena: the fore-part collision case leads to a more significant rotational yaw motion, which relatively slows down the energy conversion during the collision process and extends the collision duration.

3.3 *Structural damage and energy absorption*

The final deformation and damage of the struck structure are shown in Figure 10, and the measured

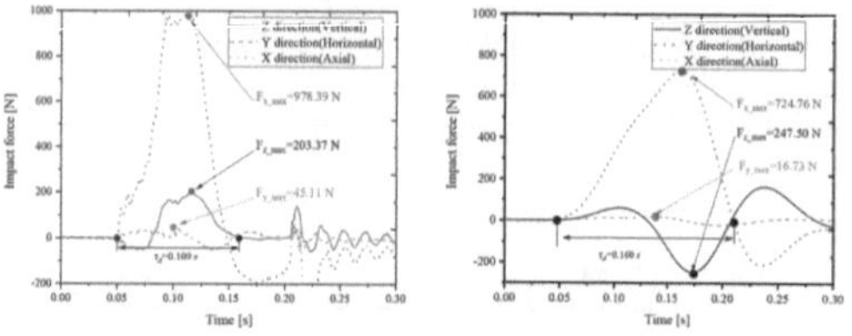

(a) Time history curves of impact forces for middle collision case (b) Time history curves of impact forces for fore-part collision case

Figure 9. Comparison of the time-history curves of the impact force during collision.

final penetration depths after the experiment were 48.33 mm for the midship collision case and 64.42 mm for the fore-part collision case. It can be observed that the final deformation in both collision cases is approximately triangular in shape. Although the collision was a perpendicular impact, the final shape is not a perfect isosceles triangle, indicating that the boundary conditions formed by the surrounding fluid during the collision significantly influence the final damage mode of the side structure.

Using the external dynamics theory of ship collisions and based on the principle of energy conservation, some energy conversion during the collision process was calculated, as shown in Table 5 and Table 6. When the impact occurred at the midship, the energy absorbed by the side structure was 16.2 J, accounting for 37.2% of the initial collision energy. In contrast, when the impact occurred at the fore part, the absorbed energy was 12.4 J, representing 24.3% of the initial collision energy. This indicates that the side structure absorbed more energy when the collision occurred at the midship. More recently, Pedersen and Zhang developed a theoretical model for calculating energy dissipation during ship collisions, which is represented by Equation (1). The corresponding energy dissipation values, $\boldsymbol{\Delta E_2}$, calculated using this method, are also listed in Table 6, where it can be seen that the results are relatively close.

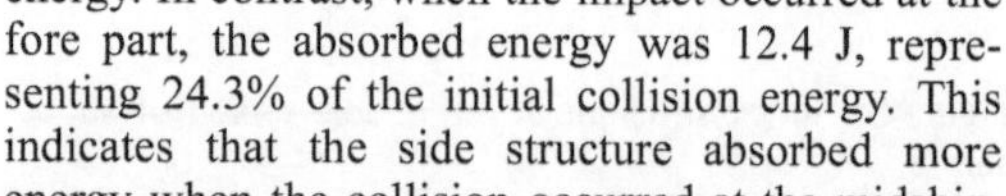

(a) Final damage mode for the case M1 (b) Final damage mode for the case M2

Figure 10. Final deformation and damage of side structure under different impact cases.

$$\Delta E_2 = \frac{1}{2}\frac{1}{D_\xi + \mu \cdot D_\eta}\left(1 - e^2\right)V_0^{\,2} \tag{1}$$

Table 5. Energy conversion during collision experiments.

	E_{Kt}					
Case	E_{AK} [J]	E_{BK} [J]	$\frac{E_{Asurge}}{E_{AK}}\times$ 100\%	$\frac{E_{Bsway}}{E_{BK}}\times$ 100\%	$\frac{E_{Broll}}{E_{BK}}\times$ 100\%	$\frac{E_{Byaw}}{E_{BK}}\times$ 100\%
M1	14.2	10.0	98.2%	72.9%	26.9%	0.02%
M2	17.1	16.0	98.5%	76.3%	6.9%	15.4%

Table 6. Energy dissipation during collision experiments.

Case	E_{K0} [J]	E_{Kt} [J]	ΔE_1 [J]	ΔE_2 [J]
M1	43.5	24.2	19.3	20.1
M2	51.1	33.1	18.0	21.6

where D_ξ and D_η are expressions derived from equation of motions; please refer to section 2.1 in Pedersen & Zhang (1998). V_0 is the initial relative velocity, and e ($0\leq e\leq 1$) is the coefficient of restitution.

4 CONCLUDING REMARKS

This study focuses on improving ship collision resistance and related issues by conducting model experiments in an experimental tank, with impact locations at the middle section and fore part. The experiments consider fluid effects, six-degree-of-freedom motion of the ship models, and structural damage simultaneously. Based on the present model experiment results, the following concluding remarks are drawn:

(1) After the collision, the striking ship exhibits a dominant surge motion, accounting for more than 98% of its residual kinetic energy in both test cases. In contrast, the motion of the struck ship is primarily characterized by sway and roll. In addition, for the bow impact, a significant yaw motion is observed, which carried up to 15% more kinetic energy than in the middle collision case.

(2) Hydrodynamic effects play a crucial role in modifying the collision response. Under equivalent impact velocity and angle, bow collisions result in up to 26% lower peak collision force but exhibit approximately 50 ms longer contact durations compared to midship collisions. This suggests that fluid resistance and local flexibility influence not only the force magnitude but also the energy dissipation timeline.

(3) Based on the measured 6-DOF motions, the energy dissipation rates were calculated using the classical principle of energy conservation. The results show that the energy dissipation rate is 37.2% for the fore-part collision and 27.3% for the midship collision. Furthermore, these results were compared with theoretical predictions using the method proposed by Pedersen and Zhang, and good agreement was observed. The close agreement validates the experimental setup and provides a valuable reference for future studies on ship collision energy absorption mechanisms.

ACKNOWLEDGEMENTS

This work is supported by the project of the Wuhan University of Technology start-up fund for Distinguished Professors (grant no. 471–40120163), and it also acknowledges the support of the project of grant no. 43220222.

REFERENCES

Calle M A G, Oshiro R E, Alves M. Ship collision and grounding: Scaled experiments and numerical analysis[J]. *International Journal of Impact Engineering*, 2017, 103: 195–210.

Ehlers S, Tabri K, Romanoff J. & Varsta P. Numerical and experimental investigation on the collision resistance of the X-core structure[J]. *Ships and offshore structures*, 2012, 7(1): 21–29.

EMSA. Annual Overview of Marine Casualties and Incidents 2023. Lisbon, *Portugal*: *European Maritime Safety Agency*; 2023.

Jones N. Some recent developments in the dynamic inelastic behavior of structures [J]. *Ships and Offshore Structures*, 2006, 1(1):37–44.

Kim S J, Taimuri G, Kujala P, Conti F, Le Sourne H, Pineau J.P., Looten, T., Bae, HS., Mujeeb-Ahmed, M.P., Vassalos, D., Kaydihan, L. & Hirdaris S. Comparison of numerical approaches for structural response analysis of passenger ships in collisions and groundings[J]. *Marine Structures*, 2022, 81: 103125.

Liu Z, Amdahl J. A new formulation of the impact mechanics of ship collisions and its application to a ship-iceberg collision[J]. *Marine Structures*, 2010, 23 (3): 360–384.

Liu K, Wang Z, Tang W, Zhang Y. & Wang G. Experimental and numerical analysis of laterally impacted stiffened plates considering the effect of strain rate[J]. *Ocean Engineering*, 2015, 99: 44–54.

Minorsky V U. An analysis of ship collisions with reference to protection of nuclear power plants[R]. *Sharp (George G.) Inc.*, New York, 1958.

Paik J K, Chung J Y, Choe I H, Thayamballi A.K. & Pedersen P.T. On rational design of double hull tanker structures against collision[J]. *In*: *The Society of Naval Architects and Marine Engineers*, 1999.Transactions SNAME. vol. 107. paper number 14.

Pedersen P.T., Zhang S. On impact mechanics in ship collisions[J]. Marine structures, 1998, 11(10): 429–449.

Ringsberg J W, Amdahl J, Chen B Q, Cho S R, Ehlers S, Hu Z, Kubiczek JM, Kõrgesaar M, Liu B, Marinatos JN, Niklas K, Parunov J, Quinton BWT, Rudan S, Samuelides M, Guedes Soares C, Tabri K, Villavicencio R, Yamada Y, Yu ZL & Zhang S. MARSTRUCT benchmark study on nonlinear FE simulation of an experiment of an indenter impact with a ship side-shell structure[J]. *Marine structures*, 2018, 59: 142–157.

Song M, Ma J, Huang Y. Fluid-structure interaction analysis of ship-ship collisions[J]. *Marine Structures*, 2017, 55:121–136.

Tabri K, Määttänen J, Ranta J. Model-scale experiments of symmetric ship collisions[J]. *Journal of Marine Science and Technology*, 2008, 13(1): 71–84.

Tabri K, Varsta P, Matusiak J. Numerical and experimental motion simulations of nonsymmetric ship collisions[J]. *Journal of marine science and technology*, 2010, 15: 87–101.

Wang G, Arita K, Liu D. Behavior of a double hull in a variety of stranding or collision scenarios[J]. *Marine Structures*, 2000, 13(3): 147–187.

Wang X, Li C B, Zhu L. Numerical Simulation of Ship Collision and Validations with Experimental Results[C]. *International Conference on Offshore Mechanics and Arctic Engineering*. American Society of Mechanical Engineers, 2021, 85123: V002T02A004.

Wevers L J, Vredeveldt A W. Full scale ship collision experiments[R]. *TNO-report 98-CMC-R1725*. The Netherlands; 1999.

Ye X, Fan W, Sha Y, Wu Q, & Ren Y. Fluid-structure interaction analysis of oblique ship-bridge collisions[J]. Engineering Structures, 2023, 274: 115129.

Yu Z, Amdahl J, Strheim M. A new approach for coupling external dynamics and internal mechanics in ship collisions[J]. *Marine structures*, 2015, 45:110–132.

Yu Z, Liu Z, Amdahl J. Discussion of assumptions behind the external dynamic models in ship collisions and groundings[J]. Ships and Offshore Structures, 2019, 14 (sup1): 45–62.

Zhang M, Liu J, Hu Z & Zhao Y. Experimental and numerical investigation of the responses of scaled tanker side double-hull structures laterally punched by conical and knife edge indenters[J]. *Marine Structures*, 2018, 61: 62–84.

Zhou H, Zhu L, Zhang L, Wang X, Zhao Y & Guo K. 2025. Oblique and normal ship collision model tests in a water tank. Proceedings of the ASME 2025 44th International Conference on Ocean, Offshore and Arctic Engineering. Structures, Safety, and Reliability. OMAE.

Zhu L, Faulkner D, Atkins A G. The impact of rectangular plates made from strain-rate sensitive materials[J]. *International Journal of Impact Engineering*, 1994, 15 (3):245–255.

Zhu L, James P, Zhang S. Statistics and damage assessment of ship grounding[J]. *Marine Structures*, 2002, 15 (4):515–530.

Zhu L, Zhou Z, Pedersen PT. Ship grounding model tests in a water tank: An experimental study[J]. *Marine Structures*, 2024, 93, 103529.

Zhu L, Wang X. Ship collision model tests in a water tank[J]. *Marine Structures*, 2025, 103: 103791.

Collision and grounding in arctic conditions

Trends in Collision and Grounding of Ships and Offshore Structures – Zhu & Guedes Soares (Eds)

Dynamic responses of orthogonal woven CFRP composite laminates subjected to ice impact

W. Cai, H.K. Chen & S.X. Li
State Key Laboratory of Advanced Technology for Materials Synthesis and Processing, Wuhan University of Technology, Wuhan, P.R. China

B. Yu, Z.H. Guo & L. Zhu
School of Naval Architecture, Ocean and Energy Power Engineering, Wuhan University of Technology, Wuhan, P.R. China

ABSTRACT: Laminated fibre-reinforced composite materials are widely applied in lightweight ships and ocean engineering fields, thanks to their excellent properties like high strength to weight and stiffness to weight ratios, reduced corrosion tendency and design flexibility. However, composite materials used in ship side structures are easy to experience obvious deformations and damage under ice impact loadings, resulting in negative effects on their work performance and structural safety. Therefore, the aim of this paper is to study the dynamic responses of composite laminates subjected to ice impacts based on the numerical methods. The impact forces and deformations of steel plates were investigated. The dynamic response mechanisms of CFRP composite laminates under ice impact are revealed in this paper, which can provide useful information for the evaluation of the structural safety of ice-classed composite ships.

1 INTRODUCTION

Arctic region has abundant oil and gas resources and acts as a lifeline for many communities with the increasing exploitation of the Arctic channel. More voyages across the Arctic waters will be possible in the future due to the global warming and declined extent of sea ice. However, the increased ship traffic and resources exploitation activities in these waters may increase the probability of impacts between ice floes/icebergs and ships and offshore structures, which necessitates the safety assessment of these events. Thus, the assessment of the dynamic structural behaviour of ship plates subjected to ice impacts is of paramount importance to improving the safety of ships and offshore structures operating in Arctic region.

Much research work has been conducted to investigate the dynamic responses of marine structures under ice collision by experimental and numerical methods. Some full-scaled or scaled ice-structure interaction tests have been reported, such as full-scaled collision tests between ice bergy bit and the CCGS Terry Fox icebreaker (Gagnon, 2008; Ritch et al., 2008), full-scaled ice crushing tests conducted in Aker Arctic test basin (Määttänen et al., 2021), full-scale ice load measurement tests of Xue Long icebreaker (Wu et al., 2021), the quasi-static laboratory model tests in which an ice cone was slowly pushed against a large scale steel stiffened plate (Manuel et al., 2013), the laboratory collision tests between a floating steel structure in a tank and freshwater ice blocks (Kim et al., 2017), the ice drop impact tests (Ince et al., 2017; Yu et al., 2023), the ship-ice glancing impact test in ice tank (Sun & Huang, 2020), the large double-pendulum ice impact tests (Gagnon et al., 2020), and the horizontal ice impact tests (Zhu et al., 2020; Bi et al., 2021; Cai et al., 2020, 2022). From these scaled and full-scaled tests, the dynamic responses of ice-structure interaction were obtained, such as permanent deformations of structures, impact forces, and ice loading pressures.

In recent years, numerical methods for predicting the plastic damage of ship structure under ice impact are widely used by the numerical method. The mechanical behaviour of ice during impact is very complicated, and a matured ice material model comprehensively describing the behaviours is still lacking due to the complexity of ice constitutive model. A few material models are available in the literature for ice modelling, such as the isotropic elastic failure constitutive model (Kim & Kedward, 2000; Zhu et al., 2016), crushable foam model (Gagnon, 2011; Kim et al., 2015), the elastic-plastic and pressure dependent constitutive model (Liu et al., 2011; Gao et al., 2015; Yu & Amdahl, 2021), the rate-sensitive constitutive model (Zhang et al., 2022; Ma et al., 2022), and the damage model (Polach & Ehlers, 2013; Cai et al., 2020). These numerical models of ice material are employed in the finite element simulations to model the ice failure in ship-ice collision.

DOI: 10.1201/9781003684404-17

From the above literature review, it can be found that current experimental and numerical simulation research on ice-structure collisions mainly focuses on metallic structures, while studies on the dynamic response of composite structures under ice impacts are severely lacking. It is imperative to conduct numerical research on the dynamic response of fiber-reinforced polymer composite structures under ice-impact loads to accurately predict structural failure modes, reveal the failure mechanisms of composite structures under extreme ice-impact loads, and provide theoretical foundations for the ice-impact resistant design of composite structures of polar ships.

2 FINITE ELEMENT MODEL

Numerical simulations for ice-composite laminates impacts are performed by using the finite element software ANSYS/LS-DYNA, which is appropriate for non-linear explicit dynamic simulations with structural failure analysis. In the finite element calculation model, the length L_1 of the composite laminate is 380mm, the width B_1 is 380mm, and the height H_1 is 2.4mm, as shown in Figure 1. The layers of composite laminates are labelled as cfrp1, cfrp2, cfrp3, cfrp4, cfrp5, and cfrp6 from top to bottom, with cohesive elements COH1 inserted between cfrp1 and cfrp2, COH2 inserted between cfrp3 and cfrp4, and COH3 inserted between cfrp5 and cfrp6, as illustrated in Figure 2. The composite laminates model is modelled by four-node shell elements with five integration points throughout the thickness, defining the Hughes-Liu shell element formulation. The ice model is modelled by solid elements. In the mesh division of this simulation model, the finite element mesh size for composite laminates model is 10×10×10 mm, and the mesh size of ice model is 25×25×25 mm.

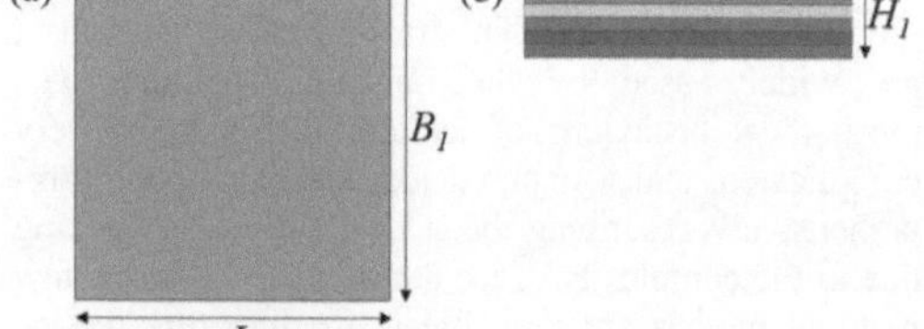

Figure 1. Geometric dimension diagram of composite laminates.

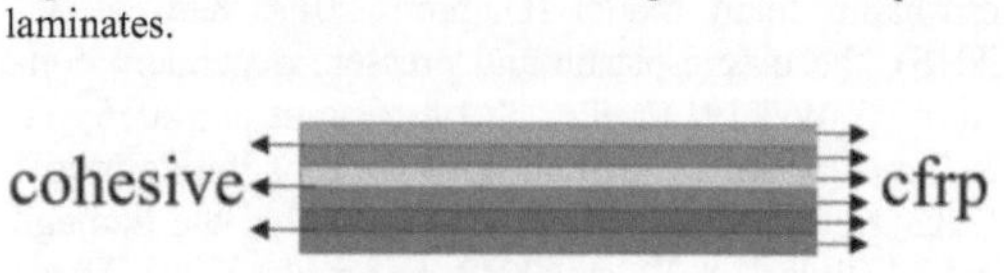

Figure 2. Schematic diagram of cohesive elements between the layers of composite laminates.

In the finite element numerical simulation process, the composite laminates model falls at a certain speed and collides with the ice model. The boundary conditions for composite laminates are Ux=Uy=0, Uz=1, Urx=Ury=Urz=0, and the boundary conditions of the ice body are Ux=Uy=Uz=0; Urx=Ury=Urz=0, as shown in Figure 3. The option *CONTACT_ERODING_SURFACE_TO_SURFACE (*ESTS) is used to define contact between plate and the newly emerged surface of ice wedge if the ice element is eroded, and the coefficient of friction is set to be 0.2. This command enables the calculation of possible contact between ice elements.

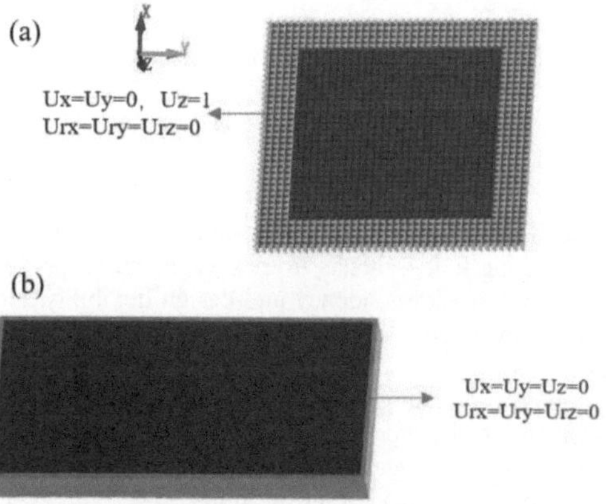

Figure 3. Boundary conditions of finite element models: (a) composite laminates model; (b) ice model.

As far as current research is concerned, due to the complexity of ice material properties, its numerical material constitutive model and failure criterion are still under continuous exploration. The elastic-plastic failure material model (*MAT_ISOTROPIC ELASTIC_FAILURE) is used in simulating the ice behaviors, which is widely used in ship-ice collision analysis. It is a non-iterative plasticity with simple plastic strain failure model; When the pressure reaches the failure pressure or when the effective plastic strain reaches the failure strain ($\varepsilon_{eq} \geq \varepsilon_f$), the element loses its ability to carry tension and will be deleted. The input parameters of ice material model are given in Table 1.

Table 1. Input parameters of ice material model (*MAT_013).

Material properties	Unit	Value
Mass density	kg/m^3	862
Shear modulus	GPa	1.628
Yield stress	MPa	0.16
Plastic hardening modulus	GPa	4.26
Bulk modulus	GPa	5.26
Failure pressure	MPa	-0.15
Plastic failure strain	-	0.0035

The material model (MAT_COMPOSITE_DAMAGE) is selected from the library of LS-DYNA to describe and simulate the material properties and failure behavior of composite laminates. The Hashin damage criterion in this composite material constitutive model can comprehensively consider the tensile and compressive strength, as well as shear strength of composite materials in different directions, allowing for in-depth analysis of the damage characteristics of CFRP laminates under impact loads, to predict the damage behavior of materials and understand the failure mechanism of materials. Therefore, the Hashin damage criterion is adopted as the basis for damage judgment of CFRP laminates, and its expression is shown in Table 2.

Table 2. Failure criteria for composite material models.

Failure mode	Failure criteria
Longitudinal tensile failure	$\left(\frac{\sigma_{11}}{X_T}\right)^2 + \left(\frac{\tau_{12}}{S_c}\right)^2 \geq 1$
Longitudinal compressive failure	$\left(\frac{\sigma_{11}}{X_C}\right)^2 \geq 1$
Latitudinal tensile failure	$\left(\frac{\sigma_{22}}{Y_T}\right)^2 + \left(\frac{\tau_{12}}{S_C}\right)^2 \geq 1$
Latitudinal compressive failure	$\left(\frac{\sigma_{22}}{2S_C}\right)^2 + \left[\left(\frac{Y_C}{2S_C}\right)^2 - 1\right]\frac{\sigma_{22}}{Y_C} + \left(\frac{\tau_{12}}{S_C}\right)^2 \geq 1$

In above formula, σ_{11}, σ_{22} and τ_{12} and respectively represent the longitudinal normal stress, transverse normal stress, and shear stress of the laminates; X_T, X_C, Y_T and Y_C are respectively longitudinal tensile strength, longitudinal compressive strength, transverse tensile strength, and transverse compressive strength; S_C is the in-plane shear strength of laminates.

The input material parameters for the constitutive model of composite materials are the mechanical performance test data of composite laminates, as shown in Table 3.

During the ice impact process, composite laminates will experience delamination. Zero thickness cohesive elements are constructed between layers to simulate the bonding behavior and delamination process between layers. Cohesive elements can define the evolution of damage and crack propagation to study the failure phenomenon between layers. The material constitutive model used for laminates is the zero-thickness cohesive model (MAT_COHESIEVE _MIXED_MODE). The initial displacement of mixed mode damage is:

$$\delta^0 = \delta_I^0 \delta_{II}^0 \sqrt{\frac{1+\beta^2}{\left(\delta_{II}^0\right)^2 + \left(\beta\delta_I^0\right)^2}} \tag{1}$$

In the equation, $\delta_I^0 = T/E_N$ and $\delta_{II}^0 = S/E_T$ are the initial displacement of single mode damage, and $\beta = \delta_{II}/\delta_1$ is Mixed mode damage initiation displacement.

The total failure displacement in the final mixed mode δ^F is

when $I_{XUM} > 0$,

$$\delta^F = \frac{2(1+\beta)^2\left[\left(\frac{E_N}{G_{IC}}\right)^{I_{XUM}} + \left(\frac{\beta^2 \bullet E_T}{G_{IIC}}\right)^{I_{XUM}}\right]^{-\frac{1}{I_{XUM}}}}{\delta^0} \tag{2}$$

when $I_{XUM} < 0$,

$$\delta^F = \frac{2\left[G_{IC} + (G_{IIC} - G_{IC})\left(\frac{\beta^2 \bullet E_T}{E_N + \beta^2 \bullet E_T}\right)\right]}{\delta^0\left(\frac{1}{1+\beta^2}E_N^\gamma + \frac{\beta^2}{1+\beta^2}E_T^\gamma\right)^{\frac{1}{\gamma}}} \tag{3}$$

In these equations, E_N and E_T are the normal and tangential stiffness, respectively; G_{IC} and G_{IIC} are type I and type II fracture energies, respectively. The input parameters of cohesive element material model are given in Table 4.

Table 3. Input parameters for constitutive model of composite materials.

Material properties	Value
$\rho/(\mathrm{t \cdot mm^{-3}})$	1.52×10^{-9}
E_a/MPa	4508
E_b/MPa	4508
E_c/MPa	1100
G_{AB}/MPa	3465.6
G_{BC}/MPa	3100
G_{CA}/MPa	3100
X_C/MPa	486.4
X_T/MPa	753.3
Y_C/MPa	486.4
Y_T/MPa	753.3
S_C/MPa	486.4
ν	0.12

Note: E_a, E_b and E_c are elastic moduli in different directions; G_{AB}, G_{BC} and G_{CA} are shear moduli in different directions respectively; ν is Poisson's ratio.

Table 4. Input parameters of cohesive element material model.

Material properties	Value
$\rho/(\mathrm{t \cdot mm^{-3}})$	1.5×10^{-9}
$E_N/(N \cdot \mathrm{mm^{-3}})$	5000
$E_T/(\mathrm{N \cdot mm^{-3}})$	5000
$G_{IC}/(\mathrm{N \cdot mm^{-1}})$	0.134
$G_{IIC}/(\mathrm{N \cdot mm^{-1}})$	0.5555
T/MPa	30
S/MPa	30

Note: T is the normal peak traction force; S is the tangential peak traction force.

There group of numerical simulations of composite laminates under ice impact were conducted, and the impact parameters, including impact mass, impact velocity and impact energy, are summarized in Table 5. The impact velocity applied to composite laminated model, and the added mass also applied to the boundary of composite laminated model in numerical simulations, as shown in Figure 4.

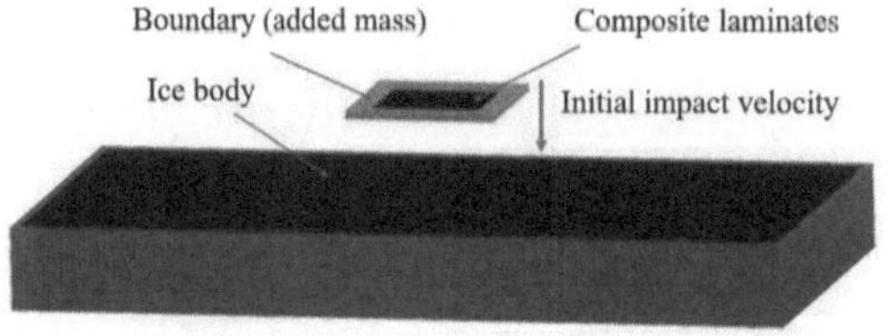

Figure 4. Impact model of composite laminates and ice body.

Table 5. Numerical calculation conditions for collision between composite laminates and ice body.

Number	Impact velocity (m/s)	Impact mass (kg)	Impact energy (J)
CLV-1.5	1.5	500	499.85
CLV-2	2	500	1000
CLV-4	4	500	4000

3 RESULT ANALYSIS

Figure 5 shows that during the falling process of the composite laminate hitting the ice body, the ice body is penetrated by the composite laminate structure, and many cracks appear on the ice body. The ice body exhibits a mixed failure mode of compression and bending.

Figure 6 shows the force time history curves of composite laminates colliding with ice layers under three numerical conditions. According to Figure 6, the maximum collision force between the composite laminates and the ice body under the numerical condition CLV-1.5 is 195kN, the maximum collision force between the composite laminates and the ice body under the numerical condition CLV-2 is 250kN, and the maximum collision force between the composite laminates and the ice body under the numerical condition CLV-4 is 473kN. The higher the collision speed, the greater the maximum collision force between the ice body and the composite laminates. During the collision process, the maximum collision force occurs when the composite material structure first comes into contact with the ice layer. As larger cracks appear in the ice, the collision force rapidly decreases and then enters the platform stage.

Figures 7-9 show the Mise stress distribution cloud maps for each layer of composite laminates under three impact conditions. Figures 10-11 show

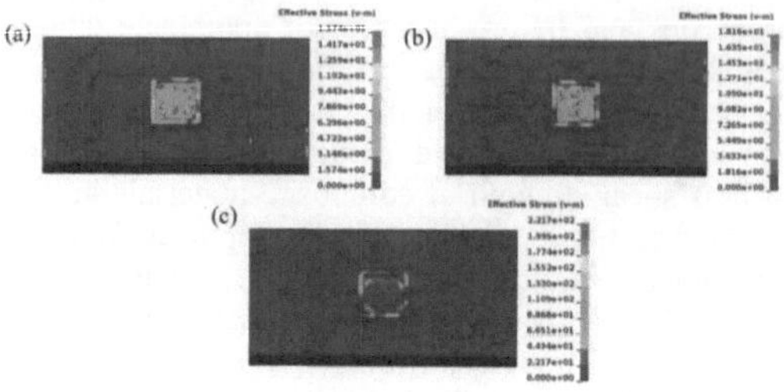

Figure 5. Failure modes of ice under three collision conditions: (a) CLV-1.5; (b) CLV-2; (c) CLV-4.

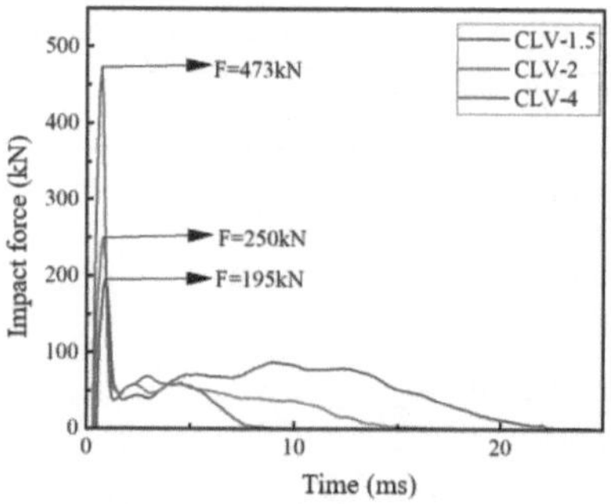

Figure 6. Force time history curves of composite laminates colliding with ice layers under three numerical conditions.

the Mise stress distribution cloud maps of interlayer cohesive elements in composite laminates under three impact conditions. At low collision speeds of 1.5m/s and 2m/s, there was no damage to the interlayer cohesive elements of the composite material. However, at a collision speed of 4m/s, significant interlayer delamination damage occurred in the composite laminates, and cohesive elements near the boundary were deleted. From Figures 9 and 12, it is seen that when there is significant damage to the cohesive elements (cohesive3) between the layers of the composite laminates, the composite material layers cfrp5 and cfrp6 experience significantly higher Mise stress. It can be inferred that when the collision speed is high, interlayer damage will occur in composite materials under ice impact, which will sharply affect the ice impact resistance and load-bearing capacity of composite laminates.

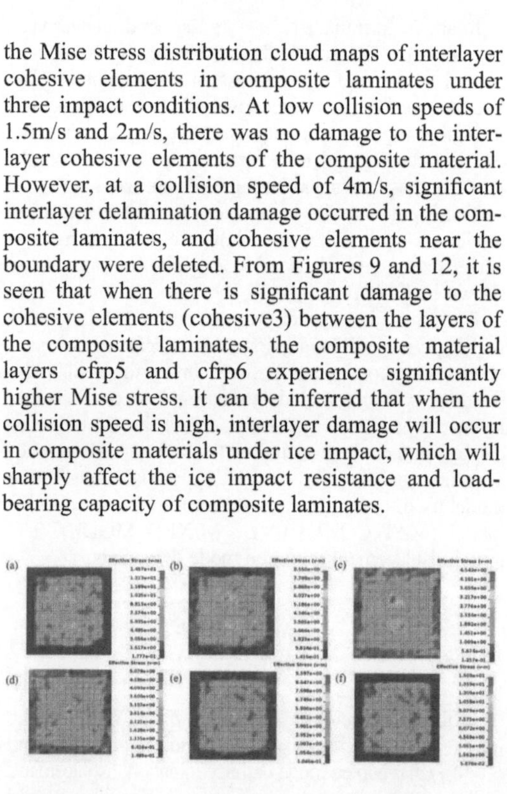

Figure 7. Cloud map of Von Mises stress distribution in each layer of composite laminates under the condition CLV-1.5: (a) cfrp1; (b) cfrp2; (c) cfrp3; (d) cfrp4; (e) cfrp5; (f) cfrp6.

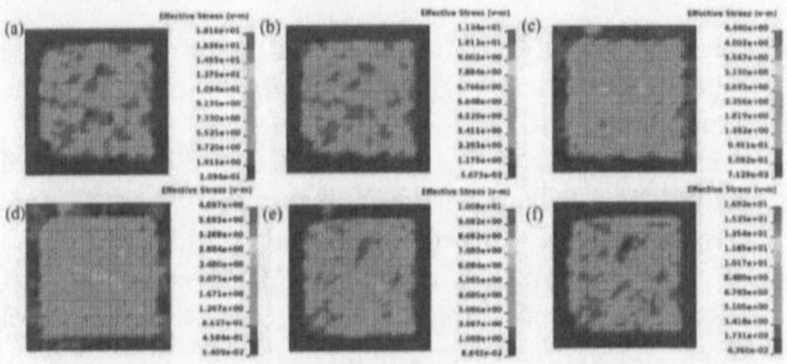

Figure 8. Cloud map of Von Mises stress distribution in each layer of composite laminates under the condition CLV-2: (a) cfrp1; (b) cfrp2; (c) cfrp3; (d) cfrp4; (e) cfrp5; (f) cfrp6.

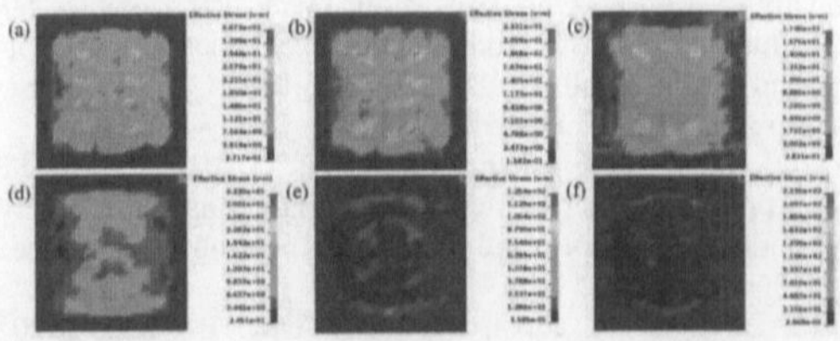

Figure 9. Cloud map of Von Mises stress distribution in each layer of composite laminates under the condition CLV-4: (a) cfrp1; (b) cfrp2; (c) cfrp3; (d) cfrp4; (e) cfrp5; (f) cfrp6.

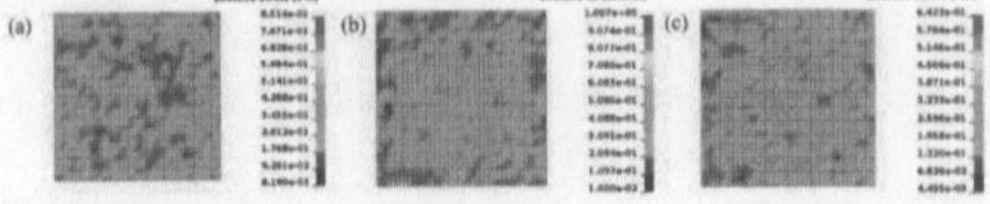

Figure 10. Stress distribution cloud map of interlayer cohesive elements in composite laminates under the condition CLV-1.5: (a) cohesive1; (b) cohesive2; (c) cohesive3.

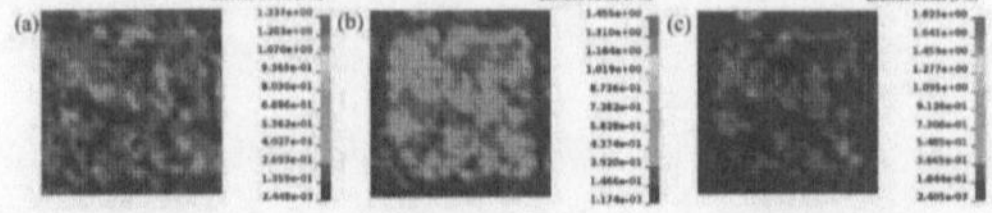

Figure 11. Stress distribution cloud map of interlayer cohesive elements in composite laminates under the condition CLV-2: (a) cohesive1; (b) cohesive2; (c) cohesive3.

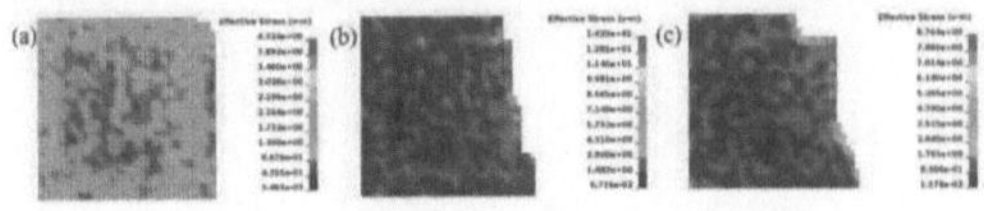

Figure 12. Stress distribution cloud map of interlayer cohesive elements in composite laminates under the condition CLV-4: (a) cohesive1; (b) cohesive2; (c) cohesive3.

4 CONCLUSION REMARKS

This paper uses numerical simulation methods to study the dynamic response of composite laminates in collision with ice bodies, and analyzes the impact force time history curves, ice body failure modes, and stress distribution characteristics and failure modes of composite materials under different collision energies. The research results indicate that during the impact process between composite laminates and ice body, the delamination damage is highly likely to occur between the laminates, which will sharply affect the ice impact resistance and load-bearing capacity of composite laminates. At present, the structural properties of composite materials are closely related to the material system and manufacturing methods. When obtaining the failure parameters of composite materials, it is necessary to maintain the stability of the material system and manufacturing method to reduce the variability of the mechanical behavior of composite materials. Moreover, it is found that how to achieve interlayer toughening of composite materials is crucial for the service safety of composite structures under ice impact.

REFERENCES

Bi, L., Zhu, L. & Cai, W., 2021. Effect of ice crushing characteristics on plastic deformation of ship plate under ice impact// *31th International Ocean and Polar Engineering Conference. International Society of Offshore and Polar Engineers (ISOPE).*

Cai, W., Zhu, L., Yu, T.X., et al., 2020. Numerical simulations for plates under ice impact based on a concrete constitutive ice model. International journal of impact engineering, 143:103594.

Cai, W., Zhu, L., Qian, X.D., 2022. Dynamic responses of steel plates under repeated ice impacts. International Journal of Impact Engineering, 162:104129.

Gagnon, R.E., 2008. Analysis of data from bergy bit impacts using a novel hull-mounted external Impact Panel. *Cold Regions Science and Technology*, 52,50–66.

Gagnon, R.E., 2011. A numerical model of ice crushing using a foam analogue. *Cold Regions Science and Technology*, 65(3): 335–350.

Gagnon, R E, Andrade, S.L., Quinton, B., Daley, C., Colbourne, B., 2020. Pressure distribution data from large double-pendulum ice impact tests. *Cold Regions Science and Technology*, 103033.

Gao, Y., Hu, Z., Ringsberg, J.W., Wang, J., 2015. An elastic–plastic ice material model for ship-iceberg collision simulations. *Ocean Engineering*, 102: 27–39.

Ince, S.T., Kumar, A., Park, D.K., Paik, J.K. 2017. An advanced technology for structural crashworthiness analysis of a ship colliding with an ice-ridge: Numerical modelling and experiments. *International Journal of Impact Engineering*, 110: 112–122.

Kim, E., Storheim, M., Amdahl, J., Løset, S., von Bock und Polach, R.U.F., 2017. Laboratory experiments on shared-energy collisions between freshwater ice blocks and a floating steel structure. *Ships and Offshore Structures*, 12(4): 530–544.

Kim, H., Kedward, K.T., 2000. Modeling hail ice impacts and predicting impact damage initiation in composite structures. Am Inst Aeronautics Astronautics, 38(7):1278–88.

Kim, H., Daley, C., & Colbourne, B., 2015. A numerical model for ice crushing on concave surfaces. *Ocean Engineering*, 106: 289–297.

Liu, Z., Amdahl, J. & Loset, S., 2011. Plasticity based material modelling of ice and its application to ship-iceberg impacts. Cold Regions Science and Technology, 65, 326–334.

Määttänen, M., Marjavaara, P., Saarinen, S., Laakso, M. 2011. Ice crushing tests with variable structural flexibility. *Cold regions science and technology*, 67(3): 120–128.

Ma, D., Li, X., Manes, A., Li, Y.L. 2022. Numerical modelling of ice: Mechanical behaviour of ice under high strain rates. *International journal of impact engineering*, 172:104375.

Manuel, M., Gudimetla, P.S.R., Daley, C., Colbourne, B. 2013. Controlled plastic deformation of a grillage using artificial freshwater ice at large scale, *Proceedings of the 22nd International Conference on Port and Ocean Engineering under Arctic Conditions (POAC 2013)*; June 9-13, Espoo (Finland).

Polach, R. B. & Ehlers, S., 2013. Model scale ice—Part B: Numerical model. *Cold regions science and technology*, 94: 53–60.

Ritch, R., Frederking, R., Johnston, M., Browne, R., Ralph, F. 2008. Local ice pressures measured on a strain gauge panel during the CCGS Terry Fox bergy bit impact study. *Cold Regions Science and Technology*, 52(1):29–49.

Sun, J. & Huang, Y., 2020. Investigations on the ship-ice impact: Part 1. Experimental methodologies. Marine Structures, 72: 10277.

Wu, G., Kong, S., Tang, W., Lei, R.B., Ji, S.Y. 2021. Statistical analysis of ice loads on ship hull measured during Arctic navigations, *Ocean Engineering*, Volume 223,108642.

Yu, T., Liu, K., Wang, G., Liu, J.J., Wang, Z.L. 2023. A tri-axial ice model for simulating ice-stiffened panel impact: Experiments and numerical modeling. *Marine Structures*, 88: 103358.

Yu, Z., Amdahl, J., 2021. A numerical solver for coupled dynamic simulation of glacial ice impacts considering hydrodynamic-ice-structure interaction, *Ocean Engineering*, 226: 108827.

Zhang, J., Liu, Z., Ong, M C., Tang, W.Y. 2022. Numerical simulations of the sliding impact between an ice floe and a ship hull structure in ABAQUS. *Engineering Structures*, 273: 115057.

Zhu, L., Qiu, X.M., Chen, M.S., Yu, T.X. 2016. Simplified Ship-Ice Collision Numerical Simulations, 26th International Offshore and Polar Engineering Conference, ISOPE 2016.

Zhu, L., Cai, W., Chen, M., Tian, Y.K., Bi, L.Z. 2020. Experimental and numerical analysis of elastic-plastic response of ship plates under ice floe impacts. *Ocean Engineering*, 218:106174.

Trends in Collision and Grounding of Ships and Offshore Structures – Zhu & Guedes Soares (Eds)

Study on the damage limit strength of stiffened plate depression caused by lower ice under crack influence

J.Y. Chen, Z.K. He & G. Liu
School of Naval Architecture & Ocean Engineering, Dalian University of Technology, Dalian, P.R. China

ABSTRACT: Global climate change has caused sea ice melt, and the probability of ship-ice collision in the polar region has greatly increased. In order to study the structural safety after ship-ice collision, this paper uses the nonlinear finite element method to select a polar ship stiffened plate structure. Considering the influence of initial cracks, LS-DYNA is used to simulate the collision between ice body and stiffened plate structure, and the ultimate strength of stiffened plate after collision damage is studied. The results show that the impact velocity is negatively correlated with the ultimate strength of the structure, and the buckling mode of the stiffened plate presents different forms with the change of the impact velocity. The location and number of the initial crack have different effects on the ultimate strength of the stiffened plate. The research in this paper can provide a reference for the structural safety after collisions.

1 INTRODUCTION

With the development of Arctic shipping routes and energy sources, polar ships will inevitably collide or squeeze with sea ice during polar navigation. Long-term navigation in harsh environments can easily cause cracks to germinate at the weld, and the crack-containing stiffened plate depression caused during the collision process will gradually weaken the ultimate strength of the ship. Since the structure of ships traveling in polar regions is difficult to be effectively repaired and strengthened in a short period of time after damage occurs, in this case, it is necessary to evaluate the remaining ultimate strength of the structure under the damaged state of the ship to ensure the safety of navigation.

For the problem of polar ice impact, Cai et al. (2021) studied the structural response and damage law of ice finite element material numerical model (Cai et al. 2021), ice shape (Ni et al., 2024; Müller et al., 2024) and impact velocity (Yan et al., 2017) after collision between hull structure and sea ice. Rahbar-Ranji et al. (2015) and Bahmani et al. (2024) used a series of nonlinear finite element methods to analyze the ultimate strength of double-span stiffened plates with cracks, and discussed the influence of transverse cracks, crack length and crack position on the ultimate strength of stiffened plates. In the finite element analysis, Xu et al. (2017) studied the influence of the ultimate strength of the double-span stiffened plate under various loads, and discussed the influence of the lateral load of the compressor (Xu et al., 2024), tensile and bending loads (Liu et al., 2023), uniaxial cyclic load (Hu et al., 2022), biaxial cyclic load (Li et al., 2025) on the ultimate strength of the stiffened plate with cracks. For stiffened plates with dents, many scholars have studied them and have carried out a lot of work. Ao et al. (2020) studied the residual ultimate strength of two-span stiffened plates with impact crater damage under compressive load by using commercial software ABAQUS, and discussed the influence of important factors such as depression depth, depression size and depression angle on the ultimate state of stiffened plates in detail. Bian et al. (2018) used the finite element simulation calculation to obtain the influence of the polar collision dent damage on the ultimate strength of the hull two-span stiffened plate. According to the material mechanical properties test of EH36 steel at low temperature, Zhang et al. (2023) used the reduction factor evaluation method and based on the empirical formula of the ultimate strength of the double-span stiffened plate under axial compression, proposed to use the flexibility of the stiffened plate and the flexibility of the stiffened plate shell as the change parameters, and used the least square method to fit the reduction factor to obtain the empirical formula of the residual ultimate strength of the double-span stiffened plate under low temperature with collision damage. Zhang et al. (2022) carried out experimental procedures and nonlinear finite element analysis to examine the ultimate strength characteristics of double-span stiffened plates with cracks and corrosion pits under uniaxial longitudinal compression loads. The research of the above scholars is based on the prefabricated depression of the stiffened plate, that is, the depressed stiffened plate is not generated by the collision between the actual ice body and the stiffened plate, which is inconsistent with the actual situation. Moreover, the stiffened plates studied are mostly double-span and single-span stiffened plate structures. The stiffened plate

DOI: 10.1201/9781003684404-18

has no ribs and other structures, which is very different from the actual structure in the hull, which may result in the lack of engineering application.

Based on the above literature review, it can be seen that a large number of scholars have used experimental and numerical methods to study the ultimate strength of cracked stiffened panels and the ultimate strength of stiffened panels after dent damage. However, because the above scholars have adopted prefabricated depression and simplified stiffened panels. The structure is a double-span single-span form, and the ultimate strength of stiffened panels with cracks in real ships after ice collision has not been studied in detail. In the actual navigation process of polar ships, the ultimate strength of cracked stiffened plates caused by ice impact will seriously affect the safe navigation of ships. The purpose of this paper is to study the ultimate strength of stiffened plates after considering ice impact on cracked stiffened plates by numerical method, which can provide reference for the safety design and evaluation of polar ship structures.

2 FINITE ELEMENT MODEL OF STIFFENED PLATE DENT DAMAGE

In the process of related research and analysis, two finite element models were used, which were stiffened plate model before impact and stiffened plate model after impact. Because the two models show a clear order relationship, that is, in the actual collision process, there is a state before the impact, and then the state after the impact is generated, so when constructing the finite element model, it is also necessary to follow this order.

2.1 *Finite element model before impact*

The nonlinear finite element analysis software LS-DYNA is used to simulate the test process of the collision between stiffened plate and ice. In this paper, the actual structural model of a polar ship bow is used to study the deformation of the hull structure during impact. The stiffened plate adopts four-node Belytschko-Tsay shell element, and five integral points are set in the thickness direction. The size of the model is 5600mm × 4500mm, and the thickness of the outer plate is 16mm. The stiffener is T-shaped, the width of the wing plate is 120 mm and the thickness is 10 mm, the height of the web is 295 mm and the thickness is 10 mm; the rib adopts an L-shaped wing plate with a length of 45 mm and a thickness of 7 mm, a web height of 155 mm and a thickness of 7 mm, a reinforcing rib span of 1500 mm, and a rib span of 800 mm. Taking into account the actual working conditions, the displacement and rotation of the edge of the stiffened plate are set to 0, and its movement is constrained to ensure the accuracy of the final calculation results.

The sea ice model selects the wedge ice that is common in the polar case. The ice body is modeled using an eight-node Solid164 element. The upper half of the model is a cuboid with a size of 3000mm × 700mm × 1500mm, and the lower half of the cone is oblique. The length is 2121mm, the long side is 3000mm, and the thickness is 700mm. The displacement and rotation of the sea ice in all directions are constrained, and only the displacement in the z direction is allowed.

The contact option (*CONTACT_AUTOMATIC _SINGLE_SURFACE) is used to define the contact between the stiffened panel and the stiffened panel, the contact option (*CONTACT_ERODING_ SINGLE_SURFACE) is used to define the contact of the ice body itself, and the contact option (* CONTACT_ERODING_SURFACE_TO_SURFACE) is used to define the contact between the ice and the plate, and the initial velocity of the ice is given. In this paper, the calculation of the collision is considered to be finished after the plastic deformation of the stiffened plate is stabilized. After the impact calculation is completed, the permanent plastic deformation, residual stress and plastic strain of the stiffened plate are stored in the output result file, which can be used as the initial condition for the next calculation.

Considering the influence of mesh size on the calculation accuracy, the convergence analysis of the impact area is carried out. Different mesh sizes are selected in the impact center area of the stiffened plate and the collision area at the lower end of the wedge ice body for convergence analysis. The impact center area of the stiffened plate is selected 100mm, 50mm, 25mm and 12.5mm for calculation, and the mesh at the lower end of the wedge ice body is selected 100mm, 80mm, 60mm, 40mm and 20mm for convergence analysis. The maximum depth and depression area generated by the impact are compared. By analyzing the calculation results, it is found that when the grid size of the stiffened plate is 25 mm, reducing the grid size can no longer cause a large change in the calculation results. Therefore, the grid size of the impact area of the stiffened plate structure is determined to be 25 mm, and the other areas of the stiffened plate are 100 mm. The grid size at the lower end of the wedge ice body is determined to be 20 mm, and the grid size at the upper end of the wedge ice body is 60 mm. In the first step of calculation, the stiffened plate is fixedly constrained around, and the motion of the constrained sea ice is only allowed to move in the Z axis. The final finite element model is as follows Figure 1.

In order to prove the reliability of the calculation, Xu et al., (2024) studied the large deformation of single stiffened plate. The results show that the finite element results are highly consistent with the experimental results, indicating that the calculation results are highly consistent with the deformation under the actual impact.

In this paper, the influence of cracks on stiffened plates after ice collision is studied. The method of

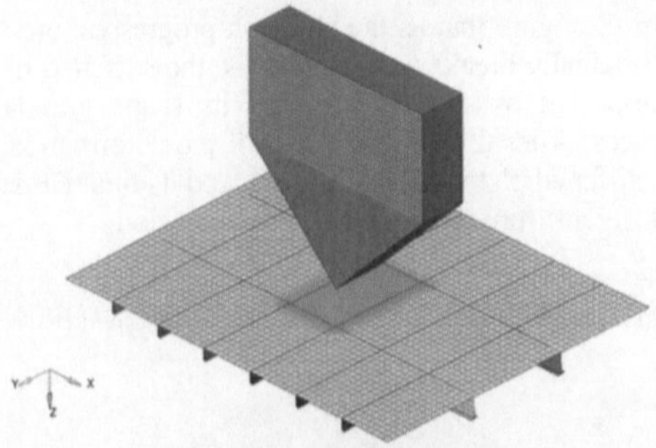

Figure 1. Finite element model before impact.

separating nodes is used to simulate cracks. For consistency, all cracks are 50mm in size. Therefore, a variety of crack forms are selected for calculation, and the location of cracks can be referred to the following Figure 2.

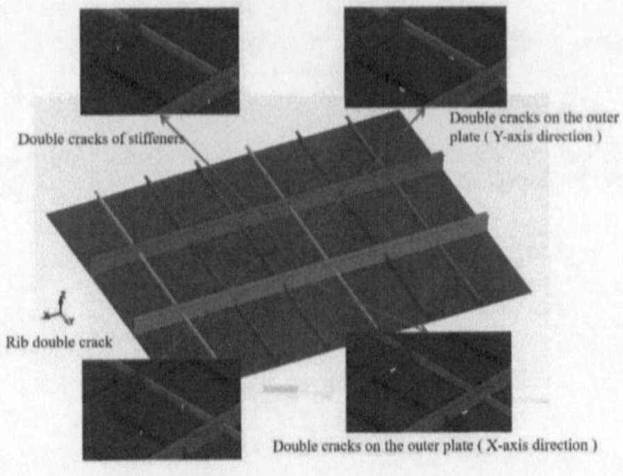

Figure 2. Crack location diagram.

2.2 *Finite element model of dented damage*

The simulation in this paper is coherent, and the damage model is the stiffened plate model after impact. The command in LSDYNA is used to import the deformation data of the stiffened plate after impact and load it on the undeformed stiffened plate as the initial model to study the ultimate strength under the dent damage. The material parameters of the stiffened plate remain unchanged, and the boundary conditions are selected according to the actual situation. The X direction is the loading compression direction. The boundary conditions of the model are set as follows Figure 3: for boundary A-C and B-D, Rx = Rz = 0, Uy = 0; for boundary A-B, Ry = Rz = 0, Uz = 0; for boundary C-D, Ry = Rz = 0, Ux = Uz = 0.

The contact option (*CONTACT _ AUTOMATIC _ SINGLE_SURFACE) is used to define the contact between the stiffened panel and the stiffened panel. The fixed displacement is applied to the loading end of the structure by (BOUNDARY _ PRESCRIBED _ MOTION _ SET) to simulate the change of the structure under the compression load. The command in LSDYNA is used to select the impact center area as the cross section of the load displacement curve.

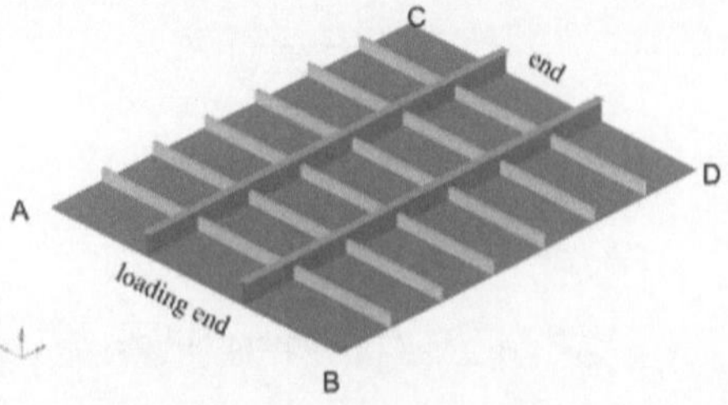

Figure 3. The boundary condition diagram of stiffened plate model.

Finally, the ultimate strength of the plate frame is obtained by the load displacement curve.

2.3 *Material model of stiffened plate and ice*

The material model used in the numerical simulation of the stiffened plate is * MAT _ 24 (MAT _ PIECEWISE _ LINEAR _ PLASTICITY) in LS-DYNA, which can take the real stress-strain curve as the input of the material. * MAT _ 24 can also consider the strain rate effect of the material under impact through the Cowper-Symonds constitutive model. The model expression is as follows (Cowper et al., 1()957):

$$\frac{\sigma_d}{\sigma_s} = \left(1 + \frac{\dot{\varepsilon}}{D}\right)^{\frac{1}{q}} \tag{1}$$

where σ_d is the dynamic flow stress corresponding to the plastic strain rate, σ_s is the related static plastic flow stress, ε is the material strain rate, and D and q are the material strain rate sensitivity coefficients. The Cowper-Symonds constants D = 40.4 and q = 5 (Kulkarni et al.2020) of low carbon steel were entered into the finite element program.

The parameters in the finite element material of the stiffened plate are shown in the Table 1 (Xu et al., 2024).

Table 1. Stiffened plate Mechanical Properties.

Stiffened plate Mechanical Properties	Input values
Density [kg/m3]	7850
Young's modulus [GPa]	206
Poisson's ratio	0.3
Yield strength [MPa]	299
Tensile strength [MPa]	484

The *MAT_13 (MAT_ISOTROPIC_ELASTIC _ FAILURE) in LS-DYNA is used to simulate the ice material. In the simulation, the constitutive model is simplified to a bilinear form, and the critical failure strain and the cut-off pressure failure criterion are combined to simulate the failure of ice. For the compression failure of ice, if the equivalent plastic strain is greater than the critical failure plastic strain, the element is deleted. For the tensile failure of ice, once the hydrostatic pressure exceeds the cut-off pressure, the separation between the units fails. The following expressions is shown (MANUAL. 2007):

$$\varepsilon_{ep}^{p} = \sqrt{\frac{2}{3}\varepsilon_{ij}^{p} : \varepsilon_{ij}^{p}} \tag{2}$$

$$P = \frac{\sigma_1 + \sigma_2 + \sigma_3}{3} \tag{3}$$

$\varepsilon_{ep}^{p} > \varepsilon_{f}^{p} or P < P_C$, element failure

ε_{ep}^{p} is the equivalent plastic strain of the element, ε_{f}^{p} is the critical failure strain, P is the hydrostatic pressure, and P_C is the cutoff pressure. The model can consider the plastic strengthening stage of ice materials and simulate the propagation of macroscopic cracks. The input parameters of mechanical properties of ice are shown in Table 2 (Yan et al., 2017).

Table 2. Ice Mechanical Properties.

Ice Mechanical Properties	Input values
Density [kg/m^3]	900
Shear modulus [GPa]	2.2
Bulk modulus [GPa]	5.26
Yield stress [MPa]	2.12
Plastic hardening modulus [MPa]	4260
Failure strain	0.02
Failure pressure [MPa]	-4

3 DISCUSS

3.1 *Effect of impact velocity on damage of stiffened plate with cracks*

In this section, the LSDYNA software is used to carry out accurate calculation and analysis of the selected model. Specifically, the structural damage characteristics of stiffened plates and stiffened plates with crack defects under various impact velocities are mainly explored. Through the accurate comparison of the maximum depth of the depression presented by the structure after the collision and the increment of the structural area change before and after the collision, the quantitative evaluation and systematic analysis of the damage degree are realized.

The Figure 4 shows the impact process of the finite element numerical calculation results. It can be seen

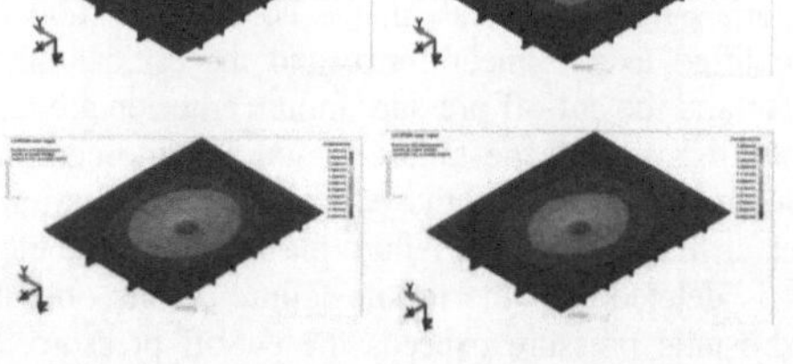

Figure 4. The deformation of stiffened plate in the impact process.

from the figure that as the collision progresses, the sea ice gradually breaks. In this process, the stiffened plate is impacted by the sea ice, and its shape gradually changes. After the collision, part of the deformation of the stiffened plate will be restored, and its final damage is different from the shape after the collision.

a) Deformation cloud diagram of complete stiffened plate un-

der 2-knot speed impact

b) Deformation cloud diagram of complete stiffened plate under 6-knot speed impact

c) Deformation cloud diagram of complete stiffened plate under 10-knot speed impact

Figure 5. Comparison of actual and prefabricated depressions.

The reinforced plate model selected in this paper is established according to the size of the actual ship reinforced plate, so it is believed that the calculation results in this paper can represent the actual ship reinforced plate used in engineering application. The simplified stiffened plate is the form of single span and double span stiffened plate commonly used in other papers. The Figure 5 is the damage cloud diagram of the actual hull stiffened plate after collision. It can be seen from the cloud diagram that there are some differences in the damage forms of stiffened plates after impact with different speeds of ice. It can also be seen from the results that there is a difference between the damage of the stiffened plate after the actual collision and the prefabricated depression of Zhang et al (Zhang et al., 2023; Ao et al., 2020). The pits after the actual damage will not show a tendency to be too localized. The depression of the stiffened plate model after the actual collision has an impact on the overall stiffened plate. The greater the impact speed, the greater the area affected by the depression.

The results of the depression deformation of the ice body after colliding with the cracked stiffened plate at different speeds are as follows Figure 6.

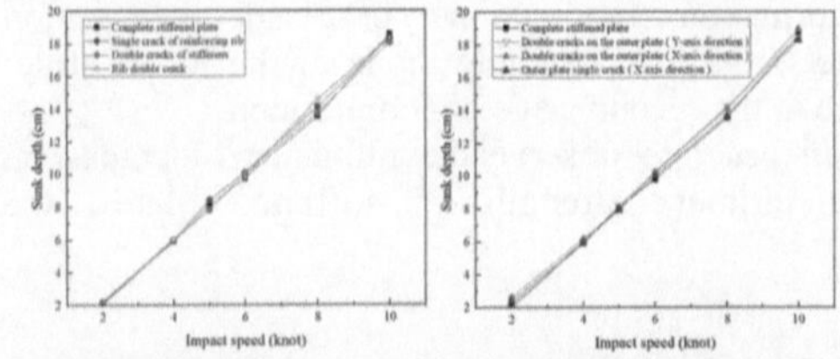

a) The variation diagram of the depth of the depression with the impact velocity

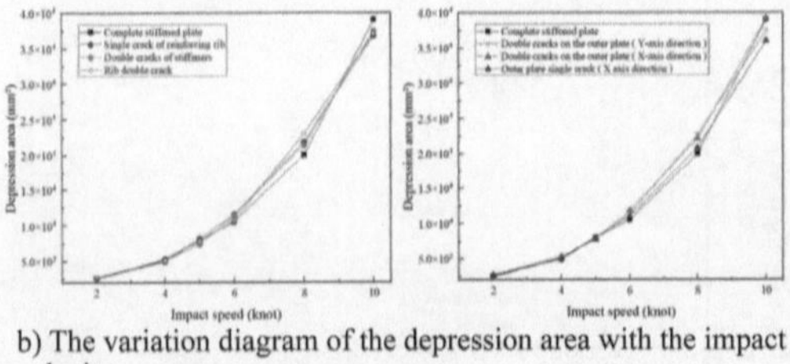

b) The variation diagram of the depression area with the impact velocity

Figure 6. Damage of stiffened plate under different impact velocities.

It can be seen from the results that as the impact velocity increases, the depth and area of the depression gradually increase. However, by comparison, it is not difficult to find that the speed and area of the depression generally show an increasing trend after the crack-containing structure is impacted. When the impact velocity reaches 10 knots, it is not difficult to find that compared with the crack-free structure, the depth and area of the depression of the crack-containing structure show a downward trend. In view of this phenomenon, it can be considered that the impact velocity is too large, and the crack opening of the crack-containing structure after high-speed impact causes part of the energy dissipation.

After increasing the crack, the dent damage at the same impact velocity generally shows an increasing trend. By comparison, it is found that the damage caused by single crack is significantly smaller than that caused by double cracks, and the effects of cracks at different positions are also different. Among them, the rib crack has the greatest influence, and its maximum depth maximum increases by 7.5 %, while the dent area maximum increases by 15 %. The corresponding stiffener crack has less influence. The reason is that the position of the stiffener crack is parallel to the loading direction of the external force, and the influence along the direction of the external force is smaller than that in other directions.

3.2 *The influence of crack location on the ultimate strength of stiffened plate*

In the process of simulation, in order to accurately obtain the residual ultimate strength of the selected structure after the impact depression, the load displacement curve of the impact area is extracted, and the curve of the force and displacement of the impact area with time is extracted by LSDYNA command. The accurate load displacement curve drawn by these data is as follows Figure 7.

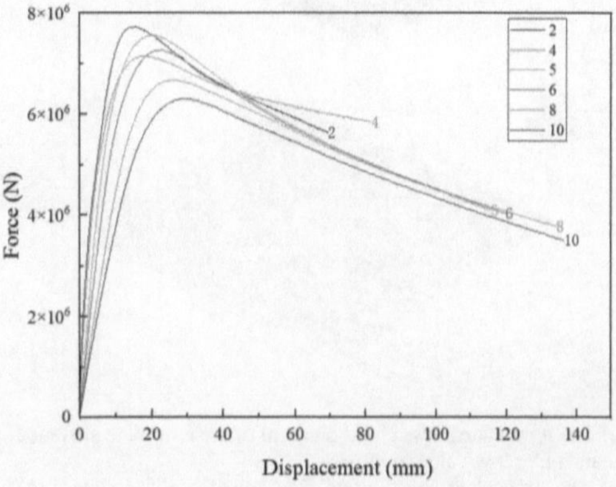

Figure 7. The load displacement curve of the intact stiffened plate after impact at different impact velocities.

The Figure 7 is the load displacement curve extracted by LSDYNA command under the compression load of the complete stiffened plate after impact at different impact speeds. From the curve characteristics, with the gradual increase of the impact load, the structural displacement also gradually increases. When the curve shows obvious nonlinear change and the slope decreases sharply, the corresponding load value is the residual ultimate strength of the structure. As the impact velocity increases, the extreme value of the load displacement curve gradually decreases and the amplitude of the curve shows an increasing trend.

It is found from the calculation results that when the impact velocity is greater than 5 knots, the final buckling mode presents a different form from that when the impact velocity is less than 5 knots. Therefore, in this paper, the impact velocity less than 5 knots is defined as the low speed state, and the impact velocity greater than or equal to 5 knots is defined as the high speed state. The Figure 8 shows the stress cloud diagram of the stiffened plate under the limit state after low speed and high speed impact.

From the Figure 8, it can be found that The buckling modes of stiffened plates after low-speed collision and high-speed collision are different. Under the limit state of the actual stiffened plate structure, due to the gradual increase of the impact velocity, the buckling mode of the final structure during lateral compression has changed, showing two different buckling modes as shown in Figure 8 a.b. The buckling modes in the literature all show the buckling modes under high-speed impact. The simplified structure fails to find the first buckling mode. The main reason is that the simplified stiffened plate has no structural constraint in the direction of the vertical stiffener, so that the first mode is not found.

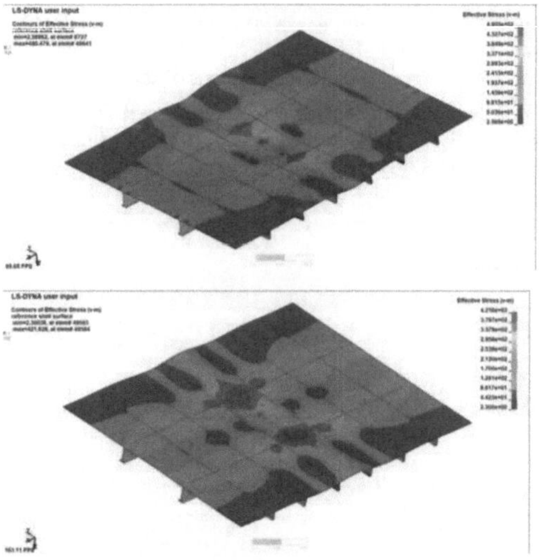

a) The limit state stress cloud diagram of the complete stiffened plate under low velocity impact
b) The limit state stress cloud diagram of the complete stiffened plate under high velocity impact

Figure 8. The comparison diagram of stress cloud diagram under the ultimate state of stiffened plate.

The Figure 9 is the limit state cloud diagram of different crack positions under compression under 10-knots impact. It is not difficult to find from the diagram that the existence of cracks leads to a certain difference in the limit state of the stiffened plate. Compared with Figure 8.b, the stress in the limit state where the crack exists is greater than that without crack. Compared with other cases with cracks, the cracks near the stiffener show a larger stress distribution, especially at the connection between the stiffener and the outer plate, the stress in the limit state is larger than that in other cases.

From the Figure 10, it can be seen that the final buckling mode presents the form of the outer plate

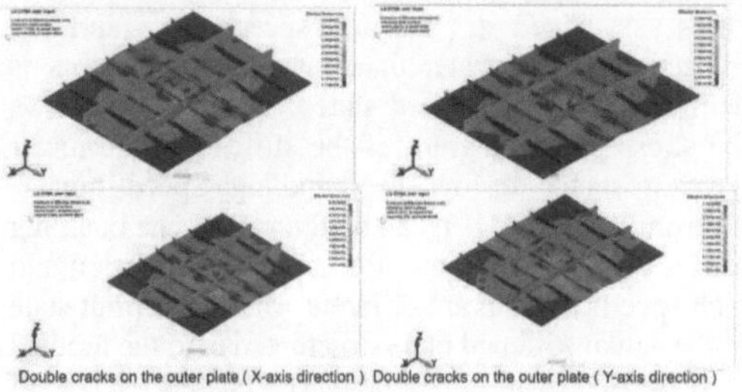

Figure 9. Stress cloud diagram of stiffened plates with cracks at different positions under limit state.

convex on both sides of the impact center depression in the case of small impact, while in the case of large impact, the final buckling mode presents the form of the impact center and both sides depression. The main reason for this phenomenon is that when the impact velocity is too large, the deformation of the single outer plate cannot resist the damage caused by the impact, and the stiffener is subjected to the external force beyond its yield limit, which causes plastic deformation, and this part of plastic deformation affects the final buckling mode of the whole structure in the second stage of compression.

In order to observe the influence of cracks on the ultimate strength of stiffened plates, the

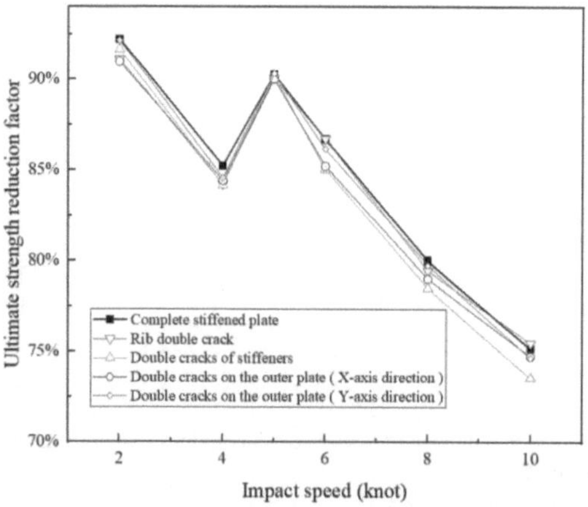

Figure 10. Ultimate strength reduction factor at different crack locations.

ultimate strength reduction factor (Zhang et al., 2023) is the ratio of the ultimate strength of the actual force to the ultimate strength of the full force. The ultimate strength reduction factors of different crack positions are as follows Figure 10.

The ultimate strength of cracked and uncracked stiffened plate ice after impact is analyzed, and the results are shown in the Figure 10. It can be seen from the figure that due to the change of buckling mode, the residual ultimate strength of the structure changes abruptly at the impact of 5 knots. The ultimate strength under the overall buckling mode is larger than the local buckling mode caused by the similar impact velocity, and the existence of cracks will not affect the change of buckling mode, but the existence of cracks leads to the increase of the residual ultimate strength under the impact of 5 sections.

It can be seen from the diagram that the influence of different crack positions is not the same. Compared with the crack-free structure, the ultimate strength of the crack-containing structure is generally smaller than that of the crack-free structure; compared with cracks at other positions, the crack at the stiffener has the greatest influence on the ultimate strength, while the rib crack has the least influence, which is contrary to the damage after impact. For the outer plate crack, the ultimate strength damage in the direction parallel to the loading end is generally smaller than that in the direction perpendicular to the loading direction. It can be seen from the figure that the residual ultimate strength of the overall structure under small impact still has a certain safety margin. In this case, the ship can still sail safely.

3.3 *Effect of crack number on ultimate strength of stiffened plate*

In order to further explore the influence of the number of cracks on the structural performance, the structural stiffeners and the joints between the outer plate and the stiffeners were selected as the research objects. Single crack and double crack conditions were applied respectively. The finite element analysis method was used to systematically study the influence of different crack numbers on the mechanical properties and bearing capacity of the structure.

The Figure 11 shows the damage and residual ultimate strength of different crack numbers under ice impact. Through the detailed analysis of Figure 11.a and Figure 11.b, it can be seen that compared with the single crack, the existence of double cracks will lead to a significant increase in the damage degree of the stiffened plate after impact. Under the action of load, the double cracks will accelerate the initiation and propagation of the micro-cracks inside the stiffened plate, so that the damage evolution process of the material is accelerated. Compared with the cracks located at the joints, after increasing the number of cracks, the damage growth after the increase of the cracks located at the joints is greater than that at the stiffeners. Through accurate data measurement and calculation, it is found that the maximum increase of the damage degree of the stiffened plate caused by the increase of the cracks at the joints can reach 7.3 %. And in the ribs increased by 1.8 %.

It can be seen from Figure11c that the ultimate strength reduction at the stiffener with increasing the number of cracks is greater than that at the connection between the stiffener and the outer plate, and the ultimate strength under the crack at the stiffener is less than that at the connection under the same number of cracks. This shows the opposite form of damage. In view of this phenomenon, the study found that it is mainly due to the fact that in the process of applying load, the external force directly acts on the stiffener, and the external force at the connection is small, resulting in the final situation opposite to the damage.

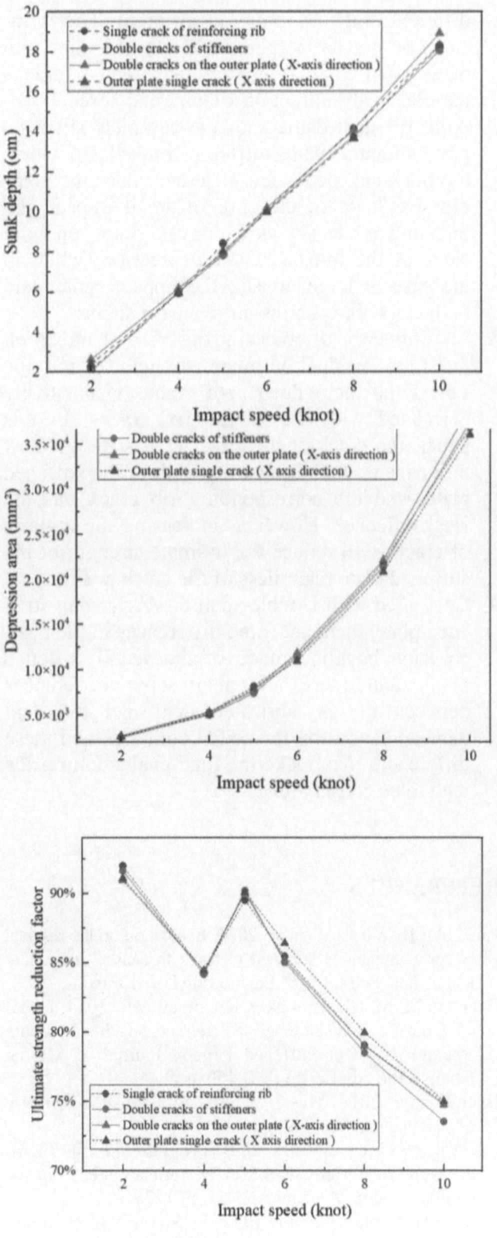

a) Sag velocity variation diagram
b) The change diagram of depression area
c) Variation diagram of ultimate strength reduction factor

Figure 11. The influence of the number of cracks on the damage and ultimate strength of stiffened plates.

4 CONCLUSIONS

In this paper, the ultimate strength of ice impact cracked stiffened plate structure after dent damage is analyzed by numerical simulation. The following conclusions can be obtained from the analysis and discussion results:

1. The existence of cracks will lead to the increase of the damage degree of stiffened plates after impact, and the influence of

different crack positions is different. The influence of cracks perpendicular to the direction of external force is greater than that of cracks parallel to the direction of external force.
2. After the dent damage of the complete stiffened plate structure under different impact velocities, the buckling modes are different under the compressive load. In the case of small impact, the buckling mode shows a convex shape on both sides of the impact center depression, while in the case of large impact, the impact center and both sides show an inward concave shape.
3. The influence of cracks at different positions on the final residual ultimate strength is not the same, and the influence of cracks is positively correlated with the number of cracks. In this paper, the crack at the stiffener has the greatest influence on the ultimate strength of the stiffened plate, and the corresponding rib crack has the least influence. However, increasing the number of cracks will reduce the ultimate strength of the stiffened plate regardless of the crack position.
4. Compared with Double-span or single-span stiffened plate, there are some differences in the post-collision buckling mode of the actual stiffened plate, and there are also some differences between the prefabricated dent and the dent damage caused by the actual collision, and these differences may make the final evaluation results tend to be conservative.

REFERENCES

Ao L, Wu H, Wang D, Wu W. 2020. Evaluation on the residual ultimate strength of stiffened plates with central dent under longitudinal thrust. Ocean Engineering 202: 107167.

Bahmani M, Morshedsolouk F, Khedmati MR. 2024. Effect of Cracked and Notched Stiffeners on the Ultimate Strength of Steel Stiffened Panels. Journal of Marine Science and Application 23: 460–469.

Bian Xuanyi. 2018, Master's thesis of Jiangsu University of Science and Technology.

Cai Wei, Zhu Ling, Bi Luze. 2021, Progress in Finite Element Numerical Methods for Ice Materials under Ship-Ice Collision. Ship Mechanics 25:1116–1126.

Cowper G.R. and Symonds 1957 P.S.Strain hardening and strain-rate effects in the impact loadings of cantilever beams, Brown University Division of Applied Mathematics, Report No. 28,.

Hu K, Yang P, Xia T, Song Y, Chen B. 2022. Numerical investigation on the residual ultimate strength of cracked stiffened plates under extreme cyclic loads. Ocean Engineering 244: 110426.

Kulkarni AA. 2020 Effects of strain rate and strain hardening on a cantilever under impact. Proceedings of the Institution of Civil Engineers - Engineering and Computational Mechanics 173: 70–82.

Li D, Chen Z. 2025. Progressive collapse response and ultimate strength evaluation of stiffened plates with welding residual stress under combined biaxial cyclic loads and lateral pressure. Marine Structures 99: 103703.

Liu G, Chen Z, Zhou J. 2023. Numerical Investigation on the Residual Ultimate Strength of Central-Cracked Stiffened Plates under Tensile and Bending Loads Using XFEM. Journal of Marine Science and Engineering 11: 302.

Müller F, Böhm A, Herrnring H, von Bock Und Polach F, Ehlers S. 2024. Influence of the ice shape on ice-structure impact loads. Cold Regions Science and Technology 221: 104–175.

Manual, K. U. S. (2007). LS-DYNA keyword User's manual. LSTC Co., Livermore, CA.

Ni B, Wang Y, Xu Y, Chen W. 2024. Numerical Simulation of Ship Collision with Rafted Ice Based on Cohesive Element Method. Journal of Marine Science and Application 23: 127–136.

Rahbar-Ranji A, Zarookian A. 2015. Ultimate strength of stiffened plates with a transverse crack under uniaxial compression. Ships and offshore structures 10: 416–425.

Xu J, Guo Y, He Z, Li Z, Liu G, He Y. 2024. Simplified analytical method for predicting the large deformation of stiffened plates with initial cracks under repeated impacts. Ocean Engineering 304: 117886.

Xu MC, Song ZJ, Pan J, Guedes Soares C 2017. Ultimate strength assessment of continuous stiffened panels under combined longitudinal compressive load and lateral pressure. Ocean Engineering 139: 39–53.

Yan Mengjiao, Zhu Ling. 2017, Research on dynamic response of hull plate structure under floating ice collision load. Journal of Wuhan University of Technology (Transportation Science and Engineering Edition) 41:268–272.

Zhang Jing, Li Mingjun, Wang Xueliang, Shi Xinghua. 2023, Parametric sensitivity analysis of ultimate strength of stiffened plates under polar low temperature and depression damage. Ship Mechanics 27: 750–762.

Zhang J, Shi XH, Guedes Soares C, Liu J. 2022. Ultimate strength of stiffened panels with a crack and pits under uni-axial longitudinal compression. Ships and offshore structures 17: 319–338.

Trends in Collision and Grounding of Ships and Offshore Structures – Zhu & Guedes Soares (Eds)
© 2026 The Author(s), ISBN 978-1-041-16361-9

Structural response of hull plates under impact of ice wedges produced at different temperatures

Z.H. Guo
Key Laboratory of High Performance Ship Technology (Wuhan University of Technology), Ministry of Education, Wuhan, P.R. China

L. Zhu
School of Naval Architecture, Ocean and Energy Power Engineering, Wuhan University of Technology, Wuhan, P.R. China

R. Sun
School of Naval Architecture, Ocean and Energy Power Engineering, Wuhan University of Technology, Wuhan, P.R. China
Sanya Science and Education Innovation Park of Wuhan University of Technology, Sanya, P.R. China

ABSTRACT: In polar regions, the temperature of ice varies with seasonal changes and geographical distribution. Changes in temperature significantly affect the mechanical properties of ice. For instance, the compressive strength of ice increases as the temperature decreases, and ice is more likely to exhibit brittle failure modes at lower temperatures. Consequently, the structural response of marine structures to ice impacts also varies with the temperature of the ice. In this study, wedge-shaped ice were prepared at temperatures ranging from -34.6°C to -2.2°C, and experiments were conducted on the impact of wedge-shaped ice on rectangular plates. During the experiments, the relationships between plate deformation, impact force, ice rebound velocity, and other relevant parameters were compared as functions of temperature. The research findings can provide valuable references for the design of polar ships operating in different ice conditions.

1 INTRODUCTION

With the opening of the Arctic route, the number of ships has increased, primarily for freight, energy development, and scientific research. However global warming has accelerated the melting of glaciers, leading to an increase in floating ice and icebergs, which poses challenges to navigation. Therefore, studying ship-ice collision performance is crucial for determining ice loads, assessing hull structures, reducing accident losses, and guiding future ship design.

In recent years, numerous studies have focused on investigating the dynamic structural behavior of ship structures under the impact of floating ice or icebergs using experimental and numerical methods. Gagnon (2008) conducted full-scale collision tests involving the CCGS Terry Fox icebreaker and floating ice floes. Gagnon & Wang (2012) also conducted scaled model collision tests between structures and ice, with the experimental results showing good agreement with numerical simulations. Ritch et al. (2008) performed small iceberg collision tests on the Terry Fox icebreaker, measuring the local impact forces acting on the hull structure during collisions. Kim et al. (2017) conducted indoor experiments on freshwater ice impacting floating stiffened steel structures. Yan & Zhu (2017) conducted numerical and experimental studies on collisions between plates and ice wedge, investigating the influence of ship-ice collision parameters on structural dynamic response and plastic deformation. The results demonstrate that the maximum deformation of the plate exhibits a linear increase with the rising impact velocity of the ice. Zhu et al. (2020) compared the differences between ice wedge and rigid wedge impacting ship plates, revealing that under the same kinetic energy, the damage caused by rigid to the plates was significantly greater than that caused by ice, reinforcing the coefficient method introduced by Zhu et al. (2016). Building on this, Cai et al. (2022) carried out repeated impact tests of ice on ship plates, proposing an energy criterion for addressing the pseudo-shakedown issue under repeated ice impact loads. New analytical method to estimate deformation of plate under repeated ice impact was proposed by Zhu et al. (2015).

In the field of numerical simulations, many scholars have proposed various ice material models, such as the crushable foam model (Gagnon, 2011; Kim et al., 2015), the elastic-plastic model (Liu et al., 2011; Gao et al., 2015; Zhu et al., 2016; Cai et al., 2020; Yu et al.,

DOI: 10.1201/9781003684404-19

2021; Mokhtari et al., 2022; Yu et al., 2023), and the viscoelastic model (Mokhtari et al., 2023).

In polar regions, ice temperatures vary with seasonal changes and geographical distribution. Temperature changes significantly affect the mechanical properties of ice. For instance, the uniaxial compressive strength of ice increases as the temperature decreases. (Gagnon & Gammon 1995; Jones 1982; Rist & Murrell 1994). Based on this, Derradji-Aouat (2003) proposed a yield equation for ice by summarizing the results of triaxial tests, which is dependent on temperature and strain rate.

Shi et al. (2017) proposed a temperature-gradient-dependent ice material model. They simulated collisions between rigid flat plates and icebergs of different geometric shapes, analyzing the effects of iceberg geometry and ice model temperature on the pressure-area curve. Similarly, Yu et al. (2023) developed a temperature-dependent ice material numerical model and conducted numerical simulations of the ice impact tests on stiffened plates performed by Kim et al. (2013). The results demonstrated that temperature has a significant influence on structural deformation. Yu et al. (2024) conducted uniaxial compression experiments on ice at different temperatures and used a temperature-dependent ice material numerical model to simulate the experiments.

Currently, research on the structural response of ship structures under ice impacts at different temperatures is primarily based on numerical simulations, with experimental studies being relatively limited. This paper presents an experimental study on the impact of wedge-shaped ice on rectangular plates under different temperature conditions. During the experiments, particular attention was paid to comparing the variations of key parameters such as plate deformation, impact force, and ice rebound velocity with temperature.

2 ICE WEDGE IMPACT TESTS

In the ice-plate impact experiments, floating ice was simplified into a wedge shape, and the geometric dimensions of the ice model are shown in Figure 1. The ice model was formed by freezing a mixture of water and crushed ice at a constant temperature, with the crushed ice particles being approximately 5 mm in size. The surface temperature of arctic sea ice is close to 0°C in summer and drops below -45°C in winter (Lindsay & Rothrock 1994). This study conducted nine impact tests at ambient temperature of 10°C, utilizing ice specimens prepared across a temperature range from -34.7°C to -2.2°C. Table 2 summarizes the test conditions and results.

The temperature of the ice specimen was determined by averaging measurements from five distinct points using an infrared thermometer as shown in Figure 2.

The dimension of rectangular plate was illustrated in Figure 3. The tensile test data for the plate are presented in Figure 4, while its geometric and material properties are summarized in Table 1.

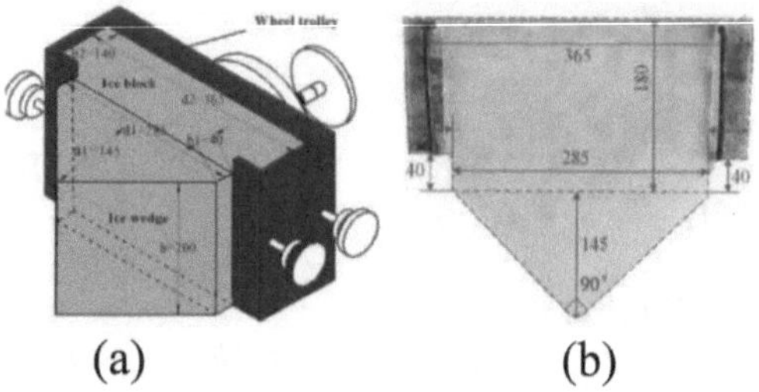

Figure 1. Ice specimen and dimensions. (a) Oblique view, (b) Top view.

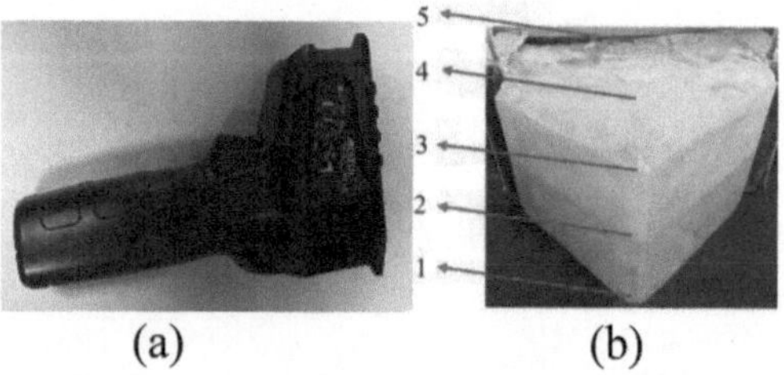

Figure 2. (a)Infrared temperature measurement equipment, (b)Distribution of temperature measurement points.

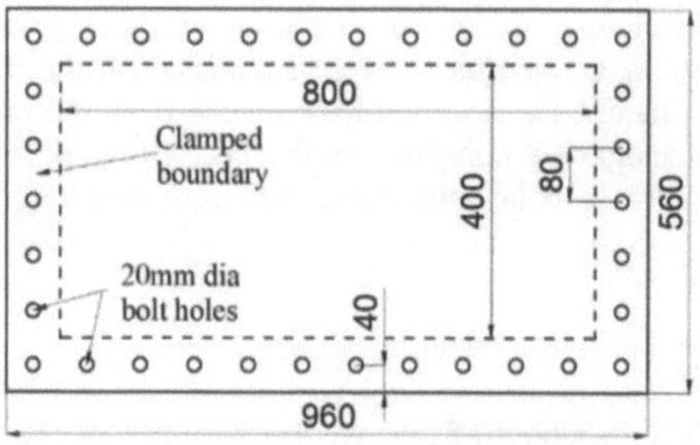

Figure 3. Plate specimen and dimensions.

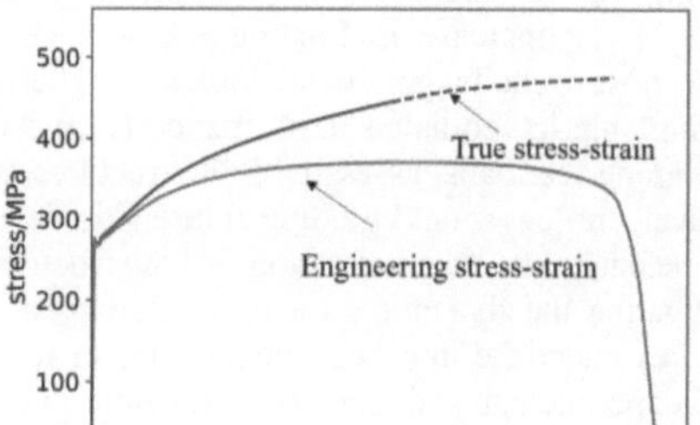

Figure 4. Plate specimen and dimensions.

Table 1. Geometric and material parameters of the plate.

Parameters	Value	Parameters	Value
Length (m)	0.8	Width (m)	0.4
Thickness t(m)	0.002	Density $\rho(kg/m^3)$	7850
Yield Stress (MPa)	275	Ultimate tensile strength (MPa)	450

The ice wedge slide down along the track of the horizontal impact experimental facility, thereby achieving the initial impact velocity, as shown in Figure 5. The impact location is at the center of the specimen. High-speed camera, accelerometer, and laser displacement sensor were used to measure the velocity of the ice wedge, the acceleration of the ice wedge, and the displacement history of the center point of the flat plate specimen, respectively. The Müller-BBM PAK (M+P) data acquisition system and the TMR data acquisition system were employed to output the signal data, as illustrated in Figure 6.

Figure 5. Horizontal impact experimental facility in WUT-IFSI lab.

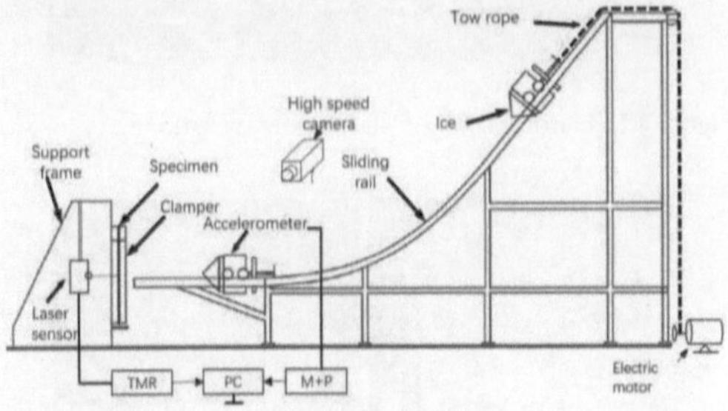

Figure 6. Schematic illustration of measurement system in impact tests.

3 EXPERIMENTAL RESULTS

Figure 7 illustrates the relationship between the permanent deformation at the center of the plate and the ice temperature. It can be observed that when the temperature is above -25°C, the permanent deformation of the plate gradually increases as the ice temperature decreases. However, when the temperature drops below -25°C, the permanent deformation of the plate does not show significant changes with further temperature reduction.

Notably, two sets of irregular data were observed in the experiments, specifically in Test No. 4 and Test No. 6. In these tests, the ice exhibited irregular splitting after impact, as shown in Figure 8. This may be attributed to defects introduced during the preparation of the ice samples. Other test conditions, such as No. 1 and No. 5, demonstrated more consistent failure patterns: the wedge tip was flattened, forming a nearly rectangular contact surface, as show in

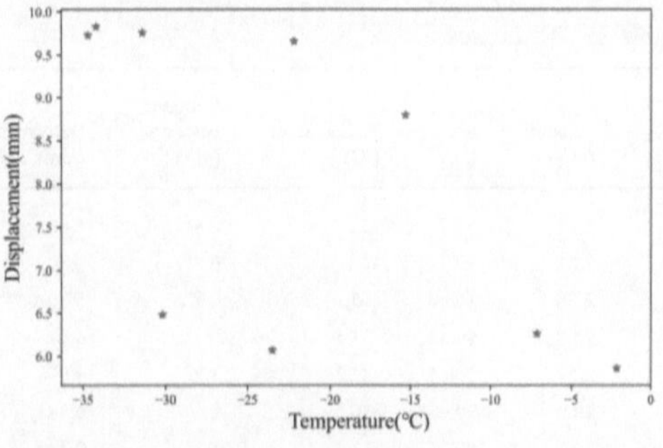

Figure 7. Plate permanent deformation.

Figure 9. Additionally, Figure 9 shows that at lower temperatures, the fracture width of the ice is shorter.

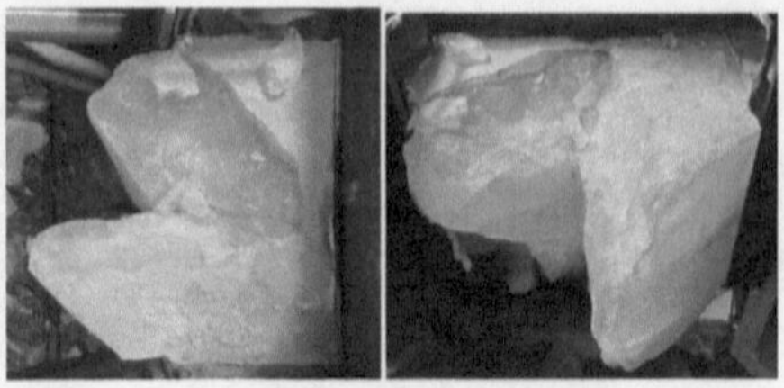

Figure 8. Irregular fracture of ice.

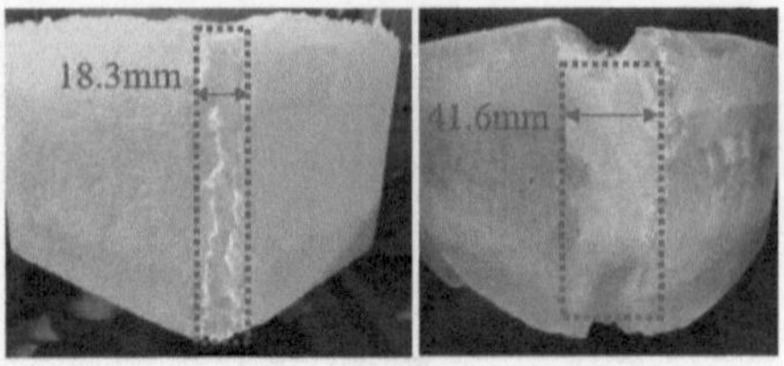

(a)Case of No.5 (b) Case of No.1

Figure 9. Regular ice failure.

Figure 10 presents the relationship between the fracture width of the ice and temperature. As the temperature decreases, the fracture width initially decreases and then stabilizes. The fracture width of the ice reflects its impact strength. Although previous studies have shown that the uniaxial compressive strength of ice increases with decreasing temperature (Gagnon & Gammon, 1995; Jones, 1982; Rist & Murrell, 1994), the stress state of ice during impact is highly complex. Therefore, the impact strength of wedge-shaped ice is not only determined by its compressive strength.

On the other hand, as the temperature decreases, ice becomes more prone to brittle failure, characterized by a sudden loss of load-bearing capacity once the compressive strength is reached (Rist & Murrell 1994). This phenomenon may also contribute to the observation that the impact strength of ice does not continuously increase with decreasing temperature.

Table 2. Test cases.

Test No.	velocity (m/s)	Mass (kg)	Temperature (°C)	Rebound velocity (m/s)	Deformation (mm)
1	2.61	99.7	-2.2	0.84	5.68
2	2.61	99.7	-7.1	0.87	7.80
3	2.58	100.0	-15.3	0.92	8.80
4	2.62	99.5	-23.5	0	6.07
5	2.61	100.1	-22.2	1.22	9.66
6	2.60	99.6	-30.2	0	6.48
7	2.61	99.7	-31.4	1.16	9.76
8	2.58	99.6	-34.7	1.10	9.73
9	2.60	99.9	-34.2	1.15	9.83

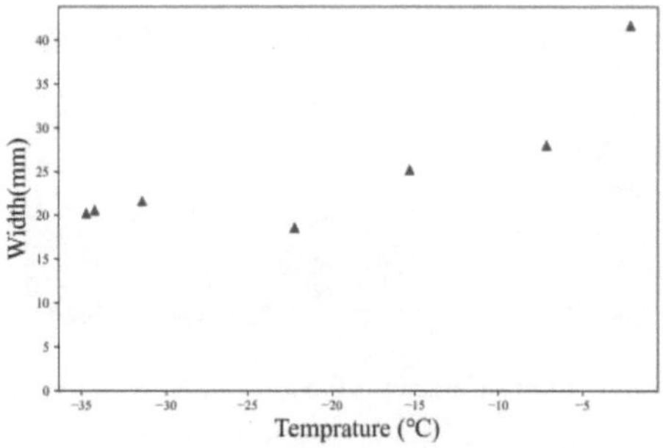

Figure 10. Fragmentation width versus temperature.

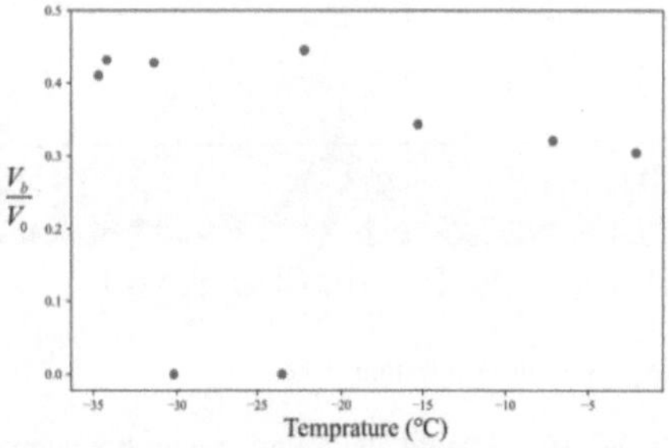

Figure 11. Dimensionless rebound velocity versus temperature.

Let the ice impact velocity be V_0, the rebound velocity be V_b, and the dimensionless rebound velocity be V_b/V_0. The relationship between the dimensionless rebound velocity and temperature is plotted in Figure 11. It can be observed that the rebound velocity of the ice initially increases with decreasing temperature and then stabilizes. When the ice undergoes splitting failure, its rebound velocity drops to zero.

After the impact tests, permanent deformation was observed on the surface of the plate, along with water stains, as shown in Figure 12. The presence of water stains indicates that the ice melted within an extremely short time during the impact. This is because the heating caused by the collision leads to the melting of ice particles at the collision interface (Nietiadi et al. 2017). Additionally, the pressure at the collision interface is high, and the melting point of ice decreases with increasing pressure, making the ice in the collision zone more prone to melting (Bejan et al.1992). Therefore, this rapid melting phenomenon also weakens the further enhancement of the ice's impact strength due to temperature reduction.

The temperature of the ice was measured immediately after the impact. Figure 13 shows the temperature distribution of the No. 8 ice sample after impact. It can be seen that the temperature of the ice at the impact area is approximately 3°C higher than at other locations. The temperature in the non-impact area remains close to that immediately after removal from the freezer, thus demonstrating that environmental temperature exerts minimal influence on the ice body's thermal state.

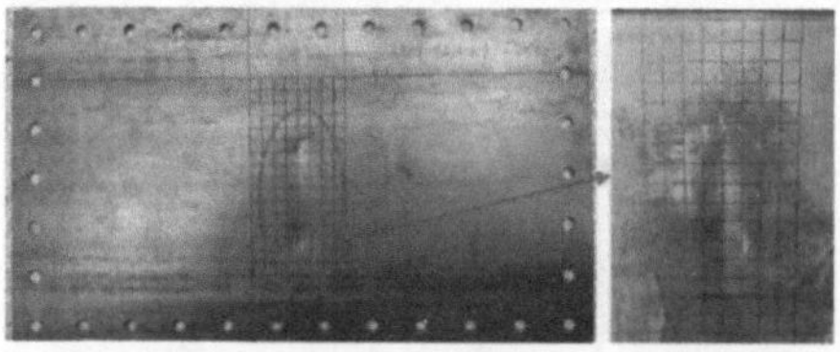

Figure 12. Deformation of plate under case no. 8.

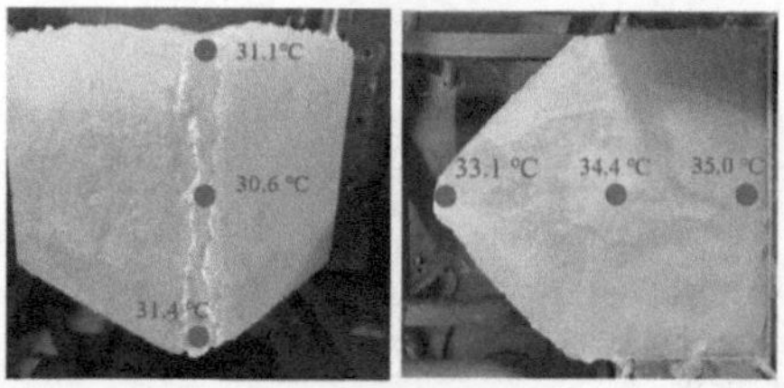

Figure 13. Temperature distribution after impact of case no. 8.

Figure 14 shows the impact force-time curves at different temperatures, while Figure 15 presents the deformation-time curves at the center of the plate under different temperatures. As previously analyzed, when the ice temperature drops below -25°C, the deformation of the plate and the fragmentation of the ice no longer change significantly with temperature. Therefore, to achieve greater visual clarity in the figures, only the data for Test No. 5 at -22.2°C is plotted for temperatures below -25°C. From Figure 14, it can be observed that as the temperature decreases, the peak force gradually increases, and the duration of the load shows a decreasing trend, indicating that the contact time between the ice and the plate also decreases.

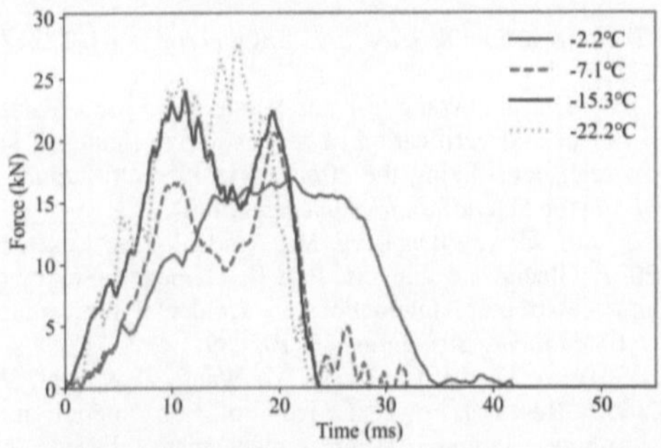

Figure 14. Load-time curve.

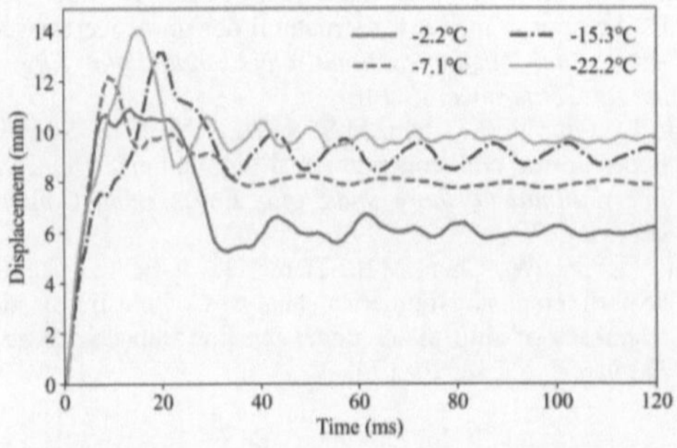

Figure 15. Deformation in the central point of plate versus time history.

4 CONCLUDING REMARKS

This article conducted experiments on the impact of rectangular plates by ice at different temperatures. The following conclusions can be drawn:

1. When the ice temperatures are between -25°C and 0°C, the permanent deformation of the plate and the rebound velocity of the ice increase as the temperature decreases, while the fracture width of the ice decreases. However, when the temperature drops below -25°C, these three parameters no longer change significantly with temperature.
2. When the ice temperatures are between -25°C and 0°C, the peak impact force gradually increases as the temperature decreases, and the duration of the load gradually decreases.
3. During the impact process, the ice absorbs heat and melts due to the collision, causing the temperature of the ice in the collision zone to rise.

ACKNOWLEDGEMENTS

This work was supported by the Wuhan University of Technology start-up fund for Distinguished Professors (grant no. 471-40120163).

The author would like to thank Mr. Y.W. Wang, Mr. B. Yu, Mr. J.S. Guo, Mr. Y.K. Liu and X. R. Chen for their assistance in the execution of the experimental work of the manuscript.

REFERENCES

Amdahl, J., Ehlers, S. & Leira, B.J. (Eds.). 2013. Collision and grounding of ships and offshore structures. *CRC Press*.

Bejan, A., & Tyvand, P.A. 1992. Pressure melting of ice under a body with flat base. *Journal of Heat Transfer, Transactions ASME*, 114(2), 529–531.

Cai, W., Zhu, L., Yu, T.X. & Li, Y.G. 2020. Numerical simulations for plates under ice impact based on a concrete constitutive ice model. *International Journal of Impact Engineering*, 143, 103594.

Cai, W., Zhu, L. & Qian, X.D. 2022. Dynamic responses of steel plates under repeated ice impacts. *International Journal of Impact Engineering*, 162, 104129.

Derradji-Aouat A.. 2003. Multi-surface failure criterion for saline ice in the brittle regime. *Cold Regions Science and Technology*, 36(1-3), 47–70.

Gagnon, R.E. & Gammon, P.H., 1995. Triaxial experiments on iceberg and glacier ice. *Journal of Glaciology*, 41 (139), 528–540.

Gagnon, R.E. 2008. Analysis of data from bergy bit impacts using a novel hull-mounted external Impact Panel. *Cold Regions Science and Technology*, 52(1), 50–66.

Gagnon, R.E. 2011. A numerical model of ice crushing using a foam analogue. *Cold Regions Science and Technology*, 65(3), 335–350.

Gagnon, R.E. & Wang, J. 2012. Numerical simulations of a tanker collision with a bergy bit incorporating hydrodynamics, a validated ice model and damage to the vessel. *Cold Regions Science and Technology*, 81, 26–35.

Gao, Y., Hu, Z.Q., Ringsberg, J. W. & Wang, J. 2015. An elastic–plastic ice material model for ship-iceberg collision simulations. *Ocean Engineering*, 102, 27–39.

Jones, S.J. 1982. The confined compressive strength of polycrystalline ice. *Journal of Glaciology*, 28(98), 171–178.

Kim, H., Daley, C. & Colbourne, B. 2015. A numerical model for ice crushing on concave surfaces. *Ocean Engineering*, 106, 289–297.

Kim, E., Storheim, M., Amdahl, J., Løset, S.,& von Bock und Polach, R. U. F. 2017. Laboratory experiments on shared-energy collisions between freshwater ice blocks and a floating steel structure. *Ships and Offshore Structures*, 12(4), 530–544.

Lindsay, R. W. & Rothrock, D. A. (1994). Arctic sea ice surface temperature from AVHRR. *Journal of Climate*, 7(1), 174–183.

Liang, H., Martin, J.M. & Mogne, T.L. 2003. Experimental investigation of friction on low-temperature ice. *Acta Materialia*, 51(9), 2639–2646.

Liu, Z., Amdahl, J. & Loset, S. 2011. Plasticity based material modelling of ice and its application to ship-iceberg impacts. *Cold Regions Science and Technology*, 65, 326–334.

Mokhtari, M., Kim, E. & Amdahl, J. 2022. Pressure-dependent plasticity models with convex yield loci for explicit ice crushing simulations. *Marine Structures*, 84, 103233.

Mokhtari, M., Kim, E. & Amdahl, J. 2023. A non-linear viscoelastic material model with progressive damage based on microstructural evolution and phase transition in polycrystalline ice for design against ice impact. *International Journal of Impact Engineering*, 176, 104563.

Nietiadi, M.L., Umstätter, P., Alabd Alhafez, I., Rosandi, Y., Bringa, E.M. & Urbassek, H.M. 2017. Collision-Induced Melting in Collisions of Water Ice Nanograins: Strong Deformations and Prevention of Bouncing. *Geophysical Research Letters*, 44(21), 10–822.

Polach, R. B. & Ehlers, S. 2013. Model scale ice—Part B: Numerical model. *Cold regions science and technology*, 94: 53–60.

Rist, M.A. & Murrell, S.A.F. 1994. Ice triaxial deformation and fracture. *Journal of Glaciology*, 40(135), 305–318.

Ritch, R., Frederking, R., Johnston, M., Browne, R. & Ralph, F. 2008. Local ice pressures measured on a strain gauge panel during the CCGS Terry Fox bergy bit impact study. *Cold Regions Science and Technology*, 52 (1), 29–49.

Shi, C., Hu, Z.Q., Ringsberg, J. & Luo, Y. 2017. Validation of a temperature-gradient-dependent elastic-plastic material model of ice with finite element simulations. *Cold Regions Science and Technology*, 133, 15–25.

Wei, P., Niu, W., Yao, C., He, Z., Zheng, Y.Y. & Ma, W. 2024. Microscopic Thermo-Mechanical Properties and Phase Transition of Bulk Ice-Ih. *International Journal for Numerical and Analytical Methods in Geomechanics*, 48(18), 4491–4505.

Yan, M.J. & Zhu, L. 2017. Study on Dynamic Response of Hull plate Structure under the Impact load of Floating Ice. *Journal of Wuhan University of Technology (Transportation Science and Engineering)*, 41(2):268–272,2017.

Yu, T.Q., Liu, K., Wang, J.X. & Wang, Z.L. 2020. Establishment and verification of a constitutive model of ice material considering the effect of temperature. *Journal of Marine Science and Engineering*, 8(3), 193.

Yu, Z., Lu, W., van den Berg, M., Amdahl, J. & Løset, S. 2021. Glacial ice impacts: Part II: Damage assessment and ice-structure interactions in accidental limit states (ALS). *Marine Structures*, 75, 102889.

Yu, Y., Zhao, D., Li, C., Liang, Y., Wang, Z. & Sun, W. 2024. Research on ship-ice collision model and ship-bow fatigue failure mechanism based on elastic-plastic ice constitutive relation. *Ocean Engineering*, 306, 118005.

Zhu, L., Shi, S.Y. & Yu, T.X. 2015. A new ice load-response model for structural design of ice classed ships. *The 25th International Ocean and Polar Engineering Conference*. ISOPE.

Zhu, L., Qiu, X.M., Chen, M.S. & Yu, T.X. 2016. Simplified ship-ice collision numerical simulations. *The 26th International Offshore and Polar Engineering Conference. ISOPE.*

Zhu, L., Cai, W., Chen, M.S., Tian, Y.K. & Bi, L.Z. 2020. Experimental and numerical analyses of elastic-plastic responses of ship plates under ice floe impacts. *Ocean Engineering*, 218, 108174.

Trends in Collision and Grounding of Ships and Offshore Structures – Zhu & Guedes Soares (Eds)
© 2026 The Author(s), ISBN 978-1-041-16361-9

Cohesive element method based ship-ice collision simulation with different bow shapes

Z.H. Song
School of Civil Engineering, Tianjin University, Tianjin, P.R. China

N.Z. Chen
School of Civil Engineering, Tianjin University, Tianjin, P.R. China
State Key Laboratory of Hydraulic Engineering Intelligent Construction and Operation, Tianjin University, Tianjin, P.R. China

ABSTRACT: The sailing resistance of polar ships is significantly influenced by bow shape. To reasonably reduce both the sailing resistance and cost of polar ships, it is essential to evaluate the impact of the bow shape on ice loading. Therefore, to examine the influence of bow shape on ice loading, a numerical model of ship moving straight ahead in level ice is developed using the cohesive element method. The cohesive zone model is employed to simulate the interactions between ices. The buoyancy of seawater is incorporated through the introduction of elastic supports. Using this model, the effects of the water plane angle and the profile angle of the bow of a polar ship on ice loading are studied. The results show that the resistance of the ship is significantly reduced when an optimal bow shape is selected. Furthermore, to facilitate a quick and straightforward assessment of the impact of bow shape on resistance, a plot to demonstrate the effects of the water plane angle and profile angle on the resistance of a polar ship is provided.

1 INTRODUCTION

Driven by climate change, the rapid decline in Arctic sea ice coverage has significantly increased the accessibility and strategic importance of polar regions. In particular, the Northern Sea Route and the Northwest Passage are drawing increasing attention due to their potential to reduce global shipping distances and transit times (Tang et al. 2022). Different from the traditional sailing, the sailing of polar ships involve interaction with sea ice, which introduces additional resistance. This polar environmental presents significant challenges for both ship design and route planning. In order to ensure the structural safety and economic viability of ships operating in ice-covered waters, accurate prediction of ice-induced resistance is essential. Therefore, a better understanding of ship-ice interaction and the use of robust numerical methods for resistance estimation are vital to improving the design and operational efficiency of polar ships.

Existing studies (Myland & Ehlers 2016) have shown that, in addition to ice properties and ship speed, the bow shape is one of the primary factors affecting resistance during ice sailing. The interaction between the bow and the ice is directly influenced by geometric characteristics of the bow. Despite the apparent similarities in the bow design of icebreaking ships, considerable differences in ice resistance may still be observed. These differences can lead to significant variation in propulsion power requirements. As the design of polar ships is constrained by time and cost considerations, an accurate estimation of icebreaking resistance at the early design stage is considered highly beneficial. Therefore, the influence of bow shape on ice-induced resistance requires systematic investigation.

Several methods were developed to predict ice-induced resistance considering ship-ice interaction, including theoretical or semi-empirical formulations, energy-based approaches, and numerical simulation-based methods incorporating fracture models.

In the category of theoretical or semi-empirical formulations, various models have been proposed to capture the physical effects of the icebreaking process. The approaches developed by Sawamura (2012), Valanto (2001) and Lindqvist (1989) are commonly referenced. Among them, Lindqvist (1989) introduced parameters to describe bow geometry and decomposed the total ice resistance into distinct components, which allowed separate estimation of each contributing force. The accuracy of semi-empirical fosrmulations was assessed by Erceg & Ehlers (2017), who identified substantial deviations across different ship sizes. Prediction errors as high

DOI: 10.1201/9781003684404-20

as 100–200% were reported, especially for larger ships.

Different from the traditional theoretical or semi-empirical formulations, energy-based methods offer an alternative by evaluating the distribution of collision energy between the ship and the ice. For instance, the study by Zhu et al. (2020) examined the energy sharing mechanism during ship-ice collisions, providing insights into the transfer and dissipation of kinetic energy in ship-ice collision.

To better describe the fracturing behavior of ice during collisions, numerical simulation based methods that incorporate explicit fracture models have gained attention (Jiang et al. 2024). These include peridynamics (Madenci & Oterkus, 2013) and cohesive zone models (Qu et al. 2023), which enable the simulation of complex fracture patterns without predefined crack paths. Such models are capable of naturally handling non-linearities associated with fracture propagation and material separation (Song & Chen 2023). As a result, they have introduced new possibilities in the analysis of ship-ice interactions by enabling direct modeling of ice failure and fragmentation.

A ship-ice collision simulation was conducted using the cohesive element method, which leverages the advantages of the cohesive zone model. By incorporating ice fracture mechanisms, the method enables the simulation of localized crack initiation and propagation. Complex bow geometries are accommodated, and the evolution of contact forces during interaction with ice is captured. The paper is structured as follows. Section 2 details the numerical model, encompassing the cohesive element method and geometric modeling of ship bows. Section 3 presents and analyzes the simulation results. Section 4 summarizes the key findings and offers concluding remarks.

2 NUMERICAL MODEL

In this section, the numerical model utilized to simulate ship-ice interaction are constructed and it consists of three parts. In the first part, the cohesive element method employed to capture the fracture and fragmentation behavior of ice is introduced, along with its finite element implementation. This method enables the simulation of crack initiation, propagation, and element separation under dynamic loading. In the second part, the bow geometries under investigation are defined, with specific emphasis on variations in water plane angle and profile angle. Finally, the other simulation settings are introduced.

2.1 *Cohesive element method for ice fracture*

The cohesive element method, based on the cohesive zone model, originates from fracture mechanics and is used to characterize material behavior near the tip of a propagating crack. Dugdale (1960) and Barenblatt (1962) observed a relationship between the extent of plastic yielding and applied external load when studying yielding at the end of a slit in a sheet. This relationship later became known as the traction-separation law. As shown in Figure 1, the region ahead of the crack tip is defined as the cohesive zone (Song et al. 2024). Within this zone, the traction perpendicular to the direction of crack propagation is determined by the separation between the opposing surfaces, in accordance with the traction-separation law (Song & Chen 2024).

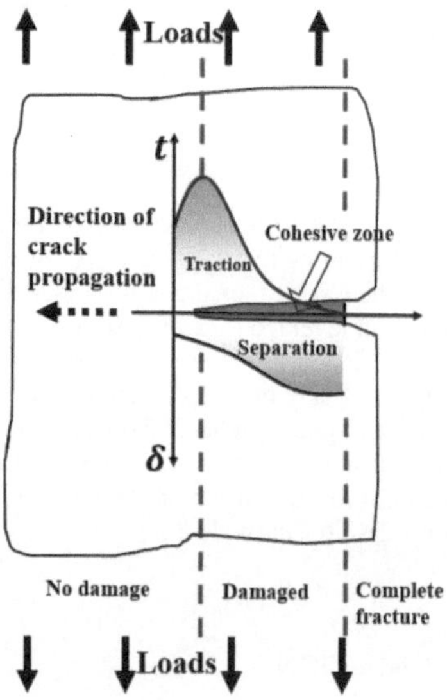

Figure 1. Schematic diagram of cohesive zone.

2.1.1 *Traction-separation law*

A bilinear traction-separation law is used to describe the constitutive behavior of ice within the fracture zone and it is defined by an initial linear elastic response followed by a softening branch based on energy representing damage evolution.

The initial linear elastic response of ice can be written as:

$$\boldsymbol{t} = \begin{bmatrix} t_n \\ t_s \\ t_t \end{bmatrix} = \begin{bmatrix} E_{nn} & E_{ns} & E_{nt} \\ E_{ns} & E_{ss} & E_{st} \\ E_{nt} & E_{st} & E_{tt} \end{bmatrix} \begin{bmatrix} \varepsilon_n \\ \varepsilon_s \\ \varepsilon_t \end{bmatrix} = \boldsymbol{E\varepsilon} \quad (1)$$

where t is traction vector and it is composed of the normal traction t_n and two shear tractions t_s and t_t. The subscript ii of E_{ii} represents the direction. ε_n, ε_s and ε_t are the strains in three directions. Based on ε, the separation δ in cohesive zone can defined as:

$$\boldsymbol{\delta} = \begin{bmatrix} \delta_n \\ \delta_s \\ \delta_t \end{bmatrix} = \frac{1}{T_{ori}} \begin{bmatrix} \varepsilon_n \\ \varepsilon_s \\ \varepsilon_t \end{bmatrix} = \frac{1}{T_{ori}} \boldsymbol{\varepsilon} \quad (2)$$

where δ_n, δ_s and δ_t are the component of δ in three directions. T_{ori} is the initial thickness of the cohesive zone.

t is increased with the increase of ε. When the traction is increased to a certain value, damage of ice will occur. The criterion of damage initiation is quadratic nominal stress criterion and it can be formulated as:

$$\left(\frac{t_n}{\sigma_{n,max,0}}\right)^2 + \left(\frac{t_s}{\sigma_{s,max,0}}\right)^2 + \left(\frac{t_n}{\sigma_{t,max,0}}\right)^2 = 1 \quad (3)$$

where $\sigma_{n,max,0}$, $\sigma_{s,max,0}$ and $\sigma_{t,max,0}$ are the maximum values of the traction that the cohesive zone of ice can reached.

After damage occurs, the cohesive zone will enter the softening stage and the stiffness of the cohesive zone of ice will degenerate (Song & Chen 2025). In the softening stage, as the separation displacement increases, the resistance and traction in cohesive zone of ice continue to decrease. In this stage, the rule of damage evolution is governed by the BK (Benzeggagh - Kenane) criterion. Based on the BK criterion, when the fracture energy in the cohesive zone reaches a critical value, the cohesive zone will be completely ineffective and the ice will be completely broken.

2.1.2 *Combination with finite element*

The aforementioned traction-separation law is implemented in a finite element framework to simulate the fracture behavior of ice. In the finite element model, two types of elements are used to represent the ice sheet. Solid elements are applied to model the bulk deformation of the ice, while cohesive elements are inserted to capture the failure process. Given that fracture in the ice sheet can initiate at arbitrary locations during ship-ice collision, cohesive elements are placed at all potential failure interfaces. As illustrated in Figure 2, cohesive elements are embedded along the interfaces between all neighboring solid ice elements. This configuration enables the simulation of spontaneous crack initiation and propagation without predefined crack paths.

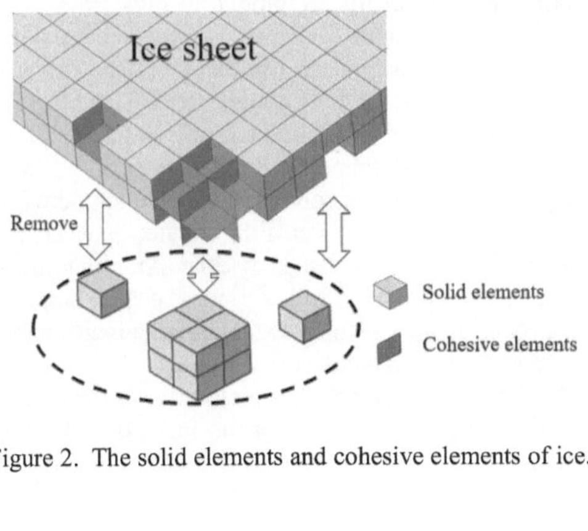

Figure 2. The solid elements and cohesive elements of ice.

2.2 *Bow shape of ships*

As discussed in the introduction, shape of bow has a significant influence on ice resistance during sailing in ice-covered waters. Therefore, the effect of bow shape variation on ice resistance requires detailed investigation.

The reference ship used in this study is a 10,000 TEU container ship, as illustrated in Figure 3. The principal dimensions of the ship are as follows. Length overall is 293 meters. Beam is 42 meters. The draft is 15 meters. The ship model was selected based on its representative geometry and size among large commercial ships operating in open and ice-covered waters.

Figure 3. The geometry of reference ship model.

To isolate and quantify the effect of bow geometry on ice resistance, modifications are applied to the bow of the baseline ship form. The adjustments are focused on two specific geometrical parameters commonly used to describe bow shape and the geometrical parameters are the water plane angle and profile angle. These parameters influence the local contact area and pressure distribution between the ship and the ice sheet during the breaking process. A series of bow variants is generated by systematically altering these angles while keeping the rest of the ship geometry unchanged. This approach allows for a focused analysis of how these shape characteristics affect the ice-breaking and crushing behavior during interaction.

2.2.1 *Water plane angle*

As illustrated in Figure 4, the water plane angle is defined as the acute angle between the longitudinal sailing direction and the bow contour at a fixed waterline plane. This angle will not exceed 90 degrees. A smaller water plane angle typically indicates a sharper bow profile, while a larger angle corresponds to a more blunt or flat-shaped bow. In the present simulations, variations in the water plane angle are introduced to investigate its influence on ice resistance. It should be noted that despite differences in water plane angle across the analyzed cases, both the draft and the beam were kept constant. This ensures that any changes in ice-induced forces can be attributed specifically to variations in bow sharpness rather than changes in submerged volume or beam-related parameters.

2.2.2 *Profile angle*

As illustrated in Figure 5, the profile angle is defined as the angle between the horizontal plane and the upward direction of the bow contour on the

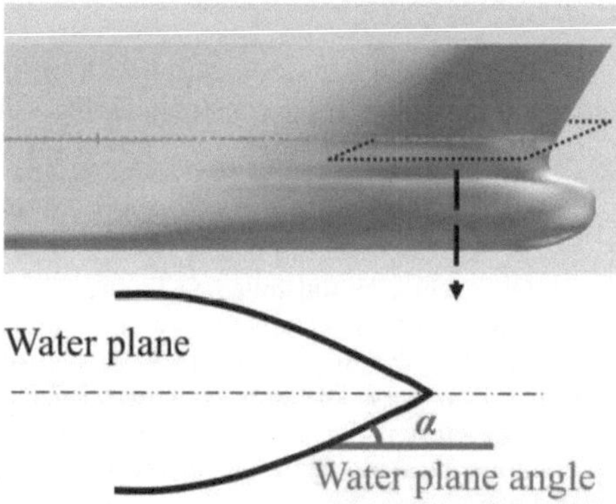

Figure 4. Water plane angle.

longitudinal centerline section. A profile angle of 90 degrees indicates a bow surface that is perpendicular to the water surface. Variations in this angle represent changes in the steepness of the bow profile in the vertical plane. In the analysis of effect of profile angle on resistance, only the profile angle was altered, while all other ship form parameters, including water plane angle, beam, and draft, were maintained unchanged. This ensures that observed differences in ice resistance can be directly attributed to the influence of profile angle rather than other geometric factors.

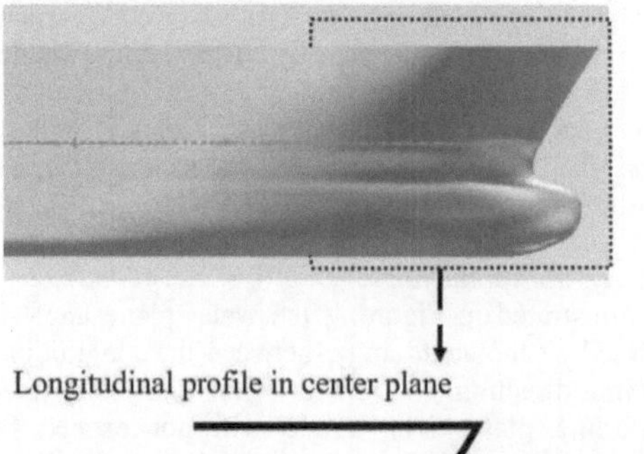

Figure 5. Profile angle.

2.3 *Numerical setting*

The ship geometry is generated in SolidWorks and modified using Python scripts to produce variants with different bow shapes. As shown in Figure 6, the models are then imported into ABAQUS for explicit dynamic analysis of ship-ice collision. An elastic foundation is placed beneath the ice sheet to simulate buoyant effects. Simply supported boundary conditions are applied to the bottom of the elastic foundation and three edges of the ice sheet.

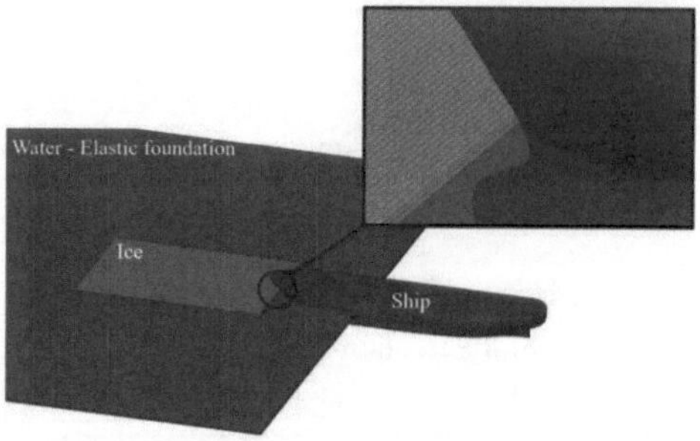

Figure 6. The finite element model of ship and ice.

The ship is modeled as a rigid body. Surface interactions are defined with hard contact in the normal direction and a friction coefficient of 0.1 in the tangential direction. Gravity acceleration is set to 9.8 m/s^2, and the forward velocity of the ship is maintained at 1.5 m/s. The average resistance force is extracted from the stabilized stage of each simulation to evaluate ice resistance for different bow shapes.

3 RESULTS AND DISCUSSIONS

In this section, the results of ship-ice collision simulations are presented. The failure modes of the ice sheet induced by ship-ice interaction are first briefly examined. Subsequently, the effects of water plane angle and profile angle on the resulting ice resistance are discussed. The resistance values are compared across different bow shapes to evaluate the influence of geometric variation on ice-breaking behavior and force response.

3.1 *Breaking of ice*

The simulation results of ship-ice collision process were analyzed to evaluate the ice breaking characteristics and resistance patterns under specific bow configurations. The results are plotted in Figs 7-8.

As shown in Figure 7, the progressive ice breaking process at four distinct time instances (20s, 40s, 60s, and 80s), in which the bow configuration is defined by a water plane angle of 45° and a profile angle of 135°.

Figure 7 demonstrates the development of ice fracture patterns during the continuous interaction between the ship bow and ice sheet. Full contact between the bow and the ice is established by 20 s. After that, fragmented ice is displaced to both sides of the ship, and the overall failure pattern exhibited relatively stable and consistent characteristics over time.

As shown in Figure 8, the variation in ice resistance over time is presented. During the first 20 s of interaction, the resistance value experienced a significant

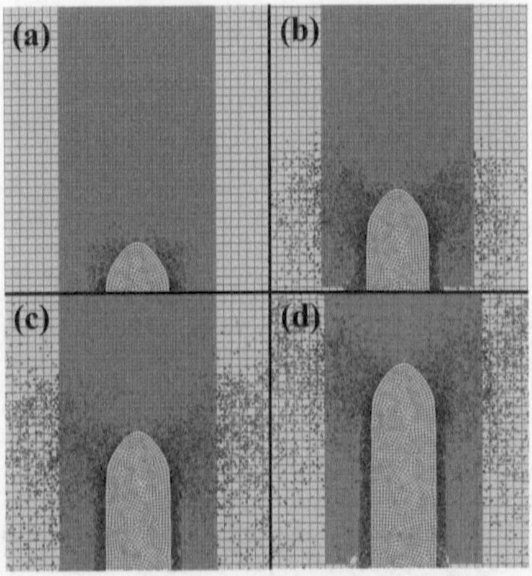

Figure 7. Ship-ice collision process at different time: (a) 20s; (b)40s; (c)60s; (d)80s.

rise, reaching approximately 6×10^6 N. Subsequently, the resistance fluctuated within a range of 4×10^6 N to 1×10^7 N, without exhibiting substantial increases. This stabilization phenomenon aligns with the visual observations from Figure 7, where the full engagement between the bow and ice sheet was achieved around 20s. The fluctuations in the resistance curve after 20s can be attributed to the cyclic nature of ice breaking processes, including crack initiation, propagation, and clearing of ice fragments.

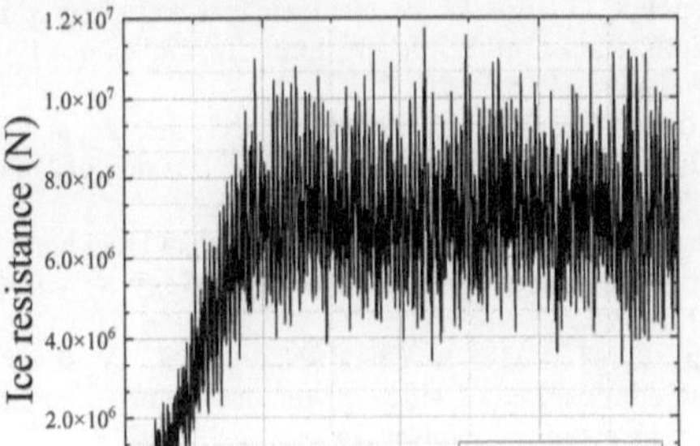

Figure 8. Relationship between ice resistance and time when α= 45° and γ = 135°.

3.2 *Water plane angle*

The influence of bow geometry on ice-breaking patterns and resistance characteristics was examined by comparing different water plane angles.

Figure 9 Displays the ice fracture patterns for two water plane angles (α = 45° and 60°), with the profile angle (γ) fixed at 45°. Both 3D and top views are provided to illustrate the spatial distribution of fragmented ice. With the increase of water plane angle from 45° to 60°, significant changes in the ice breaking patterns were observed. The larger water plane angle resulted in the formation of larger ice fragments and closer accumulation of broken ice pieces along the ship sides. This behavior is associated with a reduction in the horizontal component of the contact force.

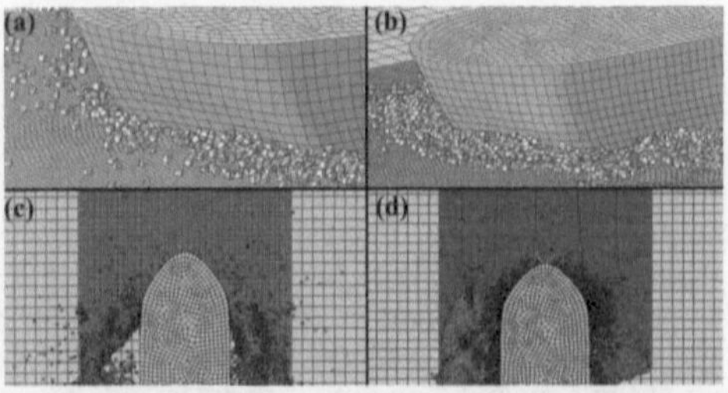

Figure 9. Fracture of ice under different water plane angle conditions: (a) α= 45° and γ =45° in 3D view; (b) α= 60° and γ =45° in 3D view; (c) α= 45° and γ =45° in top view; (d) α= 60° and γ =45° in top view.

The relationship between normalized ice resistance and water plane angle is depicted in Figure 10, where the resistance values are normalized by the resistance at α = 15°. The curve exhibits a monotonic increasing trend with a notable surge in resistance between α = 50° and 60°. The normalized resistance increased gradually from 1 to approximately 2 for α ranging from 15° to 45°, followed by a sharp rise to about 6 when α exceeded 45°. This dramatic increase in resistance correlates with the observed changes in ice breaking patterns shown in Figure 9, where larger ice fragments and more concentrated accumulation patterns were identified at higher water plane angles.

These results indicate that the water plane angle has a substantial impact on both fracture behavior and resistance, with a critical transition occurring near α = 45°.

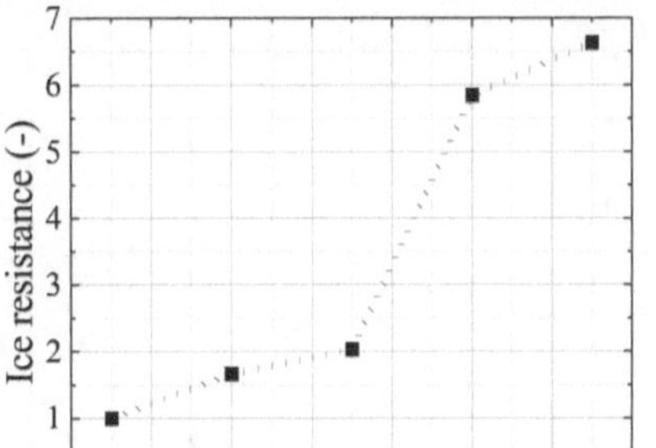

Figure 10. Relationship between water plane angle α and the ice resistance in sailing direction.

3.3 *Profile angle*

The impact of profile angle on ice-breaking mechanisms and resistance characteristics is analyzed through comparative studies with varying profile angles while maintaining constant water plane angles

Figure 11 Illustrates the ice fracture patterns under two profile angles (γ = 45° and 135°) with a fixed water plane angle (α = 45°). Both 3D and top views are presented to demonstrate the spatial distribution of ice fragments. As the profile angle increased from 45° to 135°, distinct changes in ice breaking patterns were observed. At γ = 135°, the ice fragmentation zone became more concentrated near the bow region, with significantly reduced breaking effects in areas farther from the ship sides. The upward component of the breaking force at larger profile angles resulted in ice accumulation in front of the bow and along the ship sides.

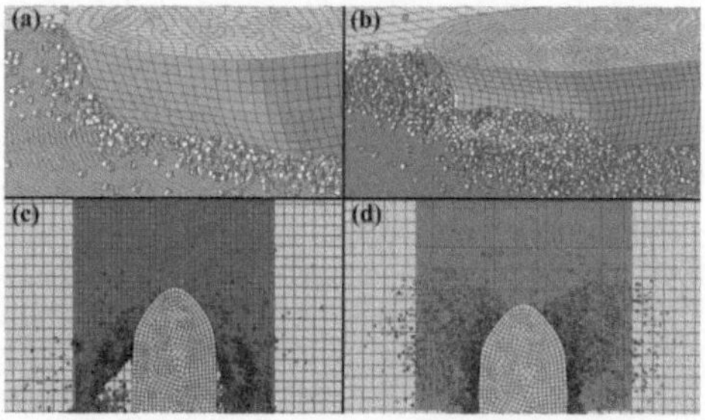

Figure 11. Fracture of ice under different profile angle conditions: (a) α=45° and γ =45° in 3D view; (b) α=45° and γ =135° in 3D view; (c) α=45° and γ =45° in top view; (d) α=45° and γ =135° in top view.

Figure 12 Shows the relationship between normalized ice resistance and profile angle, where resistance values are normalized by the resistance at γ = 120°. A nearly linear decreasing trend is observed from γ = 40° to 120°, with normalized resistance dropping from approximately 5 to 1. However, when profile angles exceeding 120°, the resistance reduction became negligible.

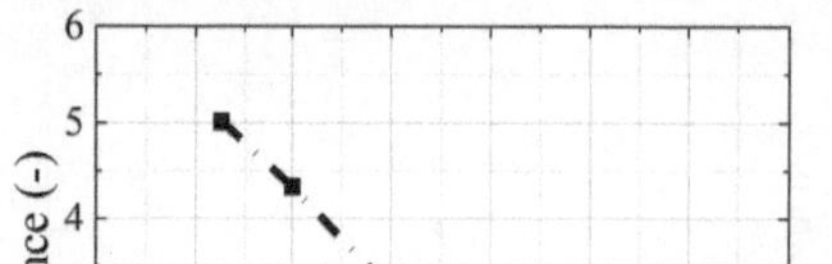

Figure 12. Relationship between profile angle γ and the ice resistance in sailing direction.

These observations indicate that increasing the profile angle generally reduces ice resistance through modified breaking patterns and ice resistance is stabilized beyond γ = 120°, likely due to the dominance of upward breaking forces at larger profile angles.

4 CONCLUSIONS

A cohesive element method based finite element framework was established to evaluate the influence of bow shape on ice resistance during ship-ice collision. Fracture behavior of ice is represented through a traction-separation law to simulate crack initiation and propagation. Buoyancy effects are modeled using elastic supports, and contact interactions are defined with normal and tangential constraints. Variations in bow shape are introduced by adjusting the water plane angle (α) and profile angle (γ).

Results show that ice resistance is significantly affected by bow geometry. Both water plane and profile angles govern the fracture pattern and resistance magnitude of ice. Ice resistance is increased sharply with α exceeding 45°.

ACKNOWLEDGEMENTS

This research was supported by the National Natural Science Foundation of China (Grant No. 52311530091). The financial support is greatly appreciated.

REFERENCES

Barenblatt, G.I. 1962. The mathematical theory of equilibrium cracks in brittle fracture. *Advances in Applied Mechanics* 7: 55–129.

Dugdale, D.S. 1960. Yielding of steel sheets containing slits. *Journal of the Mechanics and Physics of Solids* 8 (2): 100–104.

Erceg, S. & Ehlers, S. 2017. Semi-empirical level ice resistance prediction methods. *Ship Technology Research* 64 (1): 1–14.

Jiang, C., Lin, L. & Chen, N.Z. 2024. A study on auxetic-inspired side structure for enhanced crashworthiness. *Marine Structures* 93: 103545.

Lindquist, A. 1989. Straightforward method for calculation of ice resistance of ships. In Proc. *POAC'89*.

Madenci, E. & Oterkus, E. 2013. Peridynamic theory. *In Peridynamic Theory and Its Applications*: 19–43. New York: Springer.

Myland, D. & Ehlers, S. 2016. Influence of bow design on ice breaking resistance. *Ocean Engineering* 119: 217–232.

Qu, Y., Chen, N.Z. & Yuan, L. 2023. Cohesive zone model based reliability analysis for a sandwich pipe. *Ocean Engineering* 270: 113550.

Sawamura, J. 2012. Numerical investigation of ice bending failure and ice submerging force for ship maneuvering in level ice. In *Proc. 21st IAHR International Symposium on Ice*, Dalian, China: 1116–1128.

Song, Z.H. & Chen, N.Z. 2023. A modified cyclic cohesive zone model for low-cycle fatigue crack initiation

prediction for subsea pipelines under mode I loading. *Ocean Engineering* 276: 114200.
Song, Z.H. & Chen, N.Z. 2024. A modified cyclic cohesive zone model for fatigue crack initiation prediction for mooring chain steels. *Engineering Fracture Mechanics* 300: 109985.
Song, Z.H., Chen, N.Z. & Zheng, T. 2024. Cyclic cohesive zone model (CCZM)-based fatigue crack initiation prediction for a mooring chain. In *International Conference on Offshore Mechanics and Arctic Engineering*. American Society of Mechanical Engineers.
Song, Z.H. & Chen, N.Z. 2025. Prediction of fatigue crack initiation location and orientation for subsea pipelines using cohesive zone model. *In Innovations in the Analysis and Design of Marine Structures* 207.
Tang, X., Zou, M., Zou, Z., Li, Z. & Zou, L. 2022. A parametric study on the ice resistance of a ship sailing in pack ice based on CFD-DEM method. *Ocean Engineering* 265: 112563.
Valanto, P. 2001. On the cause and distribution of resistance forces on ship hulls moving in level ice. In Proc. *International Conference on Port and Ocean Engineering under Arctic Conditions*.
Zhu, L., Cai, W., Chen, M., Tian, Y. & Bi, L. 2020. Experimental and numerical analyses of elastic-plastic responses of ship plates under ice floe impacts. *Ocean Engineering* 218: 108174

Collision between ships and offshore structures

Trends in Collision and Grounding of Ships and Offshore Structures – Zhu & Guedes Soares (Eds)

Nonlinear dynamics of ship-bridge pier collisions using FSI method

Y.H. Chen & T. Wang
Huazhong University of Science and Technology, School of Naval Architecture & Ocean Engineering, Wuhan, P.R. China

J. Pan
School of Naval Architecture, Ocean and Energy Power Engineering, Wuhan University of Technology, Wuhan, Hubei Province, P.R. China

M.C. Xu
Huazhong University of Science and Technology, School of Naval Architecture & Ocean Engineering, Wuhan, P.R. China
Wuhan Lituo Bridge Protection Technology Co. Ltd., Wuhan, Hubei, P.R. China

ABSTRACT: The collision between the ship and pier is a highly complex nonlinear dynamic response process, involving the coupling between the ship and the surrounding fluid field. This study aims to validate the reliability of the Fluid-Structure Interaction (FSI) method by comparing the finite element results with experimental data. The dynamic response of ships in different weights and velocities is analyzed for the collision with bridge pier. The results show that the FSI model provides accurate predictions compared to experimental data. It is suggested that the impact force should be filtered for 50 milliseconds to account for the structural response time, rather than using the peak value. For ships with the same kinetic energy, the maximum impact force remains similar, but heavier ships cause longer collisions due to their lower speed and more gradual energy release. As the impact velocity increases, the peak force becomes larger and the collision lasts longer. Higher velocities also cause additional peaks in the force-time curves.

1 INTRODUCTION

With the development of transport industry, the number of bridges and ships gradually increase, which may lead to a high number of accidents between ships and bridges, resulting in serious impact on buildings and society (Perera & Guedes Soares, 2015). As a result, the issue of ship-bridge collisions is increasingly paid attention (Larsen, 1993). Hence, it is essential to predict the impact loads of ships for the safety assessment of collision. Carlebur (1995) conducted a series of collision tests on full-scale ships. And the results were used to verify the damage prediction of ship structures for collisions. Sha & Hao (2013) conducted impact tests to evaluate barge impact force and pier response, and then proposed the empirical formulae to predict the peak impact force and impulse.

However, due to the high cost of the experiments, the advent of the finite element method has compensated for the lack of experimentation and the reliability of the method has been proven (Wang et al. 2018). It has been widely used in the collision investigation between the bridge and ship (Gholipour et al., 2020). Sha & Amdahl (2019) established the high-resolution FE models of a container ship impacting a prestressed concrete pontoon wall for investigating structural deformation and energy absorption. Kameshwar et al. (2018) evaluated the performance of bridge columns with varying geometric and design parameters subjected to barge impact with different pile cap locations, by carrying out numerical simulation of 2000 bridges under barge impact, where the grillage modelling method was used on bridge deck. Oppong et al. (2020) established the FE models of bridge pier with the bent cap, pier columns, pile cap, and piles under barge impact to capture and compare the structural response of both intact and damaged bridge piers. Fan et al. (2020) developed a high-resolution finite element (FE) model and an associated Operational Performance Fragility Surface (OPFS) model (Fan et al., 2020) for a four-span continuous bridge, aiming to establish a methodology for vessel-impact fragility assessment of reinforced concrete (RC) bridges under corrosion-induced deterioration. Additionally, the study examined the influence of both corrosion and pile scour on the fragility of bridge structures subjected to barge impacts (Fan et al.,

DOI: 10.1201/9781003684404-21

2021). Liu et al. (2025) developed a detailed P-a model for ship-bridge collisions, focusing on two specific collision scenarios. They created a refined FE model of a ship and validated it with experimental data. The study concluded that the P-a model can accurately predict impact forces and damage in ship-bridge collisions, considering various critical parameters. Wang et al. (2022) studied how to predict the crushing force of ship bows for bridge collision safety. They tested models and used the IUM method to calculate these forces. The conclusion is that ship bow crushing force depends on both the structure's volume and its internal structure's stiffness and collapse mode.

The motion and dynamic response of the ship can be affected by the surrounding fluid. However, the fluid effect was not considered in the above numerical simulations. The fluid-structure interaction (FSI) method is important for studying ship-bridge collisions because it accounts for both the water flow and the ship's movement. Traditional models do not capture the influence of fluid on the ship and bridge, especially in complex environments. Therefore, this paper validates the effectiveness of the FSI method through experiment. Based on FSI method, A few impact velocities are selected to study their influence. It also compares the dynamic response of ships with different weights and velocities to bridge pier under the same kinetic energy.

2 COLLISION TEST BETWEEN BARGE AND BRIDGE PIER

In 2004, ship-bridge collision tests were conducted on the St. George Island Causeway Bridge (Consolazio et al., 2006), as shown in Figure 1. The bridge is located in the northwest panhandle region of Florida and extends approximately 4 miles in length. It was constructed in the 1960s and spans across Apalachicola Bay, connecting St. George Island at the south end to the mainland of Florida at the north end. Due to the bridge's structural inadequacy in terms of modern vessel collision resistance standards, impact tests were conducted on the side Pier-3. This pier is located in the portion of the South Bridge adjacent to the channel.

In order to obtain the detailed results of the dynamic response of the bridge during collision, a high-speed portable data acquisition system was installed in the tests. Each impact block was designed to match-as closely as was feasible-the shape and stiffness of the pier column, which is made of concrete with 2.44 m length, 1.83 m width and 0.66 m thickness. This was done so that interaction between the barge and impact block would closely mimic the interaction of barge struck the pier column directly. Each block was consequently designed as a heavily reinforced deep concrete slab. Sufficient stiffness was provided such that local deflections within the block would be minimal in comparison to barge deformations and pier displacements. There are reinforcement bars arranged inside. The impact block consisted of a concrete block and four force sensors to measure the bridge collision force. The principal dimensions of the impact barge in the test are showed in Table 1.

Table 1. Principal dimensions of the impact barge in the test (Consolazio et al. 2006).

Length	Breadth	Molded depth	Draught	Displacement
46.13 m	15.10 m	3.81 m	0.648 m	344 t

(a) Impact testing between the barge and bridge pier

(b) Impact block (c) Bridge pier

Figure 1. Barge impact testing of the St.George Island Causeway bridge.

3 FINITE ELEMENT MODEL

3.1 *FE model of striking barge*

The explicit FE code LS-DYNA is used to simulate the impact procedures. The finite element model of the striking ship using 4-node shell 163 elements is shown in Figure 2. The beam structures such as the bow reinforcement of the hull are simulated by BEAM161 elements. The material properties of mild steel are presented in Table 2, which is defined by the *MAT_PLASTIC_KINEMATIC. Cowper-Symonds constitutive equation that could take into account the influence of strain rate is adopted herein as following (Cower 1957):

$$\frac{\sigma'_0}{\sigma_0} = 1 + (\frac{\dot{\varepsilon}}{C})^{\frac{1}{p}} \qquad (1)$$

where, σ'_0 is the dynamic yield stress under plastic strain rate. $\dot{\varepsilon}$ and σ_0 is the corresponding static yield stress; C and p are constants for steel materials (Jones, 2011).

The velocity of the striking barge is 0.95 m/s, which is the same with the test. In order to avoid penetration at the mesh interface with different material properties and mesh sizes, the CONTACT_AUTOMATIC_ SURFACE_TO_SURFACE is defined for the ship and pier contact. And the dynamic and static friction coefficients are set as 0.3 and 0.2 respectively. During impacting, the ship would fold, and then the internal structure might contact each other, which is also considered by automatic contact method.

Table 2. Material properties of barge steel (Consolazio et al. 2006).

Properties	Units	Mild steel
Mass density (ρs)	kg/m3	7850
Young's modulus (E)	GPa	206
Poisson's ratio (v)	-	0.3
Yield stress (σy)	MPa	306
Ultimate tensile strength (σu)	MPa	370
Critical failure strain		0.3
C		40.4
P		5

Figure 2. FE models of the ship and bridge with FSI method.

3.2 *FE model of bridge pier*

The finite element model of bridge pier is showed in Figure 2. The solid 164 element is used to model the concrete of the bridge, struck block and soil. Since no damage was caused to the pier in the tests, a simple linear elastic material model *MAT_LINEAR ELASTIC is used for the simulation. It is sufficiently accurate to represent the stiffness of the pier. The concrete material properties of the bridge pier are showed in Table 3. To simplify the calculations, the same model is used for the soil material model. The range of soil is set as 11.89 m length, 6.06 m width and 19.82 m depth (Consolazio et al., 2006), and the corresponding material properties is showed in Table 4. The perimeter is set to a no-return rejection boundary condition to simulate an infinite surrounding soil body.

Table 3. Material properties of bridge concrete.

Position	Pier	Bearing platform	Superstructures
Mass density	2402.78 kg/m^3	2402.78 kg/m^3	2402.78 kg/m^3
Young's modulus	28.34 GPa	34.05 GPa	29.49 GPa
Poisson's ratio	0.23	0.14	0.13

The impact block is shown in Figure 3. It also used solid 164 element, with the same material parameter settings as the pier. The reinforcing bars are simulated by beam 161 element, in which the elastic modulus is 200 GPa, Poisson's ratio is 0.3 and the yield strength is 420 MPa. It is coupled to the concrete block using a common-node approach.

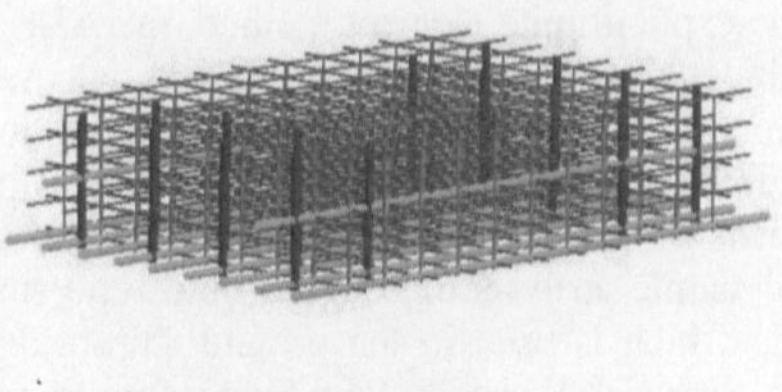

(a) Bars of the impact block

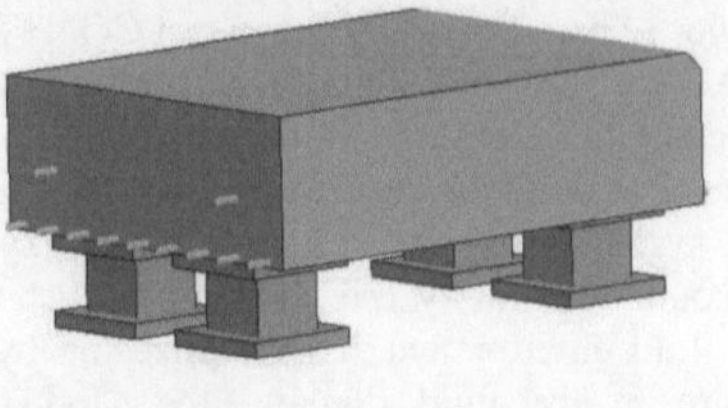

(b) Impact block

Figure 3. Impact block set on the bridge pier.

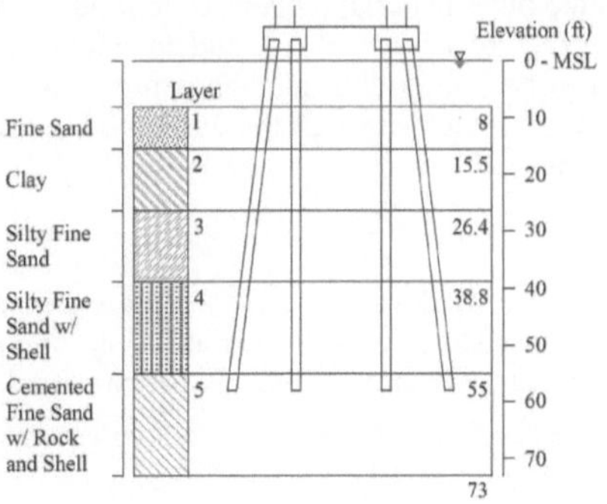

Figure 4. Soil profile in the testing field (Consolazio et al. 2006).

Table 4. Material properties of soil (Consolazio et al. 2006).

Soil layers	Depth (m)	Soil density (ρ) (kg/m^3)	Young's modulus(E) (MPa)	Poisson's ratio(ν)
Fine sand	2.44-4.73	1762.03	6.20	0.20
Clay	4.73-8.05	1601.85	2.39	0.25
Silty fine sand	8.05-11.83	1906.20	12.10	0.30
Silty fine sand	11.83-16.77	1986.29	55.84	0.35
Cmt. fine sand	16.77-22.26	1986.29	154.44	0.40

3.3 *Modelling of water surrounding ship*

In the explicit finite element method, the ALE method combines the merits of both the Lagrangian and Eulerian formulations and can overcome the numerical computational difficulties caused by the severe distortion of element (Zhao, 2003). The ALE method can simulate the dynamic analysis of coupled fluids and structures, hence which is used in the present ship-bridge collisions analysis. A penalty-based algorithm was used in the coupling to capture the fluid-structure interaction behavior (Zhao 2003). The keyword CONSTRAINED_LAGRANGE IN _ SOLID was utilized to couple the wedge model and the fluid model.

The gravitational acceleration is applied to the entire system using the *LOAD_BODY_Z keyword to account for gravity effects and establish a hydrostatic fluid environment. To initialize the hydrostatic pressure in the fluid domain, the *INITIAL_HYDROSTATIC_ALE keyword is defined, enabling rapid stabilization of the hydrostatic pressure gradient. Additionally, considering the presence of air inside the ship, the *INITIAL_VOLUME_FRACTION_GEOMETRY keyword is used to initialize the air.

Unlike other material models, fluid models in LS-DYNA require a MAT_NULL and equations of state (EOS) to be jointly defined. The expression for the deviatoric viscous stress of the MAT_NULL can be written as:

$$\sigma_{ij}' = \mu \xi_{ij}' \quad (2)$$

where $\sigma_{ij'}$ is deviatoric viscous stress, μ is the dynamic viscosity, ξ_{ij}" is the deviatoric strain rate.

The linear polynomial EOS can be expressed by:

$$P = \left[C_0 + C_1\mu + C_2\mu^2 + C_3\mu^2\right] + [C_4 + C_5\mu + C_6\mu^2]e_{ipv0} \quad (3)$$

$$\mu = 1/V - 1 \quad (4)$$

where P is the pressure value, e_{ipv0} is the initial specific internal energy, V is the relative volume, and C1~ C6 are user-defined constants.

Table 5 lists the air and water parameters of the material and the equation of state (EOS). Since the range of water surrounding ship is very wide and can be considered as infinite fluid field, which is simulated by non-reflecting boundary conditions with BOUNDARY_NON_REFLECTING key word. The fluid domain is set as 60 m length and 30 m width, the height of air and water domain are set as 4 m and 6 m which was determined based on the forward movement distance of the vessel and the size of the structural models. The flow of water is defined by inflow and outflow velocity boundaries in the longitudinal direction. The gravitational acceleration is applied on the whole FE model, which could produce the hydrostatic pressure.

Table 5. Material and EOS parameters of the fluid field (Ye 2023).

Fluid type	Material model/EOS	Parameter	Values
Air	MAT_NULL	Density	1.18 kg/m^3
		Viscosity coefficient	1.844×10^{-5} N·s·m^{-2}
		Cutoff pressure	-10 Pa
	EOS_LINEAR_POLYNOMIAL	C_4	0.4
		C_5	0.4
		E_0	2.533×10^5 Pa
		V_0	1
Water	MAT_NULL	Density	998.21 kg/m^3
		Viscosity coefficient	8.684×10^{-4} N·s·m^{-2}
		Cutoff pressure	-100 Pa
	EOS_LINEAR_POLYNOMIAL	C_0	1.01×10^5 Pa
		C_1	2.25×10^9 Pa

3.4 *Test conditions and loads conditions*

In the experiment, three series (P1, B3, P3) of 15 forward impacts were conducted on the two piers at different speeds by barges under different loading conditions (Consolazio et al., 2006). For simulation, this section selected the P3T3 collision test. In this test, the barge weight is 344 tons and the impact speed was 0.94855 m/s (3.11 ft./sec.). The sea surface was calm during the test with an average wind speed of 2.26336 m/s (4.4 knots). In order to study the influence of impact velocity, the impact velocities with 1.9, 2.5, 3.0, 3.5, 4.0 m/s are considered.

In additional, ship-bridge collisions will be simulated with different speeds and weights, while keeping the kinematic energy constant. The goal is to analyze the dynamic response of the bridge during these collisions.

Therefore, a 344 t barge with an impact speed of 1.9 m/s and a 1376 t barge with an impact speed of 0.95 m/s were selected for comparative analysis. The energy of the impacting ships remained consistent at 0.62 MJ. The loads conditions are shown in Table 6, in which M_s is the mass of the container ship, V_f is the velocity of water flow, V_s is the collision velocity of the container ship, $f_{\max}$ is the maximum impact force.

Table 6. Loads conditions.

NO.	M_s/t	V_f/(m/s)	V_s/(m/s)	$f_{\max(\mathrm{MN})}$
1	344	0	0.95	2.25
2	344	0	1.9	3.84
3	1376	0	0.95	3.90
4	344	0	2.5	4.79
5	344	0	3.0	5.12
6	344	0	3.5	5.70
7	344	0	4.0	6.32

4 ANALYSIS RESULTS

4.1 *Comparative analysis of the numerical and experimental results*

It is important to verify the FE model analysis including the setting of material properties, the boundary condition, applied method of loads and the definition of contact. Hence, the configuration setting of the FE model of ship and bridge pier will be validated by testing data to ensure the reliability of numerical results. The report of measurement data of the impact test (Consolazio et al., 2006) is used in the validation of FE model with FSI method. The impact speed of the barge is 0.95 m/s in the test, which is set as initial impact velocity in the numerical simulation. The stress distributions of the barge and bridge pier in the numerical simulation are showed in Figure 5. It can be seen that the damages of the ship bow are mainly distributed in the area of collision, and corresponding structural deformation is around 0.25 m at the moment with maximum impact force. The maximum von Mises stress on the barge and bridge pier are 300 MPa and 13.9 MPa.

Due to the spikes of data, it is necessary to remove the high-frequency part for representing the impact dynamic response of structure. An examination of pier deformation profiles shows the top of the pier moves significantly less than the point of impact as a result of the inertial effect of the superstructure

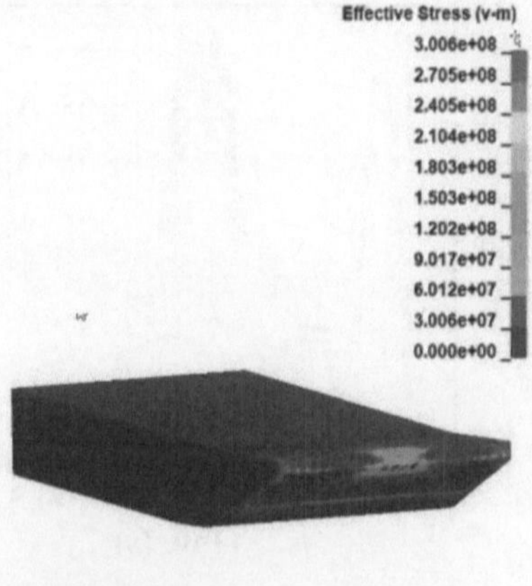

(a) Striking barge

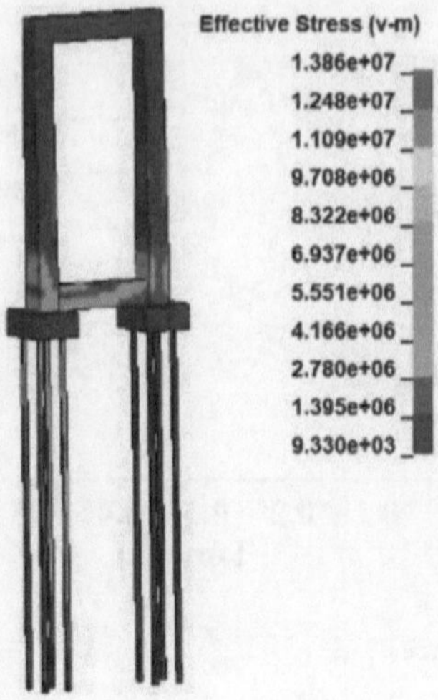

(b) Bridge pier

Figure 5. The stress distributions of the barge and bridge pier in the numerical simulation.

(Chopra, 2001). The peak dynamic force is the largest impact force computed during the simulation, which is not representative of the design structural demands that engineers need to consider, because the structure has not had time to respond to the rapid change in loading. The average filtering method is generally used to deal with the impact force data of the FE analysis, and the corresponding expression is given by.

$$F_f(t) = \frac{\int_{t-\frac{\Delta t}{2}}^{t+\frac{\Delta t}{2}} F(t)dt}{\Delta t} \tag{5}$$

where *Ff*(t) is the impact force after filter, *F(t)* is the origin impact force in the numerical simulation, and *Δt* is the moving average window of a selected time interval.

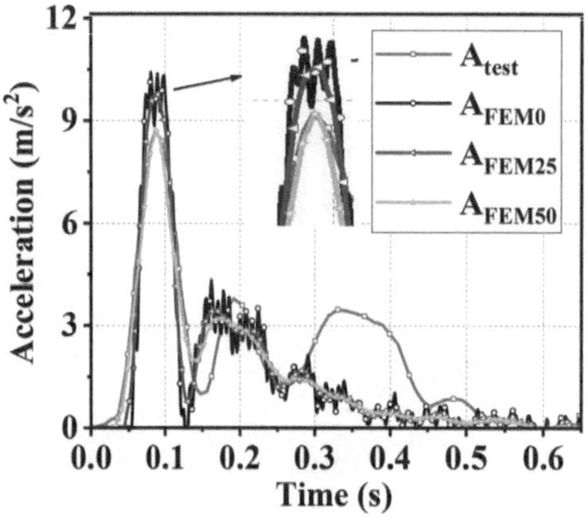

(a) Acceleration

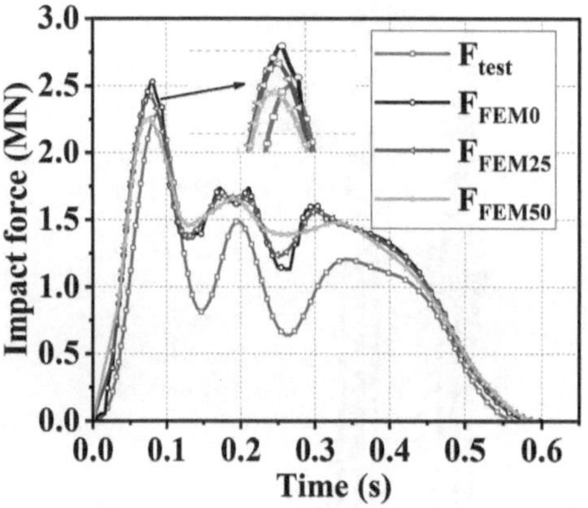

(b) Impact force

Figure 6. Time history curves of impact force and acceleration in the test and FE analysis with FSI method.

Table 7. Maximum value of the acceleration and impact force.

Data	Test	FEM Origin	FEM △t=25 ms	FEM △t=50 ms
Acceleration (m/s^2)	8.71	10.42	9.77	8.71
Force (MN)	2.29	2.53	2.42	2.25

It is important to choose reasonable time interval for the moving average window that is used to filter out high spikes for the analysis of dynamical response. However, if the selected time interval was too long or too short, the filtered data would be inappropriate that could not reflect the real dynamic response. The peak average force with twenty-five and fifty milliseconds are considered herein (Severino, 2005). The time histories of the acceleration and impact force in the FE analysis and experiment are showed in Figure 6 Time history curves of impact force and acceleration in the test and FE analysis with FSI method, in which the subscript of FEM0, FEM25, FEM50 means the numerical origin data, filtering time interval with 25 ms and 50 ms in the FE analysis. The impact durations between them are very close. Due to the simplification of the abutment and pile-soil models during the process of modeling, the response of the pier under impact loading is different from the test. There are three peak values of acceleration in the test, and two peak values in the simulations, which might be caused by the water current that is not considered in the FE analysis. The maximum value of the acceleration and impact force are showed in Table 7. It can be seen that the peak acceleration and impact force in the FE analysis without considering data filter is 19.7% and 10.5% lager than that in the test. However, the peak impact force filtered with time interval 50 ms is 2.25 MN, which very close to that in the test 2.29 MN. There is only 1.7% difference between them. Hence, it is suggested to filter the numerical data by average filtering method with 50 ms time interval.

4.2 *Effect of kinematic energy*

The time history curves of impact force at the same kinematic energy are shown in Figure 7. When the kinetic energy of the impacting ship is doubled, the collision duration increases and the maximum impact force nearly doubles as well. Additionally, there is another peak appears on the impact curve. The higher the energy of ship, the stronger and longer the collision.

In contrast, when ships of different weights and velocities collide with the same kinetic energy, the maximum impact force does not differ significantly. However, a heavier ship would result in a collision with a longer duration. According to formula 5, when a ship has a larger tonnage, its speed must be lower to maintain the same kinetic energy as a smaller ship. This slower speed results in a more gradual release of energy during the collision. As a result, the peak force occurs later and the collision lasts longer. A larger ship with a lower speed distributes its energy over a longer time, while a smaller ship with higher speed releases its energy more quickly, causing a shorter collision duration. This difference in energy release leads to longer collision duration and delayed peak impact force for larger ships.

The maximum stress distributions of barge are shown in Figure 8. Under the same kinetic energy, the maximum stress differences between barges with different tonnage and velocities are relatively small. Despite the differences in ship weight and velocity, the distribution of the impact force is similar throughout the structure. The maximum stress distributions of bridge are shown in Figure 9. The maximum stress on the bridge under different velocities with the same kinematic energy are 219 MPa and 225 MPa. The difference between them is small. Stress concentrations are observed in the area of the

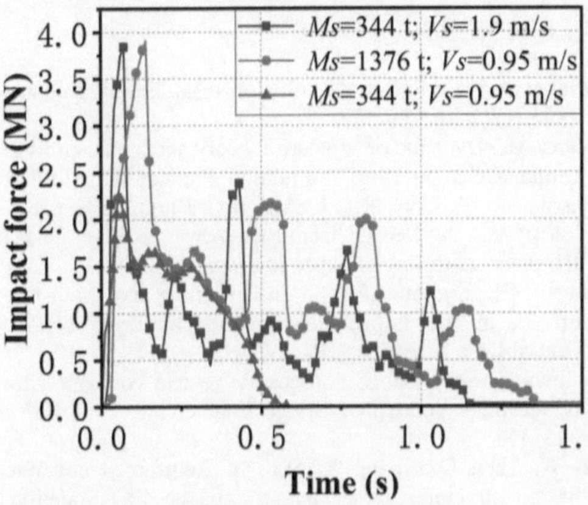

Figure 7. Time history curves of impact force.

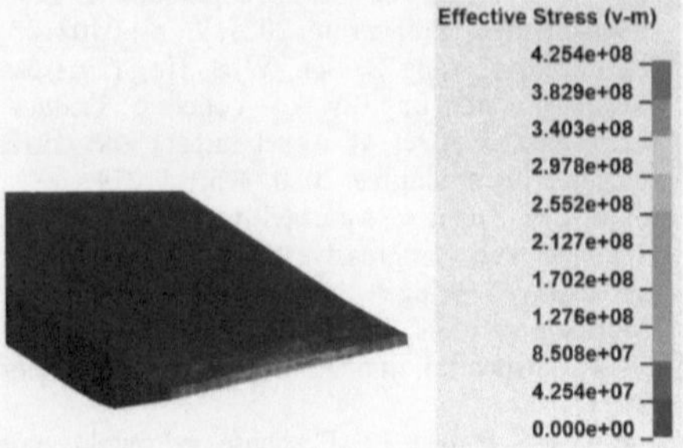

(a) M_s=344 t; V_s=0.95 m/s

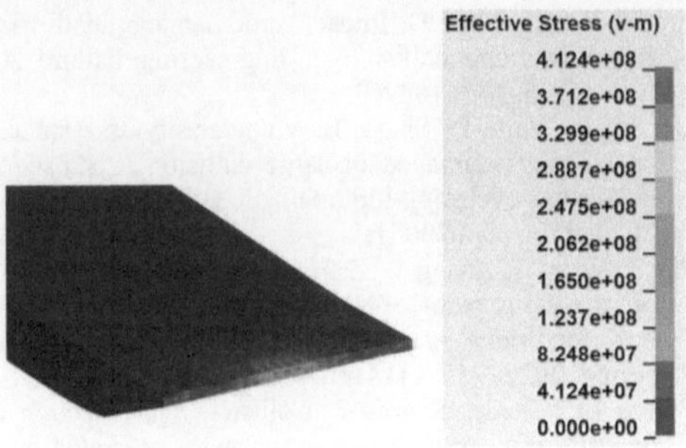

(b) M_s=1376 t; V_s=1.9 m/s

Figure 8. Maximum stress distributions of barge.

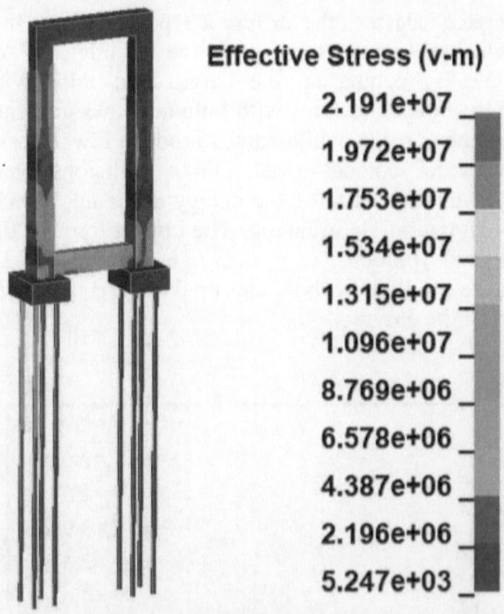

(a) M_s=344 t; V_s=0.95 m/s

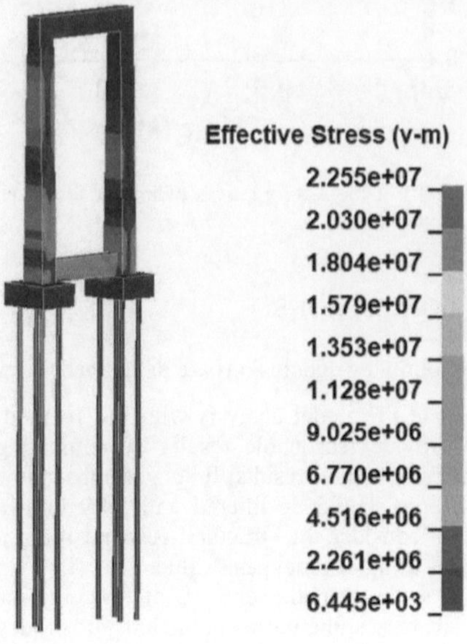

(b) M_s=1376 t; V_s=1.9 m/s

Figure 9. Maximum stress distributions of bridge.

impact block. In additional the stresses are mainly distributed in the pile on the collision side and no stress is generated on the other side.

4.3 *Effect of impact velocity*

The time history curves of impact force under different velocities are shown in Figure 10. The maximum impact forces are recorded in Table 6. As the impact velocity increases, the peak impact force also becomes larger and the duration of the collision becomes longer. When V_s=4 m/s, the peak impact force reaches 6.32 MN. In additional, as V_s>0.95 m/s, there is a fourth peak in the time history curves of impact force. The larger the kinetic energy of the impacting ship, the larger the peak impact force generated. Despite the different types of ships, this conclusion is consistent with the findings of Ye (2023). By comparing the barges and ships with bulbous bows, the ships with bulbous bows generate higher peak force, while barges produce lower peaks from wider contact areas. Barge collisions have longer duration because the energy dissipates slowly through overall deformation. The impact force of the ship with bulbous bow decreases more quickly, because the bulbous bow can be damaged to absorb most of the energy.

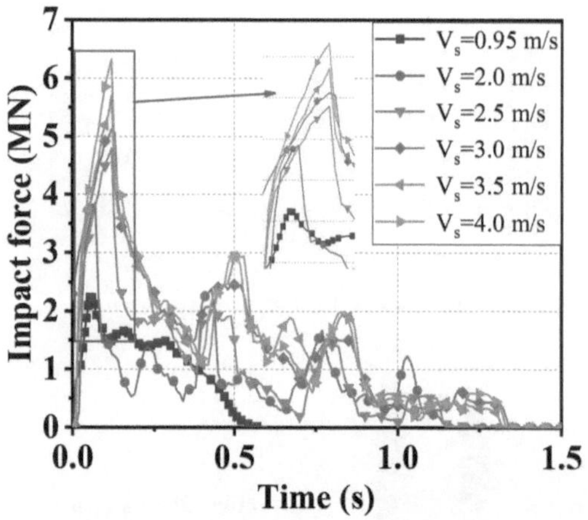

Figure 10. Time history curves of impact force under different velocities.

5 CONCLUSIONS

The following conclusions are developed.

(1) The FE model analysis with FSI method could provide reasonable results by comparing with experimental results. It suggests that the impact force should be filtered with fifty milliseconds to consider the structural respond time, instead of using original peak value.
(2) As the kinetic energy of the impact ship increases, the more energetic ship collides more violently and for a longer period of time.
(3) For ships with the same kinetic energy, the maximum impact force remains relatively consistent regardless of the ship's weight or velocity. However, heavier ships result in longer collision durations due to their lower speed which leads to a more gradual release of energy.
(4) As the impact velocity increases, the peak force becomes larger and the collision lasts longer. Higher velocities also cause additional peaks in the force-time curves. There are differences between barges and ships with bulbous bow in the time history of impact force.

REFERENCES

Carlebur A F C. Full-scale collision tests. Safety science, 1995, 19(2-3): 171–178.

Chopra AK. Dynamic of structures: theory and applications to earthquake engineering, 2nd Edition, Prentice Hall, 2001.

Consolazio GR, Cook RA, McVay MC. Barge impact testing of the St. George Island Causeway Bridge, Phase III: physical testing and data interpretation. 2006.

Cowper, GR, Symonds PS. Strain-hardening and strain-rate effects in the impact loading of cantilever beams. Brown Univ Providence Ri. 1957.

E1-Tawil S, Severino E, Fonseca P. Vehicle collision with bridge piers, Journal of Bridge Engineering, 2005,10(3): 345–356.

Fan W, Shen D, Huang X, Sun, Y. Reinforced concrete bridge structures under barge impacts: FE modeling, dynamic behaviors, and UHPFRC-based strengthening[J]. Ocean Engineering, 2020, 216: 108116.

Fan W, Sun Y, Sun W B, Huang, X. Effects of corrosion and scouring on barge impact fragility of bridge structures considering nonlinear soil–pile interaction[J]. Journal of Bridge Engineering, 2021, 26(8): 04021058.

Fan W, Sun, Y, Yang C, Sun, W B, He, Y. Assessing the response and fragility of concrete bridges under multi-hazard effect of vessel impact and corrosion[J]. Engineering structures, 2020, 225: 111279.

Gholipour G, Zhang C, Mousavi AA. Non-linear numerical analysis and progressive damage assessment of a cable-stayed bridge pier subjected to ship collision. Marine Structures, 2020, 69: 102662.

Jones N. Structural impact. Cambridge university press, 2011.

Kameshwar S, Padgett J E. Response and fragility assessment of bridge columns subjected to barge-bridge collision and scour[J]. Engineering Structures, 2018, 168: 308–319.

Larsen OD. Ship collision with bridges: The interaction between vessel traffic and bridge structures. IABSE, 1993.

Liu X, Wu H, Chen D. Impact force-damage depth model for ship-bridge collision[J]. Engineering Failure Analysis, 2025, 167: 108963.

Oppong K, Saini D, Shafei B. Vulnerability assessment of bridge piers damaged in barge collision to subsequent hurricane events[J]. Journal of Bridge Engineering, 2020, 25(8): 04020051.

Pan J, Wang T, Zhang W Z, Huang, S W, Xu, M C. Study on the assessment of axial crushing force of bulbous bow for bridge against ship collision[J]. Ocean Engineering, 2022, 255: 111411.

Perera LP, Guedes Soares C. Collision risk detection and quantification in ship navigation with integrated bridge systems. Ocean Engineering, 2015, 109: 344–354.

Sha Y, Amdahl J. Numerical investigations of a prestressed pontoon wall subjected to ship collision loads[J]. Ocean Engineering, 2019, 172: 234–244.

Sha Y, Hao H. Laboratory tests and numerical simulations of barge impact on circular rein-forced concrete piers. Engineering structures, 2013, 46: 593–605.

Wang JJ, Song YC, Wang W. Calibrations of numerical models by experimental impact tests using scaled steel boxes. Engineering Structures, 2018, 173: 481–494.

Ye X, Fan W, Sha Y. Fluid-structure interaction analysis of oblique ship-bridge collisions. Engineering Structures, 2023, 274: 115129.

Zhao HO. LS-DYNA Dynamics Analysis Guide[M]. 2003.

Trends in Collision and Grounding of Ships and Offshore Structures – Zhu & Guedes Soares (Eds)
© 2026 The Author(s), ISBN 978-1-041-16361-9

Analysis of structural crashworthiness of submarines in ship-submarine collision

S.Y. Choi, I.J. Ban & S.C. Shin
Department of Naval Architecture and Ocean Engineering, Pusan National University, Busan, South Korea

D.H. Seo
Bureau Veritas, Busan, South Korea

ABSTRACT: Submarines are exposed to the potential risk of collisions with marine structures or ships during operation, and such accidents can result in significant human and material damage, as well as cause substantial losses in military operations. While research on ship-to-ship collisions has been actively conducted, studies on submarine collisions are relatively lacking due to military confidentiality. This study analyzes structural damage by assuming a scenario where the bow of a ship collides with a specific part of the submarine. To evaluate the damage behavior of various submarine structures, design variables such as hull plate thickness, reinforcement size, and spacing were set, and several models were created based on these factors. This research aims to evaluate the crashworthiness of submarines in ship-submarine collisions based on various design parameters and scenarios. The results of the research can serve as predictive data for the initial damage assessment in the event of a real-world submarine collision. Furthermore, they can provide a fundamental basis for the establishment and improvement of structural design criteria incorporating the Accidental Limit State (ALS) concept in future collision scenarios.

1 INTRODUCTION

1.1 *Research background*

Submersible offshore structures such as submarines operate under extreme environmental conditions and are particularly vulnerable to accidental collisions due to limited visibility and operational constraints. These incidents can result in severe structural damage, casualties, and environmental harm. However, research on submarine collisions remains limited, primarily due to military confidentiality and restricted data access. While ship-to-ship collision studies are relatively abundant, few have addressed localized damage and plastic deformation mechanisms in submarine collisions.

In this study, a collision scenario was developed based on a real-world incident involving the USS Greeneville, a U.S. Navy submarine, and the Japanese fisheries training ship Ehime Maru in 2001 (Drumheller & Benoit, 2004). The scenario assumes a submarine with only its sail portion exposed above the water surface colliding with the bow of a large container ship. This type of event is characterized by significant energy transfer between two massive marine structures, inducing substantial plastic deformation in the submarine's hull. A considerable portion of the impact energy is absorbed through structural deformation of the submarine.

According to real-world damage cases and numerical simulation results, impact energy is primarily absorbed through the mechanism:

plastic tensile deformation such as denting and rupture of the shell plating, and then folding and crushing deformations of internal structures like webs, decks, and bottoms (Zhang et al., 2019). Particularly during impact events, local buckling, yielding, and eventual fracture of the hull and stiffeners are observed (Paik & Pedersen, 1996), making quantitative evaluation of deformation magnitude and absorbed energy essential.

This study focuses on plastic deformation behavior, a key factor in assessing the residual strength and structural integrity of submarines after collision. Four representative design parameters—shell thickness, frame spacing, frame size, and deep frame size—were derived from initial scantling formulae and varied independently. Finite element simulations were conducted for each case up to the point where plastic deformation stabilized.

The results highlight the effects of each parameter on deformation and structural stability, providing valuable data for future design improvements under the Accidental Limit State (ALS) framework.

DOI: 10.1201/9781003684404-22

Figure 1. USS greeneville undergoing repairs after offshore collision.

2 NUMERICAL MODELING AND METHODOLOGY

2.1 *DARPA SUBOFF (struck structure)*

2.1.1 *Geometric modeling of DARPA SUBOFF*

The submarine model was based on the DARPA SUBOFF hull form, scaled to 90 meters to represent a typical single-hull configuration. The impact was assumed at the midsection.

To facilitate the implementation of the MCOL technique (Section 2.3), the forward and aft sections were modeled as rigid bodies, and internal components such as ballast tanks, trim tanks, and torpedo tubes were omitted for simplicity.

The primary structural elements, including the pressure hull shell plating and ring stiffeners, were modeled in detail using initial scantling formulae. As shown in Figure 2, ring frames were symmetrically arranged around the impact location to ensure structural uniformity.

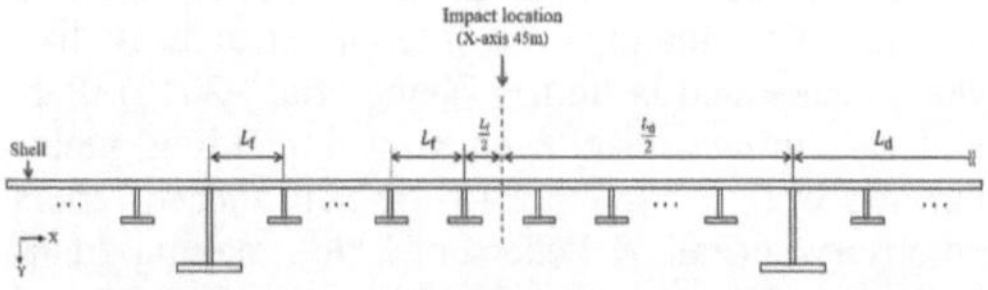

Figure 2. Ring frame layout centered on impact location.

The submarine's geometry and principal dimensions were defined based on initial design requirements and served as the baseline for all variant cases. Figure 3 and Table 1 show the main dimensions of the DARPA SUBOFF model.

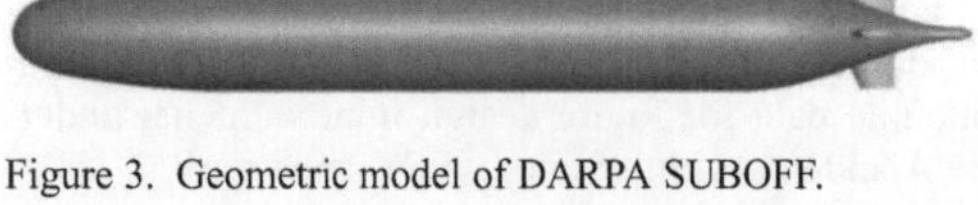

Figure 3. Geometric model of DARPA SUBOFF.

Initial scantling formulae of the struck structure are summarized in the Table. 2 below (Oh, 2010).

Table 1. Principal dimensions of DARPA SUBOFF.

Overall Length	m	90
Diameter	m	10.5

Table 2. Suggestion of initial scantling formulae.

Parameter	Symbol	Initial Scantling Formulae
Shell Thickness	h	$0.790 \times \frac{pR}{\sigma_{0.2}}$
Frame Space	L_{fr}	$1.557 \times \sqrt{R \times h}$
Web Height	h_w	$7.552 \times h$
Web Thickness	t_w	$0.645 \times h$
Flange Width	b_{fl}	$3.827 \times h$
Flange Thickness	t_{fl}	$1.414 \times h$

2.2 *Material model of DARPA SUBOFF*

HY-100 high-strength steel, known for its toughness and collision resistance, was used for the submarine's pressure hull. Its impact behavior was modeled using the Johnson–Cook constitutive model (MAT_015) in LS-DYNA, which incorporates strain-rate hardening and thermal softening. The material parameters were calibrated to match HY-100 properties, and the formulation is shown in Table 3 (Holmquist, 1987).

Table 3. Strength model constants for HY100.

Strength Constants For$\sigma = (A + B\varepsilon^n)(1 + C\ln\dot{\varepsilon}^*)(1 - T^{*m})$		
A	MPa	758
B	MPa	402
n		0.26
C		0.011
m		1.13

To capture fracture behavior, a Johnson–Cook failure model was also applied. Element erosion was triggered when the cumulative damage parameter reached a critical value. The failure strain is defined by the equation in Table 4 (Holmquist, 1987):

Table 5 summarizes the major parameters used in the MAT_015 model for HY-100 steel. (Holmquist, 1987):

2.3 *KCS (striking structure)*

2.3.1 *Geometric modeling of KCS*

To obtain conservative results, a 3600-TEU class container ship (KCS) developed by KRISO was selected as the striking body. Only the bow was modeled as deformable with detailed stiffeners, while the rest of the hull was treated as rigid. HT32 steel and a shell thickness of 19 mm were used to reflect the structural characteristics of the bulbous bow, with additional stiffeners added for fidelity.

Table 4. Fracture model constants for HY100.

Fracture Constants		
For $\varepsilon_f = [D_1 + D_2 EXP(D_3\sigma^*)][1 + D_4 \ln \dot{\varepsilon}^*][1 + D_5 T^*]$		
D_1		0.61
D_2		2.07
D_3		-0.5
D_4		0.01
D_5		0.0
Spall Stress, σ_{Spall}	MPa	5600
ε_f^{min}		0.034
$\sigma^* = \sigma_m/\bar{\sigma}$		
$\sigma^* = \sigma_m/\bar{\sigma}$		
$^* = (T - T_{Room})/(T_{Melt} - T_{Room})$		

Table 5. Material properties of HY100.

ELASTIC PROPERTIES		
Elastic Modulus, E	GPa	207
Poisson's Ratio, ν		0.3
Shear Modulus, E/2(1+ν)	GPa	79
Bulk Modulus, E/3(1-2ν)	GPa	172
THERMAL PROPERTIES		
Density, ρ	kg/m³	7746
Conductivity, k	W/mK	34
Specific Heat, c_p	J/kgK	502
Melting Temperature, T_{melt}	K	1793

To better represent the vessel's kinetic energy, the fully loaded condition was considered by applying a point mass corresponding to the ship's displacement and center of gravity. This approach enhanced the accuracy and realism of the collision simulation. The geometry and main dimensions of the KCS are shown in Figure 4 and Table 6 (Van et al., 1997).

Figure 4. Geometric model of KCS.

Table 6. Principal dimensions of KCS.

Lpp	m	230
Lwl	m	232.5
Bwl	m	32.2
D	m	19
T	m	10.8
Displacement	m³	52,030

2.3.2 *Material model of KCS*

The KCS structure employed the MAT_024 material model in LS-DYNA to simulate the elasto-plastic behavior of the bulbous bow, where significant deformation is expected during collision. HT32 steel was used for the shell plating and mild steel for the stiffeners. The stiffener layout replicated the actual structural configuration to improve the realism of the analysis. Strain-rate sensitivity was modeled using the Cowper–Symonds formulation, as given in Equation (1) (Cowper & Symonds, 1957).

$$\sigma = \sigma_0 \left[1 + \left(\frac{\dot{\varepsilon}}{C} \right)^{1/P} \right] \quad (1)$$

where: σ_d = dynamic yield stress; σ_0 = static yield stress; $\dot{\varepsilon}$ = strain rate and C, P = constants of Cowper–Symonds relation.

To accurately simulate material behavior up to final fracture, a failure criterion was defined by specifying a critical plastic strain value. Element erosion was activated once this threshold (FAIL) was reached. The key material parameters used in the MAT_024 models are summarized in Table 7 (DNV GL, 2018).

Table 7. Material properties of Mild & HT32.

Property		Mild	HT32
Yield Stress	MPa	235	315
Elastic Strain	%	0.2	0.20
Ultimate Stress	MPa	450	530
Critical Strain	%	20	16.7
Density	kg/m³	7850	7850
Young's Modulus	N/m²	2.06×10^{11}	2.06×10^{11}
Poisson's Ratio		0.3	0.3
Tangent Modulus	MPa	1085	1303
Hardening Parameter		1.0	1.0
Strain Rate (C)	s^{-1}	40.4	3200
Strain Rate (P)	s^{-1}	5.0	5.0

2.4 *MCOL-analysis software and motion equation*

MCOL is a rigid-body motion analysis technique commonly used in naval architecture and offshore engineering. It simulates vessels' six degrees of freedom (6-DOF) motion during events such as collisions or groundings by incorporating the inertial and damping effects of surrounding fluids.

The MCOL method is integrated with structural analysis software to reflect hydrodynamic forces caused by changes in the vessel's position and velocity at each simulation timestep. In this approach, the vessel is modeled as a rigid body. The overall motion response, influenced by external and hydrodynamic forces, is governed by the equation (2) (Le Sourne et al., 2003; Cui et al., 2024).

$$[M + M_\infty]\ddot{x} + G\dot{x} = F_W(x) + F_H(x) + F_V(x) + F_C$$

Where x = earth-fixed position of the center of mass matrix; $\dot{x}$ = velocity vector; = acceleration vector; M = structural mass matrix; M_∞ = added mass matrix; G = gyroscopic matrix; F_W = Wave damping force; F_H = restoring force; F_V= viscous force; and F_C = Contact force.

This governing equation is integrated over time to simulate the vessel's motion response and has been widely applied to numerical motion analyses of ship collisions and groundings (Le Sourne et al., 2003; Cui et al., 2024). The hydrodynamic parameters required for the analysis, such as the added mass matrix, hydrostatic restoring matrix, buoyancy parameters, and wave damping matrices for the body, are pre-calculated using Ansys-AQWA and imported via a .mco file. This approach enables realistic modeling of complex fluid–structure interaction (FSI) phenomena without explicitly meshing the fluid domain, thereby ensuring both computational efficiency and accuracy (Cui et al., 2024).

Furthermore, previous studies have reported significant differences between simulations that neglect hydrodynamic forces by fixing the body and those that incorporate MCOL-based hydrodynamic effects, particularly in terms of rebound motion, velocity change, and absorbed energy following a collision (Le Sourne et al., 2001; Le Sourne et al., 2003). These findings highlight that incorporating the MCOL technique is essential for producing realistic collision simulation results (Le Sourne et al., 2003).

3 FINITE ELEMENT ANALYSIS SETUP

3.1 *Element type and mesh size*

Shell elements were used for both the submarine and container ship. A 200 mm mesh was applied to the submarine and the deformable region of the KCS for contact accuracy, while a coarser 1000 mm mesh was used for the KCS rigid region to improve computational efficiency.

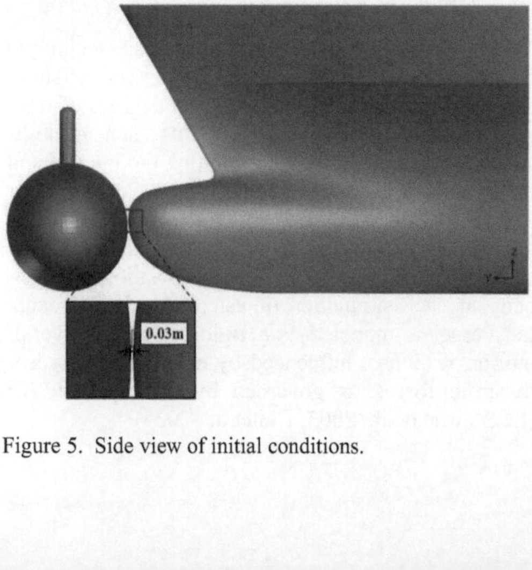

Figure 5. Side view of initial conditions.

3.2 *Initial conditions*

The collision analysis was defined by impact location and velocity. The impact occurred at the submarine's midsection, 45 meters along its length, where the bulbous bow of the KCS strikes. A point mass was applied to align the contact point with the submarine's center of gravity (COG).

A collision velocity of 12 knots and a 90° impact angle were used to represent a typical coastal operating scenario. The initial setup is shown in Figure 5.

3.3 *Contact modeling*

The physical contact phenomena between the striking and struck structures were modeled using a contact algorithm.

1. Interaction contact between the KCS and the submarine,
2. Self-contact within the KCS,
3. Self-contact within the submarine.

Both static and dynamic friction coefficients were set to 0.3.

4 DESIGN VARIABLES AND SCENARIO

4.1 *Design variable definition and classification*

In this study, key structural design variables were defined to evaluate the collision resistance of the submarine, and scenario-based analyses were conducted accordingly.

Four primary design variables were selected based on the initial scantling formulae (Oh, 2010): shell thickness, frame spacing, frame size, and deep frame size. These variables and their formulae are presented in Table 8.

Table 8. Initial scantling formulae.

Shell Thickness	h	$0.790 \times \frac{pR}{\sigma_{0.2}}$
Frame Space	L_{fr}	$1.557 \times \sqrt{R \times h}$
Web Height	h_w	$7.552 \times h$
Web Thickness	t_w	$0.645 \times h$
Flange Width	b_{fl}	$3.827 \times h$
Flange Thickness	t_{fl}	$1.414 \times h$

4.2 *Parameter conditions*

The maximum diving depth was set to 600 meters to estimate hydrostatic pressure in the initial scantling formulae, ensuring a balanced consideration of structural stability, feasibility, and design margin. Although actual operational depths are classified, this study adopted a conservative condition, reflecting typical safety factors used in submarine design.

To evaluate the influence of design diving depth variation on collision resistance, hydrostatic pressures corresponding to a ±50m deviation from the nominal depth were considered. The deep frame section was designed with a cross-sectional area three times larger and a second moment of area ten times greater than those of the ring frame section (Lee, 1996).

Reference values for the design variables, derived from the initial scantling formulae, are summarized in Table 9.

Table 9. Submarine design variables according to design.

Parameter	550m	600m (base)	650m
Shell Thickness	37	34	40
Frame Space	690	660	715
Web Depth	280	257	303
Web Thickness	24	22	26
Flange Width	142	132	154
Flange Thickness	53	49	57
Deep Web Depth	501	460	532
Deep Web Thickness	43	40	46
Deep Flange Width	254	237	270
Deep Flange Thickness	95	88	100

* Unit: mm

4.3 *Scenario*

A total of nine scenarios were constructed, as summarized in the Table 10 below.

Table 10. Design cases based on submarine design variables.

Case 1	Base
Case 2	Shell Thickness Reduced
Case 3	Shell Thickness Increased
Case 4	Frame Spacing Reduced
Case 5	Frame Spacing Increased
Case 6	Frame Size Reduced
Case 7	Frame Size Increased
Case 8	Deep Frame Size Reduced
Case 9	Deep Frame Size Increased

The scenarios consisted of a base case and a series of single-variable modified cases, where each structural design variable was individually altered. This setup was intended to analyze the influence of each variable on the collision resistance of the submarine.

5 RESULTS AND EVALUATION OF CRASHWORTHINESS

Based on the KCS–Submarine model, a total of nine simulation cases were conducted by individually varying one of four primary structural design parameters: shell thickness, frame spacing, frame cross-sectional dimensions, and deep frame cross-sectional dimensions. To assess the structure's collision resistance, two key response indicators were evaluated: indentation and absorbed energy.

Figure 6 Presents a summary of the penetration depth and absorbed energy results for the base case, which serves as a reference for comparative evaluation.

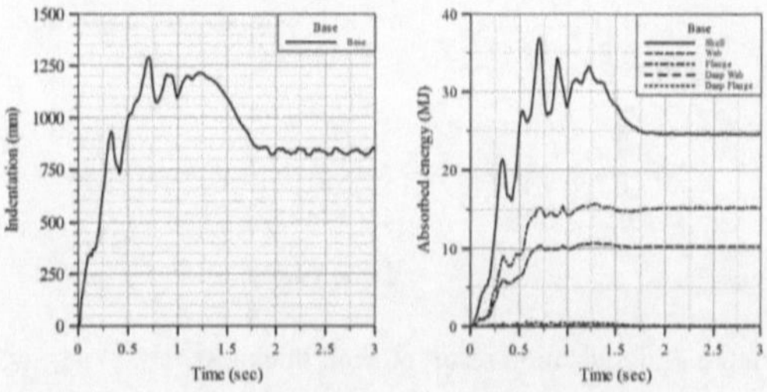

Figure 6. Indentation, absorbed energy of base case.

5.1 *Shell thickness*

The deformation response under collision loading exhibited notable sensitivity to changes in shell thickness. When the shell thickness was reduced, the maximum and permanent indentation depths increased to approximately 1429.9 mm and 1025.8 mm, corresponding to increases of 10.7% and 22.5%, respectively, compared to the baseline values of 1291.5 mm and 837.1 mm. These results indicate that a thinner shell leads to increased local deformation and more severe structural damage. In contrast, increasing the shell thickness significantly improved deformation resistance, with maximum and permanent indentation depths reduced to 670.6 mm and 347.5 mm, representing reductions of 48.1% and 58.5%, respectively, as shown in Figure 7.

These variations in indentation behavior were directly linked to changes in absorbed energy. When the shell thickness was increased, the total absorbed energy decreased markedly to 27.11 MJ, a 57.1% reduction from the base case (63.17 MJ), indicating enhanced stiffness and reduced plastic deformation. Conversely, reducing the shell thickness increased the absorbed energy to 68.47 MJ, 8.4% higher than the baseline, suggesting greater deformation and energy accumulation. As illustrated in Figure 8, a thicker shell contributed to improved energy dissipation efficiency and significantly mitigated structural damage, whereas a thinner shell promoted excessive energy absorption and local failure.

5.2 *Frame spacing*

Variations in frame spacing had a considerable influence on structural deformation under collision loading. When the frame spacing was reduced, the maximum

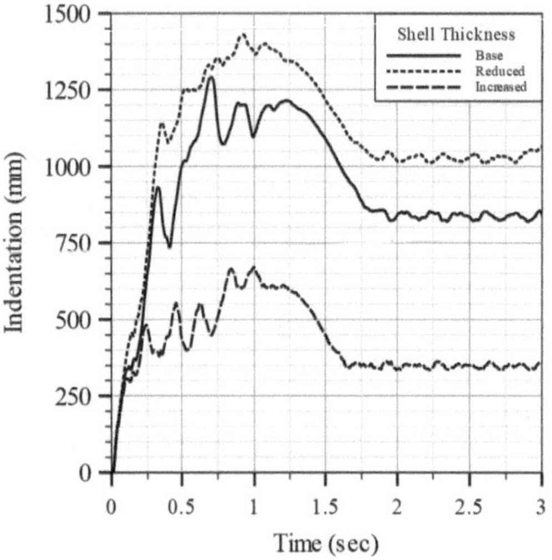

Figure 7. Indentation result of shell thickness.

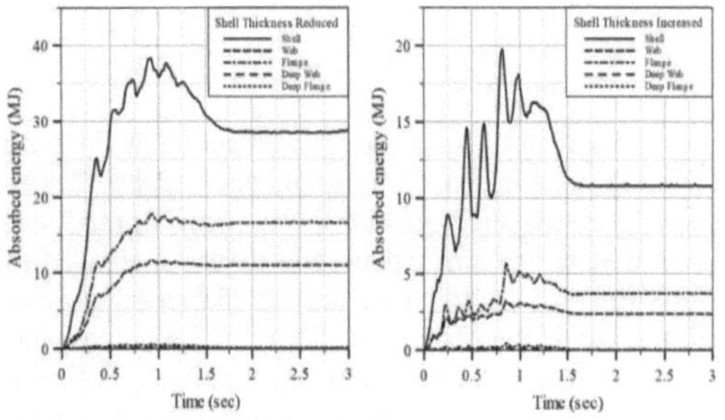

Figure 8. Absorbed energy result of shell thickness.

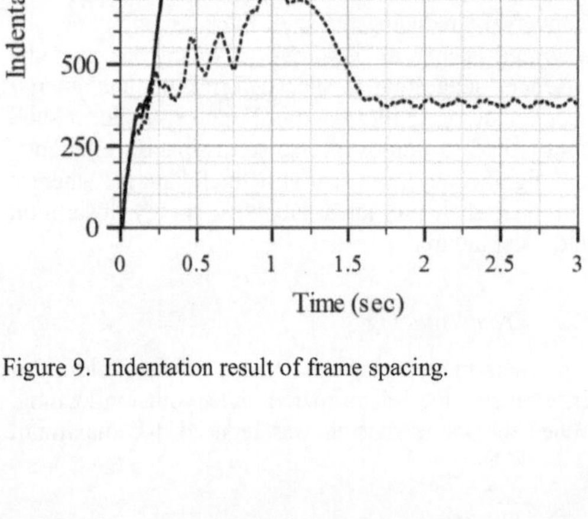

Figure 9. Indentation result of frame spacing.

and permanent indentation depths decreased to 766.6 mm and 379.9 mm, representing reductions of 40.6% and 54.6%, respectively, compared to the baseline values of 1291.5 mm and 837.1 mm. These results demonstrate that closely spaced frames effectively suppressed impact-induced deformation, thereby mitigating structural damage, as illustrated in Figure 9. In contrast, increasing the framc spacing yielded only marginal improvements, with maximum and permanent indentation depths reduced by just 4.9% (1228.0 mm) and 3.5% (807.6 mm), respectively, indicating negligible benefits in deformation resistance.

In terms of energy absorption, reducing the frame spacing resulted in a significant decrease in absorbed energy to 32.92 MJ, representing a 47.9% reduction from the base case (63.17 MJ), reflecting enhanced structural stiffness and limited plastic deformation during collision. Conversely, increased frame spacing resulted in a total absorbed energy of 57.88 MJ, showing only an 8.4% reduction and thus offering minimal improvement in energy dissipation capacity, as shown in Figure 10.

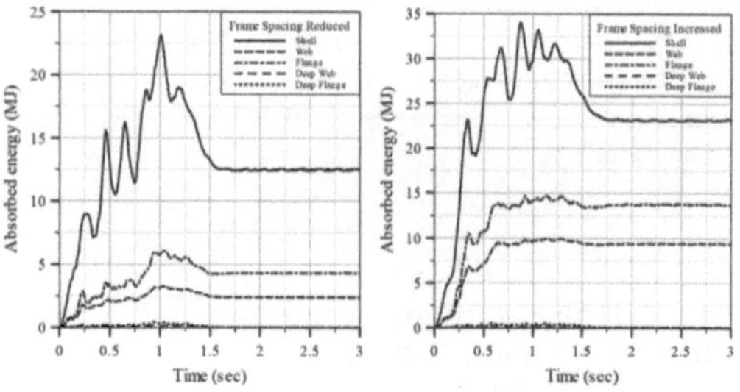

Figure 10. Absorbed energy result of frame spacing.

Overall, while narrower frame spacing markedly improved resistance to indentation and energy absorption, wider spacing neither significantly worsened deformation nor contributed meaningfully to impact load redistribution.

5.3 *Frame size*

Variations in the frame cross-sectional dimensions had the most pronounced effect on collision-induced structural responses. When the frame cross-sectional dimensions were reduced, the maximum and permanent indentation depths increased to 1610 mm and 1249.1 mm, corresponding to 24.7% and 49.2% increases relative to the base case, respectively. These results indicate a significant loss in structural stiffness and resistance. In contrast, an increase in frame cross-sectional dimensions led to a 53.2% reduction in maximum indentation (604.5 mm) and a 66.7% reduction in permanent indentation (279.3 mm), clearly demonstrating improved capacity to suppress local deformation, as shown in Figure 11.

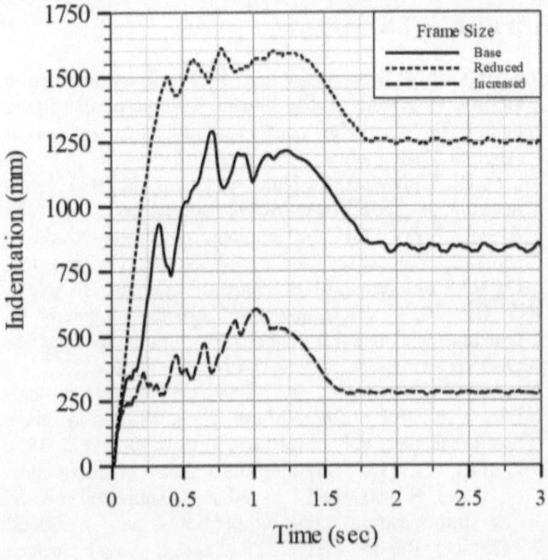

Figure 11. Indentation result of frame size.

Regarding energy absorption, reducing the frame dimensions resulted in a total absorbed energy of 76.19 MJ 20.6% higher than the baseline value of 63.17 MJ indicating increased plastic deformation due to diminished structural stiffness. Conversely, an increase in frame cross-sectional dimensions decreased the absorbed energy to 26.85 MJ, representing a 57.5% reduction, and demonstrating effective mitigation of impact energy through enhanced structural rigidity. Overall, an increase in frame cross-sectional dimensions improved structural stability by limiting indentation and reducing deformation-induced energy absorption, as clearly illustrated in Figure 12. In contrast, reduced frame dimensions led to intensified structural damage resulting from excessive deformation and greater energy accumulation.

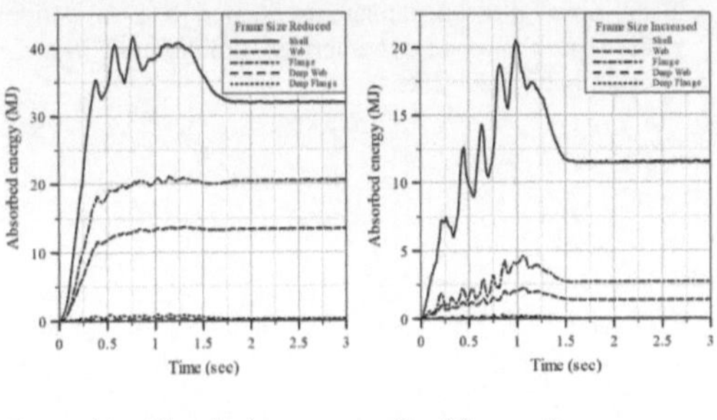

Figure 12. Absorbed energy result of frame size.

These findings underscore that frame stiffness is a key parameter governing both deformation suppression and energy dissipation mechanisms under collision conditions.

5.4 *Deep frame size*

The collision response due to variations in deep ring frame size exhibited more complex trends than other design variables. When the frame size was reduced, the maximum and permanent indentation depths decreased by 27.6% and 33.9%, reaching 935.0 mm and 553.4 mm, respectively. The corresponding absorbed energy also decreased to 41.3 MJ, a 34.4% reduction from the baseline, indicating more efficient energy distribution and reduced structural damage, as shown in Figure 13. In contrast, increasing the frame size yielded marginal improvements, with only a 0.8% decrease in maximum indentation (1281.5 mm) and a 0.5% reduction in permanent indentation (832.8 mm).

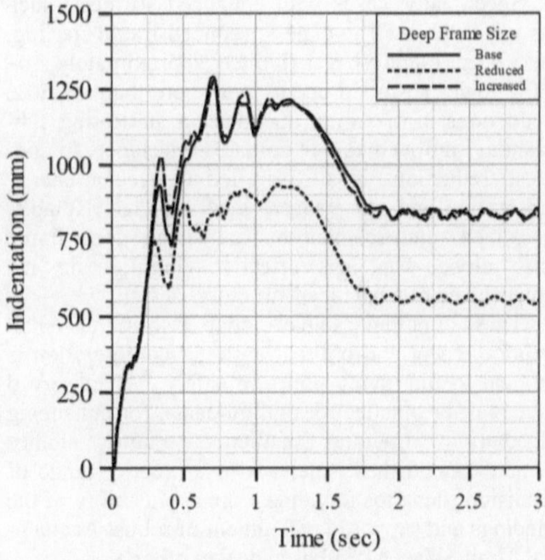

Figure 13. Indentation result of deep frame size.

The increased deep frame size resulted in a substantial reduction in absorbed energy, decreasing to 36.0 MJ a 42.9% reduction compared to the baseline. However, its influence on limiting shell indentation was negligible. This implies that while a stiffer frame improves structural resistance to deformation-induced energy dissipation, it does not significantly affect the indentation response. Accordingly, variations in frame stiffness had minimal impact on the deformation characteristics of the shell and adjacent stiffening members during collision events, as illustrated in Figure 14.

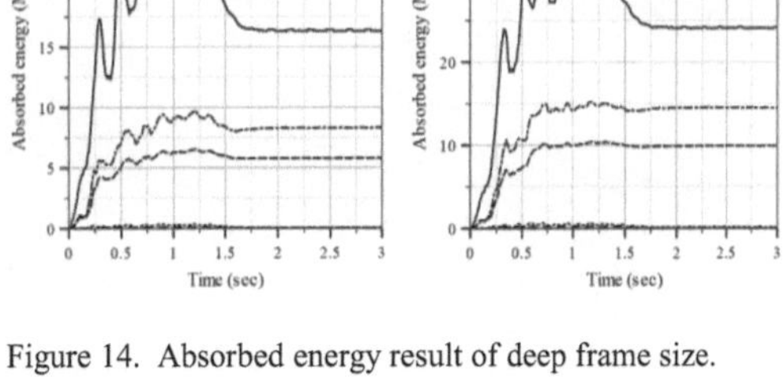

Figure 14. Absorbed energy result of deep frame size.

6 CONCLUSIONS

This study investigated the effects of single-variable changes in structural design parameters derived from initial scantling formulae at 50-meter diving depth intervals on the collision resistance of a submarine hull. The results showed a strong correlation between structural stiffness and both energy absorption and damage severity. Key parameters such as frame size and shell thickness had a pronounced effect on indentation and energy dissipation behavior, with increased structural stiffness generally resulting in reduced absorbed energy and shallower indentation depths.

Specifically, cases with enhanced stiffness, such as increased shell thickness, reduced frame spacing, and deep frame size, exhibited approximately 10–20% lower absorbed energy and more than 40–50% reductions in maximum indentation, indicating substantial improvements in collision resistance. In contrast, reductions in stiffness led to greater energy absorption and more severe local damage. Notably, localized reinforcement by increasing deep frame size alone was less effective, highlighting the importance of a balanced structural design.

These findings suggest that prioritizing shell reinforcement and optimal stiffener configuration is critical to improving collision safety. The observed trends offer practical design guidance for enhancing submarine structural resilience. Future studies should extend these analyses to a broader range of collision scenarios to validate the applicability of the findings and support development of robust Accidental Limit State (ALS)-based design criteria.

ACKNOWLEDGEMENTS

This work was supported by the Korea Institute of Energy Technology Evaluation and 438 Planning (KETEP), the Ministry of Trade, Industry, and Energy (MOTIE) of the Republic of Korea 439 (no.20224000000090).

REFERENCES

Cowper, G.R. & Symonds, P.S. 1957. Strain hardening and strain-rate effects in the impact loading of cantilever beams. Technical Report, Brown University Division of Applied Mathematics.

Cui, Z., Bi, Y., Davies, A., Du, S. & Huang, Y. 2024. Introduction of LS-DYNA MCOL solver coupling with Ansys Aqwa for the application in shipbuilding analysis. Proceedings of the 2024 International LS-DYNA Conference, Metro Detroit, Michigan, USA.

DNV GL. 2018. Determination of structural capacity by non-linear FE analysis methods. Recommended Practice DNVGL-RP-C208, Høvik: Det Norske Veritas.

Drumheller, K. & Benoit, W.L. 2004. USS Greeneville collides with Japan's Ehime Maru: Cultural issues in image repair discourse. Public Relations Review 30(2): 177–185.

Holmquist, T.J. 1987. Strength and fracture characteristics of HY-80, HY-100, and HY-130 steels subjected to various strains, strain rates, temperatures, and pressures. Technical Report NSWC TR 88-252, Naval Surface Warfare Center, Dahlgren, Virginia.

Lee, J.S. 1996. Ship and Offshore Structure. Seoul: Sejong Publishing Co.

Le Sourne, H., Couty, N., Besnier, F., Kammerer, C. & Legavre, H. 2003. LS-DYNA applications in shipbuilding. Proceedings of the 4th European LS-DYNA Users Conference, Ulm, Germany.

Le Soure, H., Donner, R. & Besnier, F. 2001. External dynamics of ship-submarine collision. Preprints of the 2nd International Conference on Collision and Grounding of Ships, Copenhagen.

Oh, D.H. 2010. A study on optimal structural design of submarine pressure hull. Master's thesis, Pusan National University, Busan, Korea.

Paik, J.K. & Pedersen, P.T. 1996. Modelling of the internal mechanics in ship collisions. Ocean Engineering 23(2): 107–142.

Van, S.H., Kim, W.J., Kim, D.H., Yim, G.T., Lee, C.J. & Eom, J.Y. 1997. Measurement of flows around a 3600TEU container ship model. Proceedings of the Annual Autumn Meeting, SNAK, Seoul: 300–304.

Zhang, S., Villavicencio, R., Zhu, L. & Pedersen, P.T. 2019. Ship collision damage assessment and validation with experiments and numerical simulations. Marine Structures 63: 239–256.

Trends in Collision and Grounding of Ships and Offshore Structures – Zhu & Guedes Soares (Eds)
© 2026 The Author(s), ISBN 978-1-041-16361-9

Numerical investigation on the anti-collision performance of fender for a spar-type floating offshore wind turbine

M.B. Du, Y.C. Zhang, Y.C. Yuan & W.Y. Tang
Shanghai Jiao Tong University, Shanghai, P.R. China

Z.Q. Hu
School of Engineering, Newcastle University, Newcastle upon Tyne, UK

ABSTRACT: The potential collision hazard between ships and floating offshore wind turbines (FOWTs) highlights the importance of researching cost-effective fenders. This paper aims to investigate the crashworthiness of a Spar-type floating offshore wind turbine equipped with a typical fender consisting of an outer plate and an inner rubber layer. Numerical simulations using LS-DYNA with user-defined subroutine are conducted to analyze the fender's dynamic response. The results demonstrate that the fender exhibits effective energy absorption and damping characteristics, reducing the maximum plastic strain by approximately 60%. Later, the protective effects of the fender with various parameters, including layer thickness, perforated hole and hole azimuthal were compared. Increasing layer thickness enhances energy absorption and reduces plastic deformation. Perforated rubber layers could perform better, although larger apertures slightly diminish the rubber's contribution. A perforated design outperforms solid fenders when the hole azimuthal is less than 60°.This study offers valuable insights for optimizing collision protection in FOWTs.

1 INTRODUCTION

In recent years, with the growing global demand for renewable energy, floating offshore wind turbines (FOWTs) have garnered widespread attention as an emerging form of wind power generation. However, FOWTs face numerous challenges in deep-sea environments, one of which is the risk of collision with ships. During routine maintenance operations, service vessels and passing ships are at high risk of colliding with the wind turbines, causing significant damage to the turbines' day-to-day operations. Furthermore, such collisions may pose serious threats to the structural integrity of the turbines and the surrounding marine ecosystem.

To address this potential risk, the design and implementation of fenders are of paramount importance. Fenders not only mitigate the impact forces on the turbine foundation and reduce structural damage, but also alleviate the dynamic response of the turbine nacelle, ensuring the normal operation of the generator system.

Han et al. (2019) designed four types of fenders: rubber, aluminum, rubber-aluminum, and aluminum-rubber. By analyzing collision forces, energy changes, and plastic strain, they studied the impact of material thickness and sequence on the fender's performance. Yue et al. (2021) proposed a novel fender using the fractal structure and compared it with traditional solid structures, examining the influence of rubber layer perforations and the number of segmented holes on impact resistance. Mozafari et al. (2022) used experimental testing and numerical simulations to investigate the impact response of aluminum foam, agglomerated cork, and polyurethane foam, comparing the impact resistance of fenders made from these three lightweight materials. Nie et al. (2024) proposed a novel steel fender assembly with quasi-negative stiffness, providing low stiffness during impact collisions, significantly reducing collision forces and top accelerations.

However, existing research and designs primarily focus on fixed offshore wind turbines, while studies on fenders for deep-sea floating wind turbines remain relatively scarce. It is important to note that there are some differences between floating and fixed offshore wind turbines in collision research. Due to the floating characteristics of floating offshore wind turbines, the way energy is transferred during collisions differs from that of fixed offshore wind turbines. Therefore, the study of collisions between FOWTs and vessels must take external dynamics into account, i.e. hydrodynamic loads and global motions of platform (Ren et al., 2022). Zhang

DOI: 10.1201/9781003684404-23

(2021) established a fluid-structure coupling calculation model for collisions between FOWTs and vessels, designing six types of anti-collision facilities and studying their protective principles. However, the full-coupling model requires significant computational time and is not suitable for the impact resistance analysis and optimization of fenders.

This paper investigates the crashworthiness performance of a Spar-type floating offshore wind turbine equipped with a typical fender comprising an outer steel plate and an inner rubber layer. Numerical simulations using LS-DYNA are conducted and both hydrodynamic as well as mooring loads are considered through a user-defined subroutine LOADUD. The crashworthiness performance of the fender is examined by varying the thickness of both the steel plate and rubber layer, as well as the aperture and angle of the rubber perforations. This study aims to explore the protective mechanisms and provide both theoretical insights and practical guidance for optimizing collision protection technologies in FOWTs.

2 NUMERICAL MODEL

2.1 *FE model of the striking vessel and the FOWT*

The FE model presented in this article consists of three components: the striking vessel, the FOWT, and the fender, shown as Figure 1.

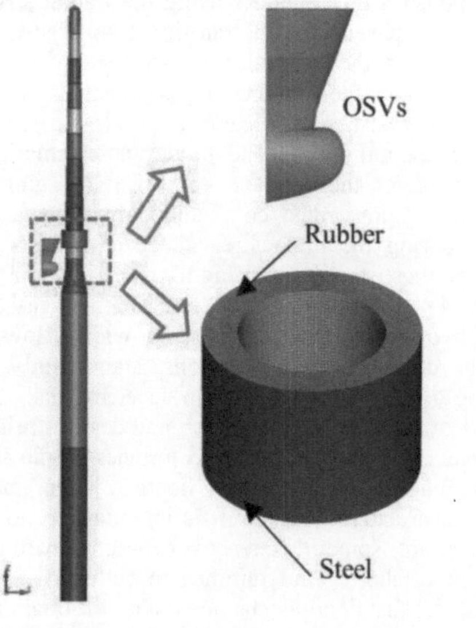

Figure 1. The finite element model.

A Hywind Spar-type floating platform with a 5 MW baseline wind turbine (Jonkman et al., 2009) were selected and the main properties are presented in Table 1.

A mesh convergence analysis for a tubular structure is conducted to decide the proper mesh size for FOWT. The tubular structure with the same

Table 1. The main properties of 5 MW Hywind Spar-type floating wind turbine.

5 MW Hywind Spar-Type Floating Wind Turbine	
Blade Length (m)	61.5
Blades Quantity	3
Nacelle CM Location Above Tower Top (m)	1.75
Rated Rotor Speed (rpm)	12.1
Elevation to Tower Base (Platform Top) Above SWL (m)	10
Elevation to Tower Top Above SWL (m)	87.6
CM Location of Tower Above SWL (m)	43.4
Total Draft (Tower Base) Above SWL (m)	120
Platform Top Above SWL (m)	10
Taper Top Below SWL (m)	4
Taper Bottom Below SWL (m)	12
Platform Diameter Above Taper (m)	6.5
Platform Diameter Below Taper (m)	9.4
CM Location of Platform Below SWL (m)	89.9
CM Location of whole FOWT System Below SWL (m)	78
Number of Mooring Lines	3
Fairleads Below SWL (m)	70
Radius To Anchors Form Platform Centreline (m)	853.87
Unstretched Mooring Line Length (m)	902.2
Total Mass (tons)	8066

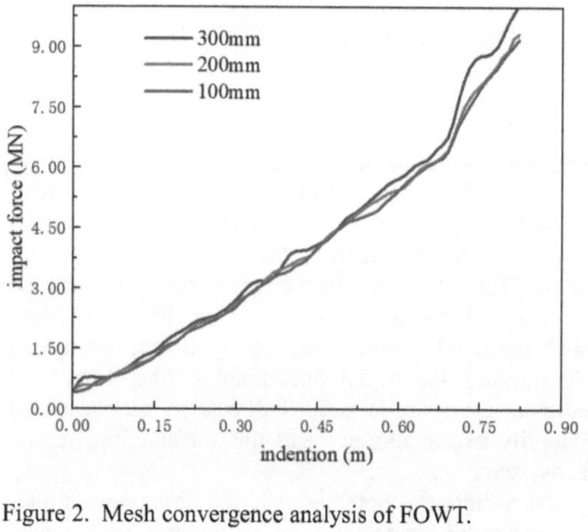

Figure 2. Mesh convergence analysis of FOWT.

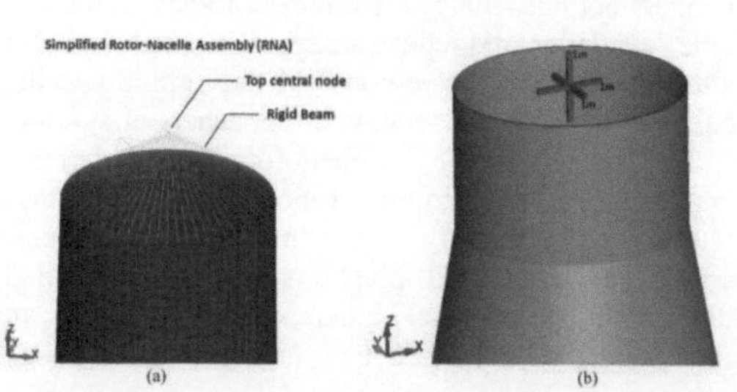

Figure 3. Simplified RNA system and rigid beam set in the FE model.

thickness as FOWT is fixed and crushed by a rigid bow. The results of different mesh sizes are compared in Figure 2. Considering the reasonable accuracy and computational efficiency, the 200 mm element sizes should be acceptable for collision area.

The FOWT was modeled using Belytchko-Tsai shell elements with 4 integration points through the thickness (Hallquist, 2013). The ballast tank was represented as a rigid part, while other components were deformable. A mesh size of 400 mm was used for parts which may not get large deformation, and a finer mesh of 200 mm was applied to the tower top and taper areas (ship collision zone). The rotor-nacelle assembly, the catenary mooring cable and the coordinate systems adopted the method proposed by Zhang (2022), shown as Figure 3.

The FOWT's plastic region below 18m is modeled as high-strength elastoplastic steel with power-law hardening, featuring a yield strength of 540MPa, a strength coefficient of 760MPa, a hardening exponent of 0.15, a mass density of 7850kg/m^3, and a Poisson's ratio of 0.3. Strain rate effects are neglected, and a constant failure strain of 0.3 is assumed. Above 18m, high-strength elastic steel is uniformly employed to prioritize motion response analysis and enhance computational efficiency.

The collision risk of offshore service vessels (OSVs) against OWTs has been highlighted (Christensen et al., 2001), prompting the selection of an OSV with a bulbous bow as the striking vessel. Only the bulb bow was modeled, with the ship's total mass of 4000 tons concentrated on this part to accurately simulate collision energy. During the simulation, the bow is treated as a rigid body, which. The shell element thickness is defined as 15mm, with a mesh size of 200mm in the collision zone and 400mm in non-critical areas.

The collision height is defined as 7 meters above the SWL (Still Water Level). The ship's initial velocity is set to 2 m/s along the X-axis, with constraints applied to the remaining five degrees of freedom. The OSV's movement induces surrounding fluid motion, modeled using added mass (Petersen et al., 1981). A constant added mass coefficient of 0.05 (Zhang et al., 2019) is applied for surge motion of OSV, which equals to 200 tons in this paper.

2.2 *FE model of the fender*

The fender is installed on the upper part of the Spar-type platform, with its center height positioned 7 meters above the SWL (Still Water Level). The overall structure has a height of 6 meters, with the inner rubber layer featuring an inner diameter of 3.25 meters. The rubber is enclosed by a steel protective shell. The mesh size for rubber and steel shell is set as 100mm, shown as Figure 1.

The rubber in the fender is modeled using the Ogden rubber constitutive model and 8 point hexahedral elements to more accurately capture its nonlinear mechanical behavior under large deformations. Additionally, the inclusion of hourglass energy in the energy balance ensures effective control over the influence of hourglass modes on the simulation results. Kim et al. (2023) presented a Ogden model through compression test data obtained from solid rubber fender models, and the data was adopted in this work, as shown in the Table 2 below.

Table 2. The parameters of the Ogden model.

μ_1/MPa	0.028
μ_2/MPa	5.814
μ_3/MPa	0.954
α_1	4.391
α_2	0.032
α_3	2.589

An elastoplastic model is adopted for the outermost shell material of the fender. The steel has a yield strength of 345MPa. The Young's modulus, mass density and Poisson's ratio are 212GPa, 7850kg/m^3 and 0.3, respectively. The tangent modulus of 1.29GPa is adopted to consider the kinematic hardening and the critical strain of 0.3 is used to consider the material failure.

In this study, automatic surface-to-surface contact is utilized to consider the structural interaction during ship collision. The static and dynamic friction coefficients for steel-steel contact are both set to 0.2 while the static friction coefficient of 0.9 and dynamic friction coefficient of 0.2 are adopted for rubber-steel contact (Yue et al., 2022). Furthermore, to prevent penetration within the rubber of the perforated hole, self-contact is applied to the rubber, with dynamic and static friction coefficients set to 0.4 and 0.6, respectively.

2.3 *External dynamics modeling*

This study models the hydrodynamics and mooring loads considering the coupled effects of internal and external dynamics in the FOWT collision. The method proposed by Zhang (2022) is utilized and implemented via the development of a subroutine LOADUD.

To simplify the analysis, this study assumes static water conditions, with hydrostatic forces comprising radiation forces and restoring forces. The added mass matrix $\mu_{jk}(\infty)$, radiation damping coefficients $B_{jk}(\omega)$, and hydrostatic restoring matrix $C^{Hydrostatic}$ were obtained through frequency-domain hydrodynamic analysis using the SESAM/HydroD. These parameters were stored in LOADUD for subsequent hydrostatic load calculations. The radiation forces were then calculated as:

$$F_{\mathrm{j}}^{Radiation}(t) = -\sum_{k=1}^{6} \left[\mu_{jk}(\infty)\ddot{x}_k(t) + \int_0^t K_{jk}(t-\tau)\dot{x}_k(\tau)d\tau\right] \quad (1)$$

Where, the first term on the right side refers to the added mass force, $\mu_{jk}(\infty)$ is the (j, k) component of the added mass matrix at infinite frequency. The second term on the right side is a convolution integral representing the wave radiation damping. $K_{jk}(t-\tau)$ is known as the retardation function and is calculated by knowing the damping coefficient in all frequency, given as:

$$K_{jk}(t) = \frac{2}{\pi}\int_{0}^{\infty} B_{jk}(\omega)\cos\omega t d\omega \tag{2}$$

Where $B_{jk}(\omega)$ is the radiation damping at frequency ω. Hydrostatic loads, including buoyancy, are calculated as:

$$F^{Hydrostatic} = \begin{pmatrix} 0 \\ 0 \\ \rho g V \\ 0 \\ 0 \end{pmatrix} - C^{Hydrostatic}x(t) \tag{3}$$

Where the first term on the right side is the buoyancy force, ρ is the fluid density, g is the gravitational acceleration constant and V is the displaced volume of fluid when the support platform is at the upright floating position. The second term indicates the change in the hydrostatic loads in 6DOF. $C^{Hydrostatic}$ is the linear hydrostatic restoring matrix and x(t) is the platform motion at time t.

The mooring system is not truly modelled in the FE model and the mooring forces are calculated in the user-defined subroutine (Jonkman, 2007). The mooring model is linear, and the calculated loads are the contribution from all mooring lines, expressed as:

$$F^{moortng} = F_0^{Lines} - C^{moortng}x(t) \tag{4}$$

Where $C^{moortng}$ is the linearized restoring matrix from whole mooring system and x(t) is the platform motion. For catenary mooring lines, F_0^{Lines} is the pretension at fairleads from the weight of the cable not lying on the seabed.

After obtaining the hydrostatic and mooring loads, the LS-DYNA structural analysis will be conducted with the calculated user loads in LOADUD. During the computational process, the loads are dynamically updated at each time step based on the current motion state, with the hydrostatic loads and mooring loads being applied to the kinematic reference point. The kinematic reference point of platform was on the SWL centre, which is also the central node of the cross beams, as shown in Figure 2b.

3 SHIP-OWT COLLISION ANALYSIS

3.1 *Energy balance verification*

The energy transfer with time during ship-FOWT collision process has been carried out. Both cases including fender and excluding fender were presented in Figure 3, respectively. The hourglass energy is found less than 5% of the total energy, and obvious energy balance is achieved, indicating that the results should be reliable.

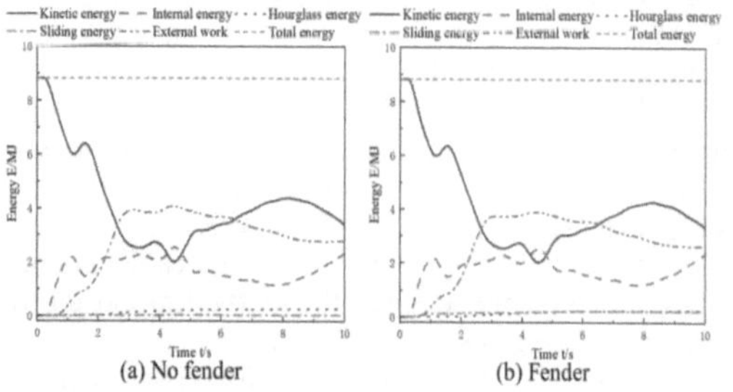

Figure 4. The energy curve of collision.

In both cases, the initial kinetic energy falls down when the ship impacts the FOWT. A considerable part of internal energy acquired for FOWT due to plastic deformation and elastic vibration. The struck body also gets some kinetic energy and starts to move around. In the meantime, external work from hydrodynamic loads and mooring loads rises due to the platform motion.

Table 3. Cases definition.

Case No	Thickness of steel shell	Thickness of rubber	Aperture diameter	Hole azimuthal
C1	No Fender	-	-	0
C2	10 mm	600 mm	-	0
C3	10 mm	900 mm	-	0
C4	10 mm	1200 mm	-	0
C5	10 mm	1500 mm	-	0
C6	30 mm	600 mm	-	0
C7	50 mm	600 mm	-	0
C8	30 mm	1500 mm	-	0
C9	50 mm	1500 mm	-	0
C10	10 mm	600 mm	70%	0
C11	10 mm	600 mm	50%	0
C12	10 mm	600 mm	30%	0
C13	10 mm	900 mm	70%	0
C14	10 mm	900 mm	50%	0
C15	10 mm	900 mm	30%	0
C16	10 mm	600 mm	70%	20°
C17	10 mm	600 mm	70%	40°
C18	10 mm	600 mm	70%	60°

In the following sections, by changing the thickness of the rubber, the steel plate, the aperture and angle of the rubber perforations, the sensitivity analysis for these parameters on the crashworthiness performance of the fender is conducted. The numerical cases are summarized as Table 3.

3.2 *Influence of fender thickness on crashworthiness performance*

3.2.1 *Influence of rubber thickness*

The collision force is depicted in Figure 5. As the thickness of the rubber component in the fender increases, the maximum collision force between the vessel and the fender also gradually increases. This could be explained that the added thickness of the rubber significantly increases the mass of the floating offshore wind turbine, resulting in a larger normal contact force during the collision.

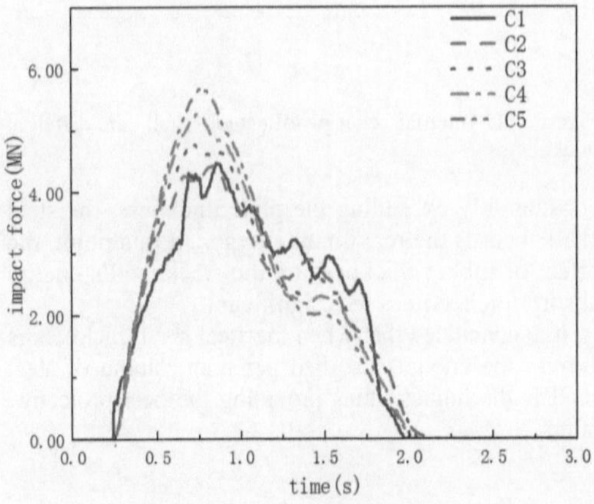

Figure 5. Contact force under different rubber thickness.

Figure 6 illustrates the structural internal energy of the FOWT. It can be seen that the maximum value of the case with the addition of a fender is consistently lower than that without a fender. This indicates that the fender effectively absorbs collision energy, thereby reducing the deformation of the tower. As the rubber thickness increases, the maximum internal energy of the FOWT decreases by 23.5%, 25.8%, 28.1%, and 29.2%, respectively, compared to the scenario without a fender. Initially, the energy absorption effect of the fender improves, but this improvement gradually slows down as the rubber thickness increases. When the reduction in maximum internal energy is normalized by the rubber volume, it becomes clear that the fender with a 600mm thickness offers the best energy absorption effect per unit volume.

At 0.85 seconds, when the collision force peaks, Figure 8 illustrates the stress-strain conditions in the collision region of the spar-type platform. It is evident that the fender efficiently safeguards the FOWT's foundation, absorbing the vessel's initial kinetic energy and mitigating the collision load, thereby preventing stress concentration. As the rubber thickness increases, the protective device absorbs more internal energy, leading to a significant reduction in both maximum stress and maximum strain on the spar-type platform. This highlights that a thicker rubber more effectively mitigates collision-induced damage.

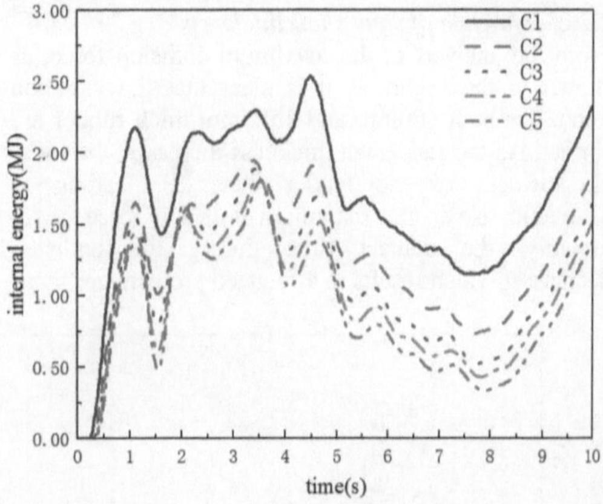

Figure 6. Internal energy of FOWT at different rubber thicknesses.

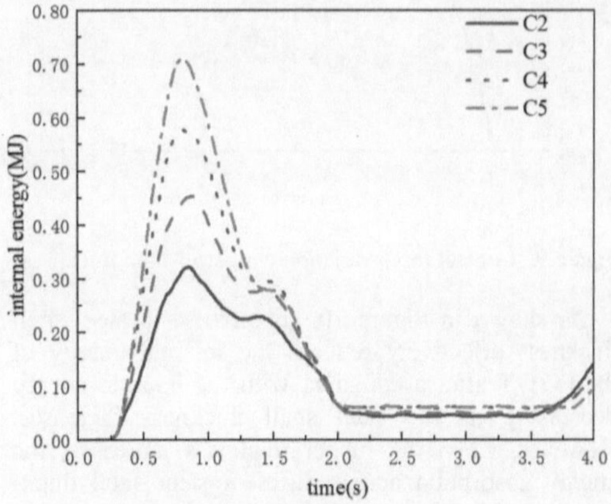

Figure 7. Internal energy of rubber at different thicknesses.

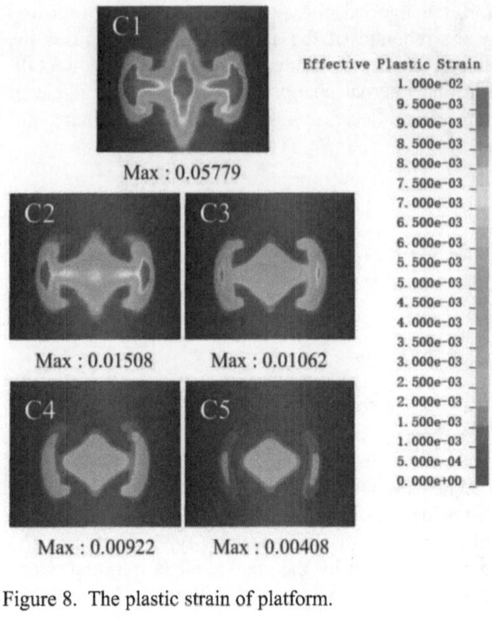

Figure 8. The plastic strain of platform.

3.2.2 *Influence of steel shell thickness*

From the analysis of the maximum collision force, as shown in the Figure 9, it is clear that the variation trends for both 600mm and 1500mm thick rubber are similar. As the steel shell thickness increases, the overall stiffness of the fender increases, leading to a notable rise in the maximum collision force. Additionally, the contact time during the collision decreases, which results in a reduced protective effect.

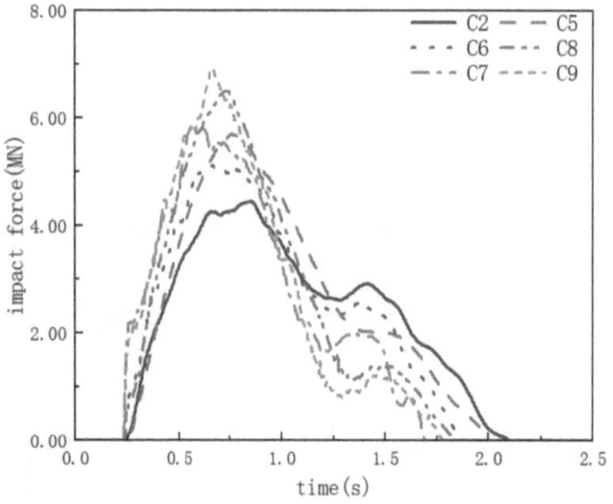

Figure 9. Contact force variation with steel shell thickness.

As shown in Figure 10, an increase in steel shell thickness effectively reduces the internal energy of the FOWT after a collision, with the internal energy decreasing as the steel shell thickness increases. However, when the rubber thickness increases, the energy absorption across different steel shell thicknesses tends to become more consistent. In comparison to the internal energy absorbed by the steel shell of the fender in Figure 11, the internal energy trend for each case is generally similar. After the first collision, the internal energy increases sharply, followed by the rebound of the steel shell, which causes the internal energy to decrease. After the second collision, the internal energy undergoes another cycle of increase and decrease before gradually stabilizing.

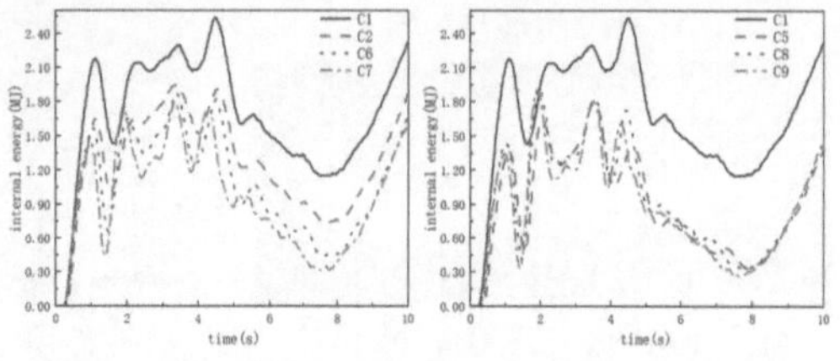

Figure 10. Internal energy of FOWT at different steel shell thicknesses.

When the steel shell thickness is 30mm (C6 and C8), which is equal to the steel plate thickness at the collision position of the spar-type platform, the energy absorbed by the steel shell is minimal. Conversely, at a thickness of 50mm (C7 and C9),

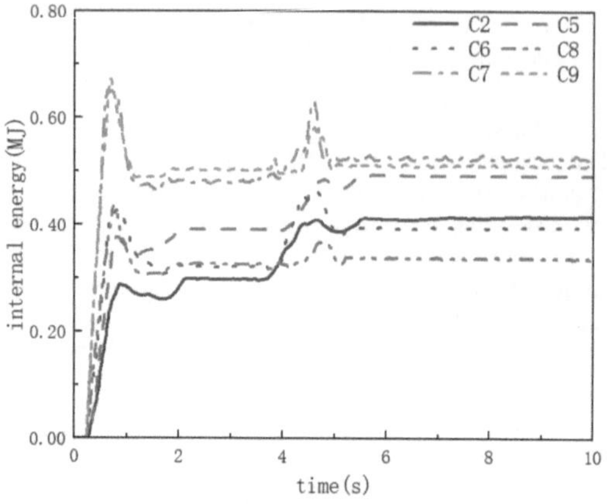

Figure 11. Internal energy of steel shell at different thicknesses.

considerably exceeding the plate thickness, the steel shell absorbs the maximum energy. At this point, the effect of rubber thickness on the steel shell's energy absorption becomes less significant.

It is concluded that when the steel shell thickness is 10mm, the energy absorbed per unit volume of steel shell is the highest, thus providing the best protective effect.

3.3 *Influence of perforated rubber layer on crashworthiness performance*

Perforated structures could provide an effective means to enhance the performance of anti-collision by extending the duration of elastic deformation and preventing the rapid application of collision forces that could otherwise cause damage. This study further examines the performance of a single-hole perforated rubber layer in the fender.

Three-hole diameters, 70%, 50%, and 30% of the rubber layer's thickness, are considered, as illustrated in Figure 12.

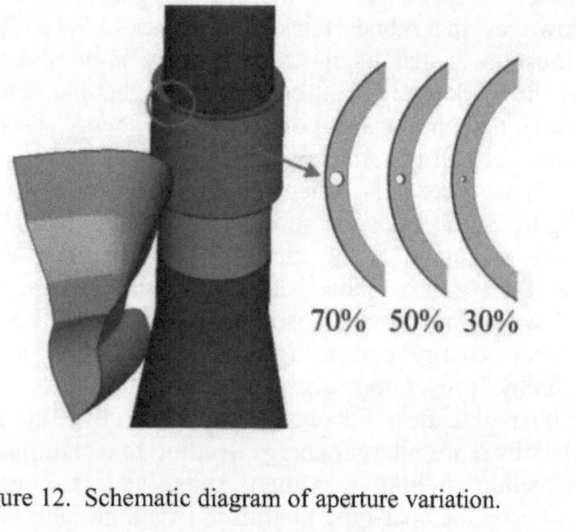

Figure 12. Schematic diagram of aperture variation.

The relationship between contact force and time is illustrated in Figure 13. Upon impact, the ship's bow initiates a rapid increase in contact force starting at 0.25 seconds. The contact force increases more gradually for the perforated rubber compared to the solid rubber. For the 600 mm thick device, the maximum force is reached between 0.65 and 0.85 seconds, while for the 900 mm thick device, it occurs between 0.7 and 0.9 seconds. As the force curve stabilizes, the fender undergoes elastic-plastic deformation, predominantly remaining within the elastic range. A comparison indicates that perforated rubber results in a 5% reduction in the maximum collision force.

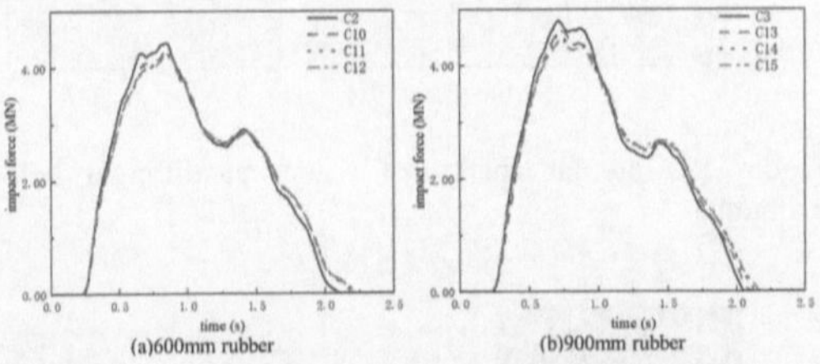

Figure 13. Contact force change with aperture.

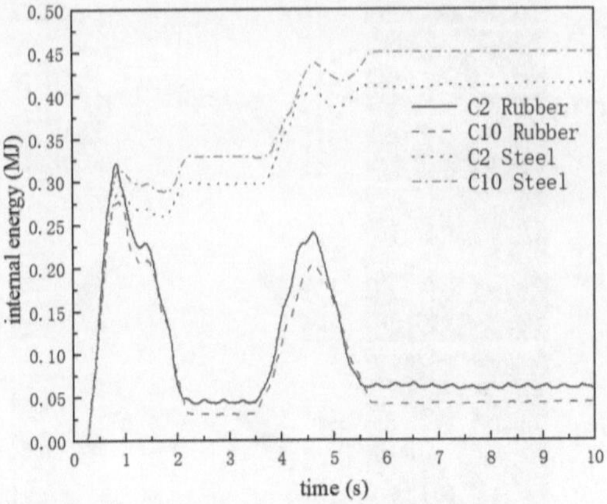

Figure 14. Internal energy of fender at different apertures.

As shown in the Figure 14 the energy absorbed by the steel shell after perforation exceeds the energy lost by the rubber. The steel shell, undergoing plastic deformation, maintains a consistent energy absorption capacity throughout the process. As the aperture size increases, the internal energy absorbed by the outer steel shell continues to rise, gradually enhancing fender's energy absorption capability. This helps protect the rubber layer from excessive damage.

Figure 15 further compares the internal energy absorbed by each fender and its proportion relative to the overall structural internal energy. Figure 16 illustrates the structural strain of FOWT and the stress of rubber when the collision force reaches the peak. It is obvious that fenders with larger diameter holes are more effective in absorbing energy during ship collisions. Additionally, as the hole diameter increases, the maximum strain at the collision area of the spar-type platform decreases, while the maximum stress in the rubber layer rises.

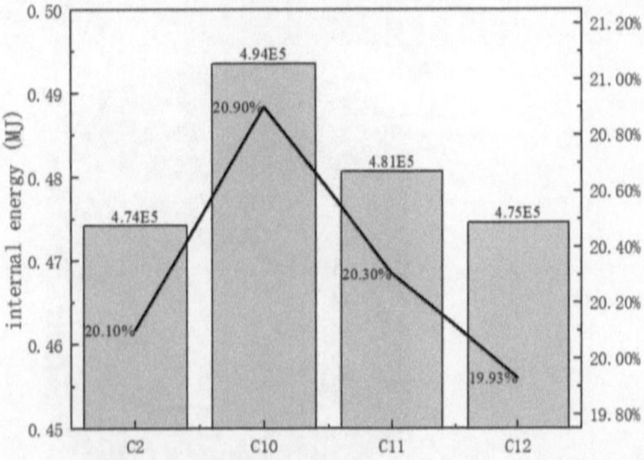

Figure 15. Internal energy and energy ratio of fender with different size of aperture.

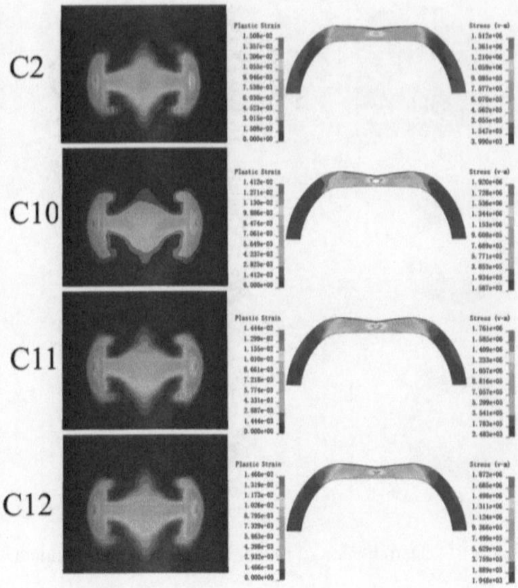

Figure 16. Plastic strain of the platform and stress of rubber.

It can be explained that perforating the rubber creates additional air gaps that help dissipate the external impact and pressure, resulting in a more uniform stress distribution. Consequently, the rubber experiences greater deformation, and its maximum stress increases. This process effectively reduces the maximum strain at the collision point of the spar-type platform, mitigating potential structural damage. It is also reasonable to speculate that a fender with more holes result in better performance.

3.4 *Effect of hole azimuthal on crashworthiness performance*

Four distinct angles 0°, 20°, 40°, and 60° were selected for analysis, as illustrated in Figure 17.

Figure 18 illustrates the deformation pattern of the fender, and it is obvious that the azimuthal of the hole significantly influence the indentation of the fender.

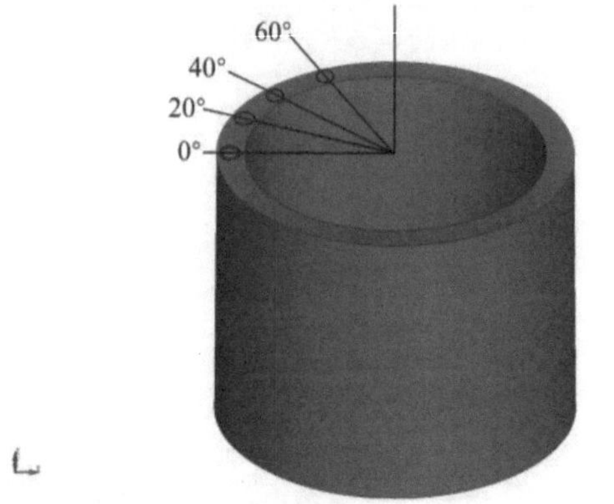

Figure 17. Schematic diagram of hole azimuthal variation.

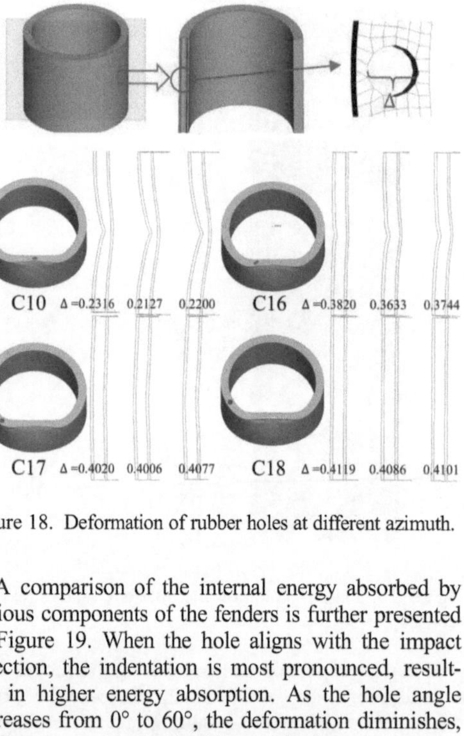

Figure 18. Deformation of rubber holes at different azimuth.

A comparison of the internal energy absorbed by various components of the fenders is further presented in Figure 19. When the hole aligns with the impact direction, the indentation is most pronounced, resulting in higher energy absorption. As the hole angle increases from 0° to 60°, the deformation diminishes, leading to a corresponding reduction in energy absorption. At 60°, the hole has minimal effect on energy absorption, with the fender's performance closely resembling that of solid rubber.

Observation of the damage at the collision site of the spar-type platform, as depicted in Figure 20, reveals that when the hole is aligned with the impact direction, the fender absorbs a greater amount of energy, and the maximum strain in the collision area is relatively lower. As the hole angle increases, the maximum strain gradually rises, approaching the behavior of solid rubber. This occurs because, at larger hole angles, the direction of energy transfer shifts, reducing the effectiveness of the rubber layer in absorbing energy. Consequently, more energy is transferred to the spar-type platform at the contact point, leading to more significant damage. Based on the current findings, it may be inferred that increasing the number of perforations in proximity to the collision zone would substantially improve the energy dissipation characteristics and impact mitigation performance of the fender system, consequently leading to a marked reduction in both peak stress concentrations and plastic strain distributions on the spar-type platform.

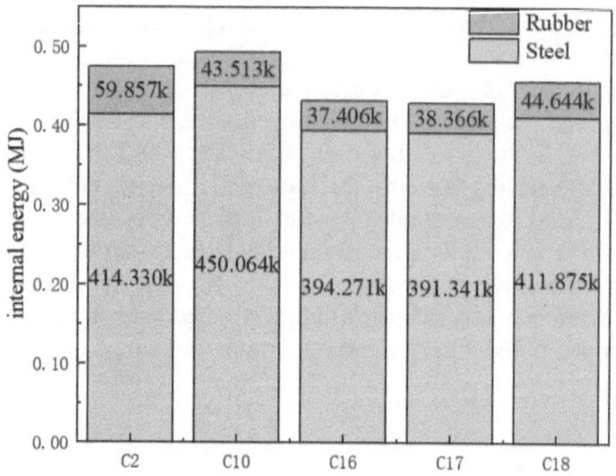

Figure 19. Internal energy of fender at different hole azimuthal.

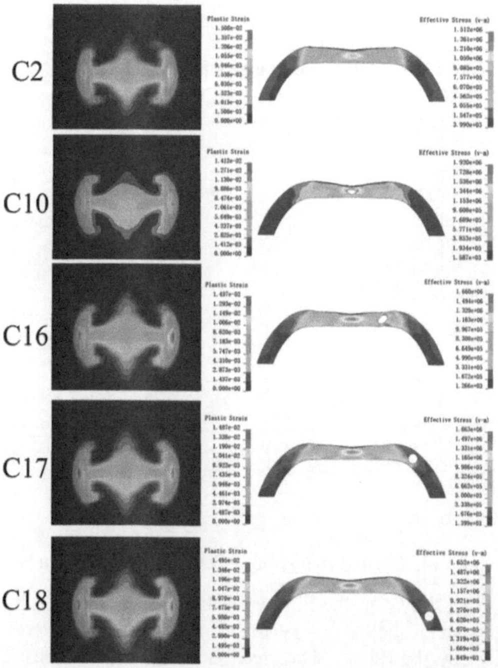

Figure 20. Plastic strain of the platform and stress of rubber.

4 CONCLUSIONS

The research focuses on evaluating the crashworthiness of a composite fender consisting of an outer steel shell and an inner rubber layer. By systematically varying the thickness of the steel and rubber layers, as well as the perforated hole and hole azimuthal in the rubber, the study examines the impact of these parameters on the fender's anti-collision performance, with a focus on local structural damage. The key findings are as follows:

1. The integration of the fender effectively reduces the structural internal energy of FOWT, thereby minimizing structural damage. As the rubber thickness increases, the absorbed internal energy rise, and the fender with the rubber thickness of 600mm obtains the best ratio of energy absorption per unit volume.

2. The perforated hole in the rubber introduces air gaps, which effectively delay the occurrence of the maximum contact force and reduce its peak value, achieving better anti-collision performance. Additionally, the energy absorption capacity of the fender improves with the increment of hole diameter.

3. As the hole azimuthal increases, the energy absorption capability of the fender starts to decrease, leading to an increase in structural damage, which gradually begins to resemble that of solid rubber, resulting in reduced protection effectiveness.

ACKNOWLEDGEMENTS

This work is funded by Guangdong Basic and Applied Basic Research Foundation (2024A151524 0008), Shanghai Pujiang Program (23PJ1406900) and China Postdoctoral Science Foundation (2024T170548). The authors would like to show their gratitude for these funding sources.

REFERENCES

Christensen, C.F., Andersen, L.W., Pedersen, P.H. 2001. Ship collision risk for an offshore wind farm. Structural Safety, 1–7.

Hallquist, J.O. 2013. LS-DYNA user's manuals version 971, vols. 1 & 2.

Han, Z.W., Li, C., Deng, Y.H., Liu, J.H. 2019. The analysis of anti-collision performance of the fender with offshore wind turbine tripod impacted by ship and the coefficient of restitution. Ocean Engineering 194: 106614.

Jonkman, J., Butterfield, S., Musial, W., Scott, G. 2009. Definition of a 5-MW reference wind turbine for offshore system development (No. NREL/TP-500-38060). Golden, CO (United States): National Renewable Energy Lab. (NREL).

Jonkman, J.M. 2007. Dynamics modeling and loads analysis of an offshore floating wind turbine. Boulder: University of Colorado at Boulder.

Kim, H.J., Park, S.M., Thomas, G., Paik, J.K. 2023. Solid Rubber Fenders to Prevent Structural Damage in a Low-Speed Collision Between a Ship-Shaped Offshore Installation and a Shuttle Tanker Working Side-by-Side in Offloading Operation. International journal of maritime engineering 165: A145–A164

Mozafari, H., Distefano, F., Epasto, G., Gu, L.X., Linul, E., Crupi, V. 2022. Design of an Innovative Hybrid Sandwich Protective Device for Offshore Structures. Journal of marine science and engineering 10: 1385.

Nie, Y.H., Fang, H., Meng, X.J. 2024. Ship collision mitigation for offshore wind turbine monopile foundations via high-general/low-collision-stiffness steel fenders. Ocean Engineering 299: 117272.

Petersen, M.J. & Pedersen, P.T. 1981. Collisions between ships and offshore platforms. Offshore technology conference. Houston, Texas.

Ren, Y.L., Meng, Q.S., Chen, C., Hua, X.G., Zhang, Z.L., Chen, Z.Q. 2022. Dynamic behavior and damage analysis of a spar-type floating offshore wind turbine under ship collision. Engineering Structures 272: 114815.

Yue, X.Z., Han, Z.W., Li, C., Zhao, X.L. 2021. The study on structure design of fender of offshore wind turbine based on fractal feature during collision with ship. Ocean Engineering 236: 109100.

Yue, X.Z., Han, Z.W., Li, C., Zhao, X.L. 2022. Study on Performance of Rubber and Aluminum Foam Protective Device for Offshore Wind Turbine. Journal of engineering for thermal energy and power 37(05): 115–123

Zhang, D.C. 2021. Research on ship impact performance of floating wind turbine structure (Master's thesis). Harbin: Harbin Institute of Technology

Zhang, S.M., Villavicencio, R., Zhu, L., Pedersen, P.T. 2019. Ship collision damage assessment and validation with experiments and numerical simulations. Marine Structures 63: 239–256.

Zhang, Y.C. & Hu, Z.Q. 2022. An aero-hydro coupled method for investigating ship collision against a floating offshore wind turbine. Marine Structures 83: 103177.

Trends in Collision and Grounding of Ships and Offshore Structures – Zhu & Guedes Soares (Eds)
© 2026 The Author(s), ISBN 978-1-041-16361-9

Research on the bottom collision of deep-sea pressure-resistant structures

T.T. Guo, L.P. Zhang, Y.J. Zhao, J. Wang, Y. Yu & H.K. Wang
China Ship Scientific Research Center, Wuxi, Jiangsu, P.R. China
National Key Laboratory of Ship Structural Safety, Wuxi, Jiangsu, P.R. China
Taihu Laboratory of Deepsea Technological Science, Wuxi, Jiangsu, P.R. China

ABSTRACT: Deep-sea equipment faces the issue of bottom collision during underwater operations, and the impact of such high-stress collisions may significantly affect the pressure-resistant structures. Addressing the problem of bottom collision in deep-sea pressure-resistant structures, this paper conducts research based on the nonlinear finite element method. With the goal of reducing collision loads and structural stresses, an open-hole rubber cushion structure was designed as a protection device for the bottom. By comparing the calculation results with and without the rubber cushions, the necessity of installing the protective device was analyzed from the perspectives of stress, load, and energy. Finally, an engineering calculation method for the peak collision load and compression of the rubber cushions during the landing collision with the rubber cushions is established based on the energy method. The results of the engineering calculation method have an error of 11.9% compared to the dynamic method—nonlinear explicit finite element method. The conclusions drawn can provide a reference for the analysis of bottom collision issues in deep-sea equipment.

1 INTRODUCTION

Entering the 21st century, countries around the world have focused their attention on the deep sea. Deep-sea equipment, represented by manned submersibles, plays a significant role in underwater operations such as seabed observation, resource exploration, and cable deployment (Liu, 2012). When performing underwater operations, deep-sea equipment may encounter two types of collision scenarios: one is a collision during horizontal navigation with vertical fixed objects like seamounts, where the impact occurs on the front part of the pressure structure. Guo et al. (2023) conducted a systematic study on the underwater collision problem of the pressure-resistant structure of titanium alloy deep-sea equipment based on simulation methods; the other is an impact from deep-sea rocks when the equipment lands on the seabed for in-situ operations. This high-stress impact may significantly influence the pressure-resistant structure. Based on an engineering example, this paper simulates the deep-sea landing collision process of a pressure hull using the nonlinear explicit finite element program LS-DYNA. The study aimed to reduce collision loads and structural stresses while designing anti-collision structures and comparing the calculation results before and after the installation of protective devices. Finally, based on the energy method, an engineering calculation method for landing collisions with open-hole rubber cushion was proposed.

2 PRESSURE-RESISTANT STRUCTURE AND DEEP-SEA BOTTOM COLLISION MODEL ESTABLISHMENT

2.1 *Parameters of pressure-resistant structure*

The total mass of the deep-sea equipment system is 700 tons. The study analyzed the structural response when the pressure hull landed directly at a velocity of 0.1 m/s. The pressure-resistant structures features a hemispherical head and a circumferential ribbed cylindrical shell structure, with specific parameters as listed in Table 1.

The kinetic energy of the deep-sea equipment was transmitted through a limiting device to the colliding pressure hull (as shown in Figure 1), resulting in direct landing collision between the bottom and granite plates. A collision model as shown in Figure 2 was established using ANSYS.

Table 1. Main parameters of pressure hull model.

Total length (m)	Shell plate thickness (mm)	Radius (m)	weight (t)	Rib spacing (m)	Outer rib size (mm)
5.7	14	0.55	700	0.2	16×90

2.2 *Setup of material models*

The material for the pressure-resistant hull is high-strength steel, employing an ideal elastic-plastic

DOI: 10.1201/9781003684404-24

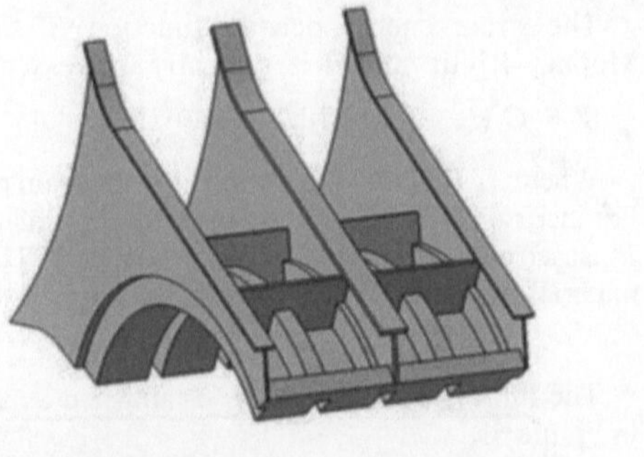

Figure 1. Limiting structure.

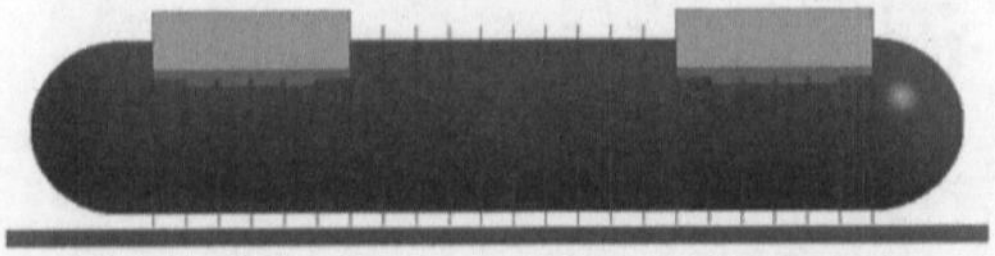

Figure 2. Landing collision model of battery cabin.

model with a modulus of elasticity of 210 GPa, yield strength of 785 MPa, and poisson's ratio of 0.3. The submarine rocks were modeled as granite, and the Holmquist-Johnson-Cook (H-J-C) model was utilized to simulate the dynamic mechanical properties of rock-soil materials, with H-J-C model parameters referenced from the reference (Wen, 2013).

Unlike metallic materials, rock-soil material's mechanical properties require consideration of static water pressure (overburden pressure on rocks). Over 20 yield criteria have been proposed; however, due to the strain rate effects exhibited by rock-soil materials under collision impacts, this study employed the H-J-C model to simulate their dynamic mechanical properties. The H-J-C model is a strain rate-dependent damage constitutive relation capable of describing material responses under high stress, large strains, and high strain rates, widely applied in numerical simulations for rock impact and integrated into LS-DYNA.

The H-J-C model consists of three components: the equation of state, yield surface equations, and damage evolution equations. The equation of state can be divided into three stages: an initial linearly elastic stage; a transitional stage representing complete internal pore compression during plastic deformation; and a fully compacted stage, modeled by a cubic polynomial, expressed as:

$$P = K_1 u + K_2 u^2 + K_3 u^3 \quad (1)$$

In the equation (1), P represents pressure and u denotes volumetric strain. Parameter K_1, K_2 and K_3 are material characteristic pressure parameters. In the figure, P_c and u_c represent the static water pressure and volumetric strain at failure due to compression, while P_l and u_l denote the static water pressure and volumetric strain during complete compaction. The yield surface equation for H-J-C is:

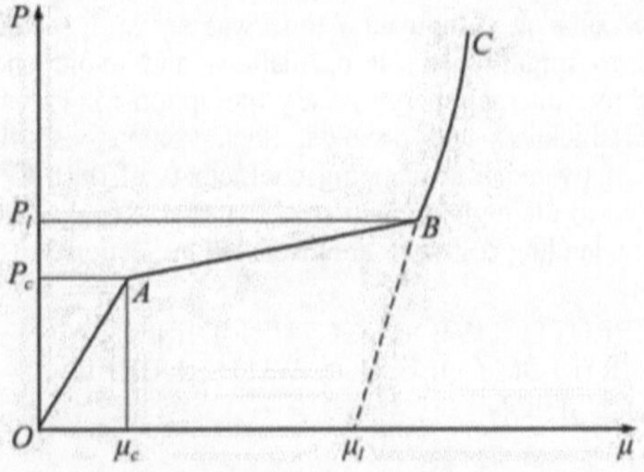

Figure 3. Equation of state.

$$\sigma^* = [A(1-D) + BP^{*N}](1 + Cln\dot{\varepsilon}^*) \quad (2)$$

where σ^* is the equivalent stress, $\dot{\varepsilon}^*$ is the equivalent strain rate, D represents damage degree, and A,B,N,C are material strength parameters.

The H-J-C model's damage evolution equation is:

$$D = \sum \frac{\Delta\varepsilon_p + \Delta u_p}{\varepsilon_f{}^p + u_f{}^p} \quad (3)$$

where $\Delta\varepsilon_p$ and Δu_p represent equivalent plastic strain and plastic volumetric strain, respectively; $\varepsilon_f{}^p$ and $u_f{}^p$ represent equivalent plastic strain and plastic volumetric strain in normal pressure.

The H-J-C model comprises 21 material parameters. Table 2 lists the parameters used in this paper:

Table 2. The H-J-C model parameters of Granite.

ρ_0(kg/m^3)	f_c(MPa)	A	B	N	C
2586	60.53	0.762	1.65	1.15	0.0013
p_i (GPa)	μ_i	K_1 (GPa)	K_2	K_3	D_1
1.2	0.101	12	25	42	0.042
D_2	T(MPa)	G (GPa)	p_C (GPa)	μ_C	SMAX
1.0	4.40	6.53	20.17	1.70×10^{-3}	7.0

2.3 *Meshing and parameter settings*

The pressure-resistant hull and circumferential ribs were modeled using the shell163 unit with a mesh size of 20 mm. The limiter structure and granite plate used solid164 units, resulting in a finite element model containing 84,280 elements (with 72,336 elements for the pressure-resistant hull). The operating water depth for the pressure structure landing was 1500 meters. In LS-DYNA, the static water pressure linearly increased to 15 MPa over 0.09 seconds before remaining constant. The initial spacing between the pressure-resistant hull and granite plate was 0.01 m to prevent collision during preloading. Surface-to-surface automatic contact algorithms were defined for the shell and thick plate.

The collision computation time was set at 1 second, and to minimize result oscillations and avoid shell negative internal energy issues, the option for varying shell thickness was disabled. Shell volume viscosity was activated, and damping coefficients of 0.01 were added to the high-strength steel material. Results from direct landing collisions are discussed in Section 3.

3 PROTECTIVE STRUCTURE DESIGN AND BOTTOM COLLISION WITH RUBBER CUSHION

3.1 *Design of protective structure*

On ships, materials such as rubber, steel, and wood are commonly used for fendering to reduce impacts during berthing or mooring operations (Ding, 2019). To mitigate the collision impact loads on the pressure hull during bottom landing, a protective structure design was conducted for the pressure-resistant hull's landing collision.

When the pressure-resistant hull lands directly without a protective structure, the large mass of the structure and small contact area may result in significant stress. To alleviate the shock load and absorb energy, protective devices are added to the bottoms of both ends of the pressure-resistant hull. Rubber materials have advantages such as low density, high stiffness, good corrosion resistance to seawater, and cost-effectiveness with easy replacement. Therefore, eight perforated rubber cushions were designed based on the pressure hull's structure and bonded to both ends of the shell. The structure of the rubber cushion is shown in Figure 4.

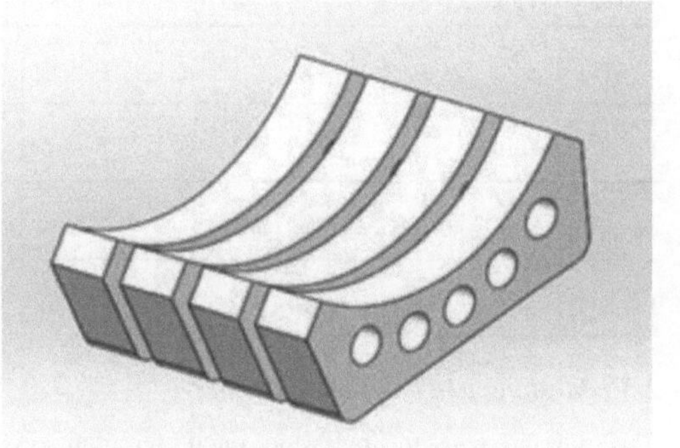

Figure 4. Open-hole rubber cushion structure.

3.2 *Bottom collision model with rubber cushion*

In the finite element model, rubber cushions are added with a hardness of 70. The material model employs the Mooney-Rivlin constitutive model, which is used to describe the hyperelastic behavior of rubber and rubber-like materials. This model belongs to the class of hyperelastic material models and is suitable for large deformations.

The strain energy density function W for the Mooney-Rivlin model is typically expressed as:

$$W = C_1(I_1 - 3) + C_2(I_2 - 3) + C_3(I_3 - 1)^2 \quad (4)$$

Where: I_1,I_2 and I_3 are the invariant of strain rensor, For incompressible materials, I_3 is typically equal to 1.

Based on literature references (Jiang, 2021), the material parameters C_1, C_2 and $C_3$are set as follows:

$$C_1 = 0.9236; = 0.2309; = 0.01 \quad (5)$$

The finite element model is established as shown in Figure 5.

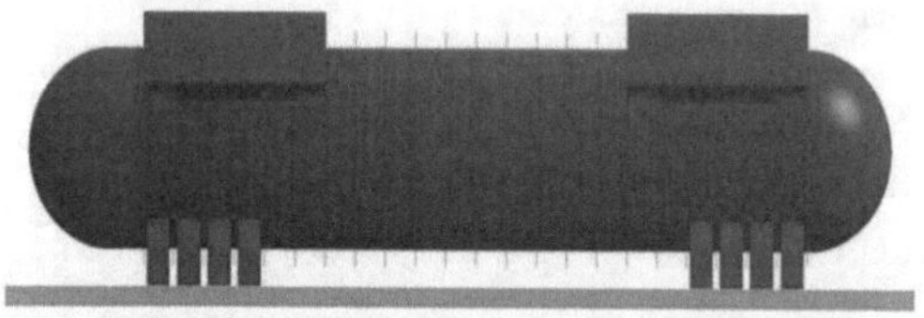

Figure 5. Landing collision model of battery cabin with protective device.

4 RESULTS COMPARISON

The results from the landing collision calculations with and without protective structures are compared, focusing on structural strength, collision loads, and energy absorption.

Figure 6 shows the stress distribution in the pressure-resistant hull under hydrostatic pressure before collision. The maximum stress is located at the shell plate between stiffeners, with a value of 526.7 MPa. Figures 7 and 8 present the structural stress distributions during maximum impact force for direct landing (without protective devices) and landing with rubber cushions, respectively. The maximum stresses occur at the contact points under maximum load, reaching 619.4 MPa and 543.2 MPa, respectively, a reduction of 76.2 MPa. Clearly, the rubber cushions significantly reduce the maximum stress.

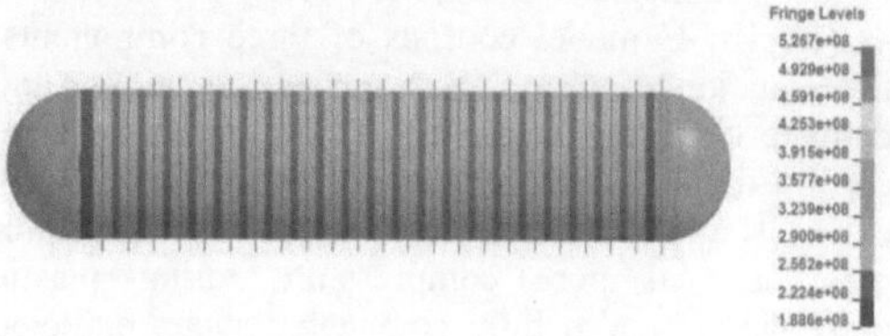

Figure 6. Stress of structure before landing.

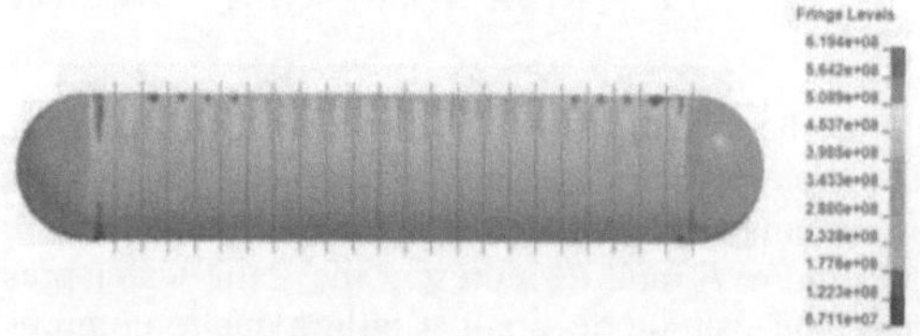

Figure 7. Stress of structure without protective device at the maximum impact force.

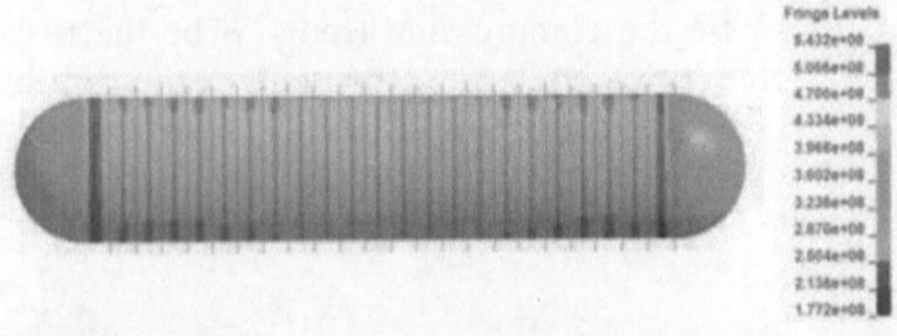

Figure 8. Stress of structure with protective device at the maximum impact force.

Figure 9 compares the collision force time-history curves for the two scenarios. Curve A corresponds to the case 1 represents the collision force during direct landing without the rubber cushion, while Curve B corresponds to the case 2 with the rubber cushion. The peak collision load for direct landing is 1526 tons, which is 4.8 times that of the case2 with rubber cushions. Figure 10 compares the impulse time-history curves under the two working conditions. The collision duration increases significantly for the model with the rubber cushion; however, both cases result in the same final impulse of 137 kN·s.

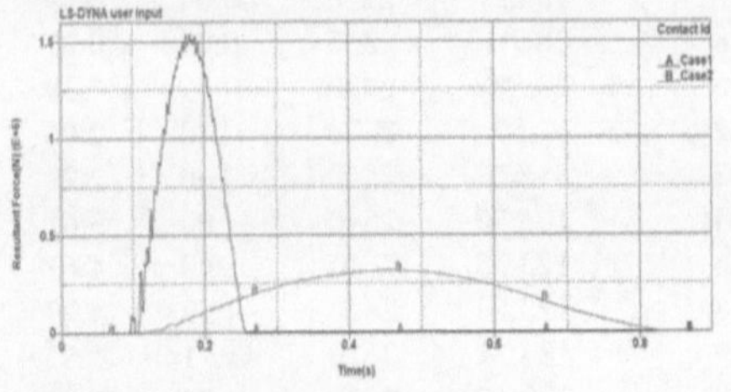

Figure 9. Comparison of time-history curves of collision force under two working conditions.

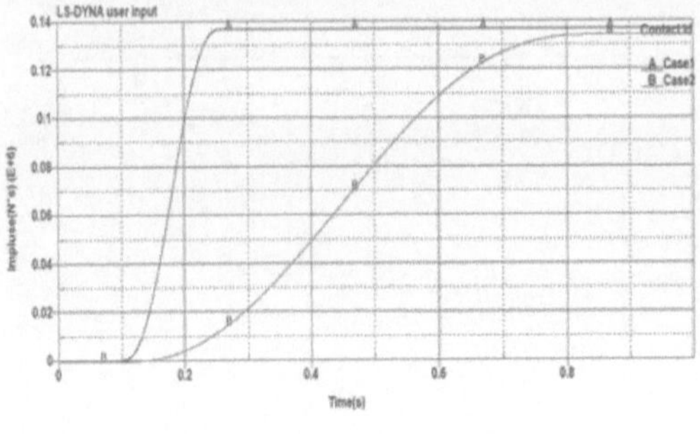

Figure 10. Comparison of time-history curves of impulse under two working conditions.

As shown in Figure 11, the external pressure is linearly increased to 15 MPa within 0.1 seconds; thus, the initial internal energy for both cases is 153.3 kJ. During the collision, the internal energy increases as kinetic energy decreases and reaches its maximum when kinetic energy drops to zero, corresponding to 159.2 kJ and 154 kJ for direct landing and landing with the rubber cushion, respectively. Since no structural failure occurs in either case, the internal energy gradually reduces back to 153.3 kJ, representing the rebound kinetic energy of the structure. Clearly, the rubber cushions significantly reduce the peak energy absorption.

Table 3. Comparison of impulse time-history curves under two working conditions.

	Peak Collision Force (kN)	Duration(s)	Inpulse (kN·s)	Average Collision Force (kN)
Direct Landing	1526	0.15	137	913
Landing with Protective Structure	318	0.7	137	196

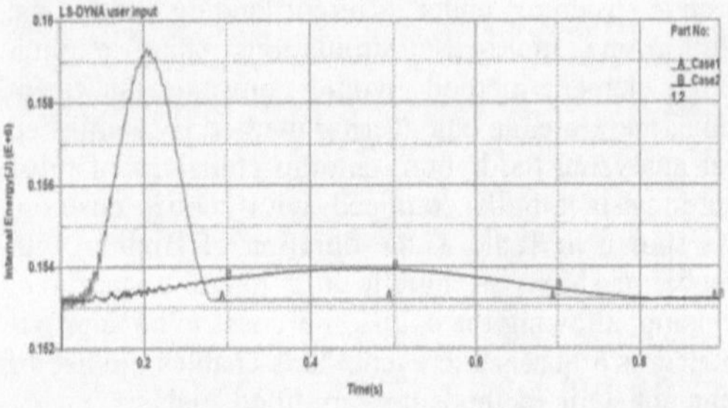

Figure 11. Comparison of energy absorption under two working conditions.

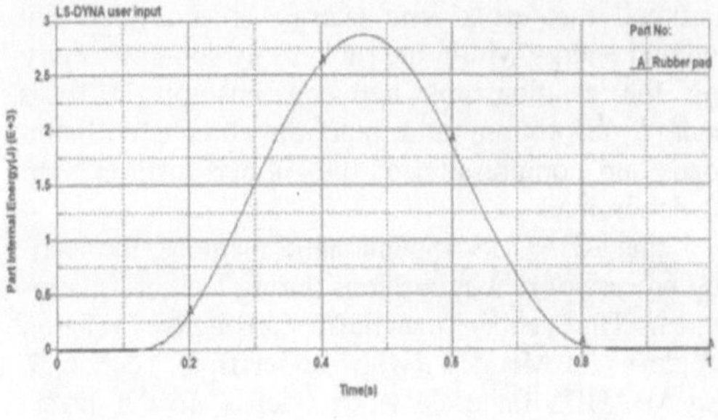

Figure 12. Internal energy time-history curves of perforated rubber cushion.

As shown in Figure 12, the rubber cushion absorbs a maximum internal energy of 2.89 kJ at the moment of maximum collision force, accounting for 83% of the total initial kinetic energy of 3.5 kJ. The pressure-resistant shell absorbed 0.49kJ of energy (approximately 14% of the total kinetic energy) at the moment of maximum impact load, while the granite absorbed 0.11kJ of energy (approximately 3% of the total kinetic energy).

Figure 13 shows the time-history curve of compression at the midpoint of the rubber cushion. Maximum deformation occurs at 0.48 seconds, with a compression of 20.5 mm.

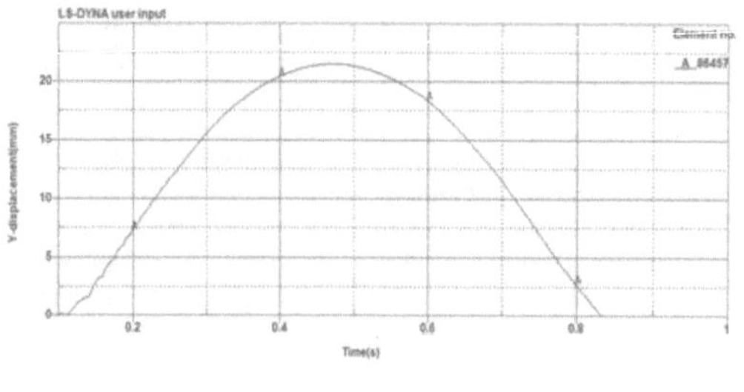

Figure 13. Time history curve of compression in the middle the perforated rubber cushion.

5 ENGINEERING METHOD FOR LANDING COLLISIONS WITH RUBBER CUSHION

In engineering design, it is necessary to quickly assess collision loads and the compression of protective structures under different landing conditions. Performing multi-case simulations using explicit finite element methods would consume significant time; therefore, an engineering method is established for analyzing the bottom landing collisions of pressure-resistant hulls equipped with rubber cushion. As shown in Table 3, the duration of landing with rubber cushions is significantly longer than direct landing, allowing the collision process to be approximated as a quasistatic event. This enables the use of implicit static methods for simplified analysis.

Previous calculations indicate that energy absorption by the pressure-resistant hull is relatively low during landing with rubber cushions, and the granite substrate can be treated as a rigid body. Thus, it is assumed that all kinetic energy is converted into internal energy within the rubber cushions. By studying the relationships between cushion stiffness, energy absorption, and reaction forces, collision loads and compression of the rubber cushions can be determined.

Using ABAQUS implicit static methods, the loading process of rubber cushions during landing is computed. The rubber material's constitutive relation employs the Mooney-Rivlin model from Section 3. An ABAQUS finite element model is created, leveraging symmetry in both the structure and loading conditions. Only half of the rubber cushion and the local region of the pressure hull interacting with it are modeled, with a mesh size of 20 mm. Fixed contact constraints are applied between the rubber cushion and the pressure hull shell plate. Loads are applied to the bottom of the rubber cushion, and results including reaction forces, stiffness, energy absorption, and maximum stress are output for different compression levels.

At a compression of 51.99 cm, the stress and deformation cloud maps of the rubber cushion are shown in Figures 14, respectively. From these calculations, performance curves (Figure 14) and parameter tables (Table 4) are derived for individual rubber cushions.

Let X be the compression (mm), F be the reaction force (kN), and E be the energy absorption (kJ). By interpolating from Figure 15, the compression and reaction force can be determined for a given energy absorption.

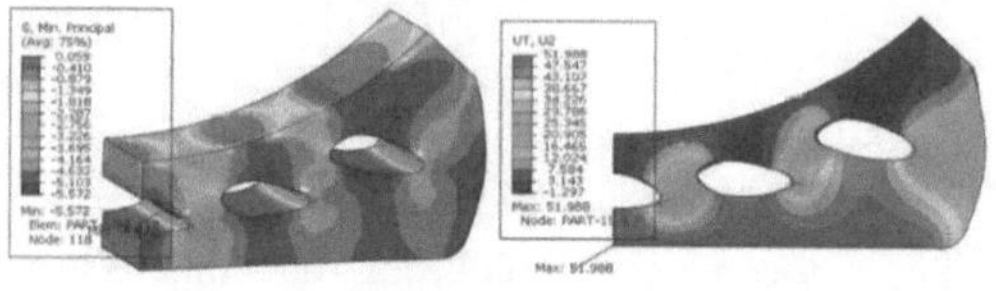

Figure 14. Stress and Deformation Cloud Map at 51.99 cm Compression.

Table 4. The relationship of rigidity, energy and force for the perforated rubber cushion.

Compression (mm)	Reaction Force (kN)	Stiffness (kN/m)	Energy (kJ)	Maximum Stress (MPa)
0.97	2.01	2070	0.00	0.07
8.03	18.23	2270	0.04	0.59
12.18	28.38	2330	0.09	0.91
18.16	43.58	2400	0.19	1.39
25.72	63.53	2470	0.40	2.06
32.92	83.62	2540	0.66	2.78
40.04	103.70	2590	1.00	3.61
47.26	123.35	2610	1.40	4.67
49.08	128.59	2620	1.52	4.99
51.99	136.21	2620	1.71	5.57
53.09	139.10	2620	1.79	5.82
55.05	143.13	2600	1.93	6.37
55.81	143.99	2580	1.98	6.69

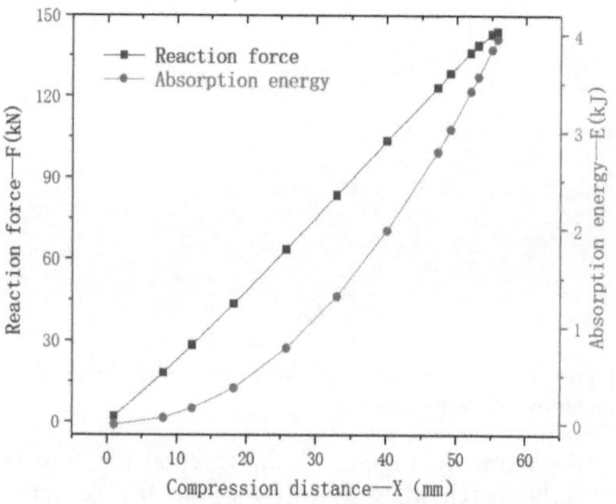

Figure 15. Curve of the relationship among force, energy absorption and compression of a single rubber cushion.

Given the total mass of the deep-sea equipment as 700 tons with an initial velocity of 0.1 m/s, the energy absorbed by the rubber cushions must satisfy:

$$E_i = E_k = \frac{1}{2}mv^2 = 3.5kJ \qquad (6)$$

where E_i is the total energy absorbed by all eight rubber cushions. For a single cushion, this amounts to approximately 0.438 kJ. Interpolation from Figure 15 yields a compression of 21.2 mm and a reaction force of 44.5 kN for one rubber cushion, resulting in a total collision force of 356 kN for all eight cushions.

This result is slightly higher than the dynamic simulation peak load of 318 kN, with a relative error of 11.9% as is shown in Table 5. The discrepancy arises because the energy method does not account for energy absorption by the pressure-resistant hull and granite substrate. Despite this, the approach provides a practical simplified method for estimating collision loads and cushion compression in rubber-cushioned landing scenarios.

Table 5. Comparison of calculation results of engineering method.

	Peak Collision Force (kN)	Compression (mm)
Nonlinear FEA method	318	20.5
Engineering method	356	21.2
Relative error	11.9%	3.4%

Based on material specifications for natural rubber, the cushion's maximum allowable stress is within the range of 8-16 MPa. As shown in Table 4, the maximum observed stress of 6.69 MPa is well below this threshold. The cushion can withstand compressions up to 55.81 mm and reaction forces exceeding 14.4 kN. During landing collision, the maximum compression of 27 mm is within acceptable limits, confirming the rationality of the rubber cushion design.

6 CONCLUSIONS

The bottom-landing collision condition of a deep-sea equipment pressure-resistant hull is studied based on the nonlinear finite element method. Aiming to reduce collision load and structural stress, an open-hole rubber cushion structure is designed as a bottom-landing protection structure. By comparing the calculation results with and without the rubber cushion, the necessity of installing the protection device is analyzed from the perspectives of stress, load, and energy. Finally, an engineering calculation method for the collision force and rubber cushion compression during landing collision is established. The conclusions are as follows:

(1) When the deep-sea equipment pressure hull directly lands on the bottom without any protection structure, it generates a significant collision load, posing a great threat to the strength of the pressure structure. Referencing the fender design of surface ships, open-hole rubber cushions are used as protective structure for the pressure-resistant hull during bottom-landing. The comparison of calculation results shows that the open-hole rubber cushions can significantly reduce structural stress, collision load, and the amount of internal energy absorbed by the hull, enhancing the safety of bottom-landing.

(2) For the landing issue of the pressure-resistant hull with rubber cushions on hard soil, the energy method can be used as a simplified calculation method, transforming the dynamic impact problem into aquasi-static process. Assuming that all the energy lost during the collision is converted into the internal energy of the rubber cushions, by studying the stiffness of the rubber cushions, a relationship between compression, reaction force, and energy absorption is established. Only simple interpolation calculations are needed to quickly estimate the maximum collision load and rubber cushions compression, and the results have an error of 11.9% compared to the dynamic method—nonlinear explicit finite element method.

(3) The rubber cushions structure has a smaller stiffness and a larger compression allowance. The calculation results show that the open-hole rubber cushions structure meets the requirements for landing collision, and the design is reasonable and reliable.

REFERENCES

Ding Z., 2019. Refined design of rubber fender for wharf. China Harbour Engineering. 39(3):42–46.

Guo T.T., Zhang L.P., Zhao Y.J., 2023. Research on deep-sea collision of titanium alloy submersible pressure shell. Advances in the Collision and Grounding of Ships and Offshore Struc-tures 347–353.

Jiang Z., 2011. Study on the Responses of Ship Fenders under Impact Load. Shanghai: Shanghai Jiao Tong University.

Liu T, Wang X, Wang S., 2012. Development status and technical progress of deep-sea manned submersible. Shipbuilding of China, 233–243.

Wen L, Li X, Wu Q., 2013. Study of Holmquist-Johnson-Cook constitutive model parameters of granite porphyry. Chinese Journal of Computational Mechanics (5):725–731.

Trends in Collision and Grounding of Ships and Offshore Structures – Zhu & Guedes Soares (Eds)
 ISBN 978-1-041-16361-9

Dynamic responses of monopile OWTs subjected to ship collision using a semi-analytical approach

A. Hammad & Z. Yu
Department of Marine Technology, Norwegian University of Science and Technology (NTNU), Trondheim, Norway

ABSTRACT: As the number of offshore wind turbines (OWTs) continue to increase and their proximity to navigational routes decreases, the risk of ship collisions with these structures has become a critical design consideration. This paper introduces a semi-analytical approach for fast and reliable analysis of the dynamic responses of monopile-supported OWTs subjected to ship collision scenarios. A Rayleigh-Ritz solution is initially adopted to determine the high-order natural frequencies and eigenmodes of monopile OWTs. The model explicitly considers tapered towers and soil-pile interactions. Dynamic responses of OWTs are then calculated using a numerical contact algorithm, where the load actions depend on the contact stiffnesses and the interaction schemes between the colliding ship and the turbine. The proposed approach is applied to the DTU 10 MW and IEA 15 MW monopile-supported OWTs subjected to head-on ship collisions. The results are discussed with respect to the turbine natural frequencies and eigenmodes, contact forces, dissipated energies (e.g., elastic and plastic) and tower top dynamic responses (e.g. displacement and acceleration).

1 INTRODUCTION

Nowadays, the offshore wind industry is rapidly expanding, with numerous wind farms set to be developed in the coming years, making offshore wind power the most advanced renewable energy technology for harnessing wind energy at sea. Monopile-supported OWTs are the most widely used substructure in wind farms, having been served as the foundation for 5181 wind turbines, accounting for approximately 78 % of all installed wind turbines up to 2023 (WindEurope, 2024). The growing number of OWTs near navigational routes increases the risk of collisions with ships, posing a safety hazard to personnel and equipment, especially when vessels are passing by or approaching for maintenance (Yu et al., 2020). By 2030, 2500 wind turbines are set to be installed in the North Sea. The Maritime Research Institute Netherlands (MARIN) estimated a risk of 1.5 to 2.5 ship collisions with OWTs per year (Professionals, 2022). For example, in 2022, a cargo ship of 37,200 tons was drifted by a heavy storm into the Hollandse Kust Zuid offshore wind farm area and crashed into a monopile foundation of an OWT causing significant dents on the monopile as shown in Figure 1(a) (Wind, 2022). Additionally, the Windfarm Support Vessel Njord Forseti struck into a wind turbine tower while traveling at a speed of roughly 20 knots in April 2020 (Administration, 2020) (see Figure 1(b)).

Regarding ship collisions, there are two main impact loads to consider: operational and accidental.

Figure 1. Damaged monopile foundation after ship collision (Wind, 2022) (left) and the Support Vessel Njord Forseti after colliding with wind farm (Administration, 2020) (right).

Operational impact loads are caused by service vessels during routine wind farm operations with an impact velocity of not less than 0.5 m/s. Although this type of impact may not cause significant permanent damage to the tower or foundation, the structures elastic response can lead to high accelerations at the rotor-nacelle-assembly (RNA), potentially affecting sensitive components like the generator, bearings and gearbox (Rasekhi Nejad & Moan, 2017). In contrast, accidental impact loads result from drifting vessels due to control failure, with an impact velocity of not less than 2.0 m/s. Operational impacts shall be accounted for in the ultimate limit states (ULS), while accidental impacts should be considered in the accidental limit states (ALS) (DNV-ST-0119, 2021).

DOI: 10.1201/9781003684404-25

Nonlinear Finite Element Analysis (NLFEA) is commonly used to study the dynamic responses of OWTs subjected to ship collisions. Bela et al. (2017) used LS-DYNA to analyse the structural response of a monopile-supported OWT during a ship collision, focusing on crushing behaviour and nacelle dynamics. In a separate study, Bela et al. (2015) conducted a parametric analysis using nonlinear numerical simulations to investigate the impact of velocity, impact point and soil stiffness on the structural behaviour of monopile-supported OWT during collisions. NLFEA with existing codes is effective for simulating complex problems but can be computationally expensive.

Analytical or semi-analytical methods are often preferred for their efficiency, accuracy and ability to provide insights into governing physics. They are commonly used in rules and design standards. Limited references are available for analysing ship collision dynamic responses and energy absorptions of monopile-supported OWTs. For instance, Pedersen (2013) proposed an analytical solution for the external dynamics of ship collision with a flexible bottom fixed OWT considering only elastic deformations based on the principle of conservation of momentum and energy. Song et al. (2021) studied the dynamic responses of a 5 MW monopile-supported OWT to ship impacts using numerical and analytical approaches. Furthermore, Ladeira et al. (2022) developed a semi-analytical method to calculate natural frequencies and mode shapes of monopile OWTs with fixed base, and analysed monopile responses using mode superposition due to a predefined collision force.

This study analyses the dynamic responses of DTU 10MW and IEA 15MW monopile-supported OWTs during ship collisions at parked condition using a semi-analytical approach. The approach adopts the Rayleigh-Ritz solution to calculate high-order natural frequencies and eigenmodes. The study also assesses the dynamic responses and energy absorption of the OWTs using a numerical contact algorithm developed by Hammad and Yu (2024) to model the interaction between the colliding bodies under hard contact loading conditions. The semi-analytical approach, including the Rayleigh-Ritz formulation and contact algorithm, was validated against nonlinear FEM (USFOS), demonstrating robust agreement in natural frequencies (less than 3% error), mode shapes, and transient collision responses. Hydrodynamic loads from wave action are excluded in this analysis, as Hammad et al. (2025) demonstrated their negligible influence (less than 5% increase in dynamic response) for parked turbines during collision events.

2 METHODOLOGY

2.1 *A Rayleigh-Ritz solution for high order natural frequencies and eigenmodes of monopile OWSs*

OWTs need to be designed to withstand accidental loads in Ultimate Limit States (ULS) and Accidental Limit States (ALS), whereby high order eigenmodes may become important and absorb significant energy. Yu and Amdahl (2023) derived a Rayleigh-Ritz solution for high-order natural frequencies and eigenmodes of monopile-supported OWTs, considering explicitly tapered towers and soil-pile interactions. The model is adopted in this paper and described briefly in this section. The transverse displacement $W(z,t)$ of a monopile OWT can be considered as a function of space and time as follows:

$$W(z,t) = w(z)q(t) \tag{1}$$

Where z indicates the vertical position along the turbine height, t is time, $u(t)$ represents the time-dependent deflection amplitude, and $w(z)$ represents structural deflection shapes that can be treated as a combination of a number of linearly independent modes:

$$\frac{w(z)}{L} = \sum_{i=1}^{n} a_i \varphi_i(z) \tag{2}$$

Where φ is the ith trial function, $a_i$is the corresponding weight and L is the characteristic length of the structure. Based on the assumed deflection shapes in Eq. (2), the Rayleigh-Ritz quotient $\omega^2 = U_{\max}/T0$ needs to be minimized, where ω is the vibration frequency, $U_{\max}$ represents the maximum strain energy and $T_{\max} = \omega^2 T0$ represents the maximum kinetic energy. This yields the characteristic equation for mode shapes with respect to the generalized stiffness matrix $\boldsymbol{K}$ and mass matrix $\boldsymbol{M}$:

$$\left(\boldsymbol{K} - \omega^2 \boldsymbol{M}\right)a = 0 \tag{3}$$

Polynomials were chosen as shape functions for structural deflections with a 6th order for the tapered tower and a 4th order for the monopile. Soil-structure interaction (SSI) plays a crucial role in the overall dynamic behaviour of the monopile-supported OWTs and should be carefully considered when assessing their responses (Jung et al., 2015). The coupled spring model, which represents the foundation flexibility using coupled translational and rotational springs at the mudline is chosen for its accuracy and simplicity in this study (see Figure 2) (Alkhoury et al., 2020). Yu and Amdahl (2023) showed that the coupled springs model suggested by Psaroudakis et al. (2020) accurately captures soil-pile interactions compared with USFOS simulations. By applying the Rayleigh Ritz principle, and enforcing continuities of displacements, slopes, bending moments and shear forces across the cross-section changes, the mass matrix $M$and the stiffness matrix K for a monopile-supported OWT with tapered tower and three coupled springs representing foundation stiffnesses can be expressed as:

$$M_{ij} = \underbrace{\int_0^{\gamma L_T} (\rho_p A_P + AM) T_{pi}\varphi_p - 1(z_2)\varphi_{p-1}(z_2) T_{qj} dz_2}_{\text{Monopile}}$$

$$+ \underbrace{\int_0^{L_T} \rho_T A_T(Z_1)\varphi_{i-1}(Z_1)\varphi_{j-1}(z_1)dz_1}_{\text{Tower}}$$

$$+ \underbrace{m_{RNA}\varphi_{i-1}(L_T)\varphi_{j-1}(L_T)}_{\text{RNA mass}}$$

$$+ \underbrace{\left[J_{RNA}\frac{d\varphi_{i-1}(z_1)}{dz_1}\frac{d\varphi_{j-1}(z_1)}{dz_1}\right]_{z_1=L_T}}_{\text{RNA rotational inertia}};$$

$$i,j \in [1,2,3,\cdots,7], p,q \in [1,2,3,4,5]; \tag{4}$$

$$AM = \begin{cases} \rho_W A_{CS} & \text{for} \quad 0<\gamma \leq h/L_T \\ 0 & \text{for} \quad \gamma>\mathrm{h/L_T} \end{cases};$$

$$K_{ij} = \underbrace{\int_0^{\gamma L_T} E_P I_P T_{pi}\frac{d^2\varphi_{p-1}(z_2)}{dz_2^2}\frac{d^2\varphi_{q-1}(z_2)}{dz_2^2}T_{qj}dz_2}_{\text{Monopile}}$$

$$+ \underbrace{\int_0^{L_T} E_T I_T(z)\frac{d^2\varphi_{i-1}(z_1)}{dz_1^2}\frac{d^2\varphi_{j-1}(z_1)}{{dz_1}^2}dz_1}_{\text{Tower}}$$

$$+ \underbrace{\begin{bmatrix} k_{hh}T_{pi}\varphi_{p-1}(z_2)\varphi_{q-1}(z_2)T_{qj} + k_{rr}T_{pi}\frac{d\varphi_{p-1}(z_2)}{dz_2}\frac{d\varphi_{q-1}(z_2)}{dz_2}T_{qj} \\ +2k_{hr}T_{pi}\varphi_{p-1}(z_2)\frac{d\varphi_{q-1}(z_2)}{dz_2}T_{qj} \end{bmatrix}_{z_2=0}}_{\text{Foundation springs}};$$

$$i,j \in [1,2,3,\cdots,7], p,q \in [1,2,3,4,5]; \tag{5}$$

where ρ_T and $A_T(z_1)$ represent the density of the tower material and area of the tower cross sections, respectively. m_{RNA} and J_{RNA} represents the mass and rotational inertia of the RNA as a lumped mass with rotational inertia at the top of the tower as illustrated in Figure 2. AM is the added mass of the surrounding water, where ρ_W and A_{CS} are water density and cross section area of submerged monopile, respectively, and h represents water depth. E_T and $I_T(z)$ are Young's modulus of the tower material and cross-sectional moment of inertia, respectively. ρ_p and A_p represent the monopile material density and the area of the monopile cross sections, respectively. T is the transformation matrix between the vector for the monopile shape function coefficients and that for the tower. k_{hh}, $k_{rr}$and k_{hr} denote the horizontal, rotational and coupled stiffness constants at the base of the monopile, respectively.

2.2 *Dynamic response (modal analysis)*

The turbine responses can be calculated by modal analysis, i.e. superposition of contributions from each mode. With the eigenmodes ψ obtained in Section 2.1 and consists of N independent modes. Given orthogonality between modes, N independent equations, one for each mode shape, can be obtained and solved separately with the following form for the ith mode:

$$\bar{M}_i\frac{d^2q(t)}{dt^2} + \bar{C}_i\frac{dq(t)}{dt} + \bar{K}_i q(t) = \bar{P}_i(t) \tag{6}$$

where $\overline{M_i}, \overline{c_i}$ and $\overline{K_i}$ are the modal mass, damping and stiffness of the ith mode, respectively, and $\overline{P}_i(t) = \int_L P(z,t)\psi_i(z)dz$ is the modal force. The terms can be normalized in a simple form as:

$$\frac{d^2q_i(t)}{dt^2} + 2\xi_i\omega_i\frac{dq_i(t)}{dt} + \omega_i^2 q_i(t) = (\overline{P}(t))_n \tag{7}$$

where ξ_i is the so-called modal damping ratio, ω_i is the eigenfrequency of the selected mode. This parameter is assumed to be constant for all modes in the present analysis. In linear systems, the differential displacements can be combined, allowing to find the displacement at a particular time t by integrating the effects of all prior load pulses.

$$q_i(t) = \frac{1}{\omega_i^d}\int_0^t e^{-\zeta_i\omega_i(t-\tau)}(\overline{P_i}(\tau))_n \sin\omega_i^d(t-\tau)d\tau \tag{8}$$

where τ is the impulse application time and ω_i^d is the damped eigenfrequency.

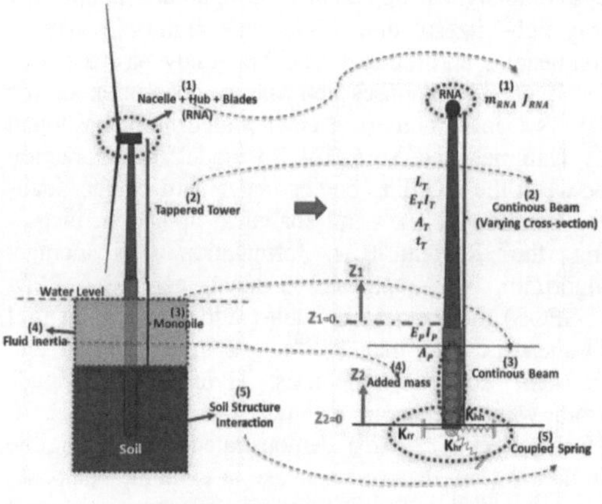

Figure 2. A simplified mathematical modelling of the monopile supported OWT.

2.3 *Contact algorithm (ship collisions)*

In 2019, the DNV-RP-C204 (2019) standard increased the design collision energy level to 50 MJ for bow/stern collisions and broadside impacts, up from 11 MJ and 14 MJ in the previous version.

However, it is still lower than the collisions reported by Kvitrud (2011), which reached up to 70 MJ. According to DNV-OS-J101 (2014), offshore wind turbines should be designed to withstand operational ship impacts at a minimum speed of 0.5 m/s and accidental ship impacts at a minimum speed of 2.0 m/s.

Hammad and Yu (2024) developed a contact algorithm for hard contact loading conditions, where the loads depend on the contact stiffnesses and responses of the interacting bodies. A contact model representing ship collision with monopile-supported OWT is shown in Figure 3. The contact stiffness between the ship and the OWT is represented by a nonlinear force-deformation curve, which can be generated using NLFEA with prescribed motions. The force-displacement curve can be simplified as piecewise linear for implementation convenience (see Figure 3). During unloading, the stiffness is assumed to be equal to the initial elastic stiffness.

A flow chart diagram illustrating the procedure of the numerical contact algorithm is shown in Figure 4. Consider the time instant t_n (the nth timestep with a constant time step being Δt), the following information is given: the ship velocity $V_S(t_n)$, the ship displacement $x_S(t_n)$, the contact force $F(t_n)$, the turbine displacement at the contact point $W_c(t_n)$, and $\delta(t_n)$ being the relative displacement between the ship and the turbine at the contact point. Upon moving to the next timestep t_{n+1}, the contact force is unknown, which depends on the turbine responses. A trial contact force is initially assumed to be equal to that at t_n, meaning that $F_{trial}(t_{n+1}) = F(t_n)$. With given trial contact force, a trial ship displacement $x_{S,trial}(t_{n+1})$ and a trial OWT contact point displacement $W_{c,trial}(t_{n+1})$ at $t_{n+1}$are derived (see Figure 4). $SR(F_{trial}(t_{n+1}))$ represents the response displacement of the monopile OWT at the contact point under given load history using modal superposition method as described in *Section* 2.2. Given the relative displacement $\delta(t_{n+1})$, assuming a trial contact force, and the contact stiffness $k(x_S(t_{n+1}))$from pre-generated force-displacement curves, a correction of the contact force $F(t_{n+1})$ at t_{n+1}can then be expressed as follows:

$$F(t_{n+1}) = F(t_n) - k(x_S(t_{n+1}))[\delta(t_{n+1}) - \delta(t_n)] \quad (9)$$

A supply vessel (UT 745) with 7500 tons displacement was selected for this study. The model has been used by Yu et al. (2022) and the corresponding force-deformation curve for bow collision was provided. A hydrodynamic added mass of 10 % for bow collision was assumed in this study as recommended by (DNV-RP-C204, 2019) with an initial impact velocity of 2 and 3 m/s, resulting in an initial kinetic energy of 16.5 MJ and 37 MJ, respectively. The impact point was set at 2 m below the water level.

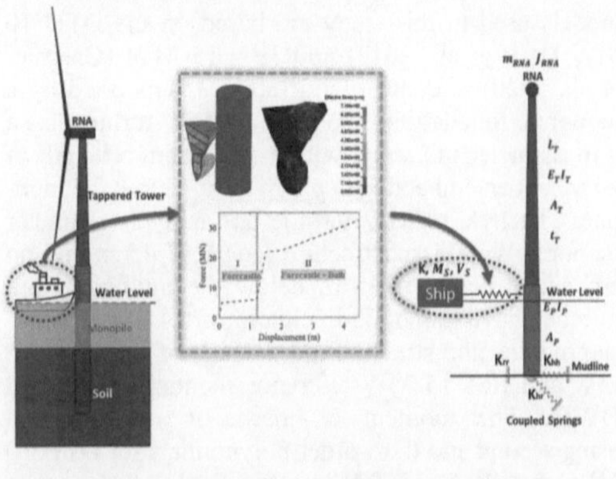

Figure 3. A contact model representing ship collisions.

It should be noted that this study assumes that only the colliding ship will experience plastic deformations represented by a nonlinear force-displacement curve to account for both elastic and plastic deformations. This is applicable to low-energy impacts from small vessels like offshore supply vessels, where the monopile remains in the elastic regime. However high-energy impacts from large vessels such as tankers or containers can cause plastic deformation in the turbine monopile, leading to energy absorption and degradation of bending stiffnesses in the damaged section.

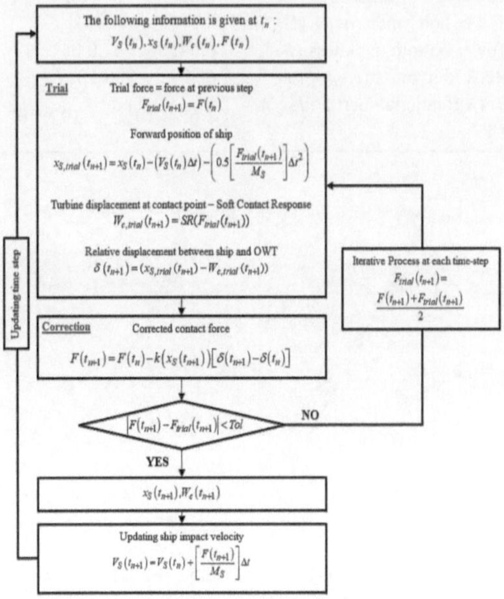

Figure 4. Flow chart illustrating the procedures of the numerical contact algorithm.

2.4 *Offshore wind turbines design*

The growing demand for increased energy production has led to the development of larger monopiles to support next-generation wind turbines. The OWT models used in this study are based on the DTU 10 MW (Bak et al., 2013) and IEA 15 MW (Gaertner et al., 2020) reference wind turbines, supported by a monopile foundation. The DTU 10 MW turbine has a 9 m diameter monopile with a substructure length of 40 m and an embedded depth of 42 m below the mudline. The IEA 15 MW turbine has an 11 m diameter monopile with a substructure length of 45 m and an embedded depth of 45 m below the mudline. Both turbines have monopile thickness of 0.11 m. Table 1 summarizes the structural parameters of the DTU 10 MW and IEA 15 MW reference monopile-supported OWTs. The moment of inertia is approximated using second and third-order polynomials for DTU 10 MW and IEA 15 MW, respectively. The cross-sectional area is fitted linearly and with a second-order polynomial for DTU 10 MW and IEA 15 MW, respectively, as shown in Figure 5.

Table 1. Structural parameters of the DTU 10MW and IEA 15 MW reference OWTs.

	DTU-10MW	IEA-15MW
Tower length (L_T) m	115.63	144.58
Tower total mass (m_T) tons	753.6	860
Tower top diameter (D_{Tt}) m	5.5	6.54
Tower top thickness (t_{Tt}) m	0.024	0.024
Tower bottom diameter (D_{Tb})m	8.3	10
Tower bottom thickness (t_{Tb}) m	0.0456	0.041
RNA total mass (m_{RNA}) tons	674	1017
RNA rotational inertia (J_{RNA}) kgm^2	1.65×10^8	3.19×10^8

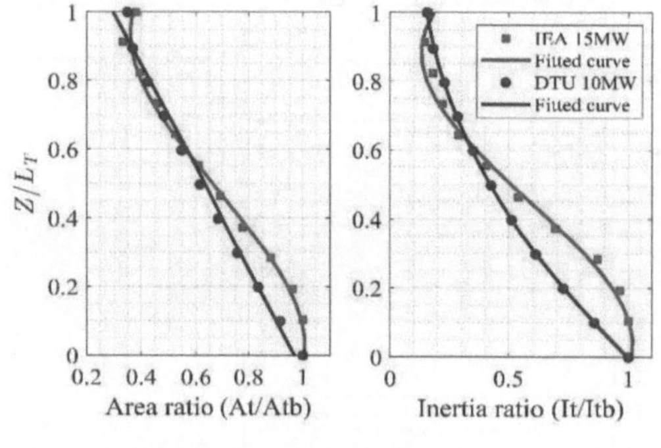

Figure 5. Fitting of DTU 10 MW and IEA 15 MW tower cross section area and moment of inertia.

3 RESULTS AND DISCUSSION

3.1 *Natural frequency and eigenmodes*

The natural frequencies of the DTU 10 MW and IEA 15 MW monopile-supported OWTs were computed up to the fourth mode using the Rayleigh-Ritz principle. Table 2 shows the natural frequencies of both turbines indicating significant differences in their dynamic characteristics. The first mode natural frequency for the DTU 10 MW turbine is 0.205 Hz compared to 0.163 Hz for the IEA 15 MW turbine, indicating the latter is more flexible due to its larger mass and dimensions. Higher-order modes also showed consistent trends, with the DTU 10 MW turbine having higher natural frequencies. Mode shapes (Figure 6) reveal that the first two modes are dominated by global bending, while the third and fourth modes show more localized deformations near the mudline and tower top.

Table 2. Natural frequencies of DTU 10 MW and IEA 15 MW monopile supported OWT.

Natural Frequencies	DTU 10 MW (HZ)	IEA 15 MW (HZ)
Mode 1	0.205	0.163
Mode 2	0.859	0.700
Mode 3	1.629	1.312
Mode 4	3.264	2.661

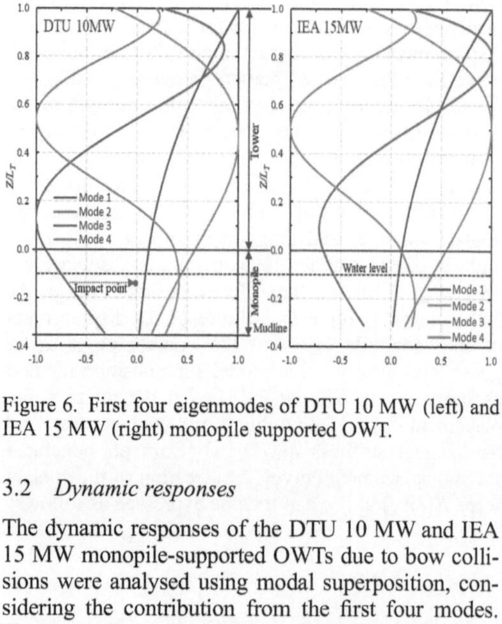

Figure 6. First four eigenmodes of DTU 10 MW (left) and IEA 15 MW (right) monopile supported OWT.

3.2 *Dynamic responses*

The dynamic responses of the DTU 10 MW and IEA 15 MW monopile-supported OWTs due to bow collisions were analysed using modal superposition, considering the contribution from the first four modes. Figure 7 shows the force and displacement time-history at the contact point of both turbines for impact velocities of 2 and 3 m/s. The peak contact force was higher at 3 m/s impact velocity. At 3 m/s, both turbines exhibited similar contact force time-history, while the IEA 15 MW turbine had a slightly higher peak force than the DTU 10 MW turbine at 2 m/s impact velocity. The displacement at the collision point is influenced by localized deformations of the structure, primarily associated with higher-order modes. This is because the collision point is typically located near the mudline (27.6 m above mudline), where higher-order modes (e.g., second and third modes) exhibit significant deformations.

The IEA 15 MW turbine has lower displacement at the collision point compared to the DTU 10 MW turbine due to its larger monopile diameter (11 m) providing greater bending stiffness near the mudline. This reduces localized deformations and results in lower displacements. At an impact velocity of 3 m/s, both OWTs show increased displacement at the collision point, but the IEA 15 MW turbine still has lower displacement than the DTU 10 MW turbine.

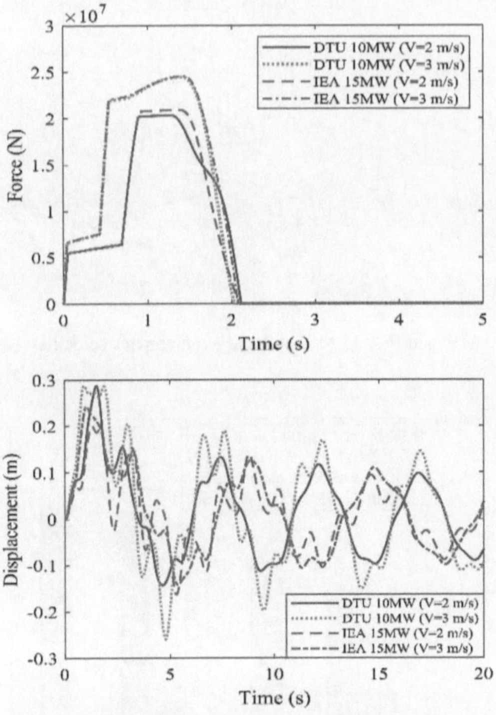

Figure 7. Contact force and contact point displacement for both DTU 10 MW and IEA 15 MW at 2 and 3 m/s impact velocities.

Figure 8 shows the tower top displacement and acceleration at impact velocities of 2 and 3 m/s for DTU 10 MW and IEA 15 MW monopile-supported OWTs. The displacement at the tower top is primarily influenced by the global bending of the structure, which is associated with the first mode. This is because the tower top is located at the free end of the structure, where the first mode (global bending) has the largest amplitude. The IEA 15 MW turbine exhibits lower displacement at the tower top compared to the DTU 10 MW turbine, despite its greater flexibility, due to its greater mass and rotational inertia, which increase its resistance to global bending. The energy absorption mechanism also plays a role as the DTU 10 MW turbine dissipates more energy through global bending (first mode) compared to IEA 15 MW turbine for both impact velocities. Increasing impact velocity leads to higher energy absorption by the first mode for both OWTs, resulting in higher tower top displacement in all cases (see Figure 9).

In OWT design, it is crucial to limit tower-top acceleration to ensure the safety of sensitive equipment in the nacelle. The maximum acceleration limit is typically set between 0.2 - 0.3 g, with g being the gravitational acceleration (Rasekhi Nejad & Moan, 2017).

The tower top acceleration of the DTU 10 MW turbine increases significantly at 3 m/s impact velocity, while the acceleration of the IEA 15 MW turbine did not show a proportional increase. This is surprising, as higher kinetic energy at 3 m/s would typically be expected to result in higher accelerations. This unexpected behaviour can be explained by examining the energy absorption curves (see Figure 9) and modal contributions to the total energy. The DTU 10 MW turbine, with its higher natural frequencies, transferred more collision energy into global acceleration compared to IEA 15 MW turbine, resulting in a significant increase in tower top acceleration at 3 m/s.

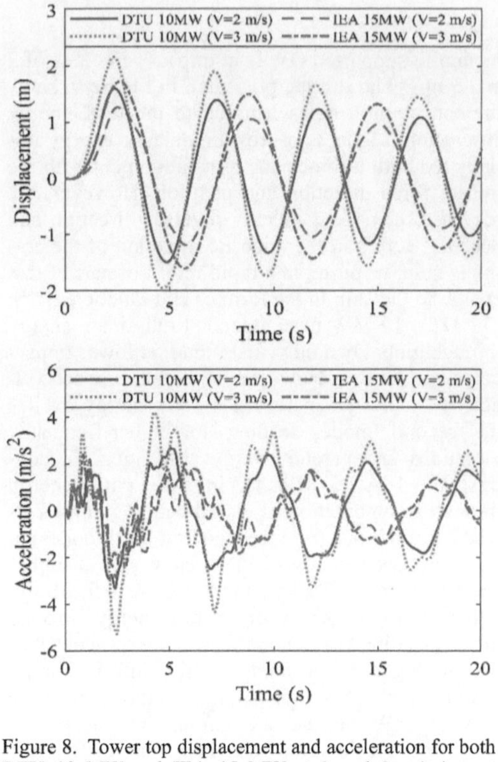

Figure 8. Tower top displacement and acceleration for both DTU 10 MW and IEA 15 MW at 2 and 3 m/s impact velocities.

3.3 *Energy absorption and modal contributions*

The energy absorption during the collision is analysed for both DTU 10 MW and IEA 15 MW

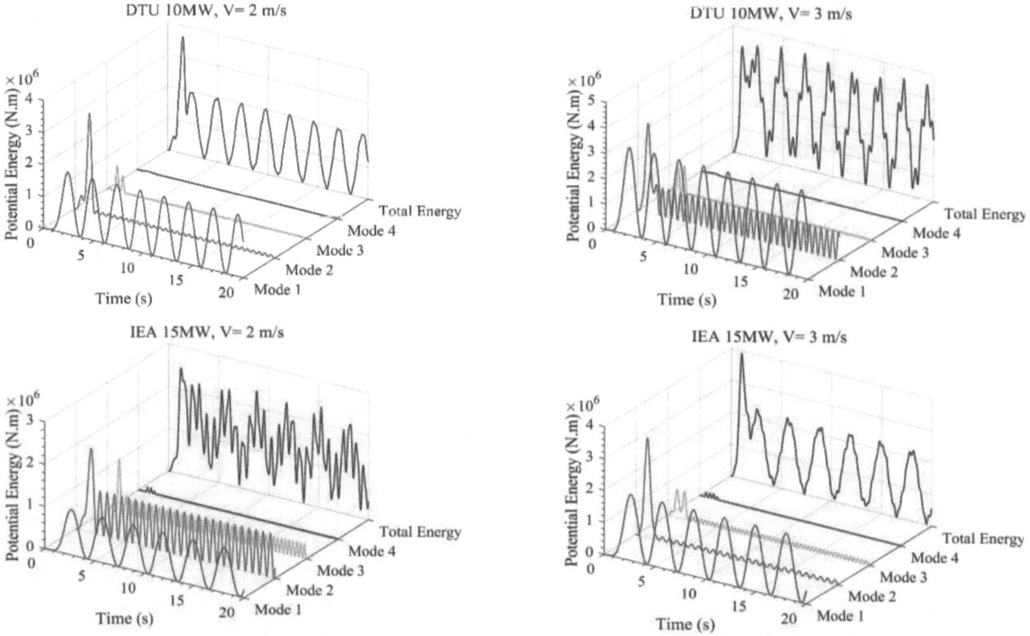

Figure 9. Elastic energy absorbed by each mode for both DTU 10 MW and IEA 15 MW at 2 and 3 m/s impact velocities.

monopile-supported OWTs at impact velocities of 2 and 3 m/s. The results, presented in Figure 9, show the contribution of each mode to the total energy absorption. Mode 2 absorbs high elastic energy initially for both turbines due to its alignment with the impact force direction and position. However, the response of mode 2 quickly reverses direction and becomes acting in the opposite direction of the colliding ship, resulting in a rapid return of part of this energy to the ship in the form of end kinetic energy. The DTU 10 MW turbine tends to dissipate energy through global bending (first mode) at lower impact velocity (2 m/s). However, at higher impact velocity (3 m/s), the increased kinetic energy excites the second mode, leading to higher-frequency vibrations and greater energy absorption. In contrast, the IEA 15 MW turbine, dissipates energy through a combination of global bending and localized deformations (first, second and third modes) at lower impact velocity (2 m/s). However, at higher impact velocity (3 m/s), the structure's flexibility allows it to absorb most of the energy through global bending (first mode), reducing the contribution of higher-order modes. This shift in energy absorption reduces high-frequency vibrations in the IEA 15 MW turbine, explaining the unexpected tower top acceleration behaviour. The differences in energy absorption can be attributed to the structural properties and natural frequencies of the turbines, highlighting the complex interplay between energy dissipation mechanisms and structural flexibility in determining their dynamic response under impact loading.

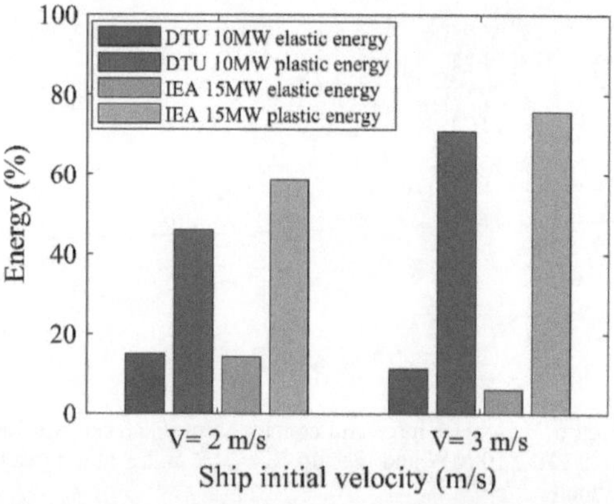

Figure 10. Percentage of elastic and plastic energy absorbed by the system for both DTU 10 MW and IEA 15 MW at different initial impact velocities.

The percentage of transferred elastic energy and dissipated ship plastic energy for DTU 10 MW and IEA 15 MW at impact velocities of 2 and 3 m/s are summarized in Figure 10. It is observed that increasing initial impact velocity generally leads to significant increase in the percentage of ship's plastic energy while reducing the absorbed elastic energy.

4 CONCLUSION

This paper investigated the dynamic behaviour and energy absorption of DTU 10 MW and IEA 15 MW monopile-supported OWTs during ship collisions using a semi-analytical approach, considering explicitly tapered tower and soil-pile interaction. The analysis revealed significant differences in their natural frequencies, mode shapes, dynamic responses and energy absorption mechanisms. The following conclusions are drawn.

- Including higher-order modes is crucial for accurately capturing the dynamic responses during ship collisions, as these modes significantly impact energy absorption.
- The IEA 15 MW turbine, despite its greater flexibility and lower natural frequencies, exhibited lower dynamic responses (displacement and acceleration) compared to the DTU 10 MW turbine. This is due to the IEA 15 MW turbine's larger monopile and tower diameter, which provides greater bending stiffness, and its greater mass and rotational inertia, which enhance its resistance to both localized and global deformations.
- The collision point displacement in both turbines is mainly affected by higher-order modes with localized deformations near the mudline, while the tower top displacement is primarily influenced by the first mode, involving global bending of the entire structure.
- At a higher impact velocity of 3 m/s, the IEA 15 MW turbine primarily dissipated energy through global bending, leading to lower tower top acceleration. In comparison, the DTU 10 MW OWT showed increased contributions from higher-order modes at higher impact velocities, resulting in larger acceleration passing the limited acceleration value of 0.3g.
- The IEA 15 MW turbine absorbs less elastic energy and dissipates more plastic energy by the ship's bow compared to the DTU 10 MW turbine.

REFERENCES

Administration, J. M. (2020). *https://www.imca-int.com/safety-events/windfarm-support-vessel-njord-forseti-hit-wind-turbine-tower-jersey-maritime-administration/.*

Alkhoury, P., Soubra, A. H., Rey, V. & Aït-Ahmed, M. (2020). A full three-dimensional model for the estimation of the natural frequencies of an offshore wind turbine in sand. *Wind Energy, 24*(7), 699–719.

Bak, C., Zahle, F., Bitsche, R., Kim, T., Yde, A., Henriksen, L. C., Hansen, M. H., Blasques, J. P. A. A., Gaunaa, M. & Natarajan, A. (2013). The DTU 10-MW reference wind turbine. Danish wind power research 2013,

Bela, A., Le Sourne, H., Buldgen, L. & Rigo, P. (2015). Numerical crashworthiness analysis of an offshore wind turbine monopile impacted by a ship. MARSTRUCT 2015 5th international conference on marine structures,

Bela, A., Le Sourne, H., Buldgen, L. & Rigo, P. (2017). Ship collision analysis on offshore wind turbine monopile foundations. *Marine Structures, 51*, 220–241.

DNV-OS-J101. (2014). Design of offshore wind turbine structures. *Det Norske Veritas: Copenhagen, Denmark.*

DNV-RP-C204. (2019). Structural design against accidental loads. *Det Norske Veritas: Norway.*

DNV-ST-0119. (2021). Floating wind turbine structures. *Det Norske Veritas: Hamburg, Germany.*

Gaertner, E., Rinker, J., Sethuraman, L., Zahle, F., Anderson, B., Barter, G. E., Abbas, N. J., Meng, F., Bortolotti, P. & Skrzypinski, W. (2020). *IEA wind TCP task 37: definition of the IEA 15-megawatt offshore reference wind turbine.*

Hammad, A. & Yu, Z. (2024). A semi-analytical approach for dynamic responses of monopile-supported offshore wind turbines subjected to accidental loads. *Applied Ocean Research, 153*, 104251.

Hammad, A., Yu, Z., Bachynski-Polić, E. & Katsikogiannis, G. (2025). Dynamic analysis of monopile-supported offshore wind turbine affected by vessel collisions. In *Innovations in the Analysis and Design of Marine Structures* (pp. 253–261). CRC Press.

Jung, S., Kim, S.-R., Patil, A. & Hung, L. C. (2015). Effect of monopile foundation modeling on the structural response of a 5-MW offshore wind turbine tower. *Ocean Engineering, 109*, 479–488.

Kvitrud, A. (2011). Collisions between platforms and ships in Norway in the period 2001-2010. International Conference on Offshore Mechanics and Arctic Engineering,

Ladeira, I., Le Sourne, H., Moran, J. & Salazar, P. (2022). A Semi-Analytical Methodology to Assess the Dynamic Elastic Response of a Monopile Offshore Wind Turbine Subjected to Low-Energy Ship Impacts. ASME 2022 41st International Conference on Ocean, Offshore and Arctic Engineering,

Pedersen, P.T. (2013). Ship collisions against wind turbines, quays and bridge piers. *Collision and grounding of ships and offshore structures. CRC Press/Taylor and Francis Group, London*, 273–280.

Professionals, M. (2022). *Crash between wind turbines and ships. https://maritime-professionals.com/crash-between-wind-turbines-and-ships/*

Psaroudakis, E.G., Mylonakis, G.E. & Klimis, N. S. (2020). Non-linear Analysis of Laterally Loaded Piles Using "p-y" Curves. *Geotechnical and Geological Engineering, 39*(2), 1541–1556.

Rasekhi Nejad, A. & Moan, T. (2017). On model-based system approach for health monitoring of drivetrains in floating wind turbines.

Song, M., Jiang, Z. & Yuan, W. (2021). Numerical and analytical analysis of a monopile-supported offshore wind turbine under ship impacts. *Renewable Energy, 167*, 457–472.

Wind, O. (2022).*https://www.offshorewind.biz/2022/12/02/monopile-damaged-by-rogue-ship-to-be-removed-contractor-wanted.*

WindEurope. (2024). Offshore wind energy 2023 statistics.

Yu, Q., Liu, K., Chang, C.-H. & Yang, Z. (2020). Realising advanced risk assessment of vessel traffic flows near offshore wind farms. *Reliability Engineering & System Safety, 203*, 107086.

Yu, Z. & Amdahl, J. (2023). A Rayleigh-Ritz solution for high order natural frequencies and eigenmodes of monopile supported offshore wind turbines considering tapered towers and soil pile interactions. *Marine Structures, 92*.

Yu, Z., Amdahl, J., Rypestøl, M. & Cheng, Z. (2022). Numerical modelling and dynamic response analysis of a 10 MW semi-submersible floating offshore wind turbine subjected to ship collision loads. *Renewable Energy, 184*, 677–699.

Trends in Collision and Grounding of Ships and Offshore Structures – Zhu & Guedes Soares (Eds)
© 2026 The Author(s), ISBN 978-1-041-16361-9

Effect of high ductility steel on damage extents and collision limit curves of a floating offshore wind turbine

M. Jeong, D.H. Yoon & J. Choung
Department of Naval Architecture and Ocean Engineering, Inha University, Incheon, Republic of Korea

E.S. Kim
Steel Solution R&D Center, POSCO, Republic of Korea

J.B. Lee
STS Engineering, Republic of Korea

ABSTRACT: The objective of this paper is to present the collision limit curve for ships in a floating offshore wind farm to be developed in the offshore area of Ulsan. To determine the collision ship, the Ministry of Oceans and Fisheries collected data on 65,000 ships that entered and left Ulsan Port over a period of about 10 years. The volume of the container ship used as the collision ship was categorized into three tiers. The collision ship was assumed to be a rigid body, and the FOWT, which was the target of the collision, was modeled as an elastic-plastic element in the collision area and as an elastic element in the rest of the ship. The surge added mass of the collision ship was considered through hydrodynamic analysis of the outer shell of the collision ship. High-strength steel (H36) and high-strength-high-ductility steel (H36-HD) with a strength of 355 MPa were applied to the sub-structure. It was assumed that the ship and FOWT collide at right angles, and collision limit lines corresponding to damage of 0.01 m2 and 1 m2 were presented.

1 INTRODUCTION

1.1 *Introduction*

Safety issues are emerging as the number of collisions between ships and offshore structures at sea is increasing. In July 2021, a jack-up barge working at a Chinese offshore wind farm tilted, causing four people to go missing, and in 2022, a cargo ship collided with an offshore substation at the Hollandse Kust Zuid offshore wind farm in the Netherlands (Durakovic, 2022). In May 2023, a cargo ship collided with a wind turbine at the Gode Wind 1 offshore wind farm in Germany (Buljan, 2023), In March 2024, a container ship collided with the Francis Scott Key Bridge in Baltimore, USA, causing the bridge to collapse (Lee, 2024). Cho & Kim (2013) performed a collision analysis of a 5MW offshore wind turbine generator, a 30,000DWT cargo ship, and an 850-ton barge. This study evaluated the vulnerability of offshore wind turbines to ship collisions. The study analyzed the collision vulnerability, which reflected uncertainties in ship weight, collision angle, and ship draft. Bela et al. (2017) analyzed the collision between a monopile-type fixed wind turbine and a 5,000-ton ship. Hao & Liu (2017) compared the impact resistance performance of various offshore structures, monopiles, and jackets. Yu et al. (2024) analyzed the collision damage of floating offshore wind turbines and assessed the risk of ship collisions. Yoon & Choung (2023) performed a collision simulation of a floating offshore wind turbine considering ductile failure and hydrodynamic effects. In the above-mentioned previous studies, no research was conducted on the collision limit curve, which indicates the ship's speed of passage in relation to the damaged extent of floating offshore wind turbines. Since FOWT farms are likely to restrict the routes of passing ships, it is likely that ships will be allowed to pass through FOWT farms in the future. This is expected to increase the likelihood of ship collisions. If the outer plating of the sub-structure is damaged due to a collision, it is likely to lead to sinking due to flooding, so it is very important to determine the extent of the damage to the outer shell caused by a ship collision. For these reasons, it is necessary to study the issue of FOWT and ship collisions based on statistical data of ships entering and leaving ports. In this study, the HC-LN model, a composite model of the Hosford-Coulomb (HC) model and the localized necking (LN) model, which has been modified to be applied to shell elements, was used to accurately predict the structural damage that occurs in a crash (Cerik et al., 2019). To consider the hydrodynamic

DOI: 10.1201/9781003684404-26

coefficient, we will conduct a frequency response analysis using Ansys/Aqwa and address the collision problem between a 22MW semi-submersible FOWT and ships entering and leaving Ulsan Port through Abaqus/Explicit (Simulia, 2023).

2 THEORETICAL BACKGROUND

2.1 *Flow of application of damage stability criteria*

Damage stability is mainly considered to ensure the safety of structures in the event of an accident such as a collision. Damage scenarios were selected in accordance with IEC 61400-3-2(International Electrotechnical Commission, 2022). Flow chart of Damage stability assessment is shown in Figure 1. The probability of collision is calculated using the time series data of the ship's position.

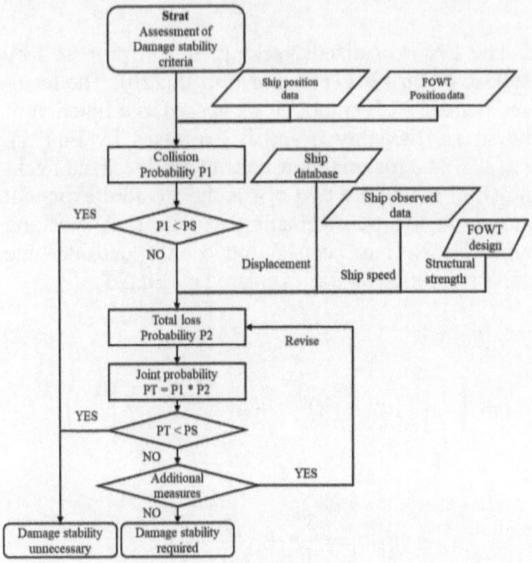

Figure 1. Flow chart of damage stability assessment.

The frequency of approach of individual ships sailing in the area where the FOWT will be installed is calculated from the position of the ship. Assuming that the variation in each navigation process of individual ships follows a normal distribution, the probability distribution of the collision range with FOWT is integrated for each individual ship as shown in Eq.(1).

$$n_{gi} = \int_{-0.5(B9+D)}^{0.5(B+D)} \frac{1}{\sqrt{2\pi\sigma^2}} \exp - \frac{(x-\mu_i)^2}{2\sigma^2} dx \tag{1}$$

The probability of approach P_g is calculated according to Eq.(3), and the ship's approach frequency n_g is calculated using Eq.(2). In this assumption, a collision is considered a disaster event and is generally treated as a Poisson process.

$$n_g = \sum_{i=1}^{n} n_{gi} \tag{2}$$

$$P_g(t) = 1 - \exp(-n_g t) \approx n_g t = n_g \ (t = 1) \tag{3}$$

The probability of total loss P2 of a structure due to a ship collision is estimated through FEM analysis and is used to determine whether the outer shell of the structure will be destroyed. PT is calculated as the product of the probability of collision P1 and the probability of total loss P2.

2.2 *Derivation process of collision limit curve*

Whether the outer skin of the FOWT is damaged or not is defined by Eq.(4) according to the amount of deformation energy.

$$E_c \leq E_d \tag{4}$$

The deformation amount E_c obtained by FEM analysis varies depending on the local structure. Therefore, it is constant and does not depend on the ship type or ship size. In Eq.(5), α is the fraction of the total deformation energy that is contributed by the deformation energy of the FOWT. If the collision ship is assumed to be a rigid body, the value of α is 1. The total absorption energy E_t can be derived through the process of Eq.(6) to (8).

$$E_d = \alpha E_t \tag{5}$$

$$M_s V_s = (M_f + M_s) V_{sf} \tag{6}$$

$$\frac{1}{2} M_f V_f^2 + \frac{1}{2} M_s V_s^2 = \frac{1}{2} (M_s + M_f) V_{sf}^2 + E_t \tag{7}$$

$$E_t = \frac{1}{2} \left(\frac{M_s M_f}{M_s + M_f} \right) V_s^2 \tag{8}$$

The reference for the ship speed according to various displacements is expressed as the representative ship speed shown in the equation, which causes damage to the FOWT outer shell. The critical speed V_s can be expressed as Eq.(9).

$$V_s = \sqrt{\frac{2E_t(M_s + M_f)}{\alpha M_s M_f}} \tag{9}$$

Figure 2 is an example of a collision limit curve, which can classify vessels around FOWT as safe or dangerous depending on the speed and displacement. The area above the total loss limit curve is considered to have sufficient energy to cause damage to the FOWT hull.

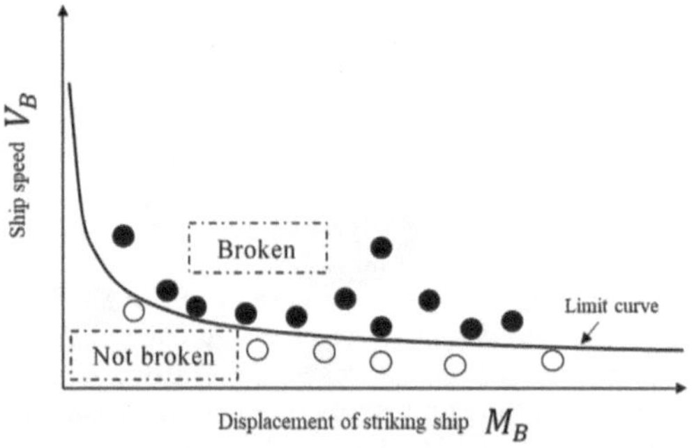

Figure 2. Example of collision limit curve.

2.3 *Flow stress and fracture model*

In order to accurately represent the hardening after necking, the combined Swift-Voce hardening model proposed by (Sung et al., 2010) was adopted in this study. The Swift-Voce hardening law is shown in Eq. (10). Here, the weighting factor α is the weighting factor for the Swift-Voce hardening law. The Swift law and the Voce law are represented by Eq.(11) and Eq.(12). The parameters K_s, ε_0, n, k_0, Q *and* β used in the equation can be obtained through tensile tests and numerical analysis, and high-quality parameters were derived in the study by Cerik & Choung (2020).

$$k\left[\bar{\varepsilon}_p\right] = \alpha k_s\left[\bar{\varepsilon}_p\right] + (1-\alpha)k_v\left[\bar{\varepsilon}_p\right] \tag{10}$$

$$k_s\left[\bar{\varepsilon}_p\right] = K_s\left(\varepsilon_0 + \bar{\varepsilon}_p\right)^n \tag{11}$$

$$k_v\left[\bar{\varepsilon}_p\right] = k_0 + Q\left(1 - \exp\left[-\beta\bar{\varepsilon}_p\right]\right) \tag{12}$$

The Hosford-Coulomb (HC) model proposed in the study by Mohr & Marcadet (2015) is shown in Eq.(15), and the stress triaxiality facto η and the load angle coefficient $\bar{\theta}$ used in the equation are expressed as functions of Eq.(13) and Eq.(14). In Eq.(19), a_{HC} is the load angle sensitivity, b_{HC} is the fracture strain coefficient, c_{HC} is the stress triaxiality sensitivity, and n_f is the transformation exponent. Functions f_1, f_2 *and* f_3 are calculated using Eq(16) to (18), and damage indicator, D, is introduced to consider the effects of stress path, as shown in Eq.(19).

$$\eta = \frac{I_1}{3\sqrt{3J_2}} \tag{13}$$

$$\bar{\theta} = 1 - \frac{2}{\pi}\cos^{-1}\left[\frac{3\sqrt{3}}{2}\frac{J_3}{J_2^{3/2}}\right] \tag{14}$$

$$\bar{\varepsilon}^{pr}_{HC}\left[\eta, \bar{\theta}\right] = b_{HC}(1 + c_{HC})^{\frac{1}{n_f}} \left[\left\{\frac{1}{2}\left(\begin{matrix}(f_1 - f_2)^{a_{HC}} \\ +(f_1 - f_3)^{a_{HC}} \\ +(f_2 - f_3)^{a_{HC}}\end{matrix}\right)\right\}^{\frac{1}{a_{HC}}} + c_{HC}(2\eta + f_1 + f_3)\right]^{-\frac{1}{n_f}} \tag{15}$$

$$f_1\left[\bar{\theta}\right] = \frac{2}{3}\cos\left[\frac{\pi}{6}\left(1 - \bar{\theta}\right)\right] \tag{16}$$

$$f_2\left[\bar{\theta}\right] = \frac{2}{3}\cos\left[\frac{\pi}{6}\left(3 + \bar{\theta}\right)\right] \tag{17}$$

$$f_3\left[\bar{\theta}\right] = -\frac{2}{3}\cos\left[\frac{\pi}{6}\left(1 + \bar{\theta}\right)\right] \tag{18}$$

$$dD = \frac{d\bar{\varepsilon}_p}{\bar{\varepsilon}^{pr}_{HC}\left[\eta, \bar{\theta}\right]} \tag{19}$$

The LN (Localized Necking) model proposed by Pack & Mohr (2017) is shown in Eq.(20). The localized necking (LN) model is expressed as a function of the stress triaxiality η and is expressed by Eq.(21), Eq.(22). The material constants of the localized necking (LN) model are two d_{LN} is the Hosford exponent and p_f is the transformation exponent. A Necking indicator, N, has been introduced to consider the stress path effect and is expressed by Eq.(23).

$$\bar{\varepsilon}^{pr}_{LC}[\eta] = b_{HC}\left(\left\{\frac{1}{2}\left((g_1 - g_2)^{d_{LN}} + g_1^{d_{LN}} + g_2^{d_{LN}}\right)\right\}^{\frac{1}{d_{LN}}}\right)^{-\frac{1}{p_f}} \tag{20}$$

$$g_1[\eta] = \frac{3}{2}\eta + \sqrt{\frac{1}{3} - \frac{3}{4}\eta^2} \tag{21}$$

$$g_2[\eta] = \frac{3}{2}\eta - \sqrt{\frac{1}{3} - \frac{3}{4}\eta^2} \tag{22}$$

$$dN = \frac{d\bar{\varepsilon}_p}{\bar{\varepsilon}^{pr}_{LN}[\eta]} \quad for \frac{1}{3} < \eta < \frac{2}{3} \tag{23}$$

3 FOWT - CONTAINER SHIP COLLISION SIMULATION SETTINGS

3.1 *Survey ships entering and leaving Ulsan Port*

As a result of collecting data on the number of ships entering and leaving Ulsan Port for 10 years (2011-2023, excluding the COVID period) from the Ministry of Oceans and Fisheries, the total number of ships was 274,416. Since the Ministry of Oceans and Fisheries data does not include the ship's specifications (displacement, speed) required for collision analysis, we obtained the gross tonnage, deadweight, and observed maximum speed of the ship from the private ship information site (*MarineTraffic.com*) using the Python web crawling module 'selenium'. The re-collected data on ships entering and leaving Ulsan Port totaled 65,069 ships. The collected data did not include the ship's displacement, so the ship's displacement was obtained from the gross tonnage or deadweight of the ship, which was aggregated using the reference (Ministry of Maritime Affairs and Fisheries, 2016) Eq.(24).

Table 1. Number of ships passing through Ulsan Port.

Type	Value	Type	Value
Chemical tanker	28607	Product Carrier	4102
General cargo ship	7117	Bulk carrier	3401
Container carrier	6616	Crude oil carrier	2607
Liquid gas carrier	6057	misc	940
Vehicle carrier	5264	Ropax/Passenger	358

$$M = \begin{cases} 1.385DWT \text{ ton} & \text{container} \\ 1.235DWT \text{ ton} & \text{tanker} \\ 1.400GT \text{ ton} & \text{LPG carrier} \\ 1.118GT \text{ ton} & \text{LNG carrier} \end{cases} \tag{24}$$

In this study, a container ship with the sharpest bow of the survey ships was selected as the collision ship. The distribution of the collision ship's displacement and speed is shown in Figure 3.

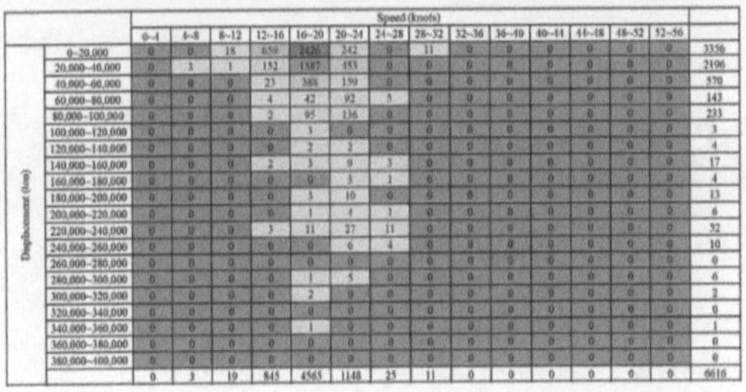

Displacement (ton)	Speed (knots) 0~4	4~8	8~12	12~16	16~20	20~24	24~28	28~32	32~36	36~40	40~44	44~48	48~52	52~56	
0~20,000	0	0	18	659	2426	242	0	11	0	0	0	0	0	0	3356
20,000~40,000	0	3	1	152	1587	453	0	0	0	0	0	0	0	0	2196
40,000~60,000	0	0	0	23	388	159	0	0	0	0	0	0	0	0	570
60,000~80,000	0	0	0	4	42	92	5	0	0	0	0	0	0	0	143
80,000~100,000	0	0	0	2	95	136	0	0	0	0	0	0	0	0	233
100,000~120,000	0	0	0	0	3	0	0	0	0	0	0	0	0	0	3
120,000~140,000	0	0	0	0	2	2	0	0	0	0	0	0	0	0	4
140,000~160,000	0	0	0	2	3	9	3	0	0	0	0	0	0	0	17
160,000~180,000	0	0	0	0	0	3	1	0	0	0	0	0	0	0	4
180,000~200,000	0	0	0	0	3	10	0	0	0	0	0	0	0	0	13
200,000~220,000	0	0	0	0	1	4	1	0	0	0	0	0	0	0	6
220,000~240,000	0	0	0	3	11	27	11	0	0	0	0	0	0	0	52
240,000~260,000	0	0	0	0	0	6	4	0	0	0	0	0	0	0	10
260,000~280,000	0	0	0	0	0	0	0	0	0	0	0	0	0	0	0
280,000~300,000	0	0	0	0	1	5	0	0	0	0	0	0	0	0	6
300,000~320,000	0	0	0	0	2	0	0	0	0	0	0	0	0	0	2
320,000~340,000	0	0	0	0	0	0	0	0	0	0	0	0	0	0	0
340,000~360,000	0	0	0	0	1	0	0	0	0	0	0	0	0	0	1
360,000~380,000	0	0	0	0	0	0	0	0	0	0	0	0	0	0	0
380,000~400,000	0	0	0	0	0	0	0	0	0	0	0	0	0	0	0
	0	3	19	845	4565	1148	25	11	0	0	0	0	0	0	6616

Figure 3. Frequency distribution table of container ship.

The surveyed container ships are mostly distributed in the 60,000 ton displacement or below. In addition, the survey shows that the speed of the container ships surveyed is mostly between 12 and 24 knots. Therefore, the representative DT of the target ship for the collision analysis was determined to be in three categories, and the maximum speed of the target ship was determined to be 20 to 25 knots.

3.2 *Dimensions of the collision ship*

Before determining the main dimensions of the container ship by its tonnage, we conducted our own research on container ships. The data collected was subjected to linear regression using a quadratic equation. The coefficients of the quadratic equation used in the linear regression and the value of the determination coefficient R^2 are shown in Table 2.

Table 2. Relationship between displacement and dimensions.

Item	ax^2	bx	constant	R^2
$Length(m)$	-5.24E-09	2.27E-03	103.35	0.949
$Breadth(m)$	-4.53E-10	2.35E-04	20.58	0.921
$Depth(m)$	-1.91E-10	1.26E-04	12.318	0.933
$T_{ful}(m)$	-2.27E-10	8.08E-05	8.15	0.963

Table 3. Dimensions of container ships surveyed on the used ship site.

No	TEU	DWT(ton)	$L_{BP}(m)$	$B(m)$	$D(m)$	$T(m)$
1	698	8157	129.58	20.6	n/a	7.4
2	704	8508	121	19.2	10.5	7.65
3	1341	17215	161.3	25	14.9	9.5
4	1345	17350	149.6	25	14.9	9.5
5	2433	34,150	188.9	32.2	n/a	11.5
6	2782	37600	186	35	17.4	n/a
7	2800	39100	225.32	29.8	16.4	11.4
8	2800	n/a	222	30	16.8	12.0
9	2689	34,150	188.8	32.2	n/a	11.5
10	2758	41,534	195.3	32.2	20.3	12.5

In this study, the results of the regression analysis were used to determine the main dimensions for the displacement. The results of the investigation of the dimensions according to the displacement were reflected in the used ship trading site as shown in Table 3 to maintain the approximate displacement. The depth was not adjusted because it has a high influence on the initial impact damage, and the target displacement was roughly satisfied by adjusting the L_{BP} and Breadth as shown in Table 4. After that, the target displacement was satisfied by adjusting the draft as shown in Table 5.

Table 4. Dimensions of container ships reflecting the survey results.

No	DT(ton)	TEU	$L_{BP}(m)$	$B(m)$	$D(m)$	$T(m)$
1	10,000	759	120*	18*	10	10
2	30,000	1867	175	26*	16	11
3	50,000	2965	204*	31*	18	12

*Adjusted dimensions

Table 5. Dimensions of the container ship after draft adjustment.

No	DT(ton)	TEU	$L_{BP}(m)$	$B(m)$	$D(m)$	$T(m)$
1	10,000	759	120	18	10	7*
2	30,000	1867	175	26	16	10*
3	50,000	2965	204	31	18	11.5*

* Adjusted dimensions

3.3 *Frequency response analysis*

The added mass has a significant impact on the damage size and collision behavior in the direction of the surge. In this study, the added mass was derived by performing frequency response analysis on the 10,000-ton, 30,000-ton, and 50,000-ton ships and FOWT. Hydrodynamic coefficient was derived using Ansys/Aqwa. Figure 4 shows the Aqwa analysis model, and Table 6 shows the mass information applied for each displacement and the number of diffraction elements used.

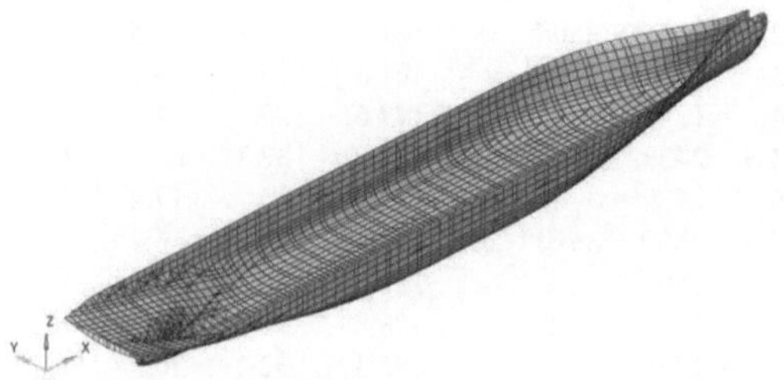

Figure 4. Example of AQWA model for displacement 10K container ship.

Figure 5 shows the AQWA analysis model of FOWT, and Table 7 shows the mass information of the analysis model and the diffraction elements used.Table 8 shows the added mass of the collision ship and FOWT obtained through the frequency response analysis.

Table 6. Container ship information for deriving the hydrodynamic derivatives.

Item	10K(ton)	30K(ton)	50K(ton)
DT(ton)	10,000	30,000	50,000
CoM_X (m)	66.19	96.53	112.53
$CoM_Y(m)$	0.00	0.00	0.00
$CoM_Z(m)$	0.53	2.05	2.05
I_x(kg-m^2)	5.61E11	3.85E12	8.69E12
I_y(kg-m^2)	1.11E13	7.17E13	1.63E14
I_z(kg-m^2)	1.12E13	7.23E13	1.65E14
Diffraction element	3252	3042	3212
Frequency range (rad/s)	0.1~2.23	0.1~1.82	0.1~1.70
Frequency	37	30	28

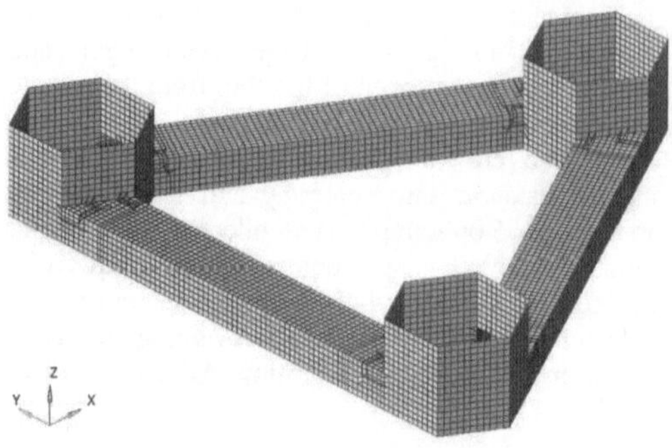

Figure 5. AQWA model for FOWT.

Table 7. FOWT information for deriving the hydrodynamic derivatives.

Item	FOWT
DT(ton)	25,299.3
CoM_X (m)	0.0
$CoM_Y(m)$	0.0
$CoM_Z(m)$	5.3
I_x(kg-m^2)	6.69E+10
I_y(kg-m^2)	6.66E+10
I_z(kg-m^2)	3.53E+10
Diffraction element	10,896
Frequency range (rad/s)	0.1~2.941
Frequency	49

Table 8. Added mass of Collision ships and FOWT.

Item	Added mass(ton)
10K ton container ship	233
30K ton container ship	680
50K ton container ship	1,145
FOWT	9,800

3.4 *FEA model for collision simulation*

Figure 6 shows the FEA modelling of a 10,000-ton container ship. The collision ship was modelled as a rigid body and a mass element and 2nd moment of inertia were placed at the center of mass. This ship was modelled with longitudinal stiffner spaced at intervals, causing the angular shape of the bow section to cause excessive collision damage. To prevent this, the bow section was divided into detailed elements as shown in Figure 9(b). The types and numbers of elements used are summarized in Table 9.

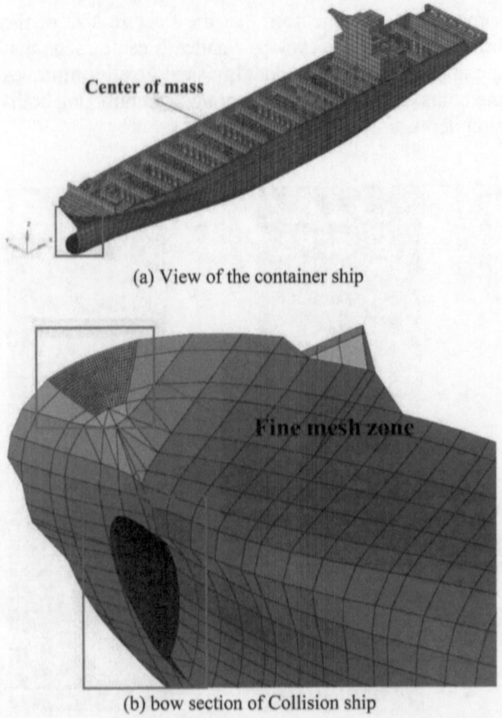

Figure 6. Container ship FEA model (Displacement 10K).

Table 9. Container ship FEA information.

Item	Element type	Material	Element
Hull	R3D4 & R3D3	Rigid	45,452
Mass	MASS	n/a	1
2nd moment of inertia	ROTARYI	n/a	1
Number of elements		45,454	

Figure 7 shows the modelling of the FOWT. The tower is a beam element, and the MASS element and ROTARYI element of the Rna are placed at the tip of the tower. The blue part at the bottom of the FOWT is the Ballst area. FOWT is modelled as a shell element. Fine mesh area of the FOWT is modelled as a elastic-plastic, and the other is modelled as an elastic. To improve the efficiency of the calculation, the longi and fine mesh element partitioning area's beam elements were changed to shell elements. The fine mesh element size is 100 mm*100 mm, and the remaining part maintains the longitude spacing element size. The types and numbers of elements used are presented in Table 10.

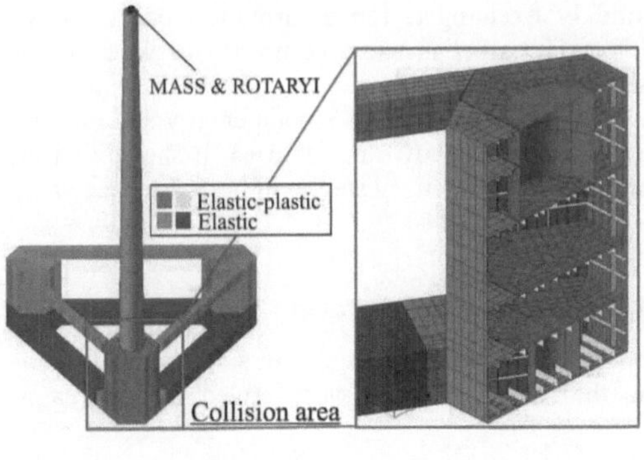

Figure 7 . FOWT FEA model.

Table 10. FOWT FEA information.

Item	Element type	Material	Element
Fine mesh	S4R & S3R	Elastic-plastic	518,509
Longi	S4R & S3R	Elastic	90,436
Tower	B31	Elastic	30
Mass	MASS	n/a	1
2nd Moment of inertia	ROTARYI	n/a	1
Number of elements			608,977

To model the contact between the collision ship and the FOWT, a penalty-based contact was defined. A friction coefficient of 0.5 was assumed to account for steel-to-steel contact.

3.5 *Material properties in finite element models*

The Young's modulus and Poisson's ratio were used as nominal values, and the final densities applied to the model are summarized in Table 11. The densities consider the ballast water in the FOWT and the added mass in the surge direction.

Table 11. Material Constant.

Item	Young's Modulus (GPa)	Poisson's ratio	Density (ton/m^3)
Ballast	210	0.3	90.463
Other	210	0.3	26.089
Tower	200	0.265	7.85

3.6 *Collision scenarios*

The initial speed used in the collision analysis was selected based on the data of Ulsan port. The collision with the tower column was assumed, and the angle of impact was determined to be 90°. The collision scenario is presented in Table 12. Figure 8 shows the collision view of the FOWT-container ship.

Table 12. Collision scenarios.

Container ship					FOWT
Ballast (ton)	Collision angle	Initial speed (knots)	Initial kinetic energy (MJ)	DT (ton)	Collision location
10,000	90	20	529.30	25,299	Center of bulkhead
30,000	90	22	1921.4	25,299	Center of bulkhead
50,000	90	25	4135.2	25,299	Center of bulkhead

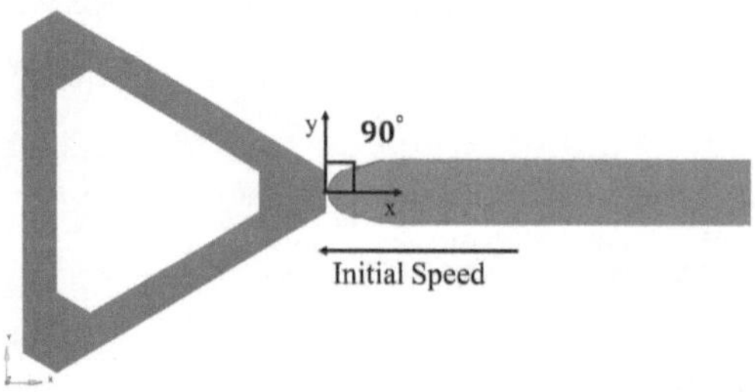

Figure 8. Collision view of FOWT-Container ship.

For conservative collision limit curve, the bulbous bow of the collision ship was adjusted to be centered at the center of the FOWT transverse bulkhead. In Figure 9, the red ship is 10,000 ton, the orange ship is 30,000 ton, and the yellow ship is 50,000 ton, and the collision positions are adjusted to have the same bulbous bow collision position. The position of the forecastle deck is adjusted accordingly.

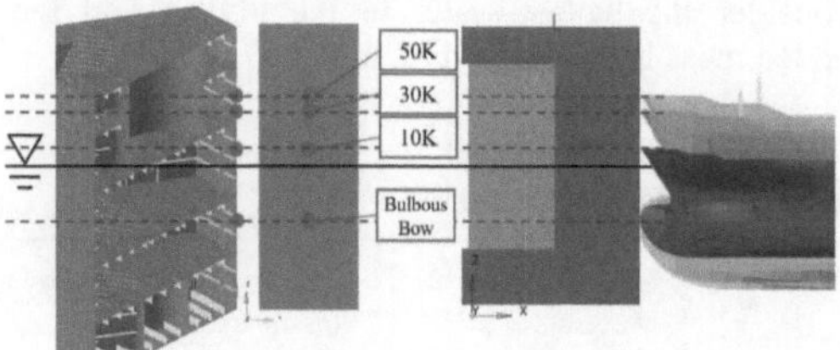

Figure 9. Relative collision position.

4 RESULTS

4.1 *Results of collision simulation*

When EH36 and EH36-HD were applied, the size of the extent was compared at the same time after the fracture occurred, and each is shown in Figure 10. As a result of comparing the size of the fracture at the same time, it can be seen that the fracture size of the high ductility steel develops smaller. It can be seen that the application of high ductility steel greatly improves the collision resistance, preventing or minimizing collision damage.

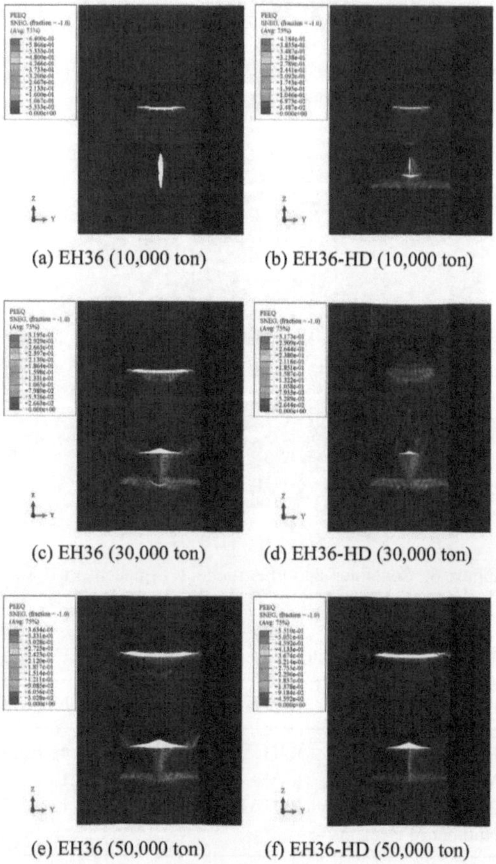

Figure 10. Comparison of damage extent (0.2s).

4.2 *Exchange of kinetic energy-deformation energy*

Figure 11 shows the history of energy exchange due to collision. The figure shows that kinetic energy is quickly exchanged for deformation energy when EH36-HD steel is used, compared to when EH36 steel is used.

The crossover time of kinetic energy and deformation energy is shown in Table 13. It shows that the effect of high ductility steel will be greater the smaller the amount of water.

4.3 *Collision limit curves*

In this study, the element size was 100 mm*100 mm, so the critical velocity was derived for two cases:

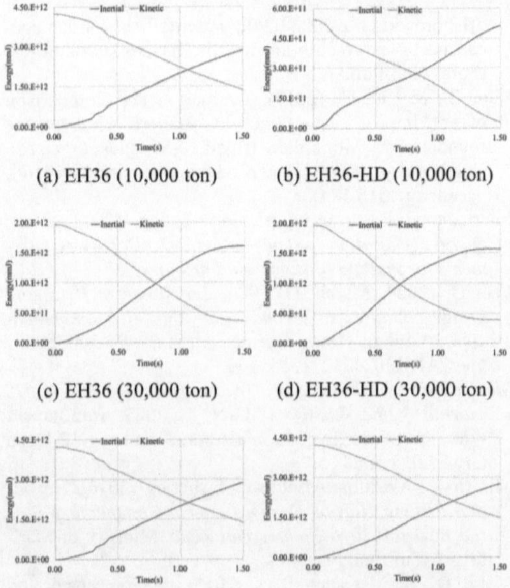

(a) EH36 (10,000 ton) (b) EH36-HD (10,000 ton)

(c) EH36 (30,000 ton) (d) EH36-HD (30,000 ton)

(e) EH36 (50,000 ton) (f) EH36-HD (50,000 ton)

Figure 11 . Comparison of kinetic energy-transformation energy exchange.

Table 13. Energy cross-time analysis (Unit: s).

Displacement (ton)	EH36	EH36-HD
10,000	0.87	0.83
30,000	0.81	0.70
50,000	1.07	1.12

Table 14. Critical speed according to the extent of the damage.

displacement (ton)	Damage extent (m^2)	Critical speed (knots)	
		EH36	EH36-HD
10,000	0.01	5.87	7.68
	1.0	6.88	9.46
30,000	0.01	3.34	5.03
	1.0	5.98	10.91
50,000	0.01	3.85	5.02
	1.0	5.61	7.03

when one element is damaged and when the damage size is 1 m*1 m. Table 14 shows the critical velocity for each case. Figure 12 shows the collision limit curve derived from the calculated critical velocity.

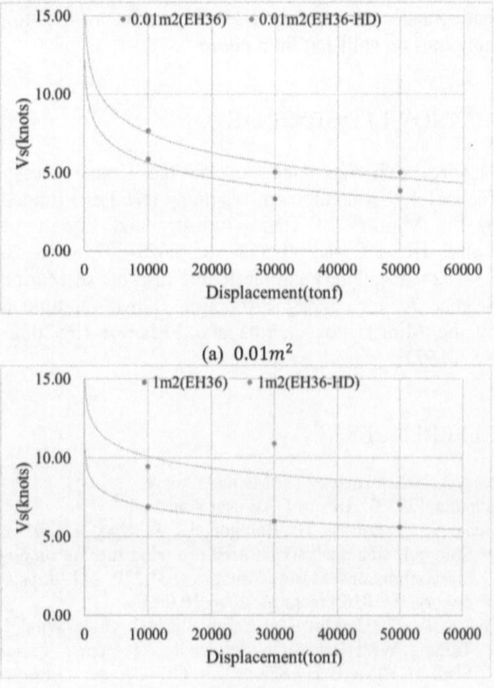

(a) $0.01m^2$

(b) $1m^2$

Figure 12. Collision limit curve by the damage extent.

5 CONCLUSION

Based on the statistical data of ships entering and leaving Ulsan Port for 10 years, representative ship types were determined and collision limit curves were presented through collision simulations. A collision simulation was conducted between a FOWT with a 22MW reference turbine and a container ship, and the displacement of the collision ship was assumed to be at level 3. The ship was assumed to collide perpendicularly with the semi-submersible floater column, and the bulbous bow was positioned at the center of the longitudinal stiffener and the center of the transverse bulkhead to eliminate the variability of the collision position depending on the displacement. When high ductility steel was applied, it was confirmed that kinetic energy was quickly converted into deformation energy, and that more critical velocity was required when the outer shell was damaged. When high ductility steel was applied to the outer shell, the collision resistance was greatly increased, and collision damage could be prevented or minimized. The extent of damage is correlated with flooding and water retention capacity. In future studies, flood analysis will be conducted to calculate flood propagation rates and post-flood damage restorability, as well as perform analyses that consider the hydrodynamic force effects (radiation damping) on collision limit curve.

ACKNOWLEDGEMENTS

This research was supported by the Korea Energy Technology Evaluation and Planning (KETEP) funded by the Ministry of Trade, Industry and Energy of Korea (RS-2021-KP002572, RS-2023-00238996 & RS-2024-00450063) and the Korea Institute of Marine Science & Technology Promotion (KIMST), funded by the Ministry of Oceans and Fisheries (RS2022-KS221571).

REFERENCES

Ansys. (2022). Ansys User Manual. Ansys.

Simulia. (2023). Abaqus/CAE user's manual.

Bela, A., Le Sourne, H., Buldgen, L., & Rigo, P. (2017). Ship collision analysis on offshore wind turbine monopile foundations. Marine Structures, 51,220–241. https://doi.org/10.1016/j.marstruc.2016.10.009

Buljan, A. (2023, May 30). Cargo Ship-Hit Gode Wind 1 Turbine Went Back into Service in 24 Hours; Vessel Said to Have Been Kilometres Off Course. *Offshore Wind*. https://www.offshorewind.biz/2023/05/30/cargo-ship-hit-gode-wind-turbine-went-back-into-service-in-24-hours-vessel-said-to-have-been-kilometres-off-course/

Cerik, B. C., & Choung, J. (2020). Ductile Fracture Behavior of Mild and High-Tensile Strength Shipbuilding Steels. *Applied Sciences*,*10*(20), Article20. https://doi.org/10.3390/app10207034

Cerik, B. C., Ringsberg, J. W., & Choung, J. (2019). Revisiting MARSTRUCT benchmark study on side-shell collision with a combined localized necking and stress-state dependent ductile fracture model. *Ocean Engineering*, *187*,106173. https://doi.org/10.1016/j.oceaneng.2019.106173

Cho, B. I., & Kim, D. H. (2013). Fragility Assessment of Offshore Wind Turbine by Ship Collision. *Journal of Korean Society of Coastal and Ocean Engineers*, *25*(4), 236–243. https://doi.org/10.9765/KSCOE.2013.25.4.236

Durakovic, A. (2022, February 2). Vattenfall Shows Damage Caused by Cargo Ship Adrift at Hollandse Kust Zuid Offshore Wind Farm. *Offshore Wind*. https://www.offshorewind.biz/2022/02/02/vattenfall-shows-damage-caused-by-cargo-ship-adrift-at-hollandse-kust-zuid-offshore-wind-farm/

Hao, E., & Liu, C. (2017). Evaluation and comparison of anti-impact performance to offshore wind turbine foundations: Monopile, tripod, and jacket. *Ocean Engineering*, *130*, 218–227. https://doi.org/10.1016/j.oceaneng.2016.12.008

International Electrotechnical Commission. (2022). *Wind energy generation systems—Part 3-2: Design requirements for floating offshore wind turbines*.

Lee, D. (2024, March 27). *Why 2.6-kilometer Baltimore bridge collapsed in seconds after ship hit (comprehensive)*. Yonhap News Agency. https://www.yna.co.kr/view/AKR20240327138551009

MarineTraffic: Global Ship Tracking Intelligence | AIS Marine Traffic. Retrieved March 14, 2025, from https://www.marinetraffic.com/en/ais/home/centerx:-12.0/centery:25.0/zoom:4

Ministry of Maritime Affairs and Fisheries. (2016). *Harbor and Fishing Harbor Design Criteria Commentary Partial Revision 2016 Commercial Zone*. Ministry of Maritime Affairs and Fisheries.

Mohr, D., & Marcadet, S. J. (2015). Micromechanically-motivated phenomenological Hosford–Coulomb model for predicting ductile fracture initiation at low stress triaxialities. *International Journal of Solids and Structures*, *67–68*, 40–55. https://doi.org/10.1016/j.ijsolstr.2015.02.024

Pack, K., & Mohr, D. (2017). Combined necking & fracture model to predict ductile failure with shell finite elements. *Engineering Fracture Mechanics*, *182*, 32–51. https://doi.org/10.1016/j.engfracmech.2017.06.025

Sung, J. H., Kim, J. hoon, & Wagoner, R. H. (2010). A plastic constitutive equation incorporating strain, strain-rate, and temperature. *International Journal of Plasticity*, Vol 26(12), 1746–1771. https://www.sciencedirect.com/science/article/pii/S0749641910000276?via%3Dihub

Yoon, D. H., & Choung, J. (2023). Collision Simulation of a Floating Offshore Wind Turbine Considering Ductile Fracture and Hydrodynamics Using Hydrodynamic Plug-in HydroQus. *Journal of Ocean Engineering and Technology*, *37*(3), 111–121. https://doi.org/10.26748/KSOE.2023.004

Yu, Y.-J., Park, S.-H., & Cho, S.-R. (2024). Collision-Damage Analysis of a Floating Offshore Wind Turbine Considering Ship-Collision Risk. *Journal of Ocean Engineering and Technology*,*38*(3),124–136. https://doi.org/10.26748/KSOE.2023.036

Trends in Collision and Grounding of Ships and Offshore Structures – Zhu & Guedes Soares (Eds)
© 2026 The Author(s), ISBN 978-1-041-16361-9

Buckling and collapse study of a floating offshore wind turbine under ship collisions

Y.L. Ren, X.G. Hua & Z.Q. Chen
State Key Laboratory of Bridge Safety and Resilience, College of Civil Engineering, Hunan University, Changsha, P.R. China
Key Laboratory for Bridge and Wind Engineering of Hunan Province, College of Civil Engineering, Hunan University, Changsha, P.R. China

Z.L. Yu
Department of Marine Technology, Norwegian University of Science and Technology, Norway

ABSTRACT: As the installed capacity of floating offshore wind turbines (FOWTs) increases in the future, the issue of ship collisions with FOWTs may become more prominent. Studying the failure mechanisms of FOWTs under ship collisions is critical for their design in the ultimate limit state (ULS) and accidental limit state (ALS). Previous studies have primarily addressed ship collisions with offshore wind turbines under moderate sea states, with limited attention to extreme conditions. Accordingly, this paper investigates the buckling and collapse of a spar-type FOWT under ship collisions during a 50-year extreme sea condition, using a co-simulation framework integrating Simulink and LS-DYNA. A series of numerical simulations, considering various ship velocities and floater poses induced by wind-wave loads, were conducted to analyze the dynamic responses and damage of the FOWT. The results reveal two distinct buckling modes: one in the lower-mid tower region and the other at the base.

1 INTRODUCTION

As offshore shallow-water wind resources become increasingly scarce, the offshore wind industry is rapidly expanding into deeper, more resource-rich areas with greater development potential (Li et al., 2024). However, with increasing water depth, traditional fixed-bottom offshore wind turbines face significant challenges, such as high costs and construction complexities, which limit their deployment in deep-sea environments (Bai et al., 2024). In this context, floating offshore wind turbines (FOWTs), known for their excellent adaptability to deep waters, have become a key area for both technological research and industrial development in the offshore wind sector. Nevertheless, both fixed-bottom and floating wind turbines are vulnerable to extreme events, such as typhoons and ship collisions (Xu et al., 2024; Niklas et al., 2023). In recent years, the frequency of ship collisions with fixed-bottom offshore wind turbines in extreme weather conditions has risen, resulting in damage to both turbines and vessels, as well as fatalities (Offshore WIND 2020; Marine Safety Investigation Unit 2023) As the number of FOWTs increases and their size expands, the risk of ship collisions under extreme weather conditions is expected to rise significantly. Therefore, to ensure the safe construction and operation of FOWTs in the future, it is crucial to study their dynamic behaviors and damage failure mechanisms under ship collisions.

Several research works have been devoted to simulate ship collisions against FOWTs using nonlinear finite element (NLFE) Tools. Echeverry et al. (2019) and Ren et al. (2022) used LS-DYNA to study the dynamic behaviors and damage of the NREL 5 MW spar-type FOWT subjected to ship collisions. In these studies, the hydrodynamic loads on the floater were calculated using the MCOL program integrated into LS-DYNA, but the wind and wave loads were not considered. Zhang et al. (2021) incorporated a theoretical model of ship collision into the FOWT tool Darwind, examining the ship collision analysis of the NREL 5 MW spar-type FOWT under wind and wave loads, with a wind speed of 11.4 m/s. This method is more efficient compared to NLFE method, but it mainly focused on the global dynamic responses of the FOWT after ship collision and some detailed transient effects cannot be captured. Zhang and Hu (2022) developed an aero-hydro-structural coupled FOWT model in LS-DYNA, using user subroutines to implement aerodynamics and hydrodynamics, and then conducted ship collision analysis

DOI: 10.1201/9781003684404-27

of the NREL 5 MW FOWT under sea conditions corresponding to a rated wind speed of 11.4 m/s. The results indicated that the wind turbine tower collapsed under high-speed impacts (5 m/s), and the structural damage may become more serious under the combined effect of wind and waves. Yu et al. (2022) used the NLFE software USFOS to study the dynamic response of a 10 MW semi-submersible FOWT under collisions from ships of 7500 and 150000 tons. This study considered both parked conditions in still water and operational conditions at a rated wind speed of 11.4 m/s. The results showed that the combined effects of collision loads, wind loads, and a taut mooring system could significantly amplify the pitch motion of the floating platform, thereby increasing the risk of overall capsizing of the floating wind turbine. Guo et al. (2022) studied the dynamic responses and damage of a tension-leg platform (TLP) floating wind turbine in still water against ship collisions using LS-DYNA. The results indicated that during a longer collision period, two distinct impacts occurred due to the pitch or roll of the TLP, with damage primarily concentrated in the contact area. Zong et al. (2023) established a bidirectional fluid-structure coupled simulation method based on Star-CCM+ and ABAQUS to study the dynamic responses and damage of a semi-submersible FOWT under ship collisions. In this study, a sensitivity analysis was conducted on parameters such as collision speed, collision angle, wind speed, and wave height, with wind speeds of 11.4 m/s, 10 m/s, and 8 m/s, respectively. The results showed that the pitch and heave amplitudes of the floating wind turbine under collision exceeded those observed without collision, and side-on collisions with the vessel had a greater impact on the structural dynamic response of the floating wind turbine. Yoon et al. (2023) developed a hydrodynamic tool HydroQus in Abaqus for calculating the hydrodynamic forces acting on the FOWT, and then studied the collisions between a semi-submersible FOWT and a ship. This study concluded that applications of suitable failure strain model and hydrodynamics into the collision simulations were of importance.

From the literature review, it is evident that current research primarily focuses on ship collision analysis of FOWTs in still-water and low-wind conditions, but with limited investigation into collision scenarios involving wind speeds greater than 11.4 m/s or extreme sea conditions. In reality, however, the initial posture and response of FOWTs at the collision moment are not the same at different sea conditions, leading to differences in dynamic behaviors and damage during and after the collision. Hence, existing research results are not comprehensive. In light of this, the present study aims to investigate the ship collision with FOWTs in extreme sea conditions. Firstly, a high-fidelity and fully coupled aero-hydro-servo-structural model of FOWT is developed using a coupling framework by Simulink and LS-DYNA, which is referred to as SimDYNA. Secondly, the reliability of the SimDYNA model is verified by comparison with OpenFAST. Finally, the present work explores the dynamic responses and damage of NREL 5 MW spar-type floating wind turbine under ship collisions in a 50-year extreme wind-wave condition.

2 METHODOLOGIES FOR THE DEVELOPMENT OF SIMDYNA

2.1 *Introduction to CO-simulation framework*

In 2022, Ansys released LS-DYNA R12, introducing bidirectional coupling capabilities with third-party software, enabling co-simulation with platforms such as Simulink and Python (Tong & Yeh, 2020). Building on this advancement, the present study develops an integrated aero-hydro-servo-structural fully coupled simulation framework for offshore wind turbines, termed SimDYNA. As illustrated in Figure 1, SimDYNA models the floating wind turbine structure in LS-DYNA, while Simulink handles the aerodynamics, hydrodynamics, hydrostatics, mooring forces, servo control systems, and rigid-body motions of the floater. Data exchange between the two platforms occurs at each time step, governed by Simulink.

2.2 *Aerodynamic modelling*

In the present work, the blade element momentum (BEM) theory is used to calculate the aerodynamic loads on the blades based on the lift and drag coefficients of the sectional airfoils and local induction velocities that are determined using the GDW model (Yang 2021). The BEM theory assumes that the forces acting on a blade element are only related to the change in momentum of the air swept by the blade element, and that the airflow between adjacent blade elements does not interact radially (Hansen 2015). Figure 2 illustrates the aerodynamic forces of an arbitrary blade element with a length of *dr*. The relative wind speed felt by the blade element is given by

$$\begin{cases} V_{rel} = \sqrt{V_{n,x}^2 + V_{t,y}^2} \\ V_{n,x} = \left(U_{0,x} - \dot{x}_{b,x}\right)(1-a) \\ V_{t,y} = \left(\Omega r + U_{0,y} - \dot{x}_{b,y}\right)\left(1+a'\right) \end{cases} \tag{1}$$

where Ω is the rotor speed; r is the local radius of the blade element; $U_{0,x}$ and $U_{0,y}$ are the wind speeds in the in-plane and out-of-plane directions, respectively; $\dot{x}_{b,x}$ and $\dot{x}_{b,y}$ denote the vibration velocities in the in-plane and out-of-plane directions, respectively; a and a' are axial and tangential induction factors, respectively, which can be determined by iterative algorithms.

The angle of attack α is obtained from

$$\alpha = \varphi - \beta - \theta \tag{2}$$

where $\varphi(r,t) = \arctan\left(\frac{V_{n,x}}{V_{t,y}}\right)$ is the flow angle; θ is the pre-twist angle at the cross secion; β is pitch angle of blade.

Then, the local lift and drag forces on the blade element with a length of *dr* can be determined, respectively, by

$$\begin{cases} dL = \frac{1}{2}\rho_a V_{rel}^2 c C_n dr \\ dD = \frac{1}{2}\rho_a V_{rel}^2 c C_t dr \end{cases} \quad (3)$$

where ρ_a is the density of air; c is the chord length of blade element; C_n and C_t are the lift and drag coefficients of the airfoil, respectively.

The total aerodynamic loads on the entire rotor is then obtained by summing the loads on all blade elements. The hub and tip losses are also considered in the aerodynamic module of SimDYNA, and the related theory can be seen in reference (Ning 2020). The wind load on the tower (Burton et al. 2011) is caculated by

$$f_t(t) = \frac{1}{2}C_{dt}\rho_a d_t|U - \dot{x}_t|(U - \dot{x}_t) \quad (4)$$

where C_{dt} is the drag coefficient of the tower; d_t is the outer diameter of the tower; U is the wind velocity including the effects of turbulent fluctuations; $\dot{d}_t$ is the velocity of the tower.

2.3 *Hydrodynamic modeling*

In this study, Morison equation is adopted to caculate the hydrodynamic loads on the slender spar platform (Lemmer 2016). The hydrodynamic force perpendicular to the a strip on the floater can be given by

$$\begin{cases} dF_M(Z,t) = C_1 - C_2 + C_3 \\ C_1 = C_M\rho_w \frac{1}{4}\pi D^2 \dot{v}_w(Z,t)dZ \\ C_2 = (C_M - 1)\rho_w \frac{1}{4}\pi D^2 \dot{v}_s(Z,t)dZ \\ C_3 = \frac{1}{2}C_D\rho_w D v_{rel}(Z,t)|v_{rel}(Z,t)|dZ \end{cases} \quad (5)$$

where C_M and C_D are the added mass and drag coefficients, respectively; ρ_w is the density of water; D is the diameter of spar platform; dZ is the length of the strip on spar platform; $v_{rel}(Z,t)$ is the relative velocity of water particle and strip.

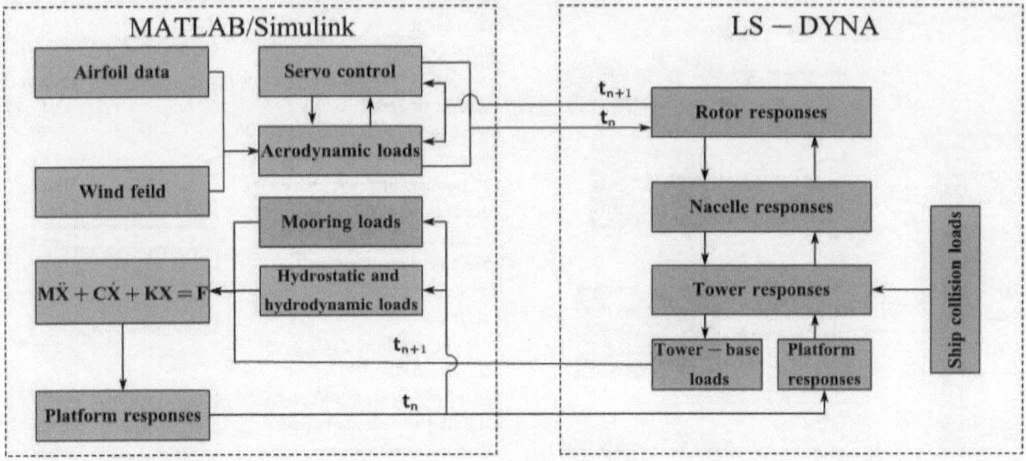

Figure 1. Coupling framework between Simulink and LS-DYNA.

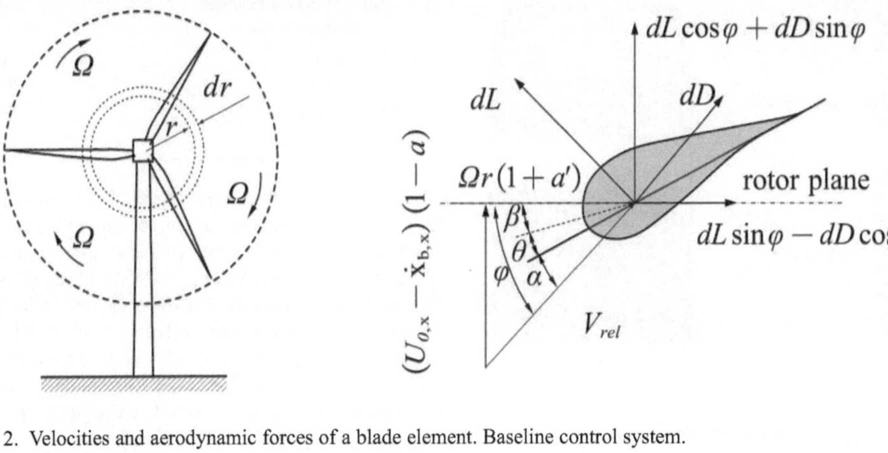

Figure 2. Velocities and aerodynamic forces of a blade element. Baseline control system.

2.4 *Baseline control system*

For the NREL 5-MW baseline wind turbine, a conventional variable-speed, variable-blade-pitch configuration was selected. The control system consists of two components: a generator-torque controller for below-rated speeds and a full-span rotor-collective blade-pitch controller for above-rated speeds (Jonkman et al., 2009). However, this study focuses on investigating ship collisions with the spar-type FOWT under extreme sea conditions, in which the wind turbine is in a parked state and the servo control system is inactive.

3 NUMERICAL MODELING

3.1 *Modeling of NREL 5 MW spar-type FOWT*

Table 1 presents main parameters of NREL 5 MW Hywind Spar-type FOWT, where CM denotes the center of mass, and SWL the still water level. Figure 3 shows the finite element (FE) model of the FOWT.

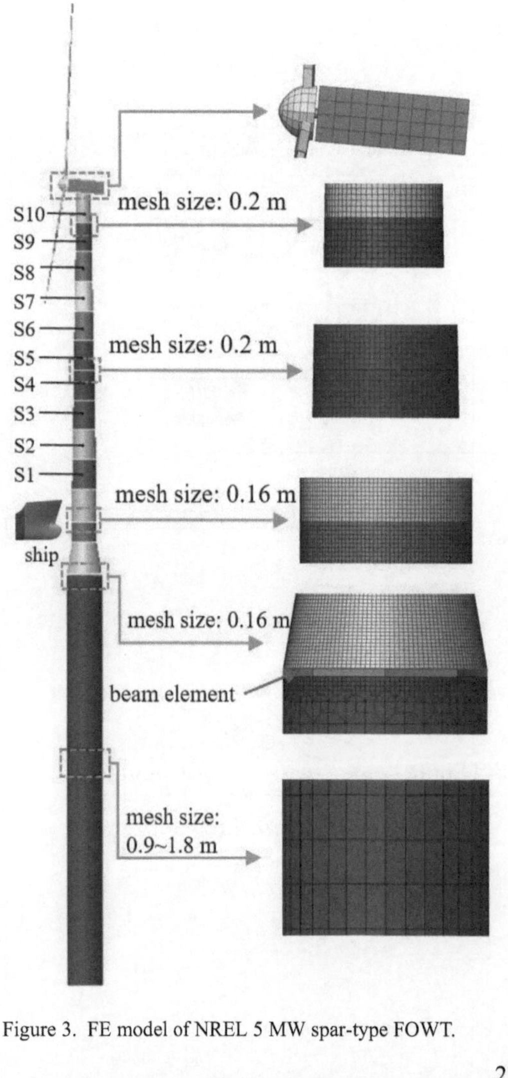

Figure 3. FE model of NREL 5 MW spar-type FOWT.

The floater and tower are modelled with Belytschko-Tsay shell elements with 5 integration points through the thickness (LSTC 2022). A beam element with a length of 0.5 m is used at the tower bottom to extract the internal forces which are then output to Simulink for solving the rigid motions of the floater. The nacelle, hub, and pitch control drivetrain are all modeled using solid elements. A general mesh size of 0.2 m was used for the tower. To capture more detailed deformation characteristics at the collision zone, a finer mesh size of 0.16 m is applied to the platform above the taper bottom. In the model, the corresponding element length to thickness ratio is smaller than 10, hence the local stress and strain fields can be captured very well (Alsos and Amdahl 2007).

The constraint methods between beam elements, shell elements, and rigid bodies can be referred to in the LS-DYNA manual (LSTC 2022).

Table 1. The main parameters of spar-type FOWT (Jonkman et al. 2009).

Description	Value
Blade length	61.5 m
Blades Quantity	3
Nacelle CM Location above Tower Top	1.75 m
Rated Rotor Speed	12.1 rpm
Elevation to Tower Base above SWL	10 m
Elevation to Tower Top above SWL	87.6 m
CM Location of Tower above SWL	43.4 m
Base Diameter of Tower	6.5 m
Top Diameter of Tower	3.87 m
Base Thickness of Tower	27 mm
Top Thickness of Tower	19 mm
Total Draft	120 m
Platform Top above SWL	10 m
Taper Top below SWL	4 m
Taper Bottom below SWL	12 m
Platform Diameter above taper	6.5 m
Platform Diameter below Taper	9.4 m
Platform Thickness above Taper	40 mm
Platform Thickness at Taper	50 mm
CM Location of Floater below SWL	89.9 m
CM Location of FOWT below SWL	78 m
Number of Mooring Lines	3
Fairleads below SWL	70 m
Radius to Anchors	853.87 m
Unstretched Mooring Line Length	902.2 m
Total Mass of FOWT	8066 tons

The elastoplastic mechanical behaviors of the tower and the platform above the taper bottom under external loads are modeled using the elastoplastic material *Mat_ PLASTIC_KINEMATIC (Mat_03). The Cowper-Symonds constitutive was adopted to consider the strain rate effect of material under impact loads (Ren et al., 2022). The elastic mechanical behaviors of the blades are simulated using the elastic material *Mat_ELASTIC (Mat_01). The structural damping parameters for the tower and blades are specified using the keyword *DAMPING_ FREQUENCY_RANGE_DEFORM. This study focuses solely on the rigid motion of the nacelle, hub, low-speed shaft, and pitch control drivetrain, which are simulated using the rigid material *Mat_RIGID (Mat_20) to capture their rigid body dynamics. Additionally, the keyword *PART_INERTIA is used to define their mass properties. The material parameters for the FOWT are shown in Table 2.

3.2 *Modeling of the ship*

Figure 4 shows the FE model of the ship established according to the Chinese ship specification. The ship has a total tonnage of 3500 tons (2500 DWT, dead weight tonnage). The length, width and height of the ship are 96 m, 15 m and 9.6 m respectively. The outer and inner plates of the ship bow are modeled by the 4-node Belytschko-Tsay shell elements with reduced integration. The mesh sizes of the outer plate and the inner plate are 0.3~0.4 m and 0.10~0.25 m respectively, with a thickness of 8~12 mm. The stiffeners inside the ship bow are simulated by Hughes-Liu beam elements for considering the nonlinear behavior from bending deformations, and the element size is 0.15~0.20 m.

The elastoplastic mechanical behaviors of the ship under collisions loads are modeled using the elastoplastic material *Mat_PLASTIC_ KINEMATIC (Mat_03), and the material parameters can be seen in Table 2.

Table 2. Material parameters of NREL 5 MW spar-type FOWT and the ship (Ren et al. 2022).

Material type	Density	Elastic Modulus	Poisson's Ratio	Tangent Modulus	Yeild stress	Failure strain
	ρ (kg/m^3)	E (Pa)	υ	E_t (Pa)	σ_y (Pa)	ε_f
Mat_01 (Blade)	1950	1.832E09	0.22	-	-	-
Mat_03 (Tower)	8500	2.06E11	0.3	1.18E09	3.45E08	0.27
Mat_03 (ship bow)	7850	2.06E11	0.3	0.88E09	2.35E08	0.34
Mat_20 (Rigid floater)	7800	2.06E11	0.3	-	-	-

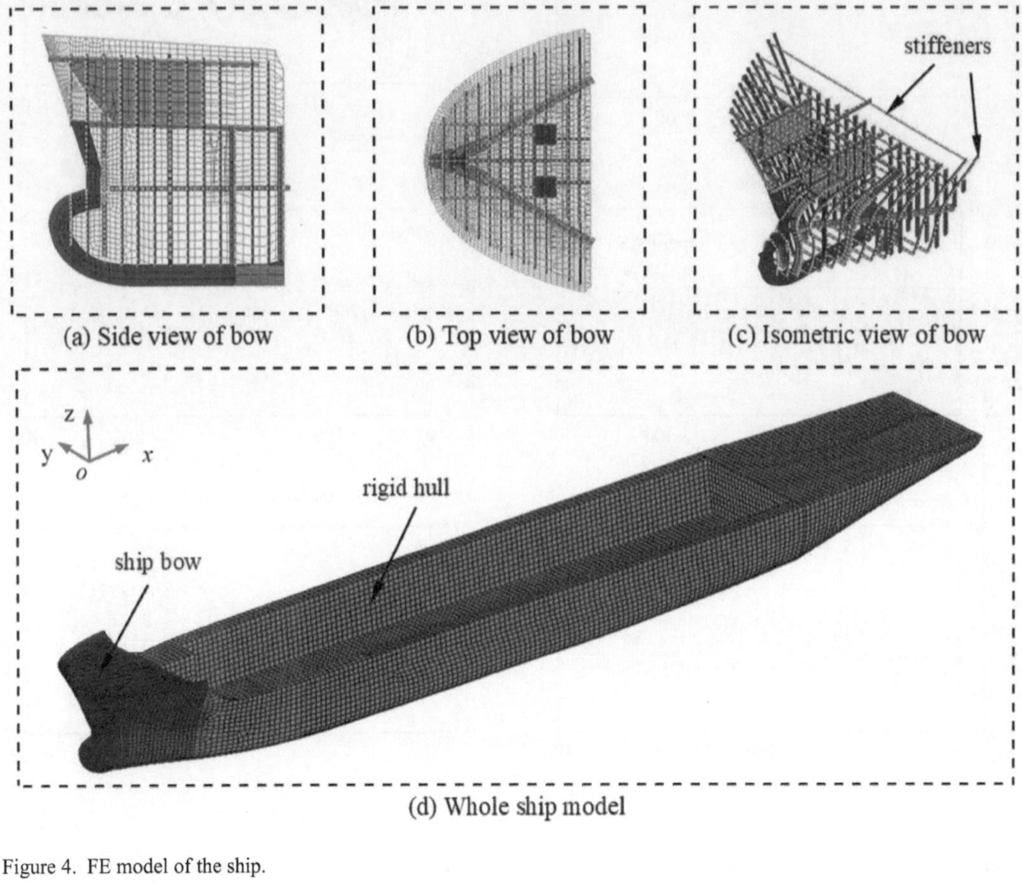

Figure 4. FE model of the ship.

4 MODEL VERIFICATION

The FE model of the ship have been verified by the authors' previous work (Ren et al., 2022). Therefore, this section focuses solely on the model verification of the FOWT.

In this test, the FOWT in operation is subjected to the combined excitations of regular wave and steady uniform wind. The wave height is 4 m and the period is 9 s, while the wind speed is 13 m/s. Figure 5 presents the time history curve of the floater displacement, tower top displacement, blade tip displacement, and rotor speed. It can be observed that the simulated results from the SimDYNA model are nearly identical to those from the OpenFAST model in the time domain. This confirms that the SimDYNA model is sufficiently accurate and can be reliably used for ship collision analysis.

5 SEA CONDITIONS AND COLLISION SCENARIOS

5.1 *Correlation between wind and wave*

In real marine environments, there is a correlation between wind and wave. This study aims to investigate the collision between a ship and a floating offshore wind turbine (FOWT) under extreme sea conditions with a 50-year return period. Based on the calculation method used in (Xu and Ishihara 2014), when the wind speed at the hub height of the wind turbine is 50 m/s, the relationship between the extreme wind speed U_{10} at a height of 10 meters above sea level, significant wave height H_s, and wave period T_p is given as follows:

$$\begin{cases} H_s = \alpha H_{0,SMB} + (1-\alpha) H_{0,swell} \\ \alpha = max\left(0.4tan^{-1}(0.34U_{10} - 1.88) + 0.39,\ 0\right) \\ H_{0,SMB} = \frac{0.3U_{10}^2}{g}\left\{1 - \left[1 + 0.004\left(\frac{Fg}{U_{10}^2}\right)^{1/2}\right]^{-2}\right\} \\ H_{0,swell} = 1.31 + \frac{(2.46-1.31)U_{10}}{12} \end{cases} \tag{6}$$

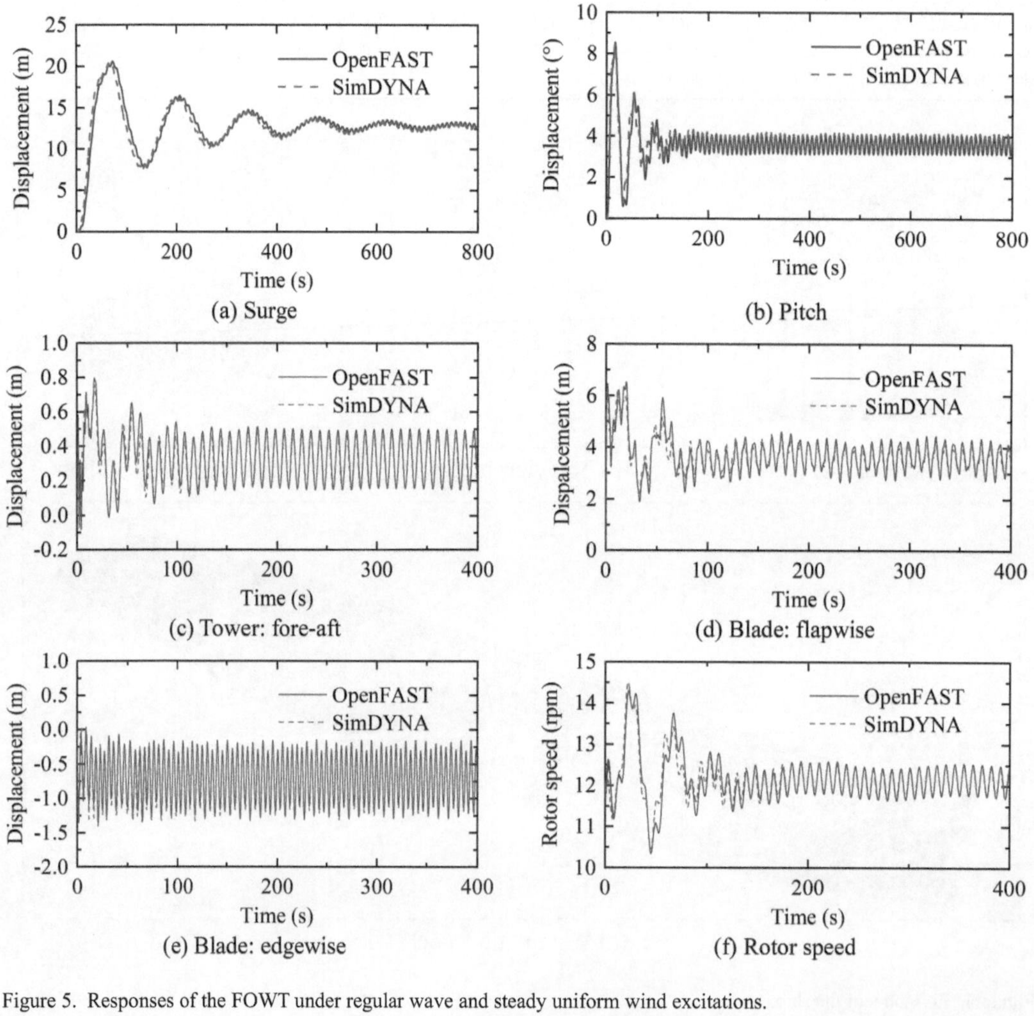

Figure 5. Responses of the FOWT under regular wave and steady uniform wind excitations.

$$\begin{cases} T_p = \alpha T_{0,SMB} + (1-\alpha) T_{0,swell} \\ \alpha = max\left(0.4tan^{-1}(0.34U_{10} - 1.88) + 0.39,\ 0\right) \\ T_{0,SMB} = \frac{1.37 \cdot 2\pi U_{10}}{g}\left\{1 - \left[1 + 0.008\left(\frac{Fg}{U_{10}^2}\right)^{1/3}\right]^{-5}\right\} \\ H_{0,swell} = 8 \end{cases} \tag{7}$$

where F=130000 m.

5.2 *Determination of collision scenarios*

Under the combined action of uniform wind and regular waves, the FOWT transitions from an initially stationary and upright position to a new steady-state oscillatory motion. However, achieving this steady-state response requires not only an extended simulation duration but also substantial computational resources, which poses challenges for the present study on ship collisions, as such events are more likely to occur during the steady-state motion of the FOWT. To address this issue, the total simulation time in this study is set to 400 seconds. An additional large damping is applied to the floater during the initial 0~95 seconds to accelerate the stabilization process, after which the actual damping properties are restored for the time from 95 to 400 seconds. Figure 6 presents the time histories of the FOWT platform's displacement and velocity at the SWL under extreme wind and wave loads, demonstrating that this approach enables the FOWT to reach a steady-state oscillatory condition within a shortened simulation time, thereby facilitating early-stage collision simulations. After 95 seconds, the floater reaches equilibrium position but continues to experience sinusoidal motion under the influence of regular waves. Hence, based on the floater's response characteristics, four typical pitch poses of P_1, P_2, P_3, and P_4, as shown in Figure 6 (b), are identified for the ship collision analysis. The corresponding pitch angles for all poses are positive, with values of 2.52°, 3.65°, 4.64°, and 3.56°, respectively. In the P_1 and P_3 poses, the floater's velocity at the SWL is close to 0, with displacements of 21.74 m and 25.80 m, respectively. In the P_2 and P_4 poses, the floater's displacements at the SWL are 23.89 m and 23.79 m, with velocities of 1.30 m/s and -1.31 m/s, respectively. Additionally, to investigate the dynamic response characteristics and failure modes of the FOWT subjected to ship collisions under extreme wind and wave loads, relatively high collision velocities of 4 m/s and 5 m/s are selected for analysis.

6 RESULTS AND DISCUSSIONS

6.1 *Collision force and local deformation of the platform*

Under the 50-year extreme sea conditions, the ship undergoes head-on collisions with the NREL 5 MW spar-type FOWT along the wind direction. Figure 7 presents the time histories of collision forces and local deformations of the platform for different collision scenarios. Significant differences are observed in the time evolution of the impact forces and local deformations across various scenarios. For collisions under the same floater pose, it is found that the impact force F_A in collision region A (ship deck impacting the floater) lags behind the impact force F_B in collision region B (ship bulbous bow impacting the floater). This is attributed to the significant pitch motion of the floater under the coupled extreme wind and wave conditions, which causes the ship's bulbous bow to contact the floater earlier than the ship's deck. Moreover, as the collision velocity increases, the impulse in region A exhibits a more pronounced growth, whereas the impulse in region B shows little variation. This is because, with deeper indentation, structural deformation and internal contact occur in the ship's deck area, leading to an increase in local stiffness and, consequently, a substantial rise in the impact force at collision region A.

The local deformation of the floater is also influenced by both the collision velocity and the floater pose. As shown in Figure 6, for the P_2 pose, the deformation in floater region A1 is minor at an impact velocity of 4 m/s but becomes significant at 5 m/s.

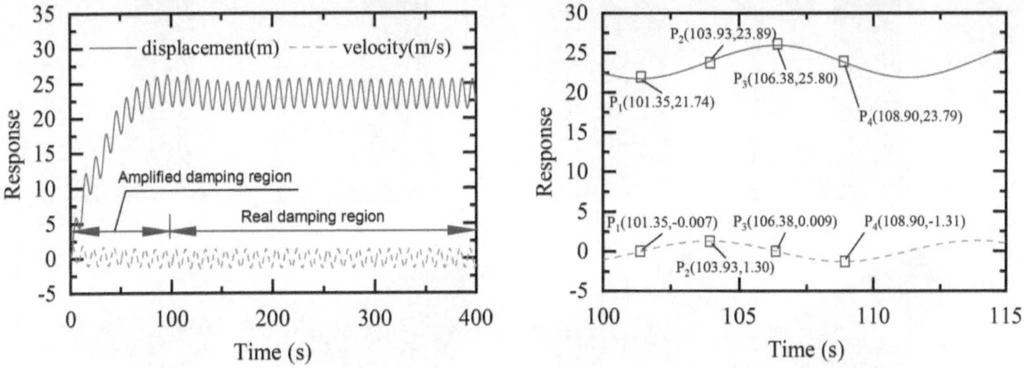

(a) Time: 0~400 s

(b) Time:100~115 s

Figure 6. Displacement and velocity time histories of the platform at SWL in surge.

Furthermore, even at the same collision velocity (4 m/s), noticeable differences in the magnitude of floater deformation are observed among different poses. These results indicate that under coupled extreme wind and wave loads, the deformation behavior of the floater caused by ship collisions is highly sensitive to variations in impact velocity and floater pose. Figure 8 illustrates the effective plastic strain contours of the platform and the ship's bow under impact velocity of 5 m/s. By comparing the damage extent in the floater and the ship, the severity of structural damage under different poses can be ranked as $P_4 > P_3 > P_1 > P_2$. A reasonable explanation for this phenomenon is that, under the P_4 pose, the ship and the floater move towards each other, resulting in the maximum relative collision velocity and, consequently, the most severe damage. In contrast, under the P_2 pose, the ship and the floater move in the same direction, leading to the minimum relative collision velocity and the least severe damage.

6.2 *Buckling and collapse of the FOWT under ship collisions*

Ship collisions not only cause severe local damage to the floater but also lead to the damage and collapse of the tower. Figure 9 illustrates the structural buckling modes and collapse poses of the FOWT after ship collisions at an impact velocity of 5 m/s. It is found that for the poses of P_1, P_2, and P_3, local buckling occurs at the bottom of the leeward side of tower section 3.

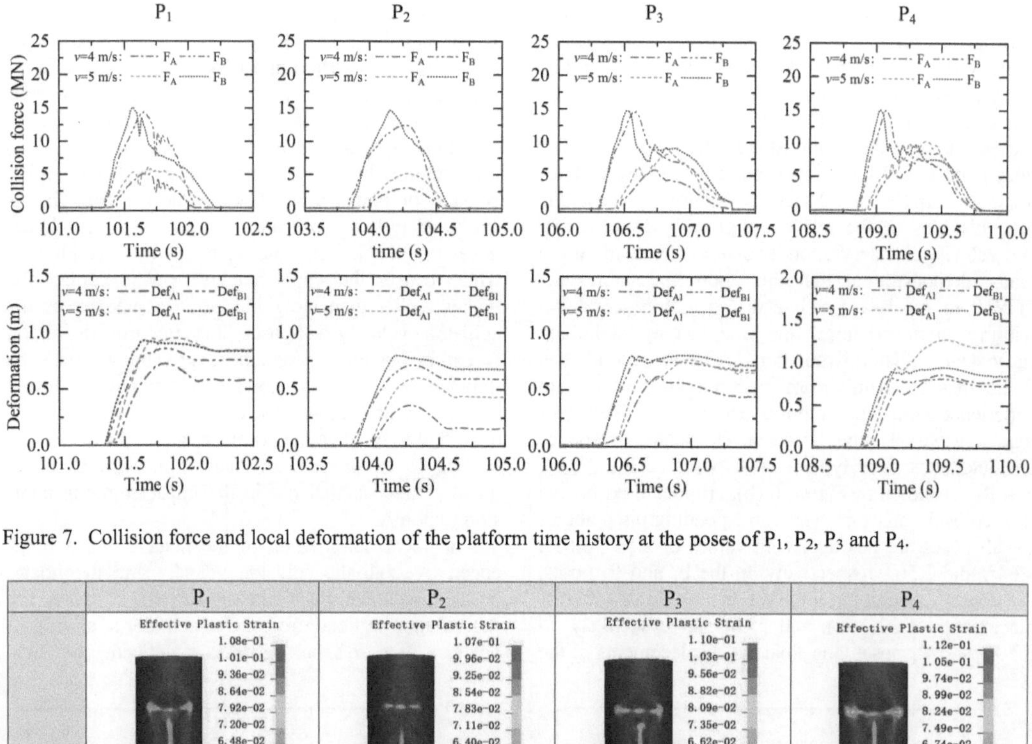

Figure 7. Collision force and local deformation of the platform time history at the poses of P_1, P_2, P_3 and P_4.

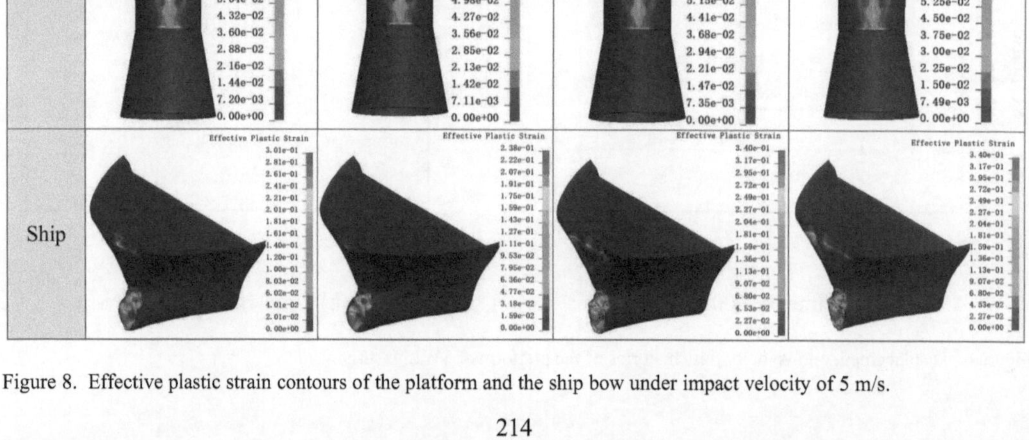

Figure 8. Effective plastic strain contours of the platform and the ship bow under impact velocity of 5 m/s.

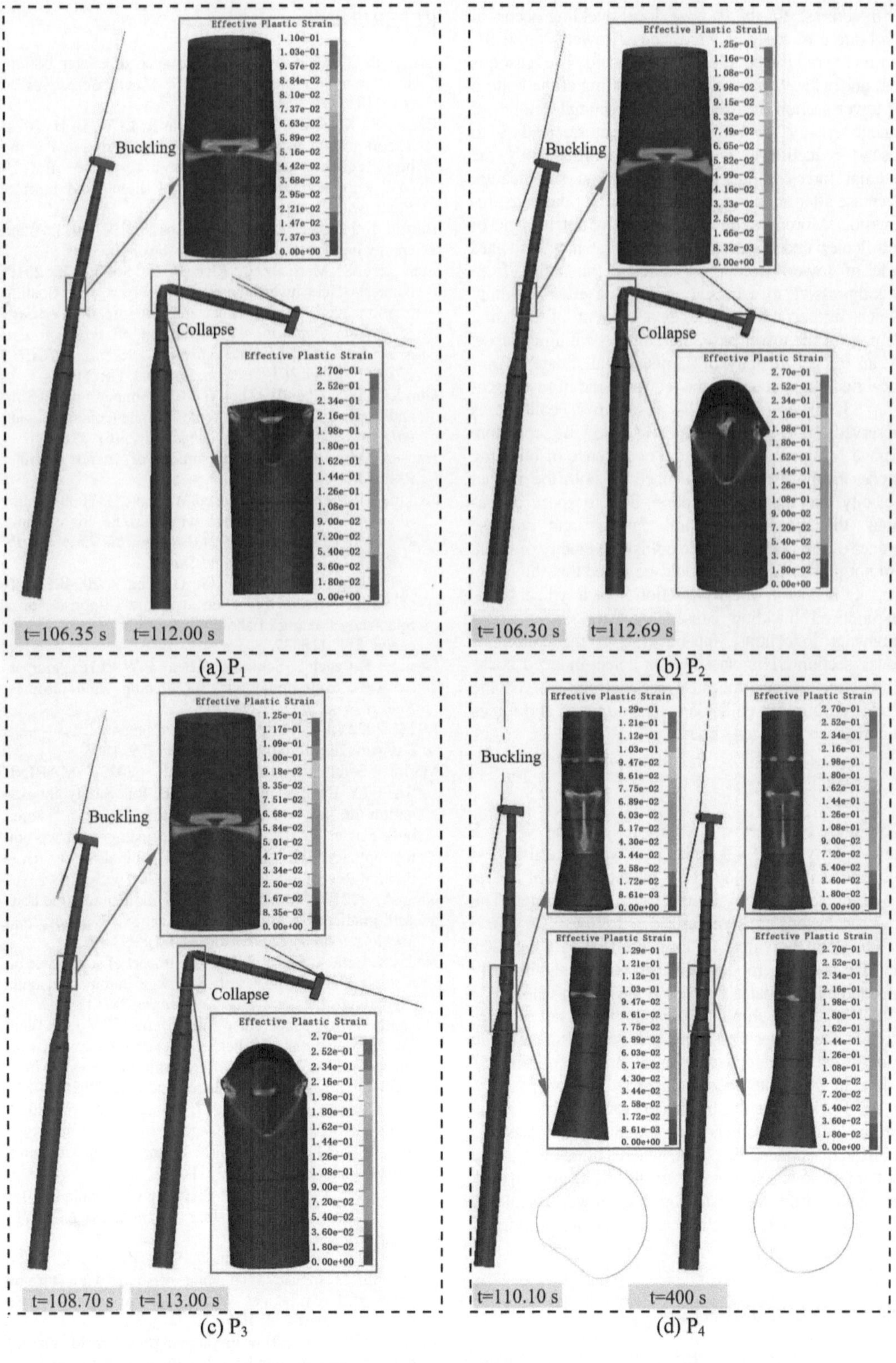

Figure 9. Buckling modes and collapse poses of the FOWT under impact velocity of 5 m/s.

In contrast, for the P_4 pose, local buckling occurs at the bottom of the windward side of tower section S1, but no overall tower collapse is observed. The reason is that under the P_4 pose, the local buckling at the bottom of tower section S1 dissipates a substantial amount of kinetic energy from the upper structure, leading to a sharp reduction in inertial effects. Consequently, the internal forces at the buckling location significantly decrease after buckling, preventing further damage progression. Moreover, due to the positive pitch angle of the floater under extreme wind loads, the windward side of tower section S1 quickly transitions from a compressive to a tensile stress state after buckling, which further inhibits the development of buckling damage at the tower base, thereby preventing collapse. At an impact velocity of 4 m/s, for all floater poses, local buckling occurs on the leeward side of tower section S3. Except for the P_2 pose, tower collapse is observed under all other poses following collision. Above results indicate that the occurrence of buckling modes in the tower is influenced by both the impact velocity and the floater's pose. Due to space limitations, the detailed structural damage and collapse modes of the FOWT under a collision velocity of 4 m/s are not presented hereIt should be noted that the buckling at the bottom of tower section S3 is a typical bending-induced buckling caused by excessive bending moments. In contrast, the buckling at the bottom of tower section S1 is classified as defect-induced buckling, which occurs because the deformation of the floater during ship collision extends upward and forces deformation at the tower base.

7 CONCLUSIONS

This study examines ship collisions with the NREL 5 MW spar-type FOWT under a 50-year return period extreme sea state, with a focus on the effects of floater pose and ship velocity on the collision outcomes. The key conclusions are summarized as follows:

(1) Both the extent of local damage to the floater and the buckling modes of the tower are significantly affected by the floater's pose and the impact velocity;

(2) When the ship's velocity is in the same direction as the floater's motion, the floater sustains the most severe damage, whereas damage is minimized when the velocities are in opposite directions;

(3) Buckling at the bottom of tower section S3 is caused by excessive bending moments and is classified as bending-induced buckling, whereas buckling at the bottom of tower section S1 results from deformation transmitted from the floater top to the tower base, leading to local instability characterized as defect-induced buckling.

ACKNOWLEDGMENTS

This research was supported by the National Science Fund for Distinguished Young Scholars (No.5202 5082).

REFERENCES

Alsos HS, Amdahl J. On the resistance of tanker bottom structures during stranding. 2007. *Marine Structures* 20 (4): 218–237.

Bai H, Xu K, Zhang M, Yuan W, Jin R, Li W, Li H. 2024. Theoretical and experimental study of the high-frequency nonlinear dynamic response of a 10 MW semi-submersible floating offshore wind turbine. *Renewable Energy* 231: 120952.

Burton T, Jenkins N, Sharpe D, Bossanyi E. 2011. Wind energy handbook. *John Wiley & Sons* 300–309

Echeverry S, Marquez L, Rigo P, Le Sourne H. 2019. Numerical crashworthiness analysis of a spar floating offshore wind turbine impacted by a ship. *Proceedings of the 8th international conference on collision and grounding of ships and offshore structures (ICCGS 2019), October 2019*. Lisbon, Portugal: CRC Press.

Guo J, Zhao Y, Chen W, Zhou G. 2022. Simulation of a Ship and Tension Leg Platform Wind Turbine Collision. *Journal of Marine Science and Technology* 29(6): 731–741.

Hansen M. 2015. Aerodynamics of wind turbine. Routledge.

Jonkman J, Butterfield S, Musial W, Scott G. 2009. Definition of a 5-MW reference wind turbine for offshore system development. *National Renewable Energy Lab. (NREL), Golden, CO (United States)*.

Li Y, Li H, Wang B, Meng H, Su O, Tang Y. 2024. Effects of various freak waves on dynamic responses of a Spar-buoy floating offshore wind turbine. *Ocean Engineering* 311: 118837.

Lemmer F, Raach S, Schlipf D, Cheng P W. 2016. Parametric wave excitation model for floating wind turbines. *Energy Procedia* 94: 290–305

LSTC. 2022 LS-DYNA keyword user's manual. *Livermore Software Technology Corporation* 1799–1808.

Marine Safety Investigation Unit. 2023. MARINE SAFETY INVESTIGATION REPORT: Safety investigation into the loss of control of the Maltese registered bulk carrier JULIETTA D. https://msiu.gov.mt/wp-content/uploads/2023/11/MV-Julietta-D_Final-Safety-Investigation Report.pdf, 2023-01-31.

Ning A. 2021. Using blade element momentum methods with gradient-based design optimization. *Structural and Multidisciplinary Optimization* 64(2): 991–1014.

Niklas K, Bera A, Garbatov Y. 2023. Impact of steel grade on a ship colliding with an offshore wind turbine monopile supporting structure. *Ocean Engineering* 287: 115899.

Offshore WIND. 2020. Three Injured after CTV Hits Wind Turbine at German Offshore Wind Farm. https://www.offshorewind.biz/2020/04/24/three-injured-after-ctv-hits-wind-turbine-at-german-offshore-wind-farm/, 2020-04-24.

Ren Y, Meng Q, Chen C, Hua X, Zhang Z, Chen Z. 2022. Dynamic behavior and damage analysis of a spar-type floating offshore wind turbine under ship collision. *Engineering Structures* 272: 114815.

Tong X, Yeh I. 2020. Cross-Platform Co-Simulation for Vehicle Safety Analysis. *16th International LS-DYNA Users Conference* 1–9.

Xu H, Wang L, Zha X, Rui S, Shen K, Guo Z. 2024. Dynamic response of floating offshore wind turbine under different stages of typhoon passage. *Applied Ocean Research* 148: 104047.

Xu N, Ishihara T. 2014. Prediction of tower loading of floating offshore wind turbine systems in the extreme wind and wave conditions. *Wind Engineering* 38(5): 463–476.

Yang Y, Bashir M, Michailides C, Li C, Wang J. 2020. Development and application of an aero-hydro-servo-elastic coupling framework for analysis of floating offshore wind turbines. *Renewable Energy* 161: 606–625.

Yoon D H, Choung J. 2023. Collision simulation of a floating offshore wind turbine considering ductile fracture and hydrodynamics using hydrodynamic plug-in HydroQus. *Journal of Ocean Engineering and Technology* 37(3): 111–121

Yu Z, Amdahl J, Rypestøl M, Cheng Z. 2022. Numerical modelling and dynamic response analysis of a 10 MW - semi-submersible floating offshore wind turbine subjected to ship collision loads. *Renewable Energy* 184: 677–699.

Zhang Y, Hu Z, Ng C, Jia C, Jiang Z. 2021 Dynamic responses analysis of a 5 MW spar-type floating wind turbine under accidental ship-impact scenario. *Marine Structures* 75: 102885.

Zhang Y, Hu Z. 2022. An aero-hydro coupled method for investigating ship collision against a floating offshore wind turbine. *Marine Structures* 83: 103177.

Zong S, Liu K, Zhang Y, Yan X, Wang Y. 2023. The Dynamic Response of a Floating Wind Turbine under Collision Load Considering the Coupling of Wind-Wave-Mooring Loads. *Journal of Marine Science and Engineering* 11(9): 1741.

Trends in Collision and Grounding of Ships and Offshore Structures – Zhu & Guedes Soares (Eds)

Modeling ship collisions against spar floating offshore wind turbines using machine learning

G. Vandegar, P. Salazar L. & P. Rigo
ANAST, Department of ArGEnCo, University of Liège, Belgium

Y.P. Sone Oo & H. Le Sourne
ICAM Engineering School - Nantes Campus, France
Nantes University, Ecole Centrale Nantes, CNRS, GeM Institute (UMR 6183), Nantes, France

ABSTRACT: With the growing deployment of offshore wind turbines (OWTs), assessing ship collision consequences is crucial for ensuring structural safety. This study presents a metamodel that combines a neural network (NN) with the MCOL external dynamics solver to efficiently predict the damage of a spar-like floating wind turbine collided by a ship. By varying parameters such as OWT geometry, impact location, and ship mass and velocity, a comprehensive dataset is first generated using finite element simulations of collisions between rigid ships and fixed monopile-supported OWTs. A Transformer-based NN learns force-penetration relationships from these simulations. To handle floating wind turbines, the model is then coupled with the MCOL solver to account for hydrodynamic forces. Results demonstrate that this method can predict the impact response of the floating offshore wind turbine (FOWT) while significantly reducing computational costs compared to conventional non-linear finite element analysis (NLFEA). It offers a promising solution for real-time collision damage assessment, with future applications including deformable ship models and multi-physics interactions.

1 INTRODUCTION

1.1 *Context*

The offshore wind energy sector is experiencing rapid growth as part of the global transition to renewable energy. Offshore wind farms offer a sustainable solution to meet rising energy demands, with Europe aiming to reach 10 GW of floating wind energy by 2030 (WindEurope, 2022). However, the expansion of offshore wind farms presents challenges related to their deployment, particularly in terms of maritime safety. Over recent years, collisions between ships and offshore wind turbines (OWTs) have already been reported. For instance, a rudderless cargo ship drifted into the Hollandse Kust Zuid wind farm in the Dutch North Sea during a storm (Figure 1), and another cargo ship collided with an OWT in the German Gode Wind farm (Figure 2). With the increasing presence of floating offshore wind turbines (FOWTs), the risk of such accidents is expected to rise. The consequences can be severe, leading to environmental hazards such as oil spills or even explosions in the case of methane or hydrogen-fueled vessels. On the wind farm side, structural damage may result in tower collapse, mooring line failure, and potential drifting of the platform, causing further collisions within the wind farm. Therefore, preventing these outcomes is essential, and the structural design of OWTs must ensure their crashworthiness based on collision accidental limit states (ALS) defined in the standards and guidelines established by different certification institutions, such as Bureau Veritas (BV) and Det Norske Veritas (DNV). These assessments follow a risk-based approach, introducing the need for rapid and accurate collision damage predictions to account for a wide range of design parameters and collision scenarios when estimating the probability of failure.

Figure 1. Damaged monopile foundation from the impact of a drifting cargo vessel (Dutch North Sea) - From (Hollandse Kust, 2022).

DOI: 10.1201/9781003684404-28

Figure 2. Cargo ship Petra L. stroke an offshore wind turbine at Gode Wind 1 wind farm (Germany), 2023 - From (Offshore-energy.biz, 2023).

1.2 *Challenges*

Modeling ship-FOWT collisions presents unique challenges due to the complexity of physics involved. These include impact loads leading to large structural deformations, hydrodynamic forces (e.g., added mass, drag, wave radiation and hydrostatic forces), mooring loads, and wind forces. In practice, non-linear finite element analysis (NLFEA) has been widely used to predict the structural damages, capturing complex phenomena such as elastoplastic deformations, ruptures, and fluid-structure interactions (Figure 3) as performed in (Echeverry, 2021), (Zhang et al., 2021) and (Yu et al., 2022). While such an approach usually provides reliable results, it is computationally demanding and requires significant expertise, making it impractical for large-scale risk assessments or real-time applications. Simplified methods are thus needed for rapid damage estimation, particularly in the early design stage, which may involve thousands of collision scenarios with different striking ships and impact conditions.

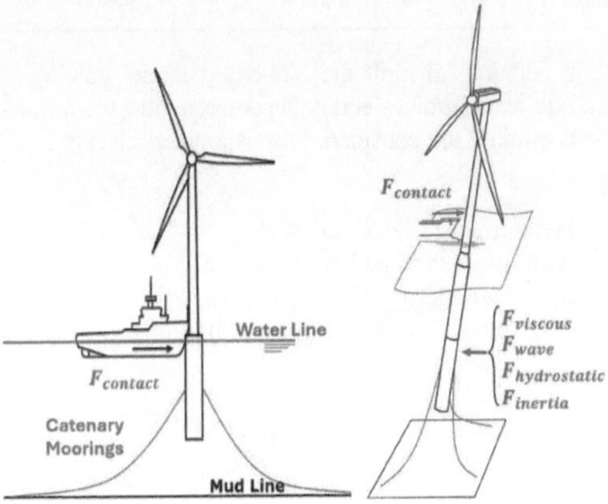

Figure 3. Forces involved in ship-FOWT collision.

1.3 *State of the art*

Despite a promising field of research, the use of AI-based metamodels in marine accidental events has been limited in the past years. Metamodels are particularly valuable in analyses requiring numerous computations, especially when analytical models lack accuracy or are unavailable. Fan et al. (2020) developed a surrogate model using response surface modelling to predict the residual compression strength of the columns of a bridge subjected to barge collisions. Das et al. (2022) compared different metamodeling techniques, consisting of neural networks (NNs), polynomial regression, and gradient-boosting regression trees, for building models capable of predicting damage extents and oil outflow in tanker collision accidents. Mauro et al. (2023) investigated both the accuracy and calculation time of surrogate models based on multiple linear regressions, NNs, and decision trees to assess the breach dimensions after passenger ship collisions. More recently, Zhang et al. (2025) proposed a NN-based model to predict ship collision damage extents under real operational conditions using Automatic Identification System data. Their metamodel, trained by considering the breach sizes obtained from super-element simulations (Le Sourne et al., 2012), allowed the computational time to reduce from 10 minutes to under 0.1 second.

1.4 *Proposed approach*

The present approach aims to provide rapid, data-driven predictions by incorporating machine learning techniques, improving collision risk assessment methodologies for floating wind farms. First, an extensive series of NLFEA is conducted to build a comprehensive database capturing various collision scenarios and structural responses. In these simulations, fixed monopile-supported OWTs are impacted by rigid striking ships. Approximately one thousand simulations are performed with varying OWT geometries, impact locations, and impactor masses and velocities. This database serves as the foundation for training an NN-based model based on Transformer architecture (Vaswani et al., 2017). The surrogate model takes the OWT particulars and collision scenario parameters as input, directly providing the corresponding force-penetration relationship, while effectively learning complex patterns from the simulated data. Then, supposing that the ballast of a spar-buoy FOWT acts like a "moving" clamped boundary condition of the tower, the NN-based model is coupled with MCOL external dynamics solver which considers the hydrodynamic forces acting on the wind turbine floater (Le Sourne et al., 2007).

2 FINITE ELEMENT MODEL

2.1 *Model description*

To generate the dataset, nonlinear finite element simulations are conducted using LS-DYNA commercial software. The OWT support structure is simplified as a cylindrical tower with overall length L uniform radius R and thickness t as done by (Ladeira et al., 2023). To represent the rotor nacelle assembly (RNA), a lumped mass m_{rna} is applied at the top while the base of the cylinder is clamped. The OWT structure is impacted by a ship of mass m_{ship} simplified as a wedge at an initial velocity v_0 and vertical location αL, while the ship movement is constrained to x-direction (Figure 4).

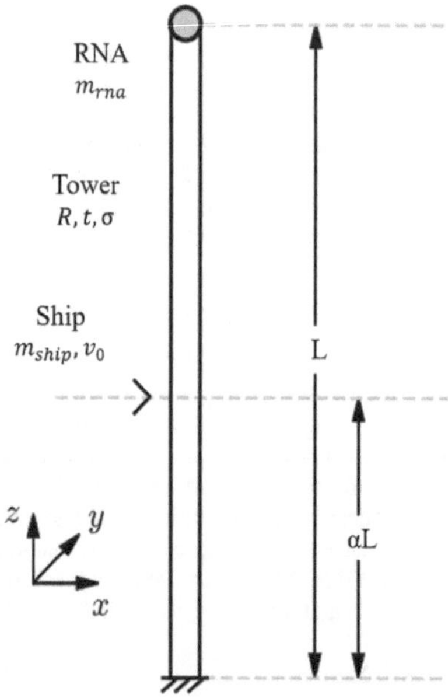

Figure 4. Simplified model of OWT and striking ship.

Fully integrated shell elements with five through-thickness integration points are chosen for both the impactor and the cylindrical tube. In accordance with the mesh convergence study performed by the authors and reported in (Ladeira et al., 2023), an element size of 25 cm is considered for the tube and the impactor (Figure 5). A bi-linear plasticity constitutive model, which properties are given in Table 1, is used to model the tower steel material, while the impactor is considered rigid. Strain-rate hardening effect is disregarded due to its inherent uncertainties and the complexity involved in accurately modeling such phenomenon (Yu and Amdahl, 2018). Finally, Germanischer Lloyd (GL) failure strain criterion is included in the model, as proposed by Zhang et al. (2004), where t is the wall thickness and l_e is the element size:

$$\varepsilon_f = 0.056 + 0.54\frac{t}{l_e} \quad (1)$$

To be consistent with the chosen criterion, only tension areas are examined, and no failures were observed in any of the simulated cases, indicating that the tower retained its local structural integrity in the impact area for the scenarios studied.

2.2 *Simulation parameters*

Ship–OWT collisions simulations are performed varying the tower geometry, the mass of the RNA, and collision scenario. To efficiently sample the parameter space and ensure broad coverage, Latin Hypercube Sampling (LHS) is used to generate the input dataset, with values selected within the ranges provided in Table 2, based on the characteristics of deployed FOWTs.

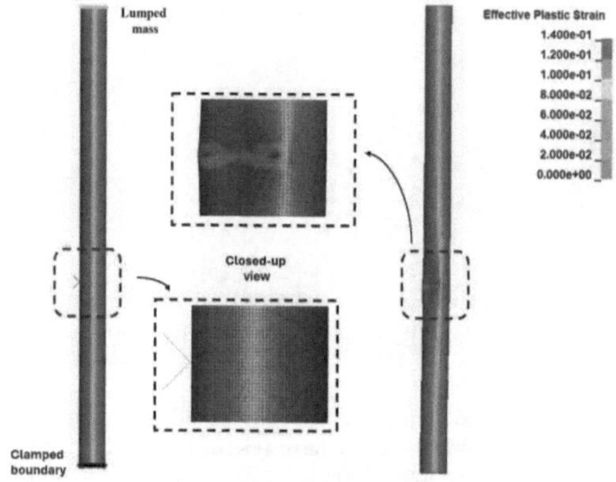

Figure 5. (Left) Initial FE model - (Right) Example of effective plastic strain distribution observed after impact.

Table 1. Material properties of the cylinder.

Property	Value	Unit
Young modulus (E)	210 000	MPa
Poisson ratio (ν)	0.33	-
Yield strength (σ)	363.7	MPa
Tangent modulus (E_T)	5 500	MPa
Density (ρ)	8500	kg/m^3

Table 2. Choice of parameters used to generate the dataset.

	R [m]	t [mm]	L [m]	m_{rna} [tons]	v_0 [m/s]	m_{ship} [tons]	α [%]
min	2.5	30	100	250	0.5	3000	20
max	5	90	200	850	5	150000	50

In addition, to limit the NN-based model training to realistic wind turbines and collision scenarios, the following conditions are considered for parameter selection:

1. Slenderness: $10 < L/2R < 25$
2. Wall thickness: $100 < 2R/t < 250$
3. Collision energy: $K_0 = \frac{1}{2} m_{ship} v_0 < 100\ MJ$
4. RNA mass scaling with tower height:

$$b - \frac{m_{rna}^{max}}{5} < m_{rna} - aL < b + \frac{m_{rna}^{max}}{5}$$

where a and b define a linear relationship between the tower length and the RNA mass:

$$a = \frac{m_{rna}^{max} - m_{rna}^{min}}{L^{max} - L^{min}}, \quad b = -aL^{min} + m_{rna}^{min}$$

3 NEURAL NETWORK BASED MODEL

3.1 *Model description*

To model the complex nonlinear force-penetration behavior observed in ship-OWT collisions, a Transformer-based architecture such as the one detailed in (Vaswani et al., 2017) is adopted. Unlike recurrent neural networks (RNNs), which process sequential data step-by-step, the Transformer leverages self-attention mechanisms to capture long-range dependencies and interactions between input features. This is particularly beneficial for learning force-penetration relationships, as the impact response is influenced by multiple factors such as material properties, structural geometry, and collision dynamics. The Transformer model takes as input the parameters listed in Table 2 and maps them to multiple output sequences representing the impact force versus ship penetration (Figure 6). The multi-head self-attention mechanism enables the model to focus on different aspects of the interaction simultaneously, ensuring that critical features such as local deformations and global structural responses are well captured. To ensure numerical stability and improve training efficiency, layer normalization and dropout are applied throughout the network (Srivastava et al., 2014).

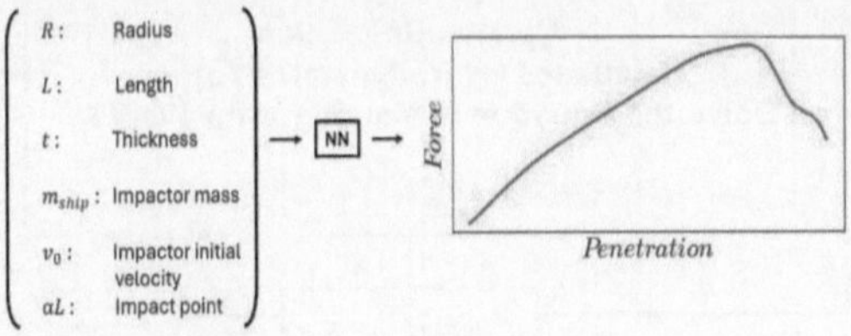

Figure 6. Inputs-Outputs pair of the NN.

3.2 *Data preprocessing*

To ensure efficient training and stable predictions, both input and output data are normalized. The input parameters, defined within the ranges provided in Table 2, are normalized as follows:

$$X_{norm} = \frac{X - X_{min}}{X_{max} - X_{min}} \quad (2)$$

For the output variables, the force is normalized using the plastic moment:

$$M_{pl} = \sigma \frac{4}{3}\left(R^3 - (R - t)^3\right) \quad (3)$$

multiplied by the arm lever $l = \alpha L$, ensuring consistency across different impact scenarios:

$$F_{norm} = \frac{F}{M_{pl} l} \quad (4)$$

To improve numerical stability and ensure energy conservation when coupling with MCOL solver, the force history is preprocessed to be strictly increasing, thus neglecting unloading effects. This approach assures convergence in the energy integration scheme. In addition, to maintain consistency across all samples, the output sequences are smoothed, ensuring well-structured input for training.

3.3 *Model training and evaluation*

The Transformer-based model is trained on the dataset generated from nonlinear finite element simulations, where each sample represents a collision scenario with varying FOWT geometry and impact conditions. The force-penetration curves obtained from these simulations serve as ground-truth data for supervised learning. To ensure robust generalization, the dataset is split into training (80%) and testing (20%) subsets. The training process minimizes the Mean Square Error (MSE) loss between predicted and simulated responses, and the R^2 score is used to measure the performance across the testing subset such as:

$$R^2 = 1 - \frac{\sum_i^n (y_i - \hat{y}_i)^2}{\sum_i^n (y_i - \bar{y})^2} \quad (5)$$

where n is the number of points in the sequence, y_i is the actual value, $\hat{y}_i$ is the predicted value and $\bar{y}$ is the mean of the sequence.

The Adam optimizer (Kingma et al., 2014) is employed for efficient weight updates, and a dynamic learning rate scheduler refines convergence. Each epoch consists of forward propagation, loss computation, backpropagation, and weight updates, with total loss monitored throughout training. To improve numerical stability and prevent overfitting, a low-pass filter is applied to the model's output before final error computation.

During evaluation, the trained model is tested against unseen LS-DYNA simulations, and performance metrics are recorded. The convergence plot in Figure 7 illustrates the stability and effectiveness of the training process. After 3000 epochs, the model achieves a R^2 score of 97.82%, demonstrating strong agreement with simulation results while significantly reducing computational costs compared to full-scale finite element models.

3.4 *Coupling of NN-based model with MCOL*

As the spar-like FOWT is concerned, its floater and tower are idealized as a uniform cylindrical column, clamped at its base to the rigid ballast tank, which serves as a "moving" clamped boundary condition - see Figure 8. This simplification is made not only in the NN-based simulations, but also in LS-DYNA /MCOL numerical simulations that are conducted to validate the proposed data-driven approach.

The contact force Fc predicted by the NN-based model is coupled with MCOL rigid-body dynamics solver to simulate the spar-like floating wind turbine response when impacted by a ship. The motion of the spar-buoy floater is influenced by hydrostatic restoring (F_H), waves (F_W), viscous damping (F_V) and water inertial forces acting on

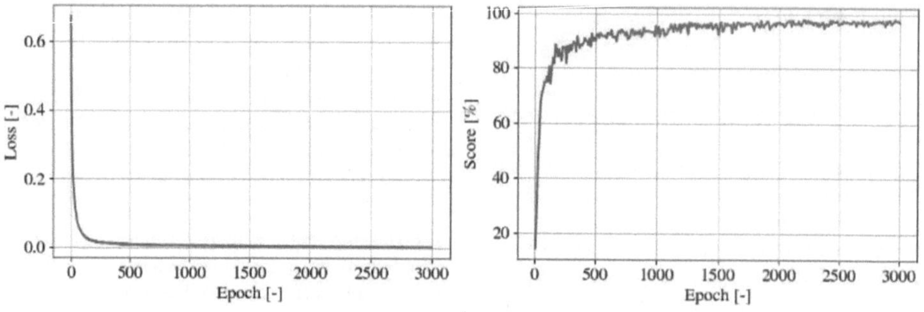

Figure 7. Convergence of the loss and of the R² score.

the floater. In MCOL, these hydrodynamic forces are calculated to solve the six-degree-of-freedom equation of motion, (Le Sourne et al., 2007):

$$M\dot{y} + G(y)y = [F_w + F_H + F_V](y, x) + F_C \quad (6)$$

where, $M = M_{FOWT} + M_\infty$ represents the total mass matrix, $G = G_{FOWT} + G_\infty$ is the total gyroscopic matrix, and x and y denote the earth-fixed position and body-fixed translational/angular velocity vectors of the FOWT's center of mass, respectively.

Beforehand, the hydrodynamic characteristics of the spar-buoy floater (i.e. water added mass M_∞, hydrostatic restoring and wave damping matrices) have been determined using HYDROSTAR seakeeping code (Bureau Veritas, 2018) and stored in a dedicated MCOL datafile (FOWT.mco in Figure9).

The NN-based predictions of the contact force F_C dynamically influence the structural response, creating a strong interaction between the tower internal mechanics and the floater external dynamics. The overall step-by-step algorithm which couples the NN-based model and MCOL solver is illustrated in Figure 9.

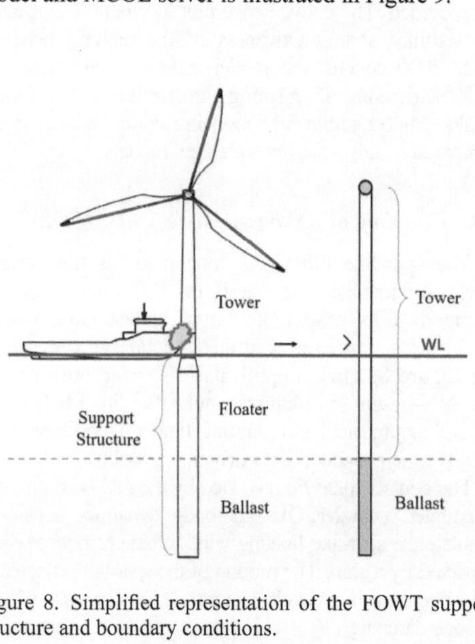

Figure 8. Simplified representation of the FOWT support structure and boundary conditions.

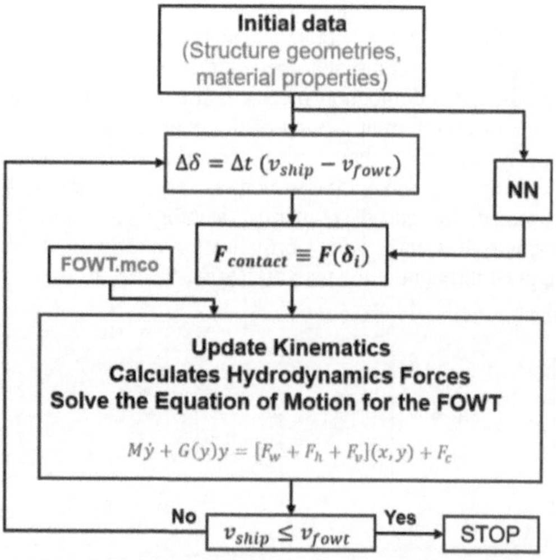

Figure 9. Scheme describing the algorithm to couple the NN-based surrogate model with MCOL solver.

4 VALIDATION OF THE DATA-DRIVEN APPROACH

4.1 *Presentation of the validation case*

To validate the developed approach, non-linear finite element simulations are performed using *BOUNDARY_MCOL card in LS-DYNA. The OC3-Hywind wind turbine, a widely documented and well-established FOWT design, is chosen as the case study. It consists of a spar-buoy structure with a cylindrical tower, a cylindrical floater including a ballast in its lower part, and a 350-tons RNA. Data used to set up the finite element model, including geometric and material properties are retrieved from (Jonkman et al., 2010) and presented in Table 3, Figure 10 and Table 4.

These values are kept constant in the simulations and only the mass and the velocity of the striking ship are varied. Finally, as the collision lasts only a few seconds, the tensile forces exerted by the catenary mooring lines used for this type of floating wind turbine do not have time to develop and are therefore neglected (Echeverry, 2021).

Table 3. Structural properties of the simplified model.

Property	Value	Unit
Mass of the RNA	350	*tons*
Cylinder radius	3.5	*m*
Cylinder thickness	30	*mm*

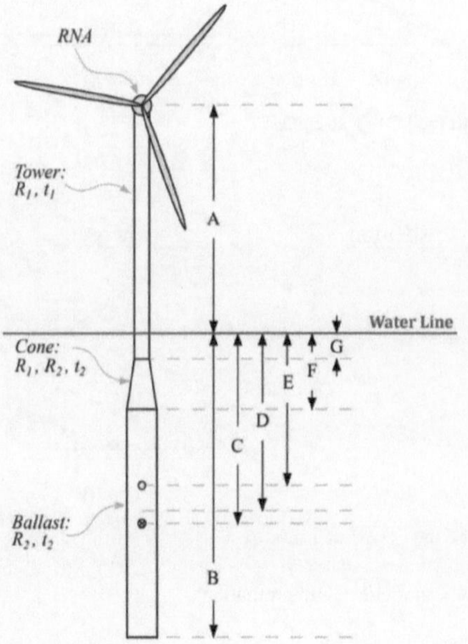

Figure 10. Diagram with the main dimensions of the OC3-Hywind.

Table 4. Main particulars of the OC3-Hywind.

Structure	Notation	Position [*m*]
Water depth	-	320
Tower top	A	90
Spar depth	B	120
Center of gravity	C	78
Mooring line location	D	70
Center of buoyancy	E	62.9
Taper bottom	F	12
Taper top	G	4

4.2 *Results*

Figure 11 compares the impact force F_C as a function of the ship penetration δ (i.e. the difference between ship and FOWT displacements at the impact point), the time history of ship penetration δ, FOWT pitch as well as energy distribution over time, including ship kinetic energy K_{ship}, FOWT kinetic energy K_{fowt}, FOWT deformation energy U_{FOWT} and FOWT hydrostatic restoring energy E_{hydro}. Two different collision scenarios are considered: a 6000-tons vessel impacting the FOWT at 2 m/s (Figure 11a) and a 24000-tons vessel colliding with the FOWT at 1 m/s (Figure 11b), both corresponding to an initial kinetic energy of 12 MJ. NN-based model/MCOL results are represented by solid lines and LS-DYNA/MCOL results by dashed lines. The discrepancies between the two approaches are calculated by Eq. (7).

$$disc. = \frac{max(FEM) - max(NN)}{max(FEM)} * 100 \quad (7)$$

The data-driven model overestimates the penetration by 5% in case (a) and 10% in case (b). The largest discrepancy is observed on the FOWT kinetic energy, which is overestimated by 27% in case (a) by the NN-based model. This suggests that although the model generally provides good predictions, its accuracy varies depending on the impact scenario.

It is also worth noting that as the FOWT pitch angle is small (less than 0.5°), the hydrostatic restoring energy is neglectable compared to the other energies. The effectiveness of the NN-based model in capturing the key physical interactions is also illustrated in Table 5, where the R^2 score (in %) is given for each result. We note that the lowest R^2 score corresponds to the FOWT kinetic energy in case (a), which agrees with the observation made above.

Table 5. Comparison of predicted and reference results for impact scenarios a) and b) using the R^2 score (in %).

	Case (a)	Case (b)
Force	98.2	98.6
Pitch	95.7	95.2
δ	99.6	97.8
K_{ship}	99.9	99.8
U_{fowt}	99.8	98.5
K_{fowt}	85.3	89.9

5 CONCLUSIONS

The proposed approach based on machine learning demonstrates significant advantages in predicting the structural response of a spar-like FOWT in case of ship collision. By leveraging a trained NN-based model, we achieve good accuracy while drastically reducing computational time compared to traditional NLFEA. Indeed, an LS-DYNA/MCOL simulation takes approximately 18 hours on a parallel 12 CPUs Intel core i7 computer and up to 1 hour if hydrodynamic effects are not accounted for. In contrast, once the training of the surrogate model is completed, an NN-based model/MCOL simulation lasts less than 5 seconds, mainly due to MCOL file reading and writing operations. If hydrodynamic effects are disregarded, as in the case of a fixed monopile supported wind turbine, the NN-based model can be used alone, providing results in less than 0.01 seconds. In both cases, the ability of the NN-based model to capture complex non-linear interactions makes it a promising alternative to conventional simulation methods, particularly for real-time assessments and rapid sensitivity analyses.

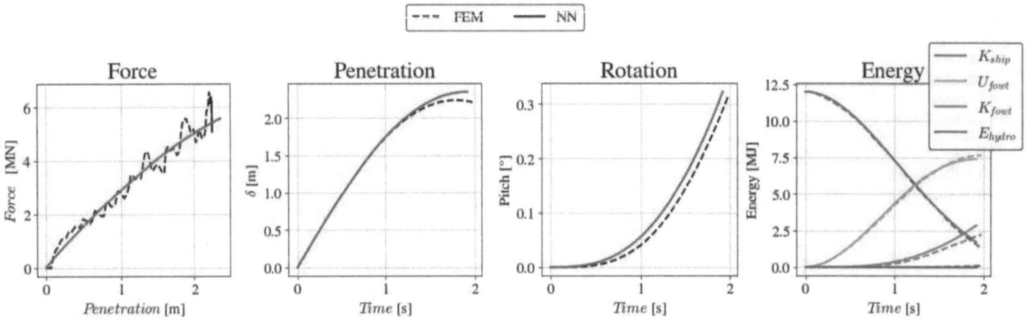

(a) 6000-tons vessel colliding with the FOWT at 2 m/s.

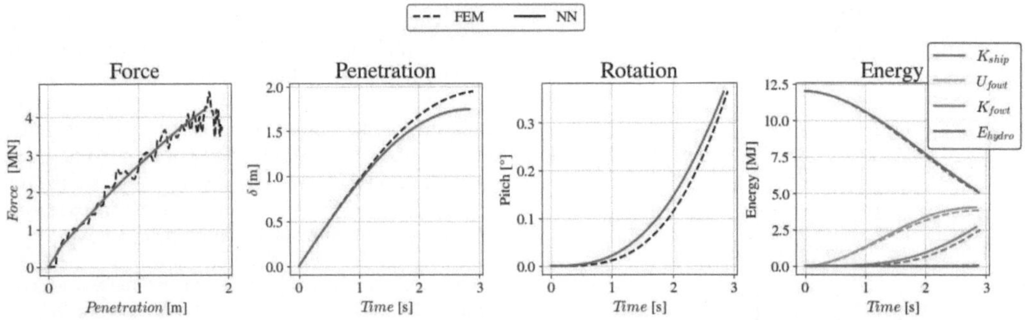

(b) 24000-tons vessel colliding with the FOWT at 1 m/s.

Figure 11. Results comparison between LS-DYNA/MCOL and the NN-model/MCOL simulations

However, some limitations remain. The NN-based model is trained from a set of numerical simulations, meaning its accuracy is inherently dependent on the reliability of the finite element models. In other words, all the shortcomings and limitations of the finite element simulations will be reflected in the NN's predictions. Moreover, the current study considers a rigid striking ship, whereas real-world scenarios involve deformable ship structures that may drastically change the energy and force distributions during the collision. Although this is a strong assumption, modeling the impactor as perfectly rigid is a first step to demonstrate the feasibility of the proposed methodology. Future research work will aim to extend the model to incorporate ship deformability and jointly evaluate ship-OWTs collision risks.

ACKNOWLEDGEMENTS

A part of this work was performed within the framework of the West Atlantic Marine Energy Community (WEAMEC) and granted by ICAM Engineering School, Pays de la Loire Region, and Europe (European Regional Development Fund).

REFERENCES

Bureau Veritas. 2018. HYDROSTAR SOFTWARE, Powerful Hydrodynamic Solver.

Das, T., Goerlandt, F., & Tabri K. 2022. An optimized metamodel for predicting damage and oil outflow in tanker collision accidents. *Proceedings of the Institution of Mechanical Engineers Part M: Journal of Engineering for the Maritime Environment*, 236(2), 412–426.

Echeverry, S. 2021. Numerical and analytical study of a spar-like floating offshore wind turbine impacted by a ship. *PhD thesis*, *University of Liege*, 1–179.

Fan, W., Sun, Y, Yang, C., Sun, W. & He, Y. 2020. Assessing the response and fragility of concrete bridges under multi-hazard effect of vessel impact and corrosion. *Engineering Structures*, 225, 111279.

Hollandse Kust. 2022. Rudderless Julietta D causes damage to foundation of wind farm Hollandse Kust. Zuid.https://tinyurl.com/ycywdkuw.

Jonkman, J., 2010. Definition of the Floating System for Phase IV of OC3. *Technical Report*, *National Renewable Energy Lab. (NREL)*, Golden, CO (United States).

Kingma, D. P., & Ba, J. (2014). Adam: A Method for Stochastic Optimization. arXiv preprint arXiv:1412.6980.

Ladeira, I., Echeverry, S. & Le Sourne H. 2023. A simplified method to assess the elasto-plastic response of standalone tubular Offshore Wind Turbine supports subjected to ship impact. *Ocean Engineering* 279 114313

Le Sourne, H. 2007. A ship Collision Analysis Program Based on Super-Element Method Coupled with Large Rotational Ship Movement Analysis. *Proceedings of the 4th International Conference on Collision and Grounding of Ships*, 131–138.

Le Sourne, H., Besnard, N., Cheylan, C. & Buannic, N. 2012 A ship collision analysis program based on upper bound solutions and coupled with a large rotational ship movement analysis tool. *Journal of Applied Mathematics* 2012.

Mauro, F., Conti, F. & Vassalos, D. 2023. Damage surrogate models for real-time flooding risk assessment of passenger ships. *Ocean Engineering*, 285, 115493

Offshore-Energy.biz 2023. Cargo ship strikes turbine at Orsted's Gode Wind 1 offshore wind farm, suffers massive damage. https://www.offshore-energy.biz/

Srivastava, N., Hinton, G., Krizhevsky, A., Sutskever, I., & Salakhutdinov, R. (2014). Dropout: A Simple Way to Prevent Neural Networks from Overfitting. *Journal of Machine Learning Research*, 15(56), 1929–1958.

Vaswani, A., Shazeer, N., Parmar, N., Uszkoreit, J., Jones, L., Gomez, A. N., Kaiser, L., & Polosukhin, I. 2017. Attention Is All You Need. CoRR, abs/1706.03762.

WindEurope. 2022. Europe can expect to have 10 GW of floating wind by 2030. https://windeurope.org/newsroom/news/europe-can-expect-to-have-10-gw-of-floating-wind-by-2030/

Yu, Z., & Amdahl, J., 2018. A review of structural responses and design of offshore tubular structures subjected to ship impacts. *Ocean Engineering*, 154, 177–203.

Yu, Z., Amdahl, J., Rypestøl, M. & Cheng, Z. 2022. Numerical modelling and dynamic response analysis of a 10 MW semi-submersible floating offshore wind turbine subjected to ship collision loads. Renewable Energy 184, 677–699.

Zhang, L., Egge, E.D., Bruhns, H, 2004. Approval procedure concept for alternative arrangements. *Proceedings of the 3rd International Conference on Collision and Grounding of Ships*, Izu, Japan, pp. 87–96.

Zhang, Y., Hu, Z., Ng, C., Jia, C. & Jiang, Z. 2021. Dynamic responses analysis of a 5 MW spar-type floating wind turbine under accidental ship-impact scenario. *Marine Structures* 75, 102885.

Zhang, M., Wang, H., Conti, F., Manderbacka, T., Remes, H. & Hirdaris, S. 2025. A hybrid deep learning method for the real-time prediction of collision damage consequences in operational conditions. *Engineering Applications of Artificial Intelligence* 145, 110158.

Trends in Collision and Grounding of Ships and Offshore Structures – Zhu & Guedes Soares (Eds)
© 2026 The Author(s), ISBN 978-1-041-16361-9

Investigation of hydrodynamic effects on scoured bridge responses during ship collisions

Z. Wang, Y.Y. Sha & J. Bogunović Jakobsen
Department of Mechanical and Structural Engineering and Materials Science, University of Stavanger, Stavanger, Norway

ABSTRACT: Bridges are subjected to various natural and man-made hazards throughout their service life, including scour and ship collisions. Over the past decades, scour has become a primary cause of bridge failures worldwide. In addition, bridges spanning navigable waterways are also at risk of accidental ship collisions. One important aspect in ship-bridge collision analysis is the hydrodynamic effect, which is typically accounted for using the constant added mass (CAM) method. This traditional CAM method has been reported to significantly underestimate collision-induced loads, especially during oblique collisions. Moreover, the hydrodynamic effects on the collision response of scoured bridges remain underexplored. This study adopts a novel coupled numerical technique to assess the hydrodynamic effects of ship collisions on a continuous reinforced concrete (RC) bridge. The finite element (FE) models of the bridge structure and ship bow are developed in LS-DYNA, incorporating soil-structure interactions (SSI) through nonlinear spring elements. The hydrodynamic forces on the ship structure during the collision event are calculated using OrcaFlex. The obtained results provide insights into the influence of frequency-dependent hydrodynamic loads on the bridge collision response considering local scour around group piles.

1 INTRODUCTION

Bridges are exposed to a wide range of natural and artificial hazards throughout their service life. Among these, scour has been recognized as one of the leading causes of bridge failures, accounting for 21.3 % of such incidents globally (Tan et al., 2020). Local scour is a natural phenomenon characterized by the erosion of sediment around bridge foundations due to flowing water, as illustrated in Figure 1. Once initiated, local scour typically becomes permanent if no retrofitting measures are implemented. The presence of local scour alters the boundary conditions at the bridge foundations, which potentially increases the vulnerability of bridges to other hazards. Therefore, it is essential to understand the impact of local scour to ensure the long-term safety and reliability of bridge structures.

Ship collisions represent another significant man-made hazard for bridges, posing a substantial threat to those across navigable waterways. With the growing demand for marine transportation, ship-bridge collision accidents have occurred more frequently in recent years. In February 2024, the port hull of a barge collided with the Lixinsha Bridge in Guangzhou, China, resulting in five fatalities and three injuries. Later in March, a container ship struck the main pier of the Francis Scott Key Bridge in Baltimore, United States, causing the complete collapse of the bridge and leading to six deaths and two injuries (Zhang et al., 2024). Figure 2 provides an overview of these two catastrophic vessel-bridge collision accidents.

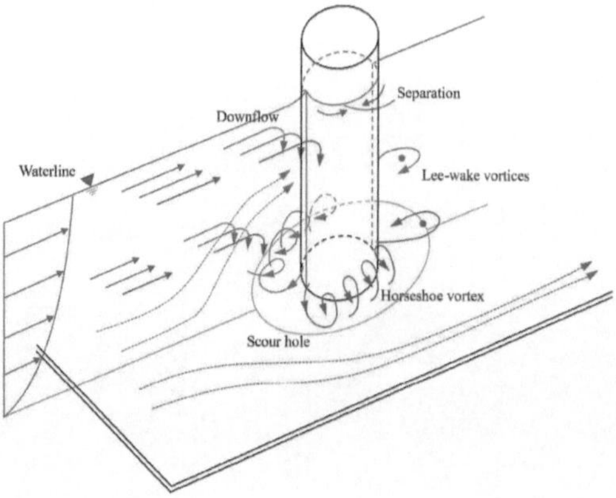

Figure 1. Schematic drawing of the local scour formation around a bridge pile (Wang et al., 2023).

With advancements in computational technology, finite element (FE) simulations have emerged as the predominant method for analyzing the highly nonlinear dynamics associated with ship-bridge collisions (Xu et al., 2024). Previous studies have attempted to model the collision interaction between a traveling vessel and a bridge, assuming that the bridge structure is in an intact condition at the time of impact (Fan

DOI: 10.1201/9781003684404-29

et al., 2020; Gholipour et al., 2018). However, this assumption overlooks the potential changes in the soil conditions caused by local scour. In response, some recent studies have explored the more complex multi-hazard events. Kameshwar & Padgett (2018) investigated bridge performance under the combined effects of barge impact and scour, revealing a clear correlation between an increasing scour depth and a higher probability of failure. Guo et al. (2020) provided further insights into the structural response of scoured bridges with group pile foundations subjected to barge collisions. Their findings indicated that while scour has a minimal effect on the maximum impact force, it significantly influences the displacement of bridge piers and pile foundations. To date, research on the dynamic behavior of bridges experiencing local scour when subjected to ship collision forces remains limited (Wang et al., 2024).

Figure 2. Collapse of (a) Lixinsha Bridge and (b) Francis Scott Key Bridge due to vessel collision accidents (Zhang et al., 2024).

The interaction between a floating structure and its surrounding water, known as the fluid-structure interaction (FSI), complicates the analysis of ship-scoured bridge collision studies. High-resolution FSI simulations using the Arbitrary Lagrangian-Eulerian (ALE) method have been developed to explicitly model the fluid (Rudan et al., 2019; Ye et al., 2023). These analyses highlight the critical role of the hydrodynamic effects in capturing ship-bridge collision interactions, particularly for accurately predicting structural damage and ship trajectories. Despite their accuracy, the ALE method demands substantial computational resources and complex modeling efforts, which limit its application in engineering practice.

A common simplification in ship-bridge collision analysis is the use of the constant added mass (CAM) approach, which considers the increased inertia forces on the ship without explicitly modeling the surrounding fluid. This approach offers a computationally efficient solution by representing hydrodynamic effects using a fixed added mass coefficient. However, the CAM method has been reported to underestimate the collision-induced loads, particularly in oblique collisions and scenarios with long collision durations (Tabri, 2012). Moreover, this method fails to capture the hydrostatic restoring forces, leading to discrepancies in predicting structural damage and motion responses (Yu et al., 2019).

To overcome the limitations of the CAM method, coupled simulations (Yu & Amdahl, 2016) were introduced to integrate external dynamics with internal mechanics. Le Sourne et al. (2012) developed a coupled approach in LS-DYNA using the large rotation rigid body solver MCOL. This method allows for the computation of hydrodynamic forces and structural deformations during a collision event through parallel simulations. However, research on the influence of hydrodynamic effects in ship collisions with scoured bridges remains scarce. To address this gap, the present study introduces a coupled algorithm that efficiently incorporates frequency-dependent hydrodynamic forces and moments into ship-scoured bridge collision analyses. The ship collision-induced forces are calculated in LS-DYNA, while the hydrodynamic forces and moments acting on the ship are computed using OrcaFlex. The soil-structure interaction (SSI) is represented by a nonlinear spring system around the pile foundations, incorporating relevant scour conditions. A series of collision scenarios, including head-on collisions with and without offset, are analyzed. The structural response is compared between the CAM method and the coupled approach in terms of collision forces, structural damage, displacement responses of the bridge, and ship motion.

2 NUMERICAL MODEL OF THE SHIP-SCOURED BRIDGE CASE STUDY

2.1 *Characteristics of ship and bridge structures*

Continuous bridges with fixed supports are among the most common types of bridge structures spanning navigable canals and rivers. In this study, a representative four-span reinforced concrete (RC) bridge is adopted as the baseline structure, with each span measuring 30 m, as illustrated in Figure 3 (Wang et al., 2023). Circular piers support the bridge deck, with each standing 25.2 m high and having a diameter of 2.8 m. The piers are reinforced with 30 mm diameter steel rebars and 20 mm diameter stirrups spaced 225 mm apart. The bridge piers rest on pile caps, with dimensions of 11.4 m in length, 10.9 m in width, and 2.5 m in height. Each pile cap distributes the loads to six steel piles with a length of 24 m and an outer diameter of 1.8 m. The group piles are arranged with center-to-center distances of 2.85 m in the longitudinal direction and 3.96 m in the transverse direction.

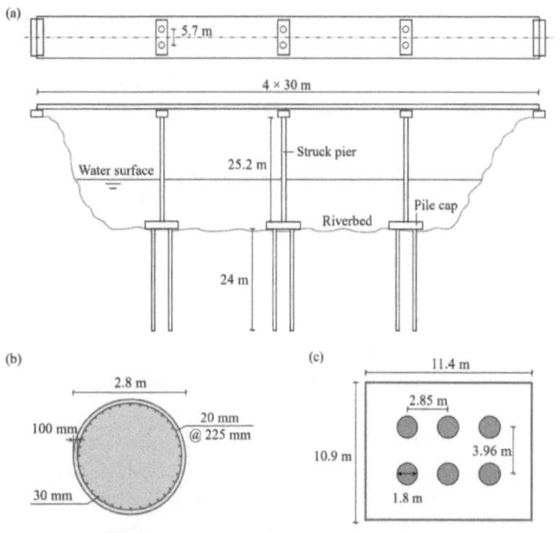

Figure 3. Geometric dimensions of the studied bridge: (a) schematic view; (b) cross-section of bridge piers; (c) layout of group pile foundation.

The ship structure considered in this study is a modern standard supply vessel with a deadweight tonnage of 9018 t and a draft of 6.66 m. It has a total length of 103.21 m, a breadth of 15.96 m, and a depth of 11.66 m, as illustrated in Figure 4.

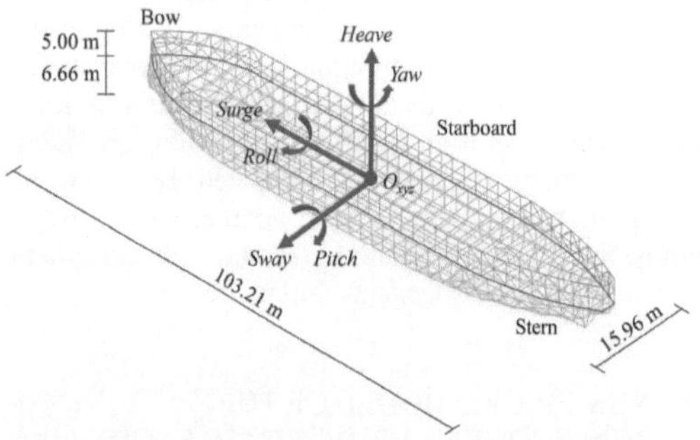

Figure 4. Panel model of the supply vessel in OrcaFlex.

2.2 *FE modeling of the ship-bridge system*

A numerical model of the ship-bridge system is developed in LS-DYNA to simulate the collision process. The key components of the model are illustrated in Figure 5. For the bridge structure, the simplified modeling approach proposed by Fan et al. (2020) is adopted. The non-impacted bridge piers are modeled as lumped masses, while the bearings are represented using translational and rotational springs. The impacted pier, which interacts with the traveling ship, is simulated using solid elements. The material model *MAT_CSCM_CONCRETE (*MAT_159) is used to account for the nonlinear behavior of concrete, including properties such as shear failure surfaces, cap hardening, and material softening. The concrete material has a compressive strength of 60 MPa, a density of 2400 kg/m^3, and a Poisson's ratio of 0.15.

The pier reinforcement and group piles are modeled as beam elements. The elastic-plastic material model *MAT_PIECEWISE_LINEAR_PLASTICITY (*MAT_024) is used to simulate the steel. Their mass density, yield stress, Young's modulus, and failure strain are provided in Table 1. The steel reinforcement beam elements are coupled with concrete solid elements by sharing common nodes. The strain rate effect for both steel and concrete is neglected in this study. Hourglass mode control is applied using the Flanagan-Belytschko formulation, with IHQ values set to 4 for steel and 5 for concrete. The corresponding hourglass coefficients are 0.05 for steel and 0.1 for concrete.

Table 1. Material properties of concrete and steel.

Component	Parameter	Value
Pier concrete	Mass density	2400 kg/m^3
	Compressive strength	60 MPa
	Poisson's ratio	0.15
Pier reinforcement	Mass density	7850 kg/m^3
	Yield stress	588 MPa
	Poisson's ratio	0.35
	Young's modulus	200 GPa
	Failure strain	0.25
Group piles	Mass density	7800 kg/m^3
	Yield stress	275 MPa
	Poisson's ratio	0.3
	Young's modulus	200 GPa
	Failure strain	0.25
Ship bow	Mass density	8580 kg/m^3
	Yield stress	275 MPa
	Poisson's ratio	0.3
	Young's modulus	200 GPa
	Failure strain	0.25

The bridge bearings supporting the superstructure are modeled using translational and rotational springs with the material model *MAT_LINEAR_-ELASTIC_DISCRETE_BEAM (*MAT_066). The stiffness of the horizontal bearing, vertical bearing, and rotational springs is 2.33×10^7 N/m, 1.87×10^{10} N/m, and 2.13×10^7 N·m/rad, respectively. The dampers are simulated using the material model *MAT_DAMPER_VISCOUS (*MAT_S02), where the damping ratios are set to 0.14 (Bi et al., 2013).

For the ship structure, the modeling approach proposed by Yu & Amdahl (2016) is employed. The ship bow is modeled using shell elements to represent the outer shell panels, decks, stiffeners, and transverse frames. The material model *MAT_024 is used to simulate the steel in the ship bow. Rigid beams are utilized to represent the ship hull girder from the bow to the stern, with their properties adjusted to accurately reflect the total mass and moments of inertia in all translational and rotational degrees of freedom of the ship structure.

The ship-bridge collision process involves contact interactions between the ship bow and the bridge pier, as well as internal contact within the ship bow itself. The contact interaction between the ship bow and the impacted bridge pier is modeled using *CONTACT_AUTOMATIC_SURFACE_TO_SURFACE, while the internal contact within the ship bow is represented using *CONTACT_AUTOMATIC_SINGLE_SURFACE. The static and dynamic coefficients of friction are 0.30 for the two-way interaction between the ship bow and the bridge pier and 0.21 for the single-surface contact within the ship bow.

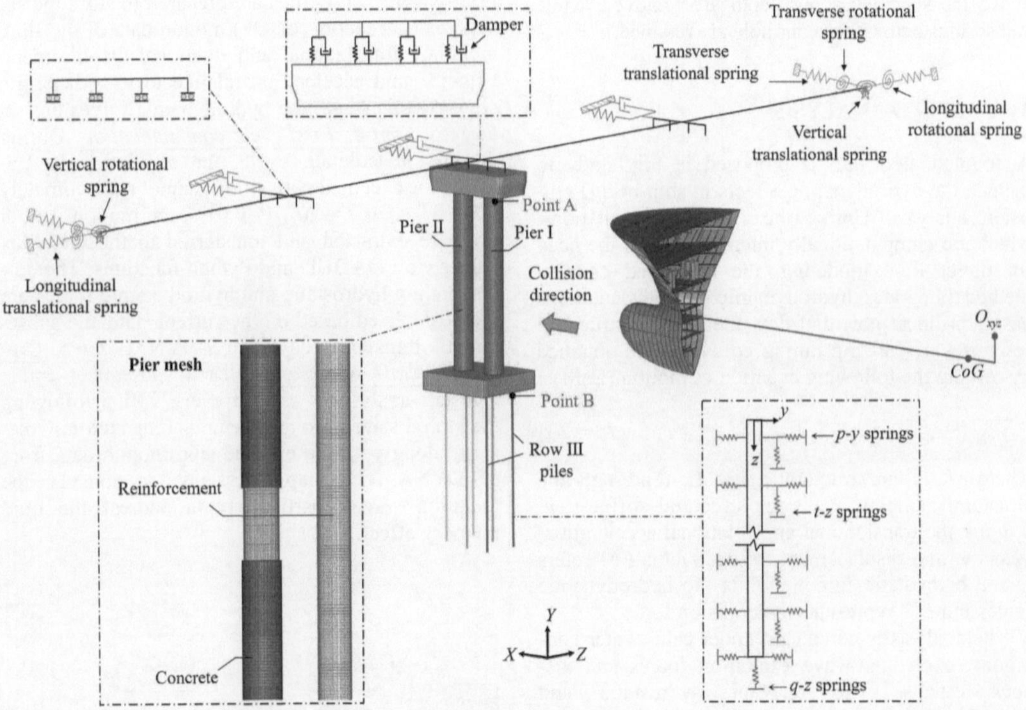

Figure 5. Finite element model of the ship-bridge system in LS-DYNA.

2.3 *Scour modeling and soil-structure interaction*

The soil conditions at actual bridge sites are often complex, typically consisting of multi-layered soils with varying properties. In this study, the soil surrounding the pile foundations is assumed to be homogeneous medium-dense sand, with a moisture unit weight of 19.6 kN/m^3 and a friction angle of 33°. To capture the scour effect on the bridge's dynamic behavior, the soil-structure interaction (SSI) is represented by a nonlinear spring system distributed along the elevation of the piles. This system includes *p-y* springs for lateral resistance, *t-z* springs for frictional resistance along the pile shaft, and *q-z* springs for end-bearing, as shown in Figure 5. The properties of these springs are determined based on the guidelines provided by the American Petroleum Institute (2007). For instance, the lateral soil resistance-deflection relations for the *p-y* springs for sand are calculated according to Equations (1) and (2):

$$p = A \cdot p_u \cdot \tan[(k \cdot d \cdot y)/(A \cdot p_u)] \quad (1)$$

$$p_u = \min[(c_1 \cdot d + c_2 \cdot l) \cdot \gamma \cdot d,\ (c_3 \cdot l \cdot \gamma \cdot d)] \quad (2)$$

where A refers to the constant accounting for static or dynamic loading conditions; p_u denotes the ultimate bearing capacity at depth d; k is the initial modulus of soil reaction; l symbolizes the pile diameter; c_1, c_2, c_3 represent the coefficients related to the friction angle of the soil; and γ is the effective unit weight.

To account for the pile group effect, the *p*-multiplier method proposed by Brown et al. (1988) is adopted. This approach modifies the *p-y* spring characteristics by reducing the lateral load capacity of the soil based on the spacing between adjacent piles. Piles nearest to the collision location are assumed to experience the greatest reduction. When subjected to significant lateral loads, the soil may undergo large plastic deformations, affecting its hysteresis behavior.

To capture this nonlinearity, different material models are used for the soil springs: *MAT_SPRING_INELASTIC (*MAT_S08) for *p-y* and *q-z* springs and *MAT_SPRING_GENERAL_NONLINEAR (*MAT_S06) for *t-z* springs. To simulate the effects of local scour on the foundation, soil springs located above the specified scour depth are removed, while soil springs below the scour hole are retained.

3 COUPLED ANALYSIS

A coupled algorithm is proposed in this study to include the hydrodynamic effects in ship-bridge collision analysis. Unlike the full FSI simulations, which are computationally intensive due to the need for direct fluid modeling, the employed coupled method integrates hydrodynamic force calculations based on linear potential flow theory. The structural responses of the ship during collisions are obtained by solving the following equation of motion (EoM):

$$\sum_{j=1}^{6} \{M_{s,ij}\ddot{u}_j + C_{s,ij}\dot{u}_j + K_{s,ij}u_j\} = F_i^{sta} + F_i^{dyn} + F_i^{col} \quad (3)$$

where M_s is the structural mass; C_s represents the structural damping; K_s is the structural stiffness; $\ddot{u}$, $\dot{u}$, u are the translational and rotational acceleration, velocity, and displacement, respectively; F^{sta} refers to the hydrostatic force; F^{dyn} is the hydrodynamic force; and F^{col} represents the collision force.

The analysis is conducted under calm water conditions, excluding wave excitation forces and viscous effects. The relevant hydrostatic and hydrodynamic forces acting on the ship structure are calculated using Equations (4) and (5):

$$F_i^{sta} = \sum_{j=1}^{6} \{-K_{h,ij}u_j\} \quad (4)$$

$$F_i^{dyn} = -\sum_{j=1}^{6} \left\{A_{h,ij}^{\infty}\ddot{u}_j + \int_{\tau=0}^{\infty} IRF_{h,ij}(\tau)\dot{u}_j(t-\tau)d\tau\right\} \quad (5)$$

where K_h denotes the hydrostatic stiffness; A_h^{∞} is the added mass at infinite frequency; IRF_h represents the impulse response function; and τ is the time lag.

Explicit FE analyses are conducted in LS-DYNA to simulate the collision interaction between the ship bow and the bridge pier, while hydrostatic and hydrodynamic forces are calculated in OrcaFlex. The coupled analysis framework is illustrated in Figure 6, and a step-by-step description is presented as follows;

Step 1: Calculating hydrodynamic properties. Diffraction analyses are performed using the panel model shown in Figure 4 to determine the hydrodynamic characteristics of the ship. The obtained hydrostatic restoring forces, frequency-dependent added mass, and damping matrices, as well as response amplitude operators (RAOs), are imported into the OrcaFlex solver. *Step 2: Generating binary executable.* The load subroutine LOADUD is modified to support key functionalities, including data reading and writing, script invocation, and the application of nodal forces. The customized subroutine is compiled with other source files using a Fortran compiler, producing a binary executable that serves as the main LS-DYNA solver. *Step 3: Initiating calculation.* The simulation is initialized in the LS-DYNA solver. Log files are prepared to store the six degrees of freedom (6DOF) motion data of the ship, including translational and rotational displacement, velocity, and acceleration relative to its axis origin O_{xyz}. At this stage, the hydrodynamic forces are set to zero. *Step 4: First-cycle communication.* During the first calculation cycle, the simulation in LS-DYNA is executed with a time step of approximately 1.7×10^{-6} s. At $t = \Delta t_{ext} = 0.01$ s, the motion data at O_{xyz} are extracted and transferred to the OrcaFlex solver via LOADUD and Python functions. The corresponding hydrostatic and hydrodynamic forces are then calculated based on the current state and subsequently transferred back to LS-DYNA. *Step 5: Creating restart analyses.* In each subsequent cycle, a restart analysis is created every 0.01 s following the stored state from the previous time step in OrcaFlex, along with the updated ship motion data from LS-DYNA. This step is essential to consider the radiation waves, particularly to address the fluid memory effect.

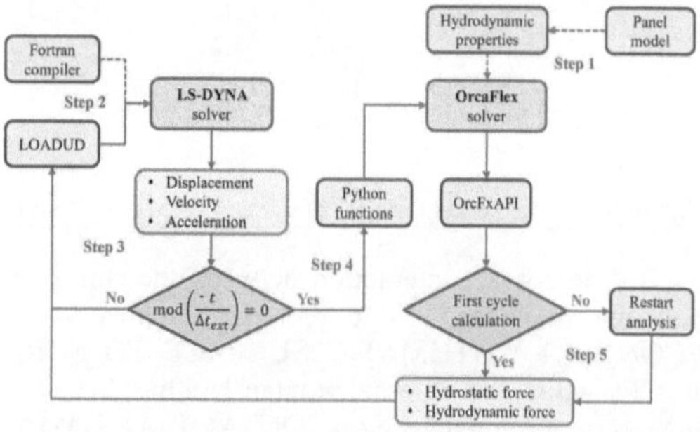

Figure 6. Coupled simulation between LS-DYNA and OrcaFlex.

4 SHIP-BRIDGE COLLISION ANALYSES

4.1 *Simulation scenarios*

The traveling speed of the supply vessel is selected as 2 m/s, corresponding to a collision energy of 18 MJ. Head-on collisions are considered in this study. To investigate the influence of hydrodynamic effects and local scour on the structural response of the ship-bridge system, six different collision scenarios are analyzed, as summarized in Table 2. The first three scenarios involve collisions on the pier center with zero offset. In Scenario C-CM-I, the CAM method is applied, assuming 10 % of the total ship mass to represent the surrounding fluid. The bridge is also assumed to be in pristine condition without local scour. This analysis is conducted solely in

LS-DYNA. In Scenario C-CP-I, the hydrodynamic effects on the ship are incorporated by activating the communication between LS-DYNA and OrcaFlex. Scenario C-CP-S introduces a scour depth of 8 m around the group piles to represent a severe scour condition (Wang et al., 2024). Another three collision scenarios with a 0.5 m offset are included. Similarly, the CAM method with a 10 % added mass is applied in Scenario O-CM-I. Scenario O-CP-I includes the hydrodynamic effects on the ship through parallel simulation between two solvers. Scenario O-CP-S further considers the scouring effect around the pile foundation of the bridge.

Table 2. Ship-bridge collision scenarios.

No.*	Collision offset	Fluid-structure interaction	Scour depth
C-CM-I	0 m	CAM method	0 m
C-CP-I	0 m	Coupled method	0 m
C-CP-S	0 m	Coupled method	8 m
O-CM-I	0.5 m	CAM method	0 m
O-CP-I	0.5 m	Coupled method	0 m
O-CP-S	0.5 m	Coupled method	8 m

* Note: C = collisions on the pier center; O = collisions with an offset; I = intact soil condition; S = local scour effect.

4.2 *Head-on collisions*

The time histories of collision forces exerted by the ship bulb and forecastle on the bridge pier are presented in Figure 7 for three head-on collision scenarios without offset. In the initial loading phase, no significant differences are observed among the three scenarios for both the ship bulb and forecastle. As shown in Figure 7(a), the peak impact force for the ship bulb reaches 17.68 MN in Scenario C-CM-I, in which the CAM method is applied. In Scenario C-CP-I, where the frequency-dependent added mass and damping are considered by implementing the coupled analysis, the peak impact force slightly increases to 18.28 MN. In Scenario C-CP-S, the peak ship bulb-bridge pier collision force is approximately 17.76 MN, where the local scour around the pile foundation is included. Although the local scour does not seem to influence the peak impact force for the ship bulb significantly, it results in a slight delay in the occurrence of the peak impact force. This can be attributed to the increased free pile length above the scour depth, which extends the natural period of the bridge structure.

The difference in the maximum collision forces induced by ship forecastle between Scenarios C-CM-I and C-CP-I, as shown in Figure 7(b), is less than 2.8 %. This is due to the relatively small contribution of hydrodynamic forces compared to the dominant collision interaction forces. In contrast, the inclusion of local scour in Scenario C-CP-S results in a noticeable increase of 9.6 % in the peak ship forecastle collision force. During the descending phase, where the collision forces drop to zero, Scenario C-CM-I with the CAM method shows a slightly slower rate of force reduction. Nevertheless, the resultant contact force between the striking ship and the struck bridge pier is primarily determined by the material strength of the ship and the impact energy.

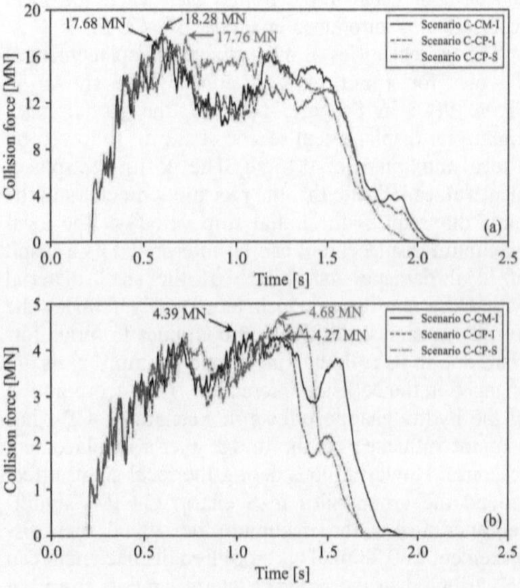

Figure 7. Time histories of (a) ship bulb- and (b) ship forecastle-bridge pier collision forces for head-on collision scenarios without offset.

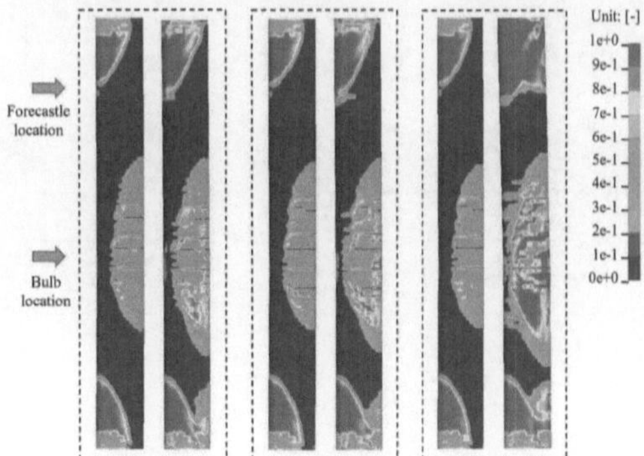

Figure 8. Effective plastic strain of the impacted pier at t = 0.5 s and t = 2.5 s for head-on collision scenarios without offset.

Figure 8 presents the effective plastic strain distribution of the impacted pier (pier I in Figure 5) at two time instants: t = 0.5 s during the initial loading phase; and t = 2.5 s during the bridge's free vibration phase after the striking ship detaches from the pier (see Figure 7). As the collision progresses, the plastic deformation expands both vertically and laterally along the pier. The damage to the struck pier is predominantly characterized by flexural failure across all three scenarios. When both the hydrostatic and hydrodynamic forces are considered in Scenario C-CP-I, the damaged area of the pier shows a slight increase compared to the CAM method in Scenario C-CM-I. More pronounced damage is observed in the impacted region and at both ends of the bridge pier when the local scour effect is introduced in Scenario C-CP-S.

The time histories of the horizontal displacement at the pier top (point A in Figure 5) are shown in Figure 9(a). In Scenario C-CM-I, the global maximum pier displacement reaches 0.22 m, followed by a local maximum of 0.15 m. The positive displacement values indicate that the pier motion occurs in the same direction as the initial ship velocity. The local maximum displacement can be interpreted as a result of local damage and failure of the steel material within the ship bow, which temporarily reduces the overall resistance. As the ship continues to move forward due to its residual kinetic energy, more parts are engaged in the collision interaction. The incorporation of the hydrodynamic effects in Scenario C-CP-I has a minor influence on the bridge pier's displacement response. However, considering the local scour effect around the group piles in Scenario C-CP-S significantly increases the maximum horizontal pier displacement to 0.28 m. This amplified displacement can be attributed to the reduced lateral support from the soil, which increases the pier's flexibility and allows for greater deflection under impact loads.

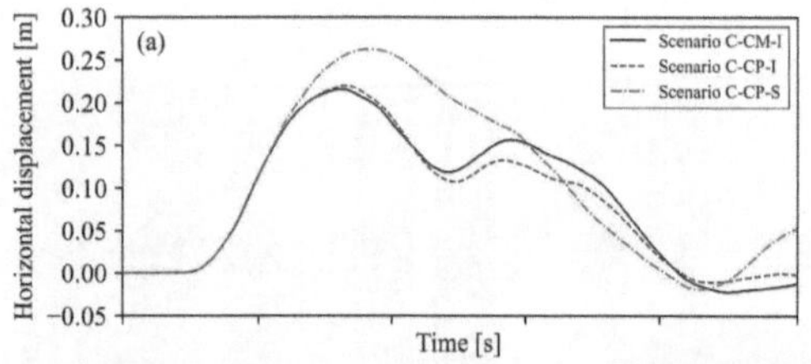

Figure 9. Time histories of the horizontal displacements at (a) point A of pier I and (b) point B of row III piles for head-on collision scenarios without offset.

Figure 9(b) presents the time histories of horizontal displacement at point B in the group piles (Figure 5). As anticipated, the hydrodynamic effects have a negligible influence on the group pile displacement response, with the maximum reaching approximately 0.2 m in both Scenarios C-CM-I and C-CP-I. In comparison, the presence of scour in Scenario C-CP-S leads to a substantial increase in the maximum pile displacement, reaching up to 0.6 m. This increase is attributed to the loss of lateral soil support above the scour hole.

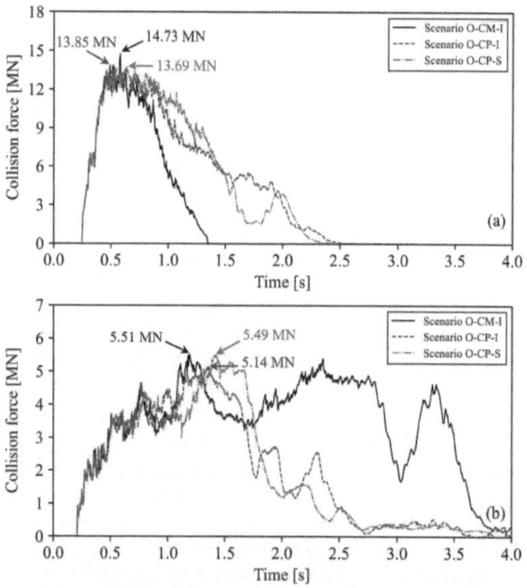

Figure 10. Time histories of (a) ship bulb- and (b) ship forecastle-bridge pier collision forces for collision scenarios with a 0.5 m offset.

4.3 *Collisions with offset*

The time histories of the collision forces for three collision scenarios with an offset of 0.5 m are presented in Figure 10. Similar to head-on collision scenarios without offset, the consideration of the hydrodynamic effects on the ship and local scour around the bridge piles exhibits limited influence on the peak impact forces for both the ship bulb and forecastle. As shown in Figure 10(a), the peak impact forces induced by the ship bulb are 13.85 MN, 14.73 MN, and 13.69 MN for Scenarios O-CM-I, O-CP-I, and O-CP-S, respectively. Notably, the peak collision forces of the ship bulb in collisions with 0.5 m offset are considerably lower than those observed in collisions without offset, as shown in Figure 7(a). The ship forecastle exhibits much higher peak forces in offset collisions, as indicated by comparing Figures 10(b) and 7(b). This difference is due to the significant roll and yaw motions of the ship using the traditional CAM method, which does not account for hydrostatic restoring forces. In Scenario O-CP-I, the incorporation of hydrostatic and hydrodynamic forces reduces the ship's roll and yaw motion, as illustrated in Figure 11. This results in a prolonged engagement of the ship bulb with the pier compared to Scenario O-CM-I, where the absence of hydrostatic restoring forces leads to earlier bulb separation from the pier. Thus, more energy is dissipated through the bulb impact in Scenario O-CP-I, explaining the earlier drop in forecastle collision force shown in Figure 10. The influence of local scour in offset collision scenarios is minimal, with negligible effect on both the peak impact force and the duration of the collision.

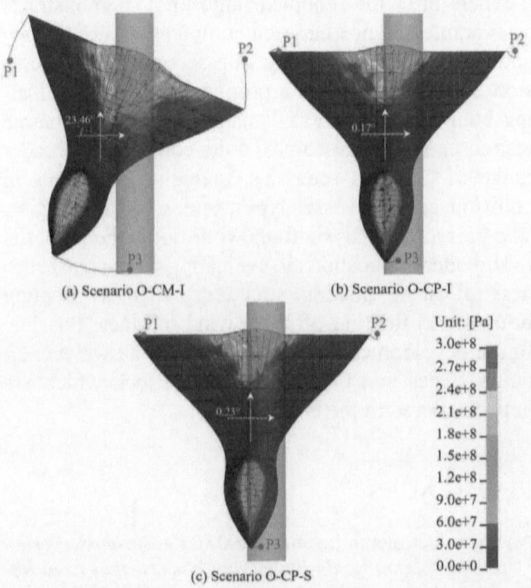

Figure 11. Effective stress in the striking ship at $t = 4$ s for collision scenarios with a 0.5 m offset. Trajectories of three reference points on the ship bow are marked in red.

Figure 11 illustrates the effective stress distribution of the striking ship at $t = 4$ s, along with the trajectories of three reference points (marked in red): P1 and P2 on the ship forecastle, and P3 on the ship bulb. In Scenario O-CM-I, where the CAM method is used, the trajectories indicate that the ship motion is mainly governed by the roll motion with moderate contributions from surge and yaw. The roll angle reaches 23.46°, in comparison to 0.17° and 0.23° in Scenarios O-CP-I and O-CP-S, respectively. This suggests that the CAM approach fails to account for the hydrostatic restoring forces, resulting in the exaggerated roll motion of the ship. Furthermore, the CAM method results in a much larger indentation and damage area in the ship forecastle, while simultaneously underestimating the structural damage in the ship bulb. The influence of local scour appears to be minor on both the ship motion and the extent of structural damage.

Figure 12 illustrates the effective plastic strain distribution of the struck pier at $t = 0.8$ s and $t = 4.0$ s. In Scenario O-CM-I, which employs the CAM method, elevated localized plastic strain is observed in the region impacted by the forecastle. In comparison, the coupled analysis used in Scenario O-CP-I exhibits a slightly more dispersed strain pattern in the bulb impact region. Additionally, the presence of local scour in Scenario O-CP-S intensifies the damage in both the bulb- and forecastle-impact regions, primarily due to the reduced lateral soil support around the piles.

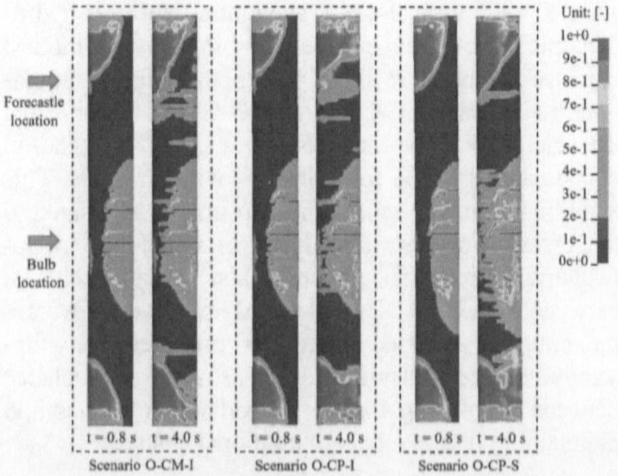

Figure 12. Effective plastic strain of the impacted pier at $t = 0.8$ s and $t = 4.0$ s for collision scenarios with a 0.5 m offset.

The time histories of horizontal displacements at the top of the impacted pier and pile foundation are presented in Figure 13. The results align with those from head-on collisions without offset, confirming that hydrodynamic effects have a limited influence on the maximum displacement response. The presence of scour notably amplifies the horizontal displacements, with a more pronounced effect on the group piles than on the pier. Furthermore, in Scenario O-CP-S, observations after $t = 2.5$ s suggest that local scour increases the free pile length of the piles, consequently resulting in a longer natural period of the bridge system.

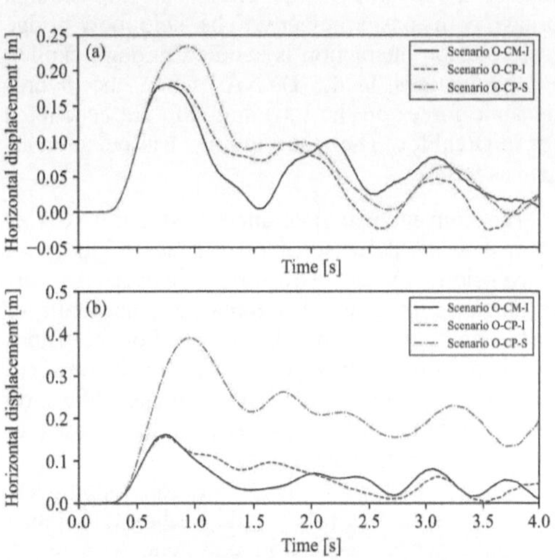

Figure 13. Time histories of the horizontal displacements at (a) point A of pier I and (b) point B of row III piles for collision scenarios with a 0.5 m offset.

4.4 *Comparison of computational cost*

A comparison of the computational demand for different simulation scenarios is presented in Table 3. The CPU time required for the coupled analyses in Scenarios C-CP-I and O-CP-I, accounting for the hydrodynamic effects on the traveling ship, shows a modest increase compared to the CAM method-based simulations in Scenarios C-CM-I and O-CM-I. Specifically, the calculation times increase by 12.8 % for head-on collisions and 8.6 % for collisions with an offset. This additional computational time is primarily attributed to the creation of restart analyses in OrcaFlex, hydrodynamic force calculations, and data exchange between the LS-DYNA and OrcaFlex solvers. Overall, the coupled analysis method is preferred for ship-scoured bridge collision analyses as it strikes a balance between accounting for the hydrodynamic effects and maintaining reasonable computational costs.

Table 3. Computational costs of different simulation scenarios.

No.	Simulation time	CPU time*
C-CM-I	2.5 s	54 h 42 min
C-CP-I	2.5 s	61 h 43 min
C-CP-S	2.5 s	54 h 57 min
O-CM-I	4.0 s	82 h 31 min
O-CP-I	4.0 s	89 h 37 min
O-CP-S	4.0 s	89 h 12 min

* Note: The CPU used is 11th Gen Intel® Core™ i9-11900K, 8 Cores, 16 Threads, 3.5 GHz.

5 CONCLUSIONS

This study presents a coupled analysis method to capture the hydrodynamic effects in ship-scoured bridge collisions efficiently. The ship bow-bridge pier collision interaction is handled through explicit FE simulations in LS-DYNA, while the hydrodynamic forces on the traveling ship are accounted for in OrcaFlex. The main findings can be summarized as follows:

- The conventional constant added mass (CAM) method yields reasonable estimates of the peak collision forces, structural damage to the impacted pier, and horizontal displacements of the pier and group piles in head-on collisions without offset. However, in offset collision scenarios, it substantially overestimates the roll motion of the traveling ship due to its omission of hydrostatic restoring forces.
- Local scour reduces the lateral soil support and increases the free pile length, leading to a longer natural period of the bridge system. As a result, the horizontal displacement responses, especially in the group piles, are amplified, along with an increase in structural damage to the impacted bridge pier.
- The computational cost of the coupled analysis, in terms of CPU time, remains comparable to that of the CAM-based simulations, while offering the additional benefit of capturing hydrostatic restoring forces and radiation wave effects induced by ship motion.

Therefore, the coupled algorithm demonstrates a favorable balance between computational efficiency and accuracy in simulating ship-scoured bridge collision events. It provides a promising tool for evaluating both local structural damage and global dynamic response. Future research could consider a broader range of collision scenarios, including variations in collision energy, vessel types, and scour conditions. The presented analysis framework holds the potential for broader applications in evaluating the crashworthiness of other floating structures, such as floating bridges and floating offshore wind turbines. Possible future developments may also incorporate wave excitation forces and the viscous fluid effect, which are not considered in the current study.

REFERENCES

American Petroleum Institute. (2007). *Recommended practice for planning, designing and constructing fixed offshore platforms: Working stress design* (Vol. 2A-WSD). American Petroleum Institute.

Bi, K., Hao, H. & Chouw, N. (2013). 3D FEM Analysis of pounding response of bridge structures at a canyon site to spatially varying ground motions. *Advances in Structural Engineering*, *16*(4), 619–640.

Brown, D. A., Morrison, C. & Reese, L. C. (1988). Lateral load behavior of pile group in sand. *Journal of Geotechnical Engineering*, *114*(11), 1261–1276.

Fan, W., Shen, D., Huang, X. & Sun, Y. (2020). Reinforced concrete bridge structures under barge impacts: FE modeling, dynamic behaviors, and UHPFRC-based strengthening. *Ocean Engineering*, *216*, 108116.

Gholipour, G., Zhang, C. & Mousavi, A. A. (2018). Effects of axial load on nonlinear response of RC columns subjected to lateral impact load: Ship-pier collision. *Engineering Failure Analysis*, *91*, 397–418.

Guo, X., Zhang, C. & Chen, Z. (2020). Nonlinear Dynamic Response and Assessment of Bridges under Barge Impact with Scour Depth Effects. *Journal of Performance of Constructed Facilities*, *34*(4), 04020058.

Kameshwar, S. & Padgett, J. E. (2018). Response and fragility assessment of bridge columns subjected to barge-bridge collision and scour. *Engineering Structures*, *168*, 308–319.

Le Sourne, H., Besnard, N., Cheylan, C. & Buannic, N. (2012). A ship collision analysis program based on upper bound solutions and coupled with a large rotational ship movement analysis tool. *Journal of Applied Mathematics*, *2012*, 1–27.

Rudan, S., Ćatipović, I., Berg, R., Völkner, S. & Prebeg, P. (2019). Numerical study on the consequences of different ship collision modelling techniques. *Ships and Offshore Structures*, *14*(sup1), 387–400.

Tabri, K. (2012). Influence of coupling in the prediction of ship collision damage. *Ships and Offshore Structures*, 7 (1), 47–54.

Tan, J., Elbaz, K., Wang, Z., Shen, J. S. & Chen, J. (2020). Lessons learnt from bridge collapse: A view of sustainable management. *Sustainability*, *12*(3), 1205.

Wang, Z., Sha, Y. & Jakobsen, J. B. (2024). Influence of local scour on the dynamic response of bridges under barge collisions. *Engineering Structures*, *308*, 117941.

Wang, Z., Sha, Y. & Ong, M. C. (2023, June 11). Numerical investigation of scoured bridge response under ship collisions. *Proceedings of the 42nd International Conference on Ocean, Offshore and Arctic Engineering*.

Xu, M. C., Wang, T. & Pan, J. (2024). A review on the assessment methods of bridge against ship collision. *Thin-Walled Structures*, *205*, 112347.

Ye, X., Fan, W., Sha, Y., Hua, X., Wu, Q. & Ren, Y. (2023). Fluid-structure interaction analysis of oblique ship-bridge collisions. *Engineering Structures*, *274*, 115129.

Yu, Z. & Amdahl, J. (2016). Full six degrees of freedom coupled dynamic simulation of ship collision and grounding accidents. *Marine Structures*, *47*, 1–22.

Yu, Z., Liu, Z. & Amdahl, J. (2019). Discussion of assumptions behind the external dynamic models in ship collisions and groundings. *Ships and Offshore Structures*, *14* (sup1), 45–62.

Zhang, W. Z., Pan, J., Sanchez, J. C., Li, X. B. & Xu, M. C. (2024). Review on the protective technologies of bridge against vessel collision. *Thin-Walled Structures*, *201*, 112013.

Trends in Collision and Grounding of Ships and Offshore Structures – Zhu & Guedes Soares (Eds)

Effect of hydrodynamics of a floating offshore wind turbine and a colliding ship on damage extents and collision limit curves

DH. Yoon, M. Jeong & J. Choung
Department of Naval Architecture and Ocean Engineering, Inha University, Incheon, Republic of Korea

ABSTRACT: IEC 61400-3-2 recommends limiting the speed of ships or restricting their passage based on a collision limit curve. This study aims to analyse the influence of hydrodynamics on these collision limit curves. HydroQus was used to account for the hydrodynamic forces acting on the colliding ship and FOWT. HydroQus, developed by the authors, is a plug-in for the commercial FEM code Abaqus/Explicit that interacts with Abaqus to apply potential theory-based hydrodynamic forces in real time. For collision simulations, 22MW-cpacity semi-submersible FOWT and a container carrier were modeled. The container carrier was assumed to be rigid, while the FOWT was modeled as elastic-plastic body. The ship was assumed to collide perpendicularly with the tower column of the FOWT. A sufficiently high initial velocity was applied to cause damage to the tower columns. The FOWT outer shell was damaged at lower ship speeds than without hydrodynamics, but the difference was not significant.

1 INTRODUCTION

1.1 *Introduction*

The development of large-scale floating offshore wind farms is expected to increase the risk of collisions between vessels navigating nearby and floating offshore wind turbines (FOWTs). For these reasons, research on FOWT-ship collision is critically needed.

Dai et al. (2013) evaluated structural damage based on collision velocity and analyzed risk factors such as human error and weather conditions in a study on collisions between monopile-type bottom-fixed offshore wind turbines (BOWTs) and vessels. Moulas et al. (2017) and Liu et al. (2022) investigated BOWT-ship collisions, observing structural damage variations with collision speed and angle. Jia et al. (2020) examined the influence of wind loads, while Bela et al. (2017) incorporated wind loads and soil stiffness in BOWT-ship collision studies. Echeverry et al. (2019) and Márquez et al. (2022) conducted collision simulations for sparand barge-type FOWTs, respectively, using the LS-DYNA (LSTC, 2023) subroutine MCOL to account for hydrodynamic forces on floating structures. Sun and Fang (2023) proposed applying composite honeycomb structures to offshore wind turbines to mitigate collision damage. Niklas et al. (2023) compared structural performance across steel grades.

DNV provides guidelines for preparing for accidental ship collisions. recommends limiting vessel speeds or restricting navigation through displacement-speed collision limit curves. Yu et al. (2022) suggested strengthening vessel operation restrictions and safety procedures to minimize the possibility of ship collisions. Storheim and Amdahl (2014) mentioned that improvements to the existing NORSOK standards are necessary due to the increasing size of supply vessels and diversification of bow designs.

The authors aim to evaluate the existing collision limit curves through FOWT-ship collision analysis. The objective is to compare cases where only the added mass of the FOWT and the colliding ship is considered, as suggested by the IEC, with cases where hydrodynamics based on potential flow theory are included. The scenario considering only the added mass of the colliding ship was investigated in the authors' previous study. For the case including hydrodynamics based on potential flow theory, radiation and restoring forces acting on the FOWT and the ship were calculated using HydroQus, developed by the authors (Yoon et al., 2023; Yoon & Choung, 2023). To predict structural damage to the FOWT during collisions, the combined necking and fracture model proposed by Pack & Mohr (2017) was applied. The colliding ship was assumed to be rigid body.

2 THEORETICAL BACKGROUND

2.1 *Equation of motion*

The governing equation of HydroQus used in this study is defined as Equation 1, according to Newton's second law. $\boldsymbol{M}$ is the mass matrix of the floating body, $\ddot{\boldsymbol{u}}$ and is the acceleration matrix at the center of mass (CoM) of the floating body. The restoring force $\boldsymbol{F}^{res}$ is calculated using the hydrostatic stiffness matrix $\boldsymbol{C}$ and the displacement $\boldsymbol{u}$ at the CoM of the floating body. The radiation force $\boldsymbol{F}^{rad}$ is calculated as shown in Equation 3 (Cummins, 1962). The added

DOI: 10.1201/9781003684404-30

mass $\boldsymbol{M}'$ is calculated as shown in Equation 4, and the impulse response function (IRF) $\boldsymbol{K}$ is calculated using the wave damping coefficient $\boldsymbol{B}$ as shown in Equation 5. Here, the integration time t_1 of the convolution integral should be sufficient for the IRF to converge to zero. These forces are calculated in real-time through HydroQus. Originally, HydroQus can also calculate first-order and second-order wave excitation forces, but these were not applied in this study. $\boldsymbol{F}^{col}$ is the impact force due to the collision between the FOWT and the ship. $\boldsymbol{F}^{col}$ is calculated by Abaqus itself.

$$\boldsymbol{M}\ddot{\boldsymbol{u}}(t) = \boldsymbol{F}^{res}(t) + \boldsymbol{F}^{rad}(t) + \boldsymbol{F}^{col}(t) \quad (1)$$

$$\boldsymbol{F}^{res}(t) = \boldsymbol{C}\boldsymbol{u}(t) \quad (2)$$

$$\boldsymbol{F}^{rad}(t) = \boldsymbol{M}'(\infty)\ddot{\boldsymbol{u}}(t) + \int_0^{t_1} \boldsymbol{K}(t-\tau)\,\boldsymbol{u}(\tau)d\tau \quad (3)$$

$$\boldsymbol{M}'(\infty) = \boldsymbol{M}'(\omega) + \int_0^{\infty} \boldsymbol{K}(\tau)\frac{\sin(\omega\tau)}{\omega}d\tau \quad (4)$$

$$\boldsymbol{K}(t) = \frac{2}{\pi}\int_0^{\infty} \boldsymbol{B}(\omega)\cos(\omega t)d\omega \quad (5)$$

2.2 *Process of fluid-structure interaction*

Abaqus is an FEA code that incorporates various material models and robust collision algorithms. However, due to the difficulty in considering hydrodynamic forces, HydroQus was employed. Figure 1 illustrates the working process of HydroQus. To calculate hydrodynamic forces based on potential flow theory, a frequency response analysis using AQWA must be conducted beforehand. The hydrodynamic coefficients obtained through this analysis are then transferred to HydroQus. HydroQus uses these coefficients to calculate hydrodynamic forces and returns them to Abaqus/Explicit. Abaqus then derives the displacements, velocities, and accelerations resulting from the transmitted forces and collisions. This procedure is repeated for all time steps in Abaqus.

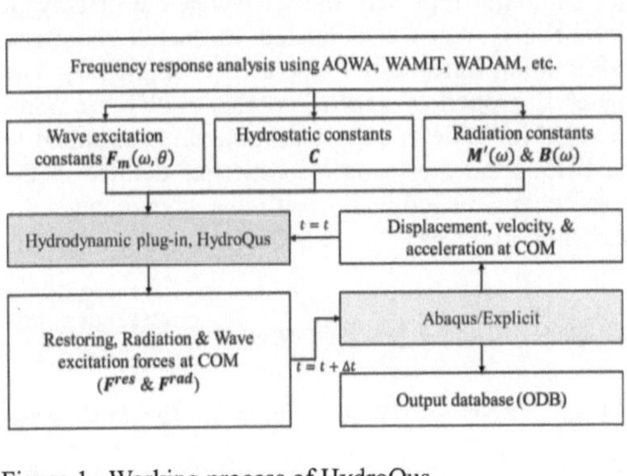

Figure 1. Working process of HydroQus.

2.3 *Limit curve*

The IEC regulations were referenced to develop the FOWT-ship collision limit curve. IEC defines the ship's velocity that causes damage to the FOWT during a FOWT-ship collision as shown in Equation 6. Here, V_B is the ship's velocity. M_A and M_B represent the masses of the FOWT and the ship, respectively, including their added masses. E_{defo} refers to the minimum energy required to cause damage to the FOWT's outer shell, which is determined through FEA. α is the ratio of energy absorbed by the FOWT, which is 1 in this study because the ship is assumed to be a rigid body.

$$V_B > \sqrt{\frac{2E_{defo}(M_A + M_B)}{\alpha M_A M_B}} \quad (6)$$

3 MODELS OF FOWT AND SHIP

3.1 *Dimensions and mass properties*

The FOWT used in this study is a semi-submersible floating structure with three columns. These three columns are connected by pontoons and decks. Figure 2 shows the substructure of the FOWT. The properties of the substructure are summarized in Table 1. For the superstructure of the FOWT, a 22MW capacity IEA public turbine and tower were applied (Zahle et al., 2024). This information is summarized in Table 2.

Figure 2. Shape of substructure.

Table 1. Properties of substructure.

Item	Unit	Value
Column height	m	30.00
Pontoon width	m	9.50
Pontoon height	m	6.00
Deck width	m	3.50
Deck height	m	4.00
Draft	m	15.00
Steel mass	ton	5,755
Ballast water mass	ton	15,749

The ship assumed to collide with the FOWT is a container carrier. In this study, the displacement of the colliding ship was assumed to be 30,000 tons and 50,000 tons. Figure 3 shows the shape of the container carrier. As a representative example, the properties of the 30,000-ton container carrier are summarized in Table 3.

Table 2. Properties of the superstructure.

Item	Unit	Value
Output power	MW	22.00
Rotor diameter	m	284.00
Tower height	m	149.39
Hub height	m	170.00
RNA mass	ton	1,216
Tower mass	ton	1,574

Figure 3. Shape of 30,000ton container carrier.

Table 3. Properties of the 30,000ton container carrier.

Item	Unit	Value
Length overall	m	175.00
Molded breadth	m	26.00
Molded depth	m	16.80
Draft	m	10.00
Displacement	ton	30,000

3.2 *Hydrodynamic properties*

Figure 4 shows the FOWT and container carrier models used in the frequency response analysis. The FOWT was modeled with approximately 11,000 diffraction panels, while the container carrier used about 3,300 diffraction panels. Non-diffraction areas were not modeled, but mass information was included. The frequency range and interval used for the FOWT's frequency response analysis were 0.1-3.0 rad/s and 0.06 rad/s, respectively. For the container carrier's frequency response analysis, the frequency range and interval were 0.1-1.8 rad/s and 0.06 rad/s, respectively. Figure 5 presents the results of the frequency response analysis. Figure 5(a) shows the heave direction RAO of both the FOWT and container carrier. Figures 5(b) and 5(c) display the surge direction added mass for the FOWT and container carrier, respectively.

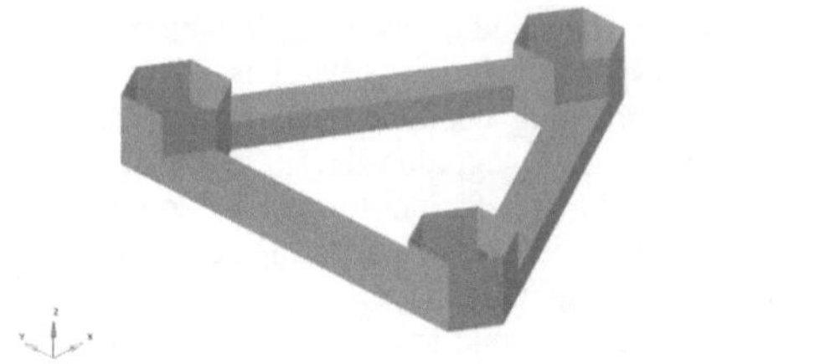

(a) FOWT

(b) 30,000ton container carrier

Figure 4. Frequency response analysis models.

3.3 *Finite element models*

Figure 6 shows the container carrier model used in the FOWT-ship collision analysis. The container carrier was assumed to be a rigid body and modeled using rigid elements. The reference node of the rigid body was positioned at the carrier's CoM. The mass information was represented at the reference node location using point mass elements and point inertia elements. The element size corresponds to the spacing of the stiffeners. The bulbous bow, where the collision with the FOWT occurs, was modeled with very small element sizes.

Figure 7 shows the FE model of the FOWT. The lower structure of the FOWT was modeled using shell elements. The tower was modeled with beam elements, and the RNA was represented using point mass elements and point inertia elements. The red and yellow areas in Figure 7 indicate the region where collision with the container carrier occurs, and elastic-plastic material properties were defined for this area. The remaining blue and green areas were defined as elastic bodies. Since the FE model does not explicitly represent ballast water, the density of the ballast area was adjusted to match the mass information (refer to the blue area in (Figure 7). The material properties applied to the FOWT are summarized in Table 4. The steel material is assumed to be EH36, and the material properties defined in the area where the collision with the ship occurs are shown in Figure 8.

4 COLLISION LIMIT CURVES

4.1 *FOWT-ship collision simulation for limit curve*

In this study, the FOWT-ship collision was assumed to be a head-on, perpendicular impact. The collision

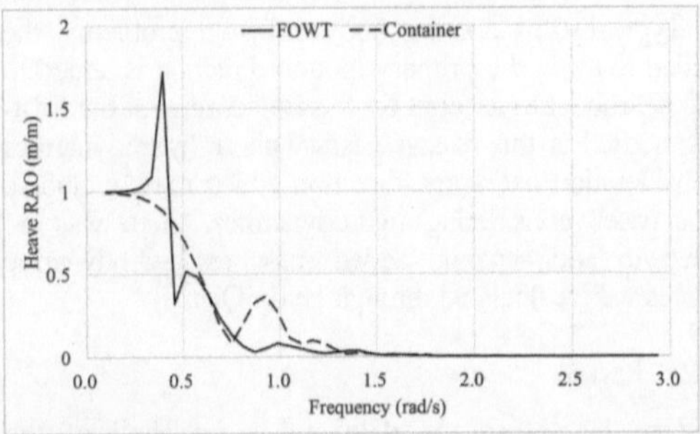

(a) Heave RAO

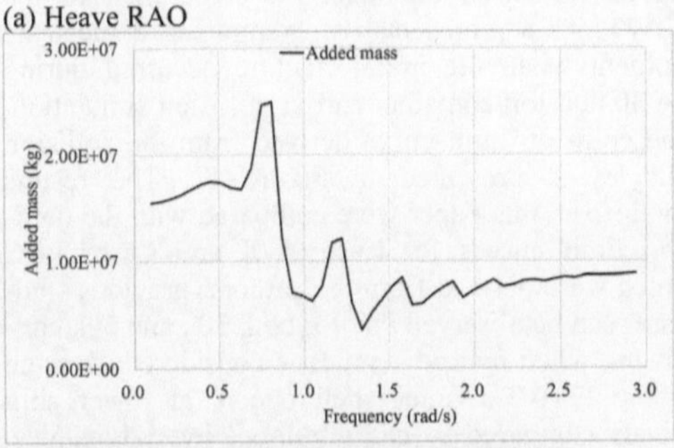

(b) Added mass in surge direction of the FOWT

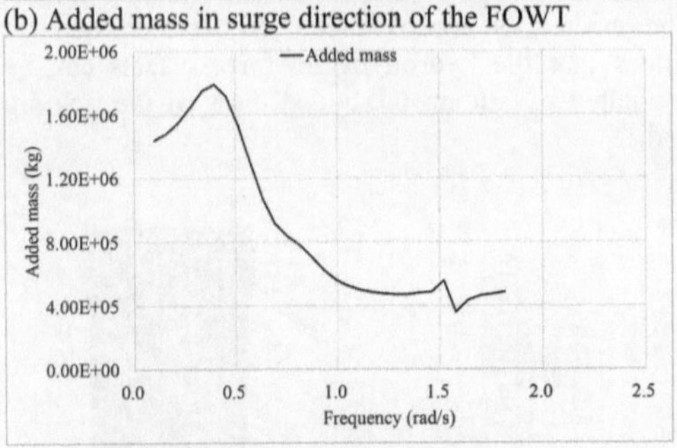

(c) Added mass in surge direction of the container carrier

Figure 5. Frequency response analysis results.

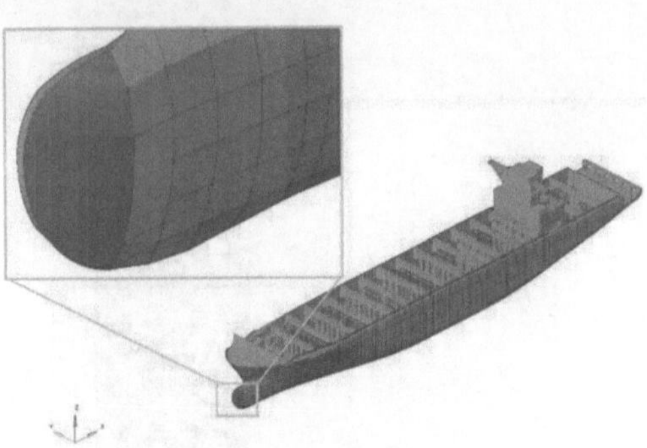

Figure 6. FE model of the container carrier.

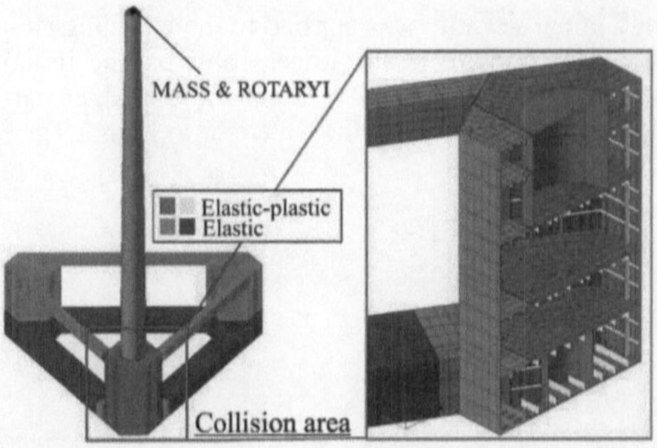

Figure 7. FE model of the FOWT.

Table 4. Material properties of FOWT.

Item	Young's Modulus (GPa)	Poisson's ratio	Density (ton/m^3)
Ballast	210	0.3	90.463
Other	210	0.3	26.089
Tower	200	0.265	7.85

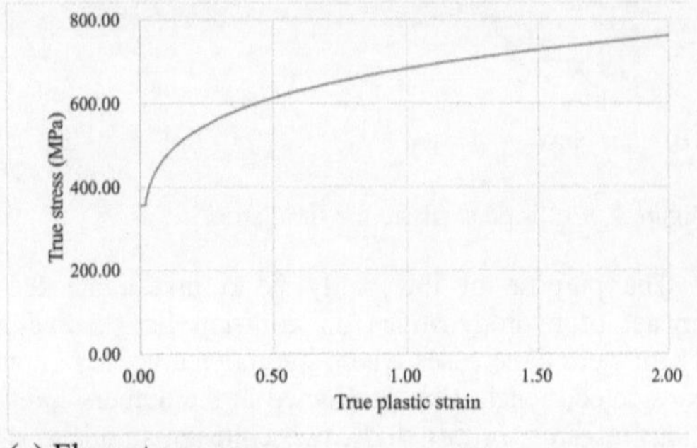

(a) Flow stress curve

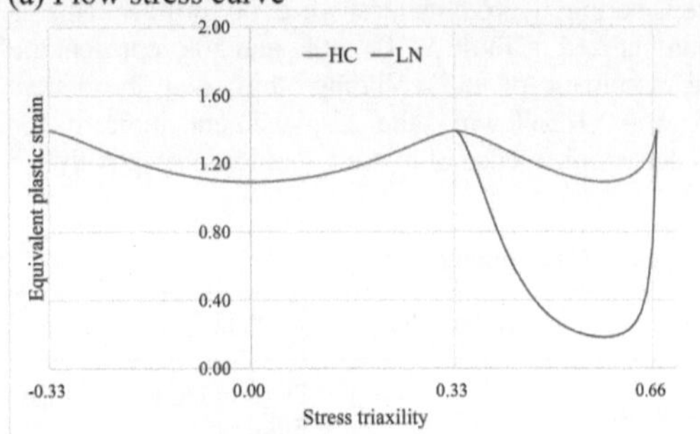

(b) Fracture strain curve

Figure 8. Material properties of collision area.

point could vary depending on the displacement of the colliding ship. To prevent this, the position of the bulbous bow's tip, where the initial contact occurs, was fixed. This position was at the center of the longitudinal stiffener and the middle of the transverse bulkhead of the tower column. Since the aim was to determine the minimum energy required for damaging the FOWT's outer shell, a sufficiently high initial velocity was applied to the colliding ship to ensure fracture in the tower column. The initial distance between the FOWT and the container carrier was approximately 100mm (refer to Figure 9).

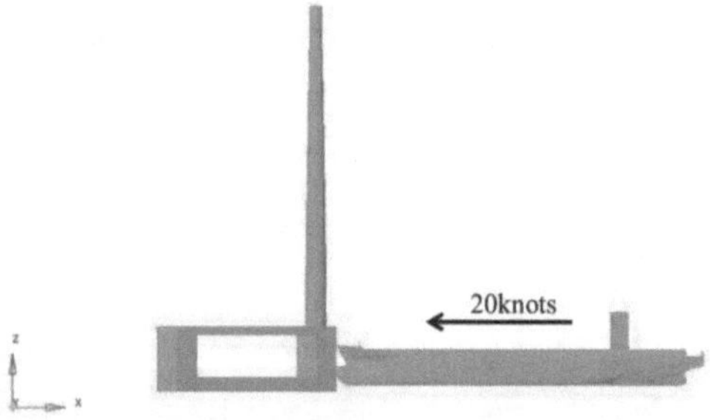

(a) Side view

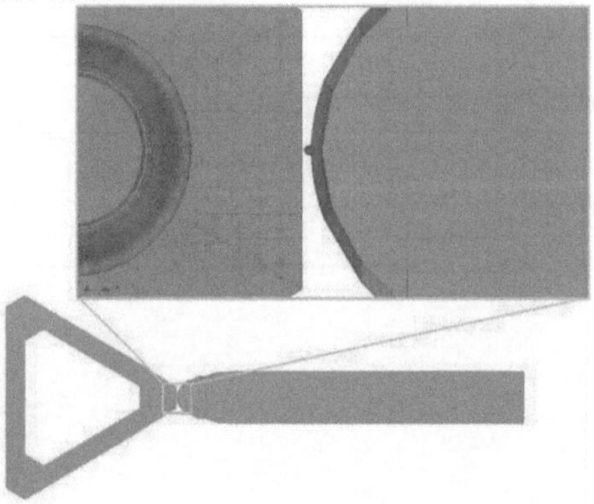

(b) Plan view

Figure 9. Collision scenario for simulation.

The purpose of this study is to investigate the impact of hydrodynamics on collision limit curves. There were three cases where hydrodynamics was not considered, which were conducted in the authors' previous research. In this study, two collision simulations considering hydrodynamics were performed. This is summarized in Table 5. 10k, 30k, and 50k represent the displacement of the colliding ship, and the letters X and O following the displacement indicate the absence or presence of hydrodynamics, respectively.

Table 5. Case summary.

Case	Hydrodynamics	Added mass by
Case10k-X	X	Density increase in FEA model
Case30k-X	X	Density increase in FEA model
Case30k-O	O	HydroQus considering $M'(\infty)$ in Equation (3)
Case50k-X	X	Density increase in FEA model
Case50k-O	O	HydroQus considering $M'(\infty)$ in Equation (3)

Typically, in floating body collision problems, the added mass in the primary motion direction is added to the FE model to account for the added mass of the floating body. For this reason, cases without hydrodynamics consideration had surge direction added mass included. For cases considering hydrodynamics, there was no need to add separate added mass as hydrodynamic forces were calculated through HydroQus.

4.2 *Results*

Figure 10 shows the damaged outer shell of the FOWT. As a representative example, it depicts the moment when the initial fracture occurred during the 30,000-ton container carrier collision simulation. The collision limit curve derived from the collision analysis is presented in Figure 11. The results obtained in this paper were compared with the collision limit curves for Case30k-X and Case50k-X, which were derived from the authors' previous studies. It can be observed that for both 30k and 50k containers, when hydrodynamics is considered, damage to the FOWT's outer shell occurs at lower ship speeds compared to cases without hydrodynamics. The error was approximately 5%, which demonstrates that the hydrodynamic force effects can be simulated by adding the added mass of the floating body.

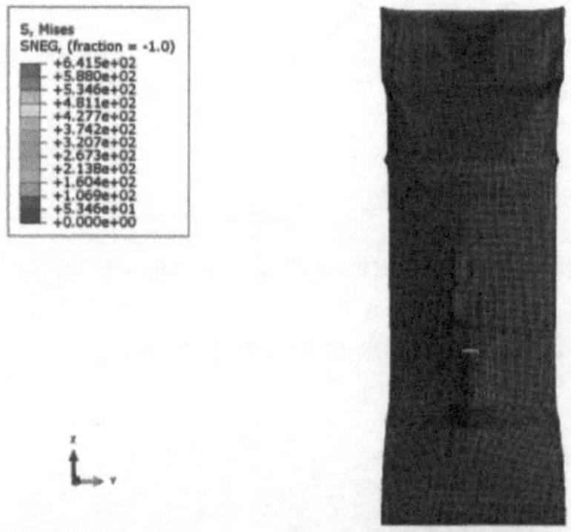

(a) Von mises stress

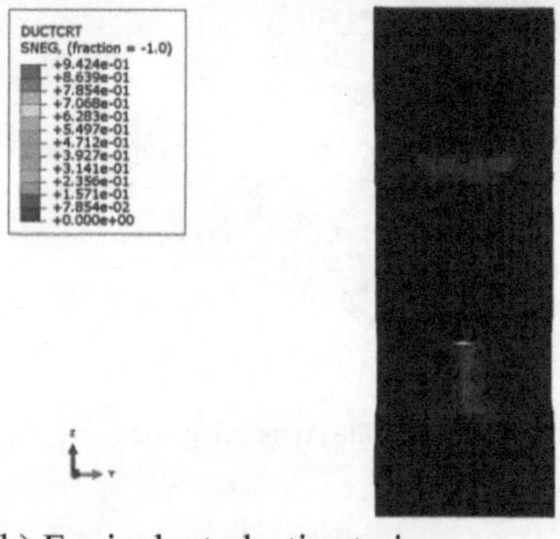

(b) Equivalent plastic strain

Figure 10. Results.

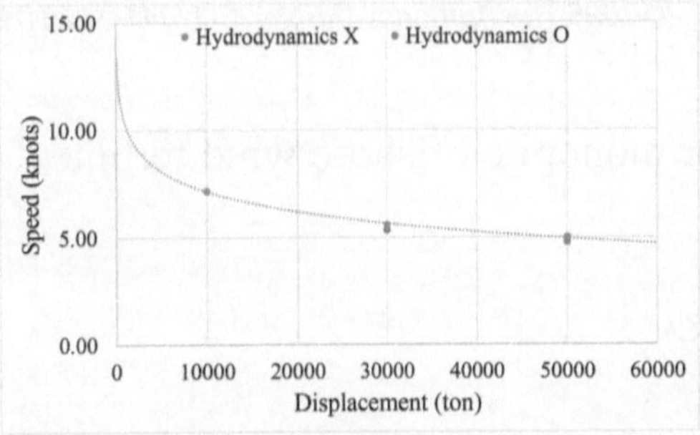

Figure 11. Collision limit curve.

5 CONCLUSIONS

To derive the collision limit curve, a 22MW class FOWT and container carriers of two different displacements were prepared. HydroQus was used to develop collision limit curves that consider hydrodynamics. The collision limit curves obtained from previous studies were compared with those considering hydrodynamics. When hydrodynamics was taken into account, damage to the FOWT's outer shell occurred at lower ship speeds, although the difference was not significant. This implies that hydrodynamic effects can be simulated by directly adding the added mass to the FE model.

The authors anticipate that the relative velocity between the FOWT and the container carrier due to wave excitation force and mooring force will be a significant factor in determining the extent of damage. Therefore, they are currently preparing collision simulations for container carriers with displacements other than 30,000 tons and 50,000 tons, simulations to analyze the effect of mooring forces, and simulations considering wave excitation.

ACKNOWLEDGEMENTS

This research was supported by the Korea Energy Technology Evaluation and Planning (KETEP) funded by the Ministry of Trade, Industry and Energy of Korea (RS-2021-KP002572, RS-2023-00238996 & RS-2024-00450063) and the Korea Institute of Marine Science & Technology Promotion (KIMST), funded by the Ministry of Oceans and Fisheries (RS-2022-KS221571).

REFERENCES

Bela, A., Le Sourne, H., Buldgen, L., Rigo, P., 2017. Ship collision analysis on offshore wind turbine monopile foundations. Marine Structures 51, 220–241. https://doi.org/10.1016/j.marstruc.2016.10.009

Cummins, W.E., United States. Navy, 1962. The Impulse Response Functions and Ship Motions, David Taylor Model Basin. United States Department of the Navy, David Taylor Model Basin.

Dai, L., Ehlers, S., Rausand, M., Utne, I.B., 2013. Risk of collision between service vessels and offshore wind turbines. Reliability Engineering & System Safety 109, 18–31. https://doi.org/10.1016/j.ress.2012.07.008

Echeverry, S., Márquez, L., Rigo, Ph., Sourne, H.L., 2019. Numerical crashworthiness analysis of a spar floating offshore wind turbine impacted by a ship, in: Guedes Soares, C. (Ed.), Developments in the Collision and Grounding of Ships and Offshore Structures. CRC Press, pp. 85–95. https://doi.org/10.1201/9781003002420-11.

Jia, H., Qin, S., Wang, R., Xue, Y., Fu, D., Wang, A., 2020. Ship collision impact on the structural load of an offshore wind turbine. Global Energy Interconnection 3, 43–50. https://doi.org/10.1016/j.gloei.2020.03.009.

Liu, X., Jiang, D., Liufu, K., Fu, J., Liu, Q., Li, Q., 2022. Numerical investigation into impact responses of an offshore wind turbine jacket foundation subjected to ship collision. Ocean Engineering 248, 110825. https://doi.org/10.1016/j.oceaneng.2022.110825.

LSTC, 2023. LSDYNA Manuals.

Márquez, L., Le Sourne, H., Rigo, P., 2022. Mechanical model for the analysis of ship collisions against reinforced concrete floaters of offshore wind turbines. Ocean Engineering 261, 111987. https://doi.org/10.1016/j.oceaneng.2022.111987.

Moulas, D., Shafiee, M., Mehmanparast, A., 2017. Damage analysis of ship collisions with offshore wind turbine foundations. Ocean Engineering 143, 149–162. https://doi.org/10.1016/j.oceaneng.2017.04.050.

Niklas, K., Bera, A., Garbatov, Y., 2023. Impact of steel grade on a ship colliding with an offshore wind turbine monopile supporting structure. Ocean Engineering 287, 115899. https://doi.org/10.1016/j.oceaneng.2023.115899.

Pack, K., Mohr, D., 2017. Combined necking & fracture model to predict ductile failure with shell finite elements. Engineering Fracture Mechanics 182, 32–51. https://doi.org/10.1016/j.engfracmech.2017.06.025.

Storheim, M., Amdahl, J., 2014. Design of offshore structures against accidental ship collisions. Marine Structures 37, 135–172. https://doi.org/10.1016/j.marstruc.2014.03.002

Sun, J., Fang, H., 2023. The analysis of crashworthiness and dissipation mechanism of novel floating composite honeycomb structure against ship-OWT collision. Ocean Engineering 287, 115819. https://doi.org/10.1016/j.oceaneng.2023.115819.

Yoon, D.H., Choung, J., 2023. Collision Simulation of a Floating Offshore Wind Turbine Considering Ductile Fracture and Hydrodynamics Using Hydrodynamic Plug-in HydroQus. J. Ocean Eng. Technol. 37, 111–121. https://doi.org/10.26748/KSOE.2023.004.

Yoon, D.H., Jeong, S.-Y., Choung, J., 2023. Collision simulations between an icebreaker and an iceberg considering ship hydrodynamics. Ocean Engineering 279, 114333. https://doi.org/10.1016/j.oceaneng.2023.114333.

Yu, Z., Amdahl, J., Rypestøl, M., Cheng, Z., 2022. Numerical modelling and dynamic response analysis of a 10 MW semi-submersible floating offshore wind turbine subjected to ship collision loads. Renewable Energy 184, 677–699. https://doi.org/10.1016/j.renene.2021.12.002.

Zahle, F., Barlas, A., Loenbaek, K., Bortolotti, P., Zalkind, D., Wang, L., Labuschagne, C., Sethuraman, L., Barter, G., 2024. Definition of the IEA Wind 22-Megawatt Offshore Reference Wind Turbine. Technical University of Denmark. https://doi.org/10.11581/DTU.00000317.

Trends in Collision and Grounding of Ships and Offshore Structures – Zhu & Guedes Soares (Eds)
© 2026 The Author(s), ISBN 978-1-041-16361-9

Auxetic-inspired anti-collision fender for monopile offshore wind turbines subjected to ship collision

T. Zheng
School of Civil Engineering, Tianjin University, Tianjin, P.R. China

N.Z. Chen
School of Civil Engineering, Tianjin University, Tianjin, P.R. China
State Key Laboratory of Hydraulic Engineering Simulation and Safety, Tianjin University, Tianjin, P.R. China

ABSTRACT: A novel anti-collision fender designed for monopile offshore wind turbines (OWTs) is proposed in this paper. The auxetic-inspired structure characterized by a negative Poisson's ratio (NPR) is employed for the proposed fender. Finite element (FE) models of the auxetic-inspired anti-collision fender, the monopile OWT and the ship are established. A series of dynamic analyses are conducted to calculate the structural responses of the OWT which is equipped with the proposed fender during collision with a maintenance ship. The improvements in anti-collision performance afforded by the auxetic-inspired anti-collision fender are discussed. The results provide reasonable insights for the novel design of anti-collision fenders that are applied to OWTs.

1 INTRODUCTION

The rapid expansion of offshore wind energy underscores the global commitment to sustainable and renewable power sources. Offshore wind turbines (OWTs), particularly those mounted on monopile foundations, harness stronger and more consistent winds compared to their onshore counterparts, making them increasingly vital in meeting energy demands. However, this growth introduces significant challenges, foremost among which is the risk of ship collisions. Such collisions can cause catastrophic damage to both turbine structures and vessels, leading to substantial financial losses, environmental contamination, and potential harm to human life.

The vulnerability of OWTs to ship collisions arises from their placement in maritime traffic routes and the increasing density of commercial and maintenance vessels in these areas. Although historical data are limited, they suggest that ship-turbine collisions, while rare, do occur and can have severe consequences. For instance, in 2003, a floating dock broke loose in Denmark, threatening nearby wind turbines, highlighting real-world risks. Recent UK accident reports indicate a rising trend in offshore wind farm incidents, including collisions, emphasizing the need for robust protection measures (Moulas et al., 2017). Consequently, developing effective collision protection systems is critical to ensuring the safety, reliability, and longevity of these essential infrastructure assets.

Current approaches to mitigating collision risks primarily involve designing and implementing fenders. Traditional fenders are typically constructed from materials such as rubber, steel, or composites, designed to absorb and dissipate the kinetic energy of impacting ships. For example, Yue et al. (2021) proposed a fractal-structured fender for OWTs, demonstrating enhanced energy absorption through its complex geometry. Zheng & Chen (2023) investigated the impact of different forms of fender on the structural safety of OWT during collision. However, existing fender designs face several limitations, including challenges in predicting and accommodating diverse impact scenarios, ranging from low-speed maintenance vessel contacts to high-energy collisions with larger ships. Moreover, reliance on simplified damage assessment models can result in either overconservative or insufficient protection, leading to unnecessary costs or inadequate safety (Moulas et al., 2017). Additionally, while various protection measures, such as crashworthy devices and anti-collision systems, have been explored, innovative materials with unique mechanical properties, such as auxetic-inspired materials, remain largely uninvestigated in this context (Ma et al., 2024).

Auxetic-inspired materials, characterized by a negative Poisson's ratio (NPR), exhibit counterintuitive deformation behavior: they expand laterally when stretched and contract when compressed. This property results in superior energy absorption,

DOI: 10.1201/9781003684404-31

indentation resistance, and fracture toughness, making them particularly suitable for impact protection applications (Luo et al., 2021). Jiang & Chen (2023, 2024) proposed auxetic-inspired structures with enhanced crashworthiness, which demonstrates superior energy absorption over traditional designs through numerical simulations. Despite their potential, auxetic-inspired materials have not been applied to the design of fenders for OWTs, reflecting a significant gap in current research and application. This lack of exploration presents an opportunity to leverage the unique properties of auxetic structures to enhance collision protection for monopile OWTs.

In this paper, an auxetic-inspired anti-collision fender for monopile OWTs subjected to ship collisions is proposed. The exceptional mechanical properties of auxetic structures to enhance energy dissipation and structural integrity during collision are capitalized for the fender. Finite element models of the fender, monopile foundation, and a representative ship are developed to simulate the dynamic response under collision scenarios. Through a dynamic analysis, the protective effects of the proposed fender are evaluated and compared with traditional designs, demonstrating its potential to mitigate collision damage more effectively.

2 METHODS

2.1 *Equation of motion in collision*

Dynamic analysis is essential for evaluating structural responses under time-varying loads, such as ship collisions with offshore wind turbines. The finite element method (FEM) is a numerical technique widely applied to simulate these dynamic events. Complex geometries, nonlinear material behaviors, and large deformations can be handled effectively. The governing equations in dynamic finite element analysis are derived from Newton's second law, expressed as the equation of motion:

$$\boldsymbol{M}\ddot{\boldsymbol{u}} + \boldsymbol{C}\dot{\boldsymbol{u}} + \boldsymbol{K}\boldsymbol{u} = \boldsymbol{F} \tag{1}$$

where $\boldsymbol{M}$ is the mass matrix, $\ddot{\boldsymbol{u}}$ the acceleration vector, $\boldsymbol{C}$ the damping matrix, $\dot{\boldsymbol{u}}$ the velocity vector, $\boldsymbol{K}$ the stiffness matrix, $\boldsymbol{u}$ the displacement vector and $\boldsymbol{F}$ the time-varying external force vector.

2.2 *Hourglass energy control*

In dynamic finite element simulations, underintegrated elements, such as reduced integration hexahedral elements, may cause hourglassing. Hourglassing is a numerical instability. Elements deform in zero-energy modes without producing strain, leading to unphysical behavior and inaccurate results. Underintegrated elements are used for computational efficiency but may fail to capture all deformation modes.

Stabilization techniques are employed to control hourglassing. Stiffness-based hourglass control involves adding an artificial stiffness matrix to the element stiffness matrix $\boldsymbol{K}_{\text{total}}$ as:

$$\boldsymbol{K}_{\text{total}} = \boldsymbol{K} + \alpha \boldsymbol{K}_{\text{hg}} \tag{2}$$

where α is a scaling factor, and $\boldsymbol{K}_{\text{hg}}$ is the hourglass stiffness matrix (Belytschko et al., 2014). Viscosity-based hourglass control introduces damping terms to dissipate energy associated with hourglass modes (Engelman et al, 1981). These techniques ensure hourglass modes are suppressed with minimal impact on the overall structural response.

In this paper, hourglass control was implemented to ensure simulation stability and accuracy. Solid elements used for the fender and monopile foundation were particularly addressed. Ship collisions involve complex contact and deformation scenarios. Effective hourglass control is critical for reliable results.

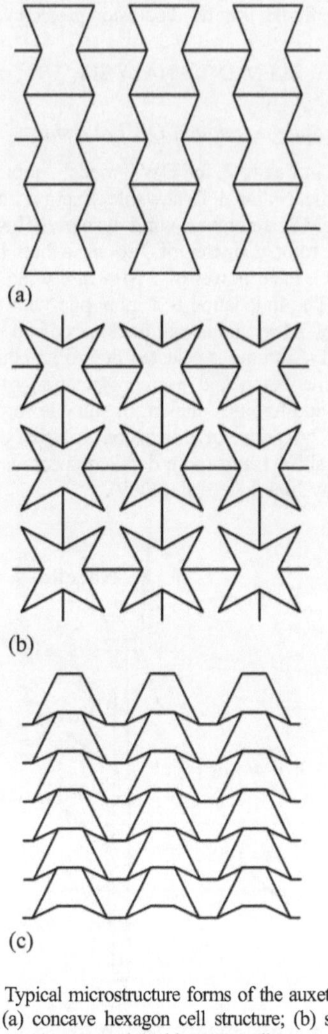

Figure 1. Typical microstructure forms of the auxetic-inspired materials: (a) concave hexagon cell structure; (b) star-shaped cell structure; (c) tetragonal double-arrow cell structure.

2.3 *Materials with negative poisson's ratio (NPR)*

Auxetic-inspired materials are defined by a negative Poisson's ratio ($\nu < 0$). Lateral expansion occurs when stretched, and contraction occurs when compressed. This property of auxetic-inspired materials is caused by their special microstructure. The microstructure forms of several typical auxetic-inspired materials are shown in Figure 1. This behavior provides superior energy absorption, indentation resistance, and fracture toughness, making them ideal for impact protection (Bollaert et al., 1993).

During collisions, compressive loads cause the fender to contract laterally. The contact area is increased, and the load is distributed more uniformly. Energy dissipation is enhanced, and the force transmitted to the monopile foundation is reduced. (Prawoto, 2012). Therefore, Auxetic properties enable larger deformations and greater energy absorption compared to conventional materials, offering improved collision protection.

3 FINITE ELEMENT ANALYSIS

3.1 *Monopile foundation OWT and ship*

As shown in Figure 2, the OWT model adopted in this paper is the National Renewable Energy Laboratory (NREL) 5MW reference wind turbine. This turbine features a rotor diameter of 126 m, a hub height of 90 m, and a rated power of 5 MW at a wind speed of 11.4 m/s. The foundation is a monopile with a bottom diameter of 7.1 m and a wall thickness of 80 mm. The NREL 5MW turbine is selected due to its comprehensive documentation and its representation of modern offshore wind turbines, making it suitable for collision analysis. The principal dimensions of the OWT are listed in Tables 1 and more detailed specifications are provided by Jonkman et al. (2009).

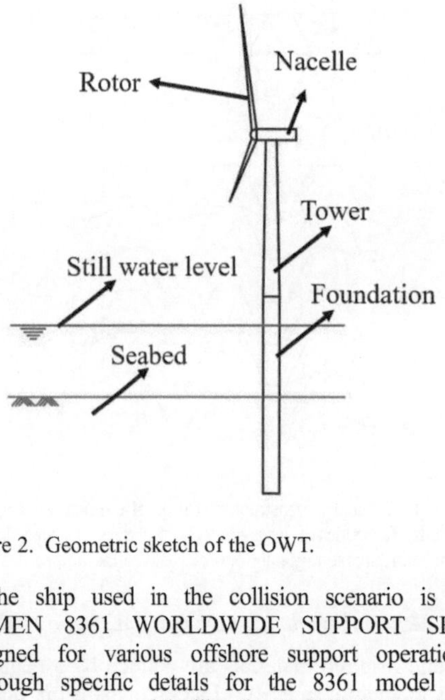

Figure 2. Geometric sketch of the OWT.

The ship used in the collision scenario is the DAMEN 8361 WORLDWIDE SUPPORT SHIP, designed for various offshore support operations. Although specific details for the 8361 model are limited, similar models like the DAMEN 8316 have dimensions of approximately 83 m in length and 16 m in beam, with a capacity for 88 personnel. These ships are equipped with generators, cranes, and other equipment for support and training tasks. This ship type is chosen for its relevance to maintenance activities in offshore wind farms, where such vessels are commonly deployed. Principal dimensions are listed in Tables 2.

Table 1. Principal dimensions of the OWT.

Parameter	Value
Operating water depth (m)	20
Nacelle height (m)	90
Depth to foundation base below SWL* (m)	98
Diameter of tower above taper (m/s)	5
Diameter of foundation (m/s)	7.1
Thickness of foundation (mm)	80
Weight of the rotor (kg)	110,000
Weight of the nacelle (kg)	240,000

* SWL=Still water level.

Table 2. Principal dimensions of the ship.

Parameter	Value
Length overall (m)	83
Beam overall (m)	16
Depth at sides (m)	7.2
Maximum draught (m)	4.25
Displacement (t)	3000

3.2 *Anti-collision fender design*

The anti-collision fender proposed in this paper consists of a sandwich structure with steel plates enclosing a honeycomb-core layer inspired by concave hexagon auxetic-inspired materials. Concave hexagon structures are widely studied for their superior energy absorption and impact resistance. Research indicates that these structures distribute loads uniformly during collisions, reducing peak stresses and enhancing toughness. For instance, concave hexagon honeycomb structures demonstrate higher specific energy absorption under in-plane impact compared to traditional honeycombs (Luo et al., 2021).

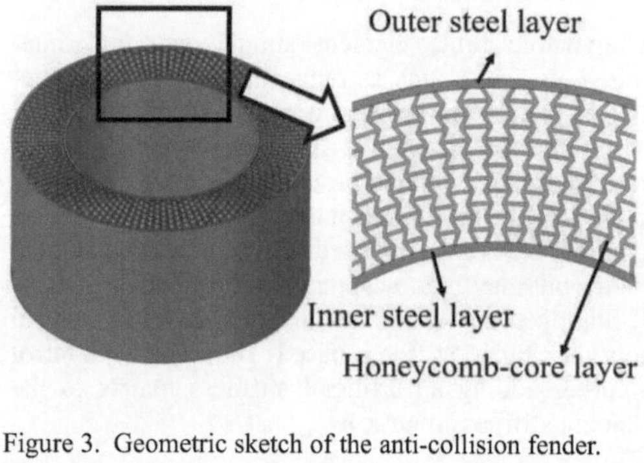

Figure 3. Geometric sketch of the anti-collision fender.

The honeycomb-core layer, integrated between steel plates, forms a fender that protects the monopile foundation from ship collisions. This design leverages auxetic-inspired properties to improve protective performance, offering a novel approach to offshore wind turbine safety. The geometric sketch of the fender is illustrated in Figure 3 and the principal dimensions of the fender is listed in Table 3.

Table 3. Principal dimensions of the anti-collision fender.

Parameter	Value
Height of the fender(m)	6.0
Thickness of steel layer (m)	0.1
Thickness of honeycomb-core layer (m)	1.8
Volume of the honeycomb-core layer (m^3)	100
Thickness of the concave hexagon (mm)	40
Central angle corresponding to a concave hexagonal cell (°)	2
Height of the fenders above the SWL (m)	1.0

3.3 *Finite element (FE) model*

The finite element model developed includes the wind turbine nacelle, monopile foundation, anti-collision fender, and ship. The wind turbine rotor is simplified as a concentrated mass at the nacelle to account for inertial effects. The nacelle and fender steel plates are made of S355 steel, modeled with the Hollomon strain hardening model, where true stress σ_t and true strain ε_t are related by (Liessem et al., 2007):

$$\sigma_t = K_H \varepsilon_t^n \tag{3}$$

where K_H is the strength coefficient and n the Hollomon hardening exponent.

The honeycomb-core layer of the fender is made of rubber. The strain energy W of the rubber are determined by incompressible Mooney-Rivlin equation (Mooney, 1940 and Rivlin, 1948):

$$W = C_1(I_1 - 3) + C_2(I_2 - 3) \tag{4}$$

Where C_i is the material parameter, I_i the deformation tensor. The properties of S355 steel and rubber are shown in Table 4.

Table 4. Material properties of s355 steels and rubber.

Properties	S355 steel	rubber
Density (kg/m^3)	7850	1350
Young's module (GPa)	206	0.10
Poisson's ratio	0.28	0.49
Yeild stress σ_y (MPa)	355	/
K_H (MPa)	700	/
n	0.2	/
C_{10} (MPa)	5	0.03
C_{01} (MPa)	0.34	0.13

This paper focus on the structural responses of the foundation. Therefore, the ship is modeled as a rigid body, with only the bow section included to reduce computational complexity. Bonded interfaces are defined between fender layers and between the fender and foundation to ensure cohesive behavior. To account for the interaction between the ship structure and water during the motion of ship that priors to collision, the added mass method is employed (Petersen, 1982). The added mass is set at 5% of the ship mass. Fixed constraints are applied at the interface between the OWT and the seabed. An initial gap of 0.5 m is set between the ship and fender to prevent initial contact. Mesh refinement is applied in the collision zone, with a minimum element size of 0.2 m, to capture deformation and stress accurately. The meshed FE model of collision area with anti-collision fender are shown in Figure 4.

Figure 4. Meshed FE model of the OWT, fender and the ship.

4 RESULTS AND DISCUSSION

To evaluate the performance of the proposed auxetic-inspired fender, its effectiveness is compared with two other configurations: no fender installation and a traditional steel fender with bulkheads, which is shown in Figure 5. The steel fender is fabricated from S355 steel and it maintains identical overall dimensions to the auxetic-inspired fender. The thicknesses of both shell plates and bulkhead specified as 0.1 m for the steel fender. All three configurations are simulated under ship impact conditions at 7 m/s.

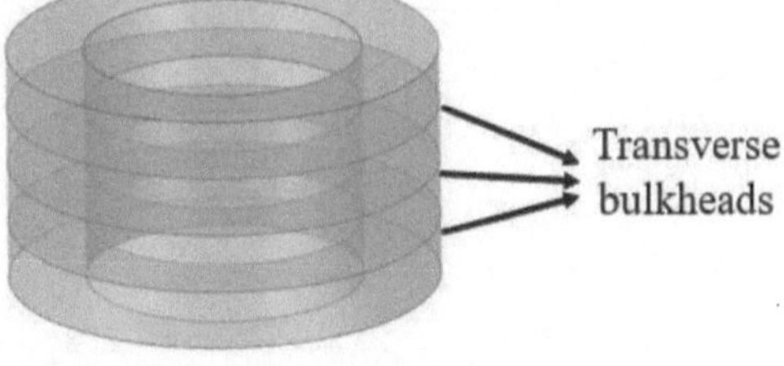

Figure 5. Steel fender with transverse bulkheads.

Figure 6 shows the time-history curves of collision forces for the three cases. The maximum impact forces recorded are 41.50 MN (no fender), 30.32

MN (steel fender), and 15.47 MN (auxetic-inspired fender). This indicates a 62.7% peak force reduction compared to the no-fender scenario and 49.0% reduction relative to the steel fender. The auxetic-inspired fender exhibits faster attainment of peak force but demonstrates more gradual force decay, resulting in the longest collision duration (1.94 s) compared to 1.37 s (steel fender) and 1.00 s (no fender). The extended collision duration suggests enhanced energy dissipation capacity through prolonged impact energy absorption, thereby reducing shock loads transmitted to the monopile.

The superior performance is attributed to the unique deformation characteristics of negative Poisson's ratio materials. These materials exhibit lateral expansion under tension and lateral contraction under compression. During ship impact, the fender's compression-induced lateral contraction facilitates greater internal deformation and energy absorption while minimizing force transmission to the foundation. This behavior aligns with existing literature demonstrating the exceptional energy absorption capacity of auxetic structures under impact loading through uniform load distribution and large deformation tolerance (Najafi et al, 2021; Gao & Liao, 2021).

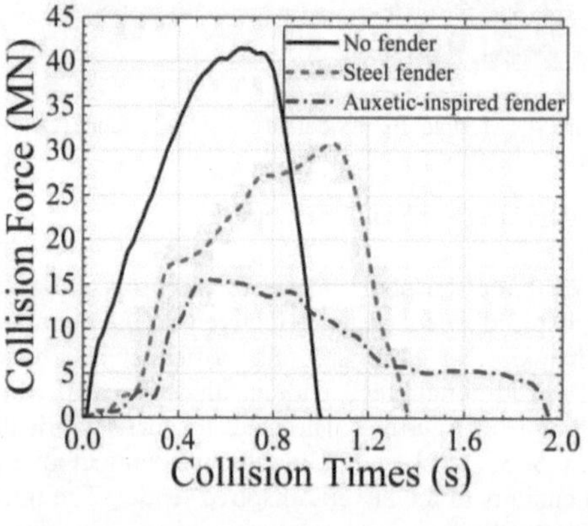

Figure 6. Collision force on the monopile foundation with different fender.

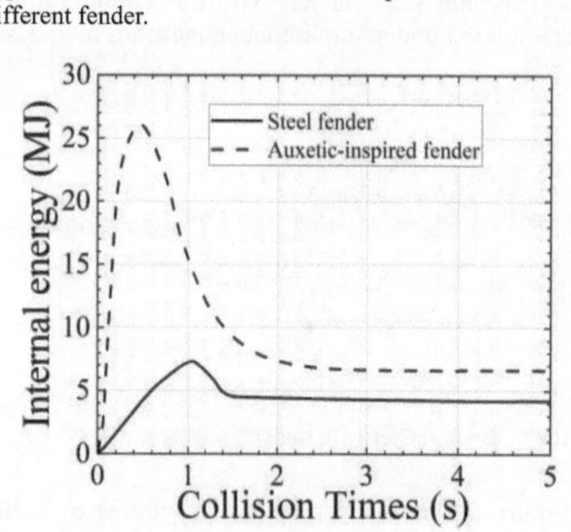

Figure 7. Internal energy of different fender.

The variation in internal energy of the auxetic-inspired fender and the steel fender during the collision is shown in Figure 7. Figure 7 shows that the internal energy of the auxetic-inspired fender is increased to a peak of 26.07 MJ rapidly at 0.48 s during the collision, which is significantly higher than the peak of 7.33 MJ observed for the steel fender. A longer duration is required for the internal energy of the auxetic-inspired fender to be decreased from the peak to a stable value compared to that of the steel fender. This indicates that the auxetic-inspired material effectively absorbs the energy generated by the collision. Therefore, the collision force transmitted to the OWT is substantially reduced.

For further discussion of the protective effect of the anti-collision fender on the monopile foundation, the distribution of von Mises stress on the monopile foundation without fender and with different fender is compared, which is shown in Figure 8.

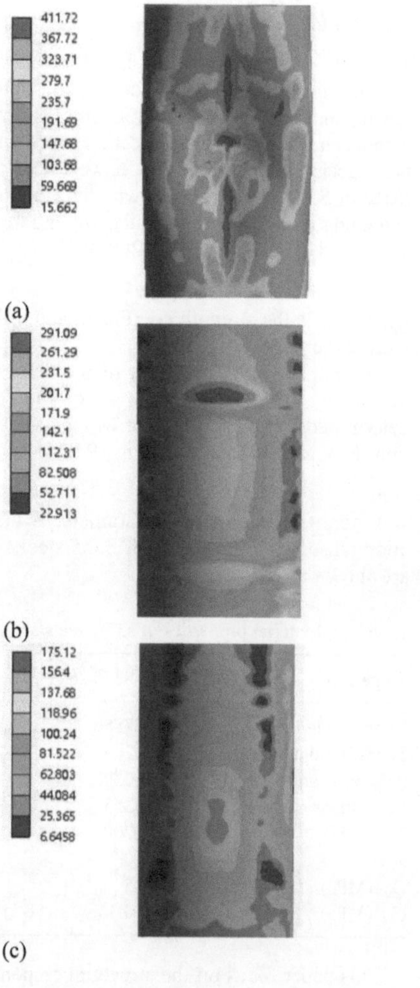

Figure 8. Distribution of mises stress on the monopile: (a) no fender; (b) steel fender; (c) auxetic-inspired fender.

Figure 8 shows that without fender protection, peak stress reaches 411.72 MPa, exceeding S355 steel's yield strength (355 MPa) and indicating plastic deformation. Stress concentrations appear as longitudinal bands at the impact point and adjacent regions. The steel fender reduces maximum stress to 291.09 MPa, but significant stress concentrations persist near fender edges and impact areas, likely caused by post-impact deformation. In contrast, the auxetic-inspired fender further decreases maximum stress to 175.12 MPa, with stress concentrations localized near the impact point and substantially reduced stress levels elsewhere. This demonstrates effective energy absorption and dissipation through the auxetic structure's lateral contraction mechanism, which concentrates deformation near the impact zone to limit stress propagation. The absence of secondary stress peaks in the auxetic system further demonstrates its structural integrity preservation during collision events. These findings corroborate previous studies showing enhanced impact resistance in auxetic structures through modified stress distribution patterns.

In a word, the proposed auxetic-inspired fender demonstrates outstanding performance in reducing peak collision forces and monopile foundation stresses. Its significantly lowered peak collision force and extended collision duration indicate superior energy absorption capacity, effectively reducing instantaneous impact loads. The reduced maximum stress and localized stress distribution further confirm its ability to protect the monopile foundation from plastic deformation and extensive damage. This advantage originates from the unique mechanical behavior of negative Poisson's ratio materials, which exhibit lateral contraction under compression to enhance energy absorption and minimize stress transmission.

In comparison with both the no-fender scenario and steel fender configuration, the auxetic-inspired fender achieves superior outcomes across all key metrics. The absence of fender protection results in excessive stresses and plastic deformation that may compromise structural integrity. While the steel fender shows moderate improvements, stress concentrations remain evident, potentially causing localized damage. The auxetic-inspired fender overcomes these limitations through its innovative design, providing an efficient and reliable collision protection solution for offshore wind turbine foundations.

5 CONCLUSIONS

In this paper, an auxetic-inspired anti-collision fender for monopile offshore wind turbines (OWTs) was proposed. The unique deformation behavior of negative Poisson's ratio (NPR) materials is considered to promote energy absorption and stress distribution of the monopile foundation during ship collision. The performance of the proposed fender under ship collision was evaluated through finite element analysis. Comparative analyses were conducted across three configurations: no fender, steel fender with bulkheads, and the proposed auxetic-inspired fender. The results show that:

- The peak collision forces and collision duration during the collision are significantly reduced by the proposed auxetic-inspired fender.
- During the collision, stress distribution is effectively confined near the collision zone by the proposed auxetic-inspired fender, which reduces widespread structural damage to the monopile foundation.

Existing research shows that the crashworthy performance of NPR structures can be influenced by the geometric parameters. As a preliminary exploration of the enhancement effect of NPR materials on the crashworthy performance of offshore wind turbines, the impact of the geometric parameters on the performance of the anti-collision fender are not considered in this paper. Systematic parametric analysis is planned for subsequent study to investigate the design considerations of the auxetic-inspired fender.

ACKNOWLEDGEMENTS

This research was supported by the National Natural Science Foundation of China (Grant No. 52311530091). The financial supports are greatly appreciated.

REFERENCES

Belytschko, T., Liu, W. K., Moran, B. & Elkhodary, K. 2014. Nonlinear finite elements for continua and structures. *John wiley & sons.*

Bollaert, V., De Schryver, F. C., Tackx, P., Persoons, A., Nusselder, J. J. H. & Put, J. 1993. A new fluorescence probe for network formation with interesting nonlinear optical properties. *Advanced materials*, *5*(4), 268–270.

Engelman, M. S., Strang, G. & Bathe, K. J. 1981. The application of quasi-Newton methods in fluid mechanics. *International Journal for Numerical Methods in Engineering*, 17(5), 707–718.

Gao, Q. & Liao, W. H. 2021. Energy absorption of thin walled tube filled with gradient auxetic structures-theory and simulation. *International Journal of Mechanical Sciences*, *201*, 106475.

Jiang, C., Lin, L. & Chen, N. Z. 2024. A study on auxetic-inspired side structure for enhanced crashworthiness. *Marine Structures*, *93*, 103545.

Jiang, C. & Chen, N. Z. 2023. Auxetic-inspired side structures for enhanced collision resistance. In *Advances in the Collision and Grounding of Ships and Offshore Structures*, Nantes, France, 313–320.

Jonkman, J., Butterfield, S., Musial, W. & Scott, G. 2009. Definition of a 5-MW reference wind turbine for offshore system development (No. NREL/TP-500-38060). *National Renewable Energy Lab, Golden, CO.*

Liessem, A., Knauf, G. & Zimmermann, S. 2007. Strain based design What the Contribution of a Pipe Manufacturer can be. In *the 17th International Offshore and*

Polar Engineering Conference, Lisbon, Portugal, ISOPE-1-07-500, 3156–3163.
Luo, C., Han, C. Z., Zhang, X. Y., Zhang, X. G., Ren, X. & Xie, Y. M. 2021. Design, manufacturing and applications of auxetic tubular structures: A review. *Thin-Walled Structures*, 163, 107682.
Ma, L., Ding, J., Zhang, X., Wang, W., Zhao, X., Sun, C. & Yang, Y. 2024. Analysis of dynamic response of offshore wind turbines subjected to ship impacts and the corresponding protection measures: a review. *Frontiers in Energy Research*, 12, 1497210.
Mooney, M. 1940. A theory of large elastic deformation. *Journal of Applied Physics*, 11(9), 582–592.
Moulas, D., Shafiee, M. & Mehmanparast, A. J. O. E. 2017. Damage analysis of ship collisions with offshore wind turbine foundations. *Ocean Engineering*, 143, 149–162.
Najafi, M., Ahmadi, H. & Liaghat, G. 2021. Experimental investigation on energy absorption of auxetic structures. *Materials today: proceedings*, *34*, 350–355.
Petersen, M. J. 1982. Dynamics of ship collisions. *Ocean Engineering*, *9*(4), 295–329.
Prawoto, Y. 2012. Seeing auxetic materials from the mechanics point of view: a structural review on the negative Poisson's ratio. *Computational Materials Science*, 58, 140–153.
Rivlin, R. S. 1948. Large elastic deformations of isotropic materials. IV. Further developments of the general theory. *Philosophical Transactions of the Royal Society of London. Series A, Mathematical and Physical Sciences*, 241(835), 379–397.
Yue, X., Han, Z., Li, C. & Zhao, X. 2021. The study on structure design of fender of offshore wind turbine based on fractal feature during collision with ship. *Ocean Engineering*, 236, 109100.
Zheng, T. & Chen, N. Z. 2023. Impact of forms of anti-collision fender on structural safety of monopile foundation subjected to ship collision. In *Advances in the Collision and Grounding of Ships and Offshore Structures*, Nantes, France, 469–474.
Zienkiewicz, O. C., Taylor, R. L. & Taylor, R. L. 2000. The finite element method: solid mechanics, vol. 2. Butterworth-Heinemann, 5, 14.

Behaviour of structures and materials under impact loadings

Trends in Collision and Grounding of Ships and Offshore Structures – Zhu & Guedes Soares (Eds)

Plastic damage predictions of impulse loaded beams

S.B. Fimreite, Z. Yu & J. Amdahl
Department of Marine Technology, Norwegian University of Science and Technology (NTNU), Trondheim, Norway

ABSTRACT: Ocean structures are subjected to a variety of loading conditions characterized by high magnitudes and short durations throughout their lifetimes. Safe utilization of the full ductile capacities of the structures requires design methods which accurately predict the elastoplastic responses of the structures. This paper investigates two simplified methods for deformation prediction. A rigid-perfectly plastic analytical method and a single-degree-of-freedom (SDOF) design method are implemented to predict the maximum responses of a plate strip subjected to lateral pressure loads. The predictions are compared against finite element simulations of elastoplastic dynamic load responses. Sixteen load cases of triangular pressure impulses of varying duration and magnitude are considered. Both simplified methods severely overestimate the magnitude of deformations. Corrections which improve the accuracy of the methods are therefore introduced. The analytical method is found to have a limited range of applicability while the corrected SDOF method gives accurate predictions of maximum dynamic response.

1 INTRODUCTION

During their lifetimes, ships and offshore structures are subjected to loads of high magnitude and short duration, e.g., impacts and collisions, wave loads, and explosions. These loads instigate dynamic responses of the structures. Metal structures typically have large ductile capacities beyond the point of first yield which may be utilized to handle severe loading cases. This will however lead to permanent deformations of the structures, requiring accurate determination of such elastoplastic dynamic effects to ensure safe and reliable design.

DNV-RP-C204 (2021) recommends a design procedure for structural components subject to explosion loads, characterized as pressure loads of high magnitude and short duration. Such loads may be categorized by load duration relative to the first natural period T of the structure: impulsive loads have duration shorter than t_d/T=0.3, quasi-static loads have duration longer than t_d/T=3 and dynamic loads occupy the intermediate interval, where t_d is the total load duration. According to the design procedure, the dynamic response of different structural components can be determined graphically for a variety of load histories. The response is determined by a single-degree-of-freedom (SDOF) analogy where an equivalent load resistance curve is created. Elastoplastic deformations may be considered as a combination of elastic and rigid-perfectly plastic response curves.

The problem of elastoplastic deformations is often simplified in analytical approaches by assuming the material behavior to be rigid-perfectly plastic, allowing determination of the permanent deformations of the structure. The elastic deformation capacity of the structure is thereby assumed to be negligible relative to the plastic deformation capacity. Hu et al. (2023) defined the ratio R between total plastic energy dissipation and maximum elastic strain energy as a limiting parameter for rigid-perfectly plastic analysis of impulsively loaded beams. For ratios of 5 and higher, the error of maximum deformation predictions was less than 10%. Xie et al. (2024) identified a dimensionless load intensity parameter from dimensional analysis of a beam subjected to a uniform pressure load. They found that the rigid-perfectly plastic model could predict the final deflection of the beam when the load intensity parameter corresponded to an energy ratio R of approximately 2.

Jones (2012) presented methods for static and dynamic rigid-perfectly plastic modelling of beams, plates and shells. For beams, all deformation was assumed to occur in plastic zones where the entire cross section was in yield, about which the remaining rigid beam sections rotated like a hinge. The plastic hinges formed close to the supports and travelled towards the midpoint of the beam for loads above a certain threshold. Furthermore, he commented on the influence of finite deformations on the rigid-plastic analysis of beams. As deformations

DOI: 10.1201/9781003684404-32

developed in an axially restricted beam, elongation of the central axis led to the development of axial forces which interacted with the internal bending moments. A simplified analysis using square interaction curves between axial forces and bending moments was included. Symonds & Mentel (1958) presented a model for the deformations of an axially restricted rigid-plastic beam subjected to an impulsive load. They considered the true non-linear interaction curve for beams with a rectangular cross section.

Numerical simulations may consider elastoplastic effects in a more rigorous manner than the simplified SDOF and rigid-perfectly plastic models. Hu et al. (2023) and Xie et al. (2024) used finite element simulations to compare the pressure load response of elastoplastic beams with predictions from rigid-perfectly plastic theory, providing limitations for rigid-perfectly plastic modelling. Development of numerical efficiency has also allowed investigations of more advanced structural problems. Ji & Wang (2014) used finite element simulations to model axial compression of stiffened panels, studying the effect of different impulsive and dynamic load histories. He & Guedes Soares (2024) numerically modelled accumulation of permanent deformations of rectangular plates due to lateral impacts.

The present paper investigates the analytical model for beam response by Symonds & Mentel (1958). The dynamic response of a beam, representative of a plate strip, is investigated using FE simulations. The results are compared with predictions from the analytical model and the SDOF method included in DNV-RP-C204. Corrections and limitations of the simplified methods are discussed further.

2 RIGID-PLASTIC ANALYTICAL METHOD

Symonds & Mentel (1958) presented a model for predicting the damage to a rectangular beam subjected to impulsive loads. The model assumed a rigid-perfectly plastic material and determined the permanent deformations of pinned and clamped beams that were axially restricted.

The deformation of a beam was divided into three stages. Firstly, plastic hinges formed at the supports and travelled towards the center of the beam as deformations developed. The central part of the beam continued to move with a constant normal velocity. As the two plastic hinges met and coalesced at the midpoint of the beam, the second stage started. The two halves of the beam rotated about the central hinge, and the velocity of the midpoint gradually decreased. In these two stages, the deformations were calculated by solving differential equations in time for the rotational angle of the beam at the support.

As the lateral deformations developed, an axial force also developed in the beam. A non-linear interaction was defined between the internal axial force and bending moment. If the energy imparted by the loading was not dissipated by the deformations in the first two stages, the axial force would at a certain point become equal to the axial capacity of the beam, and the bending moment would be reduced to zero. This was referred to as the plastic string stage, as only tensional membrane stresses were carried by the beam and the beam behaved similarly to a string. Symonds & Mentel (1958) solved the equation of motion in this stage by introducing travelling wave functions. The waves contained two contributions: one contribution from the deformation field at the onset of the string phase and one contribution from the velocity field. Permanent deformation of the beam was taken as the time of maximum deformation during this stage.

The analytical model assumed an impulsive load of sufficiently short duration. The load was then assumed to impart a uniform initial velocity to the beam before any deformations developed. The velocity corresponding to a certain load may be calculated by assuming conservation of linear momentum in the short inertial stage when the load is applied. For a triangular pressure impulse with peak magnitude p_{max} and total duration t_d acting uniformly over the beam with height H and density ρ, the corresponding velocity is given by Equation (1).

$$V = \frac{p_{max} t_d}{2 \rho H} \tag{1}$$

3 SINGLE-DEGREE-OF-FREEDOM METHOD

DNV-RP-C204 (2021) included a simplified approach for predicting the response of structural members due to time-variant loads, known as Biggs method, based on the work of Biggs (1964). Various structural components may be handled as equivalent beams. The dynamic response of the beam is determined by introduction of shape functions, reducing the dynamic equilibrium equations to a SDOF system. The static collapse mode of the beam is assumed to be an acceptable deformation shape for the dynamic response.

Static tri-linear elastoplastic load resistance curves are used as input for the dynamic response of the beam. The first linear stage of the resistance curve is defined by the elastic stiffness k_1. After the maximum capacity R_{el} is reached, the resistance remains constant, corresponding to a stiffness k_2=0. In the case of axial restraints, the developing membrane forces furthermore increase the resistance with a stiffness k_3. If the load case of the beam causes progressive formation of mechanisms, equivalent linear stiffnesses are defined.

DNV-RP-C204 (2021) further included response diagrams for different load histories. The maximum

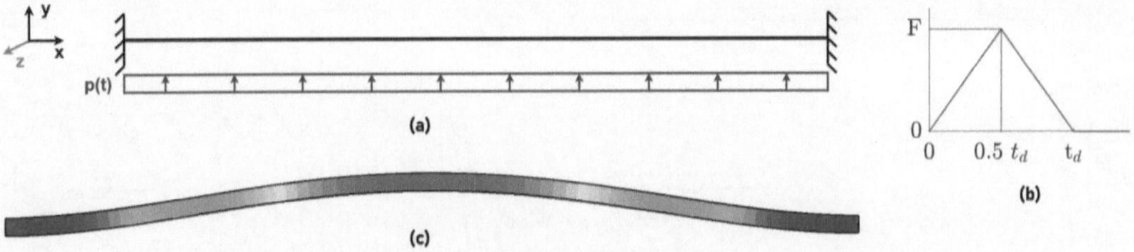

Figure 1. A) Loading case and boundary conditions of beam. b) Pressure load history. c) First eigenmode of the beam from eigenvalue analysis in Abaqus, with associated eigenperiod 7.4 ms.

dynamic response may be determined relative to the displacement at the end of the first linear part of the resistance curve, at the time when the resistance reaches R_{el} for the first time. The load duration t_d relative to the natural period T, peak load F relative to the maximum capacity R_{el}, and stiffness k_3 relative to k_1 influence the magnitude of the dynamic response.

For a clamped beam restricted against axial translations and loaded by a distributed load, DNV-RP-C204 (2021) recommends an averaged initial stiffness $k_{1,avg}$ and membrane hardening stiffness k_3 according to Equation (2) along with a maximum capacity R_{el}. E is the Young's modulus, I is the second moment of area, and L is the length of the beam. The axial capacity N_p is $\sigma_Y BH$ and the bending capacity M_p is $\sigma_Y BH^2/4$ for a beam of rectangular cross section, with width B, height H and yield stress σ_Y. Equation (2) is given in terms of lateral forces and should be divided by the area BL to convert to pressures.

$$k_{1,avg} = \frac{307EI}{L^3} \quad k_3 = \frac{4N_p}{L} \quad R_{el} = \frac{16M_p}{L} \tag{2}$$

4 FINITE ELEMENT MODEL

The dynamic response of an elastoplastic beam was investigated using the finite element software Abaqus. The beam had a square cross section with height H and width B of 15 mm and a total length L of 0.75 m. These dimensions are representative of a plate strip between stiffeners of an offshore structure. Consequently, the beam was assumed to have clamped boundary conditions including full axial restriction.

The beam was modeled with 8-noded linear brick elements with reduced integration and hourglass control. A mesh size of 3.75 mm was found sufficient for convergence of the midpoint deflections. A yield stress σ_Y of 180 MPa, Young's modulus E of 210 GPa, density ρ of 7850 kg/m^3 and Poisson's ratio 0.3 were defined to represent a steel material. Plastic hardening corresponding to a plastic tangent modulus E_t of 210 MPa was additionally included.

A uniform pressure load was applied in the positive y-direction as defined in Figure 1a. Sixteen load cases of triangular pressure histories according to Figure 1b were considered, combining four load durations t_d and four peak magnitudes F. Durations of 0.2, 0.3, 0.6, and 1.0 times the natural period T and pressure magnitudes 1.25, 2.0, 3.33, and 10.0 times the theoretical capacity R_{el} were studied. The first natural period was found to be T=7.4 ms by numerical eigenmode analysis using the linear perturbation frequency step in Abaqus. Figure 1c illustrates the associated deformation mode.

4.1 *Load resistance curve*

A load resistance curve was generated numerically for comparison against the DNV-RP-C204 recommended stiffnesses. A uniform pressure of gradually increasing magnitude was applied, sufficiently slow to limit dynamic effects. The resulting quasi-static load response curve is illustrated in Figure 2. The resistance curve generated by the DNV stiffnesses is also included in the figure.

The initial stiffness k_1 of the DNV curve was somewhat lower than the stiffness of the numerical curve, while the most significant differences between the curves were related to the membrane stiffness k_3. The resistance of the numerical simulations was then significantly higher than the DNV curve prediction.

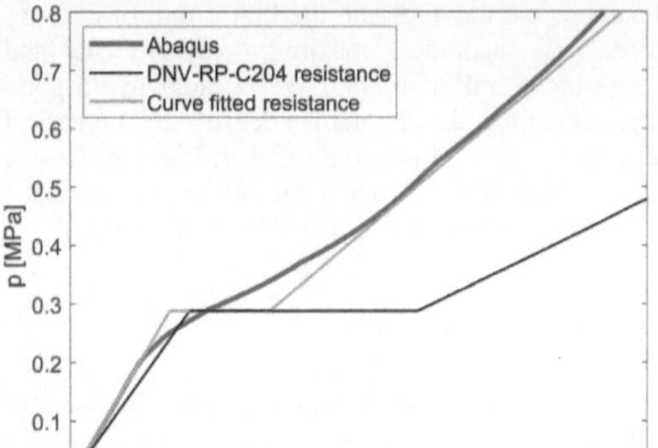

Figure 2. Quasi-static load resistance curve from Abaqus and tri-linearizations. DNV-RP-C204 curve according to Equation (2) giving k_3=0.34k_1 and curve fitted resistance k_1=67.09 MPa/m, k_3=0.5k_1.

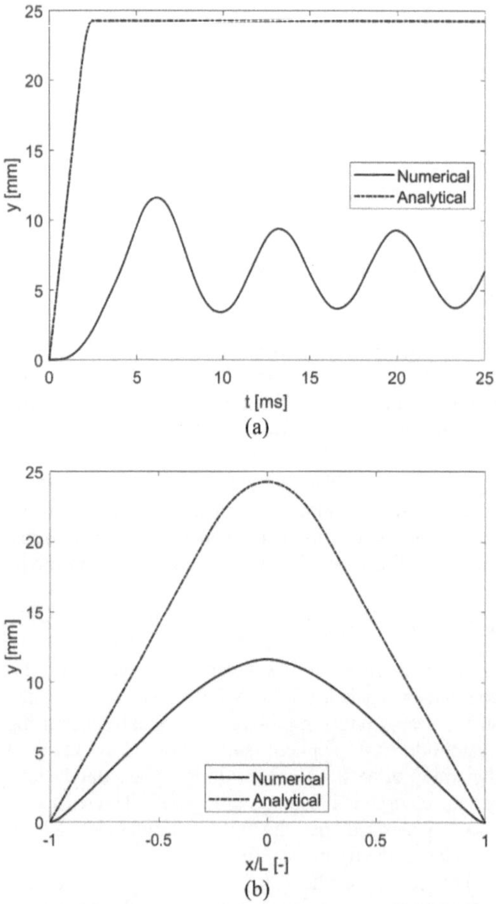

Figure 3. Response to a dynamic pressure load with duration t_d=T and peak magnitude F=1.25R_{el}. a) Midpoint deflection. b) Deformation mode at time of maximum deflection.

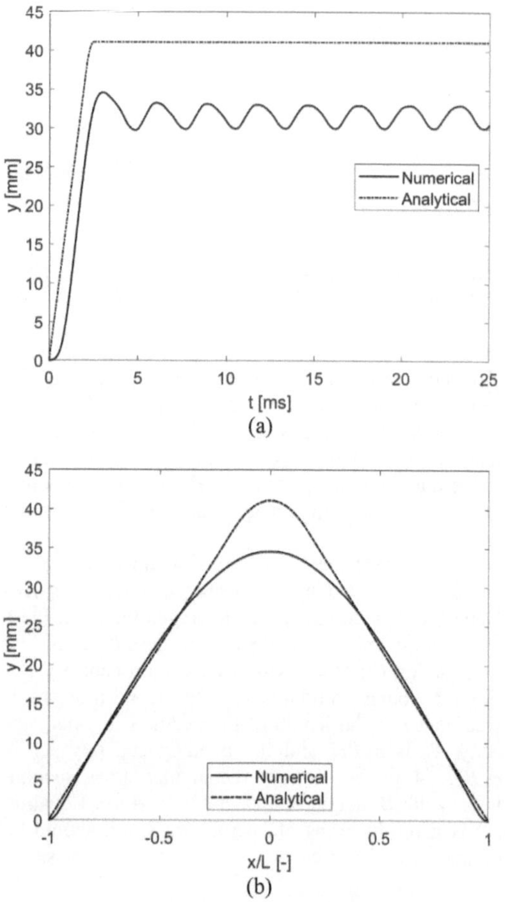

Figure 4. Response to a dynamic pressure load with duration t_d=0.2T and peak magnitude F=10.0R_{el}. a) Midpoint deflection. b) Deformation mode at time of maximum deflection.

A second tri-linearization is additionally included in Figure 2. The linear stiffness k_1 was extrapolated from the numerical curve during the first 3 mm of deformation. The theoretical maximum capacity R_{el} and a membrane stiffness k_3=0.5k_1 appeared to be good approximations for the remainder of the numerical resistance curve. This curve fitted tri-linear resistance curve retains the area under the numerical resistance curve, as recommended by DNV-RP-C204 (2021).

5 RESULTS

Figures 3 and 4 illustrate the response of the beam to a dynamic load of low magnitude and an impulsive load of high magnitude, respectively. The deformation histories of the beam midpoint and the deformation shapes at the time of maximum deformation are illustrated from the numerical simulations and the analytical method.

Figure 3 illustrates the numerical response of the beam in the dynamic loading case where the load duration was approximately 7.4 ms. The beam reached maximum deflection of 11.6 mm after 6.2 ms, before the end of the load duration. During the last part of the load duration, the beam recovered before oscillating around a permanent deformation magnitude of 6.5 mm. Simi-lar results were observed for load cases of the same duration. As the loading after 6.2 ms no longer caused increased deformations, the pressure impulse until that time is often referred to as the saturated impulse. Tian et al. (2019) recited a linear dependence on pressure magnitude of the saturated impulse and maximum deformation of a beam loaded with a rectangular pressure impulse. An identical relationship was identified for the deformations and impulses of the triangular load considered in the present study. Saturation of deformations were not observed for the shorter load durations.

The numerical maximum deformation curve in the impulsive load response in Figure 4a illustrates a maximum deformation of 34.6 mm and a permanent deformation of 31.7 mm. The relative difference between the maximum and permanent deformation was much smaller than for the load case in Figure 3. Comparing loads of constant duration and different magnitude, the relative elastic recovery decreased with increasing load magnitude. After maximum deformation, the period with which the beam oscillated around the permanent deformation magnitude was shorter than in the dynamic case in Figure 3. In all sixteen load cases, the oscillation period decreased with increasing permanent deformation magnitude.

Figures 3 and 4 also include the beam responses according to the analytical model. In the dynamic case where the load is only 25% higher than the collapse load, the rigid-plastic model severely overestimated the deformations by a factor of more than 2. Disregarding the magnitude, the deformation shape of the beam was nevertheless modelled well by the analytical model.

The analytical model also overpredicted the deformations of the high load magnitude impulsive case in Figure 4 somewhat. The magnitude of overprediction was however much smaller, with the analytical model predicting 18% larger deformations than the numerical simulations. The differences mainly appeared close to the center of the beam as illustrated by the deformation shapes. For the impulsive load case, the analytical model predicted a more triangular shape than formed by the numerical simulations. The smoother shape of the numerical simulations may be influenced by strain hardening, which is not considered by the analytical model but typically smooths the deformation shape.

Figure 5 reports the maximum deformation of the beam as determined by the numerical simulations, the analytical method and the SDOF method. Precise values are given in Table A1. Both simplified methods overestimated the deformations of the numerical simulations. The magnitude of overestimation by the SDOF method increased for increasing load duration and increasing load magnitude. Increasing load duration also increased relative overestimations by the analytical method, while no clear trend was observed regarding load level for the rigid-perfectly plastic approach. The error of the SDOF method ranged between 19-84% overestimation of the numerical deformation, while the analytical method ranged between 13-107% overestimation and a 18% underestimation of the load case of shortest duration and lowest magnitude.

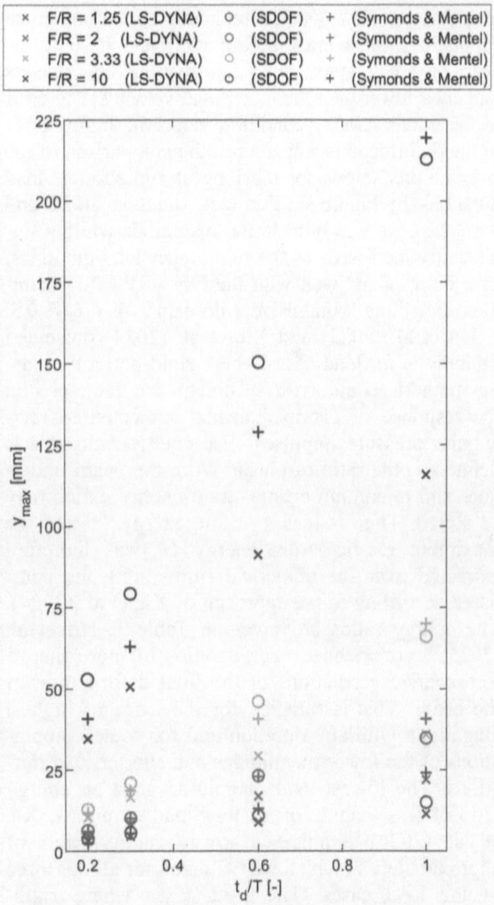

Figure 5. Maximum deformations of the beam from numerical simulations, the SDOF method and the analytical method. Precise values given in Table A1.

6 DISCUSSION

6.1 *Analytical method*

The slope of the midpoint deformation curves in Figures 3 and 4 display the vertical velocities of the beam midpoint. For the analytical curves, that was initially equal to the theoretical load equivalent velocity according to Equation (1). The agreement between the numerical and analytical curve slopes thereby reflects the suitability of the theoretical velocity. In the impulsive load case illustrated in Figure 4, the two curves had similar slopes aside from the first 0.8 ms, when the numerical deformations initiated more gradually than the analytical deformations. The acceleration of the numerical simulations lasted longer in the dynamic loading case in Figure 3, and the slope of the numerical curve did not increase to the steepness of the analytical curve. The calculated velocity is therefore a more suitable choice for the impulsive load case.

The maximum deformation velocity in the numerical simulations was lower than the corresponding calculated velocity for all the dynamic load cases of durations of 0.6*T* and 1.0*T*. For the two load durations, the calculated velocities were on average 58 and 157% higher than the maximum numerical velo-cities respectively. The calculated velocities were on average 2% higher than the maximum velocities from the numerical simulations with load duration 0.3*T*, and 18% lower than the maximum velocities generated by load duration 0.2*T*. Due to the inertia of the numerical simulations, the average deformation velocity until the beam reached maximum deformation nevertheless remained lower than the theoretical velocity. The error of the initial velocity condition induced a higher error of the deformations, apparent in the comparison of the error of predictions for the longest and shortest load durations in Figure 6. For load durations 0.2*T* and 0.3*T*, the error was of a similar magnitude while it significantly increased for the two longer load durations. This corresponds well with the DNV-RP-C204 definition of the impulsive domain of $t_d/T<0.3$.

Hu et al. (2023) and Xie et al. (2024) presented limitations for load cases where rigid-perfectly plastic methods could with sufficient accuracy predict the response of clamped beams subjected to rectangular pressure impulses. The energy ratio *R* was defined as the ratio of plastic work the beam undergoes and maximum elastic strain energy which may be stored. The dissipated plastic energy U^p and the maximum elastic strain energy U^e was therefore extracted from the numerical simulations and compared according to the approach of Xie et al. (2024). The energy ratios are given in Table 1. Hu et al. (2023) recommended energy ratios of more than 5 for accurate predictions of the final deformation of the beam. This is satisfied for the loads of highest magnitude with any duration and for some combinations of the lower magnitudes and shorter load durations. The lowest load magnitude gave an energy ratio of less than 5 for all four load durations. Xie et al. (2024) required a lower energy ratio of approximately 2, which was satisfied for all but three of the load cases. The load cases where rigid-perfectly plastic theory can be expected to be valid are indicated in Table 1.

Table 1. Energy ratios *R* extracted from numerical simulations in ABAQUS, relative to limitations of rigid-perfectly plastic theory according to Hu et al. (2023) ($R>5$) and Xie et al. (2024) ($R>\approx 2$).

Loading condition		R	R > 5	R > ≈2
t_d=1T	F=1.25R	2.9	N	Y
	F=2R	6.4	Y	Y
	F=3.33R	13.0	Y	Y
	F=10R	118.6	Y	Y
t_d=0.6T	F=1.25R	2.1	N	Y
	F=2R	4.3	N	Y
	F=3.33R	8.2	Y	Y
	F=10R	75.8	Y	Y
t_d=0.3T	F=1.25R	0.8	N	N
	F=2R	2.1	N	Y
	F=3.33R	4.1	N	Y
	F=10R	24.9	Y	Y
t_d=0.2T	F=1.25R	0.2	N	N
	F=2R	1.1	N	N
	F=3.33R	2.8	N	Y
	F=10R	11.3	Y	Y

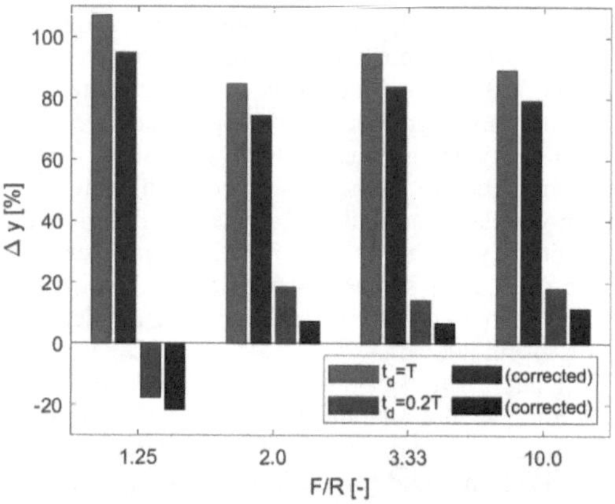

Figure 6. Error of the analytical deformations relative to the numerical simulations, including correction by increased yield stress.

Linear hardening of the material was included in the numerical model to prevent the beam from forming plastic zones at the supports which were excessively stretched. As plastic strains developed in the beam, the maximum stress therefore increased from the initial yield stress of 180 MPa. Thus, a maximum equivalent stress of 221 MPa was allowed at plastic strains of 0.2. The analytical model assumed a rigid-perfectly plastic material which conservatively underestimated the capacity of a hardening material. This discrepancy may have contributed to the analytical overestimations in Figure 5. Figure 6 illustrates the effect of increasing the yield stress from 180 MPa to 200 MPa in the analytical model. The increased yield stress reduced the magnitude of all the load responses, in most cases reducing the error of the analytical method. For the case where the analytical method initially underpredicted the response, the error was exacerbated. However, the energy ratio from this simulation was 0.2 and nearly no permanent deformations were instigated, making this an invalid application of the rigid-perfectly plastic model and thereby irrelevant for the assessment of the method.

6.2 *SDOF method*

The SDOF method overpredicted the maximum deformation of all sixteen load cases in Figure 5. However, Figure 2 illustrated that the resistance curve based on stiffnesses from DNV-RP-C204 underpredicted the resistance of the beam. The discrepancies between the two resistance curves increased for increasing deformations, particularly in the membrane resistance stage. As deformations increase with both increasing load magnitude and duration, the observed increase of error with these two factors in Figure 5 complied with expectations.

The curve fitted tri-linear curve in Figure 2 followed the numerical resistance curve much more accurately than the RP curve and conserved the area

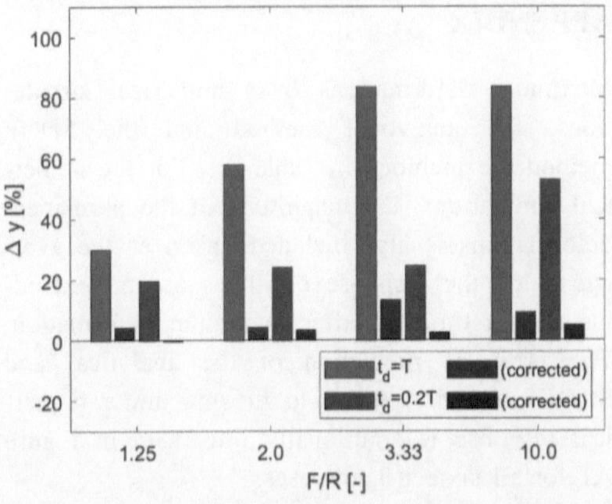

Figure 7. Error of the SDOF deformation predictions relative to the numerical simulations, including corrections from improved resistance curve.

under the curve like the RP recommends. Using this curve as input for the SDOF method slightly decreased the maximum linear response, which the dynamic response is given in terms of, and changed the graphical extraction points, which depend on the membrane stiffness k_3 relative to k_1. The effects of these corrections are illustrated in Figure 7, reducing the maximum error from 84% to 14% with errors of less than 5% in 11 of 16 load cases. The correction does however also introduce an underestimation of 2% in the load case t_d=0.3T, F=3.33R_{el}.

The membrane stiffness k_3 in DNV-RP-C204 for a beam subjected to a uniform load is the same as for a point load acting at the midpoint of the beam. For a beam with two point loads spaced symmetrically along the beam, the recommended membrane resistance is 50% higher. The error of the SDOF method was reduced to 18-36% when using the DNV resistance curve with the increased membrane stiffness.

The SDOF method assumed the eigenmode of the beam, illustrated in Figure 1, as a deformation mode throughout the dynamic response. Comparison with the numerical deformation shapes in Figures 3 and 4 illustrate that this is not an accurate representation, particularly not in the cases giving large permanent deformations. Despite this, the SDOF method was able to predict the maximum dynamic response with reasonable accuracy. When the linearized resistance curve represented the true resistance of the structure well, the SDOF method gave very good predictions of the response to both impulsive and dynamic loads.

7 CONCLUSIONS

The dynamic response of a beam subjected to uniform pressure loads has been studied by numerical simulations. Sixteen load cases of triangular pressure histories with different magnitude and duration were considered. The beam midpoint response magnitude and shape of maximum deformation was examined. Furthermore, deformation predictions from a rigid-perfectly plastic analytical method and a SDOF design method were compared with the numerical results.

The sixteen load histories instigated a range of dynamic responses. The deformations grew rapidly until maximum deformation, after which the magnitude recovered slightly. The magnitude of the recovery relative to the maximum value decreased with increasing load impulse, as did the period of oscillations for the vibration about the permanent deformations. In the load cases of longest duration, the beam deformations were saturated before the end of the load history.

The analytical method for response of an impulsively loaded rigid-perfectly plastic beam by Symonds & Mentel (1958) generally overpredicted the magnitude of the deformations. The method was most suited for the load cases of short duration relative to the natural period of the beam due to the modelling of the impulsive load as an initial velocity. Additionally, the rigid-perfectly plastic material assumption is only valid for load cases where large permanent deformations develop. The overpredictions were successfully reduced somewhat by increasing the yield stress of the model to represent the plastic hardening of the numerical material. The model predicted realistic deformation shapes of the beam.

The SDOF method as given by DNV-RP-C204 (2021) also overestimated the dynamic response of the structure. The overestimations were related to the assumed load resistance of the beam, which was underestimated by the RP. The relative error between the SDOF and numerical results therefore increased with increasing response magnitude. The accuracy was severely improved by using a more correct resistance curve. The method was then applicable to both impulsive and dynamic loading cases.

ACKNOWLEDGEMENTS

The authors gratefully acknowledge the support from high performance computation resources from the Norwegian national e-infrastructures, Project NN9585K - Accidental actions on strait crossings and offshore platforms.

REFERENCES

Biggs, J.M. 1964. *Introduction to structural dynamics*. New York: McGraw-Hill.

DNV. 2021. *DNV-RP-C204 Structural design against accidental loads*.

He, X. & Guedes Soares, C. 2024. Experimental and numerical study on the dynamic response of rectangular plates under repeated impacts. *Marine Structures* 96: 103606.

Hu, Q., Zhu, L. & Yu, T.X. 2023. Elastic effects on the dynamic plastic deflection of pulse-loaded beams. *International Journal of Impact Engineering* 176: 104550.
Ji, Z. & Wang, D. 2014. Influence of load shape on dynamic response of cross-stiffened deck subject to in-plane impact. *Thin-Walled Structures* 82: 212–220.
Jones, N. 2012. *Structural Impact*. Cambridge University Press (2nd edition).
Tian, L., Chen F.L., Zhu, L., & Yu T.X. 2019. Saturated analysis of pulse-loaded beams based on Membrane Factor Method. *International Journal of Impact Engineering* 131: 17–26.
Xie, H., Huang, X., Feng, Y. & Qui, X. 2024. Predicting the dynamic response of elastic-plastic beams by dimensional analysis. *International Journal of Impact Engineering* 193:105041.

APPENDIX

Maximum deformations from numerical simulations, the analytical method and the SDOF method are included in Table A1. For the numerical simulations, the magnitude of the permanent deformations is also included, taken as the average value the response oscillates around excluding the first period after maximum deformation. The error of prediction of the analytical and SDOF methods relative to the maximum numerical response is additionally illustrated in Figure A1 for all sixteen load cases.

Table A1. Magnitude of dynamic beam response according to numerical simulations (y^n), analytical method (y^a) and SDOF method (y^s), all given in mm. Numerical results given in terms of maximum and permanent deformation magnitude. Results for analytical results corrected by increased yield stress and SDOF method corrected by improved resistance curve in Figure 2 also included.

Loading condition		y^n	y^n (perm.)	y^a	y^a (corr.)	y^s	y^s (corr.)
t_d=1T	F=1.25R	11.62	6.47	24.10	22.67	15.09	12.11
	F=2R	22.08	18.12	40.79	38.49	36.91	23.01
	F=3.33R	36.01	32.45	70.05	66.22	66.02	40.82
	F=10R	115.97	115.31	219.52	208.04	212.88	126.55
t_d=0.6T	F=1.25R	8.92	3.58	13.06	12.20	10.97	9.14
	F=2R	17.13	11.95	23.05	21.67	23.26	17.60
	F=3.33R	29.39	25.34	40.79	38.49	48.09	31.38
	F=10R	91.21	90.00	128.99	122.09	150.42	97.36
t_d=0.3T	F=1.25R	4.71	0.23	5.64	5.12	5.71	4.76
	F=2R	8.28	3.12	10.23	9.51	10.24	8.33
	F=3.33R	16.50	11.13	18.61	17.46	20.93	16.18
	F=10R	50.67	48.20	62.92	59.47	79.26	53.40
t_d=0.2T	F=1.25R	3.27	0.08	2.69	2.57	3.91	3.26
	F=2R	5.14	0.91	6.09	5.53	6.38	5.32
	F=3.33R	10.30	5.61	11.75	11.00	12.81	10.56
	F=10R	34.56	31.65	40.79	38.49	52.89	36.27

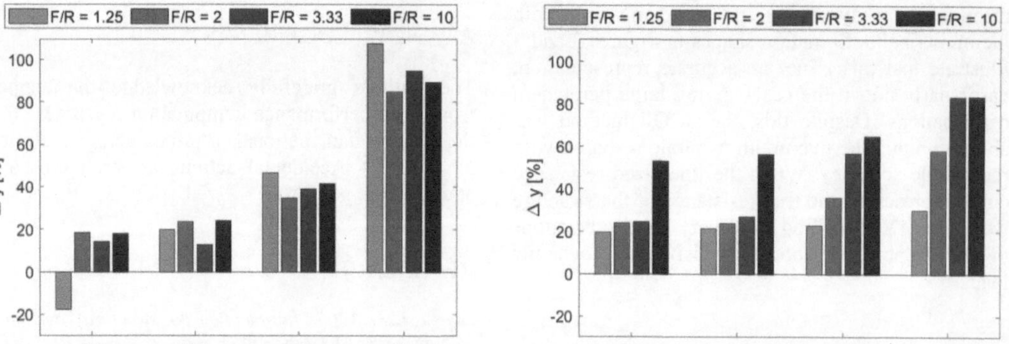

Figure A1. Error of prediction relative to maximum deformation from the numerical simulations of a) the analytical method and b) the SDOF method in all sixteen load cases.

Trends in Collision and Grounding of Ships and Offshore Structures – Zhu & Guedes Soares (Eds)

Dynamic-asymmetric model of titanium alloy and application of FEM subroutine

X.W. Jiang, J.T. Dong, Y.J. Zhao, J. Wang, W.W. Wu & L.P. Zhang
Taihu Laboratory of Deepsea Technological Science, Wuxi, Jiangsu, P.R. China
China Ship Scientific Research Center, Wuxi, Jiangsu, P.R. China
National Key Laboratory of Ship Structural Safety, Wuxi, Jiangsu, P.R. China

ABSTRACT: This paper focuses on dynamic mechanical property tests of typical α+β titanium alloy for deep-water pressure-resistant shells under different loading modes. Results show significant tension-compression asymmetric mechanical properties and strain-rate correlation in plastic deformation due to complex microstructural deformation, indicating strong dependence of α+β two-phase titanium alloy on stress state and strain rate during yield and fracture. For shells under high hydro-static pressure and underwater explosions, shock wave and hydro-static pressure coupling creates complex stress states with sudden strain-rate increases, causing varied plastic yield and damage in shell parts. Based on material's elastoplastic and damage test data, a dynamic constitutive and failure model for tension-compression asymmetric materials is constructed. A Vumat subroutine for Abaqus simulates dynamic responses and predicts safety of α+β titanium alloy shells under extreme deep-water loads.

1 INTRODUCTION

With the continuous progress of titanium alloy preparation technology in China, titanium alloys have been widely used in aerospace (Yang et al., 2014; Jiang et al., 2009), marine engineering (Meng et al., 2004; Chang et al., 2014; Fan, 2006; Fan, 2006; Zhou, 2005) and other fields due to their advantages such as light weight, high strength, stable high and low temperature toughness. For example, titanium alloy materials are selected for aerospace engine blades, rocket fuel tanks and engine brackets, satellite brackets, manned spacecraft shells, pressure-resistant shells, oil and gas pipelines and offshore platforms. Among them, the development of deep-water equipment (Ye et al., 2023; Chen et al., 2015; Wang, 2021) such as deep-water pressure-resistant shells and oil and gas pipelines needs to fully consider environmental factors such as extreme deep-sea pressure, strong seawater corrosion and low temperature, and the application demand for titanium alloys has even reached an irreplaceable position (Zhang et al., 2024). At present, there are many types of titanium alloys produced in China. There are great differences in composition and preparation technology among different types of titanium alloys. The characterization of titanium alloy material properties is not comprehensive enough. Due to the complex micro-structure of titanium alloy materials, they are extremely prone to significant stress state dependence and strain rate effects under complex conditions.

The stress of titanium alloy materials in deep-water pressure-resistant shells is relatively complex, often showing some unpredictable brittle failures (Pan & Cui, 2012). When the titanium alloy spherical shell is subjected to hydro-static pressure, after the spherical shell shows buckling room temperature deformation, the surface stress state gradually changes from compressive stress to tensile stress until failure, which indicates that with the increase of hydro-static pressure and the evolution of stress state, the unidirectional stress on the surface of the spherical shell gradually becomes complex.

Many researchers have found that titanium alloy materials have complex mechanical constitutive properties and damage rules in the plastic stage. Ahn et al. (2015) used OIM (Orientation Imaging Microscope) to analyze and study the influence of deformation twins on the strain hardening behavior of pure titanium. In the test strain rate range of 0.001-10/s, the tensile and test plastic hardening of the material increase with the increase of the strain rate, and the hardening stages of tension and compression are different, with the degree of compression hardening being greater than that of tension hardening. Based on the improved Hopkinson tension and compression bar system, Guo (2023) realized the dynamic tension and dynamic compression loading-unloading experiments of TA7 titanium alloy, and obtained the isothermal stress-strain curve under high strain rate. The experimental results show that the mechanical behavior of TA7 titanium alloy has tension-compression asymmetry,

DOI: 10.1201/9781003684404-33

and this asymmetry is strain rate-dependent. As is well known, the micro-structure of titanium alloys determines the plastic deformation behavior of the material. Generally, the plastic deformation mechanisms of titanium alloys mainly include the slip mechanism caused by dislocations and the twinning mechanism caused by twins. Jia (2015) mentioned that the titanium work of Majorell, who found that the main mechanical response behavior of α+β titanium alloys is determined by the α phase, and the β phase will only have a greater impact on the properties of titanium alloys when the temperature is high and the content is close to or exceeds 50%. Huang et al. (2007), Chichili et al. (1998) and others carried out compression tests on pure titanium and characterized the deformed microstructure using transmission electron microscopy. The results show that the twin generation rate and volume fraction of the material are positively correlated with the strain rate, and as the plastic strain approaches about 0.2, the twin volume fraction becomes stable, the material hardening reaches the maximum stress, and the material enters damage deformation.

This paper focuses on α+β titanium alloy materials, carries out research on elastoplastic and damage failure models, and develops an Abaqus finite element material subroutine model, providing a computational basis for the finite element structural model of dynamic response and failure analysis of titanium alloy structures under extreme deep-water explosion loads.

2 RESEARCH ON PERFORMANCE TESTING AND DAMAGE CONSTITUTIVE MODELS OF TITANIUM ALLOYS

2.1 *Specimen design under different loading types*

Specimen dimensions for titanium alloy were designed according to different types of loading tests to obtain material test results under varying strain rates and complex stress states (stress triaxiality). As shown in Table 1, the main test categories and key parameters are listed below. Specific specimen dimensions are illustrated in Figures 1(a)–(e). Speckle patterns were pre-sprayed on the specimens for strain measurement correction using a Digital Image Correlation (DIC) system. Figure 2 demonstrates the strain distribution contours on the specimen surface during loading.

Table 1. Geometric parameters of α+β titanium alloy specimens.

Test Type	Specimen Geometry	Loading Device	Gauge Length (mm)	Minimum Section Diameter (mm)	Strain Rate (/s)
Quasi-static Tension	Fig. 1(a)	Universal Testing Machine	40.00	6.00	0.002
Quasi-static Compression	Fig. 1(b)	Universal Testing Machine	9.00	6.00	0.002
Dynamic Compression	Fig. 1(c)	Hopkinson Tensile Bar	4.00	4.00	*2800
Notched Tensile Specimen	Fig. 1(d)	Universal Testing Machine	60.00	6.00	~0.002

Note: ~ denotes the approximate strain rate, and * indicates the average strain rate.

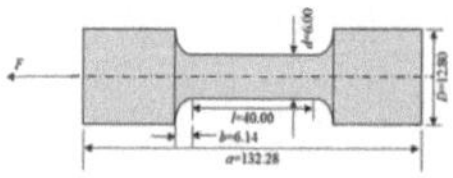

(a) Quasi-static tensile specimen

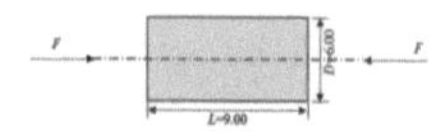

(b) Quasi-static compressive specimen

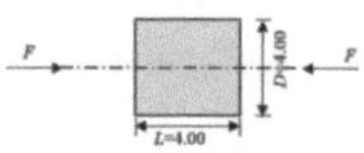

(c) Dynamic compressive specimen

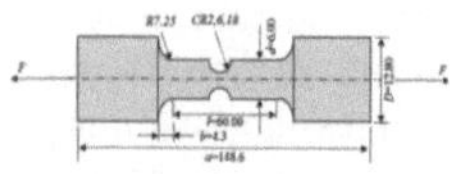

(d) Quasi-static notched tensile specimen

Figure 1. Design drawings of specimen dimensions for different types of loading tests.

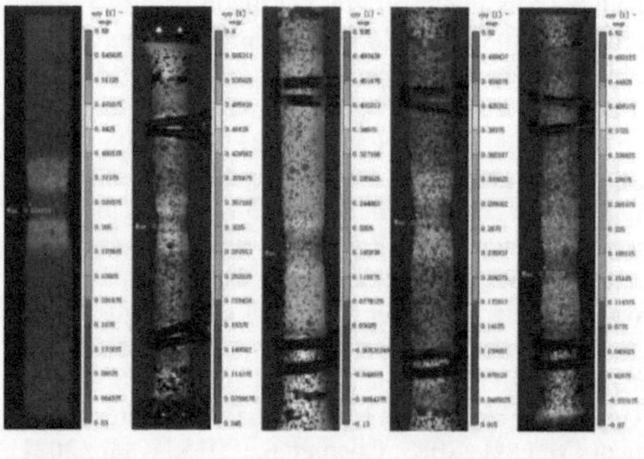

Figure 2. Surface strain nephogram of the quasi-static tensile specimen.

2.2 *Experimental results and research on elastoplastic constitutive models*

Based on the above tests, the tensile and compression test results of α+β titanium alloy under different working conditions were obtained. As shown in Figures 3(a)-(b), the load-displacement curves of the test specimens were derived from the experiments, Test 1~Test 5 are the repeated test number of compressive tests or tensile tests. By substituting the specimen area and gauge length values into the conversion formulas (1) and (2), the engineering strain and engineering stress of the material were calculated. The engineering strain and stress were then converted to true strain and true stress using the conversion formulas (3) and (4), yielding the curve results shown in Figure 4.

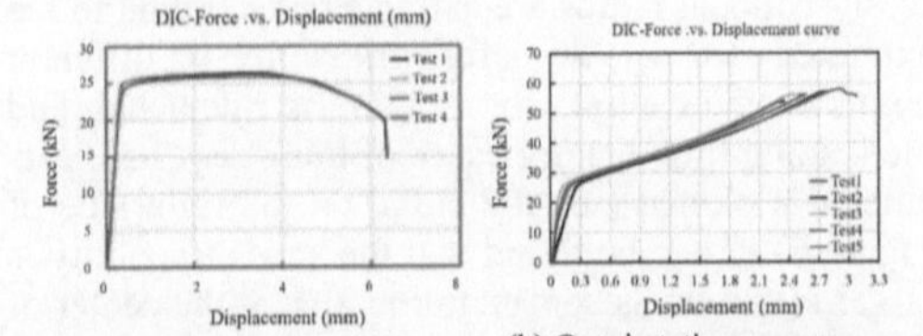

(a) Quasi-static tensile displacement-load curve

(b) Quasi-static compressive displacement-load curve

Figure 3. Load-displacement test curve of materials in quasi-static test.

$$\varepsilon_{Eng} = \frac{\Delta l}{l} \tag{1}$$

$$\sigma_{Eng} = \frac{F}{A} \tag{2}$$

$$\varepsilon_{true} = \ln(1 + \varepsilon_{Eng}) \tag{3}$$

$$\sigma_{true} = \sigma_{Eng} \cdot (1 + \varepsilon_{Eng}) \tag{4}$$

It should be noted that when using Formulas (3) and (4) to convert engineering strain and stress to true values:

I. When converting engineering strain to true strain, the compressive corresponding ε_{Eng} must be taken as a negative value. The resulting true strain value is also negative, requiring an additional sign change to obtain a positive result.
II. Similarly, during stress conversion, the compressive corresponding ε_{Eng} should be taken as a negative value to obtain the true stress value for compression.

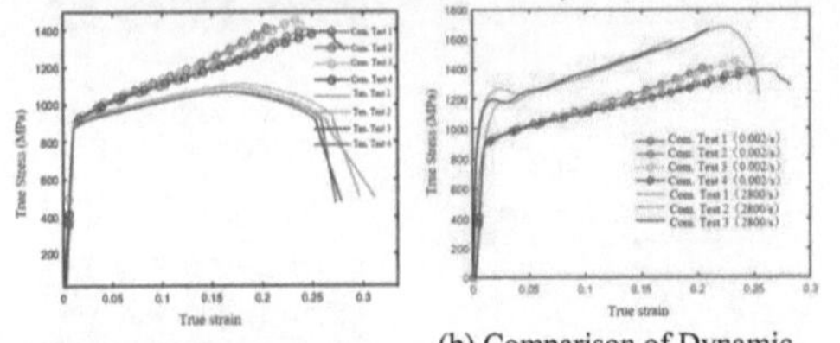

(a) Comparison of compression and tension curves

(b) Comparison of Dynamic and Quasi-static curves under the compression state

Figure 4. True stress and true strain test curve under different stress states and strain-rates.

Using the above method, the true strain-true stress test results for tension and compression at different strain rates were obtained as shown in Figure 4. From the curve results in Figures 4(a) and 4(b), it is observed that the tensile curve of the titanium alloy exhibits a distinct transition point between the elastic and plastic stages, with a strain hardening phase during plastic deformation. Additionally, Figure 4(a) reveals that the stress-strain curves of the titanium alloy show high consistency in the elastic stage, but in the plastic hardening stage, tension and compression exhibit inconsistent strain hardening capabilities. The tensile hardening degree of the titanium alloy is significantly lower than that under compression. Figure 4(b) illustrates the strain rate strengthening effect on the titanium alloy: as the strain rate increases, the overall plastic strength of the material improves.

According to the Mises equivalent stress criterion, when the Mises equivalent stress of the material reaches the initial yield limit, the material yields. At initial yielding, the yield limit values corresponding to tension and compression are equal. As plastic deformation increases, the strain hardening process is influenced by the stress state. Therefore, the subsequent yield strength can be expressed by Equation (5):

$$\sigma_s = \sigma_0 + H(\varepsilon_p, \sigma^*, \dot{\varepsilon}) \tag{5}$$

In the equation, σ_0 is the initial yield point, σ^* is the stress state parameter. When stress triaxiality η is introduced to replace σ^*, the three values of $1/3$, 0, and $-1/3$ correspond to the stress states of tension, torsion, and compression, respectively. $H(\varepsilon_p, \eta, \dot{\varepsilon})$ is the stress increment caused by strain hardening.

$$H(\varepsilon_p, \eta, \dot{\varepsilon}) = \begin{cases} H_T(\varepsilon_p, \dot{\varepsilon}), & \eta > 1/3 \\ k_1 \cdot H_T(\varepsilon_p, \dot{\varepsilon}) + k_2 \cdot H_C(\varepsilon_p, \dot{\varepsilon}), & -1/3 \le \eta \le 1/3 \\ H_C(\varepsilon_p, \dot{\varepsilon}), & \eta < -1/3 \end{cases} \tag{6}$$

By leveraging the interval continuity conditions, the following Lagrangian interpolation function is constructed to satisfy the hardening relationship within the interval region of $-\frac{1}{3} \le \eta \le \frac{1}{3}$:

$$H(\varepsilon_p, \eta, \dot{\varepsilon}) = \frac{(1+3\eta)}{2} H_T(\varepsilon_p, \dot{\varepsilon}) + \frac{(1-3\eta)}{2} H_C(\varepsilon_p, \dot{\varepsilon}) \tag{7}$$

Here, $f(\sigma_{ij})_{mises}$ denotes the Mises equivalent stress, and σ_0 represents the initial yield strength. Based on the Johnson-Cook (J-C) hardening model, the hardening functions for tension and compression under the two stress states, denoted as $H_T(\varepsilon_p, \dot{\varepsilon})$ and $H_C(\varepsilon_p, \dot{\varepsilon})$, are respectively fitted. Their expressions are given by Equation (9):

$$
\begin{aligned}
H_T(\varepsilon_p) &= 721 \cdot {\varepsilon_p}^{0.68} \\
H_C(\varepsilon_p) &= 1369 \cdot {\varepsilon_p}^{0.68}
\end{aligned}
\tag{9}
$$

Once the material enters the plastic stage, the plastic flow strain is determined using the plastic incremental constitutive method. The J_2 associated flow rule is employed for calculations, and the influence of the hardening term on the direction of plastic flow deformation is not considered. Therefore, the incremental strain is determined by Equation (10), and the flow rule potential function is given by Equation (11).

$$
d\varepsilon_{ij}^{p} = d\lambda \cdot \frac{\partial g(\sigma_{ij})}{\partial \sigma_{ij}} \tag{10}
$$

$$
g(\sigma_{ij}) = \frac{1}{2} S_{ij} : S_{ij} \tag{11}
$$

Where, $d\varepsilon_{ij}^{p}$ is the plastic strain increment, $d\lambda$ is the plastic strain increment multiplier, G is the flow potential function, and S_{ij} is the deviatoric stress tensor.

As the plastic deformation of the material increases, the material continuously enters new yield states, but there always exists a new yield surface that satisfies the new yield equation. Additionally, when the material is in a plastic state, the elastic strain still satisfies the generalized Hooke's law with the stress, and the strain increment and stress increment in the entire plastic state satisfy the series of equations in formula (12).

$$
\begin{cases}
f(\sigma_{ij}^{new})_{\text{Mises}} = \sigma_0 + \frac{(1+3\eta)}{2} H_T(\widetilde{\varepsilon_p}, \dot{\varepsilon}) + \frac{(1-3\eta)}{2} H_C(\widetilde{\varepsilon_p}, \dot{\varepsilon}) \\
\sigma_{ij}^{new} = \sigma_{ij}^{old} + d\sigma_{ij} \\
d\sigma_{ij} = C_{ijkl} d\varepsilon_{ij}^{e} = C_{ijkl}\left(d\varepsilon_{ij} - d\varepsilon_{ij}^{p}\right) \\
d\varepsilon_{ij}^{p} = d\lambda \cdot S_{ij} \\
\widetilde{d\varepsilon^{p}} = \sqrt{\frac{2}{3} d\varepsilon_{ij}^{p} : d\varepsilon_{ij}^{p}} \\
\widetilde{\varepsilon_p} = \sum \widetilde{d\varepsilon^{p}}
\end{cases}
\tag{12}
$$

2.3 *Determination of J-C Damage Parameters*

To determine the variable relationship between the damage strain of titanium alloy and the stress triaxiality, it is necessary to analyze the experimental results of the circular notched round bar specimen shown in Figure 1(e). Figure 5 displays the tensile test results of specimens with circular notches of different radii. From the DIC strain nephogram, it can be observed that stress concentration occurs at the notch of the specimen, and the maximum stress appears at the smallest radius in the center. The tensile load-displacement curve is mainly used for load input in simulation modeling to determine the stress triaxiality parameters of specimens with different notch radii.

Figure 6 shows the tensile stress nephogram of the notched specimen established by Abaqus, and the stress results of three points on the minimum cross-sectional radius are extracted. According to the extracted stresses, the stress triaxiality of different points on the central cross-section is calculated, and the loading time-history curve of this stress state parameter is plotted (see the curve on the right side of Figure 6). It can be found that the stress triaxiality is larger in the stress concentration area at the center of the specimen, and it gradually increases from the outer side to the central point. Since the stress triaxiality at the notch is not a stable value, it is necessary to determine the stress triaxiality of specimens with different circular notches through this simulation method, and extract the average value of the relatively stable interval on the stress triaxiality loading time-history curve as the evaluation value for this working condition.

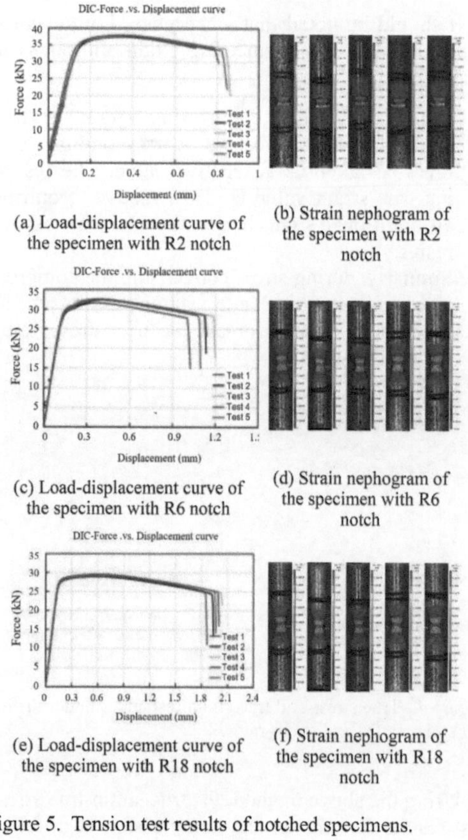

(a) Load-displacement curve of the specimen with R2 notch
(b) Strain nephogram of the specimen with R2 notch
(c) Load-displacement curve of the specimen with R6 notch
(d) Strain nephogram of the specimen with R6 notch
(e) Load-displacement curve of the specimen with R18 notch
(f) Strain nephogram of the specimen with R18 notch

Figure 5. Tension test results of notched specimens.

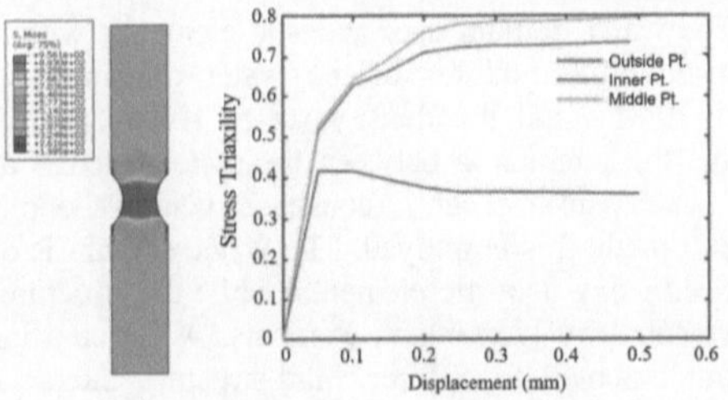

Figure 6. Stress triaxiality of the notched specimen determined by FEM simulation.

When the specimen is damaged, the load begins to decrease. The simulated equivalent plastic strain value corresponding to the moment of maximum load is taken as the initial damage plastic strain value corresponding to the working condition of the notched specimen, and the plastic strain at the point with the fastest maximum load drop rate is taken as the final fracture plastic strain (Figure 6 shows the simulation results of the notched specimen). At the same time, the initial damage strain results measured in tension and compression of the complete specimen are summarized to obtain the initial damage and final fracture plastic strain values corresponding to various different stress triaxialities. The results of each working condition are shown in Table 2.

Table 2. Initial damage strain and extreme damage strain of titanium alloy failure mode.

Dimensionless Time	η(Compression)	η (Tension)	η(R2)	η(R6)	η(R18)
0.1			1.1523	0.8028	0.5742
0.3			1.2676	0.8492	0.5912
0.5			1.2428	0.7968	0.4326
1	−1/3	1/3	1.1707	0.6478	0.4310
Ave. Triaxility			1.2084	0.7742	0.5073
Initial Damage Strain	0.2200	0. 1820	0.1402	0.1624	0.1736
Final Damage Strain	0.2400	0.2520	0.2500	0.2501	0.2502

According to the form of the J-C damage model, considering the stress triaxiality as a function of the initial damage plastic strain and the final fracture plastic strain, and fitting it in the power form of the stress triaxiality, the final result can be obtained as:

$$\begin{aligned} \varepsilon_{pf0} &= 0.0200 + 0.1800 \cdot e^{-0.3156 \cdot \eta} \\ \varepsilon_{pf1} &= 0.2500 - 0.0025 \cdot e^{-4.8270 \cdot \eta} \end{aligned} \tag{13}$$

The damage parameter D of the material can be defined as $\frac{\sum \varepsilon_p - \varepsilon_{pf0}}{\varepsilon_{pf1} - \varepsilon_{pf0}}$. Before damage occurs, the damage parameter D is 0, and when fully damaged, D is 1.

According to the above principles, the material Vumat subroutine of Abaqus was compiled to implement the description of the material plastic constitutive and damage failure mechanical model. To verify the effectiveness of the subroutine, simulations with equal load time amplitude changes were performed by establishing two types of Elements for compression and tension, respectively (as shown in Figure 7, the left model diagram is for compression and the right model diagram is for tension). It can be found that the extracted simulation curves of the Mises stress-true strain of the units are in good agreement with the test results, which can reflect that the tensile plastic hardening of the titanium alloy is weaker than its compressive plastic hardening.

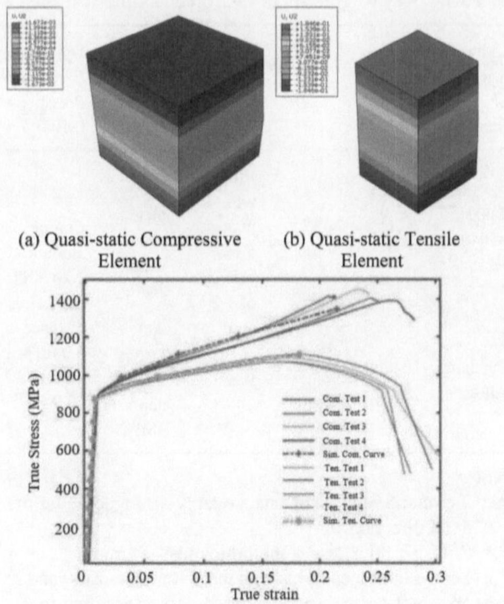

(a) Quasi-static Compressive Element (b) Quasi-static Tensile Element

Figure 7. Simulation and test verification of quasi-static tension-compression asymmetric subroutine.

3 FINITE ELEMENT SIMULATION STUDY ON DYNAMIC RESPONSE OF TITANIUM ALLOY CYLINDRICAL SHELL TO DEEP-WATER EXPLOSION

3.1 *Titanium alloy stiffened cylindrical shell model*

In this chapter, numerical simulation studies are conducted on the dynamic response of the pressure-resistant shell of a titanium alloy stiffened cylindrical shell to water depth effects under deep-water explosion using the material subroutine model developed above.

The simulation calculations in this section provide certain guiding significance for the design and safety evaluation of deep-water pressure-resistant titanium alloy shells.

The stiffened cylindrical shell is mainly composed of four structural parts: the cylindrical shell, circumferential stiffeners, flanges, and end caps. To enhance the compressive stiffness at both ends of the structure, the two ends of the cylindrical shell are designed as semi-elliptical spheres to compensate for the stiffness deficiency caused by the large span of the cylindrical shell. For convenience, the thickness of the circumferential stiffening ribs is the same as that of the shell, both designed to be 4 mm, and the stiffening ribs are evenly arranged along the axial direction of the shell at intervals. The main dimensional parameters of this pressure-resistant cylindrical shell model are shown in Table 3.

Table 3. Key parameters of reinforced cylindrical shell.

Cylindrical Shell (mm)	Stiffening Parameters	Flange	End Cap	Hydrostatic Pressure (MPa)	Dynamic Explosive Pressure (MPa)
Φ600 ×1040×4 *1	7× $T^{4\times25}_{4\times21}$ *2	Φ^{800}_{600} ×40 *3	Ellipsoid with major-to-minor axis ratio of 3:2	200 meters water depth, 2MPa	Charge amount: 1 kg TNT, explosion distance: 4.0 m, peak pressure: 10.9 MPa
Titanium Alloy	Titanium Alloy	Titanium Alloy	Steel	900 meters water depth, 9MPa	

Notes:
*1: 7 circumferential stiffeners evenly arranged along the axial direction.
*2: Stiffener thickness = shell thickness = 4mm.
*3: Semi-elliptical end caps with major-to-minor axis ratio 3:2.
*4: Hydro-static pressure conditions corresponding to 200m and 900m water depths.

3.2 *Dynamic Response Results of the Titanium Alloy Stiffened Cylindrical Shell Simulation*

For two water depths of 200m and 900m, the same explosive dynamic pressure loading is applied. The geometric structure of the model shown in Table 3 is established using Abaqus, and the developed titanium alloy material model subroutine is used for calculations of the shell, stiffeners, and flanges. Figures 8(a) and 8(b) show schematic diagrams of the geometric model and half-section mesh model in Abaqus, with a mesh size of 10mm×10mm; Figure 8(c) is the calculation assembly diagram of the overall structure of the water domain and the shell.

The water domain uses acoustic elements, with the acoustic bulk modulus of the water domain set to 2.14×10^{9} Pa and the density set to 1000 kg/m³. To model the interaction between the solid structure and the water surface, an Acoustic-Structure Coupling (ASC) method is employed. The water domain is discretized using acoustic elements, while the structure is represented by Lagrangian elements. A Tie constraint is implemented to enforce fluid-structure interaction (FSI) at the wetting surfaces, ensuring continuous transfer of loads and displacements between the fluid and solid domains. Given that the primary loads consist of explosion waves and quasi-static pressure, acoustic elements are well-suited for capturing the propagation of blast waves in the fluid. This approach also facilitates numerical convergence in fluid-structure coupling simulations, particularly for transient dynamic problems involving high-pressure loading and large deformation.

What's more, the static analysis step and explicit dynamic analysis step restart analysis in Abaqus are used to calculate the structural dynamic response of the deep-water pressure-resistant shell under the coupled action of hydro-static pressure and explosive shock waves. Figure 8(d) shows the exponential time-history curve of the explosive shock wave pressure for 1 kg TNT at an explosion distance of 4m, which is used as the dynamic load input condition for the simulation calculation model.

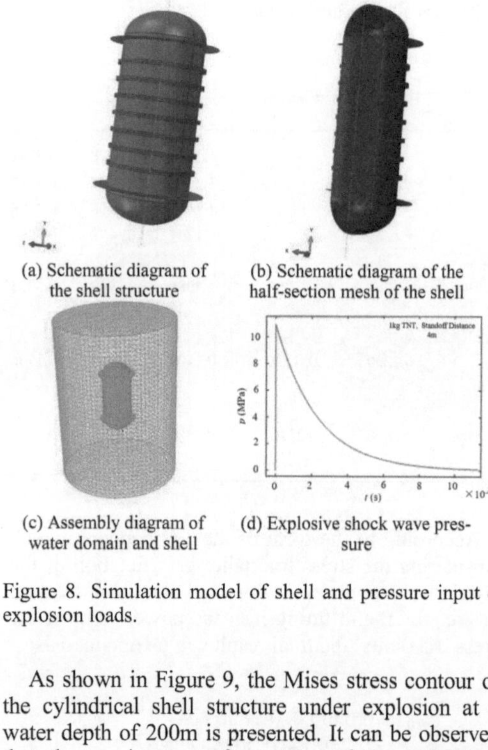

(a) Schematic diagram of the shell structure
(b) Schematic diagram of the half-section mesh of the shell
(c) Assembly diagram of water domain and shell
(d) Explosive shock wave pressure

Figure 8. Simulation model of shell and pressure input of explosion loads.

As shown in Figure 9, the Mises stress contour of the cylindrical shell structure under explosion at a water depth of 200m is presented. It can be observed that the maximum surface stress of the cylindrical shell occurs on the blast-facing surface at the center of the cylindrical shell, and a circular high-stress zone is distributed between the middle three stiffening ribs. Two stress centers are symmetrically distributed along the axial direction on the shell surface at the center of the circular arc. The stress in the cylindrical shell away from the central explosion area is relatively small. In terms of stress values, most areas of the circular high-stress zone have entered the plastic stage of the material. Additionally, from the spatial distribution of the circular high-stress zone, it can be seen that the circumferential stiffeners of the stiffened cylindrical shell exhibit a "high-stress confinement" effect, restricting high-stress loads within the three stiffened areas at the explosion center, which plays a role in stress separation on the panel and "protects" other areas of the shell.

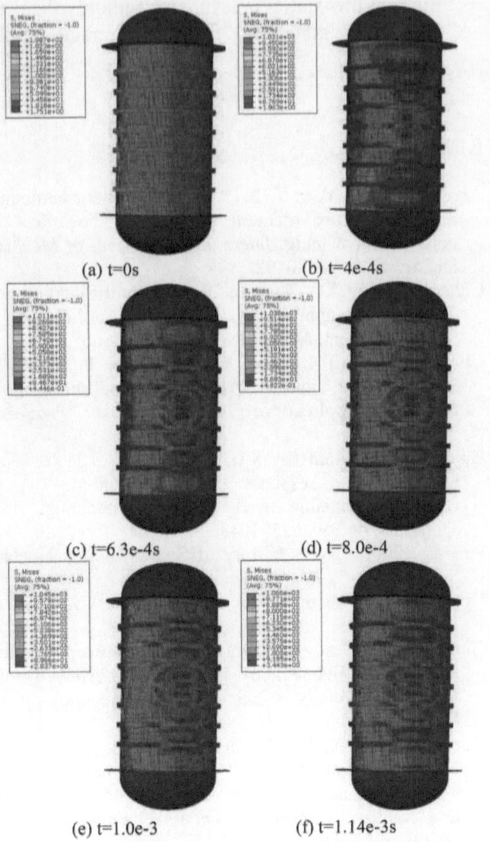

Figure 9. Dynamic response simulation results of reinforced cylindrical shell subjected the explosive loads at a water depth of 200m.

As shown in Figure 10, the Mises stress contour of the dynamic response of the titanium alloy stiffened cylindrical shell under explosion at a water depth of 900m is presented. It can be observed that when the explosive shock wave begins to propagate to 1.0×10^{-3}s, a large-area pit and breach appear on the blast-facing surface at the center of the shell structure. With the continuation of the loading time, the central pit and breach gradually expand, and fractures start to occur from the middle of the three intermediate stiffeners. The shell eventually exhibits large-area crushing deformation, and the area of the high-stress region along the axial direction of the shell is very large. Therefore, once severe fracture failure occurs in the intermediate stiffening ribs and the shell, the shell is extremely prone to forming a large-area high-stress region under the action of high hydro-static pressure, ultimately leading to instability-induced concave fracture damage of the overall structure.

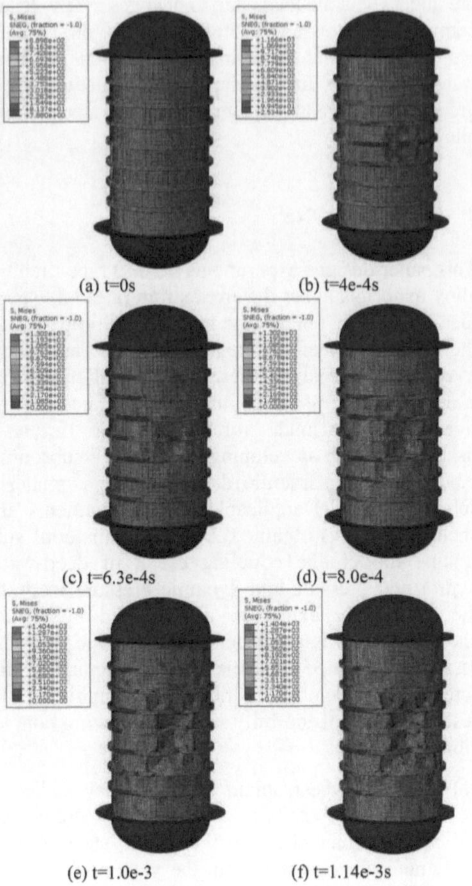

Figure 10. Dynamic response simulation results of reinforced cylindrical shell subjected the explosive loads at a water depth of 900m.

By comparing the explosive dynamic response results of the titanium alloy stiffened cylindrical shells at the two water depths, it can be found that with the increase in hydro-static pressure, the structure is more prone to severe instability damage caused by explosive impact. This implies that the hazard of deep-water explosions is far higher than that of shallow-water explosions. Therefore, when pressure-resistant deep-water structures are subjected to explosive impacts, local impact defects in the material are amplified by the coupled effect of high hydro-static pressure, ultimately leading to overall instability fracture damage of the structure.

As a comparison with steel, titanium alloy has a high strength to resist the high pressure and an corrosion resistance to withstand high-salt seawater environments. But, due to the stress-state-dependent hardening behavior of titanium alloy after plastic deformation and more significant asymmetric tension-compression properties, as well as lower fracture strain compared to traditional steel, titanium alloy shells may be more prone to asymmetric unstable failure under dynamic loads due to changes in tensile and compressive stresses. Therefore, based on titanium properties research, carrying out research on deep-water explosion for titanium alloy structures is of great significance to the safety of deep-water titanium alloy structures.

4 CONCLUSIONS

This paper designs experiments on α+β type titanium alloy materials under different strain rates and specimen forms, and obtains the mechanical properties of the material under different strain rates and stress triaxialities. Based on these mechanical properties, a tension-compression asymmetric dynamic constitutive and failure model for titanium alloy is established. Through developing a material subroutine model based on fracture damage strain, a material subroutine model applicable to solid elements and shell elements is obtained. Using this material subroutine model, the coupling effect of deep-water hydrostatic pressure and dynamic pressure loads on the deep-water explosion of stiffened cylindrical shell structures is simulated, and dynamic response results of stiffened cylindrical shell structures at different water depths are obtained. Combining the research work of the full text, the following conclusions are drawn:

(1) The α+β type titanium alloy material studied in this paper exhibits a significant strain rate strengthening effect in the plastic state. As the strain rate increases, both the yield point and the degree of hardening of the material increase. Moreover, the mechanical properties of the material are inconsistent under complex stress loading. Especially in tension and compression loading modes, the stress-strain constitutive relationship of the material has the characteristic of asymmetric plastic hardening, with compression hardening significantly stronger than tension hardening.

(2) From the deep-water explosion simulation results of the cylindrical shell structure, it can be seen that the subroutine model developed in this paper is applicable to the calculation of structural dynamic responses under complex loading conditions. This indicates that the dynamic constitutive and failure model of titanium alloy established in this paper has good mechanical calculation generalization ability.

(3) The damage degree of the cylindrical shell under the explosion condition at a water depth of 900m is more severe than that at 200m, indicating that the power of deep-water explosion is far higher than that of shallow-water explosion under the same charge amount. When pressure-resistant deep-water structures are subjected to explosive impacts in deep-water environments, local impact defects in the material are amplified by the coupling effect of high hydro-static pressure, ultimately leading to overall instability fracture damage of the structure.

REFERENCES

Ahn K, Huh H, Yoon J, 2015. Rate dependent hardening model for pure titanium considering the effect of deformation twining. *International Journal of Mechanical Sciences*. 98: 80–92.

Chang H, Wang XD, Zhou L, 2014. Current Situation and Trend of Titanium Alloys and Their Applications in Ship Equipment. *Materials China*, 9: 603–607.

Chen J, Wang TX, Zhou W, Li Q, Ge P, Xin SW, 2015. Marine Titanium Alloys at Home and Abroad and Their Applications. *Titanium Industry Progress*, 32(6): 8–12.

Chichili DR, Ramesh KT, Hemker KJ, 1998. The high-strain-rate response of alpha-titanium: Experiments, deformation mechanism and modeling. *Acta Materials*. 46(3): 1025–1043.

Fan AL, 2006. Secret History of Soviet/Russian Nuclear Submarines (5). *Modern Ships*, 5A: 41–45.

Fan AL, 2006. Secret History of Soviet/Russian Nuclear Submarines (6). *Modern Ships*, 6A: 35–39.

Guo T, 2023. Study on Dynamic Tension-Compression Asymmetric Behavior of TA7 Titanium Alloy. Hefei, Anhui: University of Science and Technology of China.

Huang W, Zan X, Nie X, Gong M, Wang Y, Xia YM, 2007. Experimantal study on the dynamic tensile behavior of a poly-crystal pure titanium at elevated temperatures. *Materials Science and Engineering*: A. 443(1-2): 33–41.

Jiang ZQ, Yang H, Zhan M, Xu XD, Li GJ, 2009. Current Situation and Prospect of Titanium Alloy Tube Development and Its Application in Aviation Field. *Journal of Plasticity Engineering*, 16(4): 44–50, 842.

Jia BH, 2015. Research on Plastic Deformation Mechanism and Constitutive Relationship of Titanium Alloy Materials. Beijing: Beijing Institute of Technology.

Meng XJ, Chen CH, Yu W, Liu YQ, 2004. Several Titanium Alloys for Marine Engineering and Their

Applications. *Annual Ocean Engineering Academic Conference*.

Pan BB, Cui WC, 2012. Experimental verification of the new ultimate strength equation of spherical pressure hulls. *Marine Structures*, 29(1): 169–176.

Wang ZT, 2021. Titanium Alloy Team of Shenyang Institute of Metal Inventions Ti62A for "Striver". *Light Metals*, 10: 18.

Yang JC, Xi JH, Yang YS, Nan L, 2014. R&D and Application of TA18 Titanium Alloy Tubes for Aerospace. *Titanium Industry Progress*, 31(4): 6–10.

Ye XW, Wang D, Qin T, Zhou J, 2023. Development of Large Manned Deep-Sea Operation Platforms in the United States and Russia. *Ship Engineering*, 45(5): 159–167.

Zhou FX, 2005. "Piranha"—Titanium Alloy Pocket Submarine. *Modern Ships*, 11A: 32–33.

Zhang, AF, Li YQ, Xu Q., Zhang B. W., Qu P., Chen W., 2024. Application Status and Prospect of Titanium Alloys for Pressure-Resistant Structures of Deep-Sea Manned Equipment. *Marine Development and Equipment Technology*, 6: 81–91.

Trends in Collision and Grounding of Ships and Offshore Structures – Zhu & Guedes Soares (Eds)

Spatio-temporal distribution identification of impact loads using orthogonal polynomials

C. Li, G.J. Shi & D.Y. Wang
State Key Laboratory of Ocean Engineering, Shanghai Jiao Tong University, Shanghai, P.R. China
Institute of Marine Equipment, Shanghai Jiao Tong University, Shanghai, P.R. China

ABSTRACT: In this paper, an inverse method that combines orthogonal polynomials and the Gaussian integration method is proposed to identify the distribution characteristics of the impact load in the spatio-temporal domain. Parabolic and trapezoidal distributions are used to simulate the spatial distribution of the tire impact loads on the deck during aircraft landings. In the time domain, the impact loads are modeled as several representative waveforms, including damped sinusoidal waves, skewed triangular waves, and post-peak sawtooth waves. By introducing singular value decomposition (SVD) with regularization in the computational process, the fitting coefficients of orthogonal polynomials for the spatial distribution of loads at different frequencies are derived, and inverse Fourier transforms are utilized to achieve spatio-temporal identification of the loads. The proposed method has been found to effectively identify the spatio-temporal distribution of tire impact loads.

1 INTRODUCTION

In modern maritime security, the capability for aircraft takeoff and landing on large offshore platforms is crucial for maintaining maritime activities and ensuring flexible scheduling. Despite advancements in constructing such platforms, accurately inputting aircraft impact loads into deck designs poses a challenge, leading to conservative design specifications. Given the limitations of current technical conditions, direct measurement of impact loads on deck structures remains difficult. Therefore, identifying the loads acting on the structure through vibration response measurement is essential. In recent years, research has explored load identification using advanced measurement techniques, signal analysis methods, and numerical computation approaches. Loads on structures are typically categorized into concentrated and distributed forces. Concentrated force identification models are simpler and can achieve reasonable accuracy with a limited number of sensors, making them widely researched and applied.

In the frequency domain, vibration force and moment determination through structural response measurement and dynamic calibration data has provided new insights for helicopter vibration analysis (Bartlett & Flannelly, 1979). Additionally, methods for locating wave sources by analyzing spectral information of dispersive waves have been proposed, allowing for simultaneous determination of spatial and temporal origins of wave sources (Doyle, 1987). Recent advancements also address stochastic dynamic load identification by integrating error analysis with weighted total least squares methods (Jia et al., 2015). Furthermore, a vehicle-bridge coupled dynamics model has been developed, combining random excitation from road surface roughness with power spectral density and deep transfer learning for identifying moving loads on bridges (Tang et al., 2022). In the time domain, a two-step iterative shrinkage/thresholding algorithm with L_1 norm minimization has been introduced to identify the locations of impact forces and their time histories. (Qiao et al., 2016). The Newmark-β method has also been applied to dynamic load identification in continuous systems, effectively reducing degrees of freedom and identifying various dynamic loads (Jiang et al., 2021). However, the accuracy of concentrated load identification is often limited in complex structural environments or varied load distributions. As engineering practices evolve, the need for improved monitoring ranges, enhanced identification accuracy, and greater system stability has become critical, emphasizing the importance of distributed load identification.

This transition poses challenges related to higher-dimensional inverse problem solving, algorithm efficiency, and noise robustness. Studies have reformulated motion differential equations as integral equations associated with dynamic loads, transforming the load identification process into a first-kind Volterra integral equation problem (Zhu & Lu, 1991). Others have utilized vibration modes and spatial orthogonal decomposition to process excitation

DOI: 10.1201/9781003684404-34

fields, integrating regularization methods to address ill-posed problems (Granger & Perotin, 1999). Force analysis methods have also been introduced to identify distributed forces on plates, combining regularization techniques to mitigate computational difficulties (Pezerat & Guyader, 2000). Advancements from the past few years have integrated the Green's function method with orthogonal polynomial fitting techniques to reconstruct distributed dynamic loads (Li et al., 2015). Additionally, methodologies combining hierarchical clustering with radial basis functions have been proposed to utilize finite acceleration response data for identifying transient loads (Liu et al., 2022). A shape function method based on moving least squares fitting has also been employed for distributed load identification (Li et al., 2017).

To improve load identification accuracy and noise robustness, experimental dynamic calibration techniques using orthogonal polynomials and Gaussian-Legendre integration methods have been proposed, demonstrating superior performance compared to conventional techniques (Jiang et al., 2021; Luo et al., 2023). However, this method primarily focuses on spatial load distribution at a single frequency, which may limit its practical application in complex loading scenarios.

This study builds on the methodology of Jiang and Luo, using SVD with regularization, to extend the application to load identification across multiple frequencies, with a particular emphasis on the spatio-temporal distribution characteristics of loads acting on beams subjected to impact loading. The improved approach provides greater simplicity compared to time-domain identification methods, thereby circumventing the challenges associated with time and spatial convolution discretization while effectively minimizing error accumulation.

2 IDENTIFICATION METHOD FOR DISTRIBUTED LOAD ON A ONE-DIMENSIONAL BEAM

2.1 *Sampling point setup*

The distributed load on the simply supported Bernoulli-Euler beam is illustrated in Figure 1.

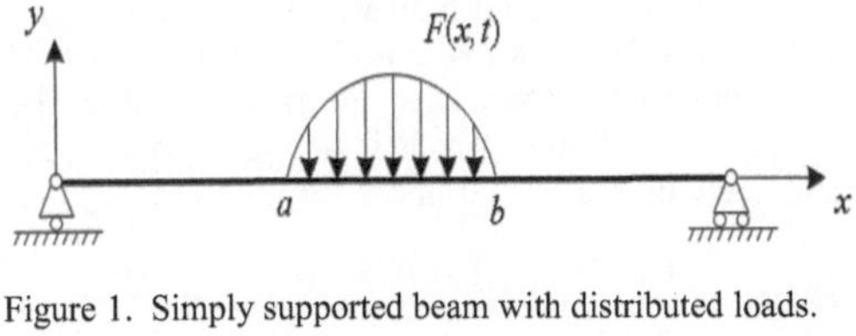

Figure 1. Simply supported beam with distributed loads.

When performing the Discrete Fourier Transform (DFT) on a finite-length signal, first set the number of sampling points as N and the time interval as Δt. Then, $t_n = n\Delta t$, where $n = 0, 1\ldots N-1$.

According to the properties of the DFT, the frequency interval is given by $\Delta f = (N\Delta t)^{-1}$. The frequency sampling points are:

$$f_k = k\Delta f = \frac{k}{N\Delta t} \tag{1}$$

where $k = 0, 1\ldots N-1$.

2.2 *Orthogonal polynomial fitting*

Legendre orthogonal polynomials are used to fit the real and imaginary parts of the distributed force in the frequency domain. Compared to the single-frequency experimental dynamic calibration methods (Jiang et al., 2021; Luo et al., 2023), this paper identifies the orthogonal polynomial coefficients related to the frequency:

$$F^r(x,f_k) = \sum_{i=0}^{m} \alpha_i(f_k)P_i(x) = \begin{bmatrix} P_1(x) \\ \vdots \\ P_m(x) \end{bmatrix}^T \begin{bmatrix} \alpha_1(f_k) \\ \vdots \\ \alpha_m(f_k) \end{bmatrix}$$

$$F^i(x,f_k) = \sum_{i=0}^{m} \beta_i(f_k)P_i(x) = \begin{bmatrix} P_1(x) \\ \vdots \\ P_m(x) \end{bmatrix}^T \begin{bmatrix} \beta_1(f_k) \\ \vdots \\ \beta_m(f_k) \end{bmatrix} \tag{2}$$

where $F^r(x, f_k)$ = the real part of the distributed force, $F^i(x, f_k)$ = the imaginary part of the distributed force, $P_i(x)$ = the i-th order Legendre polynomial, $\alpha_i(f_k)$ = the i-th order fitting coefficient at the real part frequency f_k, $\beta_i(f_k)$ = the i-th order fitting coefficient at the imaginary part frequency f_k.

For the relationship between one-dimensional beam distributed excitation and response, it can be expressed as:

$$\int_a^b H(x_\eta, x, f_k)F(x, f_k)dx = X(x_\eta, f_k) \tag{3}$$

where $H(x_\eta, x, f_k)$ = the frequency response function between the distributed load x and the response point x_η, $F(x, f_k)$ = the distributed load in the frequency domain, $X(x_\eta, f_k)$ = the acceleration response at the response point x_η, where $\eta = 1, 2\ldots u$.

This paper utilizes the Gaussian-Legendre integration discretization method to approximate the integral of the distributed force as a weighted sum of function values at Gaussian discrete points:

$$\sum_{\xi=1}^{v} w_\xi H(x_\eta, x_\xi, f_k)F(x_\xi, f_k) = X(x_\eta, f_k) \tag{4}$$

where

$$w_\xi = \frac{b-a}{(1-\lambda_\xi^2)[P_m'(\lambda_\xi)]^2}$$
$$x_\xi = \frac{b+a}{2} + \frac{b-a}{2} \cdot \lambda_\xi \tag{5}$$

with $\lambda_\xi \in [-1,1]$ being the Gaussian integration nodes, which are the roots of the Legendre orthogonal polynomials, and x_ξ is the position on the beam corresponding to the Gaussian integration nodes.

Since the signal in the frequency domain is complex, the above equation can be rewritten as:

$$\sum_{\xi=1}^{v} w_\xi \begin{bmatrix} H_{\eta k\xi}^r & -H_{\eta k\xi}^i \\ H_{\eta k\xi}^i & H_{\eta k\xi}^r \end{bmatrix} \begin{bmatrix} F_{k\xi}^r \\ F_{k\xi}^i \end{bmatrix} = \begin{bmatrix} X_{\eta k}^r \\ X_{\eta k}^i \end{bmatrix} \tag{6}$$

Substituting into Equation 2, and for the purpose of simplification, let

$$Q_1 = \left[H_{\eta k\xi}^r P_1(x_\xi) \quad \cdots \quad H_{\eta k\xi}^r P_m(x_\xi) \right]$$
$$Q_2 = \left[H_{\eta k\xi}^i P_1(x_\xi) \quad \cdots \quad H_{\eta k\xi}^i P_m(x_\xi) \right]$$
$$\theta_{\eta k\xi} = w_\xi \begin{bmatrix} Q_1 & -Q_2 \\ Q_2 & Q_1 \end{bmatrix} \qquad \Theta_{\eta k} = \sum_{\xi=1}^{v} \theta_{\eta k\xi} \tag{7}$$

$$A_k = \left[\alpha_1(f_k) \quad \cdots \quad \alpha_m(f_k) \right]^{\mathrm{T}}$$
$$B_k = \left[\beta_1(f_k) \quad \cdots \quad \beta_m(f_k) \right]^{\mathrm{T}}$$
$$\delta_k = \left[A_k \quad B_k \right]^{\mathrm{T}} \quad X_{\eta k} = \left[X_{\eta k}^r \quad X_{\eta k}^i \right]^{\mathrm{T}}$$

The final result is

$$\boldsymbol{\Theta\delta} = \boldsymbol{X} \tag{8}$$

Expand the above Equation 9 into:

$$\begin{bmatrix} \Theta_{10} & & \\ \vdots & & \\ \Theta_{s0} & & \\ & \ddots & \\ & & \Theta_{1(N-1)} \\ & & \vdots \\ & & \Theta_{s(N-1)} \end{bmatrix} \begin{bmatrix} \delta_0 \\ \vdots \\ \delta_{(N-1)} \end{bmatrix} = \begin{bmatrix} X_{10} \\ \vdots \\ X_{s0} \\ \vdots \\ X_{1(N-1)} \\ \vdots \\ X_{s(N-1)} \end{bmatrix} \tag{9}$$

2.3 *SVD with regularization*

Owing to the instability in solving the linear system $\boldsymbol{\Theta\delta} = \boldsymbol{X}$, the regularization parameter λ is introduced, resulting in:

$$\min_{\delta} \|\boldsymbol{\Theta\delta} - \boldsymbol{X}\|_2^2 + \lambda\|\boldsymbol{\delta}\|_2^2 \tag{10}$$

where the parameter λ is derived from the L-curve.

Combined with Singular Value Decomposition (SVD):

$$\min_{\delta} \left\| U\Sigma V^T\delta - X \right\|_2^2 + \lambda\|\delta\|_2^2 \tag{11}$$

where $U \in \mathrm{R}^{(sN)\times(sN)}$ and $V \in \mathrm{R}^{N\times N}$ are orthogonal matrices; $\Sigma \in \mathrm{R}^{(sN)\times N}$ is a diagonal matrix with diagonal elements as singular values $\sigma_1, \sigma_2 \ldots \sigma_N$.

Directly truncating the first N elements of $\boldsymbol{\Theta\delta} = \boldsymbol{X}$ will ignore the last $sN - N$ elements, which may lead to information loss. Thus, it is necessary to explicitly divide $U^{\mathrm{T}}\boldsymbol{X}$ into the non-zero singular value part and the zero singular value part:

$$U^T X = \begin{bmatrix} U_1^T X \\ U_2^T X \end{bmatrix} \tag{12}$$

where U_1 corresponds to non-zero singular values, and U_2 corresponds to zero singular values.

The non-zero singular value portion is solved using regularization, while the zero singular value portion is set directly to 0. For the non-zero singular value portion of regularized solution:

$$y_{1i} = \frac{\sigma_i (U_1^T \boldsymbol{X})_i}{\sigma_i^2 + \lambda}, \quad i = 1, 2 \ldots N \tag{13}$$

and for the zero singular value portion $y_2 = \boldsymbol{0}$.

Let $y_1 = [y_{11} \ldots y_{1N}]^{\mathrm{T}}$ and combining both portions yields $y = [y_1\ y_2]^{\mathrm{T}}$.

Convert to original solution as follows:

$$\boldsymbol{\delta} = V y_1 \tag{14}$$

Substituting Equation 14 back into Equation 2 facilitates the determination of the real part $F^r(x, f_k)$ and the imaginary part $F^i(x, f_k)$ of the load.

2.4 *Load frequency-time domain transformation*

The formula for the inverse DFT is:

$$F(x, t_n) = \frac{1}{N} \sum_{k=0}^{N-1} [F^r(x, f_k) + iF^i(x, f_k)] \exp(i\, 2\pi f_k t_n) \tag{15}$$

The actual load signals in the time domain are real-valued, and their Fourier transform exhibits conjugate symmetry, which indicates that the imaginary component of the load is entirely determined by its real counterpart. Therefore, when conducting the Fourier inverse transform, it is adequate to concentrate solely on the real component of the load signal:

$$\tilde{F}(x, t_n) = \frac{1}{N} \cdot \sum_{k=0}^{N-1} [F^r(x, f_k) \cos(2\pi f_k t_n) - F^i(x, f_k) \sin(2\pi f_k t_n)] \tag{16}$$

3 NUMERICAL EXAMPLES

To validate the proposed method's reliability for engineering applications, this study simplifies aircraft tire-deck impact load identification to longitudinal load analysis on a simply supported beam. Focusing on the method's longitudinal load distribution analysis, the 3D problem is reduced to 2D for effective verification. This paper employs damped sinusoidal waves (Mao et al., 2020), skewed triangular waves (Du,2021), and post-peak sawtooth waves (Zhang et al., 2023) to approximate the impact load in the time domain. The frequency components of these three typical impact loads progressively increase from left to right. This study examines the spatio-temporal distribution characteristics of three typical impact loads influenced by tire pressure, which exhibit parabolic and trapezoidal distributions (Liu, 2013), and evaluates the method's performance under the compound excitation of Gaussian white noise and impulsive interference. The noise excitation is expressed as follows:

$$F_{noise} = noise\_level \times \max F \times (randn + \sin(\pi t/T)) \tag{17}$$

Identification errors are evaluated by mean relative error (MRE), correlation coefficient (CC), and relative error (RE).

$$MRE(F,\tilde{F}) = \frac{1}{n}\sum_{i=1}^{n}\left|\frac{F(x_i,t)-\tilde{F}(x_i,t)}{F_{max}(x_i,t)}\right| \times 100\%$$

$$CC(F,\tilde{F}) = \frac{\sum_{i=1}^{p}\sum_{j=1}^{q}(F_{ij}-\overline{F})(\tilde{F}_{ij}-\overline{\tilde{F}})}{\sqrt{\sum_{i=1}^{p}\sum_{j=1}^{q}(F_{ij}-\overline{F})^2}\sqrt{\sum_{i=1}^{p}\sum_{j=1}^{q}(\tilde{F}_{ij}-\overline{\tilde{F}})^2}}$$

$$RE(F,\tilde{F}) = \left|\frac{F(x,t)-\tilde{F}(x,t)}{F_{\max}(x,t)}\right| \times 100\% \tag{18}$$

The simulation model in this paper is consistent with the experimental dynamic calibration method for load identification on beams (Luo et al., 2023).

3.1 *An investigation into the accuracy of impact load identification under various spectral characteristics*

In consideration of various impact loads and distribution forms, six cases are established, as presented in Table 1.

Six types of loads are applied at the specified positions of the simply supported beam, specifically at $x \in [400,600]$. Measurement points are set uniformly at $u = 11$with $\xi = 11$ Gauss points. To ensure identification accuracy, the fitting order should be incrementally increased while maintaining computational efficiency. When selecting the appropriate order, compare the MRE between consecutive orders. If the absolute difference falls below 0.05, select the lower order to prevent overfitting. Otherwise, continue increasing the order until this threshold is satisfied. The fitting order is established as $m = 3$ for Cases I to III and $m = 8$ for Cases IV to VI, with a total of $N = 500$ sampling points and a time interval of $\Delta t = 0.002$s.

Table 1. Different impact load forms.

Load Types	Load Forms
Case I: Parabolic-Damped Sinusoidal	$f(x,t) = [10e^{-0.3\pi t}\sin(12\pi t)](-\frac{(\chi-500)^2}{10^4}+1)$
Case II: Parabolic-Triangular	$f(x,t) = \begin{cases} \frac{10}{0.07}t(-\frac{(x-500)^2}{10^4}+1) & 0 < t \le 0.07 \\ -\frac{10}{0.33}(-\frac{(x-500)^2}{10^4}+1)\cdot(t-0.07) & 0.07 < t \le 0.4 \\ 0 & else \end{cases}$
Case III: Parabolic-Post-Peak Sawtooth	$f(x,t) = \begin{cases} 100(-\frac{(x-500)^2}{10^4}+1)\cdot(t-0.25) & 0.25 < t \le 0.35 \\ 0 & else \end{cases}$
Case IV: Trapezoidal-Damped Sinusoidal	$f(x,t) = [10e^{-0.3\pi t}\sin(12\pi t)]\cdot[-1+\tanh(0.035(x-400))-\tanh(0.035(x-600))]$
Case V: Trapezoidal-Triangular	$f(x,t) = \begin{cases} \frac{10}{0.07}[\tanh(0.035(x-400))-\tanh(0.035(x-600))-1]t & 0 < t \le 0.07 \\ \frac{10}{0.33}[\tanh(0.035(x-400))-\tanh(0.035(x-600))-1](0.07-t) & 0.07 < t \le 0.4 \\ 0 & else \end{cases}$
Case VI: Trapezoidal-Post-Peak Sawtooth	$f(x,t) = \begin{cases} 100(t-0.25)\cdot[-1+\tanh(0.035(x-400))-\tanh(0.035(x-600))] & 0.25 < t \le 0.35 \\ 0 & else \end{cases}$

The following figures present the true loads and identified loads for Cases I to III under sufficient tire pressure, demonstrating a parabolic distribution, as well as the true loads and identified loads for Cases IV to VI under insufficient tire pressure, illustrating a trapezoidal distribution, in both scenarios with and without noise excitation.

The calculation results indicate that the MRE associated with the parabolic distribution remains below 0.02%. For the more complex trapezoidal distribution requiring higher-order fitting, while its identification accuracy under noise-free conditions is lower than that of simple function distributions, the MRE stays within 1.09%. All correlation coefficients exceed 0.98, thereby confirming the effectiveness and reliability of this method for impact load identification. Under 2% compound noise excitation, the identification error associated with the trapezoidal distribution exceeds that of the parabolic distribution, ranging

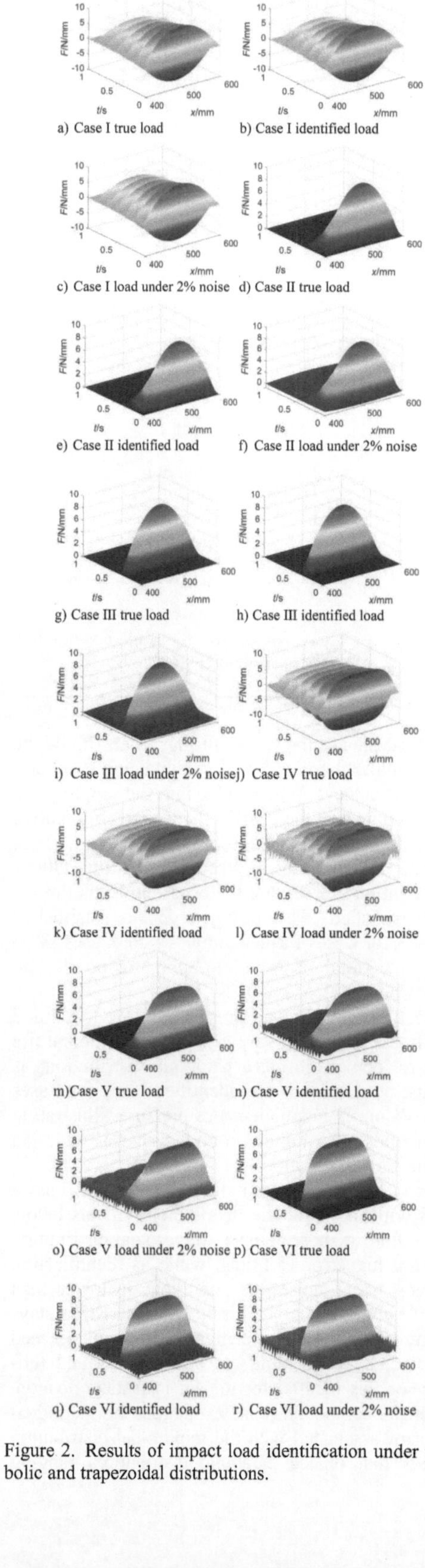

Figure 2. Results of impact load identification under parabolic and trapezoidal distributions.

Table 2. Load identification errors and correlation coefficients across various cases.

	Case I	Case II	Case III	Case IV	Case V	Case VI
MRE/%	0.0078	0.0158	0.0078	0.9682	0.9874	0.7769
MRE under 2% noise/%	1.3697	1.2857	1.3847	2.5648	2.3529	2.3880
CC	1.0000	1.0000	1.0000	0.9993	0.9981	0.9979
CC under 2% noise	0.9990	0.9978	0.9918	0.9934	0.9939	0.9821

from 2.35% to 2.56%. This finding suggests that the proposed method demonstrates a capacity to effectively manage noise and uncertainty within the input data.

Furthermore, the proposed method exhibits robust identification capabilities for loads with varying frequency components. Using Cases IV to VI as examples, Case VI encompasses the highest number of frequency components; however, its mean relative error and correlation coefficient are comparable to those of Cases IV and V, demonstrating the universality of this method in impact load identification.

3.2 *Analysis of recognition accuracy with different boundary conditions*

To validate the recognition capability of this method at different boundary conditions, excitations, as indicated in Table 3, are applied to Cases VII to X, defined as $x \in [0,1000]$. A total of $u = 11$ uniformly distributed measurement points and $\xi = 11$ Gaussian points are established on the beam. The polynomial fitting for Cases VII to X is performed at $m = 3$.

Table 3. Different boundary conditions.

Boundary Conditions	Load Forms
Case VII: simply supported beam Case VIII: fixed(L)-simply supported (R) beam Case IX: Cantilever(L) beam Case X: fixed(L)-fixed(R) beam	$f(x,t) = \begin{cases} \frac{10}{0.07}t(-\frac{(x-500)^2}{10^4}+50) & 0 < t \leq 0.07 \\ \frac{10}{0.33}(-\frac{(x-500)^2}{10^4}+50) \cdot (0.07-t) & 0.07 < t \leq 0.4 \\ 0 & else \end{cases}$

The following figures present the identification of loads with parabolic distributions under varying boundary conditions.

From the recognition results, it is evident that the errors primarily occur at the boundaries. The relative errors under symmetric boundary conditions maintain lower magnitudes, where Cases VII and X show peak values of 0.78% and 0.93% respectively. For asymmetric conditions represented by Cases VIII and IX, error escalation reaches corresponding values of 1.60% and 1.90%. The results demonstrate that the influence of boundary conditions on load identification is predominantly localized in boundary regions. Remarkably, the MRE across all tested boundary conditions remains below 0.10%, demonstrating the method's universal applicability in load identification under varying constraint scenarios.

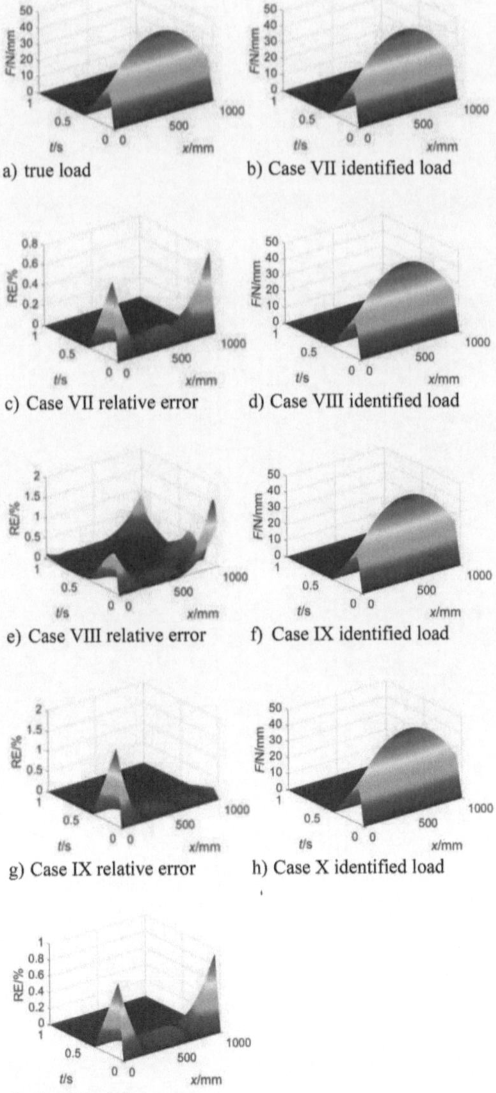

a) true load
b) Case VII identified load
c) Case VII relative error
d) Case VIII identified load
e) Case VIII relative error
f) Case IX identified load
g) Case IX relative error
h) Case X identified load
i) Case X relative error

Figure 3. Results of impact load under varying boundary conditions.

Table 4. Load identification relative errors and correlation coefficients across various cases.

	Case VII	Case VIII	Case IX	Case X
MRE/%	0.0514	0.0954	0.0608	0.0612
max RE /%	0.7782	1.6043	1.8985	0.9339

4 CONCLUSIONS

This study proposes a method for identifying the spatio-temporal load distribution of beam structures by combining SVD regularization. The effectiveness of this method is validated through simulation analysis.

The generalizability of this method in complex load identification has been substantiated through the identification of the spatio-temporal distribution characteristics of typical impact loads under parabolic and trapezoidal distributions. The introduction of the compound excitation demonstrates that the algorithm continues to yield robust identification results even under unfavorable conditions. By varying the boundary conditions, the method has been rigorously validated for its universality and effectiveness in distributed load identification.

REFERENCES

Bartlett, F.D., Flannelly, W.G., 1979. Model Verification of Force Determination for Measuring Vibratory Loads. *Journal of the American Helicopter Society* 24(2): 10.

Doyle, J.F., 1987. An experimental method for determining the location and time of initiation of an unknown dispersing pulse. *Experimental Mechanics* 27: 229–233.

Du, L., 2021. Simulation analysis on performance degradation of airfield pavement structure under aircraft load. *Southeast University.*

Granger, S., Perotin, L., 1999. An inverse method for the identification of a distributed random excitation acting on a vibrating structure part 1: Theory. *Mechanical Systems and Signal Processing* 13: 53–65.

Jia, Y., Yang, Z., Guo, N., Wang, L., 2015. Random dynamic load identification based on error analysis and weighted total least squares method. *Journal of Sound and Vibration* 358: 111–123.

Jiang, J., Ding, M., Li, J., 2021a. A novel time-domain dynamic load identification numerical algorithm for continuous systems. *Mechanical Systems and Signal Processing* 160: 107881.

Jiang, J., Luo, S., Zhang, F., 2021b. One novel dynamical calibration method to identify two-dimensional distributed load. *Journal of Sound and Vibration* 515: 116465.

Li, K., Liu, J., Han, X., Jiang, C., Zhang, D., 2017. Distributed dynamic load identification based on shape function method and polynomial selection technique. *Inverse Problems in Science and Engineering* 25: 1323–1342.

Li, K., Liu, J., Han, X., Sun, X., Jiang, C., 2015. A novel approach for distributed dynamic load reconstruction by space–time domain decoupling. *Journal of Sound and Vibration* 348: 137–148.

Liu, M., 2013. Numerical simulation of interaction between tire and asphalt pavement. *Harbin Institute of Technology*.

Liu, Y., Wang, L., Li, M., Wu, Z., 2022. A distributed dynamic load identification method based on the hierarchical-clustering-oriented radial basis function framework using acceleration signals under convex-fuzzy hybrid uncertainties. *Mechanical Systems and Signal Processing* 172: 108935.

Luo, S., Jiang, J., Zhang, F., 2023. Research on dynamic calibration experiment of distributed dynamic load identification in frequency-domain. *Journal of Vibration Engineering* 36: 706–717.

Mao, Y., Wang, Y., Chen, Y, Liu, Q., Li, M, Niu, B., Zhang, P., Yan, X., Zheng, M., Li, X., 2020. Catapult launch and arrested landing shock test methods. *Equipment Environmental Engineering* 17: 45–50.

Pezerat, C., Guyader, J.L., 2000. Force Analysis Technique: Reconstruction of Force Distribution on Plates. *Acta Acustica United with Acustica* 86.

Qiao, B., Zhang, X., Gao, J., Chen, X., 2016. Impact-force sparse reconstruction from highly incomplete and inaccurate measurements. *Journal of Sound and Vibration* 376: 72–94.

Tang, Q., Xin, J., Jiang, Y., Zhou, J., Li, S., Chen, Z., 2022. Novel identification technique of moving loads using the random response power spectral density and deep transfer learning. *Measurement* 195: 111120.

Zhang, Y., Huang, C., Li, B., 2023. A method to determine the equivalent static load of aircraft impact load. *Advances in Aeronautical Science and Engineering* 14: 157-163+177.

Zhu, J., Lu, Z., 1991. A time domain method for identifying dynamic loads on continuous systems. *Journal of Sound and Vibration* 148: 137–146.

Trends in Collision and Grounding of Ships and Offshore Structures – Zhu & Guedes Soares (Eds)
© 2026 The Author(s), ISBN 978-1-041-16361-9

Numerical study on dynamic responses of aluminum honeycomb sandwich plate under repeated slamming loadings

D. Liu & K.L. Guo
Key Laboratory of High Performance Ship Technology (Wuhan University of Technology), Wuhan, P.R. China
School of Naval Architecture, Ocean and Energy Power Engineering, Wuhan University of Technology, Wuhan, P.R. China

Y. Liao & L. Zhu
School of Naval Architecture, Ocean and Energy Power Engineering, Wuhan University of Technology, Wuhan, P.R. China

ABSTRACT: In order to study the dynamic responses of aluminum honeycomb sandwich plate (AHSP) under repeated slamming load, the numerical simulations was conducted by using ABAQUS. The slamming load obtained from the test is simplified, then loaded repeatedly, and the deformation modes of the face sheet and the core are discussed, and the time history curves of the deflection and the energy absorption are analyzed. The performance of face sheet deformation profile, permanent deflection and energy absorption of each part with the slamming number is explored. The results show that, as the slamming number increases, the deformation of the face sheets increases, but the growth rate decreases. The energy absorption of the honeycomb core and the front sheet is larger, and the energy absorption of the back sheet is the smallest, while the energy absorption of each part under each slamming load decreases with the increase of the slamming number.

1 INTRODUCTION

With the advancement of high-speed ships, ships traveling at high speeds are subjected to wave impacts, leading to significant slamming loads. These slamming loads exhibit repetitive characteristics, which can cause cumulative damage to the hull structure and may even lead to structural failure.

To improve the impact resistance of ship structures, numerous researchers have conducted studies on the dynamic responses of ship structures subjected to slamming loads. Chuang (1966) conducted free-fall impact experiments on rigid flat plates to investigate the characteristics of slamming loads. Zhu & Faulkner (1995) carried out a series of drop tests on twin-hull model, and studied the slamming pressure and structural responses. Duan (2020) and Ma et al. (2024) conducted slamming experiments on a series of wedge-shaped models with different deadrise angles, as well as scaled aluminum-stiffened plates.

Compared with traditional ship structures, lightweight sandwich structures process several advantages, including high strength-to-weight ratio, high specific stiffness, superior toughness, and excellent energy absorption capacity. Researchers have conducted investigations on the dynamic responses and energy absorption of lightweight sandwich structures under slamming loads. Zhao & He et al. (2013, 2016) conducted numerical simulations on the water entry slamming of pyramid lattice sandwich plates and corrugated hybrid sandwich plates. He et al. (2020) performed numerical simulations on corrugated composite metal sandwich plates under different deadrise angles and various slamming velocities. Guo et al. (2024) performed experimental investigations on the water-entry slamming of aluminum honeycomb sandwich plates, analyzed the slamming load characteristics and provided empirical fitting formulas for face sheet deflection and core layer compression.

As a typical type of sandwich structure, the honeycomb sandwich structure has attracted widespread attention from researchers due to its excellent mechanical properties. Zhang et al. (2009) investigated the dynamic responses of aluminum honeycomb sandwich plates under explosive loading using pendulum impact test. Xin et al. (2018) conducted low-velocity drop-weight impact tests to examine the dynamic performance of aluminum honeycomb sandwich plates.

Naval architecture and marine engineering structures are frequently subjected to repeated loads during operation, such as repeated collisions between offshore platforms and supply vessels, repeated ice impacts, and repeated wave slamming. Huynh et al. (2024) focused on the issue of repeated impulsive loading and

DOI: 10.1201/9781003684404-35

conducted parametric studies on stiffened plates of different sizes through numerical simulations. Yue et al. (2024a, 2024b) investigated structural responses under repeated impulsive loading by conducting repeated impact tests on metallic foam projectiles. They further explored the deformation mechanisms of corrugated sandwich beams and metallic monolithic plates under repeated impact loads through numerical simulations.

Based on the above analysis, research on repeated slamming remains relatively limited, particularly regarding sandwich structures. The deformation accumulation mechanisms and energy dissipation characteristics of honeycomb sandwich structures under repeated slamming loads remain unclear and require further investigation.

In this paper the dynamic responses and energy absorption characteristics of aluminum honeycomb sandwich plate (AHSP) are investigated using the LS-DYNA software. The numerical results are compared with experimental data to validate the accuracy of the numerical simulation.

2 REPEATED SLAMMING TESTS

2.1 *Experiment procedure*

The water-entry slamming experiment was conducted in the experimental water tank, measuring 29 m in length, 3.8 m in width, and a maximum depth of 1.8 m. Figure 1 illustrates the experimental setup for the water-entry slamming test of the AHSP. The device is positioned at the center of the water tank and is fixed at the bottom. The setup consists of an electromagnet, clump weights, angle-adjusting bolts, a cabinet, guide rails, and a spring damper. This setup enables water-entry slamming tests conducting at varying heights and angles.

Figure 2 shows the test specimen of AHSP, which consists of three parts: the front face sheet, the back-face sheet, and the honeycomb core. The front and back face sheets are square plates with identical size, with side length *S*=250 mm and thickness *T*=1 mm. The honeycomb core features a regular hexagonal structure with side length *l*=15 mm and height *H* =15 mm. The honeycomb cell walls are available in two configurations: single-layer walls with thickness t_a= 0.04 mm and double-layer walls with thickness t_b =0.08 mm.

Using this experimental setup, repeated slamming tests were conducted at drop height of 1.0 m, with five successive impacts. The pressure-time history curve of the front face sheet was recorded for each slamming event, and the final deflection at the center of the front face sheet was measured. The arrangement of the pressure measurement points are shown in Figure 3.

2.2 *Experimental Results*

Figure 4 presents the pressure-time history curves of the AHSP under repeated slamming tests at a drop

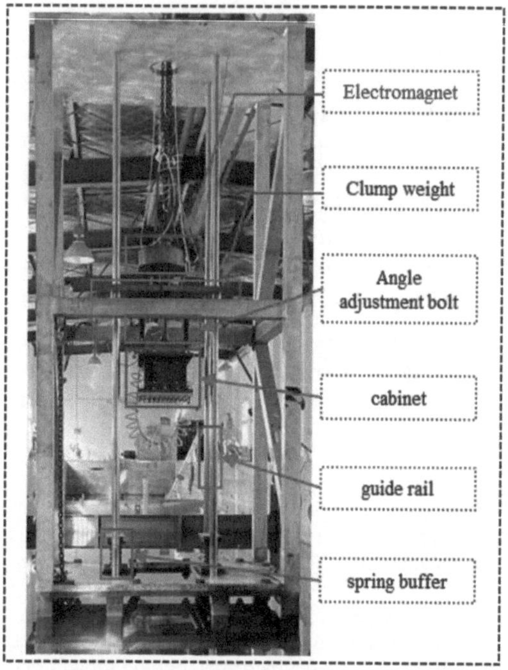

Figure 1. Slamming experimental device.

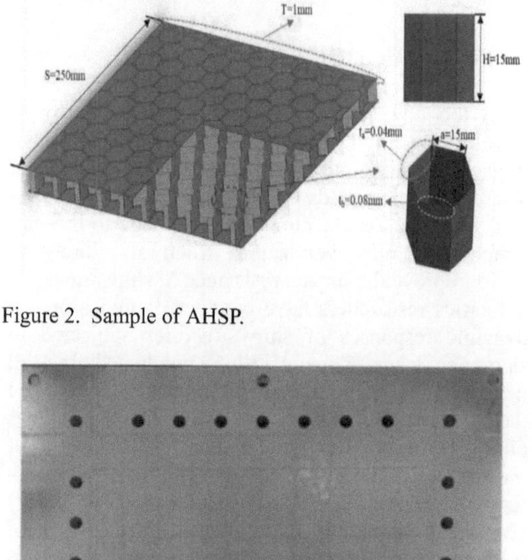

Figure 2. Sample of AHSP.

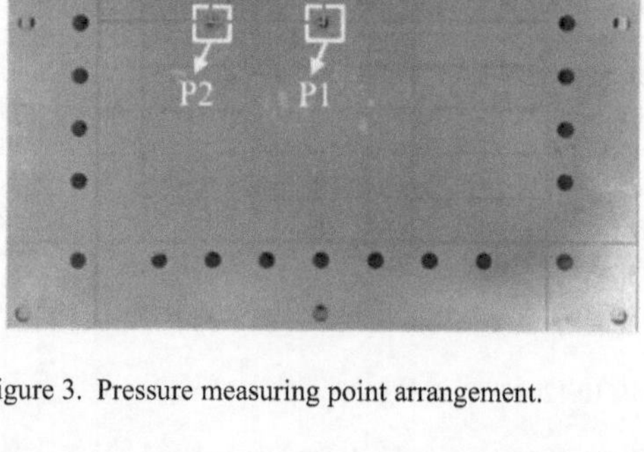

Figure 3. Pressure measuring point arrangement.

height of 1.0 m. Table 1 summarizes the permanent deflection at the center of the front face sheet after each impact.

3 NUMERICAL SIMULATION

3.1 *Finite element model*

3.1.1 *Mesh division and boundary conditions*

To validate the accuracy of the numerical simulation, the dimensions and boundary conditions of the

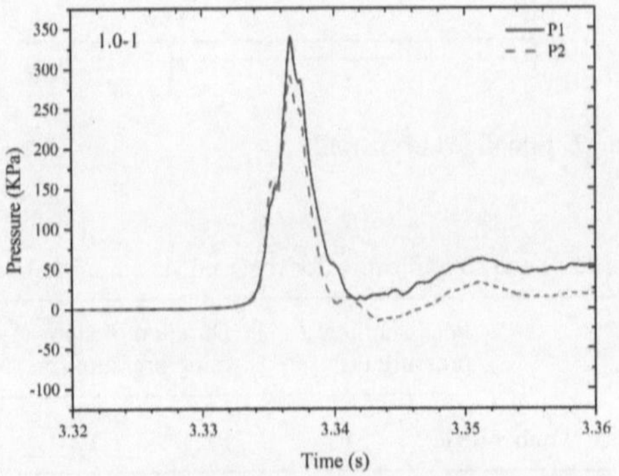

(a) First slamming experiment.

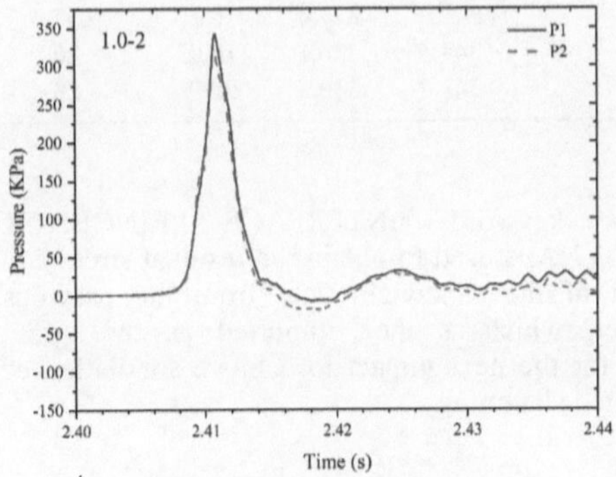

(b) Second slamming experiment

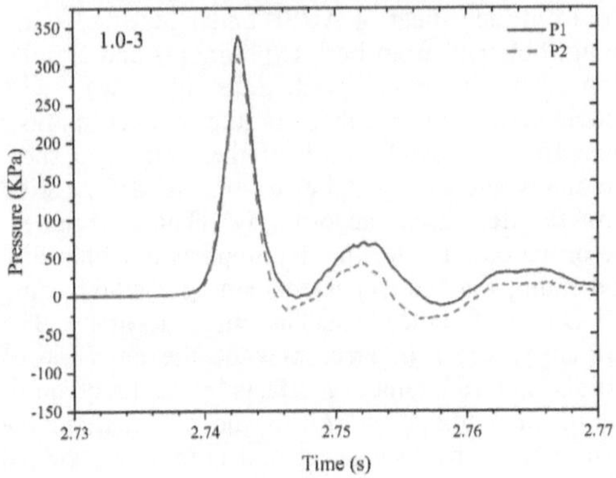

(c) Third slamming experiment

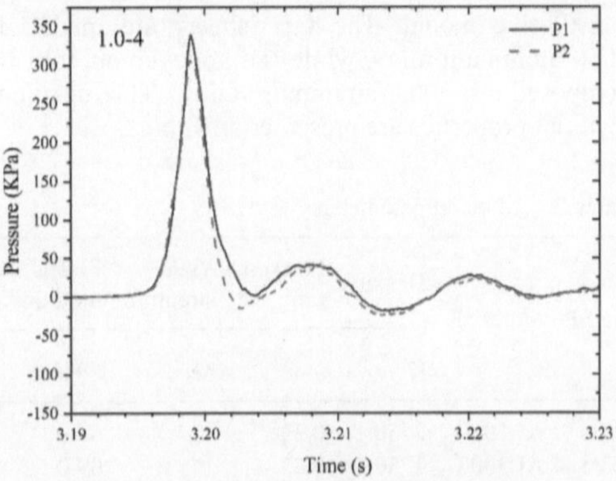

(d) Fourth slamming experiment

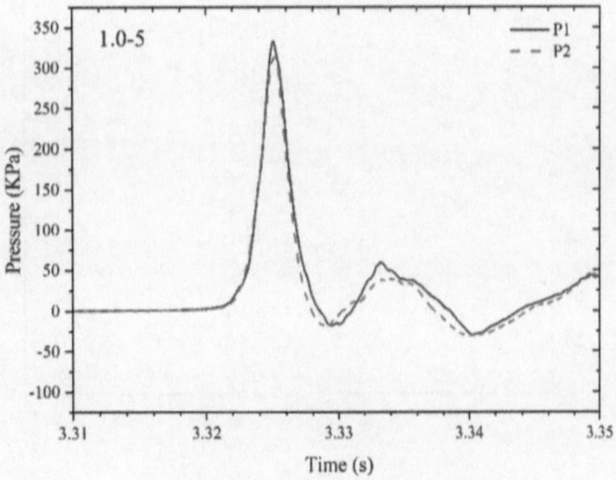

(e) Fifth slamming experiment

Figure 4. Pressure-time history curves of the AHSP.

Table 1. The permanent deflection of the front sheet.

Impact Number	Permanent deflection (mm)
1st	2.67
2nd	4.17
3rd	4.97
4th	4.99
5th	5.03

AHSP specimen in the simulation are kept consistent with those in the experiment. The AHSP has dimensions of 250 × 250 × (1 + 15 + 1) mm, and the boundary conditions are clamped. The mesh size of the core significantly affects the accuracy of the simulation. To ensure precision while maintaining computational efficiency, a mesh convergence analysis is performed. The mesh size of the AHSP face sheet is set to 5 mm, and the core mesh size is set to 2.5 mm.

3.1.2 *Material properties*

The material properties of the AHSP are modeled using *MAT_PIECEWISE_LINEAR_PLASTICITY constitutive model. The face sheets are made of 1060 aluminum alloy, while the honeycomb core is composed of 3003 aluminum alloy. The detailed material properties are presented in Table 2.

Table 2. Material parameters of AHSP.

AHSP	Materials	Density	Poisson ratio	Yield strength	Elastic modulus
		Kg/m^3	—	MPa	GPa
Plate	AL1060	2750	0.33	117.0	69.0
Core	AL3003	2750	0.33	145.0	69.0

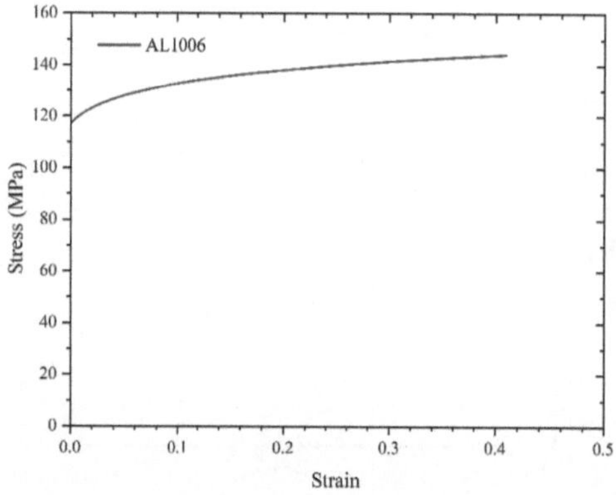

Figure 5. Plastic stress-strain curve of sheet.

3.1.3 *Simplification of load and analysis method*

To facilitate direct application of the slamming load, the method proposed by Turong et al. (2017) is adopted. The load is assumed to be uniformly distributed over the sheet surface, and the slamming pressure-time history curve is simplified as isosceles triangular pulse based on the peak pressure and peak duration. As shown in Figure 5, P represents the average of the peak impact pressures obtained from the experiment, while t represents the average duration of the slamming pressures, as summarized in Table 3.

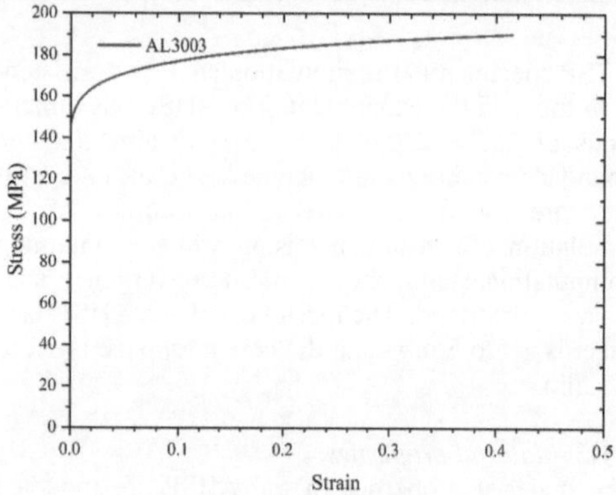

Figure 6. Plastic stress-strain curve of core.

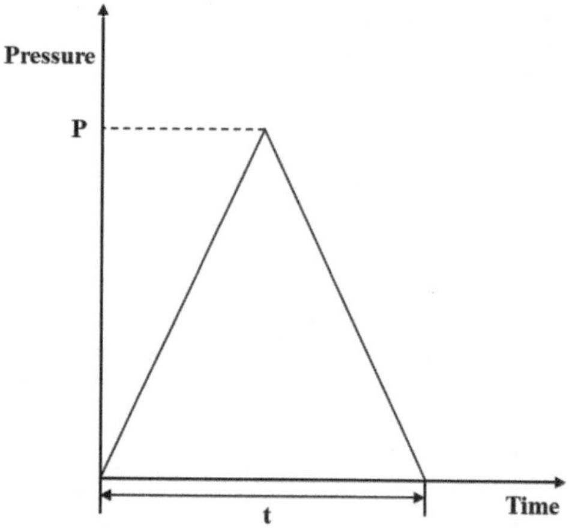

Figure 7. Simplified load profile.

Table 3. Load simplification for repeated slamming test.

	Peak slamming pressure (kPa)		Duration of slamming pressure (ms)	
Impact Number	P1	P2	P1	P2
1st	303.5	290.1	11.02	9.28
2nd	330.2	316.3	11.80	9.65
3rd	342.4	327.4	10.20	9.57
4th	344.5	326.1	10.93	8.86
5th	343.7	330.3	9.59	8.28

The keyword *INTERFACE_SPRINGBACK_LSDYNA is used to obtain the residual stress and strain of the sandwich plate from the previous impact, which is then imported as the initial state for the next impact to achieve simulation of repeated slamming.

3.2 *Verification of numerical method*

Figure 6 compares the deformation at the center of the front face sheet of AHSP under repeated slamming, obtained from both experiments and simulations. The numerical predictions agree well with the experimental results. As the impact number increases, the deformation of the front face sheet gradually increases, while the rate of deformation growth decreases accordingly. The permanent deformation error for the five impacts is within 6%, indicating that the repeated loading method using LS-DYNA for the AHSP has high accuracy. The primary sources of error may be the omission of fluid-structure interaction effects in the direct application of loading, as well as the fact that, in the experiment, the face sheet and core were bonded with polyurethane adhesive, which was not considered in the numerical simulation.

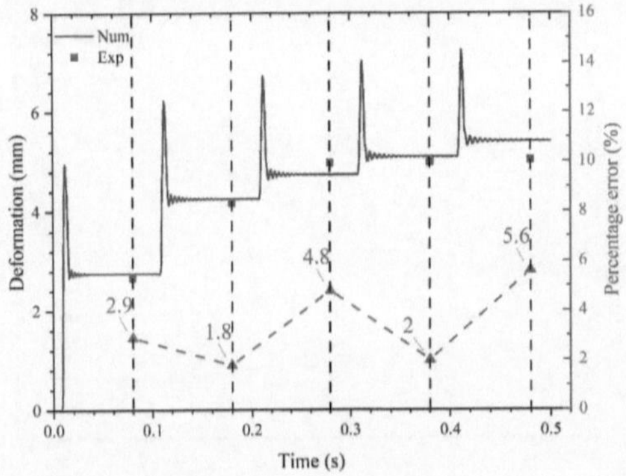

Figure 8. Comparison of deformation of the front sheet under repeated slamming.

3.3 *Dynamic responses of AHSP*

3.3.1 *Dynamic responses*

Figure 7 presents the time-history curves of the mid-point deflection for both the front face sheet and the back face sheet of the AHSP. The results indicate that under slamming loads, the face sheet initially undergoes significant instantaneous deformation. Subsequently, due to effect of the elastic recovery, the deflections decrease gradually and to be constant value, leading to permanent plastic deformation.

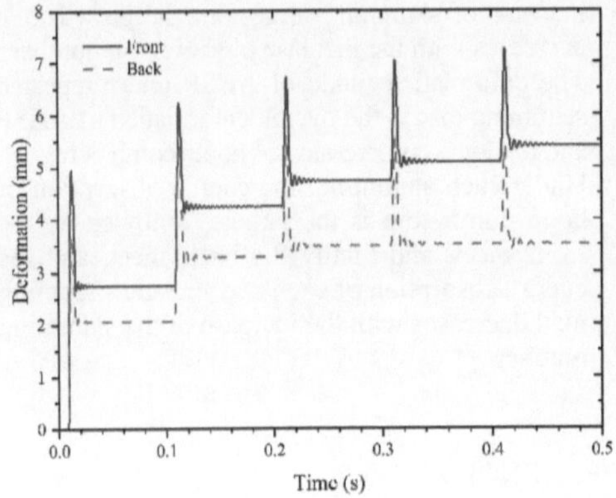

Figure 9. Deflection time history curve of the face sheet.

Figure 8 illustrates the variation of permanent deflection of the front and back face sheets of the AHSP with increasing impact numbers. Meanwhile, the deflection at the mid-point of the face sheet increased with the number of impacts, but the growth rate gradually decreased. By the third impact, the deformation of the back-face sheet tended to stabilize, whereas the front face sheet continued to experience minor increases in deformation.

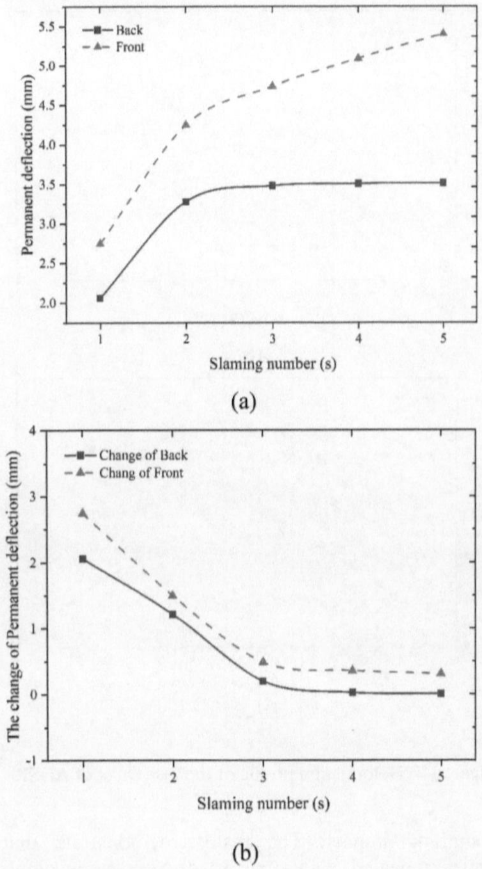

Figure 10. The permanent deflection and deflection change of the face sheet.

Figure 9 displays the cross-sectional deformation profiles of the front face sheet and back face sheet of the AHSP. The results reveal that the maximum deformation occurs at the mid-span of the sandwich plate and gradually decreases towards the edges. The deformation mode is characterized by a combination of overall bending and localized core compression.

3.3.2 *Energy absorption*

Figure 10 illustrates the time-history curves of energy absorption for different parts of the AHSP. The energy absorption increases rapidly within a very short duration after slamming, then decreases due to the rebound of the face sheets, and eventually stabilizes. However, unlike the face sheets, the honeycomb core exhibits a different energy absorption pattern. It reaches a peak energy absorption level immediately after each slamming impact and then stabilizes without experiencing rebound effects. Figure 11 presents the energy absorption of different components of the AHSP in each

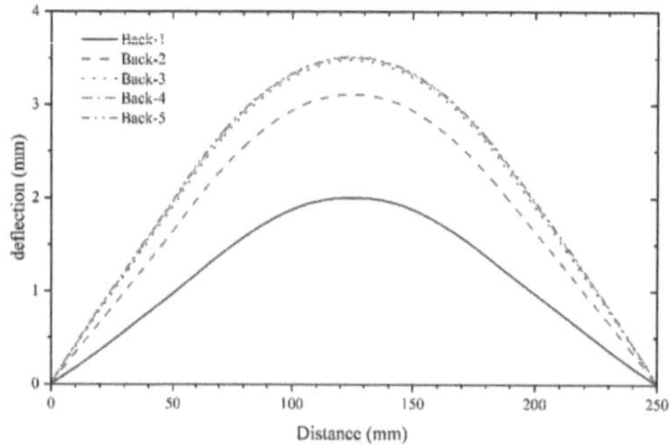

(a)

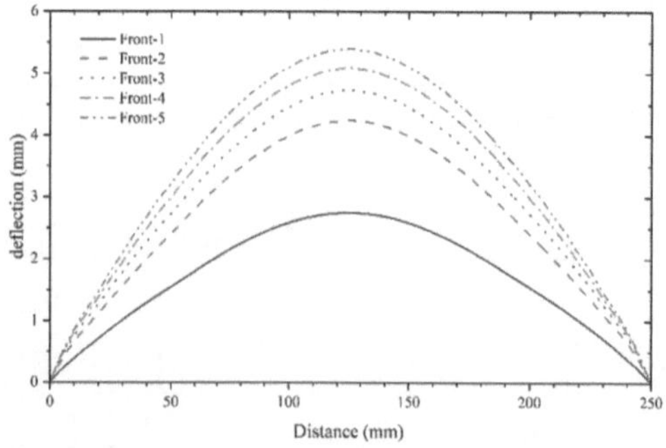

(b)

Figure 11. Deformation profile of the face sheet of AHSP.

slamming impact. The results demonstrate that under repeated slamming loads, the honeycomb core absorbs the largest amount of energy, followed by the front face sheet, while the back face sheet absorbs the least energy. Additionally, the energy absorption of all components gradually decreases with increasing slamming numbers.

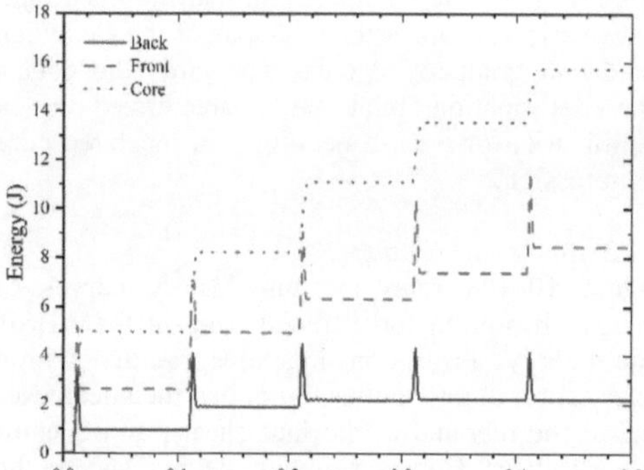

Figure 12. Energy absorption time history curve of each part of AHSP.

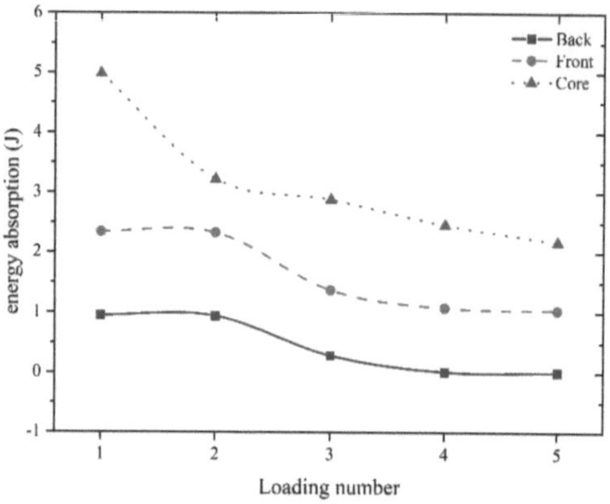

Figure 13. Energy absorption of each part of AHSP.

4 CONCLUSIONS

In this paper, numerical simulations on the dynamic responses of AHSP under repeated slamming loads are conducted using LS-DYNA, and the conclusions are as following:

(1) The results of numerical simulations agree well with those of the repeated slamming tests, indicating the proposed method is with high accuracy to analyze the dynamic responses of AHPS under repeated slamming.
(2) Under repeated slamming loads, the deformation of the front and back sheets increases with the increase of slamming times, but the growth rate decreases with the increase of slamming number.
(3) The deformation mode of AHSP under repeated slamming load is the overall deformation of AHSP and the local compression of honeycomb core.
(4) Under each slamming, the energy absorption of honeycomb core is the largest, followed by the front sheet, and finally the back sheet, and the energy absorption of each part in each slamming load decreases with the increase of the slamming number.

REFERENCES

Chuang, S.L. 1966. Experiments on flat-bottom slamming. Journal of Ship Research 10(1): 10–17.

Duan, L.L. 2020. Equivalent Design Pressure of Moving Slamming Loads and Local Response of Ship Structures. Wuhan University of Technology, PhD dissertation.

Guo, K.L., Liao, Y., Zhu, Z.K., Liu, D. & Zhu, L. 2024. Experimental study on dynamic behaviors of aluminum honeycomb sandwich plates subjected to water impact. Explosion and Shock Waves. DOI: 10.5957/jsr.1966.10.1.10.

He, M.H., Wu, H. & Cheng, Y.S. 2016. Study on the dynamic response characteristics of corrugated sandwich hybrid sandwich plate subjected to water entry slamming. Ship Science and Technology 38(15): 11–17.

He, W.T., Cui, X.F., Wang, C.Z., Deng, J.L., Hu, Q. & Zhang, Z.Y. 2021. Coupled Eulerian-Lagrangian (CEL) characterization of slamming response, and failure mechanism on corrugated sandwich structures. Applied Ocean Research, 116, 102862.

Huynh, V.V., Cho, S.R., Dang, X.P. & Truong, D.D. 2024. Practical Formula for Predicting the Residual Deflection Evolution of Steel Plates Subjected to Repeated Impacts. Journal of Marine Science and Engineering 12 (6): 956.

Ma, B., Zhu, Z.K., Zhu, L., Deng, J. & Zhao, Q. 2024. Model test and simulation study on water-entry impact of reinforced aluminum structure. Chinese Journal of Ship Research 19(S2): 100–109.

Turong, D.D., Shin, H.K. & Cho, S.R. 2017. Permanent set evolution of aluminium-alloy plates due to repeated impulsive pressure loadings induced by slamming. Journal of Marine Science and Technology 23:1–16.

Xin, Y.J., Zhang, L.W., Liu, X.M., Cheng, S.L. & Li, H.J. 2018, Impact test on aluminum honeycomb sandwich panels. Journal of Mechanical Strength 40(04): 802–809.

Yue, Z.S., Li, B.Y., Wang, X., Li, Z., Zhang, R., Wang, P. F., Cheng, L. & Lu, T.J. 2024. Elastoplastic plate shakes down under repeated impulsive loadings. Journal of the Mechanics and Physics of Solids 193, 05918, ISSN 0022-5096.

Yue, Z.S, Han, Z.Y., Wang, X., Yu, R.P., Qu, Y.L., Wang, P., Zhao, Z.Y., Zhang, Q.C. & Lu, T.J. 2024. Shakedown of metallic sandwich beams when subjected to repeated impact loadings. International Journal of Solids and Structures 295, 112793, ISSN 0020-7683.

Zhao, F., Cheng, Y.S., Liu, J. & Wang, H. 2013. Plastic dynamic response analysis of pyramid lattice sandwich plate under water slamming. Ship & Ocean Engineer 42 (03): 13–16.

Zhang, X.H., Wang, Z.H. & Zhou, L.M. 2009. Dynamic response of aluminum honeycomb sandwich plate under explosion load. Explosion and Shock Waves 29(04): 356–360.

Zhu, L. & Faulkner D. 1995. Design pressure for the wet-deck structure of twin-hull ships. Paper presented at FAST 95, Travemunde, Germany:25-27, September 1995.

Trends in Collision and Grounding of Ships and Offshore Structures – Zhu & Guedes Soares (Eds)
© 2026 The Author(s), ISBN 978-1-041-16361-9

Numerical study on dynamic properties of cosine-function multi-stable metamaterial under impact loading

X.Y. Liu & J. Pan
School of Naval Architecture, Ocean and Energy Power Engineering, Wuhan University of Technology, Wuhan, P.R. China

R.J. Xu
Marine Design and Research Institute of China, Shanghai, P.R. China

M.C. Xu
School of Naval Architecture and Ocean Engineering, Huazhong University of Science & Technology, Wuhan, P.R. China

ABSTRACT: A cosine function multi-stable metamaterial (C-MSM) for ship collision protection is proposed in this study. The C-MSM integrates a thin-walled rod with a cosine function while retaining the original unit platform, and it investigates its energy absorption performance as well as its potential application in the side protection structure of ships. A finite element analysis is conducted using ABAQUS software to examine the influence of the thickness ratio between the upper and lower rods on the performance of C-MSM. The results from quasi-static and dynamic impact simulations indicate that the energy dissipation process of C-MSM can be categorized into three distinct stages: arc deformation, layer snap-through behavior, and overall densification. By configuring five cells with varying thickness ratios, both the deformation modes and energy absorption capabilities are studied to assess the influence of thickness ratios. The results presented in this paper demonstrate the double advantages of C-MSM regarding crashworthiness and lightweight characteristics, thereby offering a novel solution for enhancing ship-side safety.

1 INTRODUCTION

Driven by the tide of globalization, the global economy has developed rapidly. International trade has become increasingly frequent, and the demand for shipping has increased greatly. As the main mode of transport for international trade, sea transport undertakes most of the global cargo transport. The rising demand makes the shipping routes constantly improve and new shipping routes constantly open up. To pursue high efficiency, ships have developed to large size and lightweight, and the number of super-large ships has increased due to the strong cargo-carrying capacity. However, the increase of ships, the high sailing density, and the differences in performance and operation, coupled with the complex marine environment, lead to an increase in the probability of ship collision accidents.

The collision accident has serious consequences, which will not only seriously damage the ship but also cause secondary disasters such as ship loss, fuel leakage, destruction of the marine ecology, and so on. With the increase in ship value and accident handling cost, economic loss increases year by year. Therefore, it is crucial to enhance the crashworthiness of ships. This not only protects the ship structure in the event of an accident, but also reduces all kinds of losses caused by accidents, and provides protection for the sustainable development of maritime transport.

Ship-side structure, as the main part of collision and energy absorption, plays an important role during the impact. However, at present, the ship-side structure is mainly composed of steel stiffened plates which has great limitations in the research of improving its crashworthiness. With the continuous study of metamaterials, researchers find that these metamaterials with special forms have the characteristics of lightweight, high energy absorption, good stability, and so on, leading to a good prospect in ship collision avoidance structure. Zhang et al. (2024) provide a comprehensive summary of the novel structures currently employed in ship-waterway structure collisions, especially, emphasizing the novel lattice structures due to their remarkably superior performance. Chen et al. (2022) design a novel three-dimensional zero Poisson's ratio structure which is applied to ship explosion-proof cabin doors. The results show that the strain energy of the new structure is 5.6 times that of the aluminum foam. Armanfar et al. (2024) introduce an approach to integrating lattice structures into shipbuilding via AM technology. The ship models with lattice structures have superior structural properties compared to conventional support elements. Jiang et al. (2023) propose a sandwich protection system against blast loads containing a graded re-entrant circular auxetic core to protect warships. The present study shows that the re-entrant circular auxetic cored

DOI: 10.1201/9781003684404-36

sandwich structure has great potential in warship blast protection applications.

Multi-stable metamaterial is a kind of cell that has multiple homeostasis and transforms homeostasis under specific external stimuli. The metamaterial can achieve recoverable energy absorption through elastic instability and deformation and has good recoverability, which has a good prospect in the field of ship protective structures. Tao et al. (2020) studied a shape-reconfigurable functionally deployable, mechanically adjustable, and reusable intelligent multi-stable metamaterial based on a 4D printing method. The metamaterial had excellent versatility and recoverability. Based on this foundational structure, Meng et al. (2021) designed a multi-step deformation metamaterial that can realize multiple stress plateaus. The theoretical, numerical, and experimental approaches all demonstrate the excellent cushion performance of this metamaterial during the quasi-static and impact process. Hou et al. (2024) proposed a novel multi-step deformation metamaterial that possessed multiple stable configurations, which can produce a unique two-step deformation mode under displacement loading. The results showed that this metamaterial had an excellent ability for energy absorption and stability deformation step. In particular, the metamaterial also had a negative stiffness effect. Winkelmann et al. (2010) discussed a composite bistable structure consisting of a material-based mechanism, which has a high ductility. Ju et al. (2024) presented a novel multi-stable metamaterial with improved and consistent energy-trapping ability based on frame-reinforced curved beams. The plate impact test shows great cushioning performance during the impact process.

This paper proposes a cosine-function multi-stable metamaterial (C-MSM) for ship collision protection. C-MSM has integrated the cosine thin wall rods while retaining the original unit platform and explored its energy absorption performance and the possibility of application in the ship-side protection structure. The FEM is established by ABAQUS, and the influence of the thickness ratio of upper and lower rods on the performance of C-MSM is investigated. The deformation mode and energy absorption performance are studied by setting five different thickness ratio cells.

2 GEOMETRIC DIMENSION AND METHOD

2.1 *Cosine-function multi-stable metamaterial structures*

Tao et al. (2020) proposed a special multi-stable metamaterial, in which the steady-state changes under different temperature changes were studied. Based on the searches before, a multi-stable metamaterial using a cosine function to control the shape of a curved bar (C-MSM) is studied as the protection of ship-side structure in this paper for its potential functional performance under ship impact loads. The 2D and 3D diagrams of C-MSM are shown in Figure 1. The C-MSM is a metamaterial composed of cosine arcs, divided into thin and thick walls, and at the highest point of the thin wall exists a vertical platform. The geometric dimensions of C-MSM cells are shown in Table 1. The shape function of the bending rod is given as following, and the coordinate system is shown in Figure 1 (a).

$$y(x) = \pm 5\cos(\frac{\pi}{10}x) \qquad (1)$$

This paper aims to optimize the multi-stable metamaterials with eye-shaped elements and enhance their energy absorption properties for future scale model impact tests, which can be used as energy absorption structures in ship collisions. Therefore, the finite elements model (FEM) is established by using ABAQUS. For the impact simulations, one 8×8 cell matrix is arrayed with a length of 272mm and a height of 168mm. The spacing arrangement method is selected to arrange five layers of cells with an interval of 10mm. The dimensions of the MSM structure are adopted as shown in Figure 2.

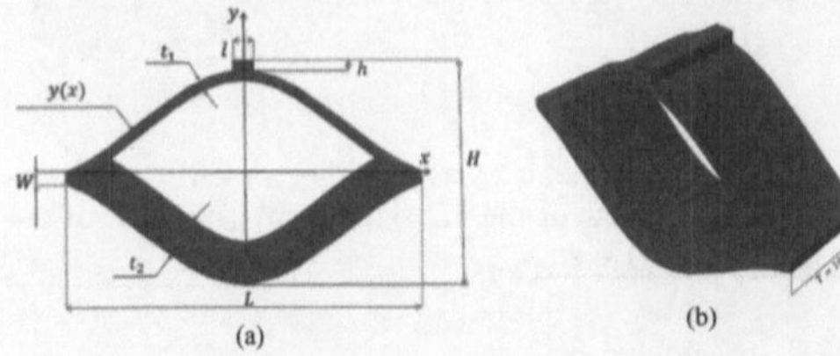

Figure 1. Geometrical configurations of C-MSM.

Table 1. Geometrical dimensions of C-MSM.

L/mm	H/mm	W/mm	l/mm	h/mm	t_1/mm	t_2/mm
34.0	21.0	1.1	2.0	1.0	1.0	4.0

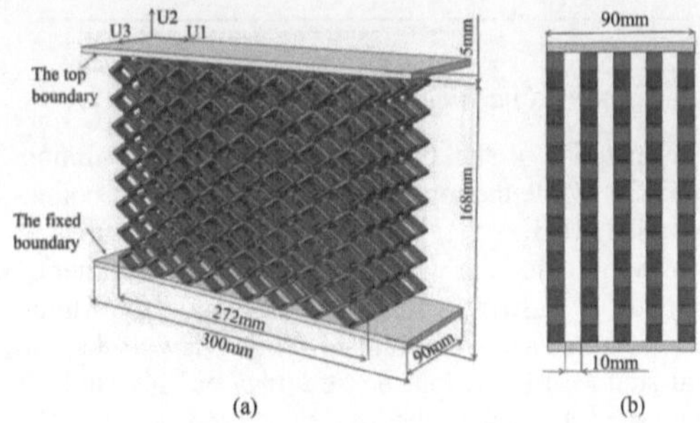

Figure 2. Geometrical arrangement of C-MSM.

2.2 *Ship bow*

The bulbous bow in the previous experiments (Pan, J. 2022) is selected as the impact head, as shown in Figure 3, set as a rigid body. Considering the small size of the C-MSM sample studied in the present paper, the similarity theory is applied to determine the geometry of the bulbous bows. According to the π criterion in the similarity theory, four key parameters during the collision are selected as the research objects under the conditions of geometric, kinematic, and dynamic similarity criteria. Assuming the material is considered consistent, time (t) and length (L) are selected as the basic physical quantities, with the relevant parameters and dimensions detailed in Table 2.

Referring to the size of the double-hull vessel's side in the previous study (Xu, 2023), the length ratio is used for the overall model, excluding the size of a single MSM. Liu et al. (2020 a) summarized the statistical values of striking ships involved in collision accidents from ships operating worldwide. A 5000T double-hull vessel is selected as the original object. The ship mass which is represented by the impact head in simulations is set to 1.865t, and the collision speed is 2.30 m/s, as shown in Figure 3.

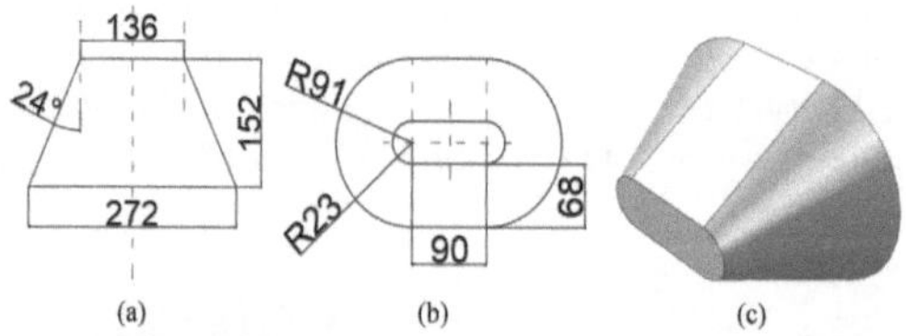

Figure 3. Diagrammatic sketches of geometric configuration: (a) Front view of the 2D sketch; (b) Top view of the 2D sketch; (c) 3D sketch.

Table 2. The scaled parameters using the similarity theory.

Parameters	Symbol	Dimension	Scaled factor	Original data	Scaled data
Length /mm	L	L	λ	3778	272
Mass /t	M	m	λ^3	5000	1.865
Density / $(kg \cdot m^{-3})$	ρ	mL^{-3}	1	7.85	7.85
Velocity / $(m \cdot s^{-1})$	v	Lt^{-1}	1	2.3	2.3

2.3 *Material parameter*

The material of the C-MSM structure is Aluminum 6061-T6, while the top boundary and the fixed boundary used AH36 steel, the basic material properties as shown in Table 3, in which the Johnson-Cook criterion is used for plasticity and failure evaluation for Aluminum 6061, as shown in Table 4. The review of the critical strain (effective plastic strain) ε_f in this study is conducted by using the criterion proposed by Ref (Hogstrom, 2010), as shown in Table 5. Universal contacts are applied to the overall FE model to avoid penetration between different parts. The dynamic friction coefficient of the model is set as 0.3 (Liu, 2022 b).

2.4 *Boundary condition*

The FEM consists of the C-MSM structure, the top boundary, and the fixed boundary, which can be shown in Figure 2. The fixed boundary is fully constrained in all Degrees of Freedom. The top

Table 3. Foundational mechanical properties of AL6061-T6 (Muniraj D,2022) and AH36 (Jiang C, 2024).

Property	Unit	AL6061-T6	AH36
Density	kg/m^3	2660	7850
Poisson's ratio	-	0.28	0.3
Young's modulus	*GPa*	70.3	206

Table 4. The Johnson-cook mode coefficient of AL6061-T6.

Material	A/MPa	B/MPa	C	n	m	D_1	D_2	D_3	D_4	D_5
AL6061-T6	270	195	0.02	0.3	1.34	0.77	1.45	0.47	0	1.6

Table 5. The plastic mode coefficient of AH36.

Material	Ultimate tensile strength /MPa	Minimum elongation	Failure uniform strainε_u	Barba constant *c* (Calle M,2015)	Critical failure strain ε_f(Deng, X. 2022)
AH36	560	0.21	0.22	0.93	0.27

boundary restricts all rotations and only the displacement in the vertical U2 direction is allowed to move. Since our research is focused on the failure mode of C-MSMs, the impact indenter is set as a rigid body.

2.5 *Load cases*

The quasi-static simulations are set up for obtaining the deformation mode and verifying the performance of C-MSMs under low load for a long time, which provides the basis for subsequent impact simulation. Under quasi-static conditions, a 70% total height (compressive height 117.6mm) compression protocol is implemented by defining the top platen as a rigid body boundary condition.

2.6 *Definition of crashworthiness*

To quantify the crashworthiness and mechanical properties of C-MSM, quantitative assessments must analyze the properties, and a series of crashworthiness indexes are defined as follows. This represents the reaction force derived from the whole surface of the fixed boundary. Energy absorption (EA) is employed here to evaluate the energy absorption characteristics of the whole C-MSMs, which collect data from the internal energy of structures.

With consideration of mass efficiency, specific energy absorption is generally used. *SEA* represents the crashworthiness index which is defined as the ratio of the absorbed energy to mass.

$$SEA = \frac{EA}{M} \quad (2)$$

Where M is the total mass of C-MSMs.

Mean crushing force (*MCF*) (Deng X,2022) is used to evaluate the crashworthiness performance of thin-walled structures.

$$MCF = \frac{EA}{\delta_{\max}} \quad (3)$$

Where δ represents the impact depth and δ_{max} is the maximum value.

The conversion rate of initial kinetic energy (λ) is used to evaluate the efficiency of MSMs in converting *EA* during the impact process.

$$\lambda = \frac{EA_{\max}}{KE} \quad (4)$$

where EA_{max} represents the maximum value of *EA*, and *KE* is the initial kinetic energy of impact indener, $KE = \frac{1}{2}mv^2 = 4.94kJ$.

2.7 *Mesh convergence analysis*

In the C-MSM model, the top and fixed boundaries are meshed using C3D8R. The maximum value of reaction force (F_R) and CPU time curves of simulations under different mesh sizes are shown in Figure 4, where CPU time represents the time required to complete the calculation under that mesh size. Considering the numerical accuracy and calculation costs simultaneously, the mesh size of 1.0 mm is selected in the subsequent simulations.

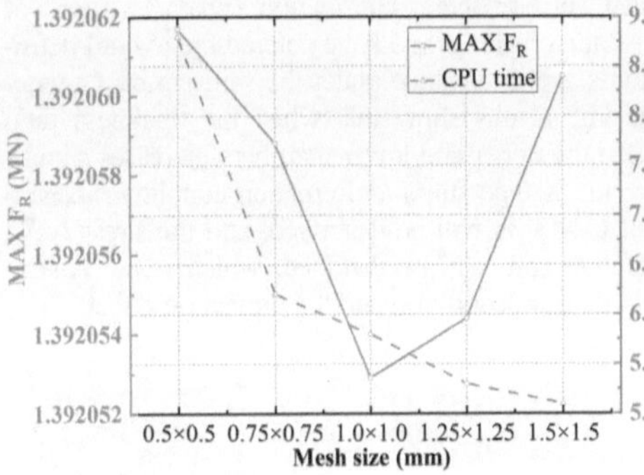

Figure 4. The F_R and CPU time curves with different mesh sizes.

3 QUASI-STATIC COMPRESSIVE RESULTS ANALYSIS

3.1 *Compressive deformation modes*

Selected six stages to show the stress contours of C-MSMs, as shown in Figure 5. C-MSMs present remarkable stage deformation characteristics, which successively go through three stages: arc deformation, stratified deformation, and total densification. As shown in Figure 5 (a), in the initial deformation stage, the specimen formed an upward and downward symmetrical curved deformation belt, in which the cell in the axial region of the arc bore greater stress. Combined with Figure 5 (b) and (c), it can be observed that snap-through instability occurs preferentially in the cells of layers 2, 5, and 6, while the cells of layers 3 and 4 maintain small shape variables. This is closely related to the deformation mechanism of C-MSMs through stress propagation along the upper and lower curved stress bands, resulting in a lower deformation response between the two curved bands. With the continuous increase of displacement load, snap-through instability begins to appear in the third layer boundary cell, which further verified the spatial distribution characteristics of the curved stress zone. When the strain reaches 80%, all cells except the first layer complete the snap-through instability, which indicates that the structure enters the whole densification stage.

It is worth noting that in the 80% strain stage, the fixed boundary only shows stress concentration in the contact area with the cell end, which effectively reduces the mechanical load in the center area of the bottom plate, showing good local protection performance.

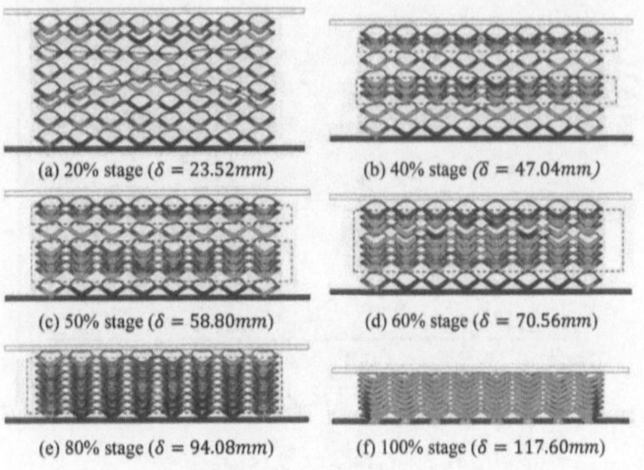

Figure 5. Deformations of C-MSMs structure under quasi-static compression.

3.2 *Crushing and energy-absorption*

Figure 6 shows the $F_R - \delta$ curve and EA curve of C-MSMs. As shown in Figure 6, the curve presents two significant inflection points at $\delta = 88.2mm$ and $\delta = 107.2mm$. The first turning point corresponds to the initial snap-through behavior of the first layer cell, and the structure is still in a low-stress state. The second turning point indicates that with the accumulation of the snap-through instability strain of the first layer cells, the structure enters the overall densification stage, which is manifested as a sharp increase in the overall stress level and a rapid increase in the total reaction of the floor. Similarly, these two inflection points also correspond to the inflection points where the rate of the EA curve increases in Figure 7.

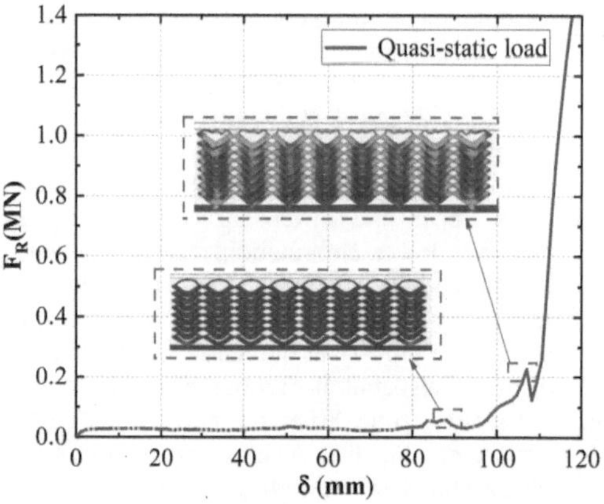

Figure 6. $F_R - \delta$ curve of C-MSMs

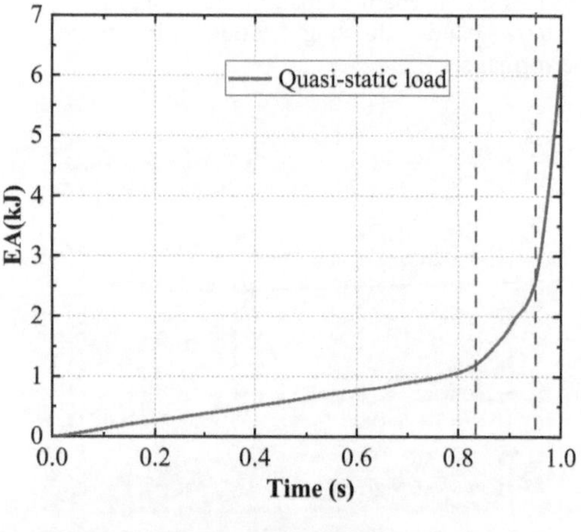

Figure 7. $EA - Time$ curve of C-MSMs

4 PARAMETER ANALYSIS WITH IMPACT LOAD

4.1 *Case introduction*

To investigate the influence mechanism of the thickness ratio between thin-walled and thick-walled rods on the mechanical properties of C-MSMs, this study introduces the thickness ratio coefficient α, whose expression is shown in formula (5). Based on this coefficient, five groups of comparison samples with different thickness ratios are designed, and the specific parameters are shown in Table 6.

$$\alpha = \frac{t_1}{t_2} \tag{5}$$

where t_1 represents the thickness of the upper rod and t_2 is the thickness of the lower rod as shown in Figure 1.

Table 6. The upper and lower rod thickness of different α

	t_1/mm	t_2/mm	α
C-1	1.0	4.0	0.25
C-2	2.0	3.0	0.67
C-3	2.5	2.5	1
C-4	3.0	2.0	1.5
C-5	4.0	1.0	4

4.2 *Results analysis*

Figure 8 shows the post-impact stress distribution of C-MSMs under different thickness ratio coefficients (α), and Table 7 shows the crushing depth of each case. C-2, C-3, and C-4 samples form a significant X-shaped curved deformation-stress coupling band. Their energy dissipation mainly depends on the intermediate cells' deformation and energy absorption mechanism. The cells in the upper and lower arc-shaped central axis region better maintain their initial configuration. With the increase of the thickness ratio α, the geometric center of the X-type stress band systematically moves down, but the structural integrity of the underlying cell can be maintained.

It is worth noting that C-1 and C-5 specimens show significantly different deformation modes: they do not form a typical X-type deformation band, but show hierarchical progressive deformation characteristics of quasi-static crushing, and have entered the overall densification stage in advance under the same impact energy input. The results show that when the thickness ratio between the upper and lower members increases significantly, the X-type stress /deformation coupling mechanism of C-MSMs will be destroyed, and the stress wave conduction path will be distorted, which is not conducive to energy absorption and the protection effect.

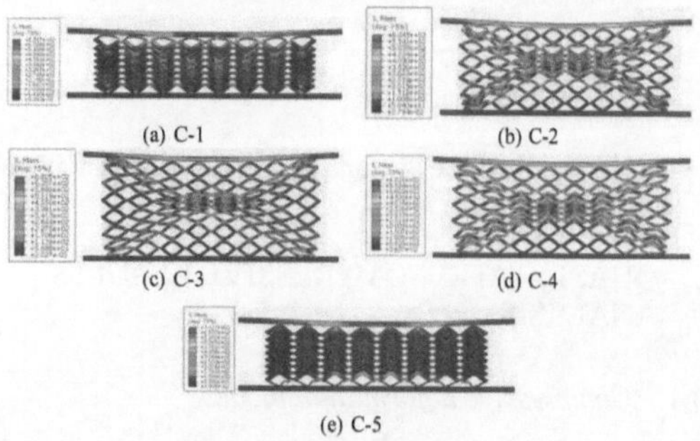

Figure 8. The dynamic collapse deformation modes with various thickness ratios.

Table 7. The crushing depth of each case.

Case	C-1	C-2	C-3	C-4	C-5
Max δ/mm	115.88	69.37	58.03	67.89	116.05

Figure 9 and Figure 10 show the evolution curves of F_R, EA, SEA and MCF of specimens with five configurations. As shown in Figure 9, each group of curves presents an elastic rebound phenomenon at the top boundary, which is represented by the reverse transition of the load curve. The C-2, C-3, and C-4 specimens have the same phase transition response sequence, showing the two-stage transition characteristics. The analysis of the dynamic stress program shows that the first turning point corresponds to the critical state of the snap-through behavior of the intermediate layer cells. The second turning point is due to the cumulative snap-through instability of cells in the cross-region of X-type stress bands. It is worth noting that the peak F_R value of the C-4 specimen is 9.09% and 8.67% lower than that of C-2 and C-3 respectively, and the curve fluctuation amplitude is reduced, forming a more stable energy dissipation platform stage. In contrast, C-1 and C-5 specimens show precocious failure characteristics. When the compression displacement reaches the critical value of 107mm, the structure enters the total densification stage. This abrupt response has dynamic similarity with the third deformation stage of the quasi-static crushing specimen.

Figure 10 and Figure 11 also illustrate the lower performance of specimens C-1 and C-5 compared with other specimens. From Figure 10, it can be found that after EA reaches the maximum, there is a small decline, which is because part of the EA is converted into the rebound momentum of the impact head. It can be found that a large or small α of the upper and lower rods hurts the deformation mode and energy absorption properties of the C-MSMs. Table 8 data shows that the energy conversion efficiency of the five groups of C-MSMs is above 93%, indicating that the structure has excellent energy absorption performance. It is worth noting that specimens C-2, C-3, and C-4 exhibit ultra-high efficiency due to the X-shaped deformation curve mechanism, and their energy conversion rate exceeds the 98% threshold. Combined with the background of the application of ship-side protection devices, R_F and SEA should be given more attention. As illustrated in Figure 9, the R_F value of C-4 demonstrates reductions of 8.67% and 9.09% compared to C-2 and C-3 respectively, while its SEA exhibits minimal decreases of merely 0.34% and 0.25% relative to these evaluation indexes.

In conclusion, C-MSM-4 ($\alpha = 1.5$)is confirmed as the optimal configuration with a suitable load response.

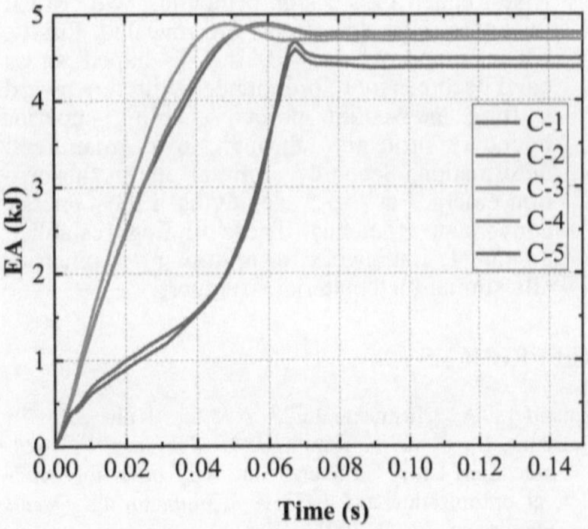

Figure 10. The curve of *EA* history.

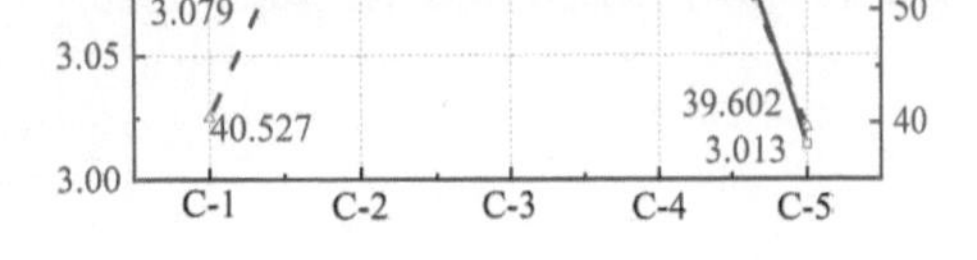

Figure 11. The curves of maximum value of *SEA* and *MCF*.

Table 8. The λ of each case.

λ	C-1	C-2	C-3	C-4	C-5
Value/%	95.02	99.05	98.95	98.72	93.04

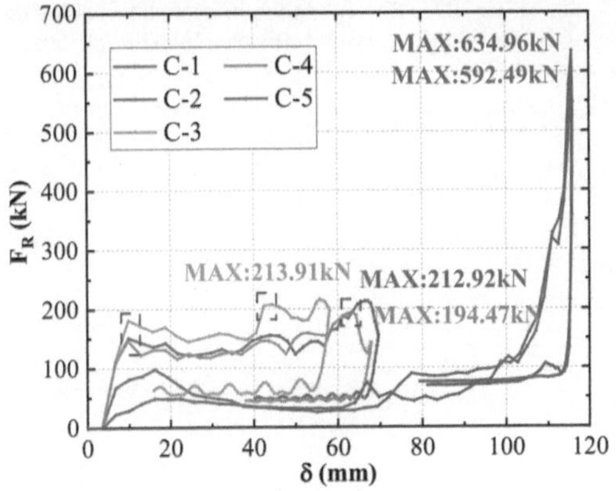

Figure 9. The curve of $F_R - \delta$.

5 CONCLUSIONS

In this study, a cosine-function multi-stable metamaterial (C-MSM) is studied for its deformation mode and energy absorption performance. Through the control of the same quality design, the influence of the ratio of the upper and lower bending rod thickness (α) on the deformation mode and energy absorption property is obtained. The main conclusions are as follows:

(1) Through quasi-static compression experiments, the deformation of C-MSMs is characterized by three distinct stages: X-type deformation, layered snap-through instability, and global densification.

(2) Under equal-mass design principles, two critical thickness ratios (α) effects are revealed. Firstly, when α approaches 1.0, the X-shaped stress band becomes more pronounced with downward shifting intersection points, enabling superior boundary protection through lower main cell deformation. Secondly, optimal energy absorption emerges at α=1.5, achieving a 93% energy conversion efficiency. These findings establish a valuable framework for optimizing geometrically similar metamaterial structures.

REFERENCES

Armanfar, A., Tasmektepligil, A. A., Kilic, R. T., Demir, S., Cam, S., Karafi, Y. & Gunpinar, E. 2024. Embedding lattice structures into ship hulls for structural optimization and additive manufacturing. *Ocean Engineering*, *301*, 117601.

Chen, H. & Li, F. 2022. Design of quadrilateral zero-Poisson's ratio metamaterial and its application in ship explosion-proof hatch door. *Ocean Engineering*, *266*, 112667.

Calle, M. A. G. & Alves, M. 2015. A review-analysis on material failure modeling in ship collision. *Ocean Engineering*, *106*, 20–38.

Deng, X., Qin, S. & Huang, J. 2022. Crashworthiness analysis of gradient hierarchical multicellular columns evolved from the spatial folding. *Materials & Design*, *215*, 110435.

Hou, X., Sheng, T., Xie, F. & Deng, Z. 2024. Mechanical logic gate design based on multi-stable metamaterial with multi-step deformation. *Composite Structures*, *335*, 118001.

Hogström, P. 2012. *RoPax ship collision–a methodology for survivability analysis*. Chalmers Tekniska Hogskola (Sweden).

Ju, X., Li, S., Zhang, Y., Wu, P. & Li, Y. 2024. Design of multi-stable metamaterial cell with improved and programmable energy trapping ability based on frame reinforced curved beams. *Thin-Walled Structures*, *202*, 112120.

Jiang, C., Lin, L. & Chen, N. Z. 2024. A study on auxetic-inspired side structure for enhanced crashworthiness. *Marine Structures*, *93*, 103545.

Jiang, F., Yang, S., Qi, C., Liu, H. T., Remennikov, A. & Pei, L. Z. 2023. Blast response and multi-objective optimization of graded re-entrant circular auxetic cored sandwich panels. *Composite structures*, *305*, 116494.

Liu, K., Zong, S., Li, Y., Wang, Z., Hu, Z. & Wang, Z. 2022a. Structural response of the U-type corrugated core sandwich panel used in ship structures under the lateral quasi-static compression load. *Marine structures*, *84*, 103198.

Liu, K., Zong, S., Li, Y., Wang, Z., Hu, Z. & Wang, Z. 2022b. Structural response of the U-type corrugated core sandwich panel used in ship structures under the lateral quasi-static compression load. *Marine structures*, *84*, 103198.

Meng, Z., Ouyang, Z. & Chen, C. Q. 2021. Multi-step metamaterials with two phases of elastic and plastic deformation. *Composite Structures*, *271*, 114152.

Muniraj, D., & Sreehari, V. M. 2022. Damage assessment of sandwich structures with Al/SiC functionally graded plasma sprayed faceplates subjected to single and repeated impacts. *Composite Structures*, *287*, 115369.

Pan, J., Wang, T., Zhang, W. Z., Huang, S. W., & Xu, M. C. 2022. Study on the assessment of axial crushing force of bulbous bow for bridge against ship collision. *Ocean Engineering*, *255*, 111411.

Tao, R., Xi, L., Wu, W., Li, Y., Liao, B., Liu, L., & Fang, D. 2020. 4D printed multi-stable metamaterials with mechanically tunable performance. *Composite Structures*, *252*, 112663.

Winkelmann, C., Kim, S. S., & La Saponara, V. 2010. Design and development of hybrid composite bistable structures for energy absorption under quasi-static tensile loading. *Composite Structures*, *93(1)*, 171–178.

Xu, S., Zhang, W., Cai, Y., Shen, W., & Wu, Y. 2023. Study on anti-penetration performance of Kevlar reinforced honeycomb liquid-filled cabin on warship: experimental verification and numerical analysis. *Ocean Engineering*, *283*, 114959.

Zhang, W. Z., Pan, J., Sanchez, J. C., Li, X. B., & Xu, M. C. 2024. Review on the protective technologies of bridge against vessel collision. *Thin-Walled Structures*, 112013.

Trends in Collision and Grounding of Ships and Offshore Structures – Zhu & Guedes Soares (Eds)
© 2026 The Author(s), ISBN 978-1-041-16361-9

Research on multi-source impact load identification of stiffened plate structure based on L1 normal regularization method

D.D. Min, G.J. Shi & D.Y. Wang
State Key Laboratory of Ocean Engineering, Shanghai Jiao Tong University, Shanghai, P.R. China
Institute of Marine Equipment, Shanghai Jiao Tong University, Shanghai, P.R. China

ABSTRACT: This paper investigates the issue of impact load identification based on structural dynamic response. Firstly, it compares the effectiveness of using different regularization methods for impact load identification, concluding that the Based on L1 norm regularization method is suitable for impact load identification. Secondly, it analyzes the impact of sensor placement schemes and sensor combinations on single-point and double-point impact load identification. It finds that the best recognition results are achieved when sensors are placed on both sides of the load in single-point load cases, the load and sensors are in a straight line with some distance; in double-point load cases, the load and sensors are perpendicular and have some distance. The optimal sensor combination involves pairing different types of sensors accelerometer is the main sensor, coupled with strain or displacement sensor. Finally, it examines the identification effects of double-point loads both same-type and different-type loads. For same-type double-point loads, the L1 regularization method can effectively identify them. However, for different-type double-point loads, due to their distinct characteristics in sparse representation, they cannot be well identified, necessitating further improvements in the load identification method.

1 INTRODUCTION

In the field of ships, identifying or monitoring environmental loads is crucial, as it helps ensure the safe operation of ships and marine engineering structures. These structures are subjected to various dynamic environmental loads, such as wave forces during navigation, wind forces on offshore platforms, ice collisions when navigating in ice zones, and impact forces on warships during explosions or weapon launches. Even though most loads cannot be directly measured to obtain their true information, they still need to be effectively identified. Depending on how impact loads affect ships through different spatial distributions, they can be categorized into point impact loads and distributed impact loads. This study focuses on the most common point impact load identification. Point impact loads include unit impact loads and multi-source impact loads.

In the monitoring of engineering structures, single-source load inversion has been extensively studied (Wang et al., 2011). This primarily involves discretizing the convolution relationship between system response and applied excitation (Wang., 2011), followed by using regularization methods such as Tikhonov regularization and L-curve criteria to mitigate ill-posed problems caused by noise and measurement errors. Regularization methods based on L2 norm Tikhonov method (Boukria et al., 2011), truncated singular value decomposition (TSVD), and function approximation are widely used in time-domain load identification(Kalhori et al., 2016). Since 2015, Qiao has made some research achievements in function approximation for impact load identification (Qiao et al., 2016). By selecting basis functions that match the shape of impact loads, new wavelet multi resolution analysis-based cubic B-spline scale functions and configuration methods based on cubic B-splines have been proposed (Qiao et al., 2015), showing better performance than TSVD methods in cantilever beam structure tests (Qiao et al., 2015). In 2018 and 2020, Wang et al. successively proposed several new regularization operators to address the accuracy of true solutions in Tikhonov regularization.

Numerical calculations and experiments have demonstrated that these regularization operators exhibit excellent filtering properties, not only approximating the truncation of extremely small singular values but also preserving larger singular values (Wang et al., 2020). For the identification of ice loads on ship hulls, Shuai Kong used three commonly regularization methods Tikhonov, TSVD, LSQR under the L2 norm. After comparison, the LSQR method was selected as the optimal regularization operator and its regularity was verified through numerical examples of ice-induced strain with noise. Finally, two load identification cases were conducted on an experimental bench, and the identified loads could accurately determine the signal characteristics of the applied loads in the time domain (Shuai et al., 2020). Wang Y compared different parameter selection methods under the L2

DOI: 10.1201/9781003684404-37

norm condition Tiknohov method, including the Generalized Cross Validation (GCV) and Ordinary Cross Validation (OCV), and TSVD, for identifying concentrated loads applied in three different directions such as vertical, lateral, and longitudinal on stiffened plates. The GCV method showed higher recognition accuracy than the TSVD and OCV methods (Wang et al., 2021).

In recent years, driven by the new theory of compressed sensing, sparse constraints based on the L1 norm have received unprecedented attention and quickly become a cutting-edge research topic in signal processing, image processing, and related fields(Tropp et al., 2010). The solution to regularization methods based on the L1 norm is typically sparse. In contrast, the solution to regularization methods based on the L2 norm, which controls the energy of the variables to be solved, is often non-sparse. The emergence of sparse theory has provided a new approach for impact load identification. From a sparse perspective, the impact load is very sparse in the time domain relative to its sampling data length, with significant values only near the impact loading zone and all zeros outside this zone.

However, in practical applications, multiple excitation sources often act simultaneously. In such cases, the high coherence caused by similar or overlapping transmission paths of various load sources makes traditional inversion methods susceptible to ill-posed problems, thereby reducing inversion accuracy. Qiao Baijie verified the advantages of the L1 norm sparse identification method over the traditional L2 norm regularization method using single-source and multi-source impact loads on thin plate structures (Qiao et al., 2019). Chen Hui conducted an identification of impact loads acting on high-speed train constraint damping plates based on the L1 norm, reconstructing both single and continuous impact loads. In view of the problem of multi-source impact load identification, CHEN Lin proposed an improved Bayesian method based on L1 norm regularization technology, and the double-point impact load identification is carried out in a typical stiffened plate structure (Chen et al., 2022). The study compared the effects of different impact conditions and response points on the identification results, as well as analyzed the impact of different hammer materials on the impact loads. In the process of impact load identification, it is necessary to analyze the influencing factors of single point and multi-point impact load identification (Chen et al., 2024).

2 IMPACT LOAD IDENTIFICATION THEORY

2.1 *Impact load identification theory*

According to the dynamic response theory, the control equation of load and response is:

$$y = Hf \tag{1}$$

In the formula, H is the transfer function of the system.

Using the least squares method to solve equation (1), we can get the objective function at this time:

$$\min_f \|y - Hf\|_2^2 \tag{2}$$

In the formula, $\|.\|_2^2$ represents the square of the L2 norm of the vector.

Expand equation (2) and then decompose the f partial derivative in it. Assuming that H^TH is a reversible matrix, we get:

$$f = (H^TH)^{-1}H^Ty \tag{3}$$

Due to the inevitable noise interference in actual measured responses, singular values in the matrix H-can be amplified by even small noise levels. Regularization methods add appropriate constraint information during the solution process, enhancing adaptability. Therefore, various regularization methods are widely used in solving ill-posed inverse problems for load identification.

Considering that L2 norm can prevent overfitting and reduce error, improve the stability of solution, L2 norm is introduced into equation (2) to get a new objective function:

$$\min_f \left\{ \|HF - y\|_2^2 + \alpha^2 \|f\|_2^2 \right\} \tag{4}$$

The objective function is the expression of Tikhonov regularization objective function. In the formula, α is the regularization parameter, which is used to control the relative size. The larger it is, the higher the stability of the result of the inverse problem, and vice versa, the higher the accuracy of the solution.

Expand equation (4) to the square, and then f decompose the partial derivative in it, we can get:

$$f_{tik} = (H^TH + \alpha^2 I)^{-1}H^Ty \tag{5}$$

The solution is the load obtained by using Tikhonov regularization method. When solving, appropriate regularization parameters should be selected. Common parameter selection methods include L curve and generalized cross validation (GCV) method, etc.

The singular value decomposition of H matrix in (5) can be obtained as follows:

$$H = (u_1, u_2, \ldots, u_m)$$

$$\begin{bmatrix} \sigma_1 & 0 & \cdots & 0 \\ 0 & \sigma_1 & & 0 \\ \vdots & & \ddots & \vdots \\ 0 & 0 & & \sigma_n \\ 0 & 0 & \cdots & 0 \end{bmatrix}_{m\times n} (v_1, v_2, \ldots, v_n)^T \tag{6}$$

In the formula, σ_i ($1 \le i \le n$) represents the singular value of the matrix H.

Substituting equation (6) into equation (5), we get:

$$f_{tik} = \sum_{i=1}^{n} \frac{\sigma_i^2}{\sigma_i^2 + \alpha^2} \frac{u_i^T y}{\sigma_i} v_i \quad (7)$$

It can be seen from equation (7) that a very small value σ_i will lead to a very large value of the solved load f. Therefore, the small singular values in matrix H are discarded and only the first k large singular values in matrix H are retained. The truncated Hk is used instead of the original H. This method is called truncated singular value regularization method (TSVD).

Therefore, the original least squares problem can be transformed into:

$$\min_f \|y - H_k f\|_2^2 \quad (8)$$

Further solution yields:

$$f_{tsvd} = (H_k{}^T H_k)^{-1} H_k{}^T y = \sum_{i=1}^{k} \frac{u_i^T y}{\sigma_i} v_i \quad (9)$$

In the above formula, k is the regularization parameter of TSVD method. In order to improve the accuracy of TSVD method, k should be taken larger.

The L1 norm represents the sum of the absolute values of all elements in a vector. Since the solution of the L1 norm is usually sparse, the L1 norm is introduced into equation (2) to obtain a new objective function:

$$\min\left\{\|Hf - y\|_2^2 + \lambda\|f\|_1\right\} \quad (10)$$

In the formula, λ is the regularization parameter, which is the same as the regularization parameter in Tikhonov method, and makes the residual term tend to balance with the load term to be sought. This method is also called sparse regularization method.

The characteristics of impact loads are as follows: within the extremely short interval at the moment of impact, the load value is non-zero and has a large magnitude; while in the non-load region during impact, the load value is essentially zero. Therefore, the impact fload signal exhibits extremely sparse properties in the time domain. Due to the introduction of the L1 norm regularization term in sparse regularization methods, the load values obtained through this regularization method also exhibit sparsity, making them suitable for identifying impact loads.

Using the truncated Newton's method to solve equation (5), the idea of this approach is: firstly, use the preconditioned conjugate gradient algorithm to calculate the search direction; secondly, employ backtracking linear search to find the minimum number of backtracks, determining the optimal computational step size; finally, use specific truncation criteria to set the termination conditions of the algorithm, thereby obtaining an approximate solution that meets the conditions. The main computational steps are as follows:

$$A = \nabla^2 \emptyset_t(f, u) \quad (11)$$

$$G = \nabla \emptyset_t(f, u) \quad (12)$$

$$A \begin{bmatrix} \Delta f \\ \Delta u \end{bmatrix} = -G \quad (13)$$

$$\emptyset_t(f, u) = t\,\|Hf - y\|_2^2 + t \sum_{i=1}^{n} \gamma u_i + \emptyset(f, u) \quad (14)$$

22 *Multi-point load identification*

Assuming the system is a linear system, when a linear system is subjected to multiple impact loads acting simultaneously, the system response is the linear superposition of the responses from each excitation source. The structural response generated by a single impact load can still be obtained through the convolution of that load with the corresponding impulse response function. Therefore, the control equations for a multiple-input multiple-output (MIMO) system can be expressed in matrix-vector form as follows:

$$\begin{bmatrix} y_1 \\ y_2 \\ \cdots \\ y_m \end{bmatrix} = \begin{bmatrix} H_{11}, H_{12}, \ldots, H_{1n} \\ H_{21}, H_{22}, \ldots, H_{2n} \\ \ldots, \ldots, \ldots, \ldots \\ H_{m1}, H_{m2}, \ldots, H_{mn} \end{bmatrix} \cdot \begin{bmatrix} f_1 \\ f_2 \\ \cdots \\ f_n \end{bmatrix} \quad (15)$$

In the formula, n represents the number of impact sources, m represents the number of response points, H_{ij} represents the transfer matrix between response point i and impact source j, f_j corresponds to the impact load, and y_i corresponds to the response point.

This can also be expressed as a compact matrix expression:

$$y = H.f \quad (16)$$

Since it is difficult to avoid measurement noise in the process of actual impact load identification, a load identification model containing measurement noise is as follows:

$$y = H.f + w \quad (17)$$

In terms of specific solution methods, the L2 norm method and the L1 norm method are included, just like the single point impact load identification. In the process of solving, the assembly transfer matrix and the structural response matrix will be newly assembled, and then the assembled load matrix will be obtained. Finally, the assembled load matrix will be divided into different load matrices, then different loads will be identified.

2.3 *Evaluation method of impact load identification*

In order to evaluate the difference between the identified load and the real load within the time range of interception, the root mean square error (RMSE) is defined, whose expression is:

$$\mathrm{RMSE} = \sqrt{\frac{\sum_{i=1}^{m}\left(f_{real}^{i} - f_{iden}^{i}\right)^2}{\sum_{i=1}^{m}\left(f_{real}^{i}\right)^2}} \times 100\% \quad (18)$$

In the formula, m represents the number of sampling $f_{real}^{i}f_{iden}^{i}$ points; represents the actual load value; represents the identified load value. The root mean square error reflects the stability and accuracy of the regularization method's identification results over the entire acquisition time range. The smaller the RMSE value, the higher the overall stability and accuracy of the identification results within the acquisition time range.

In order to evaluate the waveform similarity of the identification load results within the time interval of interception, the correlation coefficient (CC) is defined, whose expression is:

$$\mathrm{CC} = \frac{\sum_{i=1}^{m}\left[f_{iden}^{i} - E\left(f_{iden}^{i}\right)\right]\left[f_{real}^{i} - E\left(f_{real}^{i}\right)\right]}{\|f_{iden} - E(f_{iden})\|\|f_{real} - E(f_{real})\|} \times 100\% \quad (19)$$

In the formula, $f_{iden}^{i}f_{real}^{i}$ represents the average of the true load; $E(f_{real})E(f_{iden})$ represents the average of the identified load. The correlation coefficient can measure the linear relationship between the identified load and the true load. The closer the CC value is to 1, the stronger the correlation between the identified load within the selected time range and the true load, and the closer the corresponding waveform signals are.

For impact load identification, peak force is a very important index in structural health monitoring. In order to analyze the error between peak force and real peak force, peak relative error (PRE) is defined, whose expression is as follows:

$$\mathrm{PRE} = \frac{|\max(f_{real} - f_{iden})|}{\max(f_{real})} \times 100\% \quad (20)$$

In the formula, $\max(f_{iden})$ represents the peak value of the identified load; $\max(f_{real})$ represents the peak value of the real load. The smaller the PRE value, the closer the peak identification result of the impact load is to the real value, and the higher the peak accuracy.

2.4 *Sensor layout strategy based on modal analysis*

The reasonable arrangement of measurement points has an important influence on the inversion accuracy. Through modal analysis, the vibration mode information of the structure can be obtained, and then the arrangement of measurement points can be optimized to improve the inversion accuracy and reduce the problems caused by high coherence.

Modal shapes describe the deformation patterns of structures under specific modes. Different modal shapes contribute differently to responses at different locations: if measurement points are in the high-response region of the same mode, their responses are highly correlated, which can lead to high coherence in the inversion matrix, affecting the accuracy of inversion. By selecting multiple measurement points that are not arranged on the same mode, the correlation between measurement points can be reduced, improving the identifiability of the matrix and making the inversion more stable.

According to the principle of modal dynamics, the following three points can be considered in the selection of sensor measurement points:

Principle 1: Avoid high-response areas with the same mode. If a measurement point is in an area of high amplitude across multiple modes, it may become overly sensitive to specific excitation directions, leading to inversion instability. By selecting low-response areas under different modes for placement, redundant information can be reduced, enhancing the stability of the inversion.

Principle 2: Cover the main modes. Ensure that the measurement points can reflect the characteristics of different modes. In the structure, one measurement point can be located in the high response area of the first mode, and another measurement point can be located in the high response area of the second mode, so as to improve the sensitivity to different excitation sources.

Principle 3: Avoid modal nodes. At the nodes of the vibration mode, i.e. the position where the amplitude of vibration is close to zero, the measurement point may not be able to effectively capture the vibration information, so these positions should be avoided as far as possible.

3 NUMERICAL CALCULATION

In shipbuilding and marine engineering, stiffened plates are the primary structure. To effectively identify impact loads acting on ships and marine structures, a stiffened plate structure as shown in Figure 1 is selected. This stiffened plate structure measures 1.8 meters in length and 0.9 meters in width, with all four sides fixed. In numerical simulation software, the load identification performance of the stiffened plate structure under different regularization methods, varying numbers of loads, and different sensor placement schemes is compared. The stiffened plate model used in the numerical simulation is shown in Figure 2.

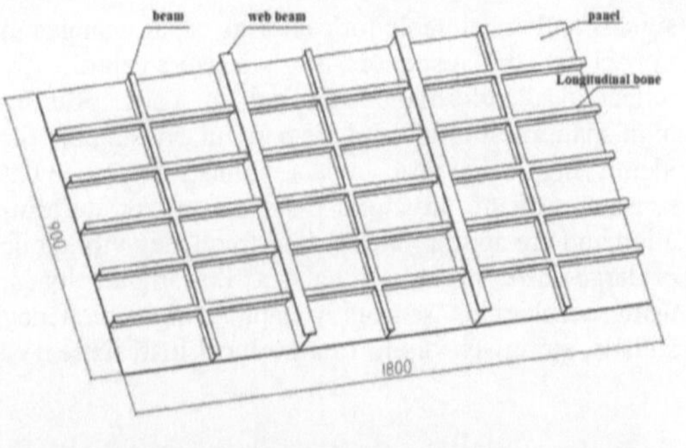

Figure 1. Typical stiffened plate structure.

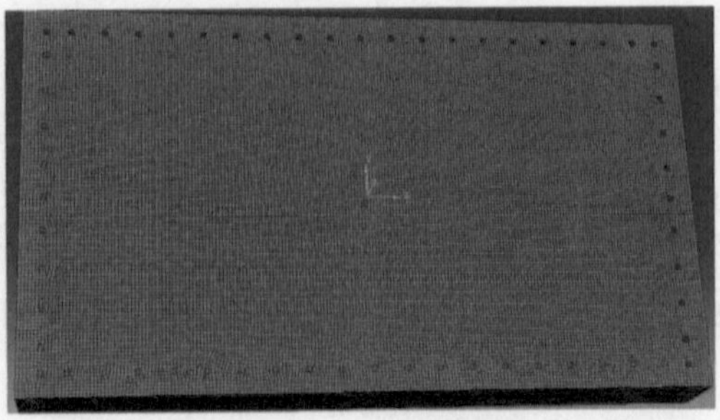

Figure 2. Numerical simulation model.

The first ten modes of the calculated structure in the numerical simulation software Abaqus are shown in the figure below:

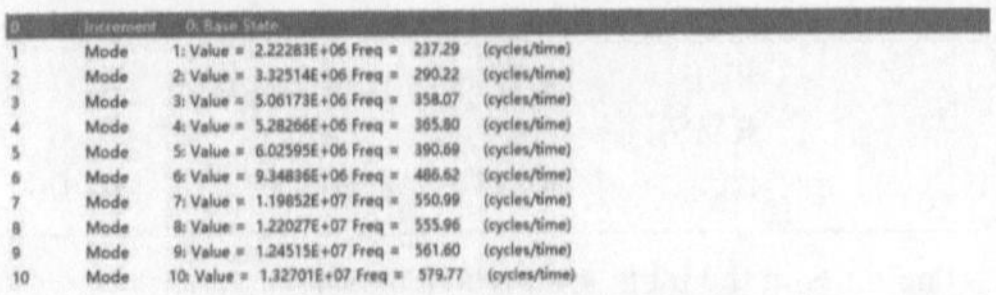

0	Increment	0: Base State			
1	Mode	1: Value =	2.22283E+06 Freq =	237.29	(cycles/time)
2	Mode	2: Value =	3.32514E+06 Freq =	290.22	(cycles/time)
3	Mode	3: Value =	5.06173E+06 Freq =	358.07	(cycles/time)
4	Mode	4: Value =	5.28266E+06 Freq =	365.80	(cycles/time)
5	Mode	5: Value =	6.02595E+06 Freq =	390.69	(cycles/time)
6	Mode	6: Value =	9.34836E+06 Freq =	486.62	(cycles/time)
7	Mode	7: Value =	1.19852E+07 Freq =	550.99	(cycles/time)
8	Mode	8: Value =	1.22027E+07 Freq =	555.96	(cycles/time)
9	Mode	9: Value =	1.24515E+07 Freq =	561.60	(cycles/time)
10	Mode	10: Value =	1.32701E+07 Freq =	579.77	(cycles/time)

Figure 3. Ten models of the calculated structure.

The first five modes of the analysis structure are shown as follows (the fourth mode is not obvious):

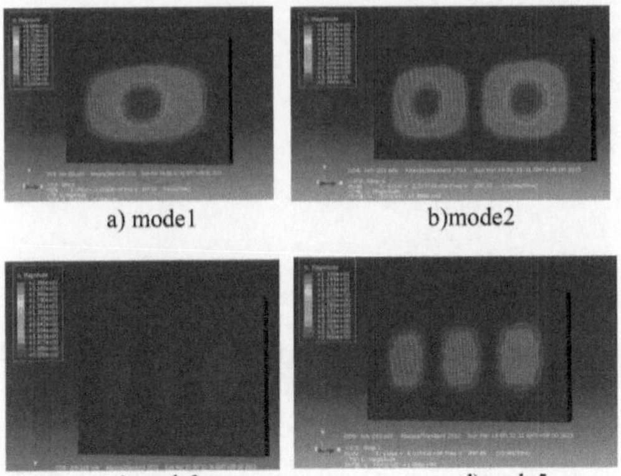

a) mode1 b)mode2

c) mode3 d)mode5

Figure 4. Five models picture.

3.1 *Single-point load*

Under the action of single point load In the numerical simulation software Abaqus, the load application points and sensor placement points are shown in Figure 6. Two acceleration sensors are used, and the loading is applied as shown in Figure 5. To simulate real-world conditions, a 10% random noise is added to the responses measured by the sensors. The three different regularization methods Tikhonov, TSVD, and L1 introduced in Part Two are used to compare the load inversion results, as shown in Figure 7.

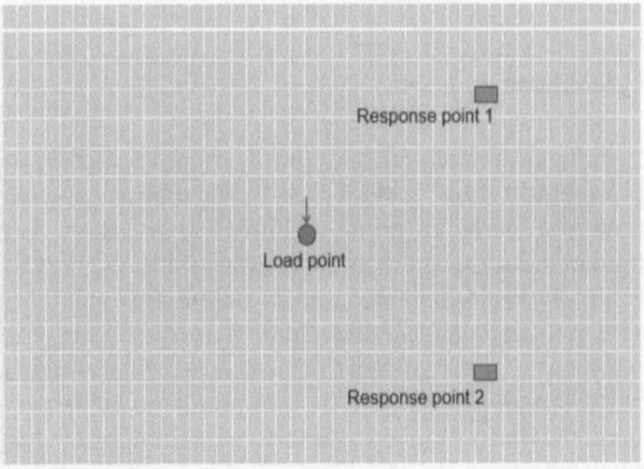

Figure 5. Impact load point and sensor layout point.

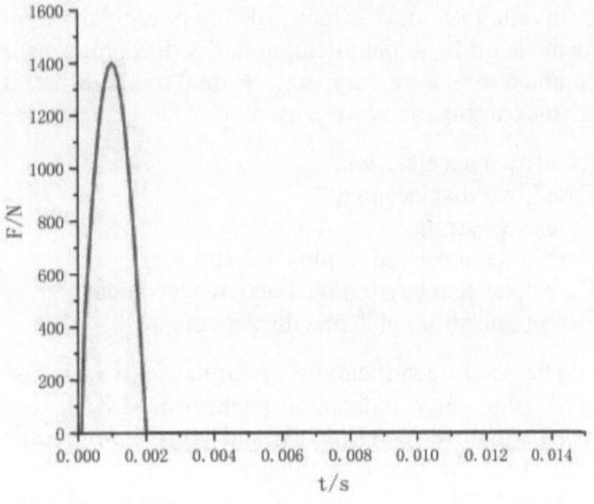

Figure 6. Single point load.

From Figure 7, it can be seen that in terms of peak error, among the three regularization methods, the L1 regularization method has the smallest error, with a peak of 1386 closest to the true load of 1400. The Tikhonov regularization method follows, while the TSVD regularization method has the smallest peak and the largest gap from the true load. The main reason is that it discards smaller singular values, leading to a smaller peak load recognition. In terms of waveform comparison, during the non-load segment 0.002s-0.015s, only the L1 regularization method identifies a straight line close to zero. Both the TSVD

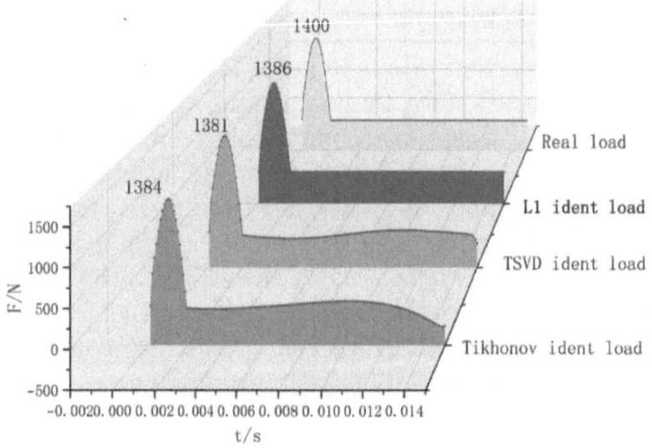

Figure 7. Effect of load identification using different regularization methods.

regularization method and the Tikhonov regularization method show curves with longer non-zero value intervals during this period, mainly because these two regularization methods are essentially L2 regularization methods, which yield more continuous solutions with non-zero values rather than sparse solutions with many zeros when solving for the load. Therefore, it can be concluded that in the identification of impact loads shown in Figure 5 where the load amplitude has fewer non-zero values, most amplitudes are zero or close to zero, the L1 regularization method can obtain sparse solutions and is also closer to the actual impact load in terms of peak load.

In addition, the effects of three regularization methods on load identification under different sensor combinations were compared. A total of six different sensor combinations were used:

Case1:two acceleration.
Case2:two displacement.
Case3:two strain.
Case4:one acceleration plus one strain.
Case5:one acceleration plus one displacement.
Case6: one strain plus one displacement.

The load identification performance was measured using three indicators: peak error (PRE), root mean square error (RMSE), and correlation coefficient (CC). As shown in Table 1.

From Table 1, it can be seen that when both sensors are of the same type, using the L1 regularization method and an acceleration sensor condition with a PRE of 0.01, RMSE of 0.031, and CC of 0.9996, all values are optimal. When the two sensors are different types, using the L1 regularization method and one acceleration plus one strain condition, the PRE is 0.0073, RMSE is 0.0275, and CC is 0.9997, all values are also optimal; however, the value for one acceleration plus one displacement condition is slightly inferior in peak error. Considering all six sensor combinations, using one acceleration plus one strain sensor significantly improves load identification compared to using two acceleration sensors.

Analyze the characteristics of different sensors: Acceleration sensors are sensitive to high-frequency signals and are suitable for capturing rapid changes in impact dynamic responses; Strain sensors can directly reflect local deformations caused by loads, offering high spatial resolution and are particularly suitable for identifying loads near impact points; Displacement sensors measure structural displacement or deformation and are appropriate for low-frequency vibrations or large-scale structural analysis. For impact loads, using acceleration sensors to capture high-frequency signals, strain sensors to capture local load forces, or

Table 1. Load recognition effect of different sensor combinations under single point load.

Method	index	Case1	Case2	Case3
Tikhonov	PRE	12.19	5.07	0.86
	RMSE	29.6	29.29	5.72
	CC	96.02	98.63	99.92
TSVD	PRE	1.06	1090.41	1.5
	RMSE	15	498.9	19.77
	CC	98.95	20.67	98.08
L1	PRE	1	47.05	1.74
	RMSE	3.1	56.43	5.84
	CC	99.96	85	99.81
Method	**index**	**Case4**	**Case5**	**Case6**
Tikhonov	PRE	501.27	1508.44	12.9
	RMSE	1121.84	3002.7	48.06
	CC	24.63	32	92.05
TSVD	PRE	4.17	6.01	233.63
	RMSE	85.39	118.5	214.27
	CC	75.74	63.45	38.9
L1	PRE	0.73	0.76	6.57
	RMSE	2.75	2.75	24.6
	CC	99.97	99.97	97.06

* The units in the table are percentages%.

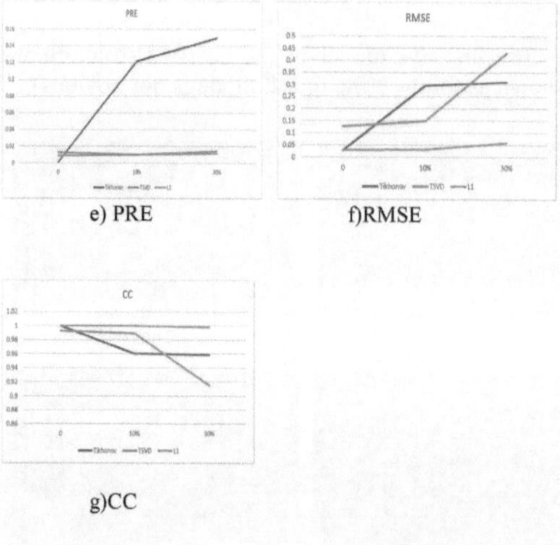

Figure 8. Load identification effect of different regularization methods under different noise levels.From top to bottom,the three indexes are PRE, RMSE, CC.

displacement sensors to capture structural deformation displacement can improve load identification accuracy. By leveraging the advantages of hybrid sensors, the effectiveness of load identification under impact load conditions can be enhanced.

Noise is an important factor affecting load identification. Under the action of single-point impact load, different noise levels are added to the response measured by the acceleration sensor, and the recognition effects of each regularization method are compared under different noise levels, as shown in Figure 8.

From Figure 7, it can be seen that under noise-free conditions, the three metrics of the Tikhonov regularization method outperform the other two regularization methods; the L1 regularization method is slightly inferior to the Tikhonov regularization method. Under 10% and 30% noise conditions, the three metrics of the L1 regularization method are significantly better than those of the other two regularization methods. Therefore, it is concluded that the L1 regularization method performs best in load recognition under noisy conditions.

The arrangement of sensors is also a critical factor affecting the effectiveness of load identification. Based on the previous structural modal analysis, four operating conditions are divided according to the line connecting two sensors with the load application point. The third condition differs from the first in that the distance between the sensor measurement point and the load application point is halved. The four layout schemes are shown in Figure 9:

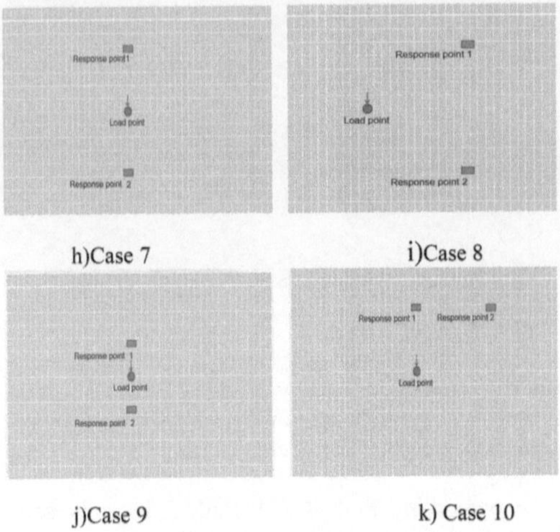

Figure 9. Four sensor layout schemes.

Under the above four layout schemes, the three indicators of load identification under impact loads were compared using acceleration sensors, as shown in Table 2. From Table 2, it can be seen that Scheme 1 has the best recognition effect, as the two sensors are precisely in the low-response region of the first five modes. In contrast, Scheme 3 has both sensors in the high-response region of the first mode. Scheme 2 has both sensors in the high-response region of the second mode, while Scheme 4 has one sensor in the high-response region of the second mode. Therefore, the optimal sensor layout is shown in Scheme 1, positioned at the normal and lower positions on the plane of the load application point.

Table 2. Load recognition effect of different sensor layout schemes under single point load.

Case	PRE	RMSE	CC
7	1.15%	2.67%	99.97%
8	1.74%	5.84%	99.81%
9	1.58%	11.35%	99.33%
10	1.69%	3.4%	99.96%

* The units in the table are percentages%.

3.2 *Dual-point load*

In the numerical simulation software, two impact loads act on the stiffened plate structure, as shown in Figure 10 and 11. Firstly, the influence of different sensor layout schemes on load identification is analyzed.

Figure 10. Double point load impact structure.

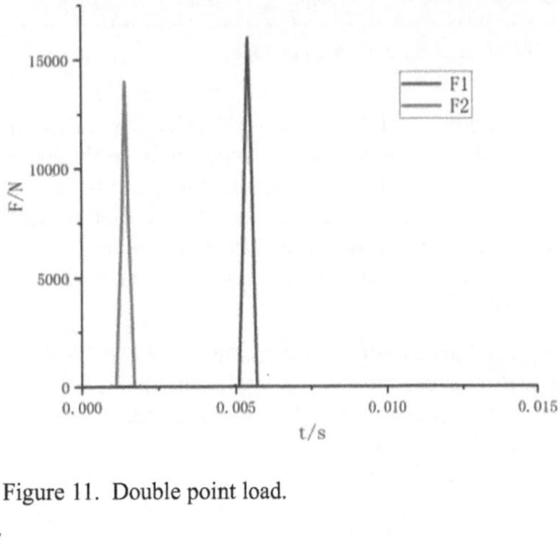

Figure 11. Double point load.

Compare the four typical sensor placement schemes shown in Figure 12: sensors placed on one side of the load application point, sensors placed at the four corners of the load, sensors and the load arranged on the same line, and sensors and the load arranged in an orthogonal shape. Firstly, analyze the case where all four sensors use the same sensor type, using an acceleration sensor with 5% noise added. Based on the research results from single-point impact loads, apply the L1 regularization method.

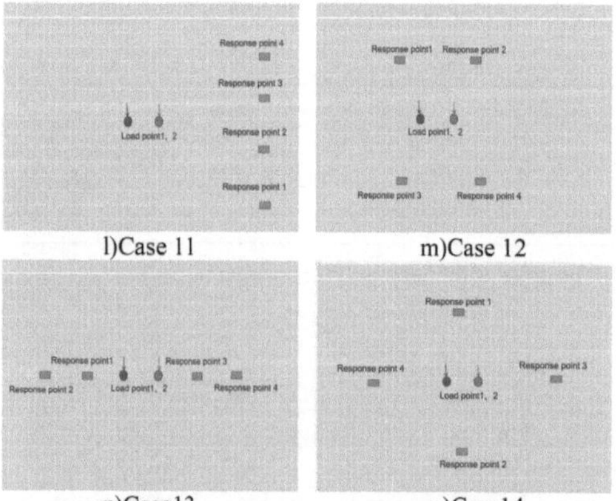

Figure 12. Four sensor layout schemes under double point load.

The peak error PRE1 and PRE2 of the two loads and the root mean square error RMSE1 and RMSE2 are compared.

Table 3. Load identification effect of different sensor layout schemes under double point load.

	PRE1	PRE2	RMSE1	RMSE2
Case 11	1.25	2.23	3.26	4.67
Case 12	0.97	1.21	3.23	2.54
Case 13	0.86	2.11	5.75	5.36
Case 14	0.64	0.62	1.25	1.11

* The units in the table are percentages%.

According to Table 3, the identification effect of scheme 4, four sensors are arranged in an orthogonal position with the load best. At this time, similar to the single point load under modal analysis, all four sensors are in the low response zone of the first five modes.

The effects of different sensor combinations on load identification were analyzed:

Case15: 2 accelerations + 2 strains
Case16: 2 accelerations + 2 displacements
Case17: 2 accelerations +1 strain+1 displacement
Case18:3 accelerations + 1 strain
Case19:4 accelerations

Table 4. Load recognition effect of different sensor combinations under double point load.

	PRE1	PRE2	RMSE1	RMSE2
Case 15	0.66	0.58	1.37	1.11
Case16	0.65	0.59	1.36	1.13
Case 17	0.66	0.56	1.38	1.08
Case 18	0.66	0.6	1.31	1.1
Case 19	0.64	0.62	1.25	1.11

* The units in the table are percentages%.

From Table 4, it can be seen that the load identification effect of the sensor combination scheme of "2 accelerations + 1 strain + 1 displacement" and "3 accelerations + 1 strain" is better than that of a single point load. Similar to a single point load, the sensor combination strategy can effectively give full play to the advantages of different sensors in capturing load information and enhance the effect of load information identification.

The recognition effect of the two loads in different forms is compared, and the double-point load is shown in Figure 13:

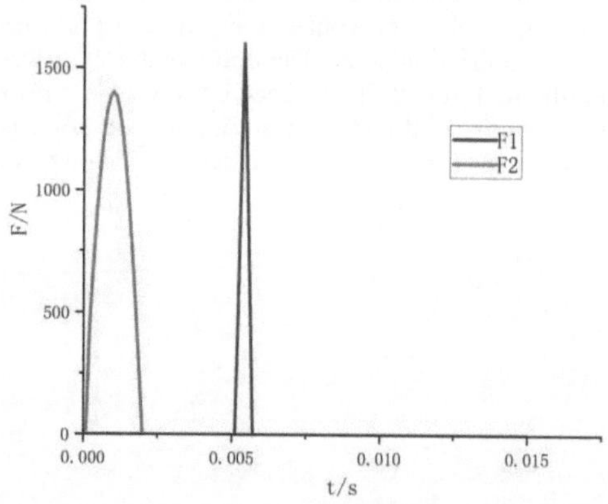

Figure 13. Two different types of loads.

Under the condition of 5% noise, the load identification was carried out by using L1 regularization method, and the influence of different sensor combination modes on the load identification effect is shown in Table 5:

By comparing Table 4 with Table 5, it can be seen that the identified load error coefficients increase under different types of loads, mainly due to overlapping or coupling between different types of loads, making it difficult to separate the observed signals. As shown in Figure 14, the load time history reconstructed using three accelerations plus one strain sensor is presented. From Figure 14, it can be observed that the identified load F1 is influenced by load F2, and during the unloading phase the loading

Table 5. Comparison of load recognition effects under different sensor combinations under different types of double-point loads.

	PRE1	PRE2	RMSE1	RMSE2
Case 15	3.8	26.6	14.5	26.05
Case 16	3.8	26.61	14.49	26.06
Case 17	3.84	26.61	14.53	26.04
Case 18	3.36	24.29	14.24	23.6
Case 19	3.36	24.29	14.24	23.6

* The units in the table are percentages%.

period of F2, an inverted load segment is identified. This indicates that different types of loads may have distinct characteristics in sparse representation. If both types are reconstructed simultaneously, more complex dictionary design or joint sparse representation might be required; otherwise, it could affect sparsity and thus impact the reconstruction results.

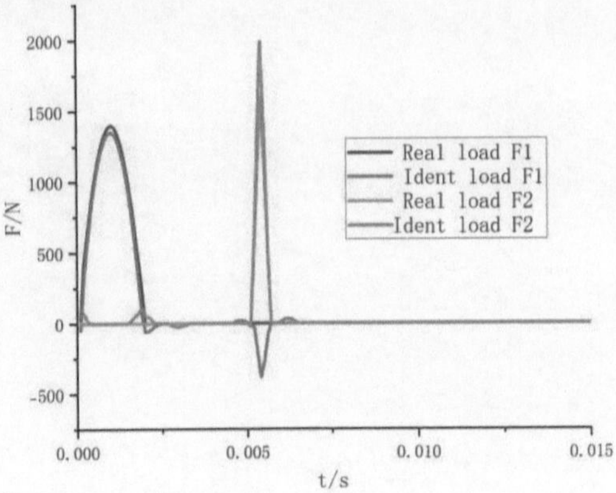

Figure 14. Identification results of different types of double point loads.

4 CONCLUSIONS

This paper investigates the issue of impact load identification based on structural dynamic response. Firstly, it compares the effectiveness of using different regularization methods for impact load identification. Secondly, it analyzes the impact of sensor placement schemes and sensor combinations on single-point and double-point impact load identification. Finally, it examines the performance analysis of double-point loads both similar and dissimilar types. From this work, preliminary conclusions can be drawn as follows:

1. In the case of single-point impact loads, different regularization methods, sensor placement schemes, sensor combinations, and noise levels are all critical factors affecting load identification. In the presence of noise, the L1 regularization method performs best compared to the other two methods. The best recognition results are achieved when sensors are placed on both sides of the load. Additionally, if the number of sensors exceeds one, a hybrid strategy, i.e "acceleration + strain" or "acceleration plus displacement" is recommended for better performance.
2. Under dual-point impact load conditions, the L1 regularization method is still employed. The sensor layout, combination form, and load type are crucial factors affecting impact load identification. When there are four sensors, they should be arranged in pairs around each load, forming an orthogonal shape, which yields the best load identification results. Additionally, a mixed strategy, i.e."2 accelerations + 1 strain + 1 displacement" or "3 accelerations + 1 strain" is more effective.
3. When comparing single-point impact loads with double-point impact loads, the factors for identifying impact loads sensor placement schemes, combination forms are consistent with those for single-point impact loads when the double-point impact loads are of the same type. However, when the two loads are of different types, due to certain coupling effects, there will be some cross-interference between the identified loads, and further regularization methods are needed to improve the load identification results.

ACKNOWLEDGEMENTS

This project is supported by National Natural Science Foundation of China (Grant No. 2241266 and Science and Technology commission of Shanghai Municipality (24ZR1436100).

REFERENCES

Boukria Z, Perrotin P, Bennani A. Experimental impact force location and identification using inverse problems: Application for a circular plate. *International Journal of Mechanics*, 2011,5(1):48–55.

Chen H. Research and engineering application of impact load identification technology based on sparse regularization method. *Southwest Jiaotong University Master Degree Thesis*, 2022.

Chen L, Wang Y, Cheng H, Liu JJ, Qiao B, Chen XF, Frequency-domain non-convex sparse regularization method for impact force identification based on spatial sparsity prior. *Journal of Vibration and Shock*, 2024, 43 (14): 148-155+188.

Kalhori H, Ye L, Mustapha S. Inverse estimation of impact force on a composite panel using a single piezoelectric sensor. *Journal of Intelligent Material Systems and Structures*, 2016,28(6).

Kong, S, Cui, HY, Tian, YK, Ji, SY. Identification of ice loads on shell structure of ice-going vessel with Green kernel and regularization method. *Marine Structures*, Vol. 74, 2020, 102820, ISSN 0951–8339.

Qiao B, Zhang X, Wang C, Zhang H, Chen, XF. Sparse regularization for force identification using dictionaries. *Journal of Sound and Vibration*, 2016, 368: 71–86.
Qiao B, Zhang X, Luo X, Chen, XF. A force identification method using cubic B-spline scaling functions. *Journal of Sound and Vibration*, 2015, 333: 28–44.
Qiao B, Chen X, Xue X, Luo, XJ, Liu, RN. The application of cubic B-spline collocation method in impact force identification. *Mechanical Systems and Signal Processing,* 2015, 64-65: 413–427.
Qiao B, Chen XF, Liu JX, et al. Study on sparse identification method of impact load in mechanical structure. *Journal of Mechanical Engineering,* 2019,*55(03)*: 81-89.
Tropp J A, Wright S J. Computational methods for sparse solution of linear inverse problems. *Proceedings of the IEEE*, 2010, 98(6): 948–958.
Wang L, Han X, Liu J, He XQ, Huang, F. A new regularization method and application to dynamic load identification problems. *Inverse Problems in Science and Engineering*, 2011,19(6):765–776.
Wang L.J. The method of positive transformation and its application in dynamic load identification. *Hunan University Master Degree Thesis*, 2011.
Wang L, Liu J, Xie Y, Gu Y.A new regularization method for the dynamic load identifcation of stochastic structure. *Comput Math Appl.*, 2018,76(4):741.
Wang L, Huang Y, Xie Y, Du Y. A new regularization method for dynamic load identification. *Science Progress*. 2020;103(3).
Wang Y, Zhou Z, Xu H, Li, S, Wu, ZJ. Inverse Load Identification in Stiffened Plate Structure Based on in situ Strain Measurement. *Structural Durability & Health Monitoring*, 2021, 15(2): 85.

Trends in Collision and Grounding of Ships and Offshore Structures – Zhu & Guedes Soares (Eds)

MV Estonia: Numerical modelling of bow arrangement collapse sequence

K. Tabri & H. Naar
Tallinn University of Technology, School of Engineering, Marine Technology Competence Center, Estonia

M. Heinvee & A. Soo
MEC Insenerilahendused, Estonia

S. Hosseinzadeh
University of Southampton, UK

T. Roosipuu
Estonian Safety Investigation Bureau, Estonia

ABSTRACT: The paper presents a numerical assessment of the progressive failure sequence of the bow visor and ramp of MV Estonia, a passenger ferry that sank on September 28, 1994. While a thorough investigation was conducted from 1994 to 1997, new investigative actives have been launched by accident investigation bureaus from Estonia, Sweden, and Finland. These new activities aim to analyze and model the sequence of events using state of the art modelling tools to gather deeper insight into the accident. Paper presents the simulation models to assess the failure sequence of the bow area, discusses the initial outcomes and compares the damages to the those reported at the visor and ramp salvaged from the sea bottom.

1 INTRODUCTION

The paper presents a numerical assessment of the progressive failure sequence of the bow visor and ramp of MV Estonia, a passenger ferry that sank on September 28, 1994. While a thorough investigation was conducted from 1994 to 1997, new investigative actives have been launched by accident investigation bureaus from Estonia, Sweden, and Finland. The goal of these new activities is to obtain a more thorough understanding of the accident through the utilization of advanced techniques and modeling software, and to replicate the series of events during the accident.

This paper presents a part of the new investigative activities and focuses on structural analysis of ship's bow visor and ramp system when subjected to harsh wave-induced forces. It is assessed whether the sea loads that were expected to occur during the accident could fail the visor locking devices and thereby eventually lead to the detachment of the visor and the consequent opening of the bow ramp.

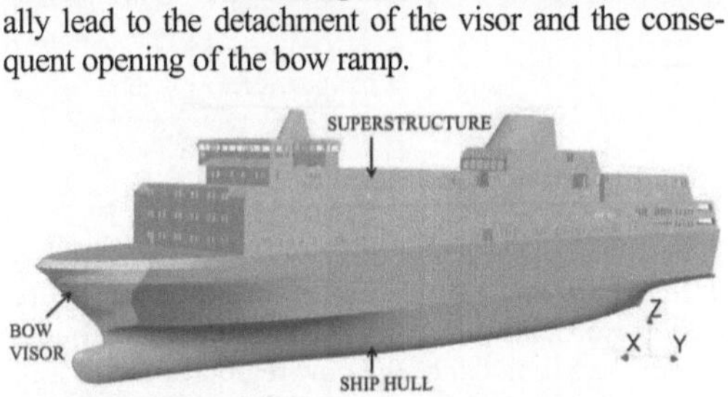

Figure 1. A view of MV Estonia with superstructure and bow visor.

The paper presents the means and simulation models that were employed to examine the physical phenomena of the failure of the bow arrangement. The wave induced forces were established based on model-scale experiments conducted at SSPA (Huss, 1995; Trägardh, 1995) and through numerical models utilizing Computational Fluid Dynamics (CFD) and Multi-Material Arbitrary Lagrangian-Eulerian (MMALE) techniques. Structural response and the failure of the components was evaluated in time-domain using explicit finite element (FE) simulations.

The simulation results are compared to the reported damages and laser-scanned models of the deformed visor and ramp, retrieved from the accident site.

2 BOW VISOR & RAMP ARRANGEMENT

Bow visor and ramp arrangement of MV Estonia is shown in Figure 1 and Figure 2. The MV Estonia had a vulnerable bow visor and ramp arrangement, in which the visor was connected into the ramp in a box-like housing. Such an arrangement was typical in the Baltic ferries in the 1970s and 1980s (Hänninen, 2007).

The visor is attached to the vessel by means of two hinges connected to the forecastle deck (C-deck in Figure 2) and through two hydraulic actuators used for opening and closing the visor. The actuators' (upper) eye fittings are attached under the visor hinge arm and actuator (lower) heel fittings welded to the B-deck.

Due to the visor's size, rounded shape with large flare angle and long hinge arms, sea forces can create

DOI: 10.1201/9781003684404-38

a significant opening moment, encouraging visor rotation by pivoting around the hinges attached to the forecastle deck. Side locks are installed on the portside (PS) and starboard (SB) side to prevent rotational movement, while a bottom lock is placed between the visor and the hull along the centerline. Additionally, there are three locating horns to mate with a recess in the visor structure to guide the visor to the correct position while closing. When at sea, the locating horns are designed for transfer of transversal sea forces only (Carlsson, 2007). Majority of the vertical sea loads are transferred to the locks and visor hinges through the visor arms.

The ramp is secured to the vessel by means of six hydraulically operated locking devices, one locking hook and two locking wedges port and starboard.

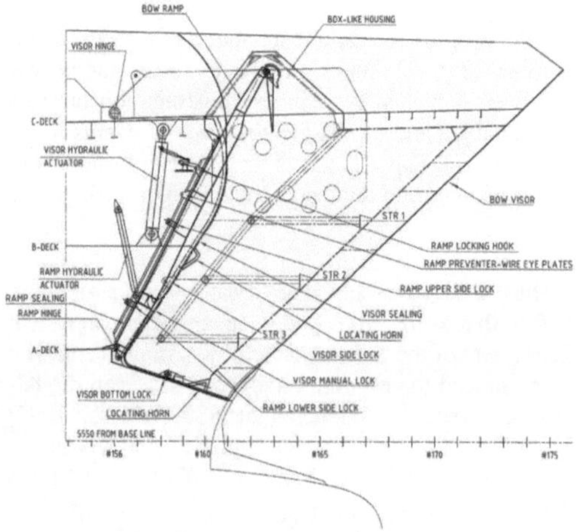

Figure 2. MV Estonia bow area arrangement with visor and a car ramp (Carlsson, 2007).

When at sea with the closed visor, the bow ramp is in its closed position and secured weathertight to form a collision bulkhead to prevent water from entering the car deck. It should be noted that the location of the ramp did not fulfill the requirements of SOLAS (Safety of Life at Sea) to ensure a safe demolition zone in case of collision.

3 ANALYSIS APPROACH & WAVE INDUCED FORCES

3.1 *Analysis approach*

The bow failure sequence is a complex process influenced by various factors such as the stochastic sea waves, ship's seakeeping characteristics, navigational choices made in challenging conditions, the condition of the locking mechanisms which have endured years of slamming impacts and numerous load cycles, potentially resulting in accumulated fatigue damage.

There are considerable uncertainties involved in establishing the wave-induced loads and the status of the ship structures at the time of the accident. Detailed numerical assessment of potential load sequences and the variations in load amplitudes is outside the scope of the current study. Therefore, this assessment partly relies on previous studies in establishing the sea state and assessing the level of the loads, see Figure 3 for the approach adopted here. The wave-induced forces and their components (forces and moments) are taken from model-scale experiments conducted at SSPA (Huss, 1995; Trägardh, 1995). CFD assessment is further conducted to validate the wave load levels and study the pressure distribution along the bow visor. CFD simulations were performed with a closed visor, while MMALE simulations were conducted to assess how the wave induced forces change as the visor locks failed. Once the locks failed, it allowed the visor to move from its closed position and thus, the load levels dropped significantly. Combining the results from model testing, CFD analysis and MMALE simulations allowed to establish the pressure distribution shape over the visor surface and the expected load levels, and to construct the characteristic load history both before and after the visor failed. Parameters used throughout the assessment are presented in Table 1.

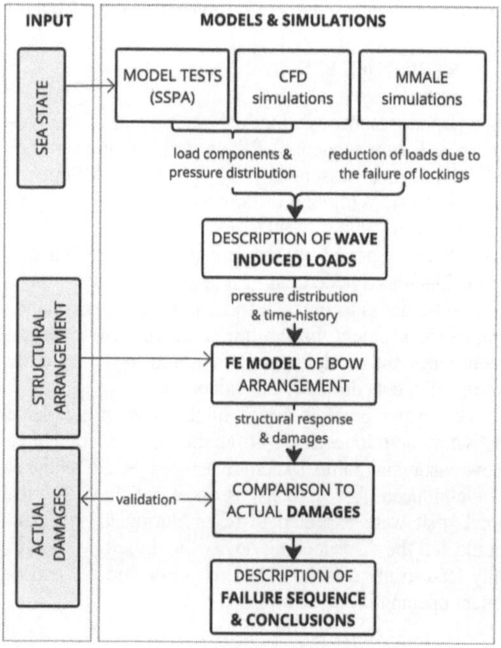

Figure 3. Adopted analysis approach.

Wave induced load distribution and its variation in time, both immediately before and after the failure of the visor locks, was applied as a distributed pressure to the FE model of the bow region of the vessel. Bow region is modelled with sufficient resolution to capture the structural failure of main components and the locking devices. FE models allow the visor to become fully detached, resulting is consequent forced opening of the bow ramp. Thus, the FE simulations cover the duration of the accident starting from the initiation of the failure of the locking devices until the visor is fully separated from the vessel and the bow ramp is torn open.

Table 1. Parameters used in the evaluation of wave-induced loads.

Parameter	Value	Unit
Length (L_{bp})	137.4	m
Breadth(B)	24.2	m
Depth(D)	16.25	m
Draft, fwd (T_{fwd})	5.17	m
Draft, aft (T_{aft})	5.61	m
Displacement	12046	tonnes
Block coefficient (C_B)	0.683	-
LCB (fwd of $L_{bp}/2$)	-4.66	m
VCG	10.62	m
Radii of gyration (k_{xx})	0.37B	m
Radii of gyration (k_{yy}, k_{zz})	0.27L_{bp}	m

3.2 *Wave induced loads from the model tests*

Due to the non-linear and random nature of the bow impact loads, the absolute quantitative loads must be judged with care. Small changes in the relative motion between the ship bow and the waves, as well as in the wave profile, can result in large differences in load values.

SSPA conducted an extensive series of the model-scale experiments to assess the load levels during the accident (Trägardh, 1995). Based on these experiments, (Huss, 1995) presented a typical combination of wave induced forces with different exceedance probability, see Table 2.

Table 2. Typical combinations of wave-induced forces constructed based on the model scale experiments conducted at SSPA, reproduced from (Huss, 1995).

Load level:	A	B	C	D	E	
Resultant force:	3.9	5.2	6.5	7.9	9.2	[MN]
Components:						
Fx	-2.7	-3.6	-4.5	-5.4	-6.3	[MN]
Fy	0.6	1	1.5	2	2.5	[MN]
Fz	-2.7	-3.6	-4.5	-5.4	-6.3	[MN]
Mx	0.6	1.7	3.1	5	7.4	[MNm]
My	4	7.5	11.3	15.5	20	[MNm]
Mz	0.5	1	1.5	2	2.5	[MNm]

The different levels represent the estimated extreme value distributions in which level A has an exceedance probability of 95%, B is the most probable maximum and E has an exceedance probability of 5%. It is not certain that these force components will act simultaneously, but it is here considered as a sufficient assumption for the average conditions (Huss, 1995). In this study, the load level B is used as a base scenario.

3.3 *CFD assessment of the wave loads and pressure distribution*

CFD model was used to assess the levels of wave-induced loads and the pressure distribution at the bow. The evaluated load levels were compared to the those from the model testing at SSPA (Table 2). Relatively short simulations were conducted that covered the duration of up to 80 seconds and <20 wave encounters. Thus, statistical assessment should be conducted to derive the expected extreme load levels for longer time durations.

A full-scale geometry of the MV Estonia is used to conduct the CFD simulations. Figure 1 illustrates the overall configuration of the vessel, including the bow visor and the superstructure. It is important to note that while the superstructure is depicted for illustrative purposes, it was not included as an active component in the computational simulation. The principal dimensions and specifications of the vessel used in the simulation are summarized in Table 1. For the purposes of this analysis, a constant ship speed of 14.5 knots (7.459 m/s) was used.

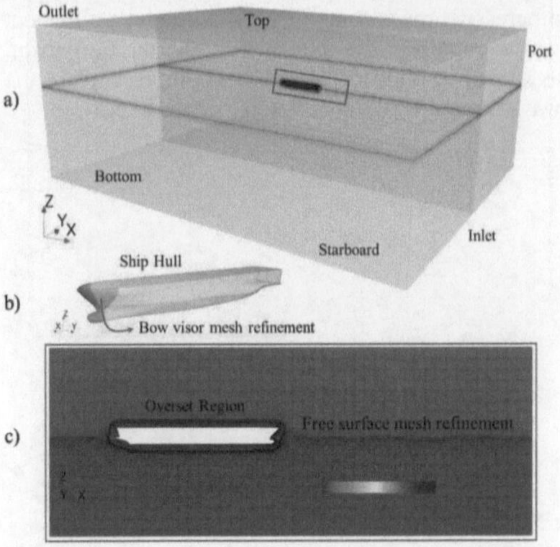

Figure 4. A general view of computational domain and mesh properties of the CFD model: a) applied boundary conditions of the domain; b) mesh on the hull and bow visor refinement; c) overset region and volume fraction of water.

Figure 4 presents the computational fluid domain along with the boundaries defined to model the problem accurately. An overset mesh technique was employed to accommodate the motion of the full-scale ship effectively. This approach allows the dynamic interaction between the moving ship and the surrounding fluid to be captured with greater fidelity.

The dimensions of the computational domain, the extent of the overset region, and the boundary conditions were selected based on the methodology proposed by (Tezdogan et al., 2015) for full-scale ship simulations. As illustrated in Figure 4, a velocity inlet boundary condition was applied at the upstream boundary in the positive x-direction, where incident irregular waves were also generated. The downstream boundary in the negative x-direction was modelled as a pressure outlet to allow fluid to exit the domain. The top and bottom boundaries were both defined as velocity inlets, ensuring appropriate flow conditions throughout the simulation domain.

A trimmed cell meshing approach was employed to produce a superior computational grid that can effectively handle intricate geometries. The background domain was constructed using a trimmed cell mesh with a base cell size of 10 meters, while the overset region, which accommodates the ship's motion, was meshed with a polyhedral grid featuring a minimum surface size of 5 cm. This setup allowed for enhanced resolution in critical regions while maintaining computational efficiency.

The mesh generation followed the guidelines outlined by the (ITTC, 2011) for CFD simulations of ships in waves. To accurately resolve the intricate flow features, including the interaction of the hull with the surrounding fluid, regions around the hull, expected free surface, and wake were subjected to progressive mesh refinement. Additional refinement was applied near the free surface and the bow visor, as shown in Figure 4b and Figure 4c, to capture the wave-structure interaction with higher fidelity.

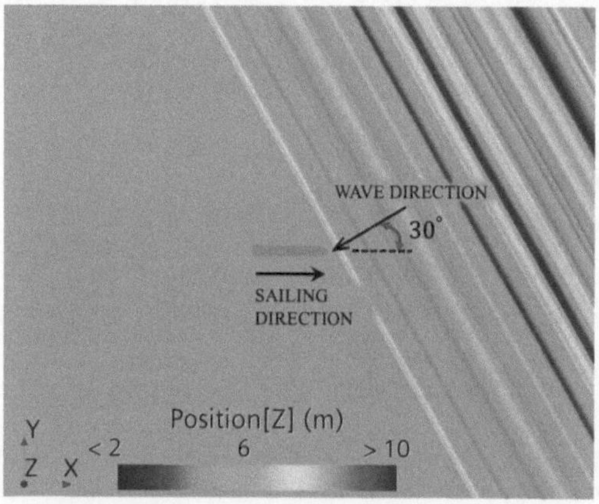

Figure 5. Top view of the domain with the initial position of the irregular wave and the direction.

Furthermore, adaptive mesh refinement techniques were employed in both the free surface and overset regions to dynamically improve resolution during the simulation, particularly in areas with steep gradients or high flow activity. The final computational mesh comprised approximately 9 million cells in the background domain and an additional 2.5 million cells in the overset region.

The Navier-Stokes equations were discretized using a finite volume method (FVM) to accurately represent the fluid flow around the ship. The unsteady Reynolds-Averaged Navier–Stokes equations were solved in conjunction with a turbulence model based on the k–ω formulation. To accurately capture the complex, nonlinear behaviour of the free surface, the Volume-of-Fluid method was used as an interface-capturing technique. The Semi-Implicit Method for Pressure-Linked Equations was adopted for pressure-velocity coupling within the FVM framework. The simulation considered a two-phase flow of water and air, with respective densities of 997.6 kg/m³ and 1.184 kg/m³.

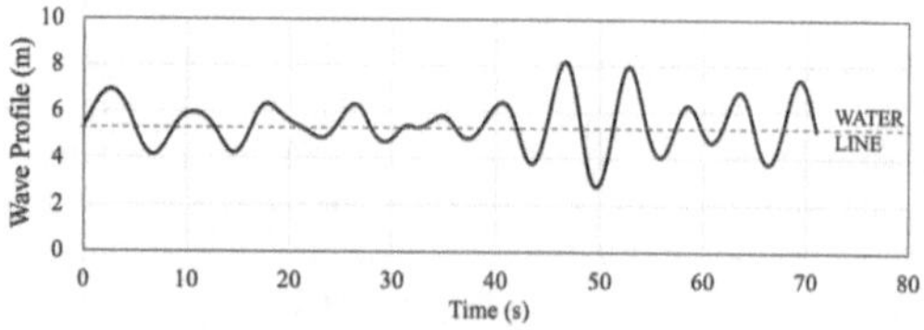

Figure 6. Time history of the oblique irregular wave generated with 30-degree angle from the port.

To realistically simulate the vessel's hydrodynamic response to wave excitation, a Dynamic Fluid-Body Interaction model was implemented, allowing the ship to move freely in the pitch, heave, and roll directions. This model enables coupling between the fluid solver and the rigid body motion solver, allowing the numerical model to dynamically compute wave-induced forces and moments acting on the hull. These loads are then used to solve the rigid-body equations of motion, resulting in time-accurate updates to the vessel's position and orientation. The simulation was carried out using the commercial CFD software Star-CCM+, which integrates both fluid and motion solvers into a single computational environment.

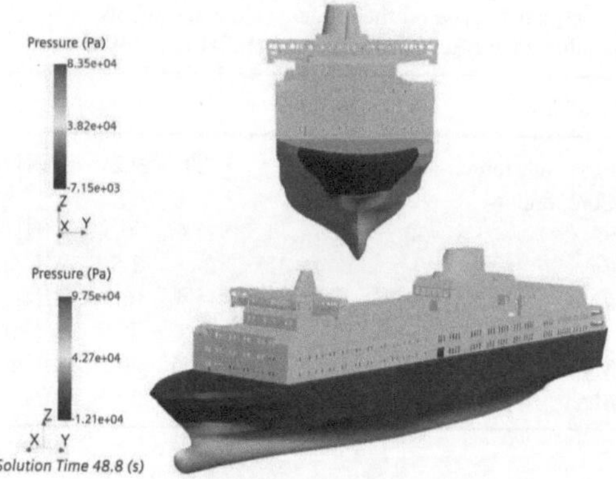

Figure 7. Pressure distribution on the ship hull and bow visor at -0.5° roll, 1° pitch, and a heave of -0.3 m.

The full-scale ship was simulated under calm water, regular wave, and irregular wave conditions. However, only the results for the irregular wave case are presented in this study. Figure 5 shows a top view of the simulated irregular wave field. The irregular waves were generated using a JONSWAP spectrum, characterized by a significant wave height of 4.5 m and a peak wave period of 8.5 seconds. To investigate the ship's response under oblique wave conditions, the incident wave direction was set at a 30-degree angle from the port side.

Figure 6 illustrates the time history of the oblique irregular wave used in the simulation. To improve numerical stability and reduce computational cost, a wave forcing method was applied at the velocity inlet and pressure outlet boundaries. This wave forcing technique also helps to mitigate issues related to surface wave reflections at domain boundaries by incorporating a gradual damping effect, which smoothly absorbs wave energy and prevents artificial disturbances (STAR-CCM+, 2024). In this study, a wave forcing zone of one ship length ($1\times L_{bp}$) was used to ensure smooth wave propagation.

All simulations were conducted with a fixed time step of 0.01 seconds and 10 inner iterations per time step to ensure convergence. Additionally, an adaptive time-stepping strategy was implemented to enhance solution efficiency during transient events. It is also important to note that a first-order temporal discretization scheme was used for the unsteady term in the Navier–Stokes equations to maintain numerical stability over long simulation durations. Figure 7 illustrates the pressure distribution on the ship hull and bow visor obtained from the CFD simulations.

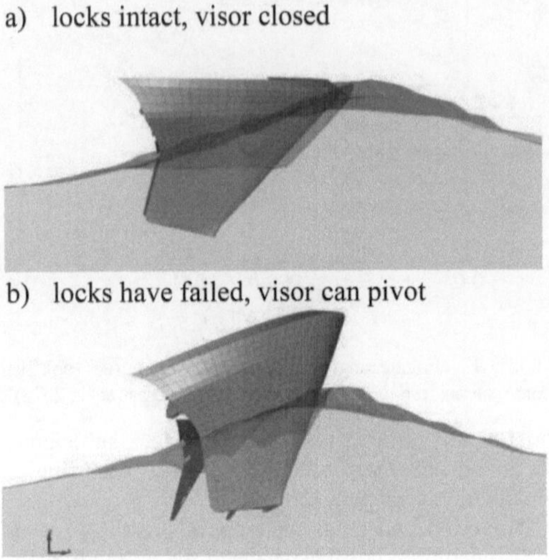

Figure 8. MMALE model developed to assess the reduction in wave-induced loads as the visor locks fail, and the visor is allowed to pivot.

3.4 *MMALE assessment of the reduction of the load levels as the visor locks fail*

MMALE simulations allow for more versatile description of the structural model and fluid structure interaction. While the viscous effects cannot be precisely modelled with MMALE, the inertial loads and wave interaction are captured well. Thus, MMALE simulations are well-suited for physical problems with dominant inertial loads, such as slamming loads (Hosseinzadeh et al., 2023)

Rather low-resolution and simplified model was prepared and simulated in simplified wave conditions due to high computational costs of the MMALE simulations, see Figure 8. In MMALE simulations the visor was either kept fixed in its locked position or was allowed to pivot as the locks were assumed failed. Only few encounters with relatively high waves were assessed. Therefore, such short simulations are not appropriate for determining maximum wave loads, but were only used to evaluate the reduction in wave-induced loads as the visor is allowed to pivot.

Table 3. Wave loads levels from MMALE simulations.

Load (MN)	Fx	Fy	Fz
Closed visor (locks intact)	9.08	3.52	10.2
Open visor (locks failed)	2.60	1.67	2.14
Load reduction	~70%	~50%	~80%

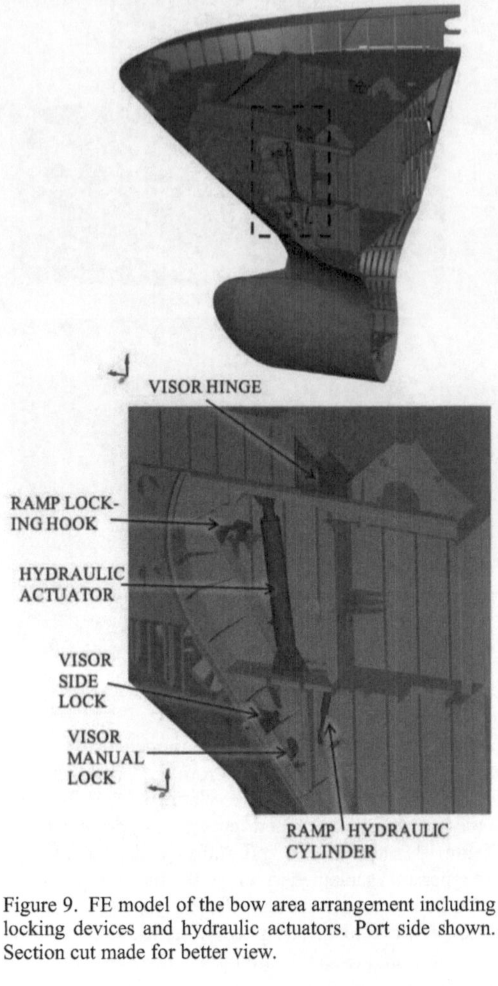

Figure 9. FE model of the bow area arrangement including locking devices and hydraulic actuators. Port side shown. Section cut made for better view.

Analysis results show that pivoting visor provides much less stiff support surface for the waves. The wave-induced loads drop by 50 to 80% depending on the load component, see Table 3. This drop is considered later in the FE simulation.

3.5 *FE model for structural response*

Fine resolution FE model of the bow was developed for structural assessment, see Figure 9 to Figure 11. Model included all the major structural components, locking devices and main hydraulic actuators. Resolution of the model was suitable to simulate the material failure and allowed modelling the response and behavior of the actuators used to manipulate the visor and the bow ramp. The simulations were conducted in time-domain using explicit integration schema in LS-Dyna.

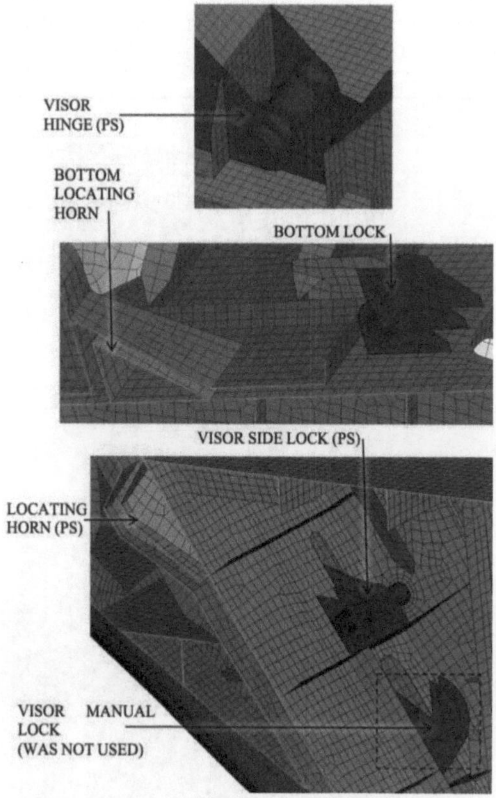

Figure 10. Visor locks, hinges and locating horns.

Steel material behavior was modelled via true stress-strain curves determined from the tensile tests of the specimens cut from the side plating of MV Estonia. Experimentally determined curves are presented in Figure 12. Element size and thickness dependent failure criteria was adopted to model the material failure.

To keep the bow structure fixed in space under the action of wave loads and gravity, the model was fully fixed at the section pointing towards the aft

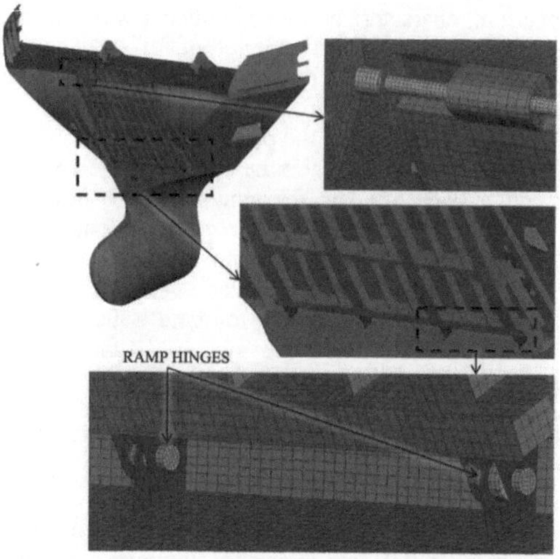

Figure 11. Bow ramp and ramp hinges.

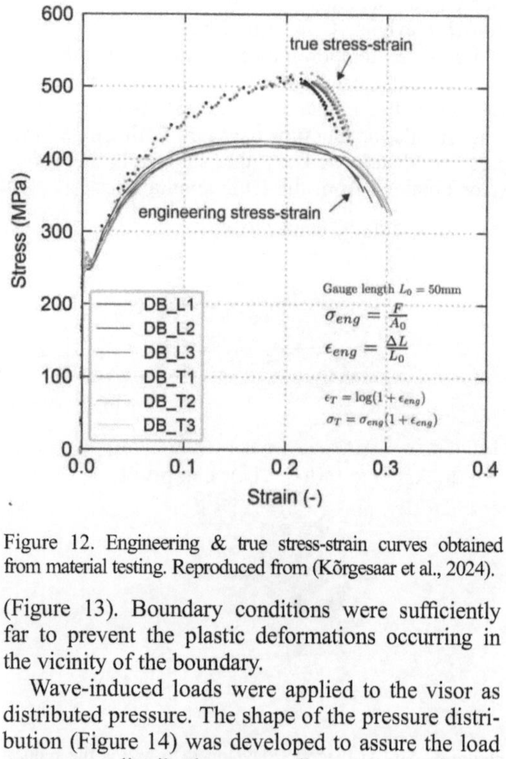

Figure 12. Engineering & true stress-strain curves obtained from material testing. Reproduced from (Kõrgesaar et al., 2024).

(Figure 13). Boundary conditions were sufficiently far to prevent the plastic deformations occurring in the vicinity of the boundary.

Wave-induced loads were applied to the visor as distributed pressure. The shape of the pressure distribution (Figure 14) was developed to assure the load component distributions according to the load level B in Table 2. The absolute load levels were achieved by scaling the magnitude of the applied pressure.

In the FE simulations the bow was loaded with several consecutive waves by prescribing the time history for the applied pressure. At the beginning of the simulation the time-varying pressure magnitude was applied to result in the desired maximum load level. After the failure of the visor locks, the level was reduced by 60% following the conclusions from the MMALE simulations, see Table 3.

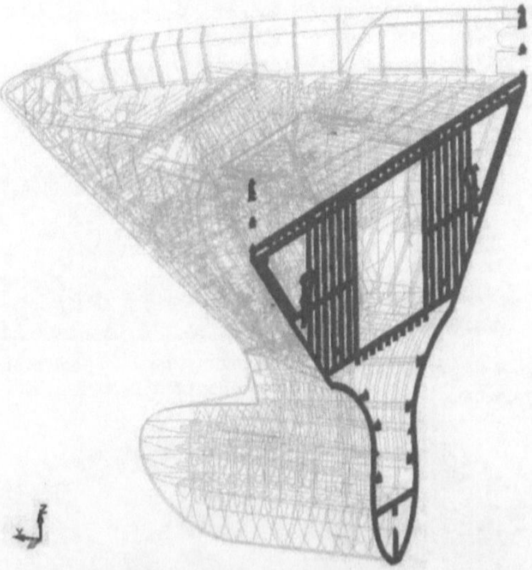

Figure 13. Boundary conditions applied to the FE model.

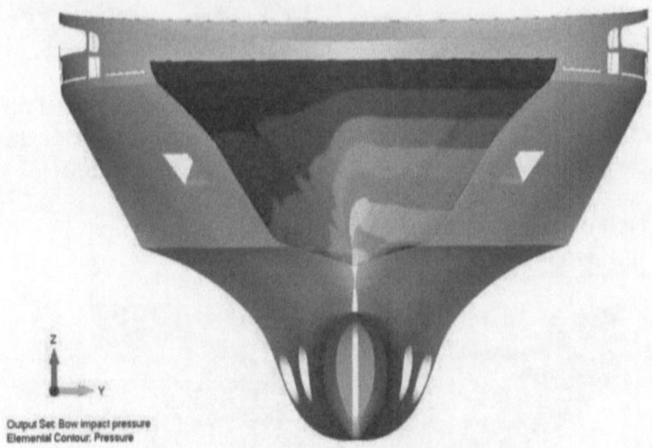

Figure 14. Shape of the pressure distribution applied to the FE model. Pressure levels vary in time to simulate the wave cycles, while the distribution shape is kept the same.

4 FAILURE SEQUENCE OF THE BOW ARRANGEMENT

The precise order in which the bow visor attachments fail is understandably difficult to predict with full confidence. Here a dynamically and kinematically admissible failure sequence is presented under the assumptions described in the above sections. The sequence is reported via the visuals from the numerical simulations and where possible, compared to the actual damages. More thorough discussion is left for final assessment report and future publications.

The time values shown in the subsequent figures serve as a reference for the order of events. Simulations are performed at an accelerated pace using a series of high waves, making it impossible to directly translate the time scale to real-life conditions. In actual wave conditions, the vessel would encounter waves of varying amplitudes, and the total duration of the failure sequence would last significantly longer.

4.1 *Visor detachment and structural damages*

As the vertical wave loads were to large extent transferred to the locking devices and the visor hinges, these experience relatively high stress levels (Figure 15). When fully intact bottom lock is assumed, its failure initiated when the wave loads were about 1.6 times the level B loads (Table 3), i.e. $F_z \cong 1.6 \times 3.6 = 5.8$ [MN]. However, if it is assumed that the cumulated fatigue damage has reduced the lock strength by 30%, the failure was initiated already at around Fz = 3.6 [MN] that corresponds to most probable maximum load level B (Table 3). Failure of the bottom lock fails at the same time together with the side locks. As the locks fail, the loads are transferred to the deck hinges, which fail as the hinge pins are torn from the eye plates connected to the hinge arms (Figure 16).

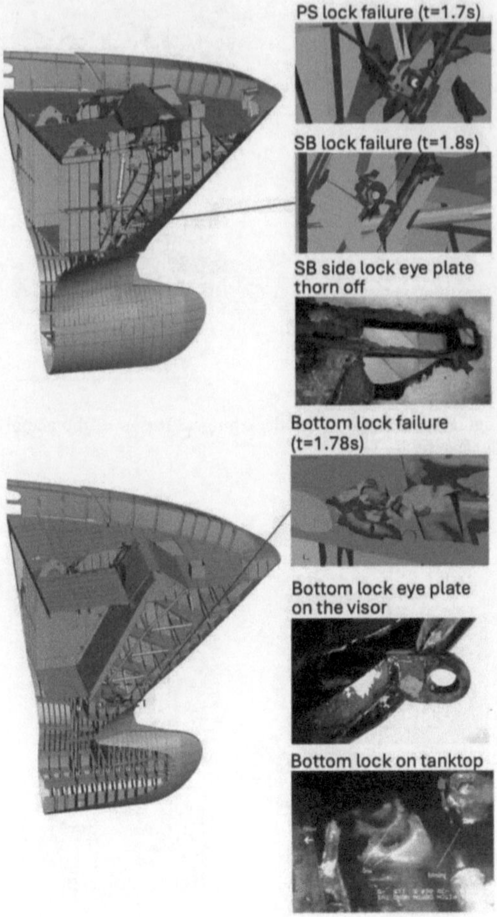

Figure 15. Failure of side locks and the bottom lock. Images of the actual damages from (Carlsson, 2007).

After the failure of the deck hinges, the visor is connected to the vessel only through the visor actuators, see Figure 17. As visor is still exposed to repeated wave loads and thereby it moves almost uncontrollably, the connection between the lower heel fitting of the actuators and B-deck fails with a part of the deck structure ripped-off (Figure 18). During this process the hydraulic actuator's piston rod on the port side is partially pulled out and bent as it becomes to contact with the front bulkhead (Figure 18). The visor bottom structure sustains significant damage from repeated impacts against the tank top.

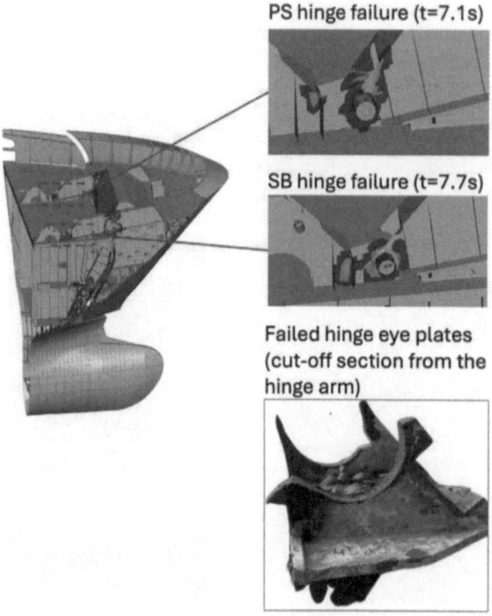

Figure 16. Failure of the deck hinges. Image of the actual damage by K. Tabri.

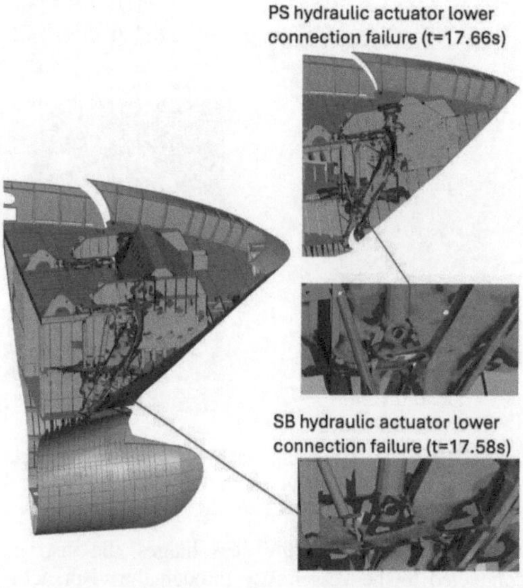

Figure 17. Tearing failure of the B-deck platting under the (lower) heel fitting of the visor hydraulic actuators.

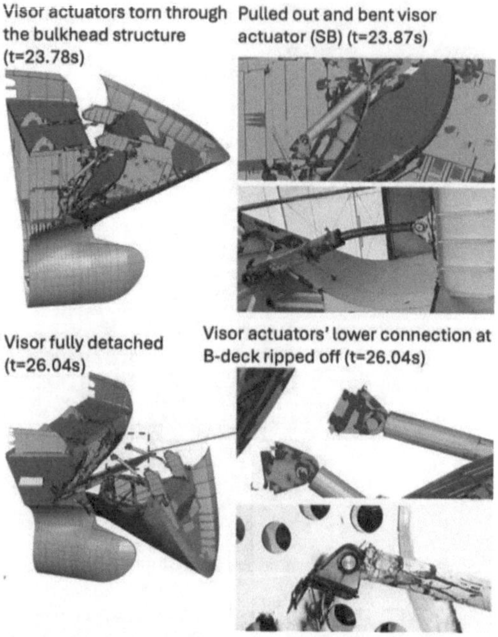

Figure 18. Visor detaching, lower ends of actuators ripped off from the B-deck and torn through the bulkhead. Images of the actual damages by K. Tabri (upper) and by (Carlsson, 2007).

Figure 19. Damage to the ship hull as the visor actuators are torn through the bulkhead.

The actuators continue to cut through the remaining part of the forecastle deck and the upper part of the front bulkhead leaving a noticeable cut (Figure 19). Soon the visor becomes completely detached from the vessel.

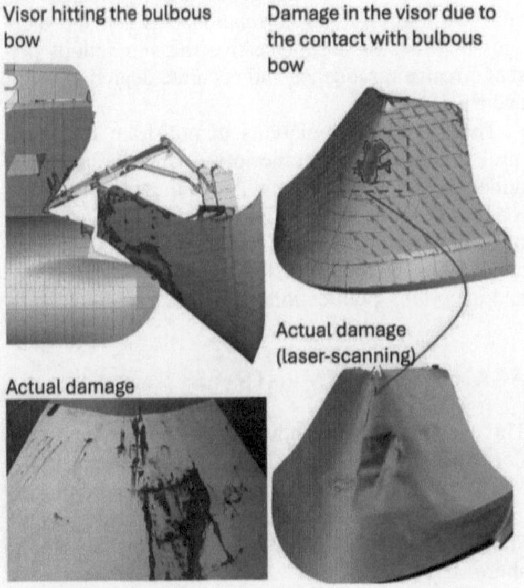

Figure 20. Damage to the ship visor due to the contact with the bulbous bow. Image of the actual damage by T. Roosipuu (left) and from the laser-scanned model of the damaged visor (right).

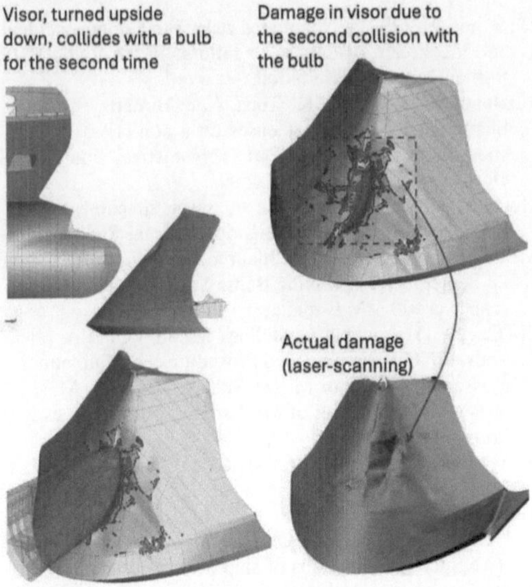

Figure 21. Visor deformations after the contact with the bulb. Image of the actual damage from the laser-scanned model of the damaged visor (right).

Visor, now fully detached from the vessel, tilts forward and collides with the bulbous bow (Figure 20). The visor flips over around the transverse y-axis and ends upside down in front of the vessel. As the vessel continues with forward

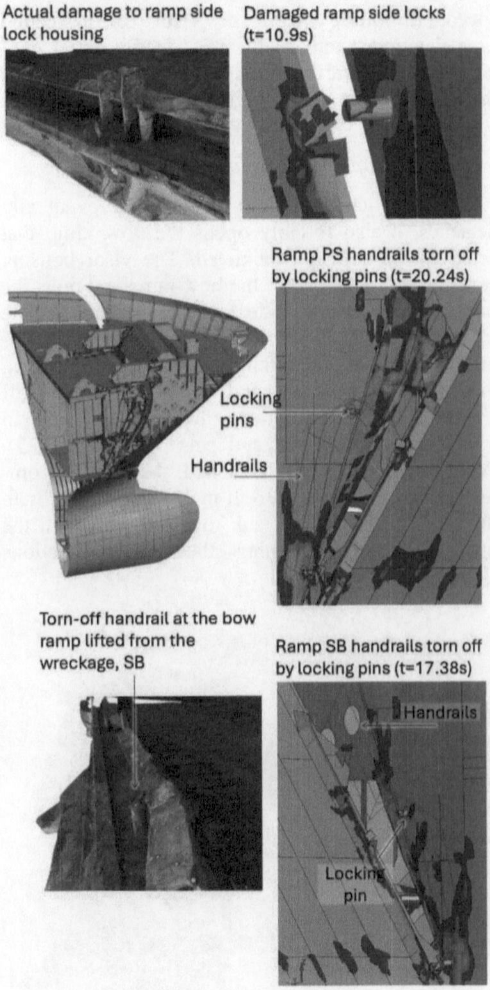

Figure 22. Damages to the bow ramp: side locks are broken, and handrails are torn off. Images of the actual damage by K. Tabri.

Figure 23. Damage in the flap shaft of the ramp. Image of the actual damage by K. Tabri.

speed, it collides with the visor for a second time, leaving a remarkable dent in the direct contact zone and partly buckles the structure around the contact zone (Figure 21).

4.1 *Bow ramp opening and structural damages*

As the visor gains freedom to move and eventually detaches, it also forcibly opens the bow ramp that acted as the collision bulkhead. The visor housing contacts the upper end of the bow ramp and pulls the ramp forward. Ramp side locks are broken as the locking pins tear the off the covers of side lock housings (Figure 22). The ramp falls forward, the locking pins get caught to the handrails and tear them off (Figure 22). Shaft for ramp flaps snags a hole in visor portside bulkhead end gets bent (Figure 23). When the visor is fully detached, the ramp is completely free to fall forward. It makes the contact with the shell plating at the tank top level and with the bulbous bow, sustaining heavy deformations (Figure 24).

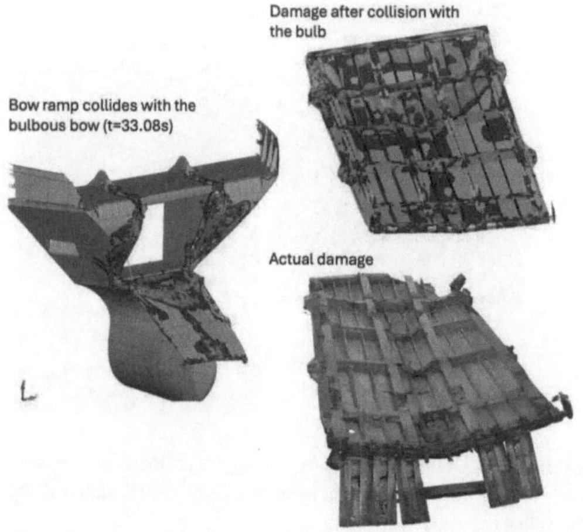

Figure 24. Damage in the flap shaft of the ramp. Image of the actual damage by K. Tabri.

5 CONCLUSIONS

Paper presents a numerical simulation of the potential failure sequence of the bow area arrangement of MV Estonia. Different data sources and state-of-the-art tools were used to construct a realistic yet accelerated wave load sequence and evaluate the ship's structural response to these loads.

Analysis revelated that the bow visor locking devices and hinges break due to forces exerted from these severe sea-loads. This led to a progressive failure sequence that commenced with the full detachment of the visor and opening of the bow ramp. Both structures suffered considerable damages that are well reported though laser-scanning and photogrammetric models. Simulated structural damages correspond well to the actual damages of the ship wreckage, visor and bow ramp. Good correlation between the damages provides an assurance that the simulations present a realistic sequence and accurate depiction of the events.

The reported simulations demonstrate the structural reaction to a specific load sequence applied to undamaged ship structures. An evaluation of various loading sequences of different magnitude and levels of fatigue degradation in locks has been conducted and will be included in the final assessment report and upcoming publications.

ACKNOWLEDGEMENTS

This work has been conducted as a part of investigation by Estonian Safety Investigation Bureau (contract no 5-4/195-1). Simulation model developments are partly supported by Estonian Research Council via the grant PRG1820 (Dynamic response of offshore structures).

REFERENCES

Carlsson, J.O. (2007). M/V ESTONIA Bow arrangement collapse-Sequence of events. Technical Report. Gothenburg, 2007.

Hänninen, H. (2007). Negotiated risks: the Estonia accident and the stream of bow visor failures in the Baltic ferry traffic. Doctoral dissertation.

Hosseinzadeh, S., Tabri, K., Topa, A. & Hirdaris, S. (2023). Slamming loads and responses on a non-prismatic stiffened aluminium wedge: Part II. Numerical simulations. Ocean Engineering, 279, 114309.

Huss, M. (1995). Wave loads on visor attachments. MV ESTONIA Accident Investigation Internal Report 1995-1997. Supplement to the final report on the capsizing on 28 September 1994 in the Baltic Sea of the RO-RO passenger vessel MV Estonia.

ITTC. (2011). Practical Guidelines for Ship CFD Applications. ITTC-Recommended Procedures and Guidelines.

Kõrgesaar, M., Saarna, M., & Šults, A. (2024). Material tests with hull plating of MS Estonia. Internal investigation report.

STAR-CCM+. (2024). STAR-CCM+ Version 19.06 Manual.

Tezdogan, T., Demirel, Y.K., Kellett, P., Khorasanchi, M., Incecik, A. & Turan, O. (2015). Full-scale unsteady RANS CFD simulations of ship behaviour and performance in head seas due to slow steaming. Ocean Engineering, 97, 186–206.

Trägardh. (1995). Model test with M/S ESTONIA. Sea loads on bow visor and yawing behaviour due to heel. Supplement to the final report on the capsizing on 28 September 1994 in the Baltic Sea of the RO-RO passenger vessel MV Estonia.

Trends in Collision and Grounding of Ships and Offshore Structures – Zhu & Guedes Soares (Eds)

Experimental and numerical investigation on the damage accumulation of aluminum alloy plates subjected to repeated impacts

D.D. Truong & V.V. Huynh
Nha Trang University, Nha Trang, Vietnam

V.V. Tuyen
Vietnam Maritime University, Hai Phong, Vietnam

S.-H. Park & S.-R. Cho
University of Ulsan, Ulsan, South Korea
UlsanLab Inc., Ulsan, South Korea

ABSTRACT: This paper presents the investigation results of the damage accumulation of plates due to repeated impacts. Three plates made of aluminum alloy in actual ships were adopted to analyze the structural responses to repeated impact loads. Five drop impact tests were conducted for each clamped models. A knife-edge striker representing severe impact cases was adopted. The plate's deflection and the striker velocity were recorded during each impact. Numerical simulations were also performed to predict the dynamic response of the plates to repeated impacts. The mechanical properties obtained from tensile coupon tests were defined in the numerical model. The impact response and deflection of the plates obtained from the tests and simulations were compared and discussed. The results showed that the deflection accumulation of the plates was noticed when subjected to repeated impact loads. The study's findings are expected to be helpful in designing marine aluminum alloy structures under repeated impacts.

1 INTRODUCTION

During their service, marine engineering structures may repeatedly contact with floating objects and ice floes, and subsequently, they can be damaged due to such repeated impacts. In addition to steel and composite, aluminum alloys are also used for the fabrication of marine vessels owing to their corrosive advantages and lightweight (Sutherland & Guedes Soares 2009). Therefore, the resistance features of the aluminum alloy structures against the mentioned loads must be fully characterized.

Regarding the single impact conditions, many scholars have focused on the structural response of the structural component made of aluminum alloys, for example, Refs. (Liu & Jones, 1987; Langseth & Larsen, 1994; Hildrum et al., 2002; Sutherland & Guedes Soares, 2009; Fagerholt et al., 2010; Jones & Paik, 2012; Mohotti et al., 2013; Liu et al., 2014; Morin et al., 2017). However, under such impact loading the damage features, including failure patterns of the aluminum structures, might differ from mass impact scenarios in reality. On the other hand, with the repeated mass impact loadings, although many studies on the response of small structural components (beams, plates) (Zhu & Faulkner, 1996; Huang et al., 2000; Chae, 2008; Jones, 2014; Cho et al., 2014; Truong et al., 2016a,b, 2018a,b, 2022a,b; Zhu et al., 2018a,b; Guo et al., 2018, 2019; Duan et al., 2019; Zhang et al., 2021; He & Guedes Soares, 2021a,b, 2022, 2025; Cai et al., 2022; He et al., 2022; Truong & Le, 2022; Huynh et al., 2024a, b; Zhu et al., 2018, 2024; Truong et al., 2024) have been reported so far, only a few studies for aluminum structures have been investigated yet (Zhu & Faulkner, 1996; Huang et al., 2000; Duan et al., 2019; He & Guedes Soares, 2021; Zhang et al., 2021; Truong & Le, 2022; Wang et al., 2024; Truong et al., 2024). Zhu & Faulkner (1996) reported the impact response of mild steel and aluminum plates and proposed analytical solutions for estimating the damage of the plates under repeated loads. Later, as a typical structural state under repeated loadings, the pseudo-shakedown was studied experimentally by Huang et al. (2000). Duan et al. (2019) also studied the repeated impact response of circular plates considering small cracks, and concluded that surface cracks considerably effect on the structural response. He & Guedes Soares (2021) analyzed the state experimentally for aluminum alloy beams under repeated mass impacts. They concluded that the beam subjected to a small impact force (small impact energy) only could

DOI: 10.1201/9781003684404-39

reach this state. Zhang et al. (2021) studied the behavior and energy absorption of aluminum honeycomb sandwich panels subjected to repeated loads. Truong & Le (2022) studied the dynamic response of plates made of aluminum alloys subjected to repeated mass impacts for various impact energies. Wang et al. (2024) also examined the fatigue and failure behaviors of 6061-T6 aluminum alloy U-notched beams. Truong et al. (2024) numerically studied the effect of different impact locations on the response of a repeatedly impacted aluminum alloy plate. The above survey revealed that those studies only focused on the simple models or the limited aluminum alloy grades and impacted certain conditions. There is still a lack of test data regarding the response of plate structures fabricating from 5083-H111, which is widely used in marine applications.

In this study, a series of repeated mass impact tests on 5083-H111 plates are conducted, and structural response is measured. Besides, a numerical analysis model is developed using the software package ABAQUS/Explicit (Abaqus 2013) to predict the damages of aluminum alloy plates due to repeated impacts. A good agreement between the predictions and test results is achieved. Discussion on the impact behavior and damage evolution is provided accordingly.

2 EXPERIMENTS

2.1 *Tensile coupon tests*

Before the main impact tests, a series of tensile tests were conducted to determine the material properties of the test model. Five tensile coupons made of aluminum alloys 5083-H111, cut from the parent plate sheets used to fabricate the test model, were tested. Tensile test coupons and test procedures were prepared and performed according to ASTM (2016). In this study, the tensile tests were performed at the University of Ulsan, Korea, as shown in Figure 1.

Figure 2 indicates the typical engineering stress-strain curve obtained from the tensile coupon tests. The average values of material properties are Young's modulus E = 70,000 MPa, yield strength σ_Y = 148.7 MPa, ultimate tensile strength σ_T = 301.5 MPa, tensile strain ε_T= 0.2351. Note that the density and the Poisson ratio are reasonably assumed as 2700 kg/m^3 and 0.33, respectively.

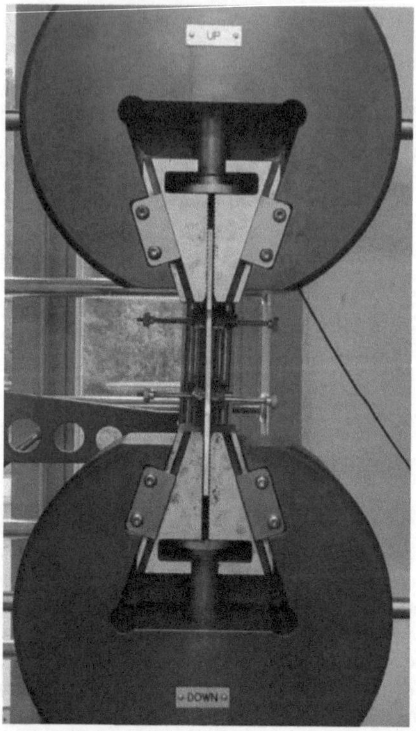

Figure 1. Tensile test machine.

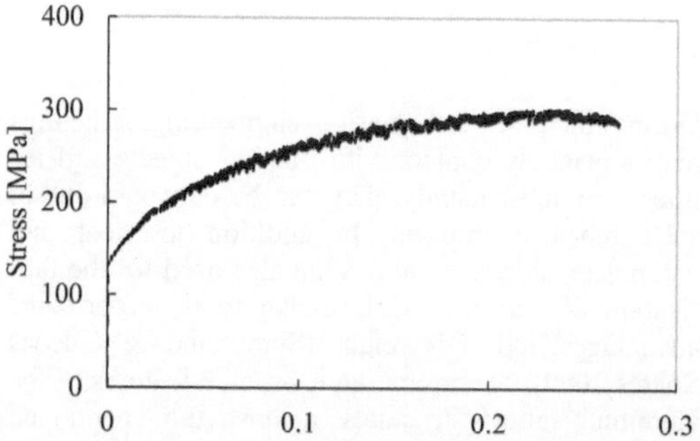

Figure 2. Typical engineering stress-strain curve.

2.2 *Repeated drop tests*

The repeated impact tests were carried out using a drop machine. The experimental set-up is shown in Figure 3. Three identical 4 mm-thick plates made of aluminum alloys 5083-H111, destinated as AUP-1, AUP-2, and AUP-3, were tested with various impact conditions. Dimensions of the test model is displayed in Figure 4. To represent the real connection of surrounding heavy structures, the plate model was bolted with a strong support frame, as shown in Figure 3. After clamping the plate on the strong base frame of the drop test frame, the striker was hung out by an electric magnet at a predefined height corresponding to a desire impact velocity. Note that the striker is fabricated as a rigid knife-edge shape having a header length of 290 mm with a radius of 17 mm, as illustrated in Figure 3.

Models AUP-1 and AUP-2 were tested with the initial impact velocity of 4.85 m/s, while model AUP-3 was tested with 4.43 m/s. Note that these velocities are associated with drop heights of 1.2 m and 1.0 m. Two striker masses of 41.8 kg and 59.82 kg were used. In this study, each plate was tested five times, in which impact velocity was kept unchanged, i.e., the same impact energy was applied for five consecutive drops. Test conditions are given in Table 1.

Table 1. Test conditions for the aluminum alloy models.

Models	No. of impacts	Mass of striker [kg]	Drop heigh [m]	Initial impact velocity [m/s]
AUP-1	5	41.80	1.2	4.85
AUP-2	5	59.82	1.2	4.85
AUP-3	5	59.82	1.0	4.43

Figure 3. Drop testing machine.

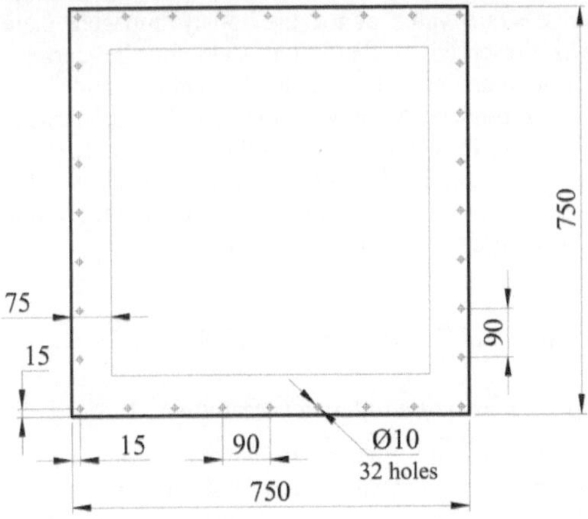

Figure 4. The 4 mm-thick AA 5083-H111 plate model.

The permanent deflection of the tested plate was measured at the center line along the x- and y-direction after each drop test. To conveniently measure the deformation, grid lines were marked with a spacing of 20 mm on both the "contact" sides and the non-contact side of the plate. Figure 5 shows the test results for the deformation profile measuring at the plate center in the y-direction for AUP-1, AUP-2, and in x-, y-directions for AUP-3. It can be seen that the plate deformed after each impact event, and the residual damage of the plate increased with the number of impacts, while its increment decreased regardless of the locations. In addition, due to the impact location being a bit different from the plate center, the deformation profile is not symmetrical from both ends.

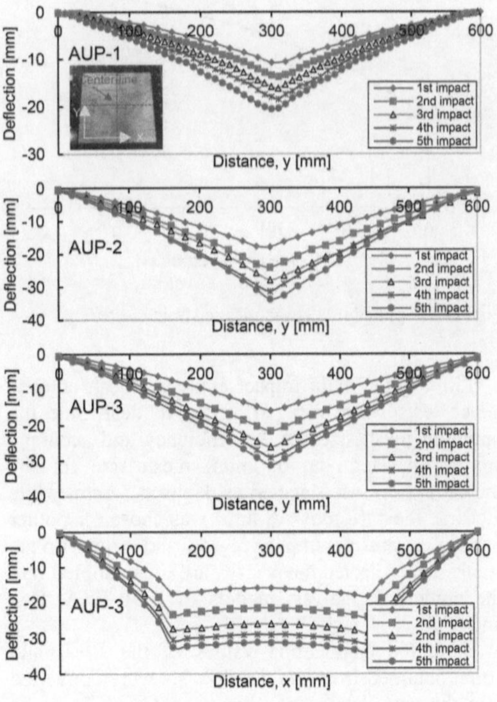

Figure 5. Deformation profiles measured at the center of the tested plates after five impacts.

3 FINITE ELEMENT ANALYSIS

The numerical simulation was performed using the finite element (FE) software package ABAQUS/Explicit (Abaqus 2013) to predict the impact response of the plates. The FE model consisted of three parts: the aluminum alloy plate, the support frame, and the striker, as displayed in Figure 6.

3.1 *Plate model*

The plate and the support-plate frame were modeled by shell elements with a reduced integration scheme (S4R) available from Abaqus library. Before

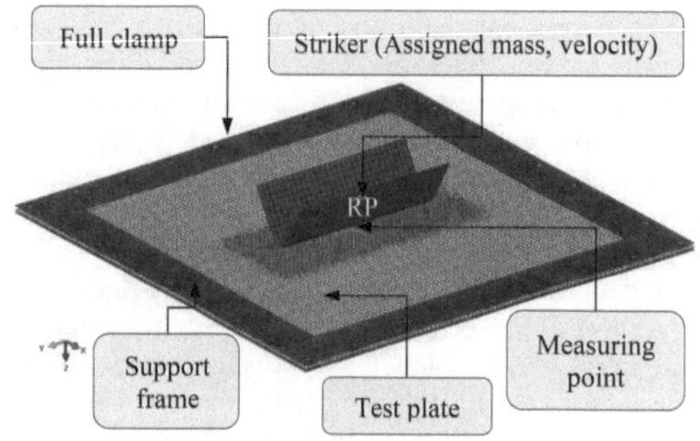

Figure 6. FE model for the impact analysis.

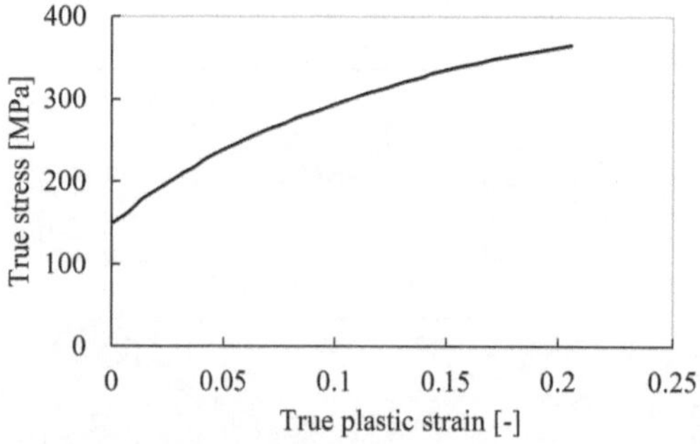

Figure 7. Applied true stress-plastic strain curve.

performing the main impact analysis, mesh convergence tests were also performed to determine the optimum mesh size for the efficiency and accuracy simulation. Here, an optimum mesh size for the impact region was selected as 4 mm × 4 mm, while a coarse mesh (8 mm × 8 mm) was chosen for other parts, i.e., outside impact region and support-plate frame. The selected mesh sizes are then adopted into the numerical analysis models of AUP-1, AUP-2, and AUP-3.

Here, the engineering values of the stress and strain obtained from the tensile tests were converted into the true data using Equations (1) – (3) to be adopted into the numerical model. The true stress-strain curve is shown in Figure 7.

$$\sigma_{tr} = \sigma_e(1 + \varepsilon_e) \tag{1}$$

$$\varepsilon_{tr} = \ln(1 + \varepsilon_e) \tag{2}$$

$$\varepsilon_{pl,tr} = \varepsilon_{tr} - \frac{\sigma_{tr}}{E} \tag{3}$$

where σ_{tr} and ε_{tr} are the true stress and true strain, respectively; σ_e and ε_e are the engineering stress and engineering strain, respectively; $\varepsilon_{pl,tr}$ are the true-plastic strain.

No strain-rate sensitivity model was defined for the plate model because the aluminum alloy is insensitive. The boundary conditions of the plate were assumed as a full clamp at the connection between the upper frame-plate and the lower support frame, in which tie constraints were simply used for bolt connections. Note that only plastic deformation mode was introduced in the numerical model because no fractures of the plate were detected in the current test in this study.

3.2 *Striker*

As the striker was certainly considered a rigid body during the tests, it was modeled using the 3-D bilinear rigid quadrilateral elements (R3D4) from the Abaqus library with a mesh size was roughly of 5 mm × 5 mm at the tip. The selected mesh size would better resemble the curvature head shape so that it ensures sufficient contact between the striker and plate, thereby providing a more realistic simulation of the contact area. Besides, a mesh size of 10 mm × 10 mm was selected for the rest of the striker, as illustrated in Figure 6. Note that, only the head part of the striker was modeled (not full geometry) for simplicity.

The striker was set to be freely translated vertically with the expected impact velocity. The initial impact velocity (V) and the mass of the striker, which are the same as those of the tests, were assigned to the referent point (RP), as displayed in Figure 6. The contact algorithm "General contact" was applied to define the contact between the striker and the plate, in which a friction coefficient was set as 0.3.

3.3 *Repeated impact scenarios*

The repeated impacts induced by the repetition of the striker's drops in the tests were simulated by repeatedly performing the calculations for identical impact energy of each impact event. In particular, after the calculation of the first impact was finished, the next calculation started with the same impact energy assigned to the striker. Note that the residual stress-strain value of the previously impacted plate was preserved as the initial state for the current-restarted analysis. The Rayleigh damping model was also introduced to remove the tested plate's elastic vibration, followed by Refs. (Truong et al., 2018a,c; He et al., 2022). To fully capture the peak and the permanent stage of the tested plate, the duration time of each impact calculation was set as 0.05 s.

4 RESULTS AND DISCUSSION

4.1 *Comparison of impact response*

The numerically obtained results for residual deflection after five impacts measured at the plate's center were compared with the test results, as shown in Figure 8. Both the results are in good agreement, with a small difference of roundly 3.8% on average, thus confirming the validity of the numerical modeling technique. However, for the later impacts the simulation produced less deflections than the tests, except for model AUP-1 tested with small mass; these discrepancies are possibly due to idealizing the boundary conditions (fully clamped, tie constraint), which would cause a difference in energy absorption from the tests. Note that this boundary condition did not allow any movements between the plate and bolts/support frame. It is apparent from Figure 8 that for both methods (test and simulation) the deflection is increased when the plate is loaded repeatedly, but its increment is also reduced accordingly. The residual deflection time histories of the plate after the 1st impact is shown in Figure 9. It is noted that the deformation progress includes three stages (i) peak deflection (Loading), (ii) elastic spring-back (Unloading), and (iii) residual deflection (Permanent) with the oscillations (see Figure 9). The plate oscillations in the test are less than that from the simulation due to the simplicity of the boundary conditions, as mentioned earlier.

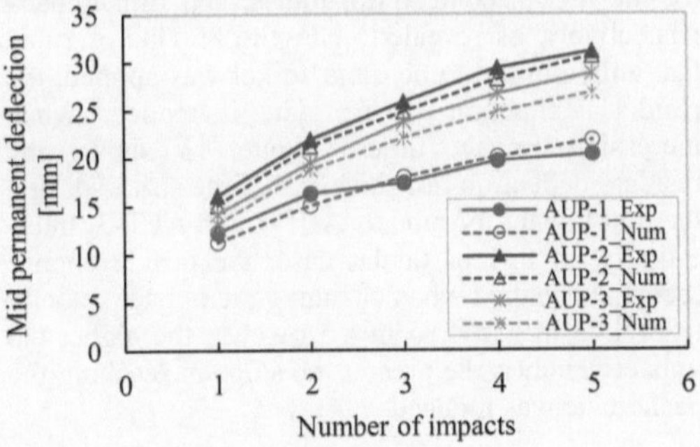

Figure 8. Residual deflection of three models for all impacts.

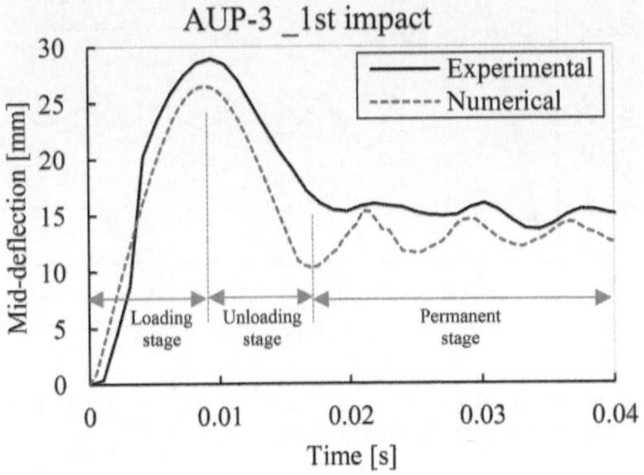

Figure 9. Deflection time histories at the center of the model APU-3 after first impact.

Figure 10 shows the typical time histories of striker velocity for the first and fifth impacts of the model AUP-2, obtained from the tests and the simulations. It can be seen that reasonable agreement between the prediction and test results was also archived. In addition, as in the tests, the numerically obtained striker velocity time history showed a decrease in the first stage, called "loading state" and an increase in the second stage, say "rebound state" with time, as displayed in Figure 10. Obviously, the rebound state is due to the release of elastic energy stored by the plate, i.e., the spring-back phenomenon. It is evident from the figure that the rebound velocity increased with the number of impacts.

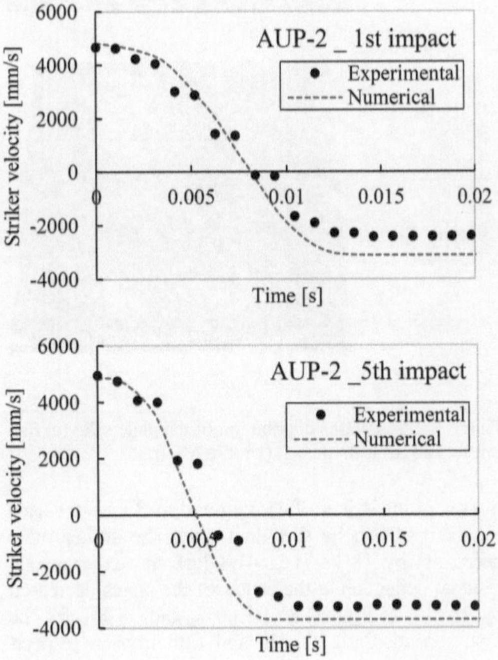

Figure 10. Velocity time histories of the striker during the 1st impact and 5th impact.

4.2 *Von-Mises distribution*

Figure 11 shows the von-Mises distribution results on the plate after the first impact and fifth impact of the model AUP-2. Due to the impacts, maximum stress occurred locally at the impact regions, indicating larger deflections were obtained along the impact line, as illustrated in Figure 5. In addition, high stress was detected near the support frame, and both holes represented the connections of adjacent structures in reality. These highly concentrated stresses can cause impact plate failure, so more attention is required to avoid fatigue damage to marine structures under repeated impact loadings.

4.3 *Effect of striking mass*

The effect of striking mass on the residual deflection of the plates under repeated impacts is investigated. Note that the models AUP-1 and AUP-2 were tested with the same impact velocity of 4.85 m/s, while two

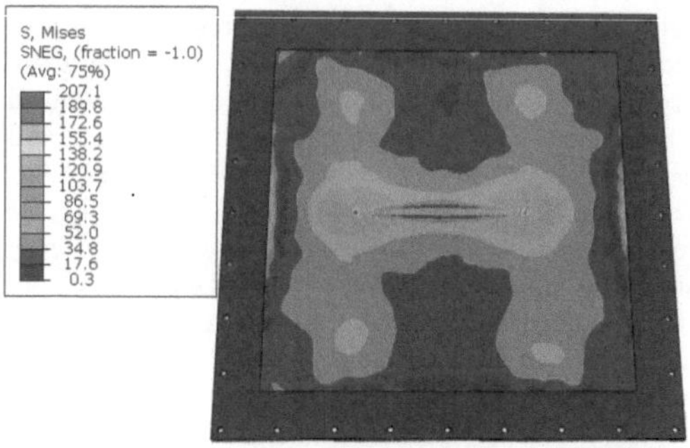

(a)

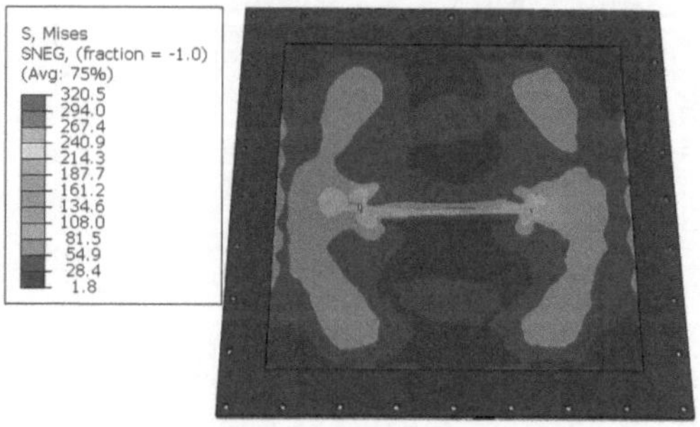

(b)

Figure 11. Von-Mises distribution of the plate after (a) first impact and (b) fifth impact. (Unit in MPa).

masses of the striker of 41.8 kg and 59.8 kg were used for them. It can be seen that when the striker mass decreased by 18 kg (i.e., 41.8 kg), the experimental residual deflection at the center of the plates decreased by 23.6%, 25.9%, 32.8%, 33.0%, and 34.4% for the first, second, third, fourth, and fifth impact, respectively; this means that although the same impact velocity applied, the reduction of residual deflection of the plate under repeated impacts is achieved, and this reduction is gradually increased with the number of impacts.

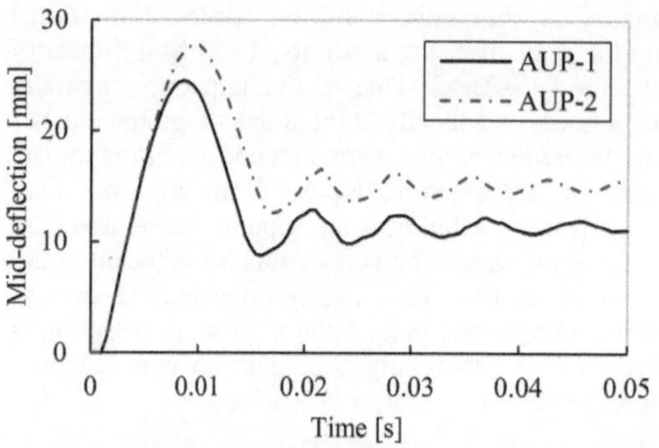

Figure 12. Deflection time histories of the models AUP-1 and AUP-2 during the first impact.

For more detail, the numerical deflection time histories during the first impact of both models are presented in Figure 12. It can be seen that both peak deflection and permanent deflection is increased when the plate is loaded with the heavier mass. Nevertheless, the more striking mass, the more time to reach the peak value was obtained, which is consistent with the observation of (Truong et al., 2018b). Besides, the slope of the loading stage in both cases is the same, suggesting the plate experiences the deflection rate for various striking masses.

4.4 *Effect of impact velocity*

Based on the numerical and experimental results of the models AUP-2 and AUP-3, the effect of impact velocity on the residual deflection of the plates due to repeated impacts is also discussed. Note that both models were tested with the same striker mass (59.82 kg), but with different impact velocities. The experimental results showed a reduction of damage to the plate for the lower impact velocity. In particular, when the impact velocity reduced from 4.85 m/s to 4.43 m/s, the residual deflection of the plate decreased by 11.8%, 11.4%, 8.1%, 9.2%, and 7.4% for the first, second, third, fourth, and fifth impact, respectively, as revealed in Figure 8. This revealed that although the same mass striker was applied, the residual deflection of the plate is reduced when increasing impact times. Figure 13 shows the residual deflection time histories of the striker during the first impact for models AUP-2 and AUP-3. Interestingly, in this particular case, the peak response seems unchanged when changing the impact velocity from 4.43 m/s to 4.85 m/s. However, the higher the impact velocity, the greater the slope in reaching the peak value was induced.

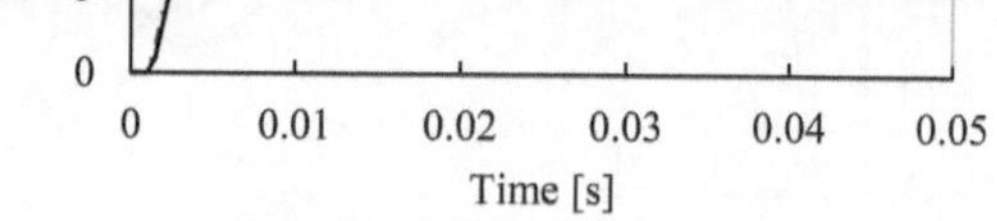

Figure 13. Deflection time histories of the models AUP-2 and AUP-3 during the first impact.

5 CONCLUSIONS

This paper examined the deflection evolution of aluminum alloy plates under repeated lateral impacts. Experiments and numerical simulations were performed in this study. The results for deflection and impact velocity time history obtained from both methods were compared, and good correlation was achieved.

It was found that the deflection of the impacted aluminum alloy plate was accumulated when the plate is impacted repeatedly, in which at the first impact the largest damage was observed, while the deflection increment was decreased. The rebound velocity of the striker was increased with impact number, implying that dissipated energy by the plate becomes reduced when the plate is loaded repeatedly.

The test results can be utilized for validations of other methods, including numerical simulations, analytical approaches, and empirical formulations.

The numerical analysis methodology developed in this study can predict the impact responses of the aluminum alloy plates. Therefore, the developed method can be adopted for simulations with large-scale models in the same manner.

Further studies on impact conditions using different striking-nose shapes, and actual-like structures seem desirable for providing more reliable data for marine design against repeated impacts.

ACKNOWLEDGMENTS

This research is funded by Nha Trang University for science and technology under grant number TR2025-13-21.

REFERENCES

ABAQUS User's Manual Version 6.13–6 (2013).

ASTM-E8: Standard Test Methods for Tension Testing of Metallic Materials. American So-ciety for Testing and Material: West Conshohocken, PA, USA, 2016.

Cai, W., Zhu, L. & Qian, X.D. 2022. Dynamic responses of steel plates under repeated ice impacts. *International Journal of Impact Engineering* 162: 104129.

Chae, G.I. 2008. Nonlinear structural behavior due to repeated interaction between ice breaking commercial ships and sea ice. *Msc thesis, University of Ulsan*.

Cho, S.-R., Truong, D.D. & Shin, H.K. 2014. Repeated lateral impacts on steel beams at room and sub-zero temperatures. *International Journal of Impact Engineering* 72: 75–84.

Duan, F., Liu, J., Wang, G. & Yu, Z.L. 2019. Dynamic behavior of aluminum alloy plates with surface cracks subjected to repeated impacts. *Ships and Offshore Structures* 14(5): 478–491.

Fagerholt, F., Grytten, F., Gihleengen, B.E., Langseth, M. & Børvik, T. 2010. Continuous out-of-plane deformation measurements of Aa5083-H116 plates subjected to low-velocity impact loading. *Int. J. Mech. Sci.* 52: 689–705.

He, X. & Guedes Soares, C. 2021a. Experimental study on the dynamic behaviour of beams under repeated impacts. *International Journal of Impact Engineering* 147: 103724.

He, X., Garbatov, Y. & Guedes Soares, C. 2022. Analysis of pseudo-shakedown of rectangular plates under repeated impacts. *Ocean Engineering* 265: 112609.

He, X. & Guedes Soares, C. 2025. Dynamic response of rectangular plates under repeated impacts from different impactors. *Int. J. Impact Eng.* 321: 120433.

Hildrum, H.G., Malo, K.A., Langseth, M. & Lademo, O.G. 2002. Numerical modelling of fracture in stiffened aluminium plates subjected to impact loading. *Structures under Shock and Impact* 7: 133–142.

Huang, Z.Q., Chen, Q.S. & Zhang, W.T. 2000. Pseudo-Shakedown in the collision mechanics of ships. *International Journal of Impact Engineering* 24: 19–31.

Huynh, V.V, Lap, C.L. & Truong, D.D. 2024a. Experimental and numerical study on the damage accumulation of stainless steel plates due to repeated mass impacts. In: *Proceedings of the 7th International Conference on Green Technology and Sustainable Development (GTSD)*, pp. 374–379, Ho Chi Minh City, Vietnam.

Huynh, V.-V., Cho, S.-R., Dang, X.-P. & Truong, D.D. 2024b. Practical formula for predicting the residual deflection evolution of steel plates subjected to repeated impacts. *J. Mar. Sci. Eng.* 12: 956.

Jones, N. 2014. Pseudo-shakedown phenomenon for the mass impact loading of plating. *International Journal of Impact Engineering* 65: 33–39.

Langseth, M. & Larsen, P.K. 1994. Dropped objects' plugging capacity of aluminium alloy plates. *Int. J. Impact Eng.* 15(3): 225–241.

Liu, B., Villavicencio, R. & Guedes Soares, C. 2014. On the failure criterion of aluminum and steel plates subjected to low-velocity impact by a spherical indenter. *Int. J. Mech. Sci.* 80: 1–15.

Liu, J. & Jones, N. 1987. Experimental investigation of clamped beams struck transversely by a mass. *Int. J. Impact Eng.* 6(4): 303–335.

Mohotti, D., Ali, M., Ngo, T., Lu, J.H., Mendis, P. & Ruan, D. 2013. Out-of-plane impact resistance of aluminium plates subjected to low velocity impacts. *Mater. Des.* 50, 413–426.

Morin, D., Kaarstad, B.L., Skajaa, B., Hopperstad, O.S. & Langseth, M. 2017. Testing and modelling of stiffened aluminium panels subjected to quasi-static and low-velocity impact loading. *Int. J. Mech. Sci.* 110: 97–111.

Sutherland, L.S. & Guedes Soares, C. 2009. Impact behaviour of GRP, aluminium and steel plates. *Analysis and Design of Marine Structures*. Taylor & Francis Group, London, U.K., pp. 293–300.

Truong, D.D., Cho, S.-R., Huynh, V.V., Dang, X.-P., Duong, H.D. & Tran, T.H. 2022b. A study on dynamic response of steel plates under repeated impacts. In: *Long, B.T., Kim, H.S., Ishizaki, K., Toan, N.D., Parinov, I.A., Kim, YH. (Eds): AMAS 2021*, LNME. *Springer, Cham.*

Truong, D.D., Dang, T.K., Vo, T.L., Dau, V.L., 2024. Effects of impact location on the dynamic response of repeatedly impacted aluminum alloy plates. *J. Fish. Sci. Technol.* 1: 134–141.

Truong, D.D. & Le, N.A.V. 2022a. Dynamic response of aluminum-alloy plates subjected to repeated impacts. In: *D. C. Nguyen et al. (Eds.): ICERA 2021*, LNNS 366. Springer, Cham.

Truong, D.D., Shin, H.K. & Cho, S.-R. 2018c. Permanent set evolution of aluminium-alloy plates due to repeated impulsive pressure loadings induced by slamming. *Journal of Marine Science and Technology* 23: 580–595.

Truong, D.D., Shin, H.K. & Cho, S.-R. 2016a. Plastic response of steel grillages subjected to repeated mass impacts. In: *Proceedings of the 7th International Conference on Collision and Grounding of Ships and Offshore Structures*. Ulsan, South Korea.

Truong, D.D., Shin, H.K. & Cho, S.-R. 2016b. Dynamic response of steel grillages under repeated mass impacts at low temperature. In: *Proceedings of the 13th International Symposium on Practical Design of Ships and Other Floating Structures*. Copenhagen, Denmark.

Truong, D.D., Shin, H.K. & Cho, S.-R. 2018a. Repeated lateral impacts on steel grillage structures at room and sub-zero temperatures. *International Journal of Impact Engineering* 113: 40–53.

Truong, D.D., Shin, H.K. & Cho, S.-R. 2018b. Response of low temperature steel beams subjected to single and repeated lateral impact. *International Journal of Naval Architecture and Ocean Engineering* 10(6): 670–682.

Wang, B., Qiu, Z., Gao, P., Xu, Y., Wang, C., Han, K., Bai, C., Yang, Q. & Zhu, Y. 2024. The experimental and numerical study of AA6061-T6 aluminum alloy under repeated impacts. *Int. J. Fatigue* 183: 108213.

Zhang, Y., Li, Y.G., Guo, K. & Zhu, L. 2021. Dynamic mechanical behaviour and energy absorption of aluminium honeycomb sandwich panels under repeated impact loads. *Ocean Engineering* 219: 108344.

Zhu, L. & Faulkner, D. 1996. Damage estimate for plating of ships and platforms under repeated impacts. *Marine Structures* 9: 697–720.

Zhu, L., Shi, S. & Jones, N. 2018a. Dynamic response of stiffened plates under repeated impacts. *International Journal of Impact Engineering* 117: 113–122.

Zhu, L., Wang, X., Guo, K., Cai, W., Jones, N. & Dai, S., 2024. Effects of indenter geometry on pseudo-shakedown of steel plates under repeated mass impacts. *Int. J. Impact Eng.* 185: 104865.

Trends in Collision and Grounding of Ships and Offshore Structures – Zhu & Guedes Soares (Eds)
© 2026 The Author(s), ISBN 978-1-041-16361-9

Study on dynamical response of improved cylindrical shell chain structure under impact load

C. Wang & J. Pan
School of Naval Architecture, Ocean and Energy Power Engineering, Wuhan University of Technology, Wuhan, P.R. China

W.Z. Zhang
Universidad Politécnica de Madrid (UPM), Madrid, Spain

M.C. Xu
School of Naval Architecture and Ocean Engineering, Huazhong University of Science & Technology, Wuhan, P.R. China

ABSTRACT: Under the condition of considerable development of water transportation, there's an increasingly high incidence of vessel-bridge collision, thus it's necessary for the study on impact-resistant structure. The periodical cylindrical shell chain (CSC) structure presents a remarkable effect on controlling the orientation of stress wave propagation, so it can be feasible to achieve the anisotropy of impact energy propagation. Moreover, impact resistance of improved cylindrical shell chain (ICSC) structure with design of heterogeneous materials could be studied. Finally, this study aims to obtain the influence of unit cell parameters (offset thickness and curvature of curved bars) in the ICSC on impact resistance for overall periodical structures, and thus optimum design form of the ICSC is determined.

1 INTRODUCTION

Thin-walled protective structures have garnered increasing attention across various applications due to their effective protection against high-energy impacts while maintaining low mass and minimal influence on the protected structures (Zhang et al., 2024). Among these protective structures, certain specialized types of metamaterials stand out due to their ability to manipulate wave propagation. This unique capability helps dissipate impact energy efficiently, reducing structural damage and enhancing reusability under repeated impact loads.

As a kind of 2D cellular structure, the CSC consists of cylindrical shell cells stacked in order (Peng, 2021). The wave propagation characteristics of the CSC are mainly affected by the material components of the single cell and the cell configuration (Kim et al., 2018). Peng et al. (2021) proposed a novel CSC with varying aspect ratios and found that the CSC with a positive aspect ratio gradient amplified the forces in wave propagation due to the dramatic increase in the apparent elastic modulus along the impact direction, whereas the CSC with a negative aspect ratio gradient showed the opposite. Furthermore, Peng et al. (2022) developed a mesoscale finite element model and a continuum-based model for CSC with a density gradient, and confirmed that the peak force transmitted to the support end increases with the increase of the density gradient parameter, thus the CSC with density gradient can control the stress pulse in a wider range. Kim et al. (2019) proposed the elliptic CSC with varying stiffness made by 3D printing and observed that the CSC with increasing stiffness was prone to accelerate wave propagation, while the CSC with decreasing stiffness showed the opposite.

Zhang et al. (2022) obtained the dispersion characteristics of the pentamode lattice-ring structure and acquired wave propagation properties. Mahmoud et al. found that particularly devised metamaterials with distinct geometry can show a superior effect of controlling elastic wave propagation (Mahmoud et al., 2024). The geometrical configuration of the CSC makes it feasible to control the wave propagation directionally, which can be the key to illustrating the property of energy diversion for the CSC.

The heterogeneous material design of phononic crystals conduces to exciting the local resonance mechanism and helps to considerably attenuate impact energy, hence providing a novel direction for designs of impact-resistant mechanical metamaterials. Based on the heterogeneous material design method, Li et al. (2024) proposed a hybrid graded origami-inspired metamaterial with multiple resonators. Then, they observed that the metamaterials can open an ultra-wide bandgap, with the amplitude of the reaction force extraordinarily reduced in the

DOI: 10.1201/9781003684404-40

bandgap range. Huang et al. (2023) added local resonant plates with heterogeneous material design to various honeycomb structures, which can trap the impact energy in the resonant plate within the bandgap range, thus contributing to the impact wave dissipation. Accordingly, in order to further enhance the impact resistance and energy dissipation of the CSC, the heterogeneous material design is utilized.

At present, it can be insufficient to study the energy diversion and conceive the improved heterogeneous material design of the CSC. In light of this, this study aims to explore the relationship between the energy diversion and the geometric configuration and apply heterogeneous material design forms to the ICSC on the basis of a comparative study for mechanical response. Besides, the function of 2 parameters (offset thickness and curvature of curved bars in ICSC) on mechanical response would be studied, hence determining the optimum design form of the ICSC. The conclusions can provide some references for the design of the new protective structure with energy diversion in the field of impact resistance.

2 NUMERICAL SIMULATION

2.1 *Geometrical parameters and working conditions*

In all working conditions, the size parameters of the shell and resonator are consistent, as shown in Figure 1, whereas the parameters of the bar can be variable. Additionally, the 2 parameters (δ_x and K_x) of the bar are relevant to the 2 factors (offset thickness and curvature of the curvature bars) in the following study.

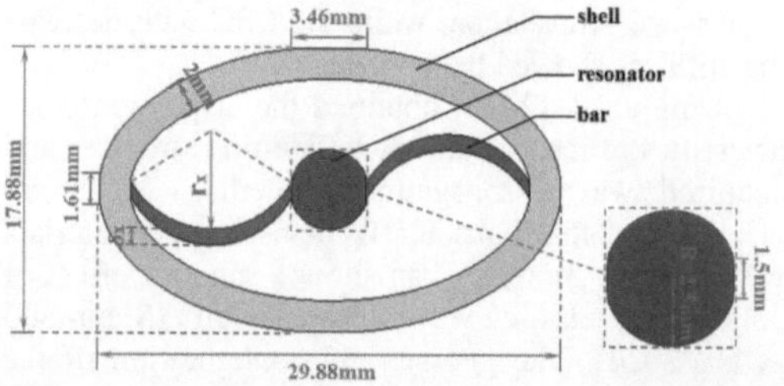

Figure 1. Configuration and size parameters of a typical unit cell.

Due to the fact that the geometries of all the curved bars are part of standard arcs, the curvature of the curved bars can be defined as:

$$K_x = \frac{1}{r_x} \tag{1}$$

where r_x (x=1,2,3) represents the radius of curved bars.

In this study, the 8×8 array is applied in all dynamic models, and the nodes of adjacent cells are merged to simulate the welds. Except for the core structure, all other components in these dynamic working conditions are shown in Figure 2.

As shown in Figure 2, the 8 sampling sites (L_y and R_y, y=1,2,3,4) are areas at the down plate that contact with the core structure. In fact, the stress distributions in all structures are highly symmetric based on the y-axis, so it can be rational to average the stress of the symmetrical sampling sites on both sides.

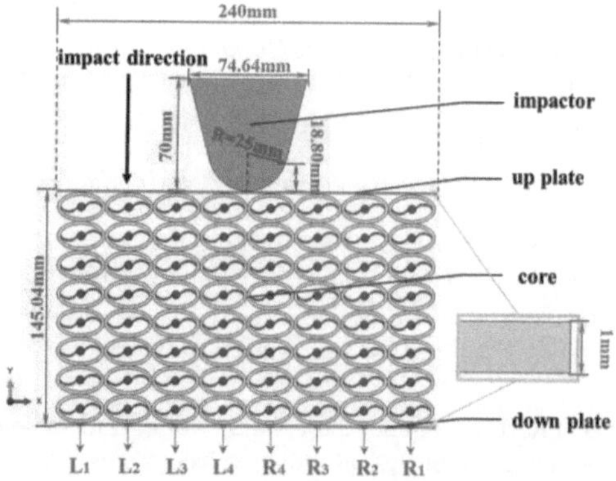

Figure 2. Diagram of typical working conditions with basic size parameters and sampling sites.

Therefore, the average stress for L_y and R_y sites can be presented as follows:

$$S_y = \frac{S_{L_y} + S_{R_y}}{2} \tag{2}$$

where both S_{Ly} and S_{Ry} (y=1,2,3,4) are the stress at the integration point with the maximum peak value among all the integration points in impact history at the corresponding left or right sampling site.

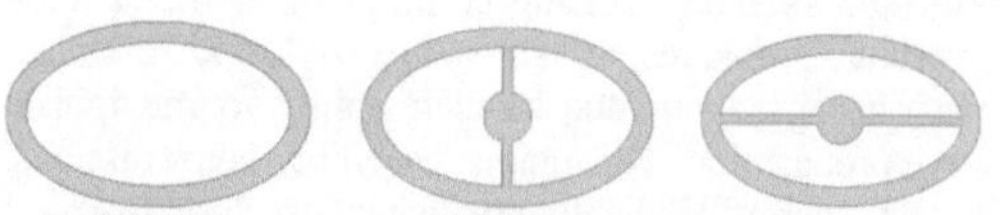

Figure 3. Unit cell of ordinary and modified forms.

The unit cell of the ordinary CSC is shown in Figure 3a, and this study aims to devise a proper modified construction for CSC. On account of this, the cylindrical shell chain with vertical bar (CSCV) and cylindrical shell chain with horizontal bar (CSCH) are designed as shown in Figure 3b and Figure 3c.

With variable parameters and configuration, all working conditions are displayed in Table 1.

2.2 *Simulation software and material parameters*

In this study, all models are set and solved by using the finite element method software ABAQUS. As to the application of materials, the material used for impactors is steel, and all up and down plates are made of aluminum. The configurations and compositions of the core structures are presented in Figure 1 and Table 1. Additionally, the basic mechanical properties of materials are shown in Table 2, and the corresponding plastic material parameters are shown in Table 3 and Table 4.

Table 1. Dynamic working conditions for all structures.

	Material				
Description	Shell	Bar	Resonator	K_x/mm^{-1}	δ_x/mm
CSC	Aluminum	-	-	-	-
CSCV	Aluminum	Aluminum	Aluminum	0	1.0
CSCH	Aluminum	Aluminum	Aluminum	0	1.0
ICSC1	Aluminum	ABS	Steel	0	0.5
ICSC2	Aluminum	ABS	Steel	0	1.0
ICSC3	Aluminum	ABS	Steel	0	1.5
ICSC4	Aluminum	ABS	Steel	0.071	0.5
ICSC5	Aluminum	ABS	Steel	0.071	1.0
ICSC6	Aluminum	ABS	Steel	0.071	1.5
ICSC7	Aluminum	ABS	Steel	0.162	0.5
ICSC8	Aluminum	ABS	Steel	0.162	1.0
ICSC9	Aluminum	ABS	Steel	0.162	1.5

Table 2. Basic mechanical properties of materials.

Material	Density (g/cm^3)	Elastic modulus (MPa)	Poisson's ratio
Aluminum	2.7	72000	0.25
ABS	1.25	3000	0.4
Steel	7.85	206000	0.3

Table 3. Plastic material parameters of ABS (left) and steel (right).

Plastic strain	True stress(MPa)	Plastic strain	True stress(MPa)
0.0000	49	0.0000	480
0.0235	58.61	0.0029	558
0.1196	59.65	0.0206	608
0.2157	60.52	0.0680	676
0.3117	61.19	0.1355	758
0.4078	61.67	0.2463	868
0.5039	61.96	0.3023	917
0.6000	62.05	2.0000	960

Table 4. JC parameters of aluminum.

Description	Value
JC constitutive parameter(A)	331.17
JC constitutive parameter(B)	576.65
JC constitutive parameter(n)	0.99
JC constitutive parameter(m)	0.945
Damage JC parameter(d1)	0.04704
Damage JC parameter(d2)	1.155
Damage JC parameter(d3)	-0.841
Damage JC parameter(d4)	-0.042
Damage JC parameter(d5)	0

2.3 *Analysis step and interaction*

For all dynamic working conditions, the analysis steps are set to the dynamic explicit analysis. To cater to the economy of calculation, the step time is finally set to 5ms.

With respect to the interaction setting of all models, it can be integral to create the contact properties at first. In the set, it is necessary to select the normal behavior as the"Hard contact" type, select tangential behavior as"Penalty contact", and set the friction coefficient to 0.3. The next operation is to set the impactor to "Face-to-face contact" with the up plate, then set the "General contact" for up(down) plate and core structures, and set the "Self-contact" for core structures. Finally, a binding connection between the up(down) plate and core structures is set.

2.4 *Loading and meshing*

The bottom of the down plate is completely fixed at each degree of freedom, and then the initial velocity of the impactor can be set at 7m/s in the pre-defined field.

Based on the analysis of mesh convergence, as for the meshing, the overall structure adopts the CPS4R integral element with seed size varying from 0.25mm to 0.5mm due to the geometry variations.

3 RESULTS AND DISCUSSION

3.1 *Comparative study on energy diversion*

In this section, the characteristic of energy diversion for modified structures, as shown in Figure 3, would be studied with the target of deciding a proper configuration for ICSC, which possesses the remarkable property of energy diversion just as CSC.

In Figure 4, both the CSC and CSCH have similar stress distribution at different moments and generate the stress path with a certain deflection angle relative to the Y-direction. The characteristic of stress diversion contributes to reducing the energy delivery along the Y-direction while boosting the energy delivery along the X-direction, thus demonstrating the excellent protective effect on the protected structure at the bottom of the down plate in the Y-direction. However, the CSCV exhibits a particularly different stress distribution, evidently generating the stress path along the Y-direction. It seems that the arrangement of vertical bars can raise the energy delivery along the Y-direction, thereby adversely affecting the protective effect on the bottom of the down plate in the Y-direction. In consequence, the CSCH shows the identical property of energy diversion just as the CSC, whereas the CSCV actually loses this property, which indicates that the design of horizontal bars instead of vertical bars added to the CSC is apparently more appropriate.

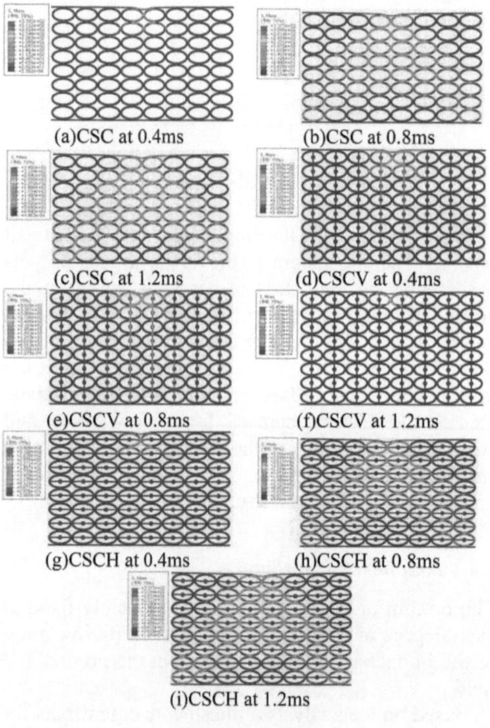

Figure 4. Stress performance of respective structures under the impact loading at different moments.

In order to more intuitively reveal the difference in the energy propagation direction due to the anisotropy of wave propagation orientation, and because the structures are symmetrical for stress distributions at both sides of the Y-direction, this study exports the average stress curves with the largest magnitude in each sampling region for working conditions in Figure 4. The larger the subscript of stress, the closer the sampling site is to the adjacent area of the Y-direction centerline. In virtue of the fact that the adjacent area of the Y-direction centerline is also usually the area where the down plate is in contact with the protected structure, it is more appropriate to have smaller stress peaks for this area.

As shown in Figure 5, despite the different magnitudes, the 4 stress curves in the corresponding

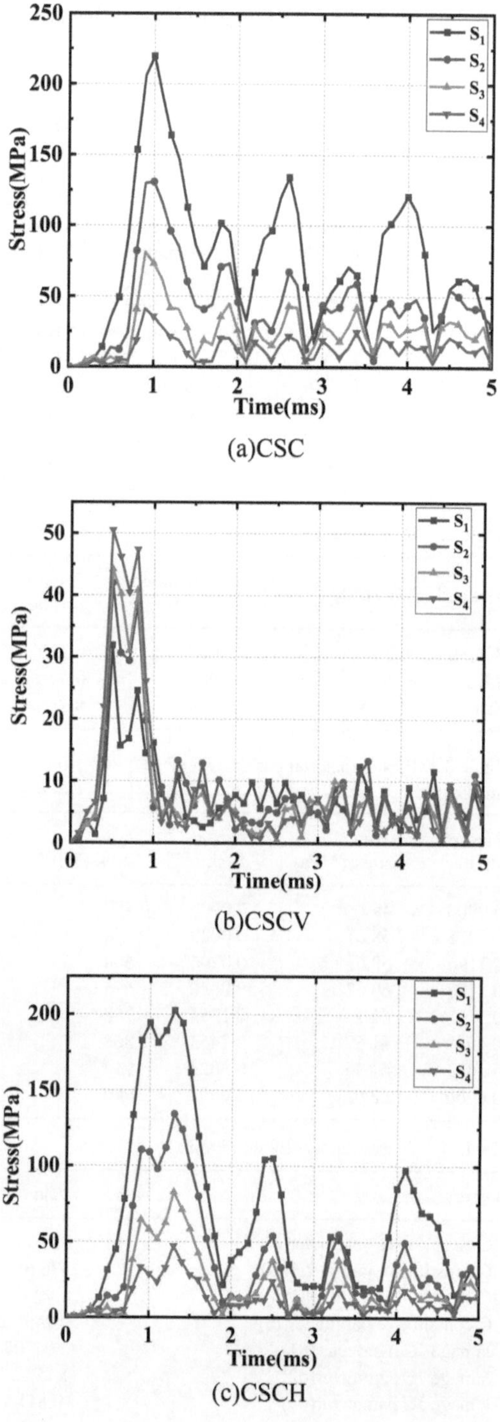

Figure 5. Average stress at sampling sites of respective structures.

structure show the analogous tendency in impact history. Furthermore, both the CSC and CSCH have the same energy propagation regularity that the peak values of stress decrease sequentially when closer to the Y-direction centerline region, whereas the regularity of the CSCV is completely opposite. As a result, the properties of stress curves explicitly demonstrate the phenomenon of energy diversion of the CSC and CSCH, and also manifest the property loss of energy diversion for the CSCV in Figure 4.

The reaction force on the down plate is a very important indicator of the effect of protection, so the reaction forces are exported at the down plate of each structure, as shown in Figure 6. It can be observed that the peak force of the CSCV substantially increased by about 74.5% compared to the CSC, while the peak force of the CSCH is close to that of the CSC.

As a result, the property of energy diversion in the CSCH is exactly the same as that in CSC, while the CSCV loses this property. Moreover, if the CSC is regarded as a control group, the reduction effect of peak reaction force for the CSCH is not optimized, and further improvement of the CSCH structure is essential.

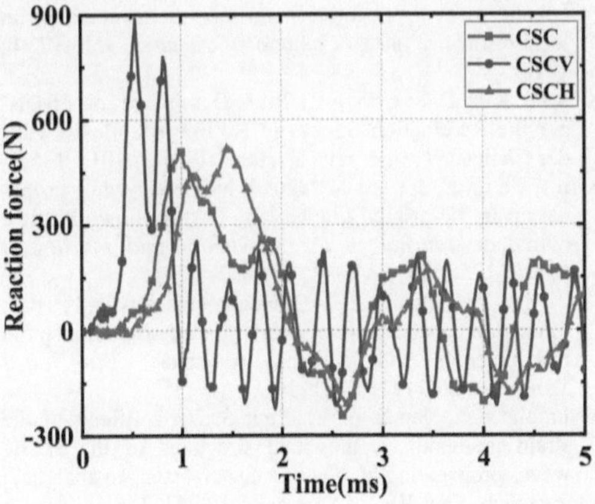

Figure 6. Reaction force on the down plate of the CSC, CSCV, and CSCH.

3.2 *Improvement for impact attenuation*

Based on the research in section 3.1, the beneficial effects of structural design for the CSCH are clarified. In conformity with the basic configuration of the CSCH, the heterogeneous material design will be introduced, and a comparative study focused on the reaction force will be conducted for the two parameters (offset thickness and curvature of the curvature bars)in this section, and an optimum form of the improved design will be obtained.

Since all the improved designs show similar stress distribution, the stress nephograms are no longer given, and the results of reaction force attenuation are mainly discussed in this section. The working condition designs of all ICSCs are shown in Table 1, and Figure 7 presents the reaction force curves of each structure under the condition that the curvatures of the curved bars are constants, with the offset thickness varying, respectively.

As shown in Figure 7a, the peak value of the reaction force sequentially increases with the increase of offset thickness when the curvature is 0. For the cases with curvatures of 0.071 and 0.162 in

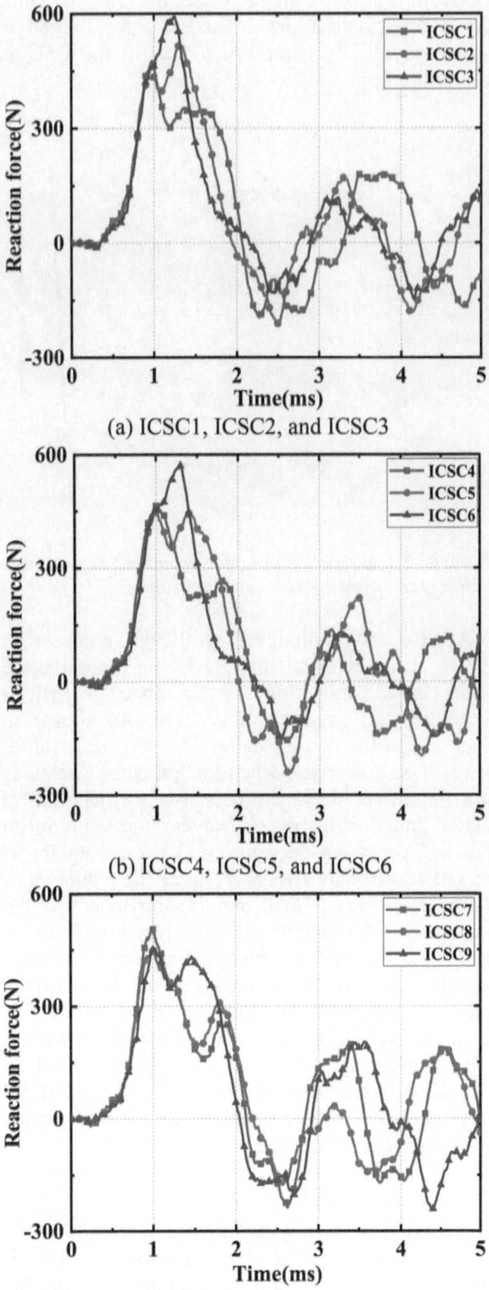

(a) ICSC1, ICSC2, and ICSC3

(b) ICSC4, ICSC5, and ICSC6

Figure 7. Reaction force on the down plate of all improved structures.

Figure 7b and Figure 7c, respectively, the ICSCs with the smallest peak reaction force are the cases with the moderate offset thickness of curved bars. Compared with the CSC, there are several improved designs with different degrees of peak force reduction, such as the ICSC1, ICSC4, ICSC5, ICSC7, ICSC8, and ICSC9, which proves that the design of large curvature is significantly beneficial for the impact attenuation. Moreover, the ICSC8 with a curvature of 0.162 and an offset thickness of 1mm shows the smallest peak reaction force, which is reduced by 11.4% compared to that of the CSC as a control group.

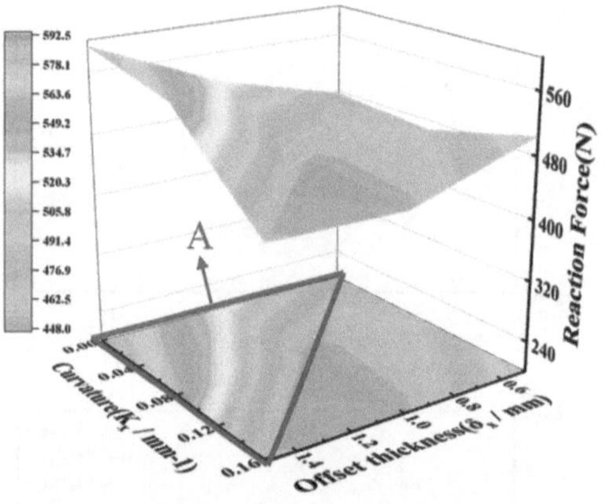

Figure 8. 3D surface plots of peak value for reaction force with varying offset thickness and curvature.

At first, in conjunction with the analysis in Figure 7, it's apparent that increasing the curvature of the curved bars within a certain range is beneficial to reduce the peak reaction force. As shown in Figure 8, in the design region "A", the peak values of the ICSCs decrease when the curvature increases and the offset thickness decreases, which almost shows linear regularity. However, it doesn't mean that the increasing curvature and decreasing thickness of curved bars always facilitate the reduction of the peak reaction force, and it's obvious that the location of the minimum peak reaction force in design regions is the center region where both the curvature and the offset thickness are intermediate values. In line with the study on the 3D surface of the peak force for ICSCs with varying curvature and offset thickness, there exists a suitable intermediate value that makes the peak reaction force of the ICSC to be optimally reduced, which provides an important avenue for optimized designs of the ICSC.

4 CONCLUSIONS

In this study, based on the comparative study of energy diversion for modified structures, a reasonable design for the ICSC is determined. Furthermore, the heterogeneous material design is utilized, and the effects of two parameters (the curvature and offset thickness of the curved bars) in the ICSC on the peak reaction force are acquired. The following are the main conclusions:

(1) The property of energy diversion in the CSCH is actually the same as that in the CSC, while the CSCV loses this property, thereby enlightening the geometrical designs for the ICSC.
(2) The heterogeneous material design is applied, and the ICSC8 with a curvature of 0.162 and an offset thickness of 1mm reduces the peak value of reaction force by 11.4% compared to the CSC. Furthermore, as for the 2 parameters (the curvature and offset thickness of the curved bars), there exists a suitable intermediate value that makes the peak reaction force of the ICSC to be optimally reduced, providing an important avenue for optimized designs of the ICSC.

REFERENCES

Huang Y., Wu J.H. & Lei Y.Z. 2023. Impact protection enhancement by negative mass meta-honeycombs with local resonance plates. *Composite Structures*, 321(11730): 1–10.

Kim H., Kim E. & Chong C. 2018. Demonstration of Dispersive Rarefaction Shocks in Hollow Elliptical Cylinder Chains. *Physical review letters*, 120(194101): 1–5.

Kim H., Kim E. &Yang J. Y. 2019. Nonlinear wave propagation in 3D-printed graded lattices of hollow elliptical cylinders. *Journal of the Mechanics and Physics of Solids*, 125: 774–784.

Li C., Sun Y. & Han Q. 2024. Bandgap tunability and impact mitigation enhancement of hybrid graded origami-inspired metamaterials with multiple resonators. *Thin-Walled Structures*, 202(112137): 1–19.

Mahmoud S.H., Jan S. & Vladimir S. 2024. Effects of the strain gradients on the band structures of the elastic waves propagating in 1D phononic crystals: An analytical approach. *Thin-Walled Structures*, 194(111316): 1–8.

Peng K.F. 2021. *Static and dynamic mechanical behavior and deformation mechanism of granular materials and chain structure*. Hefei: University of Science and Technology of China.

Peng K.F., Zheng Z.J. & Chang B.X. 2021. Wide-range control of impulse transmission by cylindrical shell chains with varying aspect ratios. *International Journal of Impact Engineering*, 158(104017): 1–16.

Peng K.F, Zheng X.J. & Zhou F.H. 2022. Elastic wave propagation characteristics of density gradient cylindrical shell chains. *Chinese Journal of Theoretical and Applied Mechanics*, 54(8): 2131–2139.

Zhang Z.H. & H b.y. 2022. Regulation Mechanism of the Shock Waves in a Pentamode Lattice-Ring Structure Subjected to Collision. *Crystals*, 12(1749): 1–21.

Zhang W.Z, Pan J & Sanchez J.C. 2024. Review on the protective technologies of bridge against vessel collision. *Thin-Walled Structures*, 201(112013):1–24.

Trends in Collision and Grounding of Ships and Offshore Structures – Zhu & Guedes Soares (Eds)
© 2026 The Author(s), ISBN 978-1-041-16361-9

Numerical study on the energy absorption effect and impact resistance of gradient honeycomb sandwich structures

T. Wang, Y.H. Chen & M.C. Xu
School of Naval Architecture and Ocean Engineering, Huazhong University of Science & Technology, Wuhan, P.R. China
Department of Naval Architecture, Ocean and Structural Engineering, School of Naval Architecture, Ocean and Energy Power Engineering, Wuhan University of Technology, Wuhan, P.R. China

J. Pan
Key Laboratory of High Performance Ship Technology, Ministry of Education, Wuhan, P.R. China

ABSTRACT: This study aims to investigate the dynamic response of gradient honeycomb sandwich structure fabricated by using bi-material 3D printing materials (PLA) under impact loads, as well as to identify the potential advantages of using sandwich structures in applications requiring impact resistance. Based on a collision system established by FE simulations, a parameter analysis is carried out to study the influence of wall thickness of unit cells, impact energy and gradient direction of unit cells on deformation mode and absorb energy capacity. The variation trend of impact force and crushing depth characteristic of the sandwich structures are investigated. Based on the structural responses from the quasi-static compression and dynamic impact, the ability to absorb energy of several core components located in different layers of cellular structure are compared, which indicates recommend structural type with specific gradient direction of being applied in protective devices with impact resistance properties.

1 INTRODUCTION

In order to effectively protect individuals and objects from impact and damage, the sandwich structure, as a widely used anti-collision structure composed of top, bottom face-sheets and the inner core, has the advantages of strong specific stiffness, outstanding impact resistance, and excellent energy absorption characteristics. Considering to its broad applicability and has garnered widespread attention in the field of energy absorption materials. In recent years, researchers have explored the energy absorption efficiency of typical periodic porous material with diverse geometric configurations including honeycomb shape, corrugated shape and negative Poisson's ratio (NPR) shape with various types of unit cells (hexagonal (Li et al., 2018), double arrow (Gu et al., 2020), star-shaped (Li et al., 2023), chiral (Hu et al., 2019), square (Kumar & Banerjee, 2020), etc.).

The conventional honeycomb sandwich structure with hexagonal core is one of the most extensively researched and widely applied protective configurations, with its mechanical behavior and failure machanism having undergone a substantial amount of experimental (Hou et al., 2011) and numerical (Ruan et al., 2003) investigations. To further enhance energy absorption performance, scholars have begun to pursue improvements to conventional honeycomb structures. Qi et al. (2020) proposed a novel re-entrant circular (REC) honeycomb structure through experimental, theoretical and numerical methods, which can dissipate extra energy due to more formed plastic angles during the quasi-static crushing process. Zhang et al. (2023) studied the dynamic response of a fully clamped foam-filled X-type sandwich beam under low-velocity impact analytically and numerically and discussed the effects of face-sheet thickness, X-type folding plate thickness and foam strength on the impact resistance of the foam-filled X-type sandwich beam. Zhang et al. investigated the dynamic responses of the sandwich panels with multi-layered corrugated core experimentally under underwater impulsive loads. The effects of impulse intensity and core configuration on the blast performance of panels are tested through designing stack orientation, layer number, core gradation and core topology. Wu et al. (2024) developed bi-material multistable auxetic honeycomb (BMAH) by using bi-material 3D printing technology. Based on the analysis of in-plane crushing, the deformation mechanism is theoretically analyzed, and the effects of geometric parameters on the performance of

DOI: 10.1201/9781003684404-41

BMAH are investigated. Liu & Li (2025) investigated re-entrant hexagonal honeycomb NPR structure added with carbon fiber reinforced Nylon material. The results show that the NPR structure reinforced with carbon fiber has 112.5% higher failure stress and 113.9% higher energy absorption efficiency than the unreinforced structure.

To increase the in-plane stiffness of lightweight honeycomb materials, the functionally graded property is proposed (Ha & Lu 2020), which can be defined as a structure in which the mechanical properties (density, elastic modulus, strength, energy absorption) change gradually over the volume due to the nature of gradient in structural design and material composition (Miyamoto et al. 2013). To explore the unique properties of gradient structures, Zhang et al. (2015) explored the crashworthiness of functionally graded thickness (FGT) tubes under axial and lateral dynamic impact loads. Grunsven et al. (2014) studied the mechanical characteristics of functionally graded lattice structures made of Ti-6Al-4V and fabricated by EBM process, in which the density was separately changed in each layer in sequence.

This paper establishes an impact system consists of indentor and honeycomb gradient sandwich structure fabricated by polylactic acid (PLA) material. A parameter analysis of impact force and deformation mode on gradient direction, wall thickness of unit cell and impact energy are carried out. Besides, based on the variation trend on energy absorption capacity on core components with different relative densities under quasi-static compression load and dynamic impact load, the impact resistance and ability to absorb energy of the honeycomb gradient sandwich structure are discussed.

2 STRUCTURE AND IMPACT RESPONSE

2.1 *FE model setup*

A collision system applying for testing structural impact resistance and ability to absorb energy is established in ANSYS/LS-DYNA, as shown in Figure 1 (a). The shape of the indentor is selected as cylinder with diameter180 mm, and its weight is 3.56kg, of which relative freedom instead of impact direction is constraint. All parts including punch and the platform are meshed by Solid164 eight-node solid element in the type of hexahedron. The *MAT_RIGID material models are adopted for the indentor and platform, respectively. The *CONTACT AUTOMATIC SURFACE TO SURFACE is employed for all Intercontact surfaces, including (i) contact between indentor and impacted structure, (ii) contact between impacted structure and platform. The static and dynamic friction coefficients were set to be 0.2 and 0.15, respectively. A FE model of sandwich structure consisting of top face sheet, bottom face sheet and unit cells with honeycomb gradient mode are modeled, where its horizontal length, horizontal width and height are set as 100 mm, 100 mm and 63 mm, respectively. The honeycomb gradient sandwich structure is meshed using quadrilateral Belytschko–Tsay shell elements (Hallquist 1998) with five integration points across the thickness. The *CONTACT AUTOMATIC SINGLE SURFACE is used in order to realize self-contacting interfaces during the crushing process. In order to investigate the changes in energy absorption efficiency along gradient direction of honeycomb structure, the sandwich structure core is divided into several parts, denoted as layer 1, layer 2 and layer 3, as shown in Figure 1 (b).

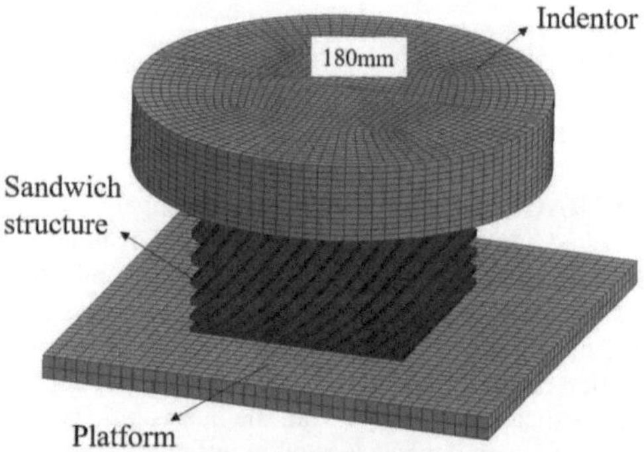

(a) Impact system

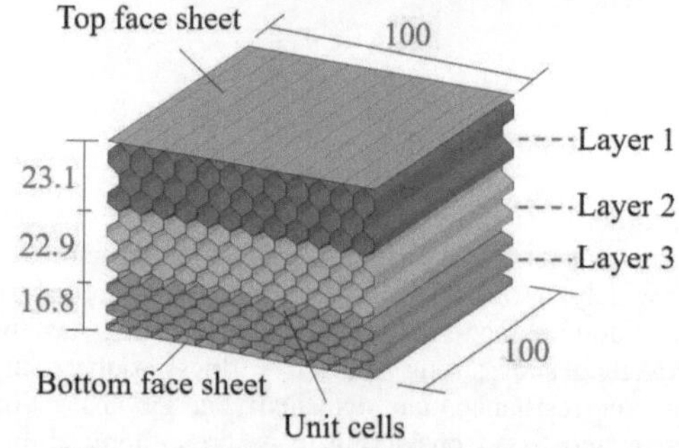

(b) Honeycomb gradient sandwich structure

Figure 1. FE model of impact system (unit: mm).

In this study, polylactic acid (PLA) as a low-cost biodegradable material with good stiffness and strength, is chosen as the materials for sandwich structures, which has been widely used in various industries. As shown in the true stress–strain curve in Figure 2, the PLA exhibits linear elasticity within the initial strain range, allowing for the calculation of the elastic modulus E as 3020 MPa. The yield stress σ_{ys} is determined to be 47.8 MPa, and density and Poisson's ratio were measured to be 1.22 g/cm3 and 0.35, respectively.

2.2 *Impact behavior analysis*

In order to investigate the influence of various parameters such as the wall thickness of unit cell, impact energy and gradient direction of unit cells on

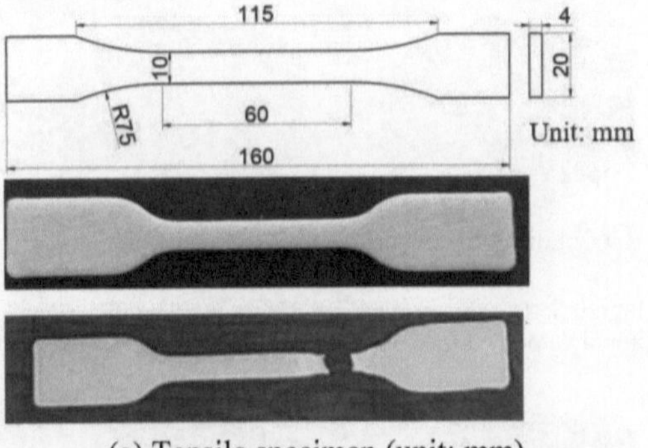

(a) Tensile specimen (unit: mm)

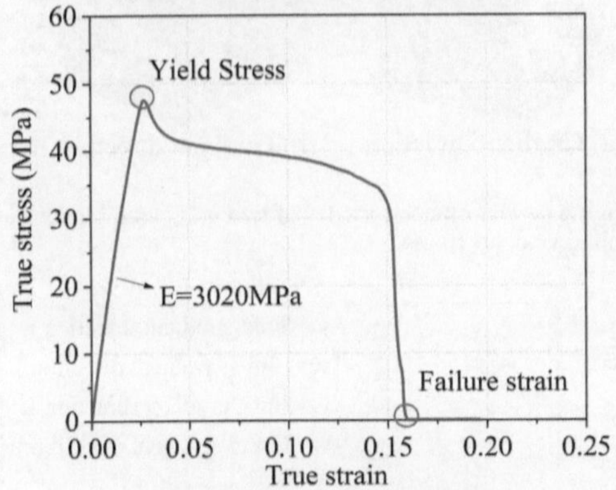

(b) True stress–strain curve

Figure 2. The analysis of PLA material model.

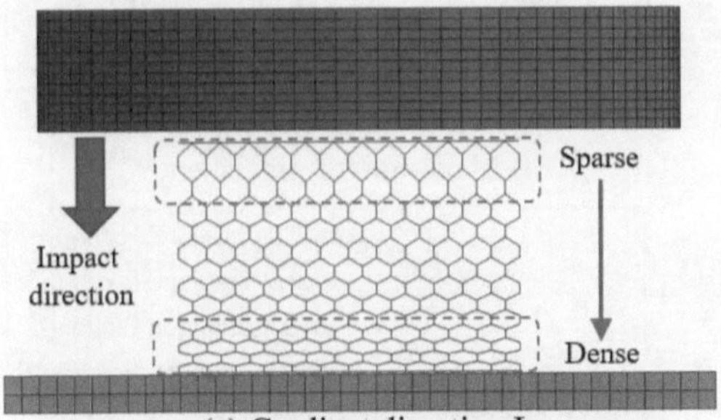

(a) Gradient direction I

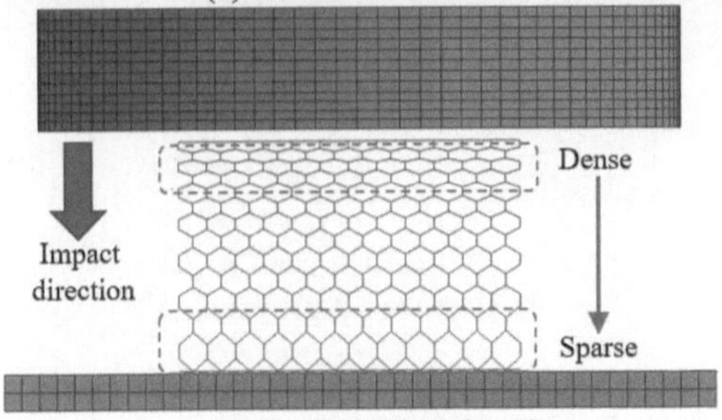

(b) Gradient direction II

Figure 3. Gradient direction of unit cells during the collision process.

the structural anti-collision performance. Impact systems are established including configuration of unit cells transforming from sparse to dense along impact direction of the indentor, denoted as gradient direction I and from dense to sparse, denoted as gradient direction II, as shown in Figure 3. Besides, two initial velocities determined impact energy size are set as 5m/s and 10m/s. The wall thicknesses of unit cell t are set as 0.7mm, 1.0mm and 1.3mm, respectively. The impact cases varying with impact energy, wall thickness and gradient direction of unit cell are listed in Table 1.

Table 1. Impact cases based on the parameter analysis.

Cases	Gradient direction	Initial velocity (m/s)	Thickness of unit cell (mm)
1	I	5	0.7
2	I	10	0.7
3	I	5	1.0
4	I	10	1.0
5	I	5	1.3
6	I	10	1.3
7	II	5	0.7
8	II	10	0.7
9	II	5	1.0
10	II	10	1.0
11	II	5	1.3
12	II	10	1.3

From the time histories of impact force shown in Figure 4, impact energy significantly affects the variation trend of impact force and impact time duration. Based on the same thickness of unit cell, comparing to load cases with high impact energy, yield force platform are observed which leads to larger duration. The main reason is that structural plastic deformation or material fracture results in longer contact duration between the indentor and sandwich structures. Besides, compression stiffness of unit cells is reduced due to decrease of its wall thickness, which also caused larger duration because of stretch and tear of wall of unit cell. Furthermore, according to contact forces between sandwich structure and platform, unit cell configuration as gradient direction II could weaken impact load more effectively, especially for the load cases with large impact energy.

2.3 *Deformation mode analysis*

Figure 5 and Figure 6 present plastic deformation states of the sandwich structures with two gradient directions of unit cells. For impact cases with low impact energy, the structural integrity is kept where no element enters into plastic fracture state.

Under large impact energy, considering to higher percentage of structural voids observed in Layer 1, lateral migration of unit cells caused by fracture of inner wall can be captured. In this case, the honeycomb structural units are subjected to stretch in lateral direction, and the junction of the inner wall of cells appears bending.

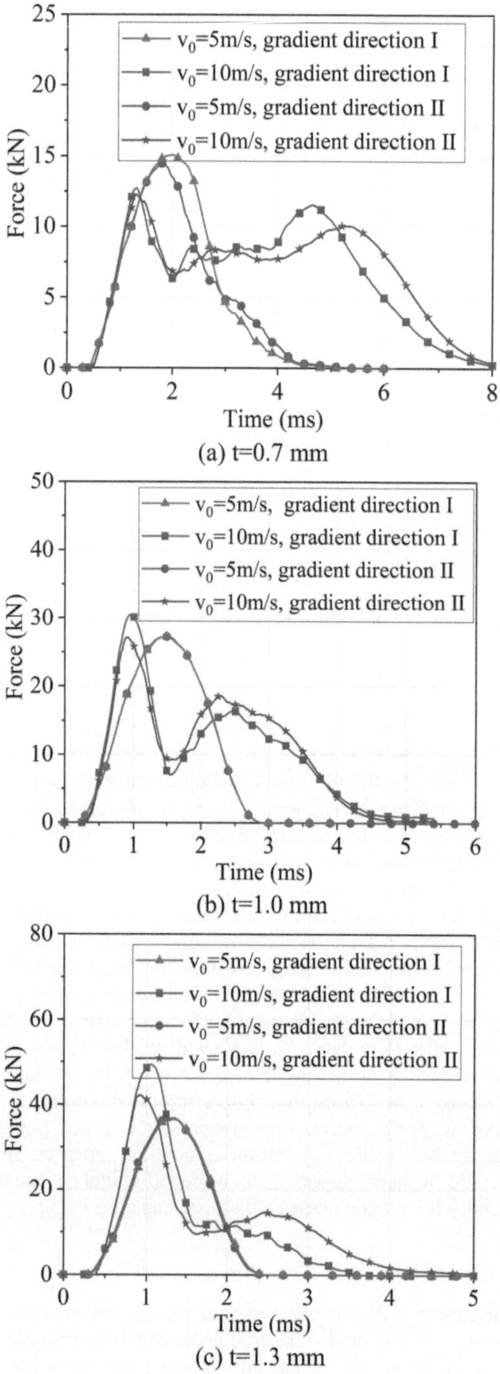

Figure 4. The impact force time histories between sandwich structure and platform with various unit cell thickness.

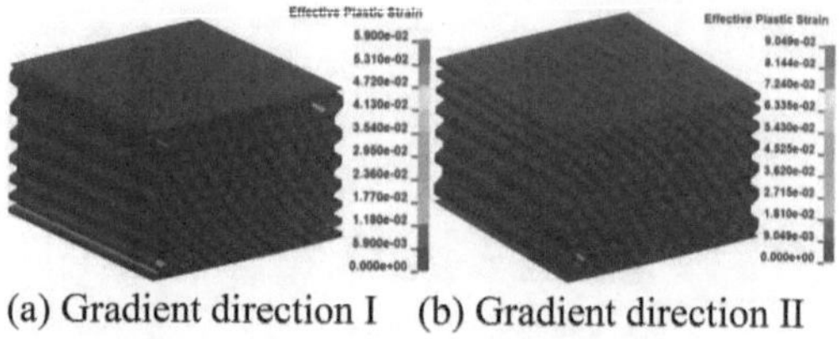

Figure 5. Deformation modes of two gradient directions at initial velocity 5 m/s.

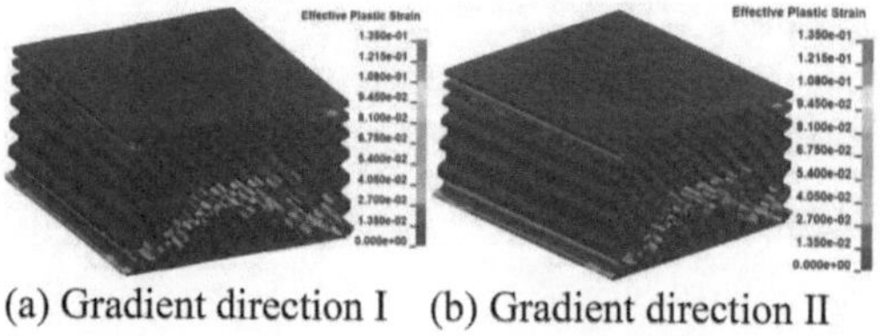

Figure 6. Deformation modes of two gradient directions at initial velocity 10 m/s.

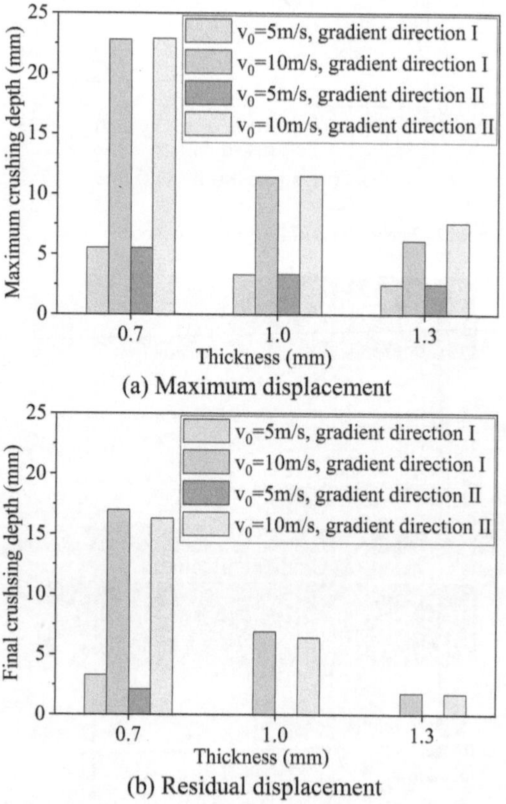

Figure 7. Displacement response of the sandwich structures under various impact energies.

Besides, damage distribution shape of the cross-section is presented as "V" mode for gradient direction I and gradient direction II. In order to further illustrate the compression deformation of sandwich structures under various impact energies and gradient directions, Figure 7 present displacement responses including the maximum displacement and residual displacement. Under the same impact energy, either gradient direction I or gradient direction II, crushing depth is basically consistent. With the wall thickness of unit cells increases, the degree of deformation decreases. From the results of residual displacement, unit cells configurated as gradient direction II have better elastic recovery performance when subjected to relatively large impact energy.

3 ANALYSIS OF ENERGY ABSORPTION EFFICIENCY

3.1 *Characteristic of energy variation trend*

During the collision process, the impact energy is mainly transformed into structural internal energy. The contribution of internal energies by each layer of unit cells under different impact energies is shown in Figure 8. Compared to the face-sheet, sandwich structure core absorbs most of the impact energy, and the energy absorption effect are more than 90%, respectively. When the unit cell thickness is thin (t=0.7 mm), unit cells of Layer 1 configurated as gradient direction II could absorb more energy under low impact energy, while the ones configurated as gradient direction I could absorb more energy under high impact energy. With the unit cell thickness increases, the more nonuniformity in the energy distribution ratios of each layer could be observed for gradient direction II, which means more plastic deformation and collapse state would occur in Layer 1.

3.2 *Discussion*

During collision process, energy absorption capacity and specific energy absorption are important criteria for evaluating the mechanical properties of sandwich structures. According to the researches by Gibson and Ashby (1997), the structural deformation of the cellular structure under the compressive loads exhibits three ideal regimes shown in Figure 9: (1) The first regime is a linear elastic region, which is characterized by the elastic modulus of cellular structure. (2) The second region of the crushing force-depth curve is the plaeau regime, where the unit cell presents plastic bending failure. (3) The third regime is the densification region, when continued compressive loads will lead to full compaction of the structure due to contact od the cell struts with the next cell struts causing a considerable rise in cellular structure stiffness.

Thus, during the quasi-static compression process, the impact energy is totally transformed into the plastic deformation energy. Through controlling the crushing depth equal to the impact case 10 m/s (gradient direction I), the variation trend of the internal under the quasi-static compression load at 5 mm/s is shown in Figure 10. Considering to strain hardening effect of material under dynamic collision, consequently, lower internal generated in the quasi-static compression. At the initial stage of crushing process, the fracture is mainly located in the transitional region from sparse to dense, and then transverse expansion deformation could be observed.

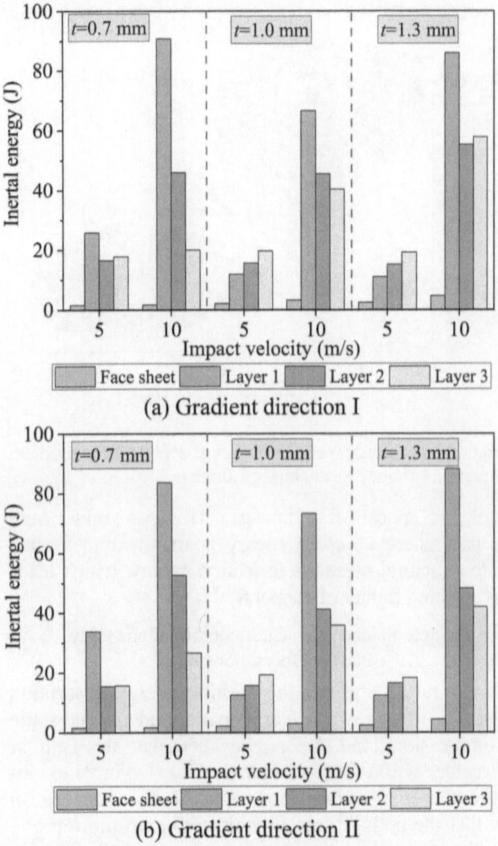

Figure 8. Distribution of internal energy on different parts.

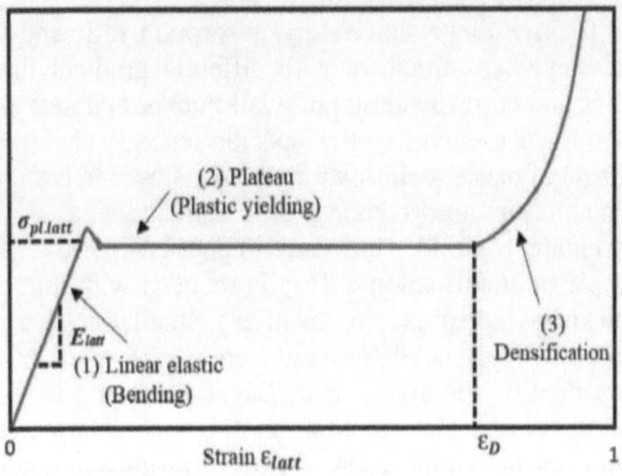

Figure 9. Typical schematic compressive stress-strain curves of cellular structures under static compression load proposed by Gibson and Ashby (1997).

Different from the quasi-static compression, the internal energy encompasses elastic deformation energy and plastic deformation energy, see Eq. (1). Based on impact cases with gradient direction II, two representative energy versus time curves of the structures with unit cell thickness 1.0 mm at initial velocity v_0=5.0 m·s^{-1} and v_0=10 m·s^{-1} are shown in Figure 11. The elastic deformation energy takes a dominate part in the internal energy under low impact energy, where the

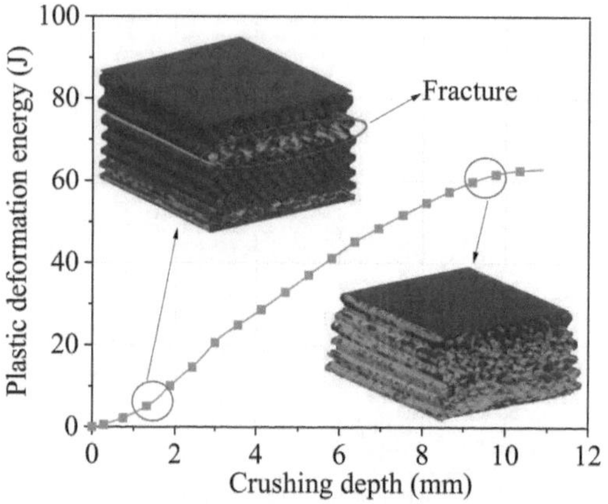

Figure 10. Energy versus crushing depth along gradient direction I (Unit cell thickness 1.0mm).

structure appears few damages. However, under high impact energy, impact energy is mainly transformed into structural plastic deformation energy, which leads to crushing failure of elements.

$$\text{Peakinternalenergy} = \text{Elasticdeformationenergy} + \text{Plasticdefoemationenergy} \quad (1)$$

Figure 12 presents the specific energy absorption (SEA) of various core components under quasi-static compression load. It can be seen that the cellular structure with relative larger density (Layer 3) exhibits its better energy absorption capacity. The main reason is that the cells in transitional region are more prone to fracture at low energy transfer velocity, leading to plastic energy gradually accumulated in cellular structure far away from the impact plane.

To investigate the energy absorption efficiency of the sandwich structures with different gradient directions and corresponding unit wall thickness under various impact energies, the specific energy absorption (SEA) of plastic deformation energy is used to compare the ability to absorb energy of the structures, as shown in Figure 13. Under the same impact energy, the SEA of plastic deformation energy is reduced with unit cell thickness increases. For relatively small thickness of unit cells, plastic deformation energy is more easily absorbed by the unit cells of Layer 1, leading to more severe plastic deformation on the area configurated with sparse unit cells. With the impact energy increases, more numbers of unit cells from Layer 3 enter plastic state for gradient direction I. Thus, for honeycomb gradient sandwich structures with larger unit cell thickness subjected to larger impact energy, unit cell configurated as gradient direction II seems to be advantages in impact resistance considering to its relatively uniform distribution of SEA of plastic deformation energy on unit cells and better reduction effect of impact force shown in Figure 4. Besides, in comparison to the SEA obtained from the quasi-static compression cases, it is found that the plastic deformation under impulse impact load concentrates at cellular structure with lower relative density (Layer 1), which could further promote the formation of the deformation band to guide stress wave, as shown in Figure 6 (b).

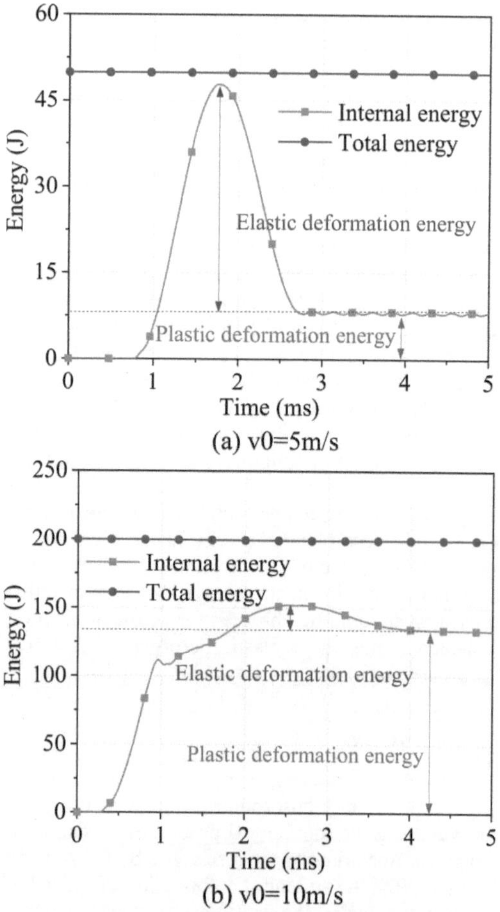

Figure 11. Energy versus time history along gradient direction I (Unit cell thickness 1.0mm).

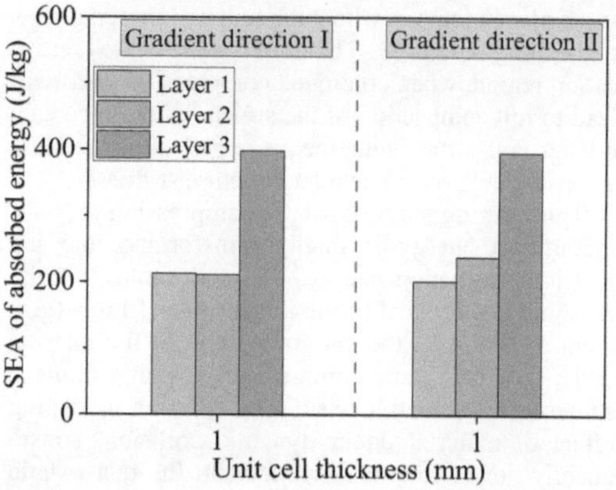

Figure 12. The energy absorption capacity of different parts under quasi-static compression load (Unit cell thickness 1.0 mm).

Additionally, in order to validate the above analytical results obtained from the FE simulations, corresponding impact tests on the graded honeycomb sandwich structures fractured by 3D-printing technique would be carried out recently.

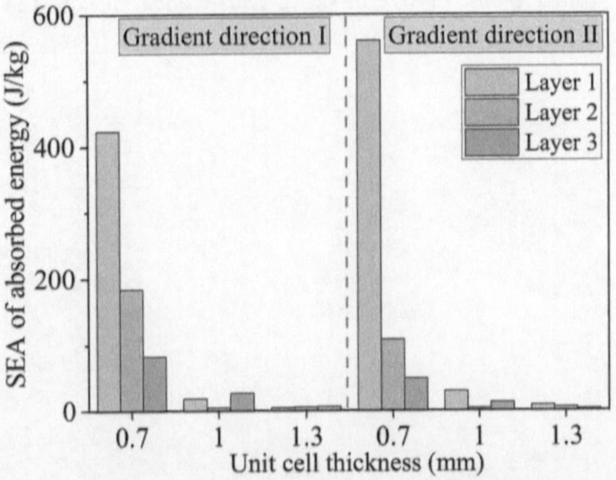

(a) Low impact energy at velocity 5m/s

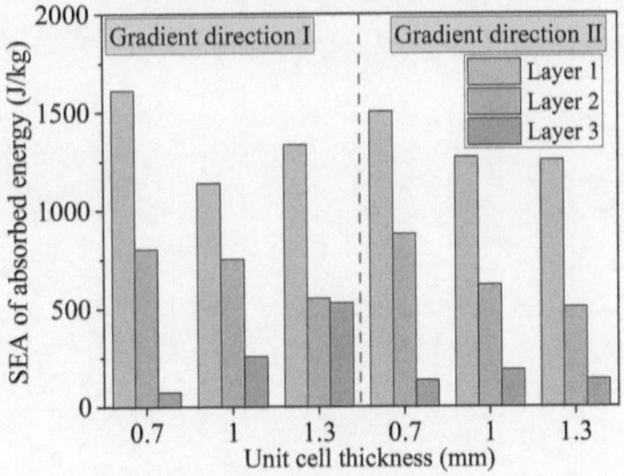

(b) High impact energy at velocity 10m/s

Figure 13. The energy absorption capacity of different parts under various impact energies.

4 CONCLUSIONS

Based on the collision system, the dynamic response and energy absorption effect of honeycomb gradient sandwich structures under impact load are studied using FE simulations. In order to investigate the impact resistance and failure mechanism, a parameter analysis on wall thickness of unit cell, impact energy and gradient direction of unit cell are carried out. The following conclusions are derived.

(1) For honeycomb gradient sandwich structures subjected to the same impact energy, the impact resistance properties are improved where unit cells are configurated as gradient direction II along impact direction;

(2) The cellular structure with relatively larger density exhibits better energy absorption capacity under quasi-static compression load, while the cellular structure with relatively lower density exhibits better energy absorption capacity under dynamic impact load.

(3) From the SEA of plastic deformation energy obtained from the cases with large impact energies, impact energy is more easily absorbed by the unit cells with larger structural voids configurated as gradient direction I, which could cause more severe structural damage compared to those one configurated as gradient direction II.

ACKNOWLEGEMENTS

This work has been supported by the National Natural Science Foundation of China (Grant No. 52371319).

REFERENCES

Ashby, M.F., Gibson, L. 1997. Cellular Solids: Structure and Properties. *Cambridge University Press*, UK, Cambridge.

Grunsven, W, Hernandez-Nava, E., Reilly, G.C., et al. 2014. Fabrication and mechanical characterisation of titanium lattices with graded porosity, Meta 4 (3) 401–409.

Gu, L X., Xu, Q L., Zheng, D M., et al. 2020. Analysis of the mechanical properties of double arrowhead auxetic metamaterials under tension. *Textile Research Journa*, 90 (21–22) 2411–2427.

Hallquist, J. LS-DYNA theoretical manual. 1998. *Livemore Software Technology Corporation*, CA: LSTC.

Hou, B., Ono, A., Abdennadher, S., et al. 2011. Impact behavior of honeycombs under combined shear-compression. Part I: experiments. *International Journal of Solids and Structures*, 48 (5) 687–697.

Ha, N S., Lu, G. 2020. A review of recent research on bio-inspired structures and materials for energy absorption applications. *Compos Part B*, 181:107496.

Hu, L L., Luo, Z R., Zhang, Z Y., Lian, M.K., Huang, L.S. 2019. Mechanical property of re-entrant anti-trichiral honeycombs under large deformation. *Composites Part B: Engineering*, 163, 107–120.

Kumar, D. & Banerjee, A. 2020. Effect of heterogeneity on crushing failure of disordered staggered-square honeycombs. *Composite Structures*, 240 112055.

Li, X.L., Li, Z.Z., Guo, Z.Y., Mo, Z.L. 2023. A novel star-shaped honeycomb with enhanced energy absorption. *Composite Structures*, 309 116716.

Li, Z., Wang, T., Jiang, Y., Wang, L M., Liu, D. 2018. Design-oriented crushing analysis of hexagonal honeycomb core under in-plane compression. *Composite Structures*, 187 429–438.

Liu, H. & Li, D. 2025. Study on the energy absorption effect and impact resistance of a composite sandwich panel with carbon fiber reinforced re-entrant hexagonal honeycomb core with negative Poisson's ratio. *European Journal of Mechanics - A/Solids*, 109: 105447.

Miyamoto Y., Kaysser W., Rabin B., Kawasaki, A., Ford, R.G. 2013. Functionally GradedMaterials: Design, Processing and Applications. Springer Science & Business Media.

Qi, C., Jiang, F., Remennikov A., Pei, L-Z., Liu, J., Wang, J.-S., Liao, X.-W., Yang, S. 2020. Quasi-static crushing behavior of novel re-entrant circular auxetic honeycombs. *Composites Part B: Engineering*, 197: 108117.

Ruan, D., Lu, G., Wang, B., Yu, T.X. 2003. In-plane dynamic crushing of honeycombs—a finite element study. *International Journal of Impact Engineering*, 28 (2) 161–182.
Wu, X W., Zhang, S., Ding, L Z., Wu, W.Q., Ma, Y.B., Deng, Z.C. 2024. Bi-material multistable auxetic honeycombs with reusable and enhanced energy-absorbing phases under in-plane crushing. *Thin-Walled Structures*, 201: 111988.
Zhang, J X., Yuan, L, Zhang, J T., Yan, J., Yuan, H. 2023. Dynamic response of foam-filled X-type sandwich beam under low-velocity impact. *Engineering Structures*, 292: 116588.
Zhang, H., Sun, G., Xiao, Z., Li, G.Y., Li, Q. 2018. Bending characteristics of top-hat structures through tailor rolled blank (TRB) process. *Thin-Walled Struct*. 123 420–440.

Trends in Collision and Grounding of Ships and Offshore Structures – Zhu & Guedes Soares (Eds)
© 2026 The Author(s), ISBN 978-1-041-16361-9

Study on penetration performance of honeycomb sandwich plate under low velocity impact

S. Zhou & K.L. Guo
Key Laboratory of High Performance Ship Technology (Wuhan University of Technology), Wuhan, China
School of Naval Architecture, Ocean and Energy Power Engineering, Wuhan University of Technology, Wuhan, China

Y.F. Zhu & L. Zhu
School of Naval Architecture, Ocean and Energy Power Engineering, Wuhan University of Technology, Wuhan, China

ABSTRACT: In this paper, the finite element model of aluminum honeycomb sandwich plate (AHSP) is established by ABAQUS, and the dynamic responses of AHSP are analyzed, including penetration and energy absorption performance. The results obtained from numerical simulations are compared with those of experiments in the previous work, and the accuracy of the numerical method has been verified. On this basis, impactor shape on the penetration performance of AHSP are determined. Results show that, the impactor shape has significate influence on the failure mode of AHSP, for instance, the sandwich plate experiences shear failure under impact from wedge impactor and cylindrical impactor, while the sandwich plate undergoes bending and tension failure under impact from spherical impactor. In addition, the impactor shape has influence on the loading and unloading process of AHSP. Under impact from spherical impactor and cylindrical impactor, the force-displacement curve appearances two peak force. While under impact from wedge impactor, more stages may occur, including the platform stage and the friction energy dissipation stage. Overall, the energy absorption of AHSP under impact from spherical impactor is the largest, followed by the cylindrical impactor, and the smallest is the wedge impactor.

1 INTRODUCTION

Honeycomb sandwich structures have been identified as having considerable application prospects in the field impact protection for ship and marine engineering, due to their excellent mechanical properties and energy absorption characteristics. A significant number of scholars have conducted related research on the dynamic response and energy absorption characteristics of honeycomb sandwich structures.

Zhang et al. (2016) conducted a low-speed impact experiment on honeycomb sandwich panels using a pendulum experimental device. The dent depth, deformation area and deformation volume of honeycomb sandwich panels under the impact of rigid balls were analyzed. Based on the principle of minimum potential energy, Zhang et al. (2017) established a rigid-plastic theoretical model and deduced the semi-analytical expressions of contact force and local dent depth. Li et al. (2017) conducted investigations on the dynamic response of gradient honeycomb sandwich panels under explosion load by combining experimental methods and numerical simulation methods. The difference of deformation modes between gradient and non-gradient honeycomb sandwich panels was compared and analyzed, and the influence mechanism of gradient on energy absorption and panel deflection was explored. Palomba et al. (2018) conducted a low-velocity impact test of a double-layer honeycomb sandwich panel, analyzed the dynamic loading and unloading characteristics of the sandwich panel, and examined the deformation and failure morphology inside the honeycomb sandwich panel using CT scanning technology. Mu et al. (2024) established a numerical simulation model to study the dynamic response characteristics of honeycomb sandwich panels under repeated impact loads with different core gradients, and analyzed the influence mechanism of core gradient on the deformation mode and energy absorption of honeycomb sandwich panels.

In the domain of practical engineering, protective structures frequently encounter high-energy impact situations, which can result in structural damage or even penetration. Recent research in this field has explored the failure and penetration response of honeycomb sandwich panels under impact load. Sun et al. (2018) conducted a breakdown experiment of

DOI: 10.1201/9781003684404-42

honeycomb sandwich panels and analyzed the local depression and penetration characteristics of honeycomb sandwich panels during the breakdown process. In a similar vein, Qin et al. (2020) conducted drop hammer impact experiments and numerical simulations, studying the plastic failure mode and penetration characteristics of honeycomb sandwich panels under hemispherical head impact. They also analyzed the influence of impact position on the deformation mode and failure failure of sandwich panels.

As stated above, there are numerous researches on the dynamic response of honeycomb sandwich panels. However, the majority of these studies concentrate on analysing failure modes and the influence of structural parameters. There is a paucity of research on the influence of panel thickness distribution ratio and impact shape on the breakdown performance of honeycomb sandwich panels, and the underlying mechanisms are not yet fully elucidated. Consequently, there is a clear need for further research in this area.

In this paper, the finite element model of the honeycomb sandwich panel is established by ABAQUS finite element software. The penetration response and energy absorption characteristics of the honeycomb sandwich panel is determined, and the comparison between the simulation results with experiments results in the literature is performed. In addition, the influences of the punch shape on the penetration responses of the honeycomb sandwich panel was explored by numerical simulation.

2 NUMERICAL SIMULATION METHOD

2.1 *Failure model*

The panel and honeycomb core are made of aluminum alloy, and the mechanical properties of aluminum alloy are related to strain rate. In this paper, the Johnson-Cook constitutive model without considering the influence of temperature is adopted. The model is described as follows:

$$\sigma=\left[A+B\left(\overline{\varepsilon_{pl}}\right)^{n}\right]\left[1+C\ln\left(\frac{\dot{\overline{\varepsilon_{pl}}}}{\dot{\overline{\varepsilon_0}}}\right)\right] \tag{1}$$

In the formula, σ is Mises equivalent stress, A, B, n and C are material parameters, $\bar{\varepsilon}_{pl}$ are equivalent plastic strain, $\dot{\overline{\varepsilon_{pl}}}$ and $\dot{\overline{\varepsilon_0}}$ are equivalent plastic strain rate and reference strain rate, respectively.

When the damage parameter D = 1, the material is considered to have failed.

$$D=\sum\left[\frac{\overline{\varepsilon_{pl}}}{\left(\overline{\varepsilon_{pl}}\right)_f}\right] \tag{2}$$

$$\left(\varepsilon_{pl}\right)_f=\left[d_1+d_2\exp(d_3\sigma^*)\right]\left[1+d_4\ln\left(\frac{\dot{\overline{\varepsilon_{pl}}}}{\dot{\overline{\varepsilon_0}}}\right)\right] \tag{3}$$

$$\sigma^*=\frac{p}{\sigma_e} \tag{4}$$

Where D is the degree of damage, $\Delta\varepsilon_{pl}$ is the increment of equivalent plastic strain, σ* is the average stress from the equivalent stress, p and σe are the impact pressure and the equivalent stress, respectively. d_1, d_2, d_3 and d_4 are the Johnson-Cook constants of aluminum alloys.

2.2 *Finite element model*

Based on the experimental framework of Reference (Qin et al., 2020), this study used ABAQUS/Explicit to establish a low-velocity impact dynamic simulation model of honeycomb sandwich structure (as shown in Figure 1). The composite structure consists of three functional components: An upper panel composed of AL2024 aluminium alloy (180 mm × 180 mm × 2 mm), a honeycomb core layer consisting of AL5052 aluminium alloy (cell inner diameter: 4.76 mm; layer height: 10 mm), and a lower panel constructed from AL2024 aluminium alloy (with geometric dimensions matching the upper panel). The impact load is applied by a rigid body punch with a geometric configuration consisting of a cylindrical substrate with a diameter of 25 mm and a hemispherical end with a radius of curvature of 40 mm.

Table 1. Material parameter.

Material	A (MPa)	B (MPa)	n	m	D1	D2	D3	D4	$\dot{\overline{\varepsilon_0}}$
AL2024	251	200	0.5	1.0	0.116	0.211	-2.2	0.12	$1.1e^{-3}$
AL5052	275	180	0.37	1.0	0.13	0.13	-1.5	0.11	$4.2e^{-4}$

Figure 1. Finite element model.

In terms of material constitutive model construction, the Johnson-Cook plasticity criterion mentioned above is used to characterize the mechanical response. The material parameters of the panel and the core are shown in Table 1. The upper and lower panels are made of aluminum alloy AL2024, and the honeycomb core layer is made of aluminum alloy AL5052.

In terms of finite element modeling, this study uses S4R reduced integral shell element to discretize the upper and lower panels and honeycomb core structure. The impact punch is defined as a discrete rigid body and discretized into R3D4 rigid shell element. The boundary conditions are set according to the actual experimental conditions. The clamped boundary conditions are established for all the nodes on the boundary of the honeycomb sandwich plate. The punch only retains the translational freedom in the vertical direction, and the punch mass is given 34.45 kg. The initial velocity of the punch is defined as 7m/s, and the impact energy is 844.025 J. The connection between the panel and the core layer adopts the Tie constraint algorithm of Abaqus/Explicit, and the complete displacement coordination between the panel and the core layer is realized by the definition of the master-slave surface. Considering the influence of mesh density on the simulation results, local mesh refinement is implemented in the contact area to ensure the accuracy of contact stress calculation and control the calculation cost.

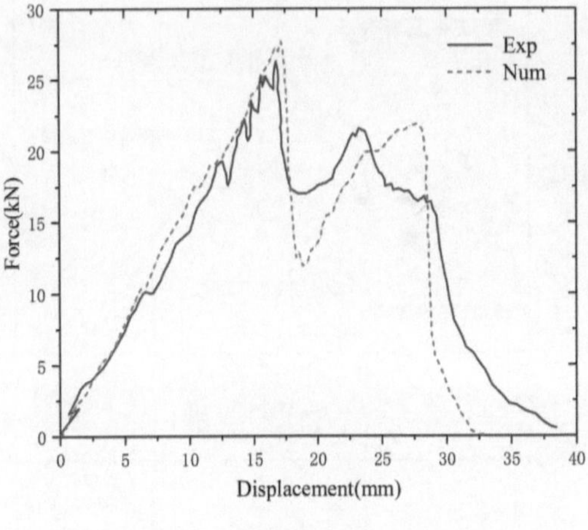

Figure 2. Comparison of force-displacement curves.

2.3 *Numerical simulation method validation*

Using numerical simulation, the breakdown process of honeycomb sandwich panel can be analyzed intuitively. In order to verify the accuracy of the simulation method, the numerical simulation results are compared with the experimental results in Reference (Qin et al., 2020).

The equipment used in the experiment is DHR-1205 drop hammer impact tester. The maximum height of the drop hammer is 28 m, the maximum impact velocity is 23.5 m/s, the mass of the drop hammer is 15 ~ 1000 kg, and the maximum impact energy is 250 kJ.

The numerical model strictly follows the experimental configuration parameters (Qin et al., 2020). The diameter of the punch is 25 mm, and the radius of the ball arc surface is 40 mm. The impact mass is 34.45 kg, the impact velocity is 7m/s, and the impact energy is 844.025 J. The effective area of the honeycomb sandwich panel is 180mm × 180mm, the thickness of the upper and lower panels is 2mm, the height of the core layer is 10mm, the inner diameter of the honeycomb structure is 4.76mm, and the wall thickness is 0.05mm.

The simulation and experimental comparison of the loading force-displacement curve of the honeycomb sandwich panel is shown in Figure 2. It can be seen from the diagram that the change trend of the impact force with the displacement obtained by the simulation and the experiment is basically the same, and the peak value of the loading force is also very similar. The comparison between the simulation results and the experimental results is shown in Table 2. It can be seen from the table that the error of the first impact force peak (F_{m1}) is 7.9 %, the error of the second impact force peak (F_{m2}) is 1.4 %, and the energy absorption error is 14.0 %. Through comparative analysis, it can be seen that the numerical simulation method can be used to predict the breakdown response characteristics of honeycomb sandwich panels, and the results have certain accuracy.

Table 2. Structural response error.

	The first peak force peak	The second peak force peak	Energy absorption
Experiment	25.5 kN	21.7 kN	426.1 J
Simulation	27.7 kN	22.0 kN	366.8 J
Error	7.9%	1.4%	14.0%

3 INFLUENCE OF PUNCH SHAPE

In order to study the influence of punch shape on the damage mode and dynamic response of honeycomb sandwich panels, hemispherical, wedge and cylindrical punches were selected to carry out numerical simulation of honeycomb sandwich panel breakdown. In the simulation, the impact velocity is 7m/s, the punch mass is 34.45 kg, the impact energy is 844.025 J, and the sandwich panel is penetrated. The size of the punch is shown in Figure 3.

3.1 *Impact force characteristics*

As illustrated in Figure 4, the time history curve of impact force under the action of a hemispherical punch can be divided into four distinct stages. In the initial stage, the upper panel exhibits overall bending and local depression, while the impact force increases linearly to a peak. Subsequently, the rupture of the upper panel results in a pronounced decline in the impact force. In the third stage, the honeycomb core

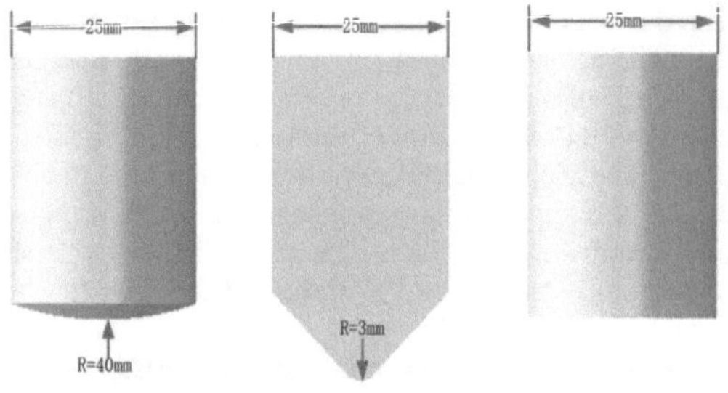

Figure 3. Punch shape and size.

layer undergoes compaction, leading to a double increase in the impact force. Finally, in the fourth stage, the bending and tensile failure of the lower panel occurs, and the impact force decreases sharply.

The impact force time history curve, under the action of a hemispherical punch, can be divided into four typical stages. In the first stage, the upper panel exhibits overall bending and local depression, and the impact force increases linearly to a peak. In the second stage, the rupture of the upper panel leads to a sharp decrease in the impact force. In the third stage, the honeycomb core layer is compacted, and the impact force rises twice. Finally, in the fourth stage, the bending and tensile failure of the lower panel occurs, and the impact force decreases sharply.

The time history curve of the impact force under the action of the wedge punch can be divided into five typical stages. In the first stage, the upper panel undergoes overall bending and local depression, forming a “one-shaped” plastic hinge line at the contact area, and the impact force increases linearly to the peak. In the second stage, the rupture of the upper panel leads to a sharp drop in the impact force. In the third stage, the honeycomb core layer is gradually compressed, and the impact force exhibits the platform phenomenon. In the fourth stage, the impact force rises twice after the core layer is compacted. In the fifth stage, the lower panel ruptures and the bearing capacity of the sandwich panel decreases. Subsequently, the lower panel is gradually torn under the action of the wedge slope, forming a ‘mouth-shaped’ break, and the attenuation rate of the impact force becomes slower. It is notable that the impact force attenuation rate increases once the wedge surface has passed completely through the lower panel.

The impact force time history curve under the action of cylindrical punch can be divided into four typical stages. In the first stage, the upper panel appears local depression and overall bending, and the impact force increases linearly to the peak. In the second stage, the upper panel breaks under the combined action of shear and tension, resulting in a sudden drop in impact force. In the third stage, the overall deformation of the lower panel occurs, and the impact force rises twice. In the fourth stage, the lower panel appears shear fracture, and the impact force decreases sharply.

From the deformation and failure diagram, it can be seen that under the action of wedge punch and cylindrical punch, the upper and lower panels of the sandwich panel mainly appear shear failure; under the action of the hemispherical punch, the upper

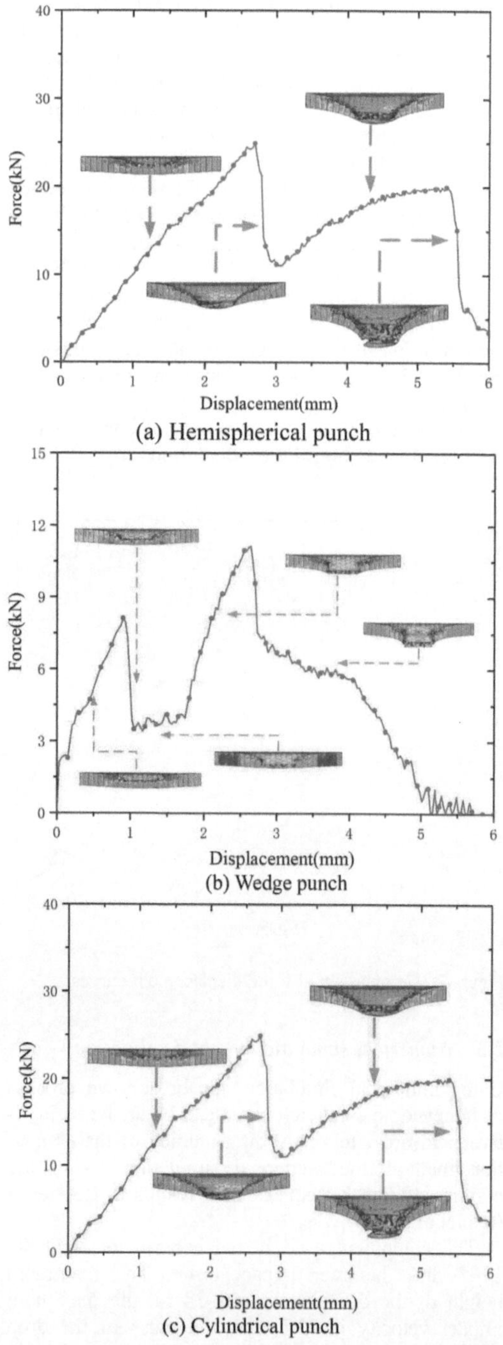

Figure 4. Impact force time history curves under different shapes of punches.

and lower panels mainly appear bending and tensile failure.

3.2 *Loading force-displacement characteristics*

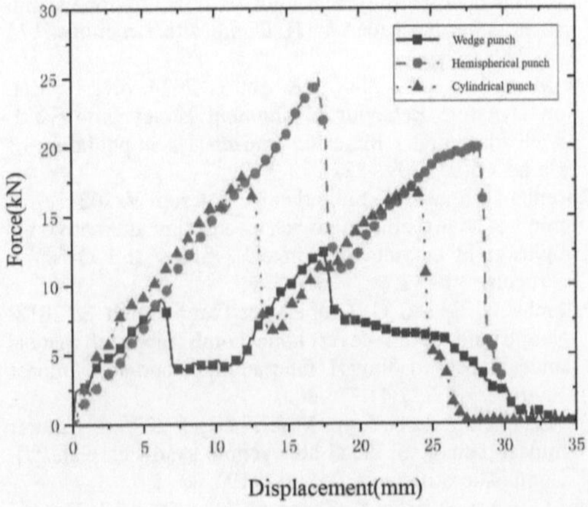

Figure 5. Loading force-displacement curves under different punch shapes.

Table 3. Response comparison under different punch shapes.

punch shape	The first stage peak force F_1 [kN]	The second stage peak force F_2 [kN]	F_1/F_2
Hemispherical punch	27.7	22.0	1.26
Wedge punch	9.0	12.3	0.73
Cylindrical punch	17.6	16.9	1.04

Figure 5 is the loading force-displacement curve of honeycomb sandwich panel under different punch shapes. It can be seen from the loading force-displacement curve that the largest loading stiffness in the first stage is the cylindrical punch, the slightly weaker is the wedge punch, and the smallest is the hemispherical punch. In addition, the curves of the cylindrical punch and the wedge punch have a process of steep increase and then slow down in the first loading stage, while the curve of the hemispherical punch is very smooth. The peak impact force data under different punch shapes are shown in Table 1.From the table, it can be seen that the largest peak impact force is the hemispherical punch, followed by the cylindrical punch, and the smallest is the hemispherical punch. From the F_{m1}/F_{m2} ratio of the first impact force peak and the second impact force peak, the F_{m1}/F_{m2} ratio of the hemispherical punch is > 1, the F_{m1}/F_{m2} ratio of the wedge punch is < 1, and the F_{m1}/F_{m2} ratio of the cylindrical punch is ≈ 1.

3.3 *Energy absorption characteristics*

Figure 6 shows the energy absorption of each part of the honeycomb sandwich panel under the action of three different shapes of punches.

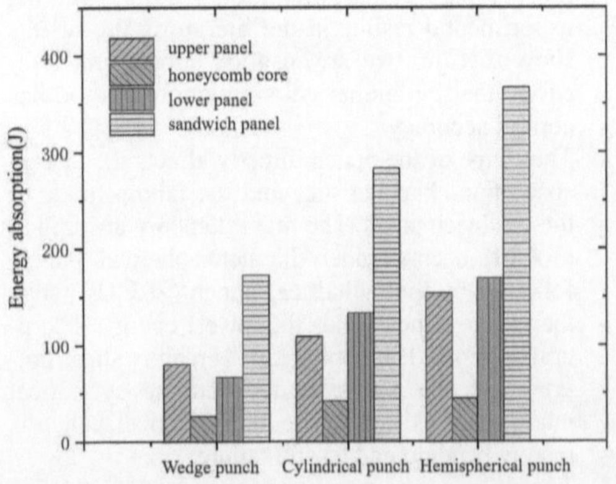

Figure 6. Energy absorption of sandwich plate under different shape punches.

For the overall energy absorption of the sandwich panel, the energy absorption under the action of the hemispherical punch is up to 366.6J, followed by the cylindrical punch is 283.3J, and the wedge punch is at least 172.4J. Under the action of wedge punch, the upper panel mainly produces local depression, and the overall deformation is relatively small. When the local depression reaches a certain degree, the upper panel appears shear failure. Shear failure also occurs after the lower panel has a certain deformation. Compared with the wedge punch, under the action of the cylindrical punch, the upper and lower panels of the sandwich panel are also dominated by shear failure, but the contact area of the cylindrical punch is larger, and the overall deformation of the upper and lower panels is also larger. Therefore, the energy dissipated by the sandwich plate under the action of the wedge punch is less than that under the action of the cylindrical punch. For the hemispherical punch, the overall deformation of the upper and lower panels is large, and the failure is dominated by tensile failure. Therefore, under the action of the hemispherical punch, the sandwich panel absorbs the most energy. Similarly, under the action of the hemispherical punch, the upper and lower panels of the sandwich panel absorb more energy than the other two punches. The honeycomb core layer is mainly compressive energy absorption, and its energy absorption is smaller than that of the panel.

4 CONCLUSIONS

In this paper, the finite element simulation method is used to study the penetration responses of honeycomb sandwich panels. The dynamic response and energy absorption characteristics of honeycomb sandwich panels are analyzed. The effects of panel thickness distribution and punch shape on the penetration performance of honeycomb sandwich structures are explored. The following conclusions can be drawn:

(1) By comparing the simulation results with the experimental results in the literature, the results show that the two are in good agreement, indicating that the numerical simulation method has certain accuracy.
(2) The shape of the punch directly affects the energy absorption characteristics and the failure mode of the sandwich panel. The highest energy absorption (366.6J) occurs under the hemispherical punch, followed by the cylindrical punch (283.3J), while the wedge punch yields the lowest energy absorption (172.4J). The upper panel is mainly shear failure when the wedge punch and the cylindrical punch are used, while the hemispherical punch is mainly bending and tensile failure.
(3) The shape of the punch also has a great influence on the loading and unloading process of the sandwich panel. Under the action of the hemispherical punch and the cylindrical punch, the force-displacement curve shows a double peak, while under the action of the wedge punch, the force-displacement curve appears more stages. In addition to the double peak stage, it also includes the platform stage and the friction energy dissipation stage.

REFERENCES

Pandey A, Singh A, Upadhyay A.K. & Shukla K.K. 2024. An experimental study of multi-core composite aluminium honeycomb sandwich panels under low-velocity impacts[J]. Journal of Sandwich Structures & Materials. 26(6): 887 –905.

Crupi V, Epasto G. & Guglielmino E. 2013. Comparison of aluminium sandwiches for lightweight ship structures: honeycomb vs. foam[J]. Marine structures. 30: 74 –96.

Li S, Li X, Wang E.Z., Wu G, Lu G & Zhao L 2017. Sandwich panels with layered graded aluminum honeycomb cores under blast loading[J]. Composite Structures. 173 (2):242 –254.

Mu M.Y., Guo K.L., Zhou S & Zhu L. 2024. Investigation on Dynamic Behavior of Gradient Honeycomb Sandwich Plate under Repeated Impacts[J]. Shipbuilding of China. 65(05):209 –221.

Palomba G, Epasto G, Sutherland L & Crupi V. 2021. Aluminium honeycomb sandwich as a design alternative for lightweight marine structures[J]. Ships and Offshore Structures. 1 –12.

Palomba G, Epasto G, Crupi V & Guglielmino E. 2018. Single and double-layer honeycomb sandwich panels under impact loading[J]. International Journal of Impact Engineering. 121(1):77 –90.

Qin Q, Chen S, Li K, Jiang M & Zhang J. 2020. Structural impact damage of metal honeycomb sandwich plates[J]. Composite Structures. 252:112719.

Sun G, Chen D, Huo X, Zheng G & Li Q. 2018. Experimental and numerical studies on indentation and perforation characteristics of honeycomb sandwich panels. Composite Structures (1):110 –124.

Yu Y., Fu T, 2025. Dynamic response of the hybrid honeycomb sandwich panel with zero Poisson's ratio under low-speed impact. Mechanics of Advanced Materials and Structures. 31(25). 7024–7038.

Zhang D, Fei F.Q. & Zhang P, 2017. Drop-weight impact behavior of honeycomb sandwich panels under a spherical impactor[J]. Composite Structures (3):633 – 645.

Zhang D, Jiang D, Fei Q & Wu Q. 2016. Experimental and numerical investigation on indentation and energy absorption of a honeycomb sandwich panel under low-velocity impact[J]. Finite Elements in Analysis and Design (1):21 –30.

Ultimate strength of ship structures and components

Trends in Collision and Grounding of Ships and Offshore Structures – Zhu & Guedes Soares (Eds)

Response of the side structure of a fishing vessel under compressive load of floating ice

Y.B. Chen
Key Laboratory of High Performance Ship Technology (Wuhan University of Technology), Ministry of Education, Wuhan, P.R. China

L. Zhu
School of Naval Architecture, Ocean and Energy Power Engineering, Wuhan University of Technology, Wuhan, P.R. China

ABSTRACT: Due to the rich fishery resources in Antarctica, more ocean-going fishing vessels navigate to the Antarctic for krill and other aquatic products, and the structural safety of fishing vessels under ice load has become increasingly prominent. In order to evaluate the strength capacity of the side structure of an Antarctic krill ship under ice load, a series of FE calculations are carried out under ice load. First, based on the requirements of IACS-PC7 ice class specification, the ice load in the floating ice compressive scene are determined, evaluation is made to check whether the structural strength met the requirements of the specification. On this basis, the ice load is increased to 4 times the elastic limit load, so that the plastic response area of the structure is further enlarged. The study provides useful information for the evaluation of the ultimate strength of Antarctic krill ship's side structure under the transverse compression of ice load.

1 INTRODUCTION

Due to the abundant Antarctic krill resources in the Antarctic, an increasing number of ocean-going fishing vessels are being deployed to conduct fishing operations in these waters. While the exploitation of Antarctic pelagic fishery resources presents substantial opportunities, it is also accompanied by significant risks and challenges. The Antarctic marine environment contains extensive floating ice, which poses severe threats to marine structures through collisions and compressive forces. Historical records of human activities in ice-covered waters reveal numerous incidents where marine structures suffered damage due to ice impacts. Consequently, it is imperative to investigate the structural load-bearing capacity of Antarctic krill fishing vessels under ice load conditions.

Regarding the structural response under ice load conditions, Qi (2014) employed a nonlinear static finite element method to evaluate the midship structural strength under both Polar Class-specified ice loads and potential ship-ice interaction scenarios. They further introduced strength acceptance criteria to analyze the ultimate load-bearing capacity of plates and stiffeners. Zhu (2015, 2016, 2017) conducted a series of ship-ice collision experiments and numerical studies, analyzing energy conversion mechanisms during impact processes. The research proposed ice load estimation methodologies and established structural dimension design curves for ice strengthened components under ice loading conditions. Geng (2019) focused on polar crude carriers, developing a direct calculation methodology for polar vessel structures under ice loads through comprehensive analysis of current ice-class specifications. Cai and Zhu (2020) performed comprehensive theoretical, experimental, and numerical simulation studies on the elastoplastic dynamic response of stiffened panel structures under ice loading. Their research revealed the effects of ice failure patterns on structural plastic deformation and damage, and further examined energy transfer mechanisms during structure-ice collision processes. It can be seen that some research on the structural response under ice loading has been carried out at present. However, there are relatively few studies on the response of ship structures with lower ice class design and weaker structural strength under ice loading, such as Antarctic krill fishing vessels.

This paper focuses on the structural response of a Antarctic krill vessel under the action of floating ice compression loads. It should be noted that the target vessel was designed with an ice notation of

DOI: 10.1201/9781003684404-43

CCS-B3, equivalent to o FSICR-IC. Previous research demonstrated that the load-bearing capacity complied with the requirements of the FSICR under IC ice class loads. To investigate the structural performance under more severe conditions, this paper further analyzed the load-bearing capacity trends of the target vessel's side structures under PC7 ice class loads. Based on the IACS ice class specifications and the vessel's parameters, the ice load panel parameters under the floating ice compression scenario were determined, and the strength assessment of the side structure under the design load specified by the regulations was conducted. Additionally, to further investigate the ultimate load-bearing capacity of the structure, the load amplitude was increased to four times the elastic limit while keeping the load panel size unchanged. The ultimate load-bearing capacity of the structure was analyzed in conjunction with the polar vessel ultimate load criteria.

2 SIMULATION

2.1 *Finite element model(FEM)*

This paper taked a krill fishing vessel in the Antarctic as the research object. Considering that the #87 and #128 positions were transverse bulkheads and there were no transverse girders or bulkheads to support the middle compartment section with a span of 40 frames (frame spacing was 0.6m), structural damage was likely to occur when subjected to ice loads. Therefore,this compartment section was selected for strength analysis in this study. Additionally, considering that the design waterline of the target vessel was positioned between the main deck and the platform deck, and ice compression scenarios typically occur near the design waterline, modeling the entire hull along the ship's depth direction would result in unnecessary computational resource consumption. Therefore, to balance computational accuracy and efficiency, the longitudinal modeling range was defined from #87 to #128, the transverse direction extended from the shipside to the hull centerline plane in longitudinal, and the depth direction covers from the upper deck to the inner bottom plate, as illustrated in Figure 1 and Figure 2. The coordinate system aligned with the hull coordinate system.

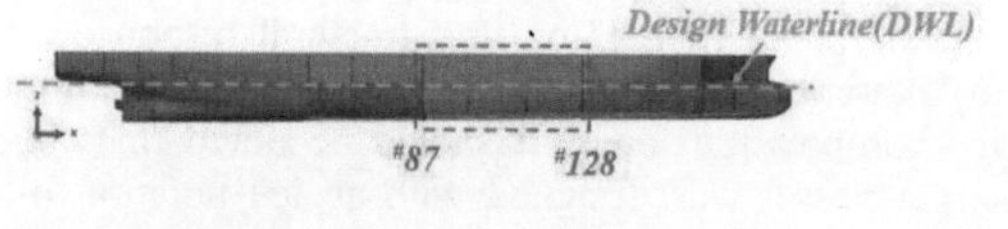

Figure 1. Hold section modeling domain.

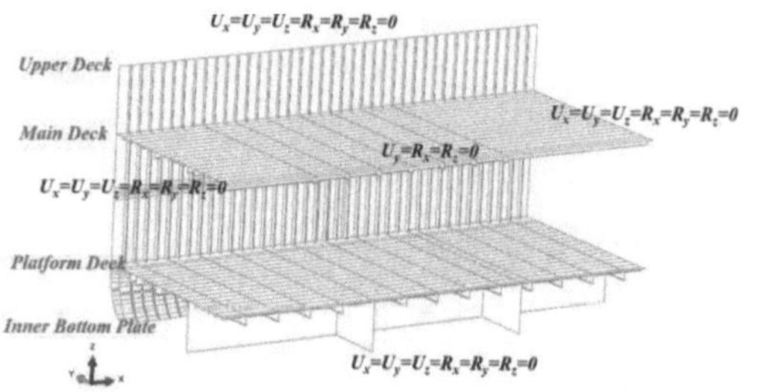

Figure 2. Hold section model schematic.

The target vessel was constructed using AH36 steel. Therefore, the Young's modulus was assigned as 206 GPa, Poisson's ratio as 0.3, material density as 7850 kg/m³, and yield strength as 355 MPa in simulation.

The side shell, frames, and deck plates were simulated using S4R elements, which enables effective capture of local deformation modes. To achieve precise compu`tational results, a base mesh size of 100mm was adopted, with local refinement to 20mm in loading zones.

Prior to conducting FE analysis, localized boundary conditions must be applied to the model. For the established FEM:

- In the X-direction: Both ends were constrained by transverse bulkheads with significant stiffness, therefore fixed support boundary conditions were applied to simulate their restraining effects;
- In the Y-direction: No constrain was imposed at the shipside, while symmetry boundary conditions were applied at the hull centerline;
- In the Z-direction: The +Z end (upper deck) and -Z end (inner bottom plate) exhibit substantial rigidity, hence both wre modeled with fixed supports.

The application positions of boundary conditions were as shown in Figure 2.

2.2 *Ice load*

For Antarctic krill vessels as ice-strengthened ships of low ice class, the ultimate bearing capacity of the side shell structure under floating ice compression scenarios was evaluated based on the IACS Polar Class 7 (PC7) requirements. In accordance with IACS Ice Class Rules (2024), the hull was longitudinally divided into four regions: bow area, bow transition area, midship area (Mi), and stern area. Each longitudinal region was further subdivided vertically into three zones: bottom zone, lower zone, and ice belt zone, as delineated in Figure 3. The computational domain in this study specifically corresponds to the Mi area.

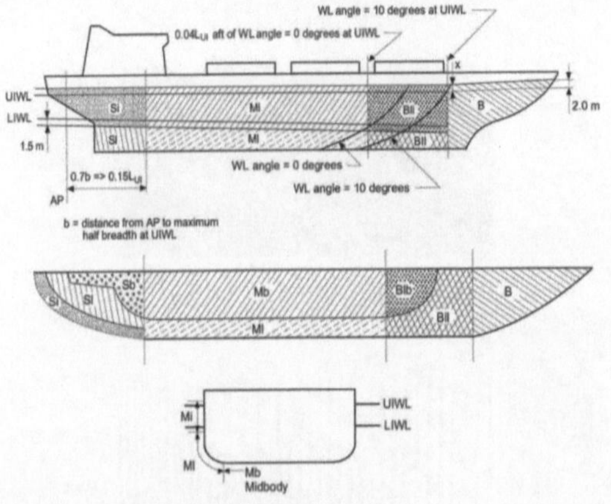

Figure 3. IACS hull zoning schematic.

The design ice load in IACS rules was represented by rectangular load patches with uniformly distributed pressure (magnitude *P*, height *b*, and length *w*). According to the requirements for the Mi area, the parameters were determined as *P* was 2.28 *MPa*, *b* was 0.587 m, and *w* was 2.115 m. Additionally, while maintaining the load patch dimensions, the pressure magnitude was further increased to 4 MPa to investigate the structural bearing capacity under more critical floating ice compression scenarios. The load patch parameters were detailed in Table 1.

Table 1. Load patch parameters.

No.	*P(MPa)*	*b(m)*	*w(m)*
Case1	2.28	0.587	2.115
Case2	4.00	0.587	2.115

Considering the higher probability of ice impact near the waterline vertically, the loading application was vertically positioned at the waterline. In the longitudinal direction, it was located at the midsection of the compartment, where the structural stiffness was minimal compared to other regions, rendering it relatively vulnerable. The loading zone configuration was as illustrated in Figure 4.

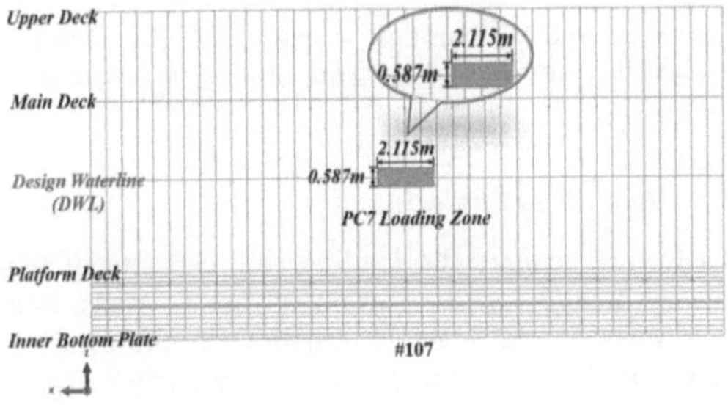

Figure 4. Loading zone schematic.

The present study employed the Riks method within the ABAQUS/Standard solver to resolve structural responses. This approach demonstrated superior convergence in static nonlinear problems. During the computational process, the solver incrementally applied the load from the specified initial magnitude to the target load, with structural response evaluation conducted at each iteration step. Consequently, while the specified ice load magnitude was 4 MPa, structural responses under progressively increasing load magnitudes below 4 MPa were concurrently obtained through this incremental procedure.

3 RESULTS AND DISCUSSION

3.1 *Stress and deformation responses*

The stress results for different loading cases were presented in Figures 5 and 6. From the stress nephograms, the following observations were derived:

1. When subjected to the design ice load magnitude of 2.28 MPa, plastic region was exclusively initiated adjacent to the loading zone for the side shell plating. As the load escalated to 4 MPa, the plastic region undergoed substantial expansion, propagating continuously until it reached the deck structures at both the upper and lower extremities.
2. Due to the presence of transverse girders in the deck arrangement, only a localized plastic region emerged at the direct interface with the side shell. Furthermore, the deck significantly enhanced the transverse load-bearing capacity of the side shell structure, thereby preventing the plastic zone from extending further into the upper deck or inner bottom plate.

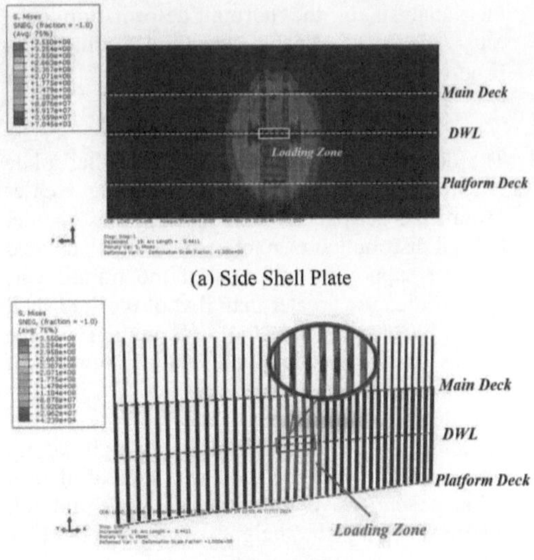

(a) Side Shell Plate

(b) Side Frames

Figure 5. The stress response of the primary load-bearing structure under case1 (loading amplitude of 2.28 MPa).

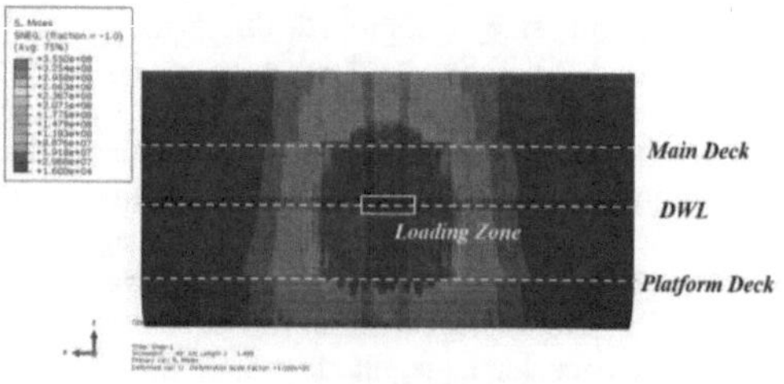

(a) Side Shell Plate

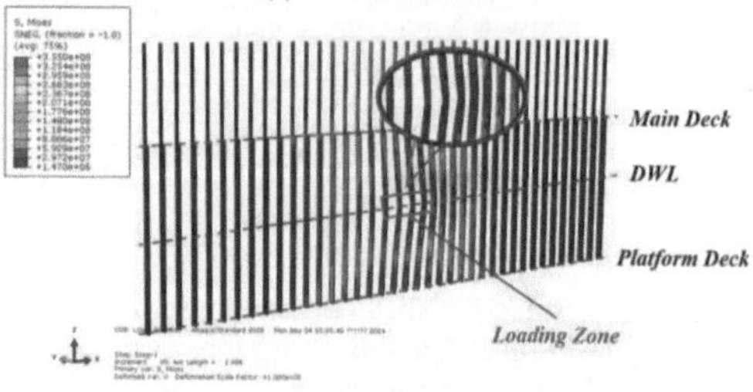

(b) Side Frames

Figure 6. The stress response of the primary load-bearing structure under case2 (loading amplitude of 4 MPa).

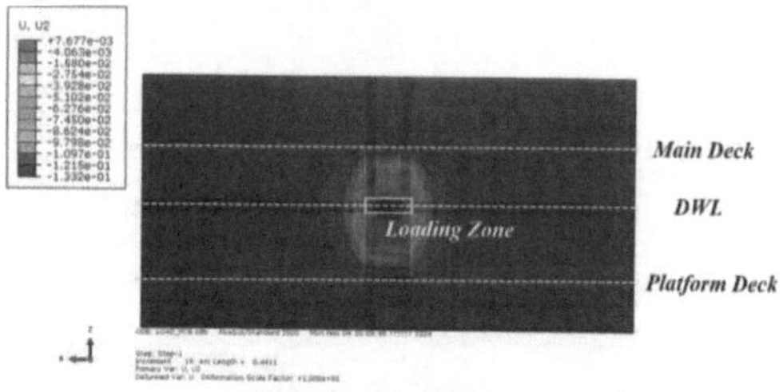

(a) Side Shell Plate

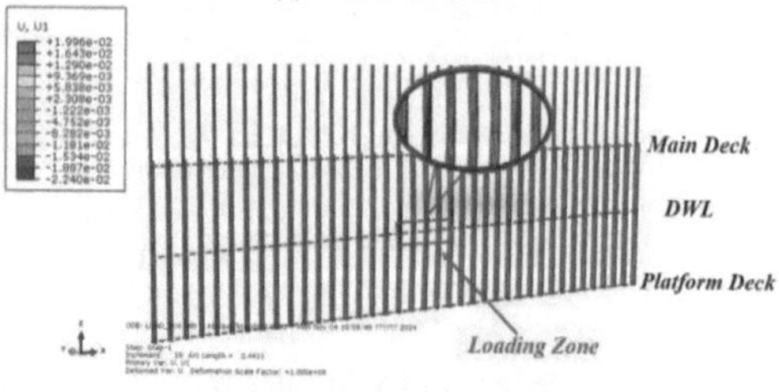

(b) Side Frames.

Figure 7. The deformation response of the primary load-bearing structure under case1 (loading amplitude of 2.28 MPa).

3. From the stress contour plot of the stiffeners, it can be observed that the frames between the main deck and the platform deck, in addition to entering the plastic response stage in the central loading region, also exhibit stress responses reaching the yield stage at their ends near the main deck and platform deck. This indicated that the decks provide significant support stiffness, resulting in substantial bending moments at the ends of the stiffeners, thereby generating high stress responses.Further analysis of the normal deformation contour plots of each structural component (Figures 7 and 8) reveals that:

1. The deformation response of the side shell plate radiated uniformly from the loading center toward the periphery, without exhibiting distinct banded distribution characteristics. This indicated that the supporting stiffness of the frames was not significantly greater than that of the side shell plating. Consequently, the deformation of the frames and the side shell plating remain at a comparable level.
2. The deformation mode of the frame web was predominantly characterized by overall lateral tilting. Even when the load was increased to 4 MPa, a semi-wave deformation mode did not manifest along the height of the web. This indicated that the ice load did not induce buckling in the frame.

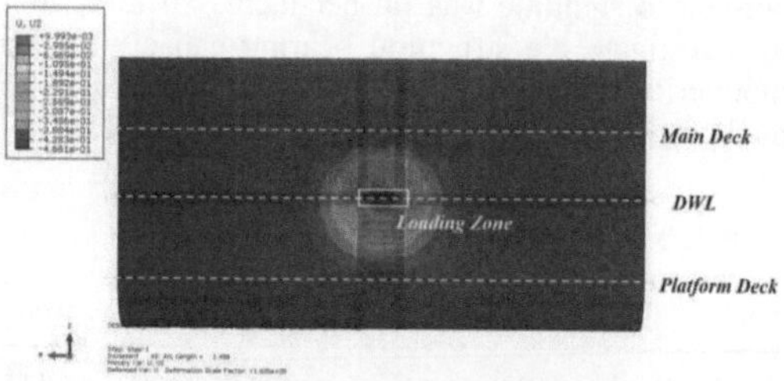

(a) Side Shell Plate

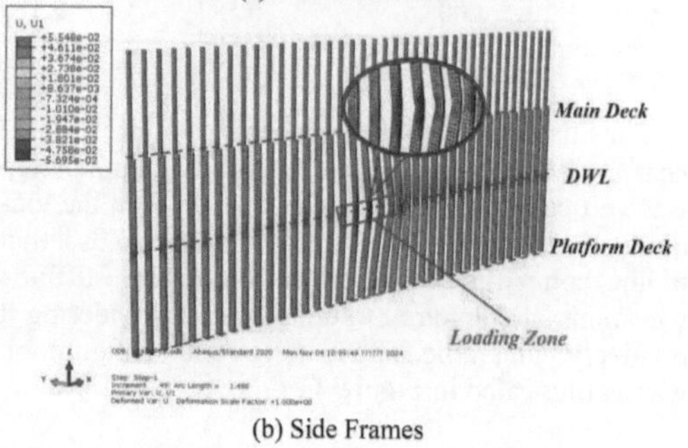

(b) Side Frames

Figure 8. The deformation response of the primary load-bearing struc-ture under case1 (loading amplitude of 4 MPa).

3.2 *Limit load analysis*

To further analyze the load-bearing capacity of the side structure under ice compression loads, the deformation and load information in the loading center region was extracted from the computational results based on the characteristics of the Riks method. The load-deformation curve was plotted, and the limit load of the structure was analyzed in accordance with the limit load criterion.

Figure 9 presented the load-displacement curve under the scope of PC7 ice load patch. It was observed from the figure that when the Y-direction deflection of the side structure reached 0.029 m, the structure entered the plastic response phase. The corresponding load at this point was 0.88 MPa, indicating that the elastic limit of the structure was 0.88 MPa. As the deformation further increased, the curve exhibited a trend of initially concave growth followed by convex growth:

1. During the concave growth phase, the load required to produce unit deformation first decreased and then increased. At this stage, the plastic response zone of the structure expanded from the loading area to adjacent panels.
2. In the convex growth phase, the load required to produce unit deformation first increased and then decreased. During this stage, all frames between the main deck and the platform deck, as well as those between #106 and #108, entered the plastic response phase, while the plastic zone continued to expand outward.

From the load-deformation curve, it was observed that even when the compressive load increased to 4 MPa, reaching four times the elastic limit, the curve did not exhibit a turning point. The side structure did not experience buckling instability, and there was no rapid decline in the load-bearing capacity of the structure. To determine the ultimate load within the safety limits, this study applied the commonly used ultimate load criterion for polar vessels: Krylov method (2009), the maximum elastic deformation method (3δ method) (2019), and the twice elastic slope method (TES method) (2007). By integrating these methods with the load-deformation curve, the limit load was determined: 1) The Krylov method, proposed by the Krylov Ship Research Institute, was a technique for determining the limit load from the load-deformation curve. The ultimate load was defined as the load corresponding to the intersection point of the tangent line in the elastic phase of the load-deformation curve and the tangent line at five times the maximum elastic deformation. 2) The maximum elastic deformation method stipulated that the deformation corresponding to the maximum load should not exceed three times the maximum elastic deformation. 3) The twice elastic slope method, introduced by ASME, involved drawing a straight line from the origin with a slope relative to the load axis that was twice the slope of the elastic phase of the load-deformation curve relative to the load axis. The load value corresponding to the intersection of this line with the load-deformation curve was termed the ultimate load.

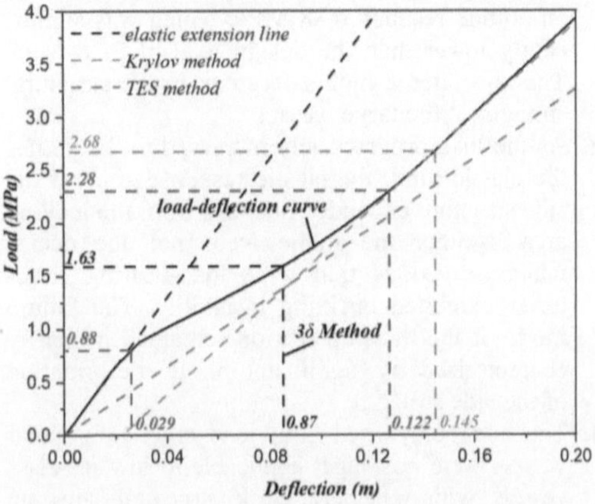

Figure 9. The load-deformation curve under the scope of PC7 ice class.

It should be noted that, as the curve initially underwent a concave growth phase upon entering the plastic region, the auxiliary lines for both the Krylov method and the twice elastic slope method did not intersect with the elastic phase straight line. Consequently, these two limit load criteria were no longer applicable. Based on the maximum elastic deformation method, the ultimate load of the target ship's side structure under the scope of PC7 ice class was determined to be 1.63 MPa. This value was relatively conservative and exhibited a significant discrepancy compared to the design load of 2.28 MPa. Therefore, the commonly used limit load criteria for polar vessels were not necessarily applicable to the strength assessment of Antarctic krill vessels under ice compression loads. Analyzing the ultimate load-bearing capacity of the structure based on the load-deformation curve in conjunction with deformation requirements was deemed more practical and meaningful.

4 CONCLUSIONS

For an Antarctic krill ship which was designed with notation of CCS-B3, the finite element model of the #87~#127 section of the vessel was created. This study defined two loading scenarios under the PC7 ice class loading area: the design load and four times the elastic limit load. Finite element calculations were conducted for these different scenarios. From the structural response results, the load-deflection curve was extracted, and the limit load-bearing capacity of the side structure was analyzed based on the limit load criteria. The main conclusions obtained were as follows:

1. Under the loading conditions of PC7 ice class, which exceeds the target vessel's design ice class of B3, the target ship's side structure entered the plastic response phase when the loading

amplitude reached 0.88 MPa, which was significantly lower than the design load of 2.28 MPa. This indicated a high safety risk in the structural strength of the target vessel.

2. As the load progressively increased to four times the elastic limit, the plastic response zone of the side structure expanded outward from the loading area. Neither the frame webs nor the decks, which served as transverse load-bearing structures, exhibited buckling instability. The failure mode of the midship section remained primarily characterized by significant plastic deformation of the side structure.
3. The commonly used limit load criteria for polar vessels were no longer applicable to low ice-class vessels with relatively weak structural strength, such as Antarctic krill ships. Analyzing the ultimate load-bearing capacity of the structure based on the load-deflection curve and deformation requirements was deemed more practical and meaningful.

REFERENCES

Aksnes V, Bonnnemaire B. 2009. Analysis of the behavior of a moored ship in variable ice drift.*The 20th International Conference on Port and Ocean Engineering Under Arctic Con-ditions*. Lulea, Sweden.

ASME. 2007. Boiler and Pressure Vessel Code.

Cai, W., Zhu, L., Yu T. X, Li, Y.G. 2020. Numerical simulations for plates under ice impact based on a concrete constitutive ice model. *International Journal of Impact Engineering*, 143: 103594.

Geng, T., Zhao, Z.Q., Xu, J., Hao, J.F. 2019. Direct calculation of polar ship ice load and structural strength evaluation. *Ship Engineering*, 41(7): 1–4.

IACS. 2024. *Requirements Concerning Polar Class.*

Qi, K.L., Liu J., Liu S.W., Wang, G. 2014. Strength Assessment of Polar Ships Under Ice Loads Using Direct Calculation Method. *Journal of Ship Mechanics*, 18 (6):662–677.

Yan, M.J., Zhu, L. Study on Dynamic Response of Hull plate Structure under the Impact load of Floating Ice. 2017. *Journal of Wuhan University of Technology (Transportation Science and Engineering)*, 41(2): 268–272.

Zeng, J., Wang, Y.W., Zheng, W.Q., Wu, J. 2019. Extreme load evaluation of ship ice belt components in polar region. *Marine Engineering*, 41(7): 5–11.

Zhu, L., Shi, S.Y., Yu, T.X. A New ice load response model for structural design of ice classed ships. 2015. *The 25th International Ocean and Polar Engineering Conference*, Hawaii, America.

Zhu, L., Qiu, X.M., Chen, M.X, Yu, T.X. 2016. Simplified Ship-Ice Collision Numerical Simulations. *The 26th International Ocean and Polar Engineering Conference*, Rhodes, Greece.

Zhu, L., Cai, W., Chen, M.S., Tian, Y.K. 2020. Experimental and numerical analyses of elastic-plastic responses of ship plates under ice floe impacts. *Ocean Engineering*, 218: 108174.

Trends in Collision and Grounding of Ships and Offshore Structures – Zhu & Guedes Soares (Eds)

Study of the ultimate strength of the ship hull considering random collisions

Y. Ding, Q. Liang & L. Zhu
Key Laboratory of High Performance Ship Technology (Wuhan University of Technology), Ministry of Education, Wuhan, P.R. China
School of Naval Architecture, Ocean and Energy Power Engineering, Wuhan University of Technology, Wuhan, P.R. China

Z. Wen
School of Naval Architecture, Ocean and Energy Power Engineering, Wuhan University of Technology, Wuhan, P.R. China

ABSTRACT: Due to the increasing ship density in the polar area, it is frequent to suffer from structural dents or small deformations caused by ship or ice floe collisions, which do not affect the continued navigation of the ship. However, when a ship is hit by a collision and further squeezed by other structures or larger floating ice, it is easy to cause structural instability. Therefore, this paper aims to study the ultimate strength of ship side structures with initial collision damage. Numerical simulations of ship side structures were performed for different impact speeds, masses, and positions to obtain collision damage across scenarios. Then, nonlinear finite element method was used to determine the ultimate strength of structures. Finally, based on extensive numerical results, a new empirical formula is developed for the residual ultimate strength of the stiffened structures, and the random distribution of ultimate strength under impact loads is determined.

1 INTRODUCTION

In recent years, ultimate limit state design methods for ships and marine structures are gaining attention in structural design. Zhu et al. (2022) studied several fitting methods for damaged ships' residual strength and proposed an optimized solution for longitudinal strength assessment. Xu et al. (2018) used regression analysis to construct an empirical formulation for combined loads. Paik & Thayamballi (1997) developed an empirical formula for the ultimate strength of longitudinally stiffened panels under uniaxial compression. Paik et al. (2001) and Harada et al. (2007) established different empirical formulas for the ultimate strength of stiffened panels under biaxial compression and lateral pressure, respectively. Zhang & Khan (2009) explored the buckling of panels and stiffened panels under axial compression and proposed a new semi-analytical formulation for ultimate strength.

Initial imperfections could occur on ship structures when encountering random collision loads, thus, the ultimate strength of stiffened panels with initial imperfections is extensively investigated. Zhang et al. (2023) developed an empirical formula for the residual strength of dent-damaged stiffened panels at low temperatures. Li et al. (2022) analyzed and proposed empirical formulas corresponding to the ultimate strength of a stiffened panel under axial compression by considering geometric imperfections. Ao et al. (2020) investigated the effect of initial imperfection parameters such as the depth, size, and angle of the dent on the ultimate strength of the structure.

These studies have improved our understanding of structural behavior under deterministic dent damage. However, collision loads in actual navigation conditions have random space and time distributions, so deterministic methods cannot effectively capture the probability distribution of random variables such as collision energy and impact location. A new formula with random collision parameters is developed to predict the ultimate strength of stiffened panels by statistical analysis, which overcomes the limitations of traditional methods and provides support for structural design under random collisions.

2 ULTIMATE STRENGTH ASSESSMENT UNDER COLLISION

A simplified numerical model is firstly established to reproduce the ship collision process by explicit dynamics method (Zhu et al., 2021), and the ultimate strength of the numerical model under collision loads is then calculated by static implicit analysis.

DOI: 10.1201/9781003684404-44

2.1 *Numerical models*

A simplified numerical model of the side structure of a krill fishing vessel is established. The investigated area of the side structure is shown in Figure 1, while the simplified stiffened structure is depicted in Figure 2. The dimensions and material properties of the stiffened structure are presented in Table 1 and Table 2, respectively.

Table 1. Geometry of the stiffened structure.

Items	Parameters	Symbols	Values
Plate	Frame spacing	l/mm	6600
	Thickness	t/mm	18
	Length	L_p/mm	6600
	Width	b_p/mm	2300
Ribs	Web height	h/mm	240
	Thickness	t/mm	10
	Panel width	b/mm	44

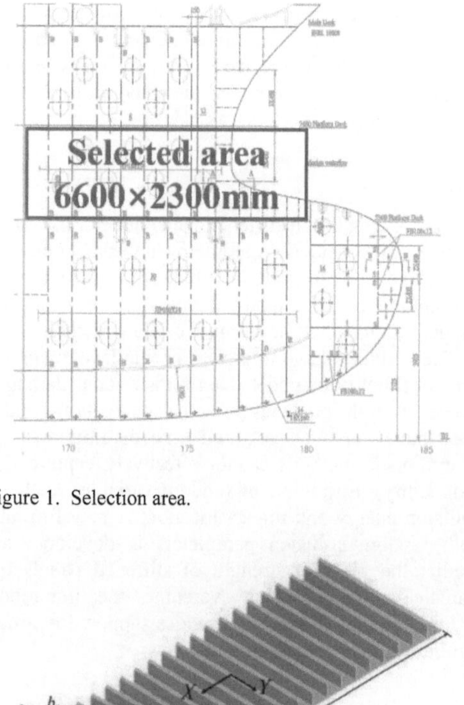

Figure 1. Selection area.

Figure 2. Stiffened structure.

2.2 *Initial imperfections by collision*

Since the boundaries of the side structure are the transverse bulkhead and the deck, the structure is modeled with solid boundary constraints. The plate is 6,600 mm length and 2,300 mm width, with twenty-two L240 × 44 × 10 angle iron. According to the mesh convergence analysis, the optimal mesh size is set to 50 mm × 50 mm, and the collision duration is 1.0-2.0 s to capture the dynamic response fully. A hemispherical punch is assumed to simulate the collision load to avoid stress concentration from geometric changes (Li et al. 2017), as illustrated in Figure 3. Figure 4 shows the deformation cloud of the structure when it is collided at its center by the punch with a speed of 3.0 m/s and a mass of 300 kg, and the maximum deformation is 26.2 cm in this collision scenario.

Table 2. Material parameters of the stiffened structure.

Density /kg·m^{-3}	Young's modulus /GPa	Yield stress /MPa	Poisson's ratio	tangent modulus /MPa
7800	206	355	0.3	206

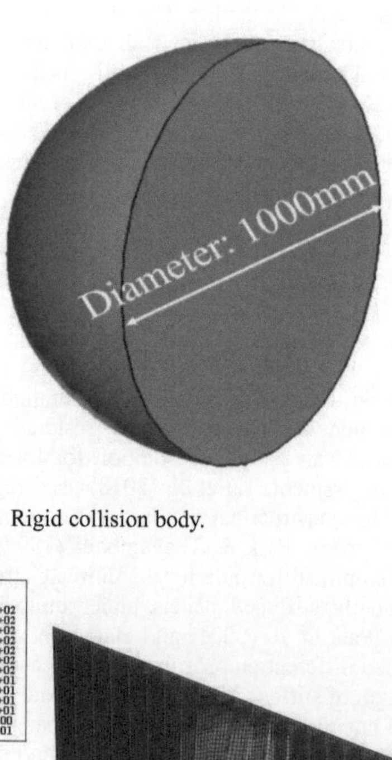

Figure 3. Rigid collision body.

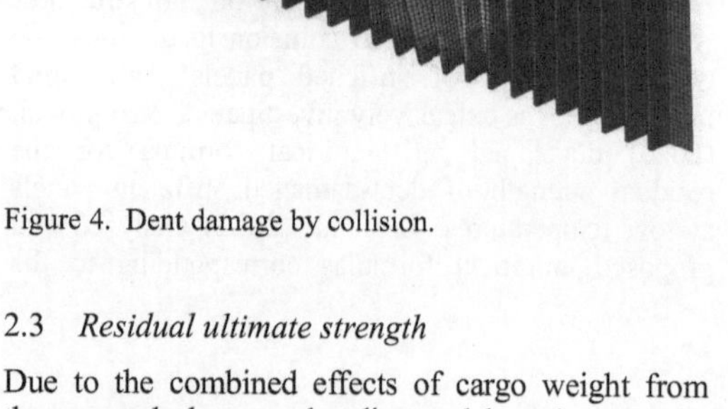

Figure 4. Dent damage by collision.

2.3 *Residual ultimate strength*

Due to the combined effects of cargo weight from the upper deck, wave bending and lateral water, the ship side is susceptible to combined loads of biaxial compression and lateral pressure. The stiffened structure on ship side only has transverse ribs, resulting in a weaker longitudinal resistance. Thus, the structure is more prone to buckling and structural failure when longitudinal compressive loads are dominant.

According to the IACS (2021) buckling capacity of overall stiffened structure, the stress multiplier factor ($\gamma_{GEB,\ bl}$) reaches its maximum value when the longitudinal to transverse loading ratio is 5:1for this structure. Therefore, in this study, the ratio of longitudinal to transverse compression is set to 5:1, and a lateral pressure of 0.066 MPa based on a draft of 6.60 m is applied.

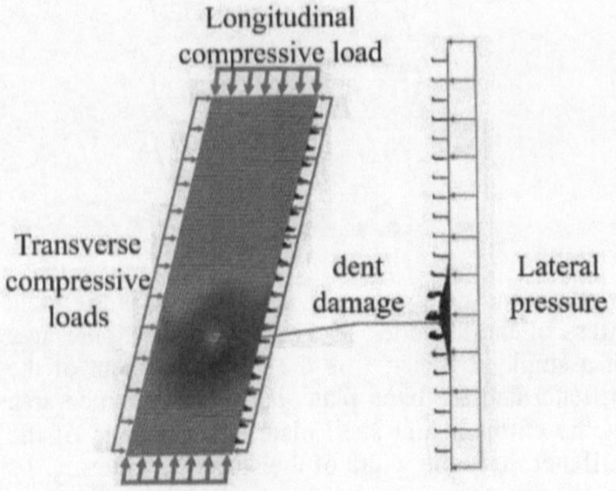

Figure 5. Load schematic.

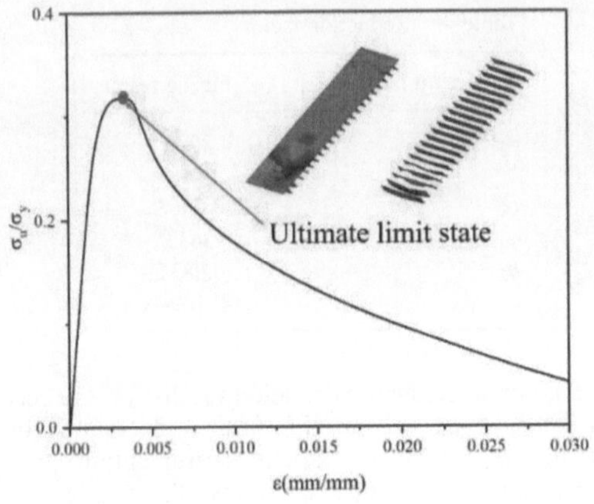

Figure 6. Load-shortening curve.

As shown in Figure 5, the longitudinal, transverse and lateral directions are parallel to the length, width and normal direction of the stiffened panel, respectively. The dent caused by the collision load is imported as an initial imperfection for the ultimate strength analysis. The load-shortening curve of the stiffened structure in a typical case subjected to axial compression and lateral pressure is attached in Figure 6.

3 ULTIMATE STRENGTH PREDICTION UNDER RANDOM COLLISIONS

3.1 *Effect of collision factors*

Figure 7 Shows the effect of collision variables on the ultimate strength of the stiffened panel. The collision speed and mass greatly reduce the ultimate strength of the structures, as shown in Figure 7(a) and Figure 7-(b), and the ultimate strength of the structures decreases with the increasing of the collision speed and collision mass. Figure 7(C) and Figure 7(d)< show the influence of the impact position on the residual ultimate strength of the structure along longitudinal and transverse directions, respectively. It is observed that the residual ultimate strength increases as the impact position moves away from the center of the structure, and it is more sensitive to the position changes in the longitudinal direction.

Table 3. Sampling range of variables.

variable	Unit	Lower limit value	Upper limit value
Speed (*V*)	mm/s	3000	5000
Mass (*M*)	kg	10	500
Position (*X*)	mm	0	3300
Position (*Y*)	mm	0	1150

The range of the collision speed in Table3 refers to the navigation speed of the krill vessel in ice-covered areas. The coordinates of *X* and *Y* indicate the longitudinal and transverse distances from the collision position to the center of the stiffened structure, respectively, as shown in Figure 2. The collision ice block is simplified as a semi-spherical rigid body, the collision mass is set between 10 and 500 kg which refers to the weight of floating ice in polar areas, while the impact speed is about the design velocity of the vessel in polar region. This approach can effectively improve sampling efficiency and provide reliable data for follow-up simulation and formula fitting.

3.2 *Prediction of ultimate strength*

An empirical formula for the ultimate strength of a stiffened panel considering dent damage is described by Zhang et al (2023), and an improved equation is proposed as in Eq.(1) in this paper. However, the measurement of the collision damage could introduce errors when assessing structural ultimate strength, so the effect of collision parameters is considered directly in the proposed formula.

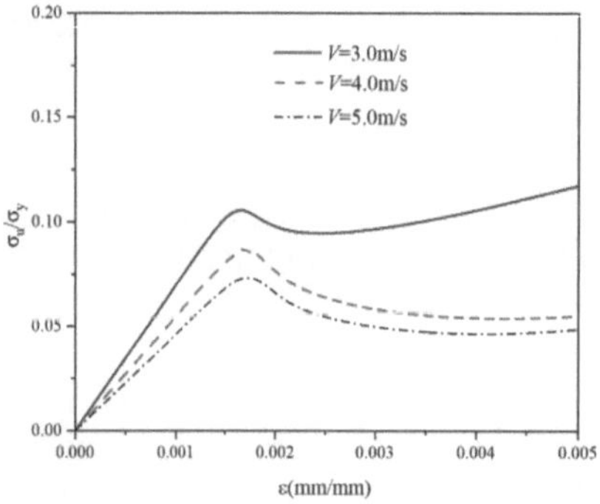

(a) Collision speeds.

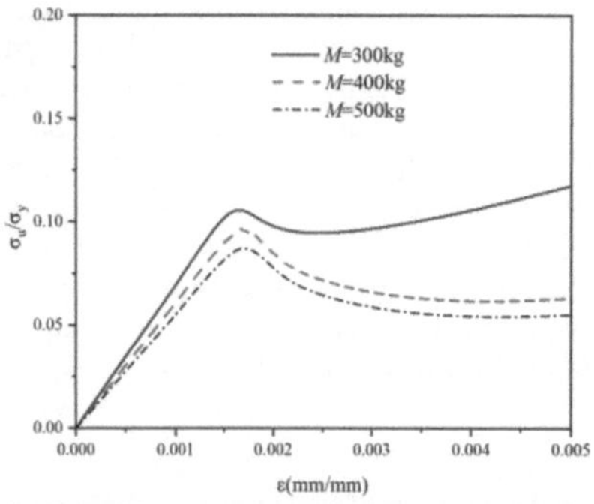

(b) Collision masses.

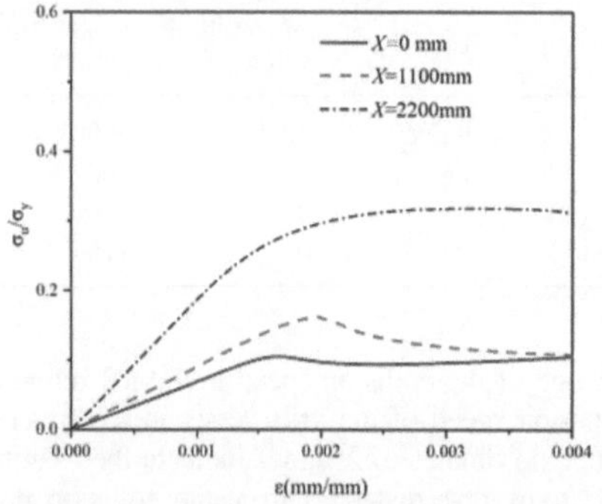

(c) Impact position along longitudinal direction(Y=0)

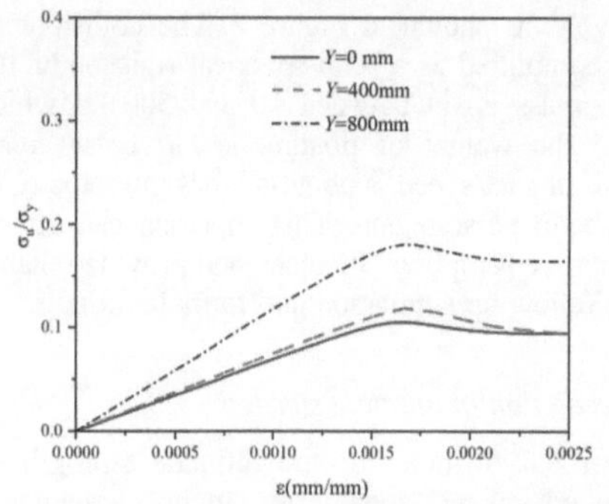

(d) Impact position along transverse direction(X=0).

Figure 7. Load-shortening curve for different collision parameters.

3.3 *Collision cases*

In this section, 250 groups combinations of the collision variables are generated randomly by the Latin Hypercube Sampling (LHS) method, and the ranges of the variables in the generated scheme are shown in Table 3.

The collision energy and the collision position have the most important influence on the collision damage as well as the residual ultimate strength, so these parameters are incorporated in Equation (1).

$$\frac{\sigma_u}{\sigma_y} = \frac{1}{\beta_c^{\ 0.28}} \frac{1}{\sqrt{1+\lambda_c^{\ 3.2}}} \tag{1}$$

where,

$$\beta_c = \beta\left(\frac{MV^2}{\sigma_y A_0}\right)^{a_1}\left(\frac{X}{0.5b_p}\right)^{b_1}\left(\frac{Y}{0.5L_p}\right)^{c_1}$$

$$\lambda_c = \lambda\left(\frac{MV^2}{\sigma_y A_0}\right)^{a_2}\left(\frac{X}{0.5b_p}\right)^{b_2}\left(\frac{Y}{0.5L_p}\right)^{c_2}$$

$$\beta = \frac{b}{t}\sqrt{\frac{\sigma_y}{E}}, \lambda = \frac{l}{\pi r}\sqrt{\frac{\sigma_y}{E}}, r = \sqrt{\frac{i_s}{a_s}}$$

$$\sigma_y = \frac{bt\sigma_{yp} + A_s\sigma_{ys}}{bt + A_s}$$

σ_{yP} is the yield stress of the plate, σ_{ys} is the yield stress of the stiffener, A_s is the cross-sectional area of a single stiffener, i_s is the inertia moment of the stiffener and the band plate, a_s is the combined area of the stiffener and shell plate, l is the span of the stiffener, b_p is the width of the stiffened structure, L_P is the length of the stiffened structure, A_0 is the cross-sectional area of the whole structure, while a_1, b_1, c_1, a_2, b_2 and c_2 are reduction factors.

Table 4. Reduction factors.

Reduction factors	Fitting results
a_1	0.2626
b_1	-2.3887
c_1	-0.0468
a_2	0.3633
b_2	-0.0831
c_2	-0.1094

The finite element calculation results for 250 sets of collision cases are used to fit the reduction factor in the Eq. (1) by the particle swarm optimization algorithm. After 1,000 iterations, the results are shown in Table 4 and the ultimate strength for stiffened panels under combined biaxial compression and lateral water pressure after random collision loads are presented in Equation (2).

$$\frac{\sigma_u}{\sigma_y} = \frac{1}{\beta_c^{\ 0.28}} \frac{1}{\sqrt{1+\lambda_c^{\ 3.2}}} \tag{2}$$

where,

$$\beta_c = \beta\left(\frac{MV^2}{\sigma_y A_0}\right)^{0.2626}\left(\frac{X}{0.5b_p}\right)^{-2.3887}\left(\frac{Y}{0.5L_p}\right)^{-0.0468}$$

$$\lambda_c = \lambda\left(\frac{MV^2}{\sigma_y A_0}\right)^{-0.0468}\left(\frac{X}{0.5b_p}\right)^{-0.0831}\left(\frac{Y}{0.5L_p}\right)^{-0.1094}$$

the meaning of each parameter is described as previous. Though the current validation is applied to transverse stiffeners, the formula allows for potential adaptation to longitudinal stiffeners when modifying the value of the coefficient for each parameters in Eq.(2) according to the arrangement of the structure as well as its loading distribution.

4 RESULTS AND DISCUSSION

4.1 *Accuracy verification*

The results calculated by Equation (2) and the ones predicted by numerical model are compared in Figure 8. It can be seen that the mean value of the correlation coefficient is 0.993, showing that the empirical formula is fitted with high accuracy.

To examine the effectiveness of the proposed empirical formulae more objectively, the coefficient of determination R^2 is calculated to assess its reliability. It is calculated by,

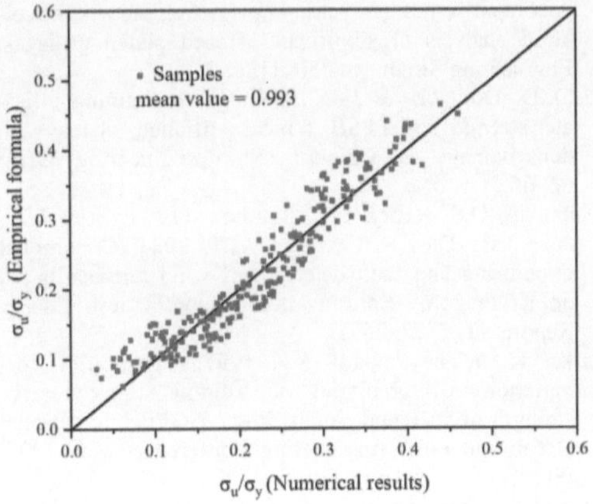

Figure 8. Deviation between numerical method and Equation (2).

$$R^2 = 1 - \frac{\sum_{i=1}^{n}(y_i - \hat{y}_i)^2}{\sum_{i=1}^{n}(y_i - \bar{y})^2} \quad (3)$$

where, y_i represents the actual values, $\hat{y}_i$ represents the predicted values by the formula and $\bar{y}$ stands for the mean value of the actual values.

The R^2 of the empirical formula is 0.9034, meeting the engineering prediction accuracy requirement ($R^2 > 0.9$; Wang et al. 2020).

4.2 *Random distribution discussion*

According to the data generated by LSH, the statistical characteristics can be estimated through the least square method as listed in Table 5, and the random distribution function of the four collision variables can be fitted. The Kolmogorov-Smirnov (K-S) test is performed to validate the distribution characteristics. The results of the K-S test are determined by the test statistic (D) and the p-value. A smaller D and a larger p-value (≥0.05)(Ohunakin et al. 2024) suggest that there is no significant difference, and the null hypothesis is not rejected. To determine the distribution type that best fits the collision variable data, p-values of different hypothetical distributions, such as normal, lognormal, and Weibull, are compared using the K-S test. The distribution with the highest p-value is selected, and the results of the distribution for the collision variables are shown in Table 5.

Table. 5. Random Distribution Characteristics of Variables.

Variable	Mean Value	Standard Deviation	COV	Distribution
V	3956.48	509.12	0.13	lognormal
M	252.13	141.56	0.56	normal
X	1542.02	794.96	0.52	normal
Y	334.47	192.21	0.57	normal

The ultimate strengths calculated by random variables and empirical formulas also have a certain degree of randomness. Therefore, the similar steps as before are used to obtain the distribution characteristics of the ultimate strength. The results demonstrate that the ultimate strength follows a Weibull distribution, as confirmed by the K-S test (statistic D = 0.0265, p-value = 0.478). The data of the calculation result shows a mean value of 0.225 with a standard deviation of 0.0817.

The fitting result has an R^2 of 0.9941, and the observed distribution curve of σ_u/σ_y almost overlaps the fitted Weibull distribution curve, as shown in Figure 9, confirming that the Weibull distribution accurately describes the random distribution characteristics of σ_u/σ_y. Using the least squares method, the estimated shape parameter k is 2.98 and the estimated scale parameter τ is 0.25, the probability density function of the ultimate strength under random collision is expressed as,

$$f\left(\frac{\sigma_u}{\sigma_y}\right) = \frac{k}{\tau}\left(\frac{\sigma_u/\sigma_y}{\tau}\right)^{k-1} exp\left[-\left(\frac{\sigma_u/\sigma_y}{\tau}\right)^{k}\right] \quad (4)$$

Based on the parameters of the Weibull distribution (k = 2.98, τ = 0.25), possible ranges for the

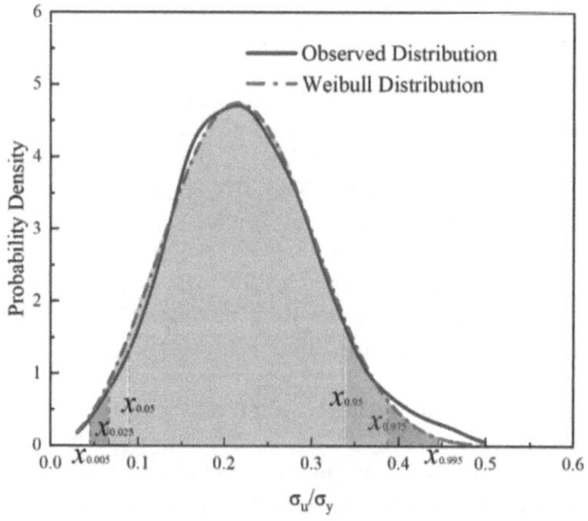

Figure 9. Observed distribution and the Weibull distribution of the ultimate strength.

ultimate strength can be constructed for different confidence levels. In Figure 9, the prediction intervals for confidence levels of 90%, 95%, and 99% are depicted, while the specific ranges of the predictions are presented in Table 6.

Table 6. Prediction intervals at different confidence levels.

Confidence levels	Lower limit value	Upper limit value
90%	0.089	0.364
95%	0.069	0.389
99%	0.046	0.443

The selection of confidence levels is essentially a question of balancing risk tolerances and costs. An analysis of prediction intervals indicates that the ultimate strength of the structure cannot exceed 44.3% of the yield strength even when considering 99% confidence level, which means that the ultimate strength under such collisions is extremely weakened and the structural strength needs to be improved in such scenarios.

5 CONCLUSIONS

This paper presents a direct prediction method for structural ultimate strength under random collision loads. In addition, the random distribution characteristics of the ultimate strength are investigated. The proposed method does not rely on the measurement of damage dimensions, and it provides a feasible theoretical basis for probabilistic safety assessment. The main results are as follows:

1. Research on the effects of collision factors suggests that the collision speed and the impact mass are negatively correlated with the ultimate capacity of the ship structure, and the ultimate capacity increases as the impact position moves away from the center of the structure.
2. The randomness research indicates that the ultimate strength of the structure follows a Weibull distribution when it encounters nearly normally-distributed collisions. The structural ultimate strength is extremely weakened under such collision scenarios, and structural reinforcement design needs to be improved.

REFERENCES

Ao, L. Wu, H. Wang, D.Y. & Wu, W.G. 2020. Evaluation on the residual ultimate strength of stiffened plates with central dent under longitudinal thrust. Ocean Engineering 202: 107167.

Harada, M. Fujikubo, M. & Yanagihara, D. 2007. Development of a set of closed-form formulae for estimation of ultimate strength of a continuous stiffened panel under combined in-plane loads and lateral pressure. ClassNK Technical Bulletin 25: 11–21.

IACS. 2021. Common structural rules for bulk carriers and oil tankers. International Association of Classification Societies.

Li, S. Georgiadis, D.G. & Samuelides, M.S. 2022. A comparison of geometric imperfection models for collapse analysis of ship-type stiffened plated grillages. Engineering Structures 250: 113480.

Li, X.D. Dou P.L. & Jing B.J. 2017. Remaining ultimate strength of FPSB typical stiffened plates with dent damage [in Chinese]. Ship Engineering 39(6): 62–66.

Ohunakin, O.S. Henry, E.U. Matthew, O.J. Ezekiel, V.U, Adelekan, D.S. & Oyeniran, A.T. 2024. Conditional monitoring and fault detection of wind turbines based on Kolmogorov–Smirnov non-parametric test. Energy Reports 11: 2577–2591.

Paik, J.K. & Thayamballi, A.K. 1997. An empirical formulation for predicting the ultimate compressive strength of stiffened panels. Proc. ISOPE International Ocean and Polar Engineering Conference, May 1997. ISOPE.

Paik, J.K. Thayamballi, A.K. & Kim, B.J. 2001. Large deflection orthotropic plate approach to develop ultimate strength formulations for stiffened panels under combined biaxial compression/tension and lateral pressure. Thin-Walled Structures 39(3): 215–246.

Wang, W. Dai, S. Zhao, W. Wang, C. Ma, T. & Chen, Q. 2020. Reliability-based optimization of a novel negative Poisson's ratio door anti-collision beam under side impact. Thin-Walled Structures 154: 106863.

Xu, M.C. Song, Z. J. Zhang, B. W. & Pan, J. 2018. Empirical formula for predicting ultimate strength of stiffened panel of ship structure under combined longitudinal compression and lateral loads. Ocean Engineering 162: 161–175.

Zhang, J. Li, M.J. Wang, X.L. & Shi, X.H. 2023. Parameter sensitivity analysis of ultimate strength of stiffened

plates under polar low temperature and depression damage [in Chinese]. Chuan Bo Li Xue/Journal of Ship Mechanics 27(5): 750–762.
Zhu, L. Cai, W. & Shi, S.Y. 2021. Review on elastic-plastic dynamic responses of ship structures under repeated impact loadings. Journal of Ship Mechanics 2(2): 256–262.
Zhang, S. & Khan, I. 2009. Buckling and ultimate capability of plates and stiffened panels in axial compression. Marine Structures 22(4): 791–808.
Zhu, Z. Ren, H. Wang, X. Zhao, N. & Li, C. 2022. Methods for fitting the limit state function of the residual strength of damaged ships. Journal of Marine Science and Engineering 10(1): 102.

Trends in Collision and Grounding of Ships and Offshore Structures – Zhu & Guedes Soares (Eds)

Damage of the deck on a modern container vessel through a dropped container

M. Radon & S. Flatt
DNV Se, Hamburg, Germany

ABSTRACT: The IMO's GHG reduction goals are driving new container ship designs, with many owners opting for LNG as a transitional fuel. For smaller existing vessels, retrofitting with Type C-tanks is a viable solution. These tanks and associated systems are often installed close to the superstructure or in the front of the ship, beneath the upper deck—with container bays above. This introduces new risks, particularly from accidental container drops during handling. This study evaluates the potential consequences of such incidents on LNG installations. Drop scenarios were developed based on the vessel's intended ports, and simulations with reasonably simplified modelling were conducted. The primary focus was on container penetration through the deck and the resulting hole size. Results showed that even extreme scenarios did not compromise the LNG tank itself. However, the study highlights the importance of positioning critical piping and equipment away from potential impact zones.

1 INTRODUCTION

1.1 *Goal of the project*

The goal of the analyses conducted within this project was to simulate the possible consequences of various accidental container drop scenarios.

The consequences were measured in possible indentation of the falling container or of the deformed structure into the areas containing critical systems. The examples of critical systems are LNG tanks themselves but also relevant piping or control equipment.

Any capacity of the critical structure to absorb deformation without producing e.g. a leak was not considered in this analysis.

An indentation far enough to come into contact in such critical equipment was directly considered as critical.

This is a reasonably conservative approach.

1.2 *Gas tanks on ships*

Container ships can be fully retrofitted with gas tanks or there is a part of the fleet that has been built as "gas ready" and can be retrofitted easier later on.

For each of the project a selection of the optimal position of the new gas tanks and corresponding infrastructure is a specific task.

It is a result of a balance including the minimisation of loss of container capacity but also the easiness of modifications of the structure and systems.

The Figure 1 shows examples of positions of the so called "gas house" on the vessel. Only positions on the deck directly under container stowage has been shown.

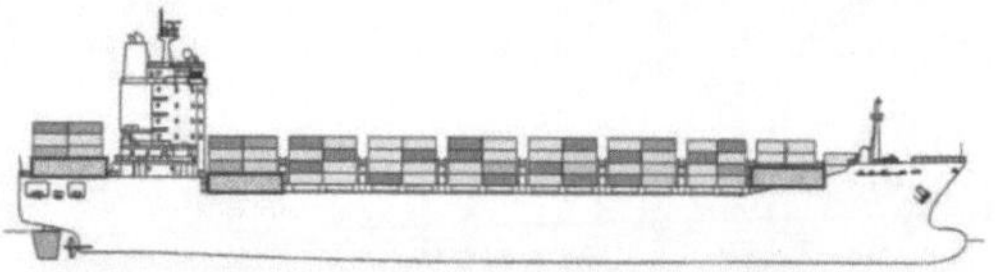

Figure 1. Examples of positions of the "gas house".

1.3 *Methodology*

The event of container drop requires a use of transient dynamic calculation methods.

To correctly account for the large deformations and finally also the breaking of the steel plates of the ship structure an explicit calculation approach was selected.

The software used was the LS-Dyna solver in combination with LS-PrePost – a pre- and postprocessing tool. The models however were prepared in various tools and imported/converted to the LS-Dyna format.

Several scenarios with different free fall height and different container masses were identified for the simulations.

The free fall part is not difficult to analyse with simple formula of the mechanics. This part of the simulation was pre-calculated and introduced as initial conditions into the explicit calculations.

DOI: 10.1201/9781003684404-45

The explicit simulation was performed for a time sufficient to completely cover the event where damages occur.

It was also verified, that if a container rebounds and re-hits the deck, the energy of this second hit is significantly smaller and does not induce damages relevant for the identification of the critical event.

The damages were quantified. In general, the depth of the indentation defines the zone not suitable for the critical installation.

Any further risk quantifications were not part of this analysis.

2 MODELLING OF THE SHIP STRUCTURE

2.1 *Model extents*

A raised structure containing the gas installation has been foreseen on the vessel. As on the beginning of the project it was not clear which scenario exactly shall be performed, the model covered the whole structure above the main deck.

Table 1. Extent of the model of the vessel structure.

Direction	Extent
Longitudinally	Full length of the Gas House
Transversally	Side to side
Vertically	Main Deck to Top of Gas House

In fact a much smaller model would be sufficient as the damages have a very local extent.

Two areas in the vicinity of the impact points were separately defined. The goal was to apply the material with failure capacity in a limited portion of the model

2.2 *FE mesh*

The mesh size was about 200 mm. Correspondingly the failure criteria as proposed in /1/ and /2/ were used. These failure criteria consider larger element sizes.

All structural elements were idealized with shell elements.

2.3 *Materials*

The Gas House is built out of steel. The upper deck and the ordinary stiffeners are made of mild steel, the girders and the vertical walls are made of HT32 steel.

The graph below shows the true stress strain material curves used for both materials.

2.4 *Boundary conditions*

The Gas House model was set as fixed for all translational degrees of freedom at the edge lines. As the effects are very local and far from the edge, this kind of boundary conditions is judged reasonable.

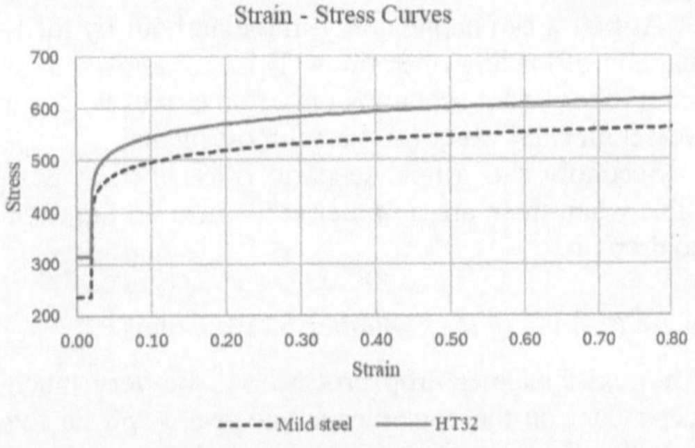

Figure 2. Material stress-strain curves.

3 MODELLING OF THE CONTAINER

The way of modelling of a container is a topic for more than one separate study. The container itself is not a very stiff structure. However, its content will very much influence the behaviour of the container during the free fall and during the crash on the deck.

The relevant parameters for the container will be:

- The mass of the container
- The centre of gravity
- The mass distribution
- The stiffness of the container with its content

For a pragmatic approach following assumptions were made:

a) The container was modelled as a rectangular bloc of a 40 ft container dimensions.
b) The container is modelled as rigid (conservative assumption).
c) The container has slightly rounded corners and edges (radius of 100mm).
d) The mass distribution is even in the whole container.
e) The center of gravity is in the geometric center of the container.

Figure 3. View of the model of a 40 ft container.

The mass of the container is discussed further in the section about the loads.

4 SCENARIOS

4.1 *General*

The accidental container drop on the deck can happen only in specific circumstances.

At first it can happen only in the harbour by loading and offloading operations. This fact allows us to limit the relevant scenarios only to the ones with the vessel on calm water (no vessel movements).

Secondly, the critical scenario is assumed as possible when there are no other containers on the deck to drop on.

4.2 *Position of the container by the impact*

The real container drop process will be very much dependent on the container actual speed and on the way the disconnection from the crane happens.

The container can hit the deck under various possible angles. In the current study it was decided to analyse 2 variants:

a) Hitting the deck with the flat bottom of the container
b) Hitting the deck with one corner so that the center of gravity of the container lies approximately above the first contact point.

It was judged that the highest penetrations shall occur when the container center of gravity is above the hit point at the time of contact. In this position the most of the kinetic energy will be transformed in plastic deformations will cause higher penetrations.

In other positions the first contact reaction will create a moment that will quickly cause a rotation of the container and a significant part of the energy will be transformed back to the kinetic energy of the container.

A tilting container might cause a larger plastic deformation area but the plastic deformation area is not a critical parameter for this accidental scenario.

4.3 *Impact point on the deck*

The impact position on the deck can also vary very much. The part of the deck close to the centerline was defined, with engineering judgment, as the most prone for critical deformations. Additionally, by an arrangement with 3 gas tanks the centerline tank has its highest part in the centerline.

It was selected, to simulate the impact on the centerline girder and in the middle between 2 girders.

Additionally, so called "swing" cases were defined. In those cases, the container swings sideways and hits the gas house wall with its edge.

4.4 *Container mass and height*

The maximum weight of a 40 ft container shall not be over 30.5 tons. Nevertheless, it happens that containers are overloaded. Even more, the overloaded containers are more likely to disconnect from the crane and fall on the deck.

On the other hand, the heaviest containers are usually not lifted to the highest crane positions as they are loaded rather in the lower tiers.

The maximum lifting height is dependent on the crane height at the container terminals and on the height of the deck itself. Here a more detailed study for every vessel is necessary to define a more specific set of container mass and drop height.

More than 20 various combinations of height and falling mass were analysed. The Mass was varying from 20 to 40 tons and the height was up to approximately 30 meters.

5 RESULTS

5.1 *Evaluation criteria*

The primary criterion that needed to be evaluated was the indentation of the container/damaged deck structure. This will allow to take further risk-based decisions on the safe distances to be kept for the critical infrastructure inside the gas house.

A second parameter was also verified, the size of the hole in the deck. This parameter has normally no further relevance but is informative.

5.2 *Results of selected scenarios*

We have selected only several scenarios to present in this paper, the most interesting ones from the analysis point of view.

Table 2. Results for selected cases.

Scenario	Container weight [t]	Drop height [m]	Penetration [m]	Hole size [m]
Swing	40	-	0.353	0.184
Flat drop	30.5	29.0	0.279	-
Drop on corner	20.5	29.0	0.783	1.445
Drop on corner	30.5	29.0	1.392	3.031

In the swing case the container can accumulate only a limited amount of kinetic energy in the horizontal displacement. The calculated penetration was below 400 mm and the hole size was abt. 200 mm.

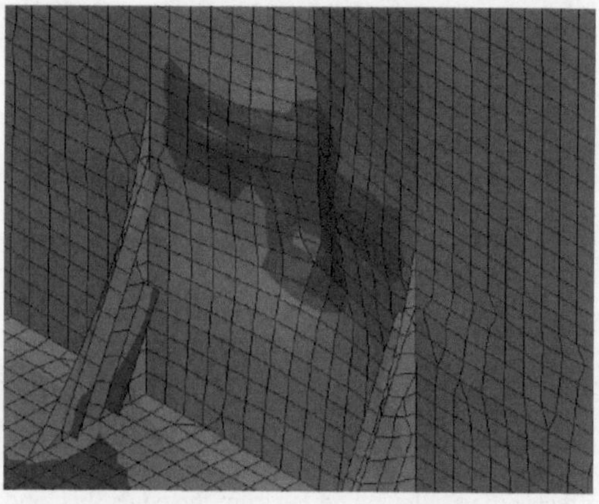

Figure 4. View of the damages in a swing case.

The vertical drop scenario with the impact on the flat bottom yields in very small damages. The impact is well distributed on the deck surface and finally no rupture occurs. The maximum penetration was below 300 mm.

The vertical drop on a corner combines a significant accumulation of kinetic energy in the vertical direction and its dissipation on a very small space. Despite the use of the container with no mass overload, this scenario is resulting with the highest damages. The calculated penetration was up to abt. 1.4 meter and the hole size above 3 meters wide.

Figure 5. View of the damages in a vertical drop on corner case.

6 CONCLUSIONS

The current investigations have shown that a dropping container can generate damages to the structure of the vessel.

In case the structure hosts vulnerable systems, like e.g. LNG tanks and related piping, such accidental drop may lead to a potentially dangerous situation.

There are many parameters of such a container drop scenario, like mass of the container, fall height, position of the container during the impact, position on the deck where the impact will occur or even the properties of the container itself and the ship structure.

Such analyses (as also this one) shall be conducted as a part of a risk analysis. This allows to select the scenarios that shall be evaluated.

As one can see from the results presented, an accidental container drop, even if unlikely to happen, can however cause critical damages. When installing critical systems close to the container cargo deck, corresponding considerations shall be made.

REFERENCES

DNV-RP-C208, *Determination of structural capacity by non-linear finite element analysis methods*, DNV, 2022.

IMO SLF 46/INF.10, *Approval procedure concept for alternative arrangements*, International Maritime Organization, 2013.

UN ADN 2013 vol 1, European Agreement concerning the International Carriage of Dangerous Goods by Inland Waterways, United Nations, 2013.

Trends in Collision and Grounding of Ships and Offshore Structures – Zhu & Guedes Soares (Eds)
© 2026 The Author(s), ISBN 978-1-041-16361-9

Experimental study on ultimate capacity of ice-strengthening structures from a heavy icebreaker

J. Zeng & Y. Wang
Marine Design and Research Institute of China, Shanghai, P.R. China

X. Yang & J. Liu
State Key Laboratory of Ocean Engineering, Shanghai Jiao Tong University, Shanghai, P.R. China

ABSTRACT: Heavy icebreakers are likely to encounter extreme ice load conditions due to severe polar environment, which makes it essential to ensure the strength of ice-strengthening structure under local ice load. In this paper, in order to research the ultimate capacity and bearing characteristics of the ice-strengthening structures subjected to lateral pressure, a quasi-static loading model test was designed. Based on the ice-strengthening structure prototypes of a heavy icebreaker and the loading areas calculated in accordance with relevant rules, the structure forms and boundary conditions of structure models and different load cases were specified. Through the experiments, the ultimate capacities and load-deflection curves under different load cases were obtained, as well as the damage modes. Nonlinear numerical simulation was performed for further comparison and verification. The results reveal a significant discrepancy in ultimate capacity among different load cases. It was also observed that the structural responses of the structure models after the loading tests mainly include bending deflection and local buckling deformation. Besides, the simulation results and experimental data are in good agreement in many aspects, such as ultimate capacity, the load-deflection curve and the damage mode.

1 INTRODUCTION

During the process of exploration of polar areas, icebreakers undertake a series of tasks such as route opening and scientific research. When it comes to heavy icebreakers, due to prolonged exposure to severe polar environment, they are likely to encounter complex and extreme ice load conditions, which set higher demands on the local strength of hull structure (Wu et al. 2020). Among various external forces acting on icebreakers which sail in ice-infested seas, ice load needs to be paid attention to particularly. Excessive ice loads may cause unpredictable damage on hull structure. Therefore, in order to ensure the normal navigation and operational safety of heavy icebreakers, it is of great importance to research the ultimate capacity of the ice-strengthening structures of heavy icebreakers under local ice loads.

The ice-strengthening region of icebreakers usually refers to the part of the hull that comes into contact and interacts with sea ice. In principle, the structure of this region should be reinforced, including increasing the size of structural components and reducing the spacing between them. Current rules have proposed detailed methods of strength assessment for ice-strengthening structures of polar ships, and many scholars also have conducted studies on the design and strength analysis of ice-strengthening structures from various perspectives (Qi et al., 2014, Xu et al., 2017). International Association of Classification Societies (IACS) recommends that when analyzing the structural strength of polar ships, ice loads can be simplified as uniformly distributed pressure acting on a specific rectangular area. Therefore, based on the above theory and considering the interaction process between hull structures and sea ice, the ultimate capacity of ice-strengthening structure of icebreakers is defined as the maximum local transverse pressure that the structure in the ice-strengthening regions is able to bear.

In recent years, when researchers study the local bearing capacity and structural responses of hull plate structures, the commonly used methods include model test, numerical simulation and a combination of both. Kim et al. (2015) demonstrated that nonlinear FEA can effectively analyze the structural responses of ice-strengthening structure under ice loads. Liu et al. (2015) put forward a simplified analytical method for the energy absorption mechanism of stiffened plates. Shi et al. (2017) carried out quasi-static loading tests for hull plates under different boundary conditions and compares results with simulations. Mikkel et al. (2018) discussed the impact of load patch length in the simulation on the bearing characteristics of ice-strengthening grillages.

DOI: 10.1201/9781003684404-46

Zhu et al. (2020) researched the nonlinear elastic-plastic structural responses of plates impacted by an ice wedge striker and an idealized rigid indenter in experimental and numerical ways. However, at present, there are few studies focusing on the ultimate capacity of ice-strengthening structures, and most researches aim at polar ships with low ice class rather than heavy icebreakers. Besides, the structural forms of the hull grillages used in the model tests are relatively simple and do not align with the features of ice-strengthening structures of heavy icebreakers.

Based on the prototypes of the bow and side ice-strengthening structures of a PC2 heavy icebreaker, this paper investigates the ultimate capacity of ice-strengthening structures through quasi-static loading model tests. The structure models, test conditions and loading method are designed and introduced. The damage and deformation modes as well as ultimate capacity of structure models under different test conditions are obtained. Numerical simulation is also performed for further verification.

2 MODEL TEST

2.1 *Ice-strengthening structure models*

The ice-strengthening structures of heavy icebreakers are usually characterized by high-strength material, thick shell plates and large strong components, which makes it hard to carry out full-scale tests for ultimate capacity due to space and loading capacity. Therefore, in this study, scale models are adopted for the experiments. The design load patch parameters of the bow and side ice-strengthening structure prototypes are calculated in accordance with IACS rules. Then, the test model ranges of the two ice-strengthening structures are specified based on the loading areas and structural forms, and the scale ratio is determined considering the practical conditions. The structure models used for loading tests are shown in Figure 1. It can be seen that both ice-strengthening structures include longitudinal girders, web frames, and frames. The sizes of the loading areas of the test models are shown in Table 1.

Table 1. Parameter table of scale load areas.

	Load patch	
Model type	Width	Height
Bow structure model	255mm	120mm
Side structure model	189mm	158mm

Based on the determined ranges of ice-strengthening structure models and adjacent structures, the boundary conditions of the structure models can be analyzed. As the longitudinal boundaries of the bow ice-strengthening structure are transverse bulkheads and the vertical boundaries are large longitudinal girders, the support in both directions are relatively strong, and its boundary conditions can be simplified as rigidly fixed on all sides. For the side ice-strengthening structure, the longitudinal boundaries are web frames while the vertical boundaries are platform decks, which means that the longitudinal support is greater than the vertical one. Its boundary conditions can be simplified as simply supported in the vertical direction and rigidly fixed in the longitudinal direction. To simulate the real boundary conditions as closely as possible during the test, it is important to design corresponding fixtures for the bow and side frame models, as shown in Figure 2. The rigidly fixed boundary is realized by the connection of bolts and ribbed channel steel with additional fixed support below, while the simply supported boundary is achieved by the connection of bolts and L-type steel without extra support.

(a) bow structure model (b) side structure model

Figure 1. Ice-strengthening structure models for loading test.

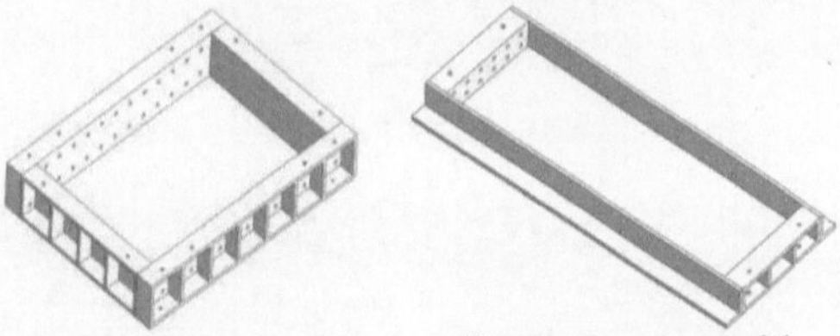

(a) bow structure model (b) side structure model

Figure 2. Boundaries design for structure models.

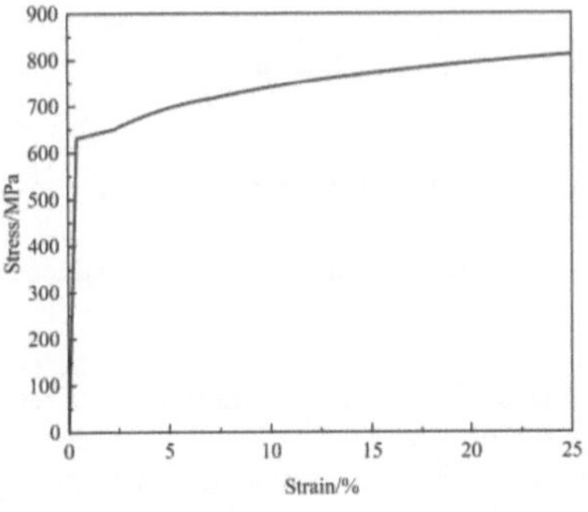

Figure 3. True stress-strain curve of EH500 steel.

The ice-strengthening structure models are made of EH500 steel, consistent with the prototype. To ensure the accuracy and effectiveness of numerical simulations in validating experimental results, it is necessary to conduct mechanical property tests on the steel of the structure models before the loading experiment. Since the engineering stress-strain curve obtained from the experiments cannot accurately reflect the true properties of the material, the engineering stress and engineering strain need to be converted into true stress and strain using specific formulas (Villavicencio & Guedes Soares, 2013). The calculated true stress-strain curve of the EH500 steel is shown in Figure 3.

2.2 *Test conditions*

According to the current guidelines, when assessing the structural strength, the simulated ice load should be applied at the position where the components are relatively weak. Therefore, for the ultimate capacity test in this study, the loading conditions should also be designed based on the distribution of strong and weak components as well as the local strength. Additionally, the loading position should be as close as possible to the center of the model to reduce boundary effects during the loading process. As a result, considering the sizes of the design load plates, the loading areas for the two structure models are specified, as shown in Figure 4.

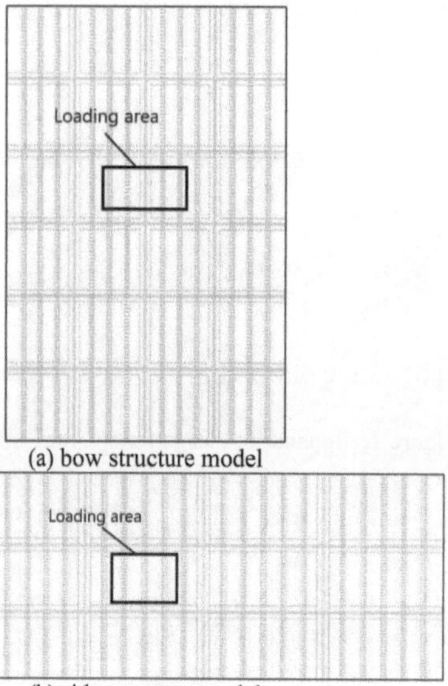

Figure 4. Loading areas for different test conditions.

In order to fully research the structural responses of the structure models in the loading process, the models should be uniformly loaded with quasi-static method until collapse, thus obtaining the ultimate capacity of the structure models.

In both loading tests and numerical simulations, the load-deflection curve is required to determine and verify the ultimate capacity of the ice-strengthening structures. Here, "load" represents the uniformly distributed ice load acting on the plate of the ice-strengthening structure, measured in MPa. "Deformation" represents the relative displacement of the main bearing component in the loading area, specifically defined as the difference between the maximum displacement of the main bearing component in the loading direction and the average displacement at the intersections between the main bearing component and its supporting components, measured in mm. To obtain the deformation responses of the ice-strengthening structures under two experimental conditions, corresponding measurement points for displacement need to be arranged. Considering the component distribution, the measurement point arrangement is shown in Figure 5. Therefore, the formula for calculating the deflection in the load-deflection curve can be derived as follows:

$$\delta = \Delta_1 - \frac{\Delta_2 + \Delta_3}{2} \tag{1}$$

Where, δ is the relative displacement used for the load-deflection curve; Δ_1, Δ_2 and Δ_3 represent the displacements of measurement points 1, 2, 3.

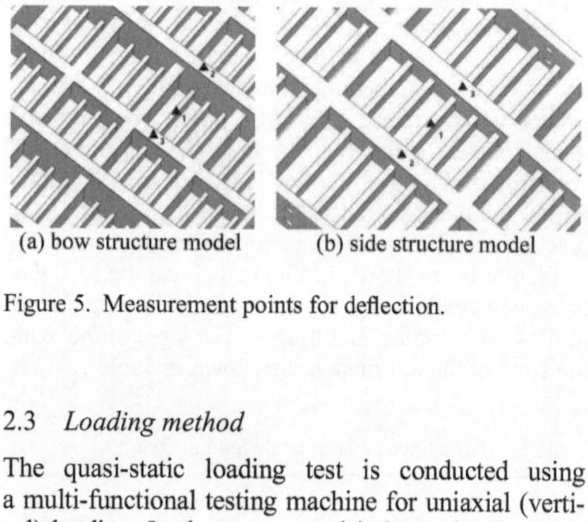

Figure 5. Measurement points for deflection.

2.3 *Loading method*

The quasi-static loading test is conducted using a multi-functional testing machine for uniaxial (vertical) loading. In the test, a steel indenter is placed on the loading area to achieve uniform loading. The load will be transferred from the testing machine to the plate surface of the structure model through the steel indenter. The steel indenter is shown in Figure 6(a), with chamfer on its bottom edges to reduce the effect of stress concentration. Due to the inevitable unevenness of the model's plate surface and the potential deformation after loading which may lead to uneven force distribution, a special sticky tape is affixed to the loading area of the structure model using double-sided tape, as shown in Figure 6(b). This sticky tape is characterized by high tensile strength and resistance to deformation.

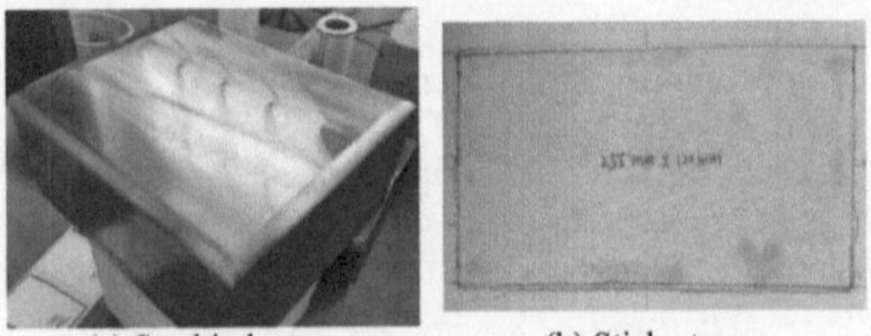

(a) Steel indenter (b) Sticky tape

Figure 6. Load application method.

In the test preparation, the ice-strengthening structure model is connected to the fixtures and then fixed on the loading platform. The loading area is marked on the plate surface of the structure model, and the steel indenter is placed accordingly. Displacement meters are installed based on the measurement point arrangement. The overall installation diagrams of the bow and side ice-strengthening structure models are shown in Figure 7.

A pre-loading is conducted to check the stability of the supports and restraint devices, and to ensure the normal functioning of the instruments and loading machine. Then, the formal ultimate capacity model test is carried out, with the load gradually increasing until the structure model collapses.

(a) bow structure model (b) side structure model

Figure 7. Structure model assembly and loading.

3 EXPERIMENTAL RESULTS

3.1 *Damage and deformation modes*

After the quasi-static loading test, the structure model can be disassembled from the fixtures. The structural response and damage modes of the bow and side ice-strengthening structure models are shown in Figures 8 and 9. Besides, a 3D scanning system is used to scan the deformation of the model after loading, as shown in Figure 10.

According to Figures 8, 9 and 10, it can be observed that the most severe damages occur in the loading area of the model. On the surface of the model, the displacement at the center of the loading aera is maximum, and the farther the point is from the center, the smaller the displacement of the point will be.

On the backside of the model, the shell plate and frames near the loading area exhibit significant bending deformation. The webs of the adjacent longitudinal girders and web frames exhibit noticeable local buckling deformation. Fractures occur at the junctions of the longitudinal and transverse strong components.

Figure 8. Damage modes of bow structure model after loading.

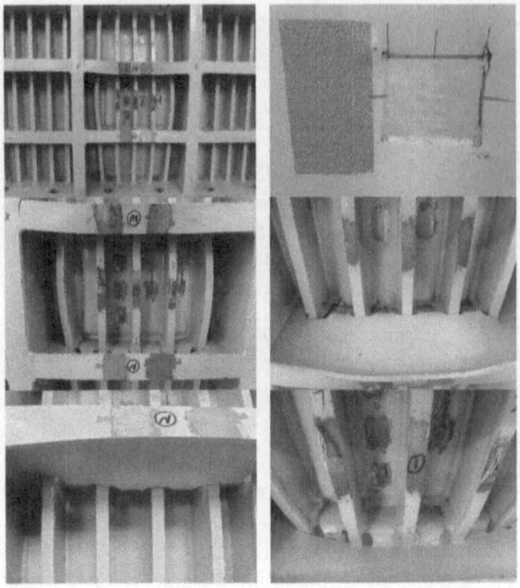

Figure 9. Damage modes of side structure model after loading.

Based on the above analysis, the structural design should prioritize the stability of strong components' webs and the stress concentration at flange connections.

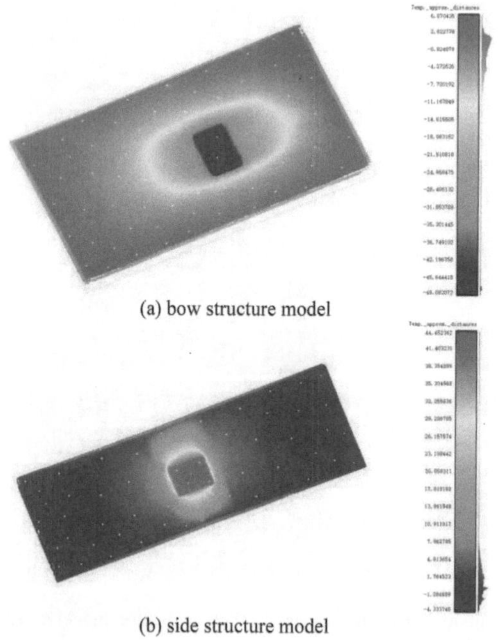

(a) bow structure model

(b) side structure model

Figure 10. Plate deformation of structure models after loading.

3.2 *Ultimate capacity of test conditions*

Based on the results of the quasi-static loading model tests, considering the load-deflection curves obtained from the experiments, the ultimate capacities of the bow and side structure models under two test conditions are presented in Table 2.

Table 2. Ultimate capacity of different test conditions.

Model type	Ultimate capacity
Bow structure model	101.69MPa
Side structure model	75.24MPa

From Table 2, it is evident that there is a significant discrepancy in the ultimate capacity of the two ice-strengthening structure models under lateral loading. In terms of structural distribution, the ice-strengthening structures in the bow region and the side region are both constructed in transverse framing systems, comprising bearing girders, web frames, frames, and shell plates. However, the spacings between strong and weak components and the scales of individual grillage in the bow ice-strengthening structure are relatively smaller, and the sizes of the components are generally bigger than those in the side structure. Therefore, the overall strength and stiffness of the bow ice-strengthening structure are greater than those of the side structure, thus enhancing its ability to resist transverse loads and boosting its ultimate capacity.

From the perspective of icebreaking modes, icebreakers usually undertake icebreaking tasks through continuous and crash icebreaking ways (Zhang et al. 2018). In both modes, the bow structure of the icebreaker directly interacts with the ice layer. For high ice-class polar ships such as heavy icebreakers, the bow structure must withstand severe impacts from thicker ice layers during crash icebreaking. Therefore, the bearing capacity of the bow ice-strengthening structure must be higher.

4 NUMERICAL SIMULATION ANALYSIS

4.1 *Plate structure modeling and test simulation*

On the basis of the design drawings, the finite element models of the ice-strengthening structure models used in the tests are established, as shown in Figure 11. The finite element models are discretized with shell elements, and the sizes and spacings of the components are consistent with the actual parameters. The applied material model is shown in Figure 3. For mesh size, after comparing relevant rules from different classification societies, 100mm×100mm is relatively applicable. Considering the scale ratio in the model test, the final mesh size is set to be 14mm×14mm.

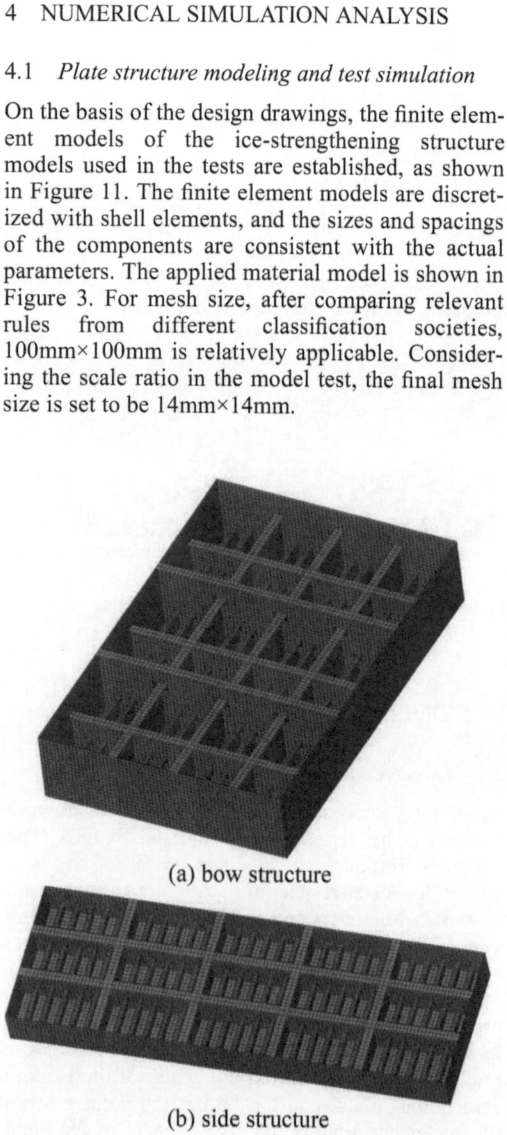

(a) bow structure

(b) side structure

Figure 11. Finite element models of ice-strengthening structures.

According to Section 2.3, the load in the model test is transferred through the steel indenter placed on the loading area. In order to simulate the actual loading process, a rectangular steel model is established. The steel model is positioned above the loading area, and a contact is defined between the bottom surface of the steel model and the plate surface of the structure model to simulate their interaction in the loading process. The simulation is performed using Abaqus explicit dynamic analysis, where a uniform downward velocity is applied to the steel model to simulate the loading process in the experiment. The assembly of the steel model and the structure model is shown in Figure 12.

(a) bow structure (b) side structure

Figure 12. Assembly between steel indenter and structure model.

4.2 *Comparisons between experimental results and numerical simulation*

According to the experimental and simulation results, the load-deflection curves of the bow and side ice-strengthening structure models during the loading process are plotted. The structural responses and corresponding ultimate capacities are compared as follows:

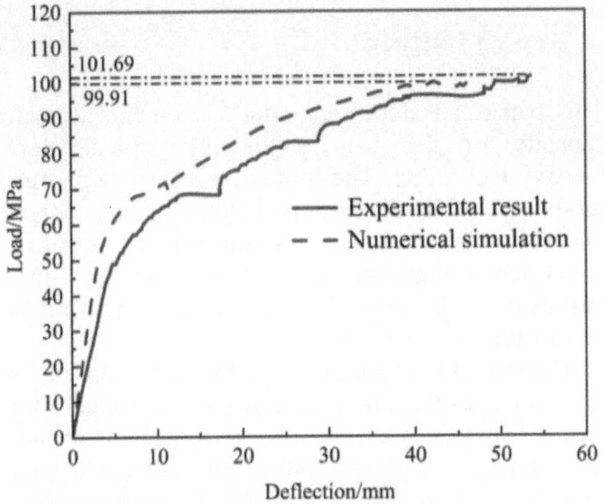

Figure 13. Load-deflection curves of bow structure under specified loading condition.

(1) The load-deflection curves of the bow ice-strengthening structure model obtained from the experiment and numerical simulation are compared in Figure 13. It can be observed that the curves obtained from these two methods are very close during both the elastic and plastic stages, showcasing a good consistency.

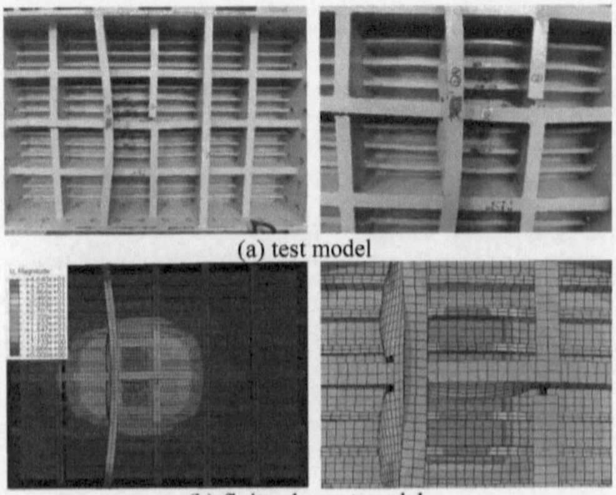

(a) test model

(b) finite element model

Figure 14. Deformation of bow structure under specified loading condition.

The structural responses and deformation modes of the damaged test model and the finite element model are compared in Figure 14. It can be observed that the finite element model exhibits significant buckling deformation in the webs of the longitudinal girders and the web frames. Besides, the panel of the longitudinal girder exhibits noticeable lateral tilting, which aligns well with the experimental results.

(2) The load-deflection curves of the side ice-strengthening structure model obtained from the experiment and numerical simulation are compared in Figure 15. It can be observed that before the load reaches 60 MPa, the trends of the load-deflection curves of the model test and the numerical simulation are relatively consistent, including the elastic stage and the start of yielding stage. After this, the load calculated by the numerical simulation remains almost unchanged, with only the displacement continuing to increase, indicating that the finite element model is unable to bear greater load. However, in the actual test, both the load and displacement show a slowly increasing trend until the ultimate stage.

The structural responses and deformation modes of the damaged test model and the finite element model are compared in Figure 16. It can be observed that the structural response of the finite element model is symmetrically distributed. The webs of adjacent longitudinal girders exhibit significant buckling deformation, and the panels exhibit slight lateral tilting, which is similar to the experimental results.

To further investigate the reasons for the differences in the latter part of the curves, the structural responses of the webs of the longitudinal girders in the model test and numerical simulation during the loading process are compared using the test video. It is revealed that after the actual model's girders

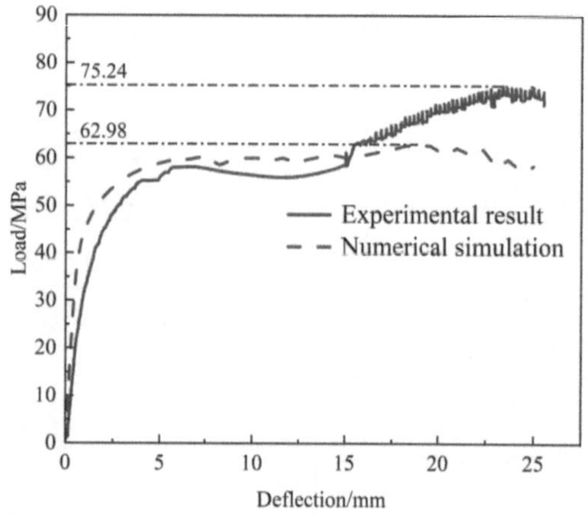

Figure 15. Load-deflection curves of side structure specified loading condition.

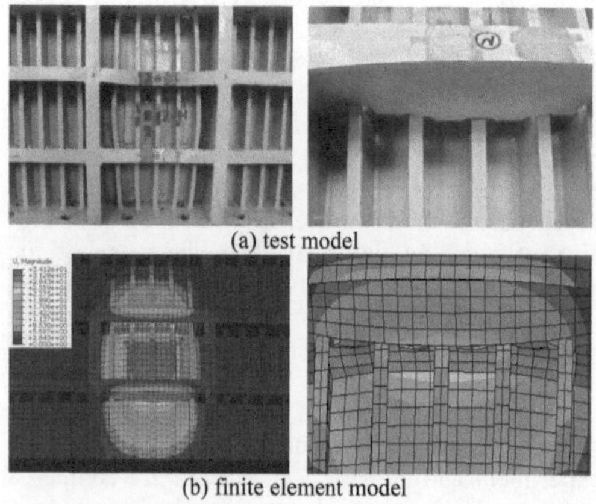

Figure 16. Deformation of side structure under specified loading condition.

exhibit significant buckling deformation, due to the through-holes, the webs of the longitudinal girder come into contact with the panels of the frames. This contact slows down the deforming trend of the components to some extent, allowing the whole structure model to continue bearing the load, thus increasing the ultimate capacity. However, after the girders of the numerical simulation model buckle, the web elements around the through-holes cannot come into "contact" with the panel elements of the frames. As a result, the webs of the longitudinal girders cannot help to bear the load, making it difficult for the bearing capacity of the numerical simulation to continue increasing. Efficient structural design should pay attention to the details where side frames intersect with longitudinal girders.

(3) Based on the load-deflection curves and the structural responses of the finite element models, the simulated ultimate capacities of the bow and side

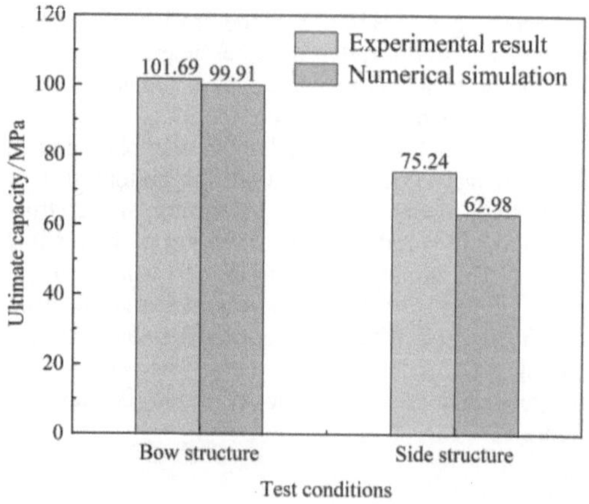

Figure 17. Ultimate capacity of different structures from model test and numerical simulation.

ice-strengthening structure models are obtained and compared with the experimental results in Figure 17.

The ultimate capacity errors between experiment and simulation of the bow and side ice-strengthening structure models are both less than 20%, showcasing a relatively satisfactory consistency. Specifically, the error of the bow structure model is smaller, while the error of the side structure model is bigger.

In conclusion, the errors between the experimental and simulation results of the ultimate bearing capacity of the frame models are acceptable. Moreover, the load-deflection curves obtained from model tests and numerical simulations are very close, and the structural responses of the structure models are also quite similar. These results demonstrate that the model tests and numerical simulations have achieved mutual verification.

5 CONCLUSIONS

This paper introduces a model test on the ultimate capacity of the ice-strengthening structures of a heavy icebreaker. The ultimate capacity and structural response of the bow and side structure models under transverse loading conditions are obtained. Numerical simulations are performed and the similarities and differences between the test and simulation results are analyzed.

The test results indicate that the ultimate capacity of the bow ice-strengthening structure is significantly greater than that of the side ice-strengthening structure. The damage modes of these two ice-strengthening structure models mainly include noticeable bending deformation of the plates and frames, as well as buckling deformation of the longitudinal girders and web frames.

The ultimate capacity, load-deflection curves, and structural responses of the structure models obtained from numerical simulations show good consistency with the test results, thus validating that the quasi-static nonlinear finite element method can accurately and effectively analyze the ultimate capacity of the ice-strengthening structures.

Based on the test and numerical results, structural design should pay attention to the stability of strong components' webs, stress concentration at flange connections, and details at intersections between side frames and longitudinal girders.

This study can provide references for the structural design of icebreakers and heavy icebreakers, as well as other polar ships.

REFERENCES

Kim, H., Dolny, J., Daley, C. 2015. An experimental study of the design and overload capacity of structural grillages subjected to ice loads. *Proceedings of the International Conference on Port and Ocean Engineering under Arctic Conditions*.

Liu, B., Villavicencio, R., Guedes, S.C. 2015. Simplified method for quasi-static collision assessment of a damaged tanker side panel. *Marine Structures* 40: 267–288.

Mihkel, K., Pentti, K., Jani, R. 2018. Load carrying capacity of ice-strengthened frames under idealized ice load and boundary conditions. *Marine Structures* 58: 18–30.

Qi, K.L., Liu, J., Liu, S.W. 2014. Strength assessment of polar ships under ice loads using direct calculation method. *Journal of Ship Mechanics* 18(3): 280–290.

Shi, S., Zhu, L., Karagiozova, D. 2017. Experimental and numerical analysis of plates quasi-statically loaded by a rectangular indenter. *Marine Structures* 55: 62–77.

Villavicencio, R., Guedes, S.C. 2013. Impact response of rectangular and square stiffened plates supported on two opposite edges. *Thin-Walled Structures* 68: 164–182.

Wu, G., Wang, Y.W., Zhang, D.J. 2020. Research status and prospect of general and structural design technology of polar icebreaker in china. *Shipbuilding of China* 61 (01): 194–203.

Xu, Y.G., Wang, Y.W., Song, Y. 2017. Study on standard design of ice framing system of polar vessel. *Ship & Boat* 28(05): 43–48.

Zhang, J., Wang, K.M., He, W.X. 2018. Study on ice load calculation method of polar icebreaker bow under continuous ice breaking. *Shipbuilding of China* 59(03): 155–163.

Zhu, L., Cai, W., Chen, M. 2020. Experimental and numerical analyses of elastic-plastic responses of ship plates under ice floe impacts. *Ocean Engineering*, 218.

New designs for structural improvement

Trends in Collision and Grounding of Ships and Offshore Structures – Zhu & Guedes Soares (Eds)
© 2026 The Author(s), ISBN 978-1-041-16361-9

Windshield optimization for drag reduction using CFD modelling and genetic algorithm

Q.P. Cheng, J.Y. Shao, X.J. Ma, Q. Jiao, X.S. Zhao & Y.O. Zhang
School of Naval Architecture, Ocean and Energy Power Engineering, Wuhan University of Technology, Wuhan, P.R. China

ABSTRACT: Wind resistance of the superstructure on container ships has a significant impact on their energy consumption. Adding windshield can effectively reduce the wind resistance of container ships. However, traditional drag reduction design methods based solely on CFD require extensive computational resources and lengthy design cycles, making it difficult to efficiently design high drag-reduction windshield shapes. To address this issue, this work proposes a drag reduction optimization design method that combines a genetic algorithm with computational fluid dynamic analysis. This method enables efficient optimization of windshield shapes, effectively reducing wind resistance and thus lowering energy consumption. The proposed method achieves a 15.97% drag reduction with the installation of the windshield, and a further improvement to 17.02% drag reduction after CFD-GA optimization. This work contributes to energy-saving and emission-reduction in the shipping industry.

1 INTRODUCTION

Shipping is a crucial component of international trade, with more than 80% of global trade goods transported by sea. As a result, shipping has also become a significant source of greenhouse gas emissions, drawing considerable attention to the industry's carbon footprint. With the development of the shipping industry, the stacking height of deck containers on large container ships continues to increase, and the trends of ship diversification and large-scale construction are deepening. Against this backdrop, green development and energy conservation and emission reduction have increasingly become core considerations in ship design. The International Maritime Organization (IMO, 2011) introduced Energy Efficiency Design Index (EEDI) requirements for newly built ships, which have been deeply integrated into all aspects of ship design and serve as key guidelines for industry development. In response to this standard, the industry has undertaken technical exploration, with researchers discovering that air resistance accounts for 5% to 15% of a container ship's total resistance. Consequently, technologies aimed at reducing wind resistance have become a focal point of energy-saving and emission-reduction research, attracting much.

Grlj & Degiuli (2023) studied different container configurations and found that compared to the even-keel condition, bow trim reduces air resistance, while stern trim increases it. Van (2019) conducted research on Wood Chip Carriers (WCC) and discovered that side shields can be used to reduce wind resistance acting on WCCs, thereby decreasing fuel consumption in maritime transport. Ikeda (2017) investigated the wind resistance when gaps existed between stacked containers. Through numerical methods, they studied the impact of these gaps on wind resistance and found that covering 50% of the gaps could reduce wind resistance by 19% at a 30° wind angle. Hassan et al. (2012) optimized container stacking arrangements using CFD numerical simulations, and their results showed that the optimized stacking method could reduce wind resistance by 30%. Without altering the hull configuration or container stacking arrangement, Deng (2022) studied different shapes of bow windshields and found that a semi-circular windshield design reduced wind resistance by 20.85%.

Currently, windshields are increasingly adopted in the industry due to their easy installation and significant drag reduction effects. However, most existing design methods still rely on empirical trial and error, requiring construction of numerous geometric models and individual CFD validations, resulting in long design cycles and high computational costs. To address this issue, this work proposes an optimization method that couples Genetic Algorithm (GA) with Computational Fluid Dynamics (CFD), enabling efficient windshield shape optimization through parametric modeling and global search.

This paper is structured as follows: Section 2 introduces parametric surface modeling for windshield, Section 3 explains the CFD numerical

DOI: 10.1201/9781003684404-47

simulation and optimization process, Section 4 analyzes the results and Section 5 concludes the present work.

2 PARAMETRIC MODELING OF WINDSHIELD

Parametric modeling of the windshield establishes a relationship between geometric parameters and model features, allowing for rapid adjustment of design parameters. This approach ensures geometric consistency among components during the optimization process while allowing the direct generation of multiple design variants through parameter-driven modeling. By integrating CFD simulation, a multidisciplinary collaborative optimization is achieved, significantly enhancing design iteration efficiency while ensuring model reliability under engineering constraints. This method not only shortens the development cycle but also optimizes aerodynamic performances.

During the parametric modeling process, the bottom hull structure remains unchanged, and a 1:100 KCS container ship model is used as the hull model. The specific model dimensions are shown in Table 1.

Table 1. KCS Ship Model dimensions and scaled model dimensions (1:100).

Parameter Name	KCS Model		Scaled Model (1:100)
	Symbol	Value	Value
Length Overall	L	232m	2.32m
Beam	B	32.2m	0.312m
Design Draft	T	10.8m	0.108m
Projected Area	A_x	638m^2	0.0638m^2

The structure of windshield is shown in Figure 1. It is formed by the combination of the side wall top line, front curve, two symmetric end curves, and spline line. Since the top line of side wall and end curves are fixed, and two of the three endpoints of the spline line are also fixed, the design of windshield treats the front curve as a cubic curve for parametric modeling. The controllable parameters of the front curve are the horizontal distance from the end point to the front and the angle between the end tangent and the horizontal line. By adjusting these two parameters, the shape of windshield can be modified.

The front curve is modelled by a cubic polynomial, with the controllable parameters being the horizontal distance from the end point to the front and the angle between the end tangent and the horizontal line.

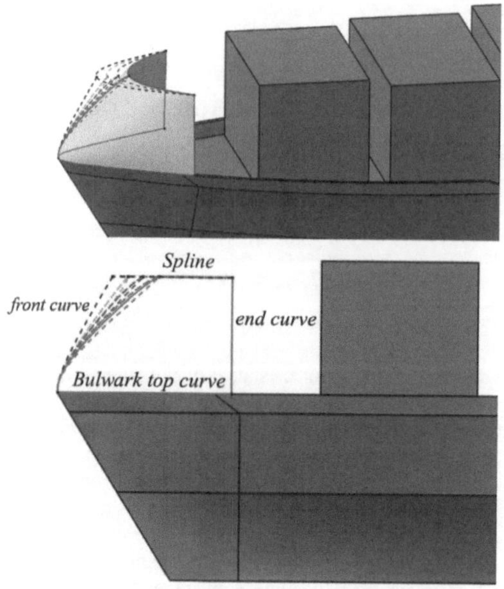

Figure 1. Parametric modelling of windshield for KCS hull.

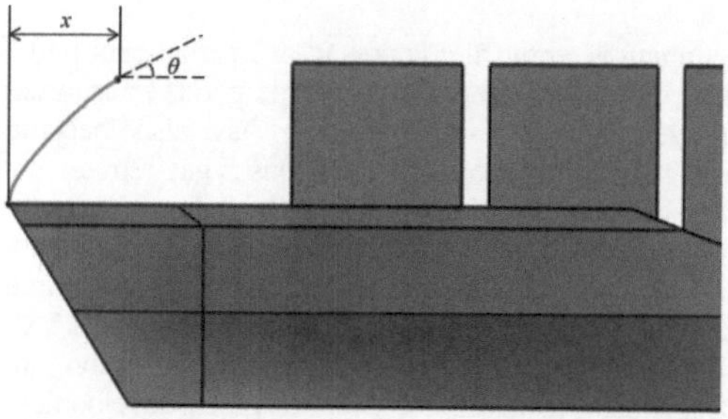

Figure 2. Optimization control parameters of windshield.

3 NUMERICAL METHODS

3.1 *Governing equations*

RANS model is adopted to simulate airflow around the superstructure of container ship:

$$\nabla \cdot \bar{u} = 0 \tag{1}$$

$$\frac{\partial \bar{u}}{\partial t} + (\bar{u} \cdot \nabla)\bar{u} = -\frac{1}{\rho}\nabla P + \nu \nabla^2 \bar{u} - \nabla \cdot (\overline{u'u'}) \tag{2}$$

Where $\bar{u}$ represents the averaged velocity vector, P is the pressure, $\overline{u'u'}$ is the Reynolds stress caused by turbulence fluctuations, ν is the dynamic viscosity, and ρ is the fluid density.This study uses the k-ε turbulence model (Rodi, 1991) to close the governing equations. For the flow around the superstructure of the container ship. This model has been extensively validated and widely used for various engineering applications. The computational domain is set as shown in Figure 4. The computational domain is established based on the ship length L as the reference length. The domain extends $1.5L$ forward from the bow and $4.5L$ behind the stern, providing sufficient space for flow development and wake formation. Considering the symmetry of the hull, a half-ship model is used to reduce computational costs. The side boundary is located $1.5L$ from the mid-section, and the top boundary is set $1.5L$ above the waterline. Boundary conditions are defined as follow: The velocity inlet is set $1.5L$ in front of the bow, while the pressure outlet is positioned $4.5L$ behind the stern. The side boundary is defined as a symmetry plane, while the waterline plane and top boundary are set as no slip wall with inlet velocity. The hull surface is defined as a stationary no-slip wall.

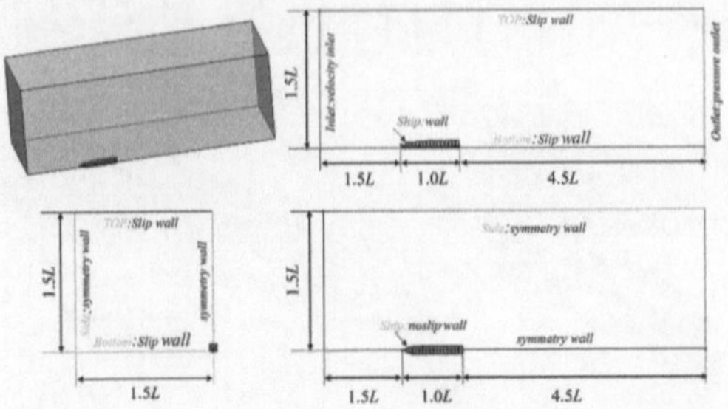

Figure 3. Computational domain and boundaries conditions.

The mesh is generated using the cutting-body method. To ensure an accurate simulation of the flow field, the boundary layer mesh satisfies $y+<1$. Additionally, the mesh is refined near windshield and wake regions to capture the complex flow field characteristics (Figure 5). The inlet velocity is set as 20m/s, corresponding to a Reynolds number of 2.96×10^6.

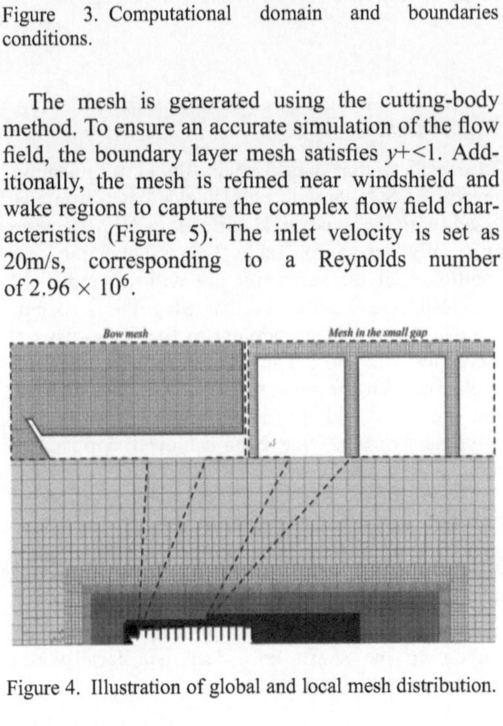

Figure 4. Illustration of global and local mesh distribution.

3.2 *Numerical validations*

Mesh convergence verification is conducted before systematic numerical analysis. Four mesh schemes are set up: coarse mesh, relatively coarse mesh, medium mesh, and fine mesh. The mean drag coefficient is used to evaluate the results between different meshes. The drag coefficient is calculated through:

$$C_d = \frac{2F_d}{\rho U^2 A_x} \tag{3}$$

Where F_d is the hull drag force.

Table 2. Grid independent results.

Mesh Type	Cell number	C_d	Error
Coarse Mesh	1.7 million	0.414	8.09%
Relatively Coarse Mesh	2.3 million	0.394	2.87%
Medium Mesh	3.5 million	0.382	0.26%
Fine Mesh	5.2 million	0.383	

Table 2 shows that the medium mesh can accurately simulate the air resistance in this case and it is selected for following simulations.

3.3 *Genetic algorithm optimization*

Figure 5 presents optimization flowchart for windshield of a container ship. CFD calculations are performed on the initial design of windshield to obtain the initial dataset. This dataset is then imported into the optimization system, where the Genetic Algorithm (GA) is used to find the optimal drag reduction rate. To facilitate the conversion of discrete data points into continuous data for the search process, the relationship between windshield parameters and the drag coefficient is constructed using Radial Basis Function (RBF) interpolation, ensuring that each data set has an estimated value. Genetic algorithm parameters are determined through preliminary sensitivity analyses, a population size of 100 is selected to leverage the diversity advantages of large populations, effectively avoiding premature convergence, 200 generations ensured sufficient iteration for stable optimization results. A crossover rate of 0.8 and mutation rate of 0.02 balanced global exploration and local exploitation. Design variables, including the horizontal distance of the windshield front endpoint($x \in [0.02, 0.1]m$) and the tangent angle ($\theta \in [20°, 90°]$), are encoded with 10-bit binary strings, achieving resolutions of$\Delta x = 0.008m$and $\Delta\theta = 0.07°$, which met the precision requirements for engineering applications. The selected data sets are then verified through CFD calculations, and the data is added to the dataset. Iterative optimization continues until the drag reduction requirements are met, forming a closed-loop optimization process. The RBF interpolation method is as follows:

$$s(x,y) = \sum_{i=1}^{n} \lambda_i \phi(||(x,y) - (x_i,y_i)||) \quad (4)$$

Where x_i represents the windshield design parameters, y_i is the drag coefficient output from CFD calculations, λ_i is the weight (coefficient), and $\phi(r)$ is the basis function. The basis function used in this study is $\phi(r) = \sqrt{r^2 + c^2}$, where c is the shape parameter.

The GA search process is as follows:

$$x' = x_{\min} + (x_{\max} - x_{\min}) \frac{\sum_{i=0}^{n-1} b_i 2^i}{\sum_{i=0}^{n-1} 2^i} \quad (5)$$

$$pop = y_{\max} - y \quad (6)$$

$$P = \frac{pop}{pop_{\max}} \quad (7)$$

Where $b = [b_0, b_1, \cdots, b_{n-1}]$ is a binary array that enables switching between binary and decimal, discretizing the continuous search space to improve search efficiency. $x_{\max}$ and $x_{\min}$ are the maximum and minimum values of windshield design parameters, and x' represents the windshield design parameters. y is the estimated drag coefficient, and $y_{\max}$ is the maximum drag coefficient, pop is the fitness, which is the difference between the maximum drag coefficient and each estimated value. $pop_{\max}$ is the maximum fitness, and P is the retention probability, used to select the minimum drag coefficient.

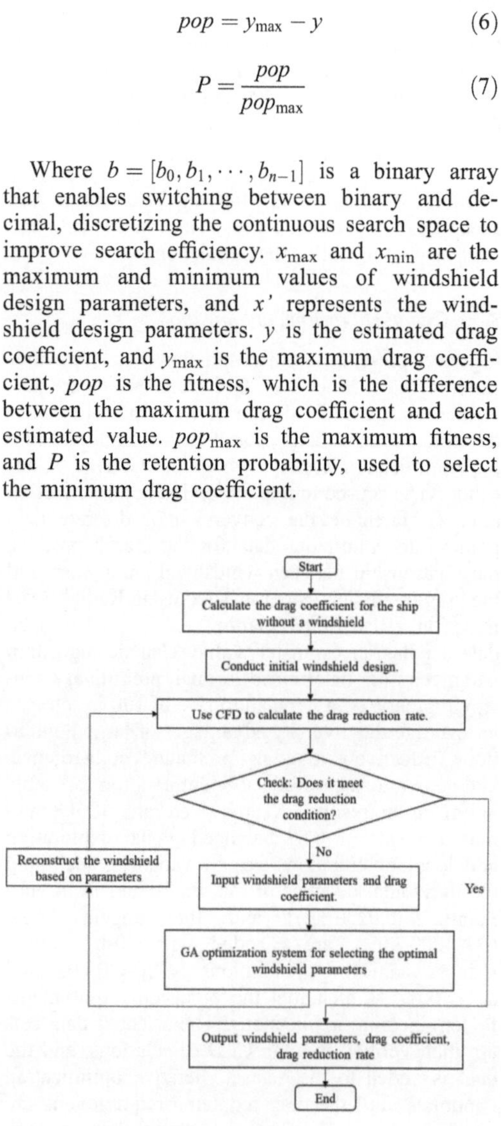

Figure 5. Optimization flowchart.

4 RESULTS AND DISCUSSION

This section analyzes drag coefficient and flow field characteristics of the container ship without/with an optimized windshield.

4.1 *Analysis of container ship without windshield*

CFD simulations have been performed for the container ship without a windshield, yielding a drag coefficient of 0.382. Figure 6 has presented the flow field, showing that the incoming flow has impacted the first container, forming a high-pressure zone in front of it. Due to the blunt shape of the container, part of the airflow has separated after striking the ship, creating a large separation zone. Another part of the airflow has formed a vortex at the bow, resulting in a pressure concentration around the ship's bow and increasing the drag coefficient.

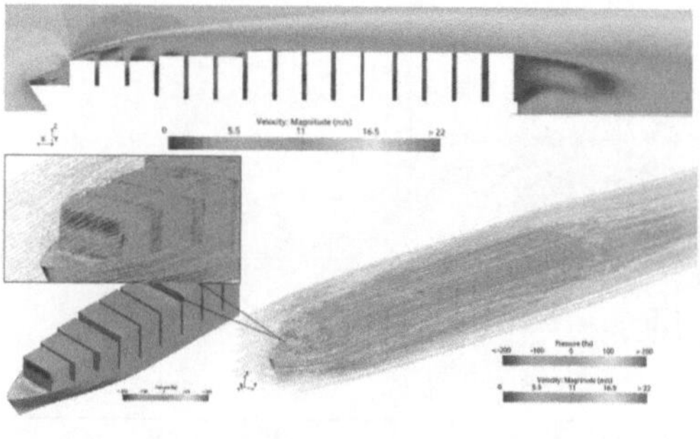

Figure 6. Flow field of KCS without windshield.

4.2 *Drag reduction of a container ship with windshield*

To provide initial data for windshield optimization, several KCS hulls with windshields have been analyzed (with the windshield surface model parameters artificially designed). Table 3 has shown the drag coefficient of the superstructure with a windshield. The results have indicated that when the horizontal distance between the endpoint of the front curve of the windshield and the ship's bow is 0.10m, and the angle between the tangent of the windshield's end and the horizontal line is 10°, the minimum drag coefficient of 0.321 has been achieved, corresponding to a drag reduction rate of 15.97%.

Comparing Figure 6 and Figure 7, with a windshield installed, the airflow has passed through windshield smooths over the first container, and the airflow behind the container has risen over its top, reducing the vortices in the middle of the container. Furthermore, the installation of the windshield has prevented vortex formation on the ship's bow deck and local vortex breakup over the first container.

Table 3. Calculation results of windshields with different parameters.

Parameter x(m)	θ(deg)	Cd	Drag Reduction Rate
0.02	*10*	*0.329*	*13.87%*
0.06	10	0.324	15.40%
0.10	10	0.321	15.97%
0.02	50	0.327	14.62%
0.06	50	0.327	14.40%
0.10	50	0.329	14.10%
0.02	90	0.325	14.92%
0.06	90	0.329	14.10%
0.10	90	0.334	12.57%

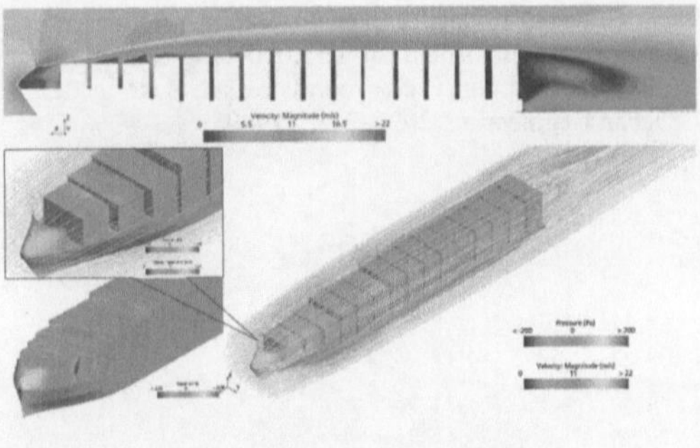

Figure 7. Flow field of KCS with windshield.

4.3 *Optimization of windshield for a container ship*

The drag coefficient data set obtained from the calculated windshield parameters has been imported into the optimization system. With GA algorithm, the optimal windshield parameters have been found to be x=0.979m, and θ=20.62°. The estimated drag coefficient from the optimization system has been 0.318, and the CFD-calculated drag coefficient has been 0.317, with an estimation error of 0.32%.

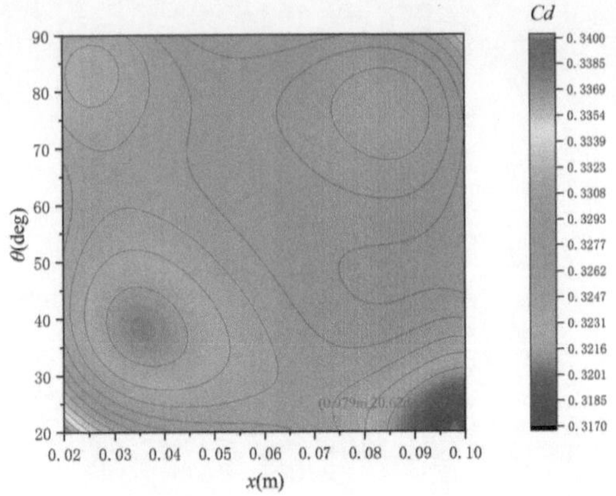

Figure 8. Optimization contour map.

By comparing the flow fields in Figures 9 and 7, it can be seen that, after optimization, the separation zone behind the first container has been further reduced, the fluid has flowed more smoothly along the surface of the container, reducing flow separation and pressure drag. The optimized windshield has provided more precise control over the flow field separation, significantly reducing the drag coefficient.

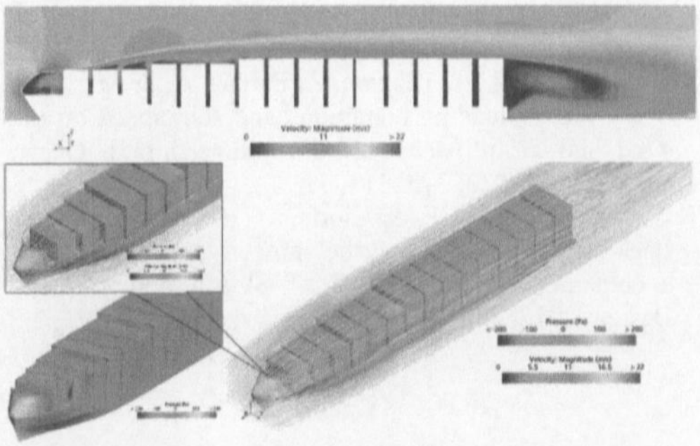

Figure 9. Flow field of KCS with windshield optimization.

Based on the CFD-GA drag reduction optimization design method, the container ship has achieved a drag reduction rate of 17.02%. Compared to the non-optimized windshield, the optimized design has improved the drag reduction effect by 6.57%. The CFD-GA based drag reduction optimization design method has efficiently provided drag-reduction solutions, reducing fuel consumption and greenhouse gas emissions during navigation.

5 CONCLUSIONS

This study proposes a CFD-GA framework for designing of windshield for container ship. The drag coefficient and flow field of container ship without and with windshield under head-on wind conditions have been analyzed. Without a windshield, obvious flow separation occurs, which leads to a high drag coefficient 0.382. Installing a windshield reduces the drag coefficient to 0.321, achieving a 15.97% drag reduction. With CFD-GA optimization, the drag coefficient is further reduced to 0.317, improving the windshield's drag reduction effect by 6.57%. This work can help reduce air resistance of container ship and promote development of the green shipping industry. The present work demonstrates the potential of CFD-GA method for efficient windshield design, to further improve the drag reduction, more design parameters can be involved to expand the design space in future work.

REFERENCES

Deng, R, Song, Z, Ren, H, Li, H, & Wu, T. Investigation on the effect of container configurations and forecastle fairings on wind resistance and aerodynamic performance of large container ships[J]. Engineering Applications of Computational Fluid Mechanics, 2022, 16(1): 1279–1304.

Deng, R, Wang, Y Q, Song, Z J, Li, H, & Wu, T C. Analysis of the characteristics of the spectral orthogonal decomposed flow fields: Numerical and experimental investigations of the air flow field around a simplified container ship model[J]. Ocean Engineering, 2022, 266: 112979.

Grlj, C G, Degiuli, N, Tuković, Ž, Farkas, A, & Martić, I. The effect of loading conditions and ship speed on the wind and air resistance of a containership[J]. Ocean engineering, 2023, 273: 113991.

Hassan, K, White, M F, & Ciorta, C. Effect of container stack arrangements on the power optimization of a container ship[J]. Journal of Ship Production and Design, 2012, 28(01): 10–19.

International Maritime Organization (IMO). Amendments to the Annex of the Protocol of 1997 to Amend the International Convention for the Prevention of Pollution From Ships, 1973, as Modified by the Protocol of 1978 Relating Thereto[J]. Amendments to MARPOL Annex VI. MEPC, 2011, 70: 18.

Rodi W. Experience with two-layer models combining the k-epsilon model with a one-equation model near the wall[C]//29th Aerospace sciences meeting. 1991: 216.

Van H N, Mizutani K, Ikeda Y. Effects of side guards on aerodynamic performances of the wood chip carrier[J]. Ocean Engineering, 2019, 187: 106217.

Van Nguyen T, Ikeda Y. 2017 S-GS4-4 Vortex Control in Gap Flow by Small Appendages to Reduce Air Resistance Acting on Deck Containers of a Ship[C]//Conference Proceedings The Japan Society of Naval Architects and Ocean Engineers 24. The Japan Society of Naval Architects and Ocean Engineers, 2017: 335–338.

Wang, G, Feng, Y, Dai, Y, Chen, Z, & Wu, Y. Optimization design of a windshield for a 12,000 TEU container ship based on a support vector regression surrogate model[J]. Ocean Engineering, 2024, 313: 119405.

Trends in Collision and Grounding of Ships and Offshore Structures – Zhu & Guedes Soares (Eds)
© 2026 The Author(s), ISBN 978-1-041-16361-9

Dynamic stiffened plating responses under green water loads for an FPSO LQ structure

Z.H. Liu
Entr-Engineering Consultancy, Aker Solutions, Trondheim Norway

ABSTRACT: A recent incident involving green water damage to the living quarters (LQ) of a Floating Production Storage and Offloading (FPSO) unit in Norway has highlighted significant concerns about the impact of green water loads. This paper presents a comprehensive method for assessing the structural integrity of stiffened steel panels subjected to potential green water loads. The green water loads were determined using advanced Computational Fluid Dynamics (CFD) analysis (FLOW-3D) through a one-way coupled approach. The resulting pressure histories were then mapped to a structural solver to evaluate the structural responses. Nonlinear finite element analysis (NLFEA) was conducted using Abaqus Explicit. Additionally, a simplified Single Degree of Freedom (SDOF) analysis was developed and executed. A representative green water impulse pattern was proposed, showing good agreement. Two simulation scenarios were chosen and examined. The proposed simplified method can be extended to any impulse load type, offering potential benefits for design applications.

1 GENERAL INTRODUCTION

FPSO ships (floating production, storage and offloading) may be exposed to extreme weather conditions during their service time. The relative motion between ship and wave will cause the water flow onto the deck, known as the 'green water loads. Consequently, the LQ (living quarter) structure could be exposed to such loads as it is normally placed towards the fore part of the ship. The living quarter (LQ) structures need to be designed with enough strength to resist this kind of load during FPSO's lifetime. Several related incidents happened on the Norwegian continental shelf, e.g., in January 2000, Petrojarl Varg FPSO was hit by green water at its fore ship (Ersdal & Kvitrud, 2000). The significant wave height is estimated to be about 12.5 m, and the peak period is estimated to be 14.5 s. Recently, in January 2024, the Åsgard FPSO slamming accident occurred (Havtil, 2024), see also in Figure 1. Luckily, no one was harmed during the incident. However, this reminds us that the design and the integrity check of the LQ on FPSO remain a challenge for the offshore industry. And this might be particularly important for those FPSOs converted from conventional oil tankers, where the LQ is vulnerable for green water loads due to relative lower freeboard comparing to purpose built FPSOs.

Buchner et al. (2002a, 2002b & 2003) discussed design aspects of green water loading on FPSOs, in which a semi-empirical design evaluation method was proposed. Due to its simplicity, this method has been utilized in engineering, especially in the early phase.

Figure 1. Damaged cabin 812 after green water loads to Åsgard (Havtil, 2024).

Faltinsen et al. (2002) performed a comprehensive study of the green water phenomenon with 2-D potential flow theory. Greco et al. (2004) did both experimental and numerical modelling focusing on understanding the hydrodynamic behaviours of water on deck. The process of the water on the deck is complicated and challenging for numerical methods. However, from the structure response side, the loads are typically demanded to be determined to evaluate the deformation of the structures. The green water loads

DOI: 10.1201/9781003684404-48

are impulse based, with short duration of rise time less than 1 second, high peak force and a gradually decreasing load period in seconds. Typically, the green water loads are similar as the Figure 2 shows. The irregular shape of the impulse type loads presents a challenge on the evaluation to the structures. Alternatively, equivalent impulse might be used instead, as discussed by Zhu et al. (2021). For quite a mount of loading panels receive the pressure loads almost at the same time, it is not straightforward to find the equivalent impulse type.

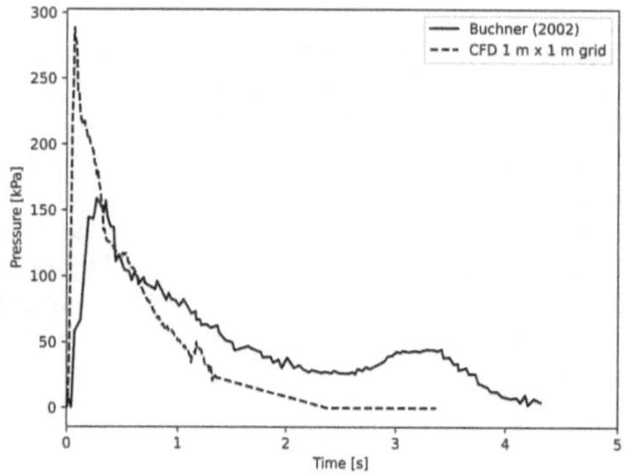

Figure 2. Illustration of typical green water loads from CFD and experimental tests.

The single degree of freedom (SDOF) analogy has been widely used to evaluate stiffened plating under impulse loads. Biggs initially proposed it in the 60s (Biggs, 1964) without considering the damping and the membrane effects in the resistant curve. DNV-RP-C204 (2019) has extended the method by including several types of membrane effects in the resistance curve. Troung et al. (2021) follows the same method and emphasizes the importance of considering strain-rate effects in the design formulations. In the '90s, Shubak R.B. proposed a simplified rigid-plastic analysis method for evaluating the stiffened plating's response under blast loads using the moving hinge hypothesis. Yu et al. (2021) further utilized this method in dealing with the stiffened panels under extreme water-slamming loads. Unlike the traditional approach, the proposed method does not need explicit pressure loads, while the water travelling velocity is required. The required water travelling velocity for the bow flare slamming scenario might be available, but it is not straightforward for the green water loads on deck since it depends on various factors, e.g., the wave crest, the bow shape, the bulwark and as well as the equipments on the deck. One argument for not using the pressure as design loads is the fact that the flexibility of the structure may play a role in the pressure loads obtained from CFD or experimental tests. Simplifying rigid structures can somewhat exaggerate the loads, and the size of the measuring strain gauge may also influence the measurements, as Faltinsen (2005) and other similar studies discussed.

This study focuses on structured responses. The green water loads are obtained from advanced CFD simulations, which is a one-way coupled analysis. The framework set-up is well-suitable for engineering purposes, both in the early phase and the detailed phase of a project. The occurrence probability of green water scenarios is not within the scope of this study. The goal is to provide an efficient analytical and numerical method to evaluate the stiffened plating under green water loads. Figure 3 shows the typical stiffened plating dimensions (with L shaped stiffener).

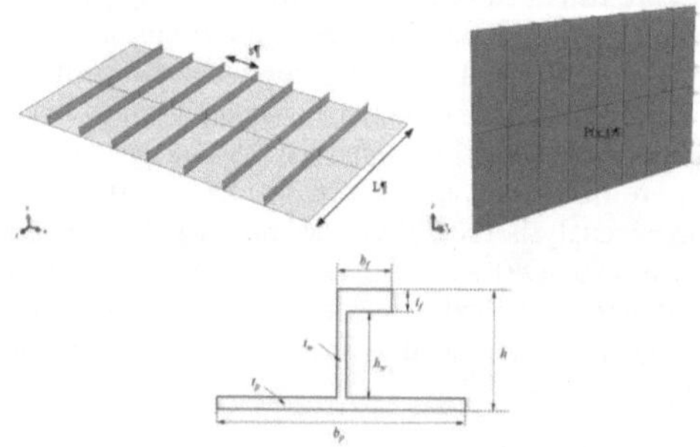

Figure 3. Stiffened plating under green water loads (HP profile stiffener as an example).

2 METHODOLOGY

In this study, the first step is to screen the design waves under ULS (within 100 years return period) condition. And use in-house tools to make irregular waves with given Hs and Tp. The extreme wave combination will be adopted and used in the CFD simulation. The load history will be extracted from the CFD simulation for each case. To obtain the structural behaviours, either the analytical or the numerical methods could be used. For the numerical method, the obtained load history will be mapped to the finite element model of the LQ wall structure. Figure 4 illustrates the procedures in detail. In the meanwhile, the SDOF (single degree of freedom) method has been also investigated, aiming to provide a trustful reference to the numerical results. The SDOF is not included in the flowchart as shown in Figure 4.

2.1 *CFD simulations*

As an important step in this evaluation process, the CFD simulation will give the design slamming impulse (pressure in time domain). In CFD simulations, the ship and LQ structures are assumed to be rigid. 3D motion of the ship is enabled with specified mass distribution and inertia properties of the ship. The actual motions of water particles are calculated in a 3D domain surrounding the ship, as well as the complex shape of the wavy water surface. The time evolution of water particle velocities and pressure is computed everywhere in the fluid. Integration of pressure and shear stresses on the body surface provides hydrodynamic forces.

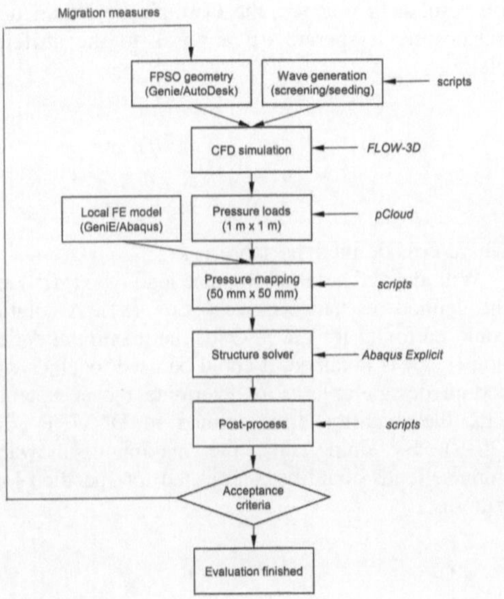

Figure 4. The flowchart for green water evaluation for LQ structures (SDOF method is done separately and not included here).

The commercial software FLOW-3D (2023). The program solves the three-dimensional Navier-Stokes equations on a fixed cartesian (rectangular) grid. The general moving object (GMO) in FLOW-3D is particularly relevant for marine applications. Objects can have prescribed velocities, forces and torques, or can move freely under the influence of hydrodynamic forces (coupled 6-DOF motion). Furthermore, objects can collide, and springs and ropes may be attached to objects, enabling the user to model mooring systems. The ability to generate waves (Airy, Stokes 5th, Fourier series method, solitary, random etc.) and let them propagate though the computational domain without excessive damping or reflection is another strong point of FLOW-3D, making the code very suited for many types of offshore simulations. Figure 5 shows a screenshot of the simulated green water loads to this FPSO, the impact of wave breaking over ship's deck, resulting in a significant volume of water surging on board. Note that the mitigation wall has been included in this case. The FPSO is imported from geometry file in STL format. A simplified model of LQ (rectangular box in the figure) was added. The pressure on the front surface of the LQ will be recorded and output for structure analysis.

Figure 6 visualized the pressure recorded in one of the simulated seeds. These recorded pressures are obtained by using the in-house developed tool called pCloud (pressure point cloud). The pressure history is recorded and output to selected panels. The panels could be moving or stationary objects. And it can cover all or part of the CFD geometry, enabling high spatial resolution in areas of interest. The panel model was created in GeniE, consists of 1 m x 1 m elements and covers the front face of the simplified LQ model. Figure 7 shows the pressure history for all panel elements under different smoothing time (left, 20 ms, middle, 50 ms and right, 100 ms). It could be seen that the smoothing interval plays a role on the pressure outputs. This is particularly relevant for numerical methods, while the pressure loads shall be unique and not dependent on the way how we measure it. With this respect, sensitivity studies to the structure responses under different smoothing time have been performed. The 50 ms smoothing time interval was selected and further used in this study since it gives most reasonable structure responses. And it also must be pointed out that with the one-way coupled analysis method, the structure flexibility has been overlooked, and the pressure could be exaggerated. The selection of 50 ms might be a good compromise for approaching the real pressure level.

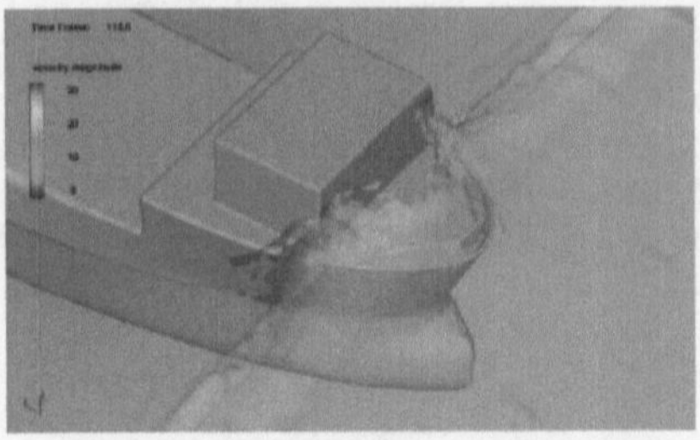

Figure 5. Simulated FPSO with green water loads on LQ wall with mitigation structure included.

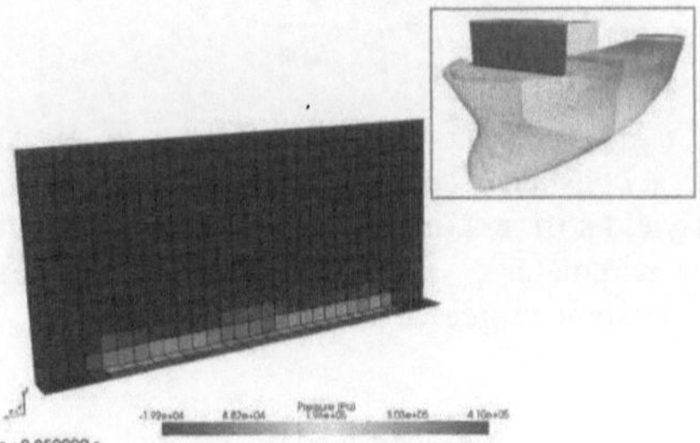

Figure 6. Visualization of the pressure time history. The grid size is 1m x 1 m representing the front wall of the LQ (the pCloud panel model is shown in the right top).

2.2 *SDOF method*

The SDOF method originated from the need to simplify the analysis of complex dynamic systems. It models a structure as a single mass-spring-damper system, where the mass represents the structure's inertia, the spring represents its stiffness, and the damper represents any energy dissipation mechanisms (Biggs,

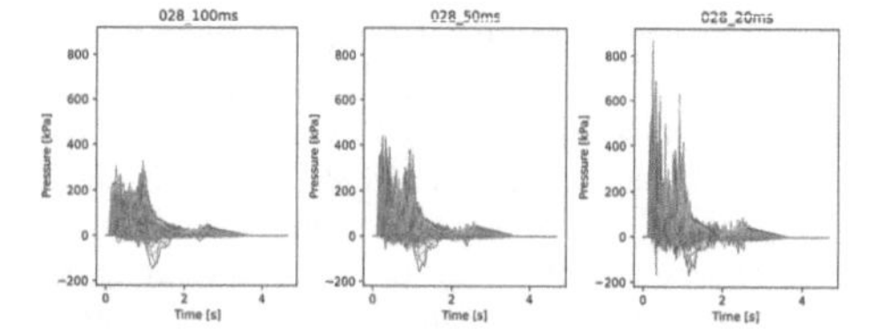

Figure 7. Pressure history for each grid (1m x 1m) under different smoothing intervals. Left: 20 ms, Middle: 50 ms, Right: 100 ms.

1964). This approach allows for the analysis of the structure's dynamic response to impulsive loads, such as those caused by explosions, slamming, and green water impacts.

To present the improved SDOF method, we include the same notations as used in Biggs (1964). Nondimensional equations of motion are as follows:

$$M\ddot{y} + ky = F_1[f(t)] \quad (1)$$

where the M is the mass, k is the resistance (stiffness), y is the displacement, $\ddot{y}$ is the velocity, $F_1[f(t)]$ is the external force in time history.

Letting $\xi = t/T$ and $\eta = y/y_{el}$ ($y \leq y_{el}$), the following equations are obtained:

$$\frac{My_{el}}{T^2}\ddot{\eta} + ky_{el}\eta = F_1[f(\xi)] \quad (2)$$

where T is the system's period, y_{el} is the maximum elastic displacement.

And since $ky_{el} = R_m$, and $T^2 = 4\pi^2 M/k$:

$$\frac{1}{4\pi^2}\ddot{\eta} + \eta = \frac{F_1}{R_m}[f(\xi)] \quad (3)$$

If $\eta > 1$, then it is in the plastic range, ky is then either a constant (perfect plastic assumption) or with post-yield stiffness (elastic-plastic response). We can use a parameter $\zeta = R/R_m$ to describe the normalized resistance force.

$$\zeta = \begin{cases} \eta & if \eta \leq 1 \\ 1 & if 1 < \eta \leq 1/k_3 \\ k_3\eta & if \eta > 1/k_3 \end{cases} \quad (4)$$

where k_3 is a normalized factor that describes the post-yield stiffness of the SDOF system. It is often described with respect to the elastic stiffness K, $k_3 = K_3/K_1$, and normally we have $k_3 \in [0, 1]$. Eq. 3 is solved by using 4th order Runge-Kutta scheme with vectorized inputs. With the current framework, the k_3 could be customized to reflect the structure resistance in a more reasonable way. For example, the true bi-linear curve as illustration in Figure 8 (right), where the flat plateau is removed from the resistance curve, and the second curve starts from the yield point. And additionally, it could be also a customized function, e.g., $k_3 = f(\eta)$, to reflect the true resistance curve either from analytical or from FE results. In practice, the k_3 might be obtained by performing a sperate FE analysis to the stiffened plating.

$$\zeta = \begin{cases} \eta & if \eta \leq 1 \\ 1 + (\eta - 1)k'_3 & if \eta > 1 \end{cases} \quad (5)$$

where k'_3 is defined the same as k_3.

With defined external impulse load type ($f(\xi)$) and the defined resistance curve ($k_3 = f(\eta)$). A solution could be found for Eq. 3. Once the maximum deflection (y_{max}) is obtained, it could be used together with certain design criteria to determine the acceptance, e.g., the ductility limit defined in DNV-RP-C204 (2019). By doing this, the maximum allowable impulse loads could be established for specified load patterns.

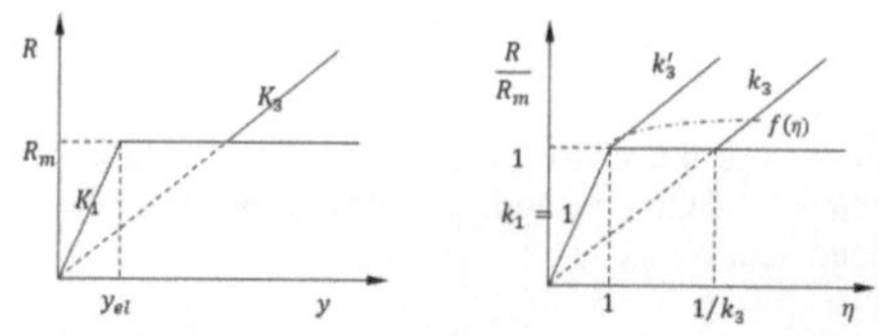

Figure 8. Simplified resistance curve (left) and normalized simplified resistance curve (right) including the improved resistance curve.

2.3 *Numerical method (FEA)*

Abaqus Explicit is chosen as the structural solver in this study. The output from the CFD analysis is the pressure time history, which is based on a 1 m x 1 m grid (selected in this study). A local model for the front wall of LQ (excluding windows) was built and meshed by GeniE/SESAM. The front wall is a stiffened panel, with 780 mm stiffener spacing and 3200 mm height between each deck. The major stiffener type is HP180x8. To capture the nonlinear failure mechanism, finer mesh (normally in the range of 10 mm to 50 mm) shall be used, see Figure 9 (a). For simplicity, the HP profile has been replaced with flat stiffener using the smearing area method, e.g., the cross-section area is kept the same. The connections are all assumed to be ideal, and no local welds are included in this model. Due to the size difference (1 m vs. 50 mm), a mapping between the Abaqus fine-meshed elements (50 mm mesh size) and the pressure grid (1 m mesh size) is needed. This was achieved using the CDKtree method within Python language (Python, 2020, 2022). A user script was made according, and it could be directly run within Abaqus CAE GUI, maximizing the work efficiency on setting up the model. The elements within each pressure grid are grouped into one surface, and the corresponding pressure history is applied by using a defined amplitude, see Figure 9 (b). The windows are not modelled in the FE model, and the corresponding pressure loads are ignored. The main reason for this is that the loads level dealt here are much higher than the design loads for the windows. As seen from Figure 10 (a, b), the defined amplitude curves are the same as the pressure history curves from the CFD outputs. The plotted pressure curves show that the pressure loads have considerable loading time in seconds.

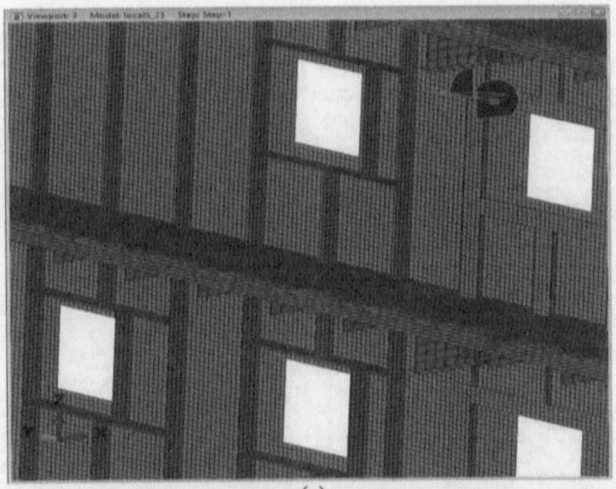

(a)

(b)

Figure 9. (a): Local view of the element mesh, general mesh size is 50 mm. (b): the pressure groups are defined by mapping the Abaqus elements with the pressure data grid. For example, five pressure groups are selected and highlighted.

Nonlinear material properties have been defined for the steels. The material grade is S235, and the plastic material properties are based on DNV-RP-C208 (2023). To simulate the damage due to high slamming loads, the ductile steel fracture model was used and defined, the gross yielding of the 1st principal plastic strain is set as 5.26%, which was implemented into Abaqus by converting to triaxiality based ductile damage, see Liu (2022) for more information. The added mass and the strain rate influences are both ignored in current study, the main argument for doing this is to simplify the process and obtain conservative results. Sensitivity studies have been performed by using different parameters with respect to the added mass and the strain rate model. The results show that both have the tendency to increase the capacity of the structure. There is a large uncertainty regarding how to quantify the added mass and the strain rate influences in the slamming events (Abrahamsen et al., 2020; Zhu et al., 2021). And it is not within the scope of this study.

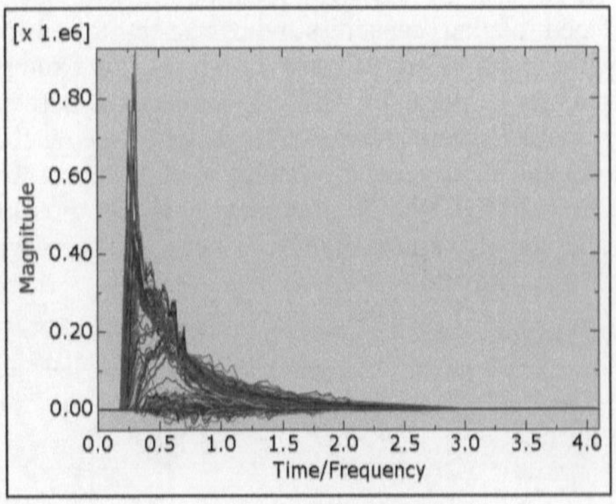

(a)

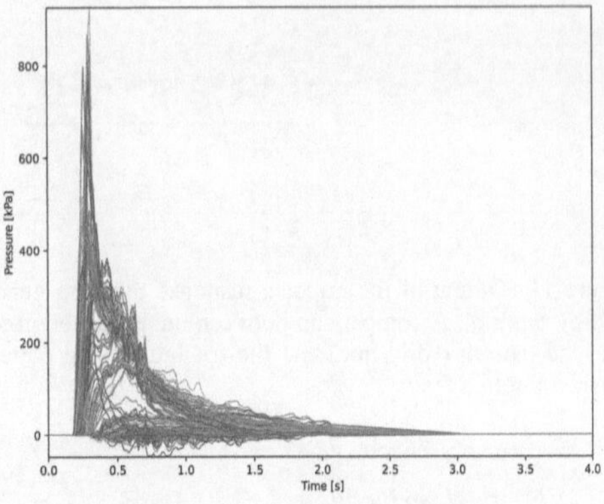

(b)

Figure 10. Defined amplitude curves for each element group, case 018_50ms. (a): Amplitudes plot in Abaqus. (b): Pressure history plots by using python from the pressure file.

3 SDOF AND FE MODEL VERIFICATION

In this section, the verification of the implemented SDOF and FE model methods will be presented.

3.1 *SDOF method verification*

The SDOF method developed in this paper has been implemented into an in-house developed tool Splast. The tool is made on general basis and be able to handle various structure types and impulse loads. The resistance curve could be either default (empirical curves from such as DNV-RP-C204, or other more advanced curve from user or literature). The stiffener profile could be standard shape or even arbitrary shapes. The damping factor could be also included. However, it might be not of interest as the maximum deflection is always the major concern in this method. The impulse loads could be also customized (theoretically could be any types), rather than the triangular load types. The developed in-house SDOF tool was tested against the DNV-RP-C204, and the same results are obtained. This tool provides a good basis for further using the SDOF method on finding out the stiffened plating capacities with respect to various impulse based loads. Figure 11 shows an example output from Splast, in which a comparison between a proposed green water loads pattern (see section below) and the triangular impulse load, both are with rise time 20 ms. With such outputs, it is then possible to find the maximum allowable peak pressure with given y_{max}/y_{el} and t_d/T.

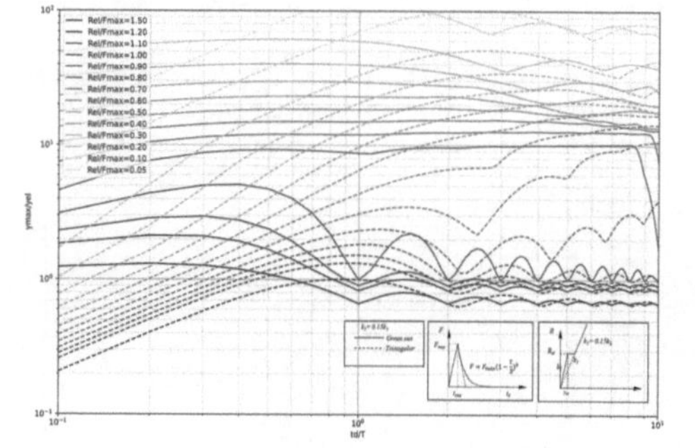

Figure 11. Output of the impulse response diagram generated by the tool. A comparison between the proposed green sea load pattern (solid line) and the triangular load pattern (dashed line).

3.2 *FE model verification*

Small-scale blast tests for stiffened panels were simulated to check the validity of the Abaqus Explicit solver. Houlston et al. (1987) presented a blast test for the stiffened panel. The stiffened panel tested was mounted flush with the ground surface 278 m from the center of the explosion for a peak side/on overpressure of 345 kPa. Figure 12 shows the comparison, present Abaqus results generally give conservative results. A comparison with Troung et al. (2021) was also done for three different scantlings (No. 1-3), see Figure 13. Still, present results give slightly larger deformations. The agreements are generally very good.

4 CASE STUDIES

Based on the established framework, we can perform the assessment to the LQ structures. In this section, both the SDOF results and the NLFEA results will be

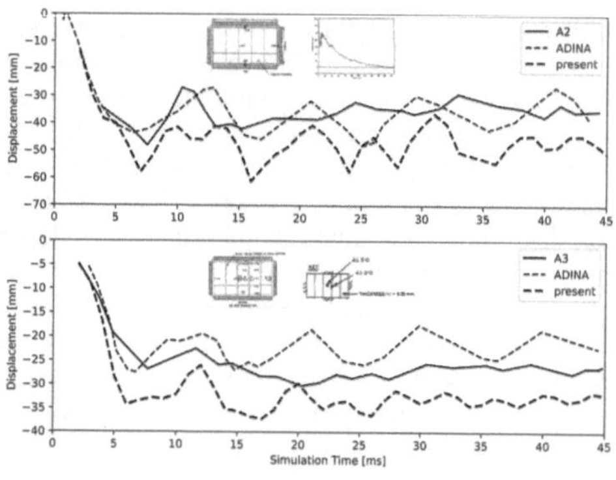

Figure 12. Comparison of different predictions with the measured motion-time history at center of panel (upper) and midspan of beam stiffener (down).

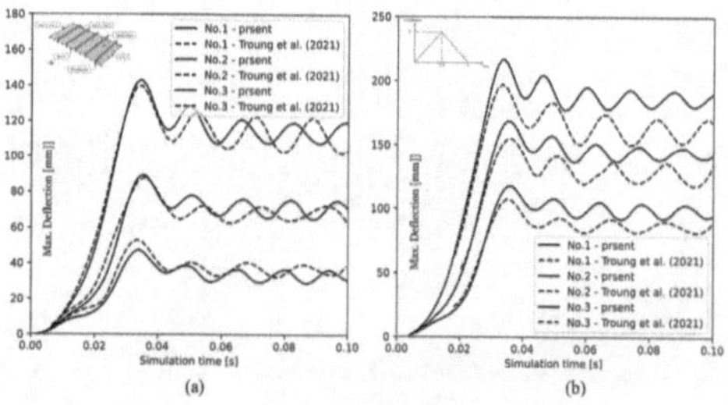

Figure 13. Numerical test comparison with Troung et al. (2021).

presented. Only two scenarios are picked from simulation case matrix and are presented here. Generally, a good agreement between SDOF and NFLEA results has been achieved. The LQ structures considered in this study shows substantial deformations under green water loads caused by irregular waves.

4.1 *SDOF results*

The front wall of LQ is mainly made of stiffened panels and openings (windows). The stiffened panel is either 8 mm or 10 mm thick and made of S235 material grade. The stiffeners are HP180x8, HP160x7 and HP140x7. The spacing is mainly 780 mm. The distance between each deck level is 3200 mm. The criterion used is the tensile fracture in yield hinges, which is defined as critical deflection (w_{crt}) (DNV-RP -C204 2019). The calculated critical deflection for the stiffened panel with HP180x8 is 142 mm (rigid constraint) to 173 mm (free constraint). The elastic deflection (w_e) is 5.9 mm. This is a conservative value for detecting the stiffener's start of fracture; the plating has a higher capacity before it starts to crack. To find out the maximum allowable dynamic pressure (peak value). The impulse curve is approached with a linear rising part, e.g., $p = p_{max} \cdot t/t_{rise}$ and a decreasing part with an empirical exponential expression, e.g., $p = p_{max} \cdot (1 - (t - t_{rise})/3)^8$.

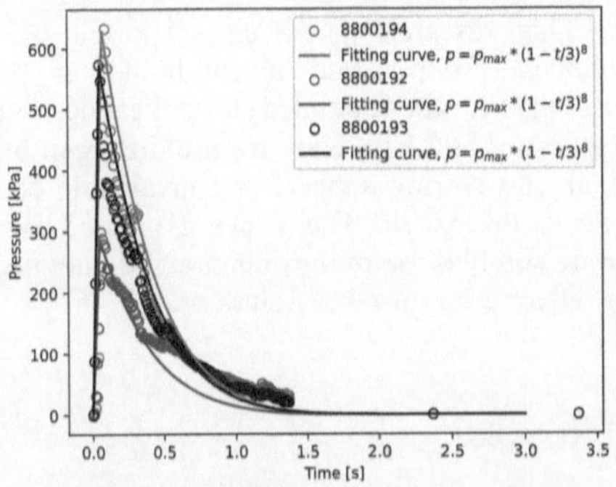

Figure 14. Typical slamming loads (selected from amplitude curves from one seeding cases in CFD simulations).

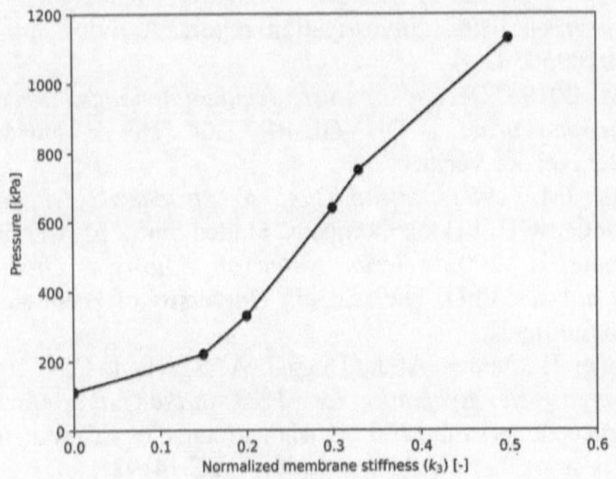

Figure 15. Peak pressure capacity with normalized membrane stiffness.

The agreement to several recorded pressure curves is quite good; see Figure 14. Implementing this idealized pressure history into the equation for the SDOF system can yield the maximum dynamic pressure; see Figure 15. The k_3 is the normalized membrane stiffness to the elastic stiffness of the SDOF system and is usually not straightforward to obtain. Based on a separate FE analysis, it could be estimated that k_3 is approx. 0.3 in this case, consequently the max. capacity (peak pressure value with assumed pattern) is then 638 kPa. Although the calculation is very simplified, it provides an intuitive impression of the structure's ultimate capacity and agrees well with the numerical results in the following section.

4.2 *NLFEA results of seed02 (Case 20)*

More than 20 cases have been simulated based on 7 different seed number and one regular wave scenario. Among all simulated results, the analytical capacity obtained (638 kPa) shows very good correlation with the numerical results. More specifically, no critical failure was observed if the peak pressure is less than 638 kPa in all simulations. This is very useful for establishing mitigation methods in CFD, no structural analysis is required.

Two of them are presented here. Figure 16 shows the screenshots from the Seed 02 in the CFD simulations. Relatively high peak pressure has been recorded, see Figure 17. Figure 18 presents the structure responses under the pressure loads. It is seen that the lower center part of the LQ structure was exposed too high-pressure impulse and the LQ structure starts to fail. Significant deformation (deck buckling) is also observed at the stair room walls on both sides.

Figure 16. Seed02 CFD simulation screenshots.

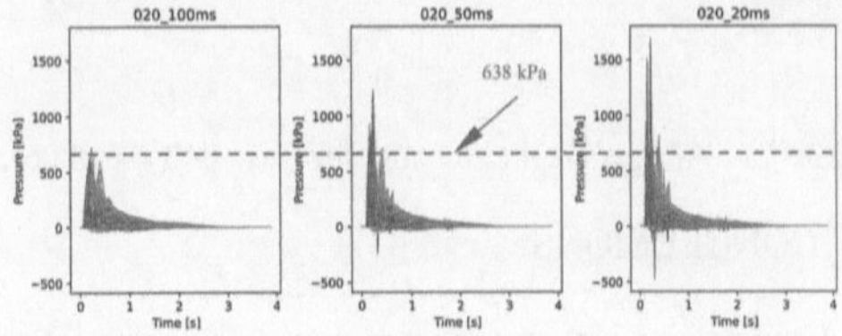

Figure 17. Pressure history under different smoothing time for Case 020 (Seed 02).

Figure 18. Abaqus Explicit results for Case 020 (Seed 02).

4.3 *NLFEA results of seed 06 (Case 30)*

Another case as shown in Figure 19, the pressure loads are within the analytical calculated maximum peak pressure (50 ms), see Figure 20. It could be expected that significant deformation may happen, but no crack will be initiated, as shown in Figure 21. The maximum deflection is located on the stair-room walls. The lower deck of the LQ structure has up to 65 mm deflection. In such case, the windows are probably will be smashed and present serious risk for the people living in LQ.

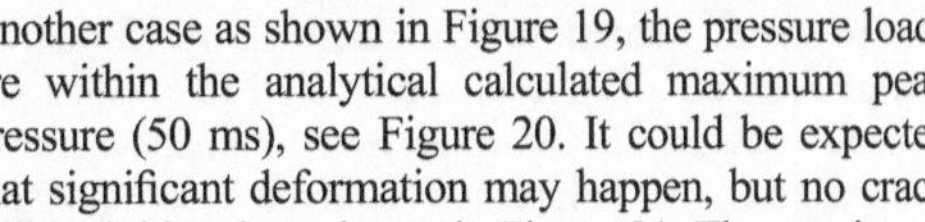

Figure 19. Seed 06 CFD simulation screenshots.

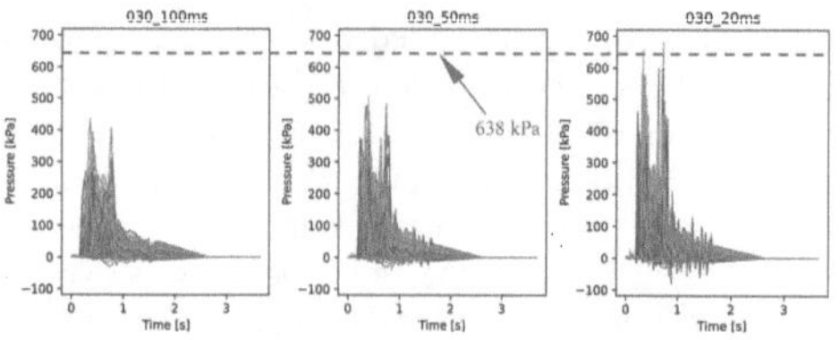

Figure 20. Pressure history under different smoothing time for Case 030 (Seed 06).

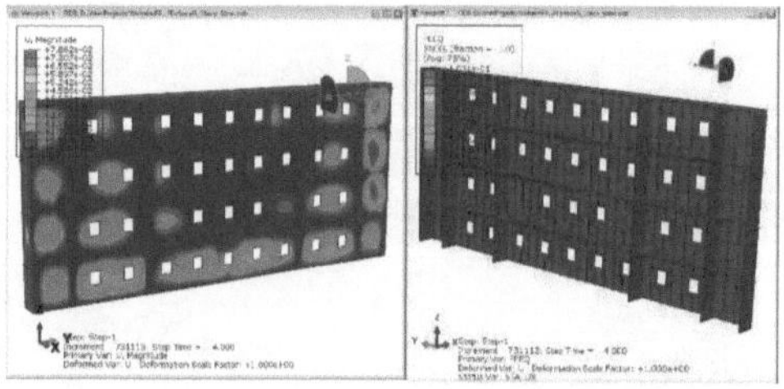

Figure 21. Abaqus Explicit results for Case 030 (Seed 06).

5 CONCLUSIONS

A Comprehensive framework has been developed for evaluating the LQ structure under green water loads in this study. It is comprised of two major parts, the CFD simulation and the dynamic structure analysis part by Abaqus Explicit. One-way coupling between the CFD and the structure solver was used. To achieve this, engineering tools have been developed to facilitate the seamless integration of different interfaces. Based on the results obtained, following conclusions could be made:

1. The framework of one-way coupled evaluation of LQ structure under green water loads has been established with the help of a series in-house engineering tools.
2. The SDOF method has been further developed with customized resistance curve.
3. An empirical pressure impulse is proposed for the green water loads and it is further applied in the SDOF method to calculate the maximum peak pressure the LQ structure can take. The results agree well with the FEA results, e.g., the FEA results for peak pressure less than the maximum peak pressure, no critical failures are observed. Consequently, the calculated maximum peak pressure could be used for optimizing the CFD mitigation measures without running new structural analysis.
4. The current analysis shows that the LQ structure consider in this model is not able to take the green water loads. Strengthen of the LQ or mitigation of green water loads must be implemented.

ACKNOWLEDGEMENTS

The author expresses his sincere gratitude to Aker Solutions for granting permission to publish this work. This research has been supported by the consultancy group, Entr, a sub-brand of Aker. Special thanks are extended to colleagues Kjetil K. Heggernes and Rolf Eide from the Bergen office for their unwavering support and invaluable contributions to this work. The views expressed in this paper are solely those of the author and do not necessarily rcflcct thosc of Aker Solutions.

REFERENCES

Ersdal G. & Kvitrud A. 2000. *Green water on Norwegian Production Ships*. Proceedings of the tenth (2000) International Offshore and Polar Engineering Conference (ISOPE). Seattle, USA, May 28-June 2, 2000.

Havtil. 2024. *Investigation of incident on 31 January 2024 involving a cabin window on Åsgard A being forced in by green water*. Investigation report. Activity number 001095054.

DNV. 2019. *Design against accidental loads. Recommended practice* DNVGL-RP-C204. Hovik, Norway: Det Norske Veritas.

Biggs JM. 1964. *Introduction to structural dynamics*. McGraw-Hill Book Company, United States of America.

Buchner B. 2002. *Green water on Ship-type Offshore structures*. PhD. Thesis. Delft University of Technology, Netherlands.

Buchner B., Voogt, A. J., Duggal, A. S., Heyl, C.N. 2002. *Green Water Evaluation for FPSOs in the Gulf of Mexico*. Proceedings of the 2002 Offshore Technology Conference. Houston, Texas U.S.A., 6-9 May. OTC 14192.

Buchner B., Lopez-Cortijo Garcia, J. N. 2003. *Design aspects of green water loading on FPSOs*. Proceedings of the 22nd International Conference on Offshore Mechanics & Arctic Engineering. Cancun, Mexico, 8-13 June. OMAE2003–37162.

Faltinsen O.M., Greco M., Landrini M. 2002. *Green Water Loading on a FPSO*. Journal of Offshore Mechanics and Arctic Engineering. Vol. 124/97.

Greco M., Landrini M., Faltinsen O.M. 2004. *Impact flows and loads on ship-deck structures*. Journal of Fluids and Structures. 19(2004), pp. 251–275.

Schubak R.B. 1991. *Nonlinear rigid-plastic analysis of stiffened plates under blast loads. Doctoral thesis*. The University of British Columbia.

Yu Z., Amdahl J., Greco M., Xu H. 2021. *Hydro-plastic response of beams and stiffened panels subjected to extreme water slamming at small impact angles*, Part I: An analytical solution.

Faltinsen O.M. 2005. *Hydrodynamics of High-Speed Marine Vehicles*. Cambridge University Press.

Dassault Systèmes. 2023. *Abaqus 2023: Abaqus Explicit* [Computer software].

Python Software Foundation. 2022. *Python Language Reference, version 3.11*. Retrieved from https://docs.python.org/3.11/.

Houlston, R. & Desrochers, C.G. 1987. *Nonlinear structural response of ship panels subjected to air blast loading*. Defence Research Establishment Suffield, Ralston, Alberta, Canada.

FLOW-3D® Version 2023R1 [Computer software]. (2023). Santa Fe, NM: Flow Science, Inc.

Virtanen, P., Gommers, R., Oliphant, T. E., Haberland, M., Reddy, T., Cournapeau, D., Burovski, E., Peterson, P., Weckesser, W., Bright, J., van der Walt, S. J., Brett, M.,

Wilson, J., Millman, K. J., Mayorov, N., Nelson, A. R. J., Jones, E., Kern, R., Larson, E., Carey, C. J., Polat, İ., Feng, Y., Moore, E. W., VanderPlas, J., Laxalde, D., Perktold, J., Cimrman, R., Henriksen, I., Quintero, E. A., Harris, C. R., Archibald, A. M., Ribeiro, A. H., Pedregosa, F., & van Mulbregt, P., & SciPy 1.0 Contributors. 2020. *SciPy 1.0: Fundamental Algorithms for Scientific Computing in Python*. Nature Methods, 17(3), 261–272.

Python Software Foundation. 2022. *Python Language Reference, version 3.11*. Retrieved from https://docs.python.org/3.11/.

DNV. 2024. *GeniE 64 version 8.9-04*, SESAM Package. Feb. 2024.

DNV-RP-C208. 2013. *Determination of structural capacity by non-linear finite element analysis methods*. Hovik, Norway: Det Norske Veritas.

Liu, Z. 2021. *Numerical simulation of dropped container impacts with an offshore platform deck in the North Sea*. Proc. JMechE Part M: J. Engineering for the Maritime Environment 1–10.

Abrahamsen B.C., Alsos H. S., Aune V., Fagerholt E., Faltinsen O. M., Hellan Ø. 2020. *Hydroplastic response of a square plate due to impact on calm water*. Phys. Fluids. 32 082103.

Zhu L., Tian L., Chen F., Yu T.X. 2021. *A new equivalent method for complex-shaped pulse loading based on saturation analysis and membrane factor method*. Int. J. Impact Eng. 158 104018.

Truong, D.D., Jang, B.-S., Ju, H.-B. 2021. Development of simplified method for prediction of structural response of stiffened plates under explosion loads. Marine Structures, 79, 103039.

Trends in Collision and Grounding of Ships and Offshore Structures – Zhu & Guedes Soares (Eds)

Impact of single-feature and multi-feature missing data on ship motion prediction with LSTM networks

P. Ni, J. Gong, Y. Qin & L. Wang
Shanghai Maritime University, Shanghai, P.R. China

ABSTRACT: Accurate ship motion prediction is critical for maritime safety. LSTM networks effectively capture nonlinear time series patterns but are sensitive to sensor failures. In this study, we analyze the impact of missing sensor data on LSTM-based prediction. Based on six degrees of freedom (6DOF) motion data, we simulate sensor failure by introducing both single and multiple feature missing scenarios. Missing Pitch and Yaw together causes R^2 to drop below 0.90 for all predicted features, and the combination of Surge and Roll leads to the lowest observed R^2 of 0.4863 for Sway. A supplementary test shows that even missing Surge alone causes performance degradation (R^2 = 0.5208), underscoring its importance. These findings highlight the vulnerability of LSTM to feature incompleteness and emphasize the need for robust modeling under realistic sensor failure conditions.

1 GENERAL INSTRUCTIONS

As the navigation environment and working conditions of ships become increasingly complex, accurate prediction of ship motion plays a vital role in marine engineering and shipping safety (Liu et al,2020). Traditional physical and linear models often struggle to maintain robustness under variable sea conditions. In contrast, recent deep learning methods, such as Long Short-Term Memory (LSTM) networks, have demonstrated strong capabilities in capturing nonlinear characteristics in time series data (Zhang et al., 2021).

Although some studies have used LSTM, RNN or neural networks to model ship motion time series, most of the work assumes that the data features are complete. For missing features, simple mean and zero padding often destroy the original dynamic pattern of the time series and cannot provide sufficiently accurate input for LSTM (Atoum et al, 2025). In addition, Due to the strong interdependence among ship motion features, missing data can propagate errors, leading to inaccurate predictions of rolling, pitching, and other motion components in LSTM-based models. (Xu et al., 2023).

In intelligent ships, reliable sensor data is crucial for real-time prediction and subsequent autonomous decision-making. However, real-world sailing conditions are highly complex, and onboard systems frequently encounter transient sensor failures and intermittent data corruption, primarily due to harsh marine environmental conditions. Sensor failures and missing data can significantly impact the prediction performance of LSTM networks to varying degrees.

Similar challenges caused by missing data have also been extensively studied in other time-series domains. For example, (Zhang et al., 2024) demonstrated that missing chlorophyll data had a greater negative impact on LSTM-based algal bloom prediction than other environmental features, due to its strong relevance to biomass variation. Similarly, (Boujoudar et al., 2024) discussed the impact of missing data in in meteorological data by comparing various imputation methods. These studies underscore that the effectiveness of LSTM models depends heavily on the completeness of critical input variables, reinforcing the need to evaluate model robustness under realistic missing data scenarios.

Based on the above background, this paper aims to address the impact of LSTM on ship motion prediction due to missing data. Specifically, the main contributions of this paper include: discussing the inability of trained LSTM to predict ship motion when a single sensor or multiple sensors fail, and quantifying the impact on prediction accuracy.

2 THEORY AND METHOD

Long Short-Term Memory (LSTM) is a type of recurrent neural network (RNN) designed to model temporal dependencies in sequential data. In this study, the LSTM is employed to predict future ship motion based on historical motion sequences. Specifically, the input to the LSTM consists of six degrees of freedom (6DOF) motion features—surge, sway, heave, roll, pitch, and yaw—recorded over the previous T time steps. The model is trained to predict the values of the same motion features at the next time step. This formulation enables the LSTM to learn the

DOI: 10.1201/9781003684404-49

temporal evolution patterns of ship behavior under dynamic sea conditions. LSTM learns temporal dependencies from historical sequential data to predict future values, making it suitable for time-series forecasting tasks such as ship motion prediction.

In time series forecasting, the input of LSTM is usually a two-dimensional matrix, represented as:

$$X = R^{T*F} \quad (1)$$

T is the time step and F is the number of features. For each time step t, the input to the LSTM unit is a feature vector: $x_t = [x_{t,1}, x_{t,2}, \ldots, x_{t,F}]$, Each column $x_{t,j}$ represents the change of a specific feature at all time steps (Hochreiter,1997). The structure of LSTM is in the following Figure 1:

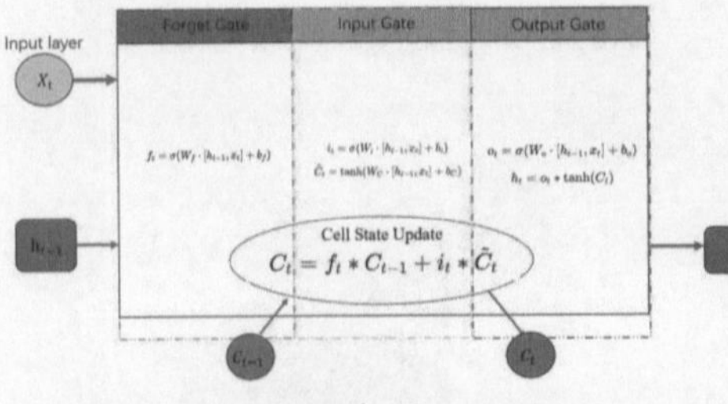

Figure 1. Structure of LSTM cell.

The feature vector x_t of each time step is the input of all gate operations. If the feature is missing, the matrix operation of calculating x_t will be disturbed, which will affect the results of each forget gate, input gate and output gate. When a column of features is missing, LSTM may not be able to capture the global relationship of the time series because the key features cannot participate in the current and future hidden state updates. C_t is the cell state, which also depends on x_t. Therefore, missing data will affect all parts of the LSTM architecture, resulting in inaccurate prediction results.

To measure the average discrepancy between the predicted values and the actual values, this study employs the Mean Squared Error (MSE) as one of the primary evaluation metrics. MSE is defined as follows:

$$MSE = \frac{1}{n}\sum_{i=1}^{n}(y_i - \widehat{y}_i)^2 \quad (2)$$

A smaller MSE value indicates higher prediction accuracy.

R^2:The coefficient of determination measures the fit of the model and its value range is [-∞,1]. The closer it is to 1, the better the fit.

$$R^2 = 1 - \frac{\sum(y_i - \widehat{y}_i)^2}{\sum(y_i - \bar{y})^2} \quad (3)$$

where n denotes the total number of samples, $\widehat{y}_i$ is the predicted value for the i th sample, and y_i is the corresponding ground truth, $\bar{y}$:The mean of the true values.

When R^2=1, the model perfectly fits the data, with predictions matching the true values exactly; when R^2=0, the model performs no better than using the mean of the target variable.; when R^2<0, the model performs worse than the mean, indicating poor predictive ability and failure to capture underlying data patterns.

3 DATA OF SHIP MOTION AND WAVE

The data source of this article is the motion data of the trimaran under regular waves with a wave direction of 30 degrees. In this study, the ship's autopilot system is utilized to collect various feature data of the trimaran over time in oblique stern waves. The differences between different working conditions are shown in Table 1. Vs is the projection of the initial ship speed, $V_w$is the wave propagation speed, λ is the wavelength, H is the wave height, and WD is the initial wave direction feature.

Table 1. Initial working conditions of different cases.

case	Vs	V_w	λ	H	WD
1	2440	2052	270	116	30°
2	2440	2258	327	146	30°
3	2440	2466	390	167	30°
4	2440	2649	450	193	30°
5	2440	2820	510	218	30°

The motion dataset under each working condition contains motion features, describing the six-degree-of-freedom motion of the ship in three-dimensional space, and LSTM is used to predict the state of the ship's motion features. They are used to simulate the dynamic behavior and response of the ship. A summary of the key variables is provided in Table 2, categorizing the features into displacement, rotation.

Table 2. Motion data feature classification.

Category	Features	Description
Translational motion	Surge, Sway, Heave	Ship displacement in X, Y, Z directions
Rotation	Roll, Pitch, Yaw	Rotation angles

LSTM relies on time series data for prediction, and the real-time operating status of the ship must be collected through sensors. Without sensors, there is no high-quality input data, and the model cannot make accurate predictions.

The selection of LSTM hyperparameters in this study is based on the discussion presented in the previous work (Xu et al., 2024), with adjustments to fit the characteristics of our dataset. The model architecture consists of two hidden LSTM layers, each with 64 units, which balances model complexity and training efficiency. A dropout rate of 0.2 is applied to mitigate overfitting by randomly disabling neurons during training. The learning rate is set to 0.001, allowing for stable convergence under the Adam optimizer. The model is trained for 30 epochs, which was found to be sufficient for loss stabilization in our experiments. ReLU is used as the activation function for nonlinearity, and input features are normalized using Min-Max scaling to accelerate gradient-based optimization.

4 RESULTS AND CONCLUSIONS

This article aims to study the impact of sensor failure on the prediction accuracy of the ship's six-degree-of-freedom motion when the input data is abnormal and thus affects the LSTM prediction. Therefore, this article first trains the LSTM model with complete data. At the same time, because null values cannot appear in LSTM, 0 is used to replace the missing data at the failure time node. The following are the model training and test results:

LSTM model training process

To compare the accuracy of LSTM in the case of missing data and complete data, LSTM is first trained with complete data and then tested with both types of data. The training file is case1 2 4 5, and case 3 is processed as a new test file for prediction. Because the training data are the same, only the missing data in the test data is different, so the loss during the training process is the same, that is, it is just testing the performance of data sets with different data on the same model training results. The training loss is shown below in Figure 2 :

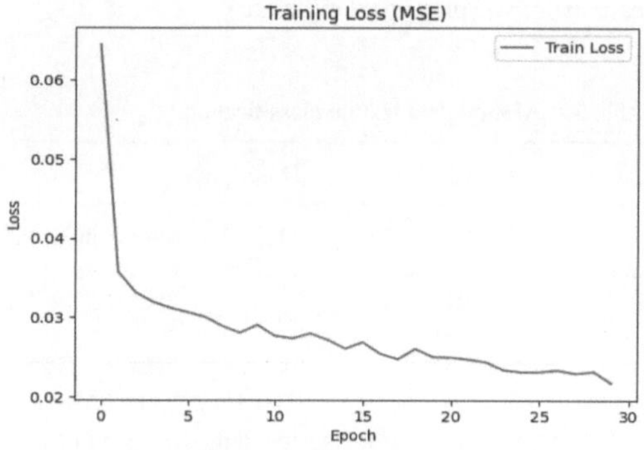

Figure 2. Comparison of R^2 with missing data.

It shows that the initial training loss is high (>0.06), but as the Epoch increases, the loss decreases rapidly and stabilizes after 10-15 rounds. Finally, after 30 rounds, the loss stabilizes at around 0.02, indicating that the model has fully converged. According to the loss curve analysis, further increasing the Epoch may not bring obvious performance improvement, so 30 rounds of training are sufficient.

Single feature data missing

Figure 3 is a correlation heatmap, which is used to show the correlation between different features. When it is close to 1, it means the two features are strongly positively correlated; otherwise close to -1: it means the two features are strongly negatively correlated; color close to 0: it means the two features have almost no correlation.

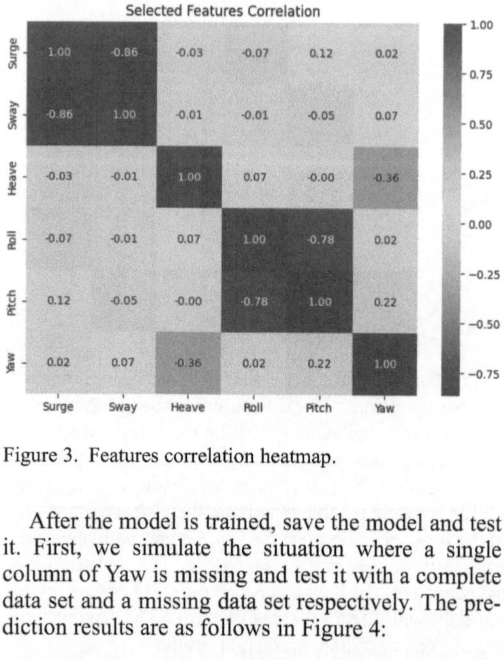

Figure 3. Features correlation heatmap.

After the model is trained, save the model and test it. First, we simulate the situation where a single column of Yaw is missing and test it with a complete data set and a missing data set respectively. The prediction results are as follows in Figure 4:

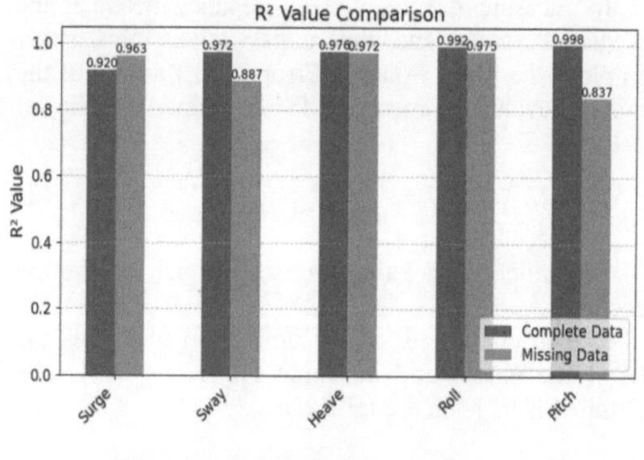

Figure 4. Comparison of R^2 with missing data.

As shown in Figure 4, the R^2 values obtained with complete data are consistently high and close to 1 across all motion components, indicating strong model performance under full feature availability.

When data is missing, the R^2 values drop significantly, especially for Pitch and Sway, suggesting that these components are more sensitive to Yaw feature loss. When R^2 is closer to 1, it means the accuracy better. In contrast, Heave and Roll show minimal R^2 degradation, indicating greater robustness to missing inputs. Notably, the R^2 value for Surge is slightly higher under missing-data conditions compared to the complete-data case. This may be due to reduced input redundancy or noise when one feature like Yaw is excluded, allowing the model to focus on more relevant signals for Surge prediction. Figure 5 shows the comparison between the predicted values and the true values of two features with large differences.

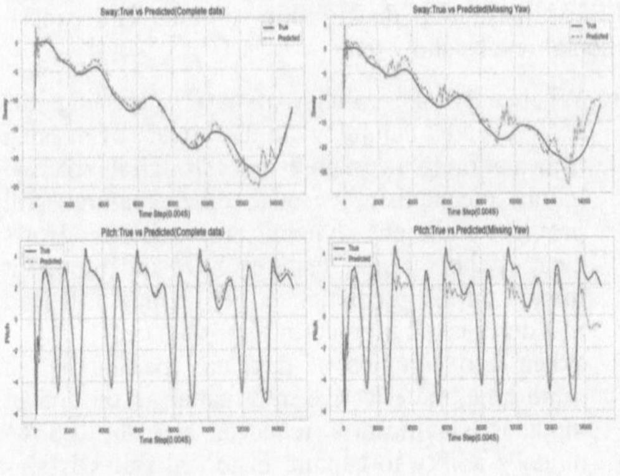

Figure 5. The effect of missing yaw data on the predictions of Sway and Pitch.

Multiple features data are missing

This study further analyzed the prediction performance of the LSTM model when two columns of input features were missing at the same time. Taking actual sensor failure as the background, the feature missing situation was simulated by artificially setting zeros, and quantitative evaluation was performed.

Since there are a lot of missing experimental data in the two columns, the typical features are selected here, that is, the combination of the two missing columns that has a particularly large impact on the prediction accuracy of other columns, as shown in the following Table 3.

Table 3. Typical impact of missing data feature combinations.

number	Missing features	Most affected features	R^2
Test A	Surge, Heave	Sway	0.5056
Test B	Surge, Roll	Sway	0.4863
Test C	Surge, Pitch	Sway	0.5076
Test D	Roll, Yaw	Pitch	0.8066
Test E	Heave, Yaw	Pitch	0.8324
Test F	Pitch, Yaw	All features	>0.90

It shows that when the correlation between the two missing columns of features and the features to be predicted is strong (such as Test A B C), the accuracy of some prediction features (Pitch) decreases significantly, with the lowest value dropping to 0.4863. When the missing feature combinations are Test D and E, the prediction performance of some features decreases moderately: R^2 is 0.8066 and 0.8324 respectively. When the correlation between the features in Test F is weak, the prediction accuracy of the model is basically unaffected: R^2 is higher than 0.9. The above results show that the impact of missing features on the model prediction performance mainly depends on the correlation between the missing features and the target prediction features.

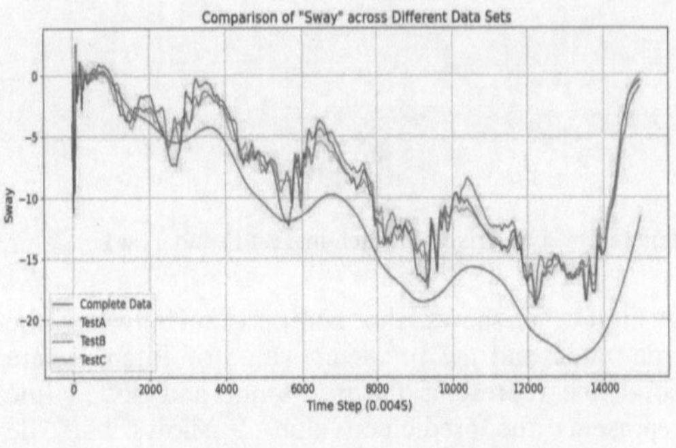

Figure 6. Comparison of sway across different data sets.

The comparison between the predicted value and the true value can be seen in Figure 6. The figure illustrates the displacement in the Sway over time across different data sets. The complete data follows a pronounced oscillatory pattern with significant shifts, especially after time step 12,000. Other data sets show similar trends but with less variation. In the early stages, all data sets behave similarly with minor fluctuations. As time progresses, discrepancies emerge, with some trajectories deviating more noticeably. After time step 12,000, the complete data shows a sharp decline and rise, while the other data sets exhibit more stable variations. This suggests that missing or modified data may have affected consistency.

Considering the strong correlation between Surge and Sway observed in Figure 3, an additional test was conducted in which only the Surge feature was removed. The resulting R^2 value was 0.5208, which is higher than the values observed in three two-feature missing cases involving Surge—namely (Test A: 0.5056), (Test B: 0.4863), and (Test C: 0.5076). This comparison suggests that while missing Surge alone significantly affects model performance, the combined effect of missing Surge with another feature further amplifies the prediction degradation. This supports the critical role of Surge and its correlation with Sway.

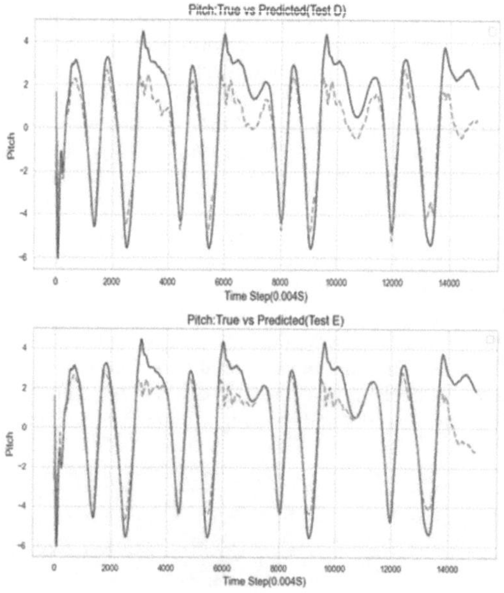

Figure 7. Comparison of Pitch in Test D and Test E.

Figure 7 shows the comparison between the true value and the predicted value of Pitch, where solid line represents the true value and dotted line represents the predicted value. It shows that the prediction performance of some features in Test D and E cases has moderately decreased, but the model can still capture the state relationship of the time series, and the impact on the prediction accuracy is medium.

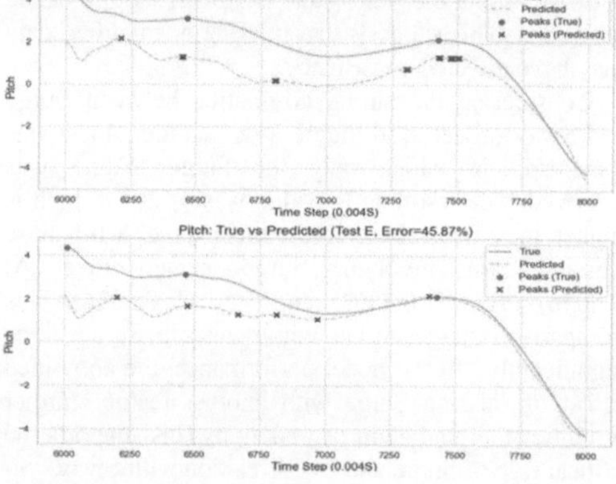

Figure 8. Impact of missing features on pitch peak error.

This paper evaluated the impact of missing features on pitch peak prediction error in Figure 8. When only the Yaw feature was missing, the peak error was 33.53%. However, when two features were missing (Test D & E), the error increased significantly by 36%-45%, indicating that the additional missing feature plays a crucial role in pitch prediction. These results highlight the necessity of preserving key features for accurate motion forecasting.

5 CONCLUSIONS

This paper studies the impact of sensor failure on LSTM prediction during ship motion. The experimental results show that:

1. When only one column of data is missing, the LSTM model can alleviate the impact of missing information to a certain extent through its internal gating mechanism, because other features still provide sufficient dynamic information. However, if the missing features have an important impact on the prediction target, it may still lead to a decrease in prediction performance.
2. When two columns of data are missing at the same time, the effective information in the model input is significantly reduced, and the LSTM model's ability to capture historical states is significantly weakened, making the prediction accuracy drop more significantly. The severity of this effect depends on the importance of the missing features and their relevance to the prediction target.

To enhance prediction accuracy under incomplete input conditions, a promising direction is to explore data completion techniques. Approaches such as mean-value substitution or MLP-based feature reconstruction may help restore missing inputs prior to prediction, thereby improving model robustness. These strategies will be further investigated in subsequent work.

ACKNOWLEDGEMENTS

This project was supported by National Natural Science Foundation of China (Grant number 52401385), Open Research Project of State Key Laboratory of Maritime Technology and Safety, China (Grant number W25CG000073).

REFERENCES

Atoum I. The critical role of evaluation metrics in handling missing data in machine learning[J].

Boujoudar M, El Ydrissi M, Abraim M, et al. Comparing machine learning algorithms for imputation of missing time series in meteorological data[J]. Neural Computing and Applications, 2024: 1–15.

Hochreiter S. Long Short-term Memory[J]. Neural Computation MIT-Press, 1997.

Liu Y, Duan W, Huang L, et al. The input vector space optimization for LSTM deep learning model in real-time prediction of ship motions[J]. *Ocean Engineering*, 2020, 213: 107681.

Xu J, Gong J, Wang L, et al. Integrating k-means clustering and LSTM for enhanced ship heading prediction in oblique stern wave[J]. *Journal of Marine Science and Engineering*, 2023, 11(11): 2185.

Xu J, Gong J, Li Y, et al. Surf-riding and broaching prediction of ship sailing in regular waves by LSTM based on the data of ship motion and encounter wave[J]. *Ocean Engineering*, 2024, 297: 117010.

Zhang C, Ding W, Zhang L. Impacts of Missing Buoy Data on LSTM-Based Coastal Chlorophyll-a Forecasting[J]. *Water*, 2024, 16(21): 3046.

Zhang T, Zheng X Q, Liu M X. Multiscale attention-based LSTM for ship motion prediction[J]. *Ocean Engineering*, 2021, 230: 109066.

Trends in Collision and Grounding of Ships and Offshore Structures – Zhu & Guedes Soares (Eds)
© 2026 The Author(s), ISBN 978-1-041-16361-9

Numerical study of fluid impact on Mark-III type flat panel in wet drop

C.Y. Wang, G.J. Shi & D.Y. Wang
Shanghai Jiaotong University, Shanghai, P.R. China

C.F. Zou
Jiangsu Ocean University, Lianyungang, Jiangsu, P.R. China

ABSTRACT: Based on the CFD method, the wet drops of the Mark III type flat panel are numerically carried out to analyze distribution characteristic of the peak impact pressure and the mechanism. This paper compares the distribution of the peak pressure between the flat region and the corrugation region of the model, and analyze the effect of the incident angle on the distribution of peak impact pressure. The results show that the effect of corrugations on distribution of the peak impact pressure of the Mark III type flat panel cannot be neglected. In addition, it also elaborates mechanisms of the air cushion effect in vertical water impact and the duel-peak pressure characteristic in inclined water impact. The study provides a reference and basis for the design of the model tests and the assessment of test results.

1 INTRODUCTION

In partially filling conditions, the external excitation causes both hull motion and free surface motion of the liquid in the tank. When the external frequency is close to the natural frequency of the liquid in the tank, violent liquid sloshing produces high-velocity movement of the free surface as well as breaking waves, which generates impact pressure on the containment system and even causes failures (Zhang et al., 2023; Wu et al., 2020). Many researchers have made contributions to the prediction and study of fluid impact pressure in tanks (Bae et al., 2020; Chen et al., 2025). Due to the nonlinearity and randomness of liquid sloshing, it is difficult to standardize the fluid impact used for strength assessment of the structure, especially for structures with local concavities such as Mark-III type flat panel. It is common to simplify or equivalent sloshing pressure to other load forms (Xie et al., 2024). Given the fluid-structure coupling when the sloshing liquid impacts the containment system, the wet drop is considered to be a reasonable research method, which is easy to carry out and repeat (Xie et al., 2024; Pan et al., 2024). Noh et al. (2009) compared the impact resistance of the Mark-III containment system with that of the KC-1 containment system by the wet drop tests. They focused on impact pressure at four positions on the model surface under different operating conditions. But for the Mark-III type flat panel, the effect of complex geometrical features on the local flow around the model before fluid impact and the evolution of gas and liquid phases cannot be neglected (Yang et al., 2023; Fang et al., 2022). Therefore, it is not reasonable to assess the fluid impact on the Mark-III type flat panel by the impact pressure at a specific position, and the distribution of the impact pressure on the model is very important (Wei et al., 2025). Due to limitation of computational and experimental cost, most studies focused on the pressure at a specific position and the pressure distribution at a specific time. There are few studies on distribution of the global fluid impact pressure.

In this paper, the wet drops of a large full-scale Mark-III type flat panel is numerically carried out based on the CFD (Computational Fluid Dynamics) method. The evolution of gas and liquid phases as well as the pressure characteristics under vertical and inclined water impacts are elaborated and the distribution characteristics and mechanism of the peak pressure on the model at different incident angles are analyzed. These works inform design of wet drop tests for flat panels with local concavities.

2 NUMERICAL APPROACH

2.1 *Governing equation*

The wet drops are numerically conducted using the CFD method in this paper. The governing equation (URANS equation i.e. unsteady Reynolds time-averaged Naiver-Strokes equation) of fluid are as follows:

DOI: 10.1201/9781003684404-50

$$\frac{\partial \bar{u}_i}{\partial t} + \frac{\partial}{\partial x_j}(\bar{u}_i \bar{u}_j) = -\frac{1}{\rho}\frac{\partial \bar{p}}{\partial x_i} + v\frac{\partial^2 \bar{u}_i}{\partial x_j \partial x_j} - \frac{\partial \overline{u_i'' u_j''}}{\partial x_j} \tag{1}$$

$$\frac{\partial \bar{u}_i}{\partial x_i} = 0 \tag{2}$$

where $\overline{u_i}$ and u_i' are the average velocity and the fluctuating velocity, respectively. $\overline{P}$ is the total pressure. Collisions between fluids and structures with complex geometries are often accompanied by large pressure gradients, flow separation, and jets. In the Realizable k-ε model, a modified transport equation for dissipation rate and a novel formulation for turbulent viscosity are applied. The dissipation rate is derived from an exact equation for transport of the mean-square vorticity fluctuation. The Realizable k-ε model satisfies certain mathematical constraints on the Reynolds stresses, consistent with the physics of turbulent flows. This makes the Realizable k-ε model suitable for numerical simulations of wet drops. In the Realizable k-ε model, turbulent kinetic energy k, dissipation rate ε and turbulent viscosity μ_t are calculated as follows:

$$\begin{gathered}\frac{\partial}{\partial t}(\rho k) + \frac{\partial}{\partial x_j}(\rho k \bar{u}_j) = \\ \frac{\partial}{\partial x_j}\left[\left(\mu + \frac{\mu_t}{\sigma_k}\right)\frac{\partial k}{\partial x_j}\right] + G_k + G_b - \rho\varepsilon - Y_M + S_k\end{gathered} \tag{3}$$

$$\begin{gathered}\frac{\partial}{\partial t}(\rho\varepsilon) + \frac{\partial}{\partial x_j}(\rho\varepsilon\bar{u}_j) = \frac{\partial}{\partial x_j}\left[\left(\mu + \frac{\mu_t}{\sigma_\varepsilon}\right)\frac{\partial \varepsilon}{\partial x_j}\right] + \\ \rho C_1 S\varepsilon - \rho C_2 \frac{\varepsilon 2}{k + \sqrt{v\varepsilon}} + C_{1\varepsilon}\frac{\varepsilon}{k} C_{3\varepsilon} G_b + S_\varepsilon\end{gathered} \tag{4}$$

$$\mu_t = \frac{\rho k^2}{A_0 \varepsilon + A_s k \sqrt{S_{ij} S_{ij} + \widetilde{\Omega}_{ij}\widetilde{\Omega}_{ij}}} \tag{5}$$

$$\begin{cases}\widetilde{\Omega}_{ij} = \Omega_{ij} - 2\varepsilon_{ijk}\omega_k \\ \Omega_{ij} = \overline{\Omega_{ij}} - 2\varepsilon_{ijk}\omega_k\end{cases} \tag{6}$$

where G_k is the turbulence kinetic energy for the mean velocity gradients, G_b is the turbulence kinetic energy for the buoyancy, Y_M is the contribution of the fluctuating dilatation in compressible turbulence to the overall dissipation rate. σ_k and σ_ε are the turbulent Prandtl numbers for k and ε, respectively. S_k and S_ε are source terms. S_{ij} is the rate-of-strain tensor and $\overline{\Omega_{ij}}$ is the mean rate-of-rotation tensor. ω_k is the angular velocity. A_0, C_2 and $C_{1\varepsilon}$ are the constants (4.04, 1.90 and 1.44).

The air in this paper satisfies the compressible ideal gas law:

$$\rho_a = \frac{P_{op} + P}{\frac{R}{M_w} T} \tag{7}$$

where ρ_a is density of air. P_{op} is operating pressure and P is local static pressure. R is universal gas constant. T is temperature and M_w is molecular weight. For the compressible ideal gas in the flow state, the flow characteristics are mainly determined by total pressure and total temperature:

$$\frac{P_0}{P} = \exp\left(\frac{\int_T^{T_0} \frac{C_p}{T} dt}{R}\right) \tag{8}$$

where P_0 is the total pressure, T_0 is the total temperature and C_p is the specific hear factor. For air, C_p is 1006.43 J/(kg·K) and P_0/P is 1.89.

2.2 *Free surface tracking*

For a typical multi-phase flow problem, tracking the two-phase interface is key to ensuring the accuracy of numerical simulation. The CLSVOF (Coupled Level-Set/Volume-of-Fluid) method combines the advantages of the Level-Set method and the VOF method. The VOF method is applied to reconstruct the two-phase interface and Level-Set method is applied to determine the normal motion of the two-phase interface. The details are as follows:

$$\frac{\partial \varphi}{\partial t} + \frac{\partial \varphi \bar{u}_i}{\partial x_i} = 0 \tag{9}$$

where φ is a continuous signed distance function that distinguishes the water phase from the gas phase within the computational domain. In the water phase $\varphi > 0$, while in the gas phase $\varphi < 0$, and at the interface of the two phases $\varphi = 0$. The curvature of the two-phase interface can be calculated as follows:

$$\kappa(\varphi) = -\frac{\partial}{\partial x_i}\left(\frac{\partial \varphi}{\partial x_i} / \sqrt{\frac{\partial \varphi}{\partial x_j}\frac{\partial \varphi}{\partial x_k}}\right) \tag{10}$$

When the cell contains both phases, the equivalent density ρ' and the equivalent viscosity μ' can be calculated by the following equation:

$$\rho' = H(\varphi)\rho_w + [1 - H(\varphi)]\rho_a \tag{11}$$

$$\mu' = H(\varphi)\mu_w + [1 - H(\varphi)]\mu_a \tag{12}$$

where ρ_w and ρ_a are the densities of water and gas phases, respectively. μ_w and μ_a are the viscosity of water and gas phases, respectively. $H(\varphi)$ is the Heaviside function, which is utilized to mitigate the discontinuities of physical properties in the vicinity of the two-phase interface.

2.3 *Dynamic mesh model*

In this study, the linear elastic body method within the framework of the smooth model is used to update the dynamic mesh. The linear elastic body method is particularly advantageous for improving mesh quality when dealing with significant boundary movements. The governing equations of the linear elastic body method are as follows:

$$\nabla \cdot \sigma(y) = 0 \tag{13}$$

$$\sigma(y) = \lambda(tr\varepsilon(y))I + 2\mu\varepsilon(y) \tag{14}$$

$$\varepsilon(y) = \frac{1}{2}\left(\nabla y + (y)^T\right) \tag{15}$$

where σ and ε are the stress and strain tensor, respectively. y is the shape variable of mesh and I is the unit matrix. λ and μ are the first and second Lamé constants, respectively. μ can also be defined as the shear modulus in these equations. The ratio of μ to λ is more important in the governing equations, and it is usually defined by the Poisson's ratio.

3 MODEL OF MARK-III FLAT PANEL

3.1 *Computational model*

As shown in Figure 1, the model has the dimensions of 2.4×2.4 m (*L*×*B*). There are some large and small corrugations intersecting with each other on the Mark-III type flat panel (Kim et al., 2025). These corrugations have gaps of 0.34 m and divide the surface into 36 separate complete area units.

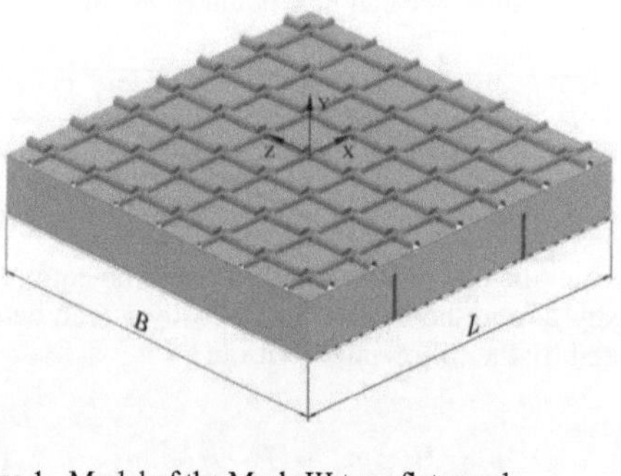

Figure 1. Model of the Mark-III type flat panel.

As shown in Figure 1, the computational domain has the dimensions of 4.0×4.0 m (*l*×*b*). The water depth *d* is 2.0 m and the height above the free surface *h* is 0.8 m. The model can only move in Y-direction. The incident angle of the model is θ. The initial gap *H* between the model and the free surface is 0.1 m. Acceleration of gravity is 9.8 m/s^2. The overlapping mesh technique is used to attain efficient computation without compromising the mesh quality. The component mesh uniformly covers the flow field around the model within 200 mm, i.e. δ=0.2 m. Through sensitivity analysis, the temporal resolution is 5×10^{-5} s and the spatial resolution of the component grid is 5.0 mm.

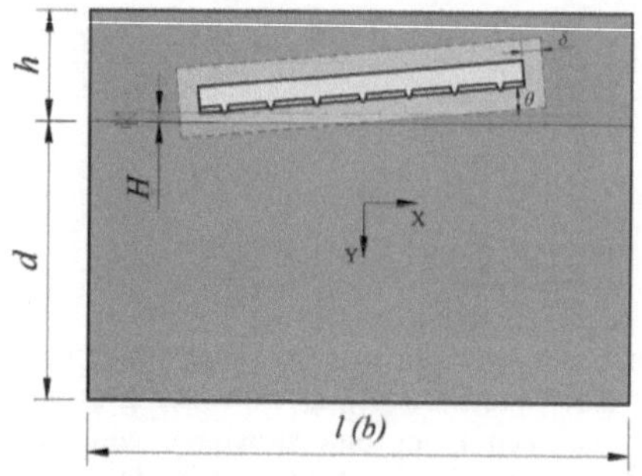

Figure 2. Cross section of the computational domain.

3.2 *Validation of numerical model*

The numerical model in this paper is first verified by the wet drop test for rigid body proposed by Yang et al. (2007). As shown in Figure 3, the wedge has the dimensions of 0.6×0.4×1.0 mm with a deadrise angle of 10 °. The total mass is 60.0 kg and the initial height is 0.5 m. The model can only move in the Y direction.

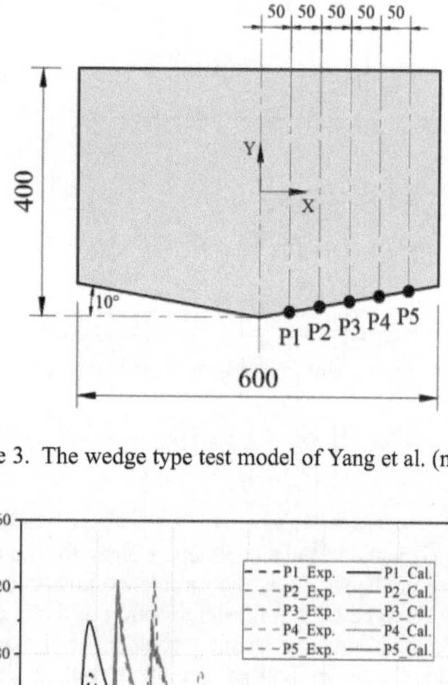

Figure 3. The wedge type test model of Yang et al. (mm).

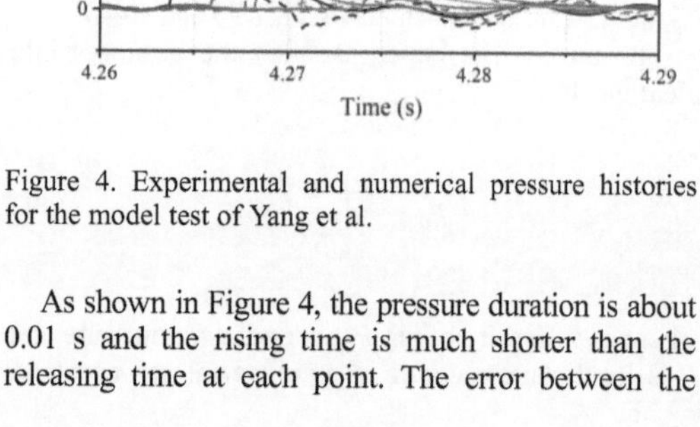

Figure 4. Experimental and numerical pressure histories for the model test of Yang et al.

As shown in Figure 4, the pressure duration is about 0.01 s and the rising time is much shorter than the releasing time at each point. The error between the numerical occurrence time of the peak pressure and the experimental results is within 0.002 s. These indicate that the numerical model can accurately reflect the real pressure characteristics. In the model tests, P2 has the highest peak pressure followed by P3. It can be concluded that the maximum pressure is located between P2 and P3. In the numerical simulation, position of the maximum pressure is also supposed to be located between P2 and P3, which is consistent with the experimental results. Thus, it can be seen that the numerical model can accurately forecast the pressure distribution.

4 RESULTS AND DISCUSSIONS

4.1 *Peak pressure in vertical water impact*

Vertical water impact is the ideal condition in which the bottom of the structure is parallel to the static free surface and the incident angle is 0 °. For a rigid body with a flat bottom, the entire surface impacts the water almost simultaneously. Consequently, air trapping is inevitable for structures with concave features. In this section, a numerical simulation of a wet drop with a total mass of 2.5 t and an initial height of 2.0 m is carried out. Since that the model is a centrosymmetric structure in vertical water impact, Figure 5 shows distribution of the peak pressure and its occurrence time on the 1/4 model to display more details about the distribution characteristics. The Y-axis serves as symmetry axis of the model. In vertical water impact, the maximum pressure occurs at the model edge. In complete area units, a higher impact pressure is noticed at the model center.

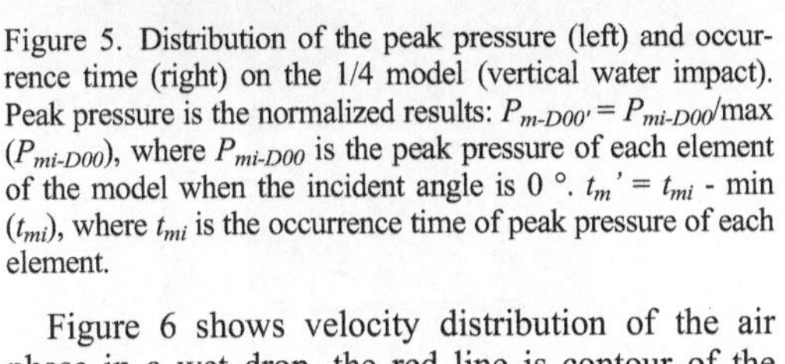

Figure 5. Distribution of the peak pressure (left) and occurrence time (right) on the 1/4 model (vertical water impact). Peak pressure is the normalized results: $P_{m\text{-}D00'} = P_{mi\text{-}D00}/\max(P_{mi\text{-}D00})$, where $P_{mi\text{-}D00}$ is the peak pressure of each element of the model when the incident angle is 0 °. $t_m' = t_{mi} - \min(t_{mi})$, where t_{mi} is the occurrence time of peak pressure of each element.

Figure 6 shows velocity distribution of the air phase in a wet drop, the red line is contour of the free surface. In comparison to the main region at the model center, the edge region has incomplete area units, so an air cushion cannot form here. At 1.15 ms, air under the edge region is compressed and quickly exhausted through the gap between the model and the free surface. At 2.45 ms, the edge region impacts the free surface, and the maximum pressure is generated. In the main region, air is still trapped between the corrugations, forming an air cushion, and air in this air cushion tends to move towards the edge region. Such gas escape persists until after peak pressure in the main region occurs, and air cushion under the model remains present throughout. In the edge region, direct fluid impact on the model led to higher impact pressure. Under the air cushion effect, peak pressure at the model center is reduced by about 70% and has a lag of about 2.0 ms in occurrence time compared with that at the model edge. Additionally, peak pressure in the corrugation region is close to that in the main region.

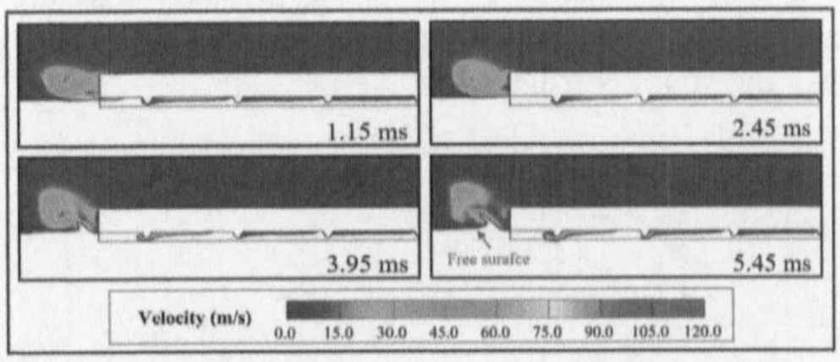

Figure 6. Contours of the free surface and velocity distribution of the air phase in vertical water impact. Referring to the occurrence time of the peak pressure in Figure 5(b), the maximum peak pressure in the edge region occurs at 2.45 ms and that in the main region occurs at 3.95 ms.

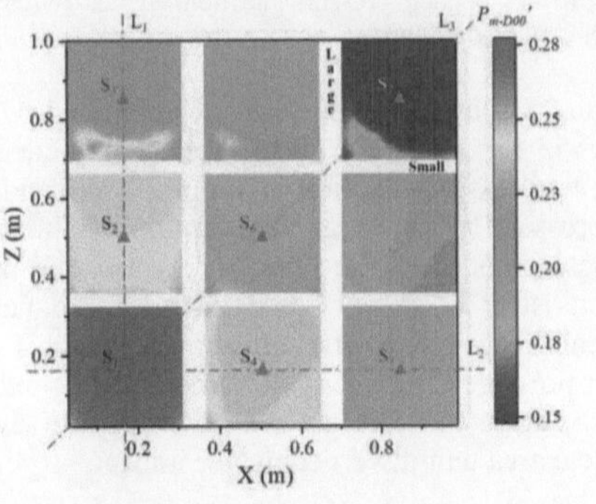

(a) Distribution of peak pressure

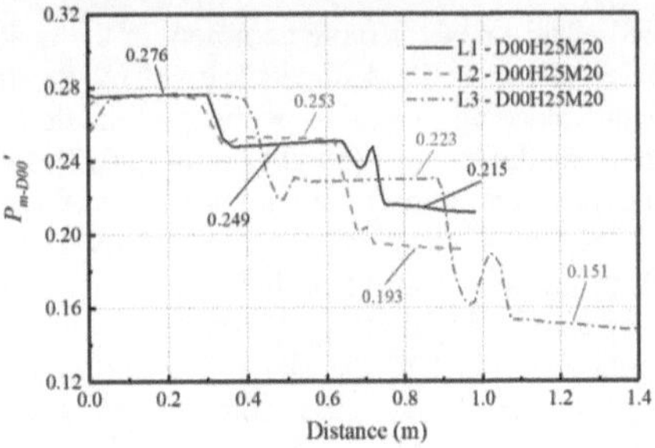

(b) Variation of peak pressure

Figure 7. Distribution and variations of the peak pressure in complete area units of the 1/4 model and configuration of pressure sensors (vertical water impact). L1, L2, and L3 all pass through the center of the area units, and S1-S7 are all located in the center of the units along the lines.

Figure 7(a) shows distribution of the peak pressure within the complete area units. The maximum peak pressure occurs at the model center and the peak pressure decreases as the distance from the model center increases. Variations of the peak pressure along L1, L2 and L3 are shown in Figure 7(b). Here, L1 and L2 are parallel to the corrugations, and L3 connects the model center to the corrugated knots. Distribution of the peak pressure on the Mark-III type flat panel is similar to the pressure distribution in the Wagner-type impact processes (Okada et al., 2000). But for each passing corrugation, there is a leap-type decay in the peak pressure and the large corrugations have stronger suppressing effect than the small corrugations.

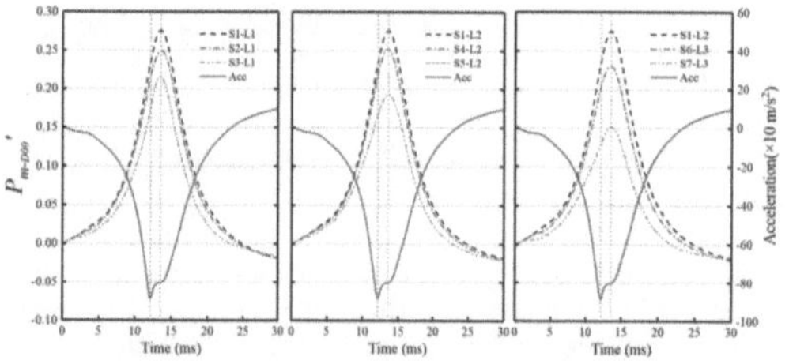

Figure 8. Time histories of impact pressure for S1-S7 and acceleration of the model (vertical water impact). The duration of the impact pressure is defined as twice the half-peak pulse duration.

Figure 8 shows pressure histories obtained by the sensors in Figure 7. For all different complete area units, the pressure duration is about 20 ms, and the peak pressure lags the peak acceleration by about 1.5 ms. Besides peak of the impact pressure, the other characteristics of the pressure histories are nearly identical. The rising and releasing times of the impact pressure are almost the same at all positions. These suggest that the air cushion exists in every complete area unit in vertical water impact.

4.2 *Peak pressure in inclined water impact*

Inclined water impact is more common. In the tanks of LNG vessels, most fluid impact caused by the three elementary loading processes or the combinations has non-zero incident angles (Lafeber et al., 2012). Given that the maximum impact pressure was observed in the model test for an incident angle of 4° (Noh et al,. 2009), this section focuses on the conditions for incident angles of 5 ° and 10 °. The model has the small corrugations parallel to the static free surface. Total mass of the model is 3.0 t and the initial height is 1.5 m.

To present more details about the distribution characteristics, Figure 9 shows distribution of the peak pressure and its occurrence time on the 1/2 model. The YZ plane is the symmetry plane of the model. The fluid flows through and impacts the corrugated surface in the positive direction of the Z-axis. In inclined water impact, the maximum pressure is observed near the root of the corrugations instead of at the model edge, which is different from the results of the vertical water impact.

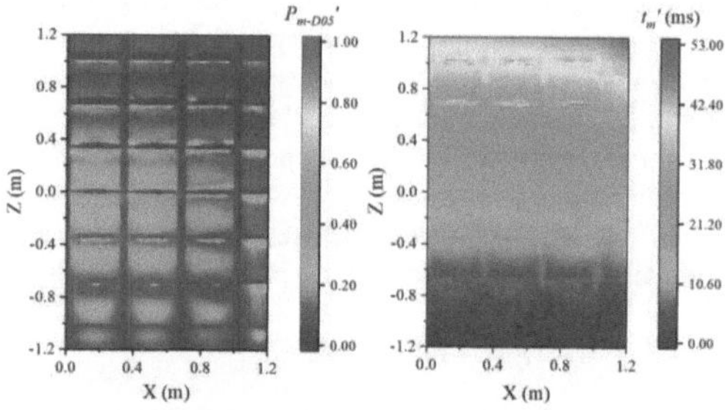

(a) Result for incident angle of 5 °

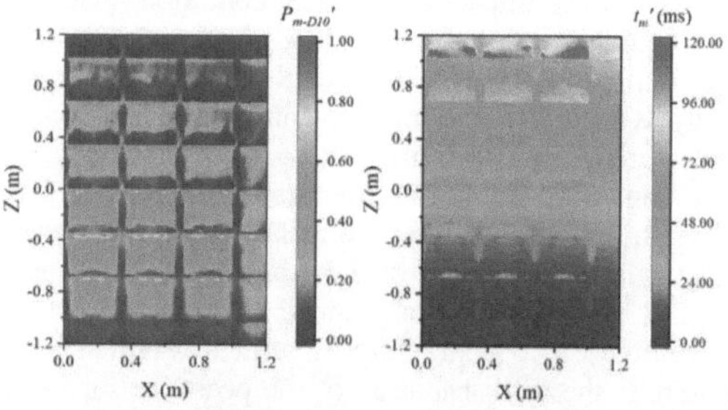

(b) Result for incident angle of 10 °

Figure 9. Distribution of the peak pressure (left) and the occurrence time (right) on the 1/2 model (inclined water impact). Peak pressure is the normalized results: $P_{m\text{-}D05'} = P_{mi\text{-}D05}/\max(P_{mi\text{-}D05})$, $P_{m\text{-}D10'} = P_{mi\text{-}D10}/\max(P_{mi\text{-}D10})$, where $P_{mi\text{-}D05}$ and $P_{mi\text{-}D10}$ is the peak pressure of each element of the model when the incident angle is 5 ° and 10 °, respectively.

(a) Result for incident angle of 5°

The peak pressure at the top of the corrugation is close to that in the main region. The peak pressure for the same height (same Z-coordinate) is observed almost simultaneously, indicating that air cushion in inclined water impact is much weaker than that in vertical water impact. However, on the upper side of the model, the gas escaping causes a lag in the occurrence time of the peak pressure. It can be assumed that a little gas tripping still exists and affects the fluid impact.

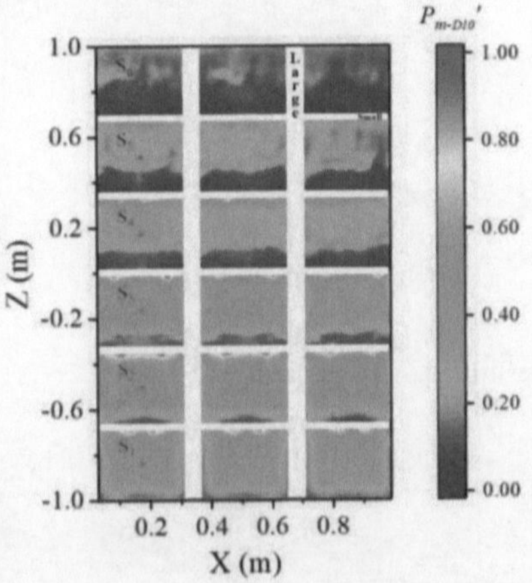

(b) Result for incident angle of 10 °

Figure 10. Distribution of peak pressure in complete area units of the 1/2 model and configuration of pressure sensors (inclined water impact). S1-S6 are all located in center of the area units.

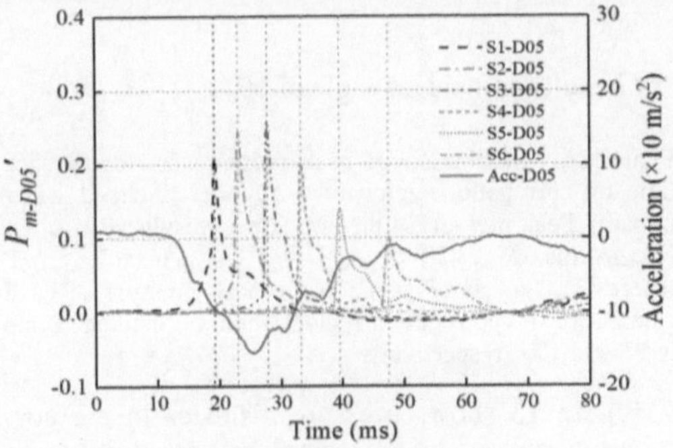

(a) Result for incident angle of 5 °

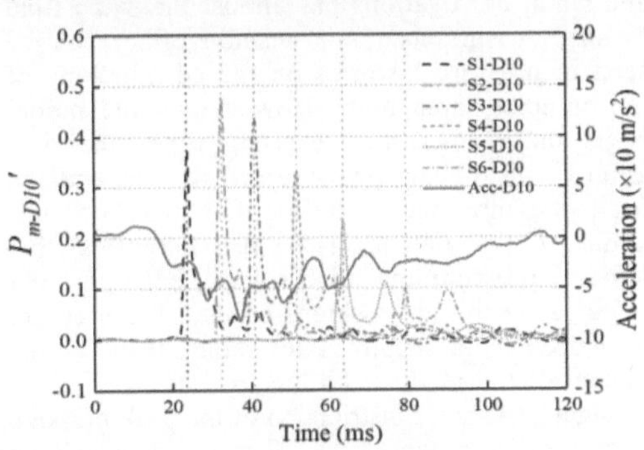

(b) Result for incident angle of 10 °

Figure 11. Time histories of impact pressure for S1-S6 and acceleration of model (inclined water impact).

Figure 10 shows distribution of the peak pressure within the compete area units. Due to the non-zero incident angle, the peak pressure at the roots of some corrugations is about 2-3 times higher than that in the main region. Because of the weak air cushion, the peak pressure at the same height (same Z-coordinate) is nearly the same. In the flow direction, the maximum pressure on the model is observed near the model center. However, the position of the maximum pressure changes significantly with the incident angle. As the incident angle increases, the position moves toward the lower side of the mode, and the maximum pressure on the model occurs earlier. Additionally, the region near the downstream face of the small corrugation, which is not directly impacted by water, undergoes expansion.

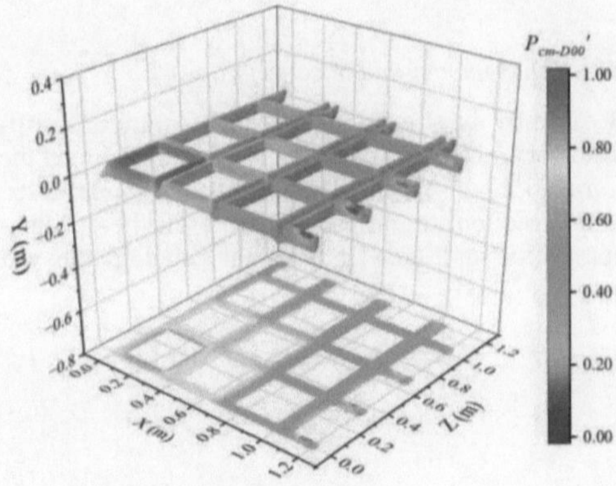

(a) Normalized peak pressure

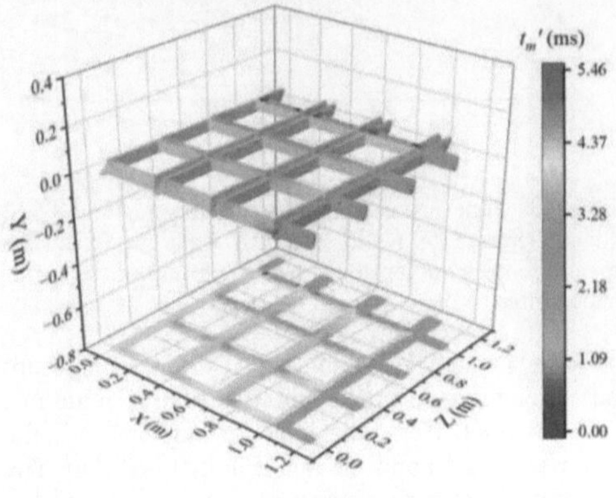

(b) Occurrence time of peak pressure

Figure 12. Distribution of peak pressure and occurrence time in corrugation region of the 1/4 model (vertical water impact). Peak pressure is the normalized results: $P_{cm\text{-}D00'} = P_{cmi\text{-}D00}/\max(P_{cmi\text{-}D00})$, where $P_{cmi\text{-}D00}$ is the peak pressure of each element of the corrugation region when incident angle is 0 °.

Figure 11 shows pressure histories obtained by the sensors S1-S6 (in Figure 10). In inclined water impact, distribution of the peak pressure and characteristics of the pressure history are similar to those in Wagner-type impact processes on the wedge (Okada et al., 2000). In this case, rising time of the impact pressure is not equal to releasing time. Moreover, as the incident angle increases, rising time of the impact pressure further decreases. This also implies that air cushion is not significant when incident angle is 5°, but there is still a small amount of gas trapping which affects the fluid impact. At an incident angle of 5°, the peak pressure at each point lags the corresponding negative acceleration peak by 2-3 ms. This difference increases to 3-4 ms when the incident angle is 10°. It can be considered that the corrugations are first subjected to the fluid impact, which causes variation in the model's acceleration before the fluid impacts the flat area units. This is the reason why the maximum pressure on the model is observed at the roots of the corrugations.

4.3 *Fluid impact on the corrugations*

Because of the special geometrical features, is difficult to accurately measure the impact pressure on the corrugation and its distribution through model tests. In this section, the impact pressure on the corrugation region of Mark-III type flat panel in the wet drop is studied.

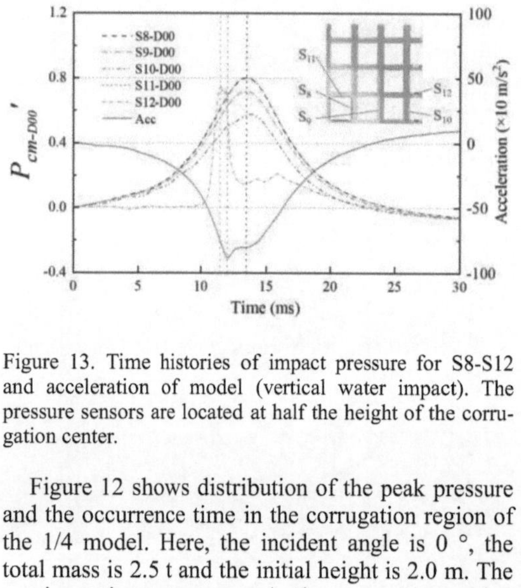

Figure 13. Time histories of impact pressure for S8-S12 and acceleration of model (vertical water impact). The pressure sensors are located at half the height of the corrugation center.

Figure 12 shows distribution of the peak pressure and the occurrence time in the corrugation region of the 1/4 model. Here, the incident angle is 0 °, the total mass is 2.5 t and the initial height is 2.0 m. The maximum impact pressure in the corrugation region occurs at the model center and also in the middle of the small corrugations near the model edge. In the same area unit, under the effect of the air cushion, the peak pressure at the root is nearly the same, and the occurrence time is also nearly the same. Moreover, peak pressure at the top of the corrugations is not greater than that at the root. At the model edge, which is a region without an air cushion, the jet generated by the impact of the top of the corrugations with free surface directly impacts on the roof of the corrugations. This causes higher pressure to occur earlier in this region.

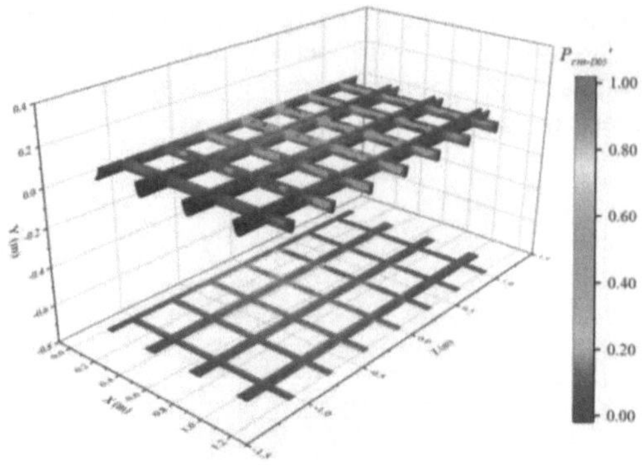

(a) Result for incident angle of 5 °

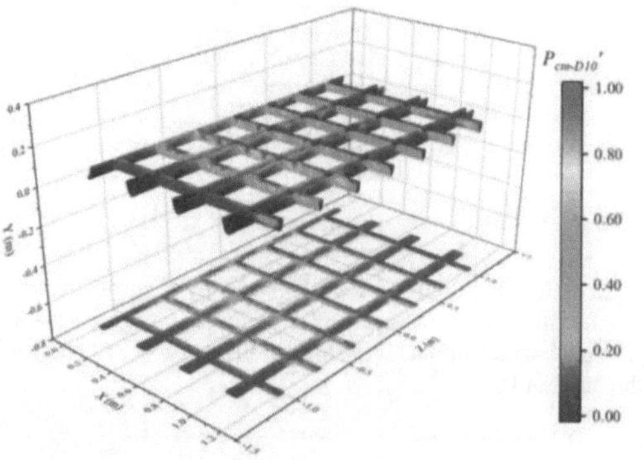

(b) Result for incident angle of 10 °

Figure 14. Distribution of peak pressure and occurrence time on corrugation region of 1/2 model (inclined water impact). Peak pressure is the normalized results: $P_{cm\text{-}D05'} = P_{cmi\text{-}D05}/\max(P_{cmi\text{-}D05})$, $P_{cm\text{-}D10'} = P_{cmi\text{-}D10}/\max(P_{cmi\text{-}D10})$, where $P_{cmi\text{-}D05}$ and $P_{cmi\text{-}D10}$ is the peak pressure of each element of the corrugation region when the incident angle is 5 ° and 10 °, respectively.

Figure 13 shows pressure histories in the corrugation region. In the complete area unit at the modl center, the impact pressure on both large and small corrugations has almost the same time history. A significant air cushion effect can be seen in pressure histories on the corrugations of the complete area units. However, at the model edge, the rising time, releasing time, and duration of the impact pressure are shorter, and the peak pressure occurs before the peak acceleration of the model. This also indicates that impact pressure at the model edge is not affected by the air cushion, instead, the pressure is caused by a localized jet rather than by the impact of the model with the free surface.

Figure 14 shows distribution of the peak pressure in corrugation region of the 1/2 model. The incident angle is 5 ° and 10 °, total mass of the model is 3.0 t and initial height is 1.5 m. In the inclined water impact, there is a significant difference between peak pressure on the upstream face and that on the

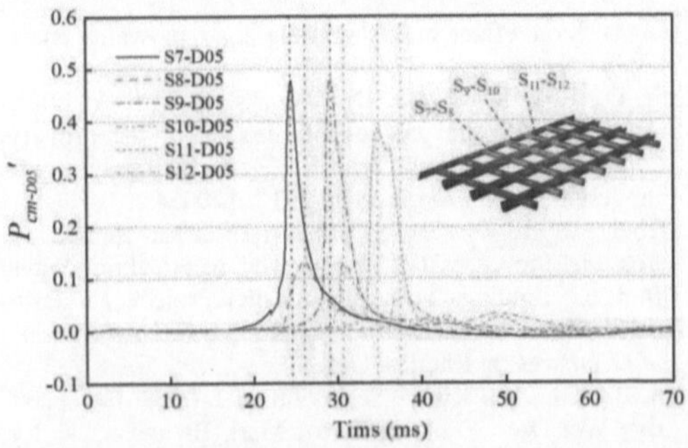

(a) Result for incident angle of 5 °

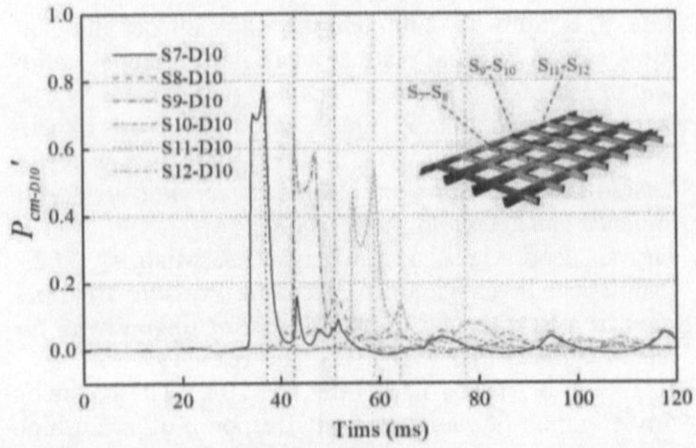

(b) Result for incident angle of 10 °

Figure 15. Time histories of impact pressure for S7-S12 (vertical water impact). The pressure sensors are located at half the height of the corrugation center. S7-S8, S9-S10 and S11-S12 are located at the upsteam face and downsteam face of the 3_{rd}, 4_{th} and 5_{th} small corrugations along the water flow direction, respectively.

downstream face of the corrugation. The maximum pressure is found on the upstream face of the corrugations. In the direction perpendicular to the flow, the distribution of peak pressure is nearly consistent, and the maximum pressure is at the model center. When the incident angle is increased from 5 ° to 10 °, the maximum pressure moves toward the free surface and the gradient in the spatial distribution of peak pressure decreases.

Pressure history on the corrugations is shown in Figure 15. It is observable that the peak pressure on the downstream face is about 20% of that on the upstream face. As the incident angle increases, the rising time of impact pressure on the upstream face decreases, while that on the downstream face increases. This implies that increase in the incident angle reduces the air cushion near the upstream face, but more gas is trapped near the downstream face. Under the condition of an incident angle of 5 °, a dual-peak pressure history is observed near the upstream face of the 5th corrugation. This dual-peak pressure history becomes more obvious under the condition of an incident angle of 10 °. As can be seen from Figure 11, the dual-peak pressure history is also observed in the center of the flat units, and it is more obvious on the upper side of the model in inclined water impact. Interestingly, this dual-peak pressure history is also observed near the static free surface when the resonant sloshing liquid impacts the tank walls (Wang et al., 2023).

In Figure 16, the air phase is hidden, position of the model, free surface, and velocity distribution are shown. The jet rapidly climbs along the surface of the model, which causes the first peak pressure in the flat and corrugation regions. By impacting the corrugation, the jet changes direction and then quickly contacts the free surface to form a closed area. This leads to a localized air trapping near the upstream face of the corrugations. As the model drops, it impacts the free surface, causing the second peak pressure in the flat region. Subsequently, the air near the upstream face of the corrugation is exhausted, the corrugations impact the free surface, and the second peak pressure in the corrugation region is observed. Although dual-peak pressure history is also observed in wet drops, its mechanism differs from that in sloshing. In sloshing, the first peak pressure results from a direct fluid impact on the wall, and the second peak pressure is caused by the falling fluid. Thus, the first peak is often greater than the second one. For wet drops, the second peak pressure, which is due to the impact of the model with the free surface, may be greater than the first peak pressure caused by the jet. Additionally, the time difference between the two peak pressures for full-scale wet drops is less than 20 ms, much shorter than that for full-scale sloshing.

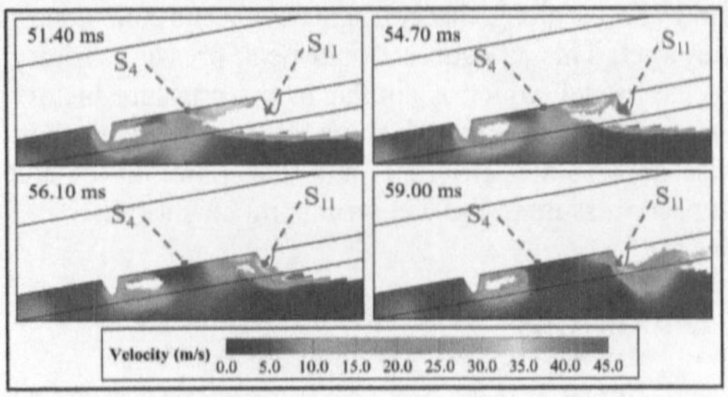

Figure 16. Contours of the free surface and velocity distribution of the water phase in inclined water impact (incident angle of 10 °). The peak pressure of S4 occurs at 51.40 ms; the first and second peak pressure of S11 occurs at 54.70 ms and 59.00 ms, respectively. The local minimum between the two peaks occurs at 56.10 ms.

5 CONCLUSIONS

In this paper, wet drops of the Mark-III type flat panel are numerically calculated using CFD method. Distribution of the peak pressure at different incident angle are shown to analyze the mechanism of fluid impact and the evolution of gas and liquid phases in vertical and inclined water impact. The conclusions drawn from this study are summarized as follows:

1. In the vertical water impact, the air cushion on the model is obvious. It can not only reduce the peak pressure and increase the rising time, releasing time and duration of the pressure, but also delay the occurrence time of the peak pressure. The distribution of peak pressure at the model center is similar to that of Wagner-type impact processes. However, the corrugations cause a leap-type decrease in the spatial distribution of peak pressure.

2. In an inclined water impact, the maximum peak pressure is observed at the root of the upstream face surface of the corrugations, which is about 2-3 times the peak pressure at the center of the flat regions. As the incident angle increases, the position of this maximum peak pressure moves toward the lower side of the model. The maximum peak pressure on the top of the corrugations is close to that within the flat regions. Air cushion in the inclined water impact is less obvious, but a small amount of gas trapping can still be observed.

3. For the corrugations of the model, the maximum peak pressure in the vertical water impact is observed at the roots of the corrugations at the model edge. At the model center, the distribution of peak pressure in the corrugation region is nearly the same as that in the flat region. In the inclined water impact, the maximum peak pressure on the corrugations is observed near the upsteam surface of the corrugations at the model center. Near the downsteam surface of the corrugations, air trapping is observed, and the peak pressure there is about 20% of that near the upsteam surface.

4. When the model impacts the free surface, the fluid climbs along the corrugated surface and generates a jet. This jet causes a dual-peak pressure history on the model, which is similar to the pressure history caused by the resonant sloshing liquid. However, their mechanisms are different, and the duration of the impact pressure in the wet drop is much shorter.

REFERENCES

Bae, J., Hwang, B., Kim, S. & Lee, J. 2020. Statistical estimation of extreme sloshing impact pressure and its validation based on dynamic behavior of insulation materials, *Functional Composites and Structures*. 2: 015001.

Chen, H., Wu, T., Wan, Z., Wang, H., Xu, P., Yang, G., & Wu, J. 2025. Numerical analysis of LNG rollover in large membrane tank under sloshing excitations, *Energy*. 315: 134351.

Fang, X, Ming, F., Wang, P., Meng, Z. & Zhang, A. 2022. Application of multiphase Riemann-SPH in analysis of air-cushion effect and slamming load in water entry, *Ocean Engineering*. 248: 110789.

Kim, D., Roh, M., Chun, D., Woo, S. & Kim, Y. 2025. A method for the profile optimization of the primary barrier of a membrane-type LNG tank using surrogate modeling, *Ocean Engineering*. 317: 120054.

Lafeber, W., Bogaert, H. & Brosset, L. 2012. Elementary Loading Processes (ELP) Involved in Breaking Wave Impacts: Findings From the Sloshel Project, *Proceedings of 22nd International Offshore and Polar Engineering Conference*. Rhodes, Greece.

Noh, B., Kim, W., Kwon, S. & Chung, J. 2009. Comparative Wet Drop Experiments of Mark III and KC-1 for Membrane Type LNG Carriers, *Proceedings of 28th International Conference on Ocean, Offshore and Arctic Engineering*. Hawaii, USA.

Okada, S. & Sumi, Y. 2000. On the water impact and elastic response of a flat plate at small impact angles, *Journal of Marine Science and Technology*. 5 (1): 31–39.

Pan, J., Zhang, W., Sun, Z., Qu, X. & Xu, M. 2024. Experimental study on the dynamical response of elastic trimaran model under slamming load. *Journal of Marine Science and Technology*. 29: 20–35.

Wang, C., Zou, C., Zhu, J., Tang, J. & Miao, Q. 2023. Numerical Investigation of Sloshing Pressure Distribution in a Rectangular Tank, *Journal of Engineering for the Maritime Environment*. 237(1): 108–119.

Wei, J., Xu, S., Yuan, Y. & Tang, W. 2025. The inversion investigation of wave slamming response of air cushion vehicle skirt airbags based on mode superposition method, *Marine Structures*. 103: 103821.

Wu, W., Zhen, C., Lu, J., Tu, J., Zhang, J., Yang, Y., Zhu, K. & Duan, J. 2020. Experimental study on characteristic of sloshing impact load in elastic tank with low and partial filling under rolling coupled pitching, *International Journal of Naval Architecture and Ocean Engineering*. 12: 178–183.

Xie, H., Li, J., Shi, G., Wang, D. & Tang, H. 2024. Characterization on the impact load of a local corner region of a liquid tank entering water, *Ocean Engineering*. 312: 119030.

Xie, H., Dai, X., Liu, F. & Liu, X. 2024. Experimental study on the slamming pressure distribution of a 3D stern model entering water with pitch angles, *Ocean Engineering*. 291: 116404.

Yang, J., Sun, Z. & Liang, S. 2023. A Numerical Study on the Water Impact of the Rigid/Elastic Box-Like Structure, *China Ocean Engineering*. 37: 333–342.

Yang, S., Lee, H., Park, T., Lee, I. & Lee, Y. 2007. Experimental and numerical study on the water entry of symmetric wedges and a stern section of modern containership, *Proceedings of Practical Design of Ships and Offshore Structures*. Houston, USA.

Zhang, Z., Wu, Q., Xie, Y. & Yu, H. 2023. Experimental and numerical investigations on the liquid tank sloshing in regular waves, *Ocean Engineering*. 271: 113668.

Risk assessment and innovative navigation systems

Trends in Collision and Grounding of Ships and Offshore Structures – Zhu & Guedes Soares (Eds)
© 2026 The Author(s), ISBN 978-1-041-16361-9

Risk coupling analysis for a ship: A case study from grounding in the west waters of Shenzhen Port

C.L. Fan & X.Y. Chen
College of Transport & Communications, Shanghai Maritime University, Shanghai, P.R. China

Y.K. Chen
Key Laboratory of Marine Environmental Survey Technology and Application, Ministry of Natural Resource, Guangzhou, Guangdong, P.R. China

B. Wu
Intelligent Transportation Systems Research Center, Wuhan University of Technology, Wuhan, Hubei, P.R. China

ABSTRACT: This study aims to measure the Risk Coupling Value (RCV) of maritime grounding accidents using the N-K model. Focusing on accidents in the waters west of Shenzhen Port, China, the framework quantifies the coupling effects of multiple risk factors, including human, organizational, environmental, and technical elements. Historical accident data from 2003 to 2022 were analyzed, revealing that the upper bound of RCV can effectively indicate potential grounding risks, with 92.5% of cases overlapping with risk priority ranges. The study highlights the importance of considering both the type and quantity of risk factors in different operational contexts, such as seasons, time windows, navigation areas, and ship types. The results indicate that this study provides a valuable tool for maritime risk coupling analysis.

1 INTRODUCTION

With the continuous growth of global trade, shipping plays a vital role in international trade. However, the occurrence of ship accidents not only causes huge economic losses, but can also cause serious environmental problems. As a common type of marine accident, ship grounding accidents occur frequently, especially in confined waters, which poses a major threat to the safety of ship navigation. Therefore, it is of great significance to analyze the risk coupling characteristics and evolution mechanism of ship grounding accidents to improve the safety of ship navigation and formulate effective risk mitigation strategies.

In the field of maritime collision and grounding analysis, Bayazit and Kaptan conducted a dynamic risk analysis of allision in port areas using a Dynamic Bayesian Network (DBN), providing insights into the complex interactions between human factors and operational conditions. (Bayazit et al., 2024) Chen et al. proposed a combination of fuzzy logic and fuzzy comprehensive evaluation methods to calculate collision risk. (Chen et al., 2014) Yu et al. developed an integrated dynamic ship risk model using Bayesian Networks and Evidential Reasoning, offering a comprehensive framework for real-time risk assessment. (Yu et al., 2021) Aydin et al. validated risk analysis using fuzzy Bayesian networks. The proposed method was applied to risk modeling of ship collisions in narrow waters to identify and minimize the potential risks associated with ship collisions in such areas. (Aydin et al., 2021) Abaei et al. employed hierarchical Bayesian analysis to estimate the probability of bottom contact based on the dynamic under keel safety margin results obtained from time-domain hydrodynamic simulations. (Abaei et al., 2018) Akhtar and Utne proposed a generic method for developing Bayesian Networks to model the risk of maritime accidents. (Akhtar et al., 2014) Akyuz proposed a hybrid accident analysis method to enhance safety in the maritime industry. The hybrid method integrates the Accident Analysis Mapping and Analytic Network Process approaches to analyze the causes of maritime accidents. (Akyuz., 2015) Fu et al. developed a framework for the quantitative analysis of grounding accident causation in Arctic shipping, considering the unique challenges posed by polar environments. (Fu et al., 2022) Youssef and Paik conducted a study on hazard identification and scenario selection for ship grounding accidents. (Youssef et al., 2018) Xi et al. investigated the human reliability of ship convoy operations in Arctic ice-covered waters and proposed recommendations to reduce collision risks by integrating the Information, Decision, and Action (IDAC) framework with the Dynamic Bayesian Network (DBN) model. (Xi et al., 2024)

DOI: 10.1201/9781003684404-51

Wang et al. proposed a hybrid methodology based on the HFACS-BN model to investigate the human and organizational factors in collision accidents between merchant ships and fishing vessels. (Wang et al., 2024) These studies enhanced the understanding of maritime collision and grounding risks, offering valuable tools and methodologies for risk assessment and management in the maritime industry.

In the field of risk analysis and assessment, Yildiz et al. demonstrated the feasibility of the modified Human Factors Analysis and Classification System for Passenger Vessels (HFACS-PV) for other types of accidents. (Yildiz et al., 2021) Fan et al. proposed a risk matrix based on the Fuzzy Analytic Hierarchy Process (AHP) for visualizing risk assessment. (Fan et al., 2024c) Zhang et al. conducted a quantitative analysis on risk influencing factors in the Jiangsu segment of the Yangtze River, offering insights into the specific regional risks and potential mitigation strategies. (Zhang et al.,2021) Li et al. incorporated a global perspective into data-driven analysis of maritime collision accident risk, highlighting the importance of integrating diverse datasets for comprehensive risk assessment. (Li et al.,2024) Guo et al. focused on risk coupling analysis of road transportation accidents involving hazardous materials in a complicated maritime environment, revealing the interconnected nature of risks in such complex scenarios. (Guo et al.,2023) Huang et al. combined the N-K model with the Functional Resonance Analysis Method to quantitatively calculate variability in risk factors, providing a novel approach to dynamic risk assessment. (Huang et al., 2022) Jiang et al. proposed an analytical model incorporating Bayesian Networks that comprehensively considered the characteristics of ship properties, organizational factors, hydrological conditions, and human factors to estimate the likelihood of ship bottom contact in fluctuating backwater zones. (Jiang et al., 2021) Liu et al. performed risk coupling analysis of subsea blowout accidents using a dynamic Bayesian network and the N-K model. (Liu et al.,2022) Qiao proposed a multi-factor risk measurement model based on coupling theory. (Qiao., 2021) Yıldırım et al. evaluated collisions and grounding accidents using the Human Factors Analysis and Classification System (HFACS) and statistical methods, offering valuable insights into the role of human factors in maritime accidents. (Yıldırım et al., 2019) Fan et al. proposed a navigation risk assessment framework for remotely controlled Maritime Autonomous Surface Ships (MASS) based on Bayesian methods. (Fan et al., 2024a) Hu et al. proposed an Enhanced Risk Assessment (ERA) framework in the field of risk analysis and assessment, combining the Systems Theoretic Accident Model and Process (STAMP) with Stochastic Petri Nets (SPN) to dynamically quantify risks in complex maritime traffic systems. (Hu et al., 2025)

In the field of ship safety and accident analysis, Bulian et al. proposed a probabilistic framework to assess the survivability of ships following bottom contact and side collision/contact accidents. (Bulian et al., 2020) Fan et al. proposed a framework for assessing the navigational risks of Maritime Autonomous Surface Ships in different operational modes (OM). This framework provides risk-based decision-making support for selecting the appropriate operational mode for autonomous ships, thereby reducing navigational risks and enhancing maritime safety. (Fan et al., 2022) Ganco examined the applicability of the NK model in the innovation process and found that the model performs well in predicting the average innovation performance but has limitations in predicting variance. (Ganco., 2017) Graziano et al. conducted a systematic analysis of human errors in ship accidents using the TRACEr taxonomy, revealing the main task errors, technical equipment, cognitive domains, and factors affecting operator performance that lead to accidents. (Graziano et al., 2016) Fan et al. proposed a framework for examining the coupling effects in the operational modes (OM) of Maritime Autonomous Surface Ships. (Fan et al., 2024b) Wang et al. identified and assessed risk influencing factors related to facilities and navigation by incorporating failure mode and effects analysis (FMEA) and risk matrix. (Wang et al., 2023)

In recent years, the application of risk coupling theory in complex systems has gradually attracted attention. The coupling effect refers to the process by which multiple risk factors interact and work together to affect the safety of a system. In ship navigation, risk factors may include human factors, organizational factors, the ship is own condition, environmental factors, and technical factors. The complex interaction between these factors can lead to a significant increase in the probability and severity of accidents. Therefore, quantifying and analyzing the risk coupling effect is essential to understand and predict the mechanism of ship accidents. In addition, the occurrence of ship grounding accidents is affected by a variety of factors, including season, time window, navigation area, and ship type. These factors not only affect the strength of risk coupling, but can also lead to uncertainty in the coupling effect. Therefore, it is of great significance to construct an analytical framework that comprehensively considers these factors to comprehensively evaluate the risk coupling characteristics of ship grounding accidents.

This study aims to quantify the Coupled Risk Value (RCV) of ships based on the N-K model, taking the ship grounding accident in the western waters of Shenzhen Port in China as a case study. The results can not only provide theoretical support for ship navigation safety management, but also provide new perspectives and methods for future related research.

2 FRAMEWORK

2.1 *Methodology: The N-K model*

The N-K model was originally introduced by Kauffman and Weinberger (1989) in the field of organism evolution and was developed further by Kauffman

(1993). The N-K model is commonly used to analyze the evolution of correlative effects of risk factors in complex networks. In this study, N reflects the number of risk factor types, and K reflects the number of coupled relations among different types of risk factors. If the number of risk factor types in a complex system is N, and each type of risk factor has n kinds of states, then N represents the number of all possible couplings.

2.2 *Data source*

The historical accident data used for this study include grounding accidents that happened in the waters west of Shenzhen Port, China, collected from the Shenzhen Maritime Safety Administration (MSA). The data from January 2003 to July 2014 were collected from the Vessel Traffic Service (VTS) center of Shenzhen MSA, while the data from August 2014 to December 2022 were downloaded from the website of Shenzhen MSA. In total, 98 grounding cases occurred over the last 20 years, which formed the data sample for this study.

The analysis of the data distribution shows that there are no obvious extreme values or anomalies in the sample, and the data distribution is relatively uniform. Therefore, although the sample size is not particularly large, the impact of contingency is minimal.

2.3 *RCV of a ship*

To calculate RCV for a specific condition, e.g., season, time window, area, and ship type, an equation is proposed as follows:

$$RCV_{ij}^{k} = \left[min\left\{T_i^j(y|x)\right\},\ max\left\{T_i^j(y|x)\right\}\right] \quad (1)$$

where RCV_{ij}^{k} is the risk coupling value of case k involving i risk types, j indicates the combination of risk types. $T_i^j(y|x)$ is the risk coupling value of risk y under the condition of x, and y is the risk set of the j-th combination of i risk types. In this study, if x is a season, $x \in$ {spring, summer, autumn, winter}. If x is a time window for OOW, $x \in$ {time window for CO, time window for SO, time window for TO}, where the time window for CO includes the two sections of 04:00–08:00 and 16:00–20:00, the time window for SO includes the two sections of 00:00–04:00 and 12:00–16:00, the time window for TO includes the two sections of 08:00–12:00 and 20:00–24:00. If x is an area, $x \in$ {channel, anchorage/berth}. If x is a type of ship, $x \in$ {container ship, bulk/general ship, liquefied cargo ship, barge/suppy}.

3 CASE STUDY

In the data sample, 80 cases involve more than one risk type. These 80 cases are used to quantify their RCV in this study. For example, case 1 represents a barge that ran aground in an anchorage during winter and the time window was 00:00–04:00. In the accident analysis, this case involved two risk types, i.e., organization and environment. Hence, y = {O, E}, i = 2, and j = 6. Based on these values and the RCVs in four conditions, $T_2^6(\mathrm{O,E}|\mathrm{winter}) = 0.0202$, $T_2^6(\mathrm{O,E}|\text{time window for } SO) = 0.0004$, $T_2^6(\mathrm{O,E}|\mathrm{anchorage/berth}) = 0.0085$, and $T_2^6(\mathrm{O,E}|\mathrm{barge/supply}) = 0.0718$. According to Eq. (1), $RCV_{26}^{1} = [0.0004,\ 0.0718]$. This process is repeated, and the RCVs of the other 79 cases are obtained. The results of all 80 cases are shown in Figure 1.

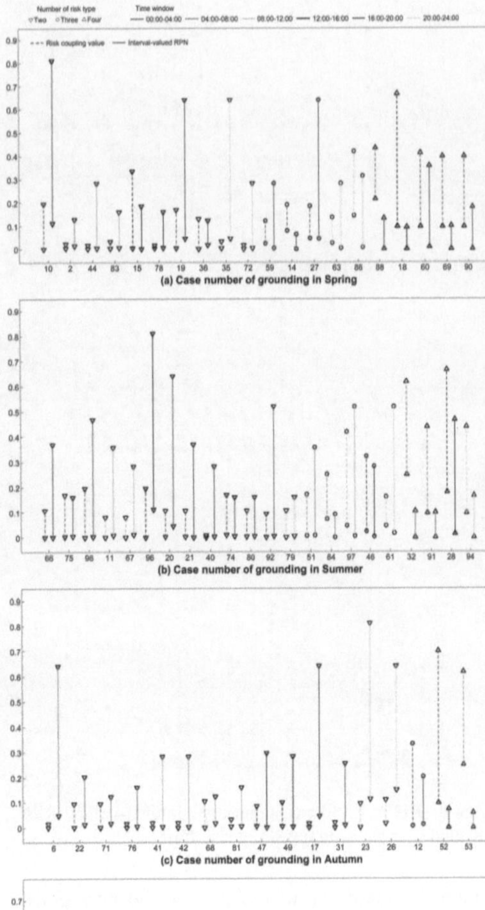

Figure 1. RCVs of 80 cases and their inteval-valued Risk Priority Number (RPN), presented by season, time window, and risk types.

Overall, when considering the number of grounding accidents across all four seasons, the quantities are nearly equal. Among the six-time windows, most grounding accidents involving two types of risks are distributed relatively evenly. However, the distribution of grounding accidents involving three or four types of risks is more sporadic and less uniform.

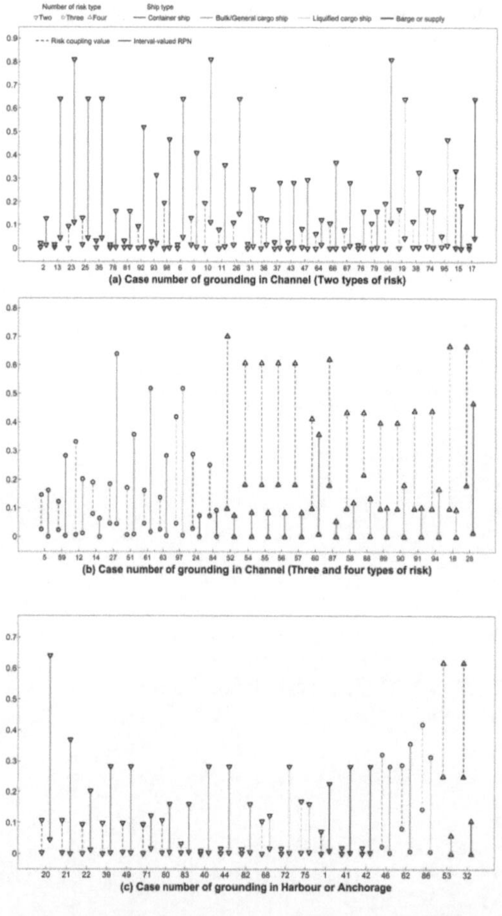

Figure 2. RCVs of 80 cases and their inteval-valued RPN, presented by area, ship type, and risk types.

Figure 2 analyzes all cases from the perspective of accident area and ship type. Figure 2 reveals that the majority of cases involved two risk types in the channel, anchorage, or berth. Figure 2(a) and (b) indicate that all four ship types experienced two to four risk coupling scenarios in the channel. Specifically, among the two-risk coupling scenarios, bulk or general cargo ships were most common, followed by container ships, liquefied cargo ships, and barge or supply vessels. In the three-risk coupling scenarios, bulk or general cargo ships again dominated, while liquefied cargo ships had the fewest occurrences. For the four-risk coupling scenarios, container ships and bulk or general cargo ships were the most frequent, with nearly identical numbers; liquefied cargo ships and barge or supply vessels were the least common, with equal counts. Figure 2(c) highlights that many container ships encountered only two risk types in the anchorage or berth. However, not all four ship types experienced three or four risk types in the anchorage or berth.

4 RESULTS COMPARISON

The RCVs of a case were compared with its risk, generated using the method of Fan et al. (Fan et al., 2022) The RCVs of all cases and their interval-valued RPN are shown in Figures 1 and 2. By comparing the upper bound of risk coupling values with the lower bound of interval-valued RPN in each of the four seasons, Figure 1 shows that the interval of RCV overlaps the interval-valued RPN in most cases involving two, three, or four risk types. Furthermore, there are certain cases where the upper bound of RCV is larger than the upper bound of interval-valued RPN, e.g., cases 14, 15, 18, 36, 60, 86, 88, 89, and 90 in the spring, cases 28, 32, 74, 75, 84, 91, and 94 in the summer, cases 12, 52, and 53 in the autumn, and cases 24, 54, 55, 56, 57, 58, and 87 in the winter. Particularly, there are cases with lower bounds of RCV that are larger than the upper bounds of its interval-valued RPN, e.g., cases 14 and 88 in the spring, case 32 in the summer, cases 52 and 53 in the autumn, and cases 54, 55, 56, 57, and 87 in the winter. However, there are several cases where the upper bound of RCV is less than the lower bound of interval-valued RPN, e.g., case 35 in the spring, cases 6, 17, 23, and 26 in the autumn, and case 13 in the winter.

By comparing the upper bound of RCVs with the lower bound of interval-valued RPN for each of these six-time windows, Figure 1 also shows that the interval of RCV overlaps the interval-valued RPN in most cases involving two, three, or four risk types. Furthermore, in certain cases, the upper bound of RCV is larger than the upper bound of interval-valued RPN, e.g., cases 52, 54, 75, 87, and 88 in the first time window, cases 32, 53, 55, 56, 57, 84, and 91 in the second time window, case 58 in the third time window, cases 15, 18, and 28 in the fourth time window, cases 12, 14, 36, 60, 74, 89, and 90 in the fifth time window, and cases 24, 86, and 94 in the sixth time window. Particularly, there are cases with lower bounds of RCV that are larger than the upper bounds of its interval-valued RPN, e.g., cases 52, 54, 87, and 88 in the first time window, cases 32, 53, 55, 56, and 57 in the second time window, and case 14 in the fifth time window. However, there are several cases where the upper bound of RCV is less than the lower bound of interval-valued RPN, e.g., cases 6 and 13 in the second time window, case 17 in the fifth time window, and cases 23, 26, and 35 in the sixth time window.

The comparison between RCV and risk values across two, three, and four coupling scenarios is summarized in Table 1. When comparing the upper bound of RCV with the lower bound of interval-valued RPN for each case, 74 cases were identified, accounting for 92.5% of the total 80 cases, where the upper bound of RCV was equal to or exceeded the lower bound of RPN. This finding suggests that the majority of RCV values overlap with the risk range, thereby validating the proposed method. Additionally, in 35 cases (43.75% of the total 80 cases), the lower bound of RCV was found to be equal to or greater than the lower bound of interval-valued RPN. If the RCV is used as an indicator of risk, in this study, 92.5% of all 80 cases could be judged as a scenario with grounding risk from the perspective of the upper bound of RCV; however, there are at least 43.75% of all 80 cases that could be judged as a scenario with grounding risk from the perspective of the lower bound of RCV. Therefore, the upper bound of RCV could be an effective leading indicator of risk.

Table 1. Comparison between RCV and risk value in three types of coupling scenarios.

	Number of cases			
	Two risk types	Three risk types	Four risk types	In total (percentage in 80 cases)
Upper bound of risk coupling value ≥ Lower bound of risk value	43	14	17	74 (92.50%)
Lower bound of risk coupling value ≥ Lower bound of risk value	6	12	17	35 (43.75%)

5 CONCLUSION

This study presents a method for measuring the Risk Coupling Value (RCV) of maritime accidents, specifically focusing on grounding incidents in the waters west of Shenzhen Port, China. The results demonstrate that the proposed method effectively captures the complexity of risk interactions in maritime operations. The RCV, represented as an interval value, provides a robust measure of risk uncertainty, reflecting the dynamic nature of maritime environments. The comparison between RCV and interval-valued RPN reveals significant overlap in most cases. Notably, the upper bound of RCV serves as a strong leading indicator of risk, with 92.5% of cases indicating potential grounding risk.

The analysis also highlights the importance of considering both the type and quantity of risk factors in maritime accidents. The distribution of RCV across different seasons, time windows, areas, and ship types reveal distinct patterns, suggesting that targeted risk management strategies could be developed for specific operational contexts. For instance, bulk and general cargo ships are more frequently involved in risk coupling scenarios, indicating a need for enhanced safety measures for these vessel types.

ACKNOWLEDGEMENTS

This work is supported by funding from the National Natural Science Foundation of China (Grant No. 52301419, 52272422, 52372416), Fund of National Engineering Research Center for Water Transport Safety (Grant No. A202404).

REFERENCES

Abaei, M.M., Arzaghi, E., Abbassi, R., Garaniya, V., Javanmardi, M., Chai, S., 2018. Dynamic reliability assessment of ship grounding using Bayesian Inference. Ocean Engineering 159, 47–55.

Akhtar, M.J., Utne, I.B., 2014. Human fatigue's effect on the risk of maritime groundings – A Bayesian Network model-ing approach. Safety Science 62, 427–440.

Akyuz, E., 2015. A hybrid accident analysis method to assess potential navigational contingencies: The case of ship grounding. Safety Science 79, 268–276.

Aydin, M., Akyuz, E., Turan, O., Arslan, O., 2021. Valid-ation of risk analysis for ship collision in narrow waters by using fuzzy Bayesian networks approach. Ocean Engineer-ing 231, 108973.

Bayazit, O., Kaptan, M., 2024. Dynamic risk analysis of allision in port areas using DBN based on HFACS-PV. Ocean Engineering 298, 117183.

Bulian, G., Cardinale, M., Dafermos, G., Lindroth, D., Rupo-nen, P., Zaraphonitis, G., 2020. Probabilistic assessment of damaged survivability of passenger ships in case of ground-ing or contact. Ocean Engineering 218, 107396.

Chen, S., Ahmad, R., Lee, B-G., Kim, D., 2014. Compos-ition ship collision risk based on fuzzy theory. Journal of Central South University 21, 4296–4302.

Fan, C., Bolbot, V., Montewka, J., Zhang, D.,2024a. Advanced Bayesian study on inland navigational risk of re-motely controlled autonomous ship. Accident Ana-lysis and Prevention 203, 107619.

Fan, C., Montewka, J., Bolbot, V., Zhang, Y., Qiu, Y., Hu, S., 2024b. Towards an analysis framework for oper-ational risk coupling mode: A case from MASS navigat-ing in restricted waters. Reliability Engineering and System Safety 248, 110176.

Fan, C., Montewka, J., Zhang, D., 2022. A risk comparison framework for autonomous ships navigation. Reliability Engineering and System Safety 226, 108709.

Fan, C., Montewka, J., Zhang, D., Han, Z., 2024c. A frame-work for risk matrix design: A case of MASS navigation risk. Accident Analysis and Prevention 199, 107515.

Fu, S., Yu, Y., Chen, J., Xi, Y., Zhang, M., 2022. A frame-work for quantitative analysis of the causation of grounding accidents in arctic shipping. Reliability Engineering & System Safety 226, 108706.

Ganco, M., 2017. NK model as a representation of innova-tive search. Research Policy 46, 1783–1800.

Graziano, A., Teixeira, A.P., Guedes Soares, C., 2016. Classification of human errors in grounding and colli-sion acci-dents using the TRACEr taxonomy. Safety Science 86, 245–257.

Guo, J., Luo, C., Ma, K., 2023. Risk coupling analysis of road transportation accidents of hazardous materials in complicated maritime environment. Reliability Engineering & System Safety 229, 108891.

Hu, S., Fang, C., Wu, J., Fan, C., Zhang, X., Yang, X., Han, B., 2025. Enhanced risk assessment framework for complex maritime traffic systems via data driven: A case study of ship navigation in Arctic. Reliability Engineering and System Safety 260, 110991.

Huang, W., Yin, D., Xu, Y., Zhang, R., Xu, M., 2022. Using N-K Model to quantitatively calculate the variability in Functional Resonance Analysis Method. Reliability Engineering & System Safety 217, 108058.

Jiang, D., Wu, B., Cheng, Z., Xue, J., van Gelder, P.H.A.J. M., 2021. Towards a probabilistic model for estimation of grounding accidents in fluctuating backwater zone of the Three Gorges Reservoir. Reliability Engineering & System Safety 205, 107239.

Li, H., Celik, C., Bashir, M., Zou, L., Yang, Z., 2024. Incorporation of a global perspective into data-driven analysis of maritime collision accident risk. Reliability Engineering & System Safety 249, 110187.

Liu, Z., Ma, Q., Cai, B., Shi, X., Zheng, C., Liu, Y., 2022. Risk coupling analysis of subsea blowout accidents based on dynamic Bayesian network and NK model. Reliability Engineering & System Safety 218, 108160.

Qiao, W., 2021. Analysis and measurement of multifactor risk in underground coal mine accidents based on coupling theory. Reliability Engineering & System Safety 208, 107433.

Wang, H., Chen, N., Wu, B., Guedes Soares, C., Wang, J., 2024. Hu-man and organizational factors analysis of collision acci-dents between merchant ships and fishing vessels based on HFACS-BN model. Reliability Engineering and System Safety 249, 110201.

Wang, X., Ma, J., Fan, C., Fu, S., Cai, Y., 2023. Towards nav-igational risk identification on a remotely controlled ship with seafarers onboard. In: The 7th International Confer-ence on Transportation Information and Safety, Aug 4-6, 2023, Xi'an, China. IEEE, 979-8-3503-0853-2/23.

Xi, Y., Zhang, X., Han, B., Zhu, Y., Fan, C., Kim, E., 2024. Advanced Human Reliability Analysis Approach for Ship Convoy Operations via a Model of IDAC and DBN: A Case from Ice-Covered Waters. Journal of Marine Science and Engineering 12, 1536.

Yıldırım, U., Başar, E., Uğurlu, Ö., 2019. Assessment of col-lisions and grounding accidents with human factors analysis and classification system (HFACS) and statistical methods. Safety Science 119, 412–425.

Yildiz, S., Ugurlu, O., Wang, J., Loughney, S., 2021. Application of the HFACS-PV approach for identification of hu-man and organizational factors (HOFs) influencing marine accidents. Reliability Engineering & System Safety 208, 107395.

Youssef, S.A.M., Paik, J.K., 2018. Hazard identification and scenario selection of ship grounding accidents. Ocean Engineering 153, 242–255.

Yu, Q., Teixeira, P., Liu, K., Rong, H., Guedes Soares, C., 2021. An integrated dynamic ship risk model based on Bayesian Networks and Evidential Reasoning. Reliability Engineering & System Safety 216, 107993.

Zhang, J., He, A., Fan, C., Yan, X., Guedes Soares, C., 2021. Quan-titative analysis on risk influencing factors in the Jiangsu segment of the Yangtze River. Risk Analysis 41(9), 1560–1578.

Trends in Collision and Grounding of Ships and Offshore Structures – Zhu & Guedes Soares (Eds)

A Bayesian network framework for rapid estimation of the damage extent after hard grounding events

D.G. Georgiadis & D. Straub
Engineering Risk Analysis Group, Technical University of Munich, Germany

M.S. Samuelides
School of Naval Architecture & Marine Engineering, National Technical University of Athens, Greece

ABSTRACT: A reliable assessment of hull-girder residual strength after a severe grounding event is critical for making informed decisions, such as determining a ship's ability to safely reach a nearby yard. A significant source of uncertainty in hull-girder residual strength assessment is the extent of structural damage. Classification societies have developed dedicated tools for rapid response damage assessment, typically relying on 2D Smith-based models. However, these models are based on deterministic methods and conservative estimates of the damage extent. To enhance this assessment, we propose a probabilistic framework for near-real-time damage assessment using Bayesian networks (BNs). The employed BN integrates multiple information sources, including underwater inspections, hydrostatic and bathymetric data, crashworthiness models, and hydraulic flow models. By formalizing these inputs and their associated uncertainties in a compact and causal framework, the BN allows for parameter updating as new evidence becomes available after the incident. We demonstrate the proposed BN model using a hypothetical grounding scenario on a Bulk Carrier.

1 INTRODUCTION

Ship grounding remains a significant threat to maritime safety, often leading to potential environmental disasters, economic losses, and human risks. The extent of damage is one of the most critical parameters in determining the residual strength of the ship after the accident. In turn, the reliable assessment of hull-girder residual strength is critical for making informed decisions, such as determining a ship's ability to safely reach a nearby yard.

During the past years, various methods for rapid damage assessment after grounding have been proposed, the majority of which are based on a deterministic approach. Nguyen et al. (2011) and Pineau et al. (2022) developed numerical analysis tools to estimate the extent of damage after grounding based on energy considerations, i.e., crashworthiness models. Ruponen et al. (2012) used a simplified hydraulic model, based on Bernoulli's principle, to assess the extent of damage extent based on the measured flooding rate. A probabilistic method to estimate the damage extent following the grounding of a passenger ship was proposed by Taimuri et al. (2023). This method, however, does not allow for belief updating on input uncertain parameters using evidence from the incident site.

Classification societies provide tools for rapid response damage assessment, such as the RRDA tool of ABS (Lee et al. 2012) or the ERS software of DNV (DNV 2025). In these tools, the estimation of the damage extent typically relies on site information acquired from the master's judgment and survey reports. A deterministic approach is usually followed; the final damage extent assessment is based on conservative estimates, e.g., by increasing the damage estimate approximately two to five times the reported value (Lee et al. 2012).

The uncertainty in the damage extent prediction along with various sources of information and available models for damage assessment (e.g., inspection outcomes, crashworthiness models, hydraulic models, onboard measurements, and water depth data) motivate the use of Bayesian networks (BNs) for damage extent assessment.

BNs provide an efficient and intuitive way to express the dependence of various uncertain parameters using a probabilistic graphical structure. BNs have been used for various applications in the maritime industry (e.g., Friis-Hansen 2000; Animah, 2024). With respect to ship grounding, BNs have been mostly applied for the evaluation of the occurrence probability of a grounding event and its associated risk assessment (Mazaheri et al. 2016; Jiang et al., 2021). A work towards hull-girder residual strength assessment was presented by Li & Tang (2019), who developed a structural reliability

DOI: 10.1201/9781003684404-52

analysis model based on BNs for quantifying the collapse risk of damaged hulls after grounding. None of the aforementioned studies, however, addresses the damage assessment after grounding.

In this paper, we present a BN model for predicting the extent and location of damage after a severe grounding scenario on a rocky seabed. The present work is based on (Georgiadis et al. 2025). The proposed BN uses existing physics-based models (i.e., crashworthiness and hydraulic models) and on-site information (i.e., flooding rate monitoring, draft measurements, hydrostatic and stability data, bathymetric data, and underwater inspection outcomes). We present a demonstration of the proposed framework on a hypothetical grounding scenario of a Bulk Carrier.

2 BAYESIAN NETWORKS

A Bayesian network is a probabilistic graphical model based on directed acyclic graphs (Jensen 2007; Pearl & Russel 2000) The nodes represent random variables and the directed links capture dependencies. The link between two nodes is based on a causal relation, but this is not necessary. Each random variable X_i is characterized by its conditional probability distribution $p(x_i|pa(x_i))$ given its parents $pa(x_i)$, which is the set of nodes with links pointing toward X_i. In the case of discrete random variables, this conditional distribution takes the form of conditional probability tables (CPTs). In the case of continuous random variables, the conditional distribution is characterized by a conditional probability density function (PDF). An BN in which all its random variables are discrete is termed *discrete BN*. If there is a mix of discrete and continuous random variables, the BN is called *hybrid BN*.

Any BN specifies the joint probability distribution $p(\mathbf{x})$ of a set of random variables as the product of the conditional distributions:

$$p(x) = \prod_{i=1}^{n} p[x_i | pa(x_i)] \tag{1}$$

As an example, consider the BN of Figure 1. The BN represents a simplified version of the crashworthiness module (defined in more detail in Section 3). A ship characterized by its mass M and speed V strikes a rock. A damage D is caused on its bottom structure. An additional node Z is included representing the outcome of an inspection. The nodes in this example represent discrete random variables whose joint PMF is:

$$p(m, v, d, z) = p(z|d)\, p(d|m, v)\, p(m)\, p(v) \tag{2}$$

As soon as a set of observations or evidence e is available, inference can take place in the network. For instance, given an inspection outcome $Z = e$, the (joint) probability distribution of the random variables M, V and D conditional on the observed value of Z is calculated as:

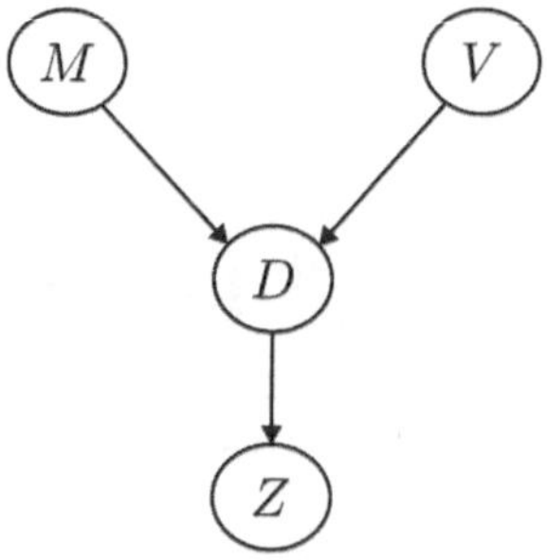

Figure 1. BN simplified example.

$$p(m, v, d|Z = e) = \frac{p(m, v, d, e)}{p(e)} \tag{3}$$

A variety of algorithms exists for inference, which can be broadly classified into two main categories: *exact* and *sampling-based* algorithms. In this work, we employ an exact inference method after discretizing the continuous random variables of the initial hybrid BN into specified intervals, thereby converting it into a fully discrete network. This discretization procedure induces an error in the final prediction of posterior probabilities. The approximation error is reduced with finer discretization but that comes with a higher computational cost.

3 BN FOR DAMAGE ASSESSMENT

In the present study, the damage is represented by a 2D rectangular shape and is mathematically described by three (3) independent continuous random variables as shown in Figure 2: the transverse location of the center of the damage Y_D, the transverse (or width) extent of the damage D_t, and the vertical extent of the damage D_v. The interest is in the prediction of 2D damage which is used as input in the evaluation of hull-girder residual strength using the 2D Smith's model of CSR (IACS 2024). It is here assumed that the critical damage location lies within the midship region of the ship, which is the most vulnerable area in terms of longitudinal bending.

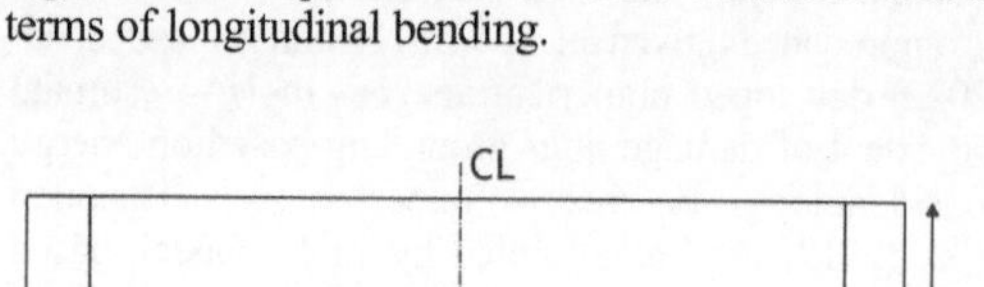

Figure 2. Two-dimensional (2D) grounding damage schematics.

The BN model illustrated in Figure 3 is proposed to probabilistically estimate the 2D damage extent and location following a hard grounding event on a rock. The BN is presented in detail in Georgiadis et al. (2025). Here, only a brief overview is given.

The BN of Figure 3 consists of discrete (double ovals) and continuous (single ovals) random variables. The green nodes represent potential evidence (reported values or calculated values outside the network) and the yellow nodes represent the output parameters of interest. The BN incorporates four (4) basic modules: (i) a crashworthiness module, which is based on energy consideration principles, (ii) a hydraulic module, which is based on Bernoulli's principle and flooding rate monitoring, (iii) a hydrostatic and bathymetric module, which includes moments equilibrium condition and water depth data, and (iv) an inspection module.

Exemplarily, we present the crashworthiness module that consists of the following nodes: mass M, mass reported M_r, impact speed V, impact speed reported V_r, damage length L_d, damage length reported $L_{d,r}$, horizontal grounding force F_h, inner hull breach IHB and damage width D_t.

The BN framework establishes probabilistic dependencies between the above parameters, allowing dynamic updates as evidence becomes available after the incident. For example, the following relationship connects the nodes *impact speed* V and *reported impact speed* V_r:

$$V_r = V + \varepsilon_V \tag{4}$$

where ε_v denotes the uncertainty in the reported value of impact speed, which is modelled by a normal distribution with zero mean and standard deviation 0.24 kn (Dalheim and Steen, 2021). As soon as evidence becomes available on the node V_r, the prior distribution of V is updated using Bayes' rule:

$$f''(v) = \frac{f(v_r|v)f'(v)}{\int_0^\infty f(v_r|v)f'(v)dv} \tag{5}$$

Probabilistic relationships, similar to those in Equation 4, are established between the true and reported values of mass and damage length.

The kinetic energy of the ship is calculated as:

$$E = \frac{1}{2}(M + M_a)V^2 \tag{6}$$

where M is the total mass (or weight) of the ship, M_a is the added mass (5% of M), and V is the impact speed. We assume that all the kinetic energy of the ship is dissipated by destroying the bottom structure (Pedersen and Zhang, 2000). Thus, the total horizontal grounding force is given as:

$$F_h = \frac{0.5(M + M_a)V^2}{L_d} \times \varepsilon_{f_h} \tag{7}$$

where L_d is the damage (stopping) length and ε_{fh} is a multiplicative error term that follows the lognormal distribution with a median equal to one and coefficient of variation (cov) 0.10 (Cerup-Simonsen et al., 2009). This error reflects the uncertainty in the horizontal grounding force prediction.

Several empirical formulas have been suggested to connect the horizontal grounding force with the damage width extent (Pedersen and Zhang, 2000; Zhang, 2002; Hong and Amdahl, 2012). Among them, the formula proposed by Cerup-Simonsen et al. (2009), which has been calibrated with a series of finite element analyses and validated against a real grounding scenario, is adopted:

$$F_h = 0.77\sigma_0 e_f^{0.71} t_{eq}^{1.17} D_t^{0.83} \tag{8}$$

where σ_0 is the flow stress of the bottom structure, ε_f is the fracture strain, t_{eq} is the equivalent thickness plate. Equation 8 is solved for D_t and inserted into the node *Damage width extent* of the BN of Figure 3. The parameters σ_0, ε_f, t_{eq} are treated as deterministic quantities. However, they can be also considered as random variables. If the inner hull is breached, Equation 8 is applied both for the outer and inner bottom; this is controlled by the parent node *Inner hull breach* which has two discrete states, *yes* and *no*.

A validation of the crashworthiness module against a real grounding accident on a single-hull oil tanker is presented in Georgiadis et al. (2025).

4 NUMERICAL INVESTIGATION

We present a numerical example to demonstrate the application of the proposed BN model to a hypothetical Bulk Carrier grounding case. The main particulars of the ship along with basic double bottom structural characteristics are listed in Table 1. A grounding scenario is assumed, which results in the damage pattern illustrated in Figure 4. More details are listed in Table 2.

Table 1. Principal particulars and double bottom structural characteristics of bulk carrier.

Main particulars	Value	Units
Length	197.00	m
Breadth	32.24	m
Depth	18.60	m
Draft design	11.30	m
Service speed	14.50	kn
Double bottom height	1.8	m
Outer (Inner) bottom equivalent thickness	38 (48)	mm
Outer (Inner) bottom flow stress	458 (458)	MPa
Outer (Inner) bottom fracture strain	0.25 (0.25)	-

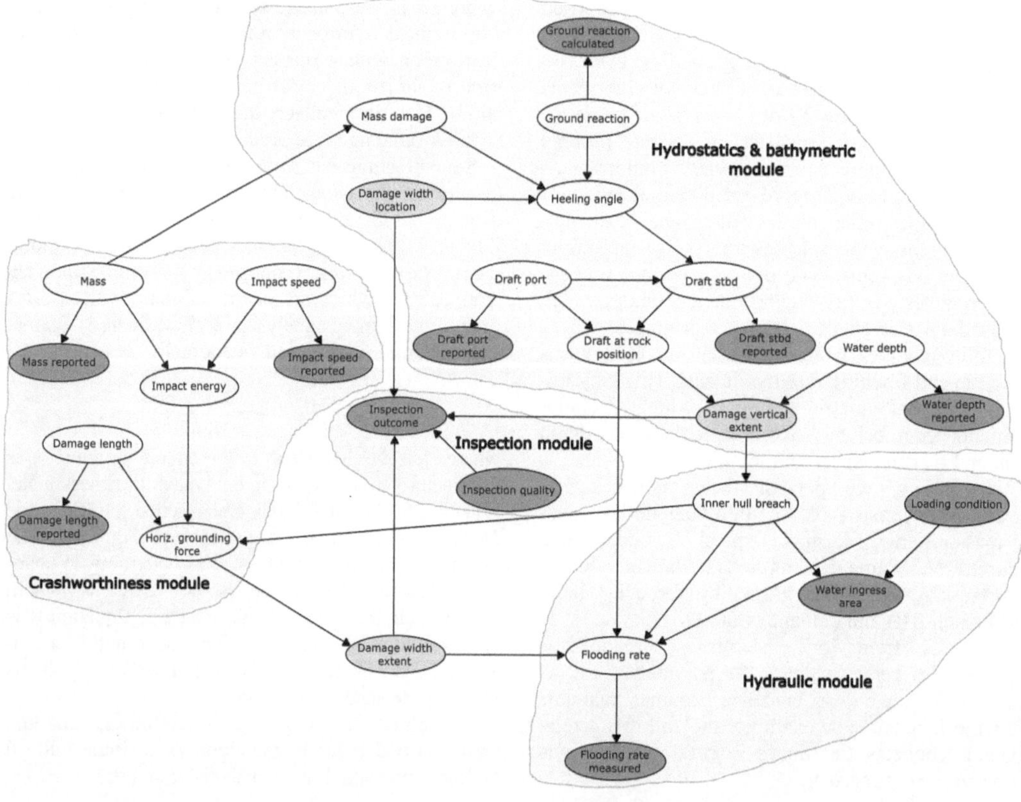

Figure 3. BN model for 2D grounding damage assessment of Bulk Carriers, from (Georgiadis et al. 2025).

Table 2. Data for the hypothetical grounding scenario.

Pre-grounded conditions	
Loading condition	Loaded
Displacement (t)	62,148
Impact speed (kn)	9.5
Draft aft (m)	11.9
Draft mid (m)	11.3
Draft fwd (m)	10.7
Water depth	10.0
Grounding conditions	
First intact long. position (m)	FE
Damage length (m)	85
Cargo hold damage length (m)	32
Flooding area	Ballast tank
Flooding rate (m^3/s)	1150
Draft port (m)	11.7
Draft stbd (m)	14.8
Heeling angle (deg)	5.5
Mass in (t)	3,509
Ground reaction (t)	3,900

(*Continued*)

Table 2. (Cont.)

Pre-grounded conditions	
Metacentric height (m)	2.4
Inspection outcome	
Damage width extent (m)	3.5
Damage vertical extent (m)	1.6
Damage width location (m)	4.5 (port side)
True values (assumed)	
Damage width extent (m)	3.1
Damage vertical extent (m)	1.8
Damage width location (m)	4.7 (port side)

4.1 *BN modeling*

The BN modeling and inference were implemented in the GeNIe software (BayesFusion, LLC 2025). Evidence is provided in the green nodes of Figure 3 based on the data listed in Tables 1 and 2. The prior distributions of the root nodes are presented in Table 3.

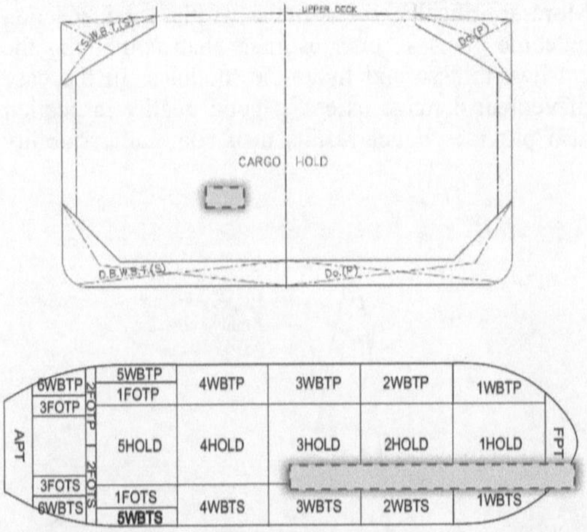

Figure 4. Ship view damage schematics.

Table 3. Prior distribution models for the root nodes of the BN.

Parameter	Distribution	Mean	St. dev	Bounds
Mass (t)	Uniform	38,025	15,574	11,050 to 65,000
Speed (kn)	Beta	9.67	3.4	0 to 14.5
Damage length (m)	Custom*	69.5	67.7	0 to 197
Ground reaction (t)	Uniform	5,000	2,887	0 to 10,000
Water depth (m)	Uniform	10	5.8	0 to 20
Draft port ground (m)	Uniform	10	5.8	0 to20
Damage center loc. (m) Uniform		0	9.3	-16.12 to 16.12

* Based on IMO (2003).

4.2 *Results*

The results from the analyses are presented in Figures 5-7. It is observed that for all cases the true value lies within the predicted range. If a single deterministic value is desired from the analyst, the 90% or 95% quantile of the posterior density function can be used, which provides a reasonable conservatism on the damage extent.

The transverse damage extent prediction is presented in Figure 5. The crashworthiness module can provide a first-cut rapid assessment of D_t. However, the uncertainty in the estimate is rather high. Integrating the hydraulic module can significantly reduce the uncertainty in the prediction. In fact, the combination of these two modules may even replace underwater inspection, which presents similar probability estimates.

The vertical damage extent prediction is illustrated in Figure 6. Using the hydrostatics and bathymetric module (i.e., information related to the damage stability of the ship, moments equilibrium, and water depth measurements), a relatively high degree of uncertainty is obtained. A significant improvement of the estimate is achieved by observing that no water has inserted in the cargo hold, i.e., no inner hull penetration. Note that even though the inner hull has not been penetrated, a probability mass is still observed above the double bottom height. This indicates damage in the inner hull plate and the attached longitudinal stiffeners. Following a residual strength analysis, the corresponding structural members should be removed from the structural analysis model. Finally, we observe that the updating with underwater inspection outcome provides similar results.

The prediction of the transverse center location of the damage is shown in Figure 7. Initially, the distribution is uniformly distributed through the entire breadth of the vessel. Information from the hydrostatics and stability of the damaged ship – moments equilibrium principle – provide a very good estimate of Y_D. However, this is achieved because in the BN model only the

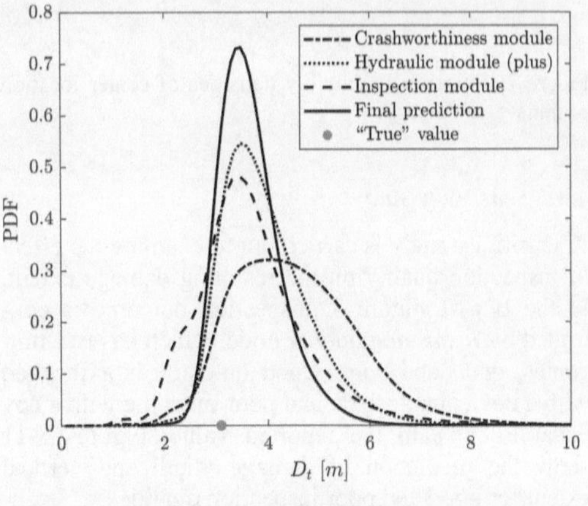

Figure 5. Posterior probability densities of damage width extent.

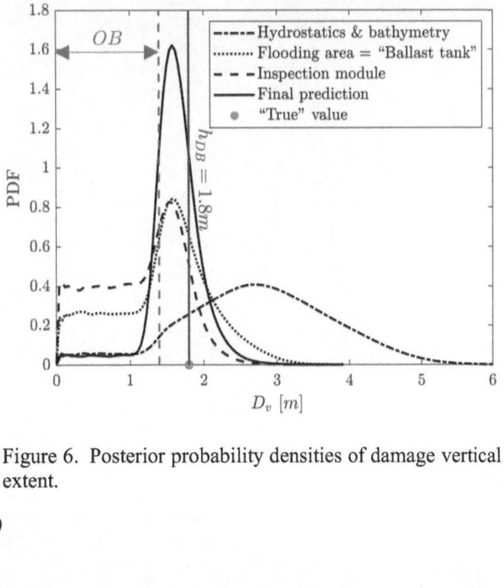

Figure 6. Posterior probability densities of damage vertical extent.

uncertainty in ground reaction estimation is considered; the metacentric height in the damage condition as well as the inflow sea water mass have been set as deterministic quantities. The mean prediction of the inspection is close to the true value. The final prediction gives a good estimate of the true location.

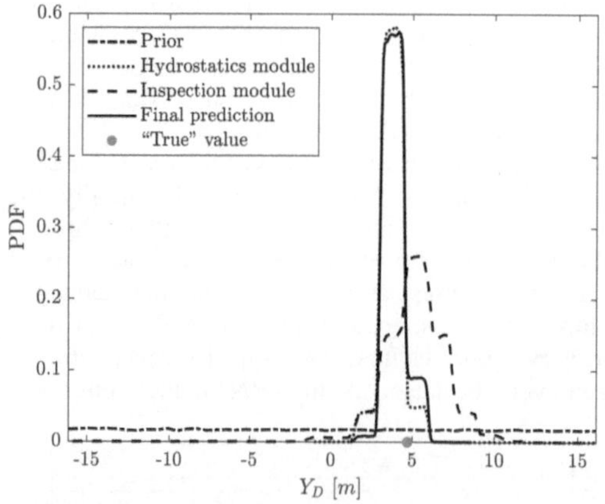

Figure 7. Posterior probability densities of center location of transverse damage.

4.3 *Sensitivity study*

A sensitivity study is carried out to examine the effect of inspection quality on the resulting damage extent. In the BN of Figure 3, inspection outcome is controlled by *Inspection quality* node, which has two outcomes, *good* and *poor*. Good outcome is associated with a cov equal to 10% and poor outcome with a cov equal to 30% in the reported value. Figures 8-11 show the prediction of damage width and vertical extent for good and poor inspection qualities.

As shown in Figures 8 and 10, a good quality inspection has a strong impact on the final prediction. More specifically, for damage width, the inspection outcome gives a better estimate than combining the crashworthiness and hydraulic modules. In the case of vertical damage extent, a good quality inspection also provides better results than combining stability

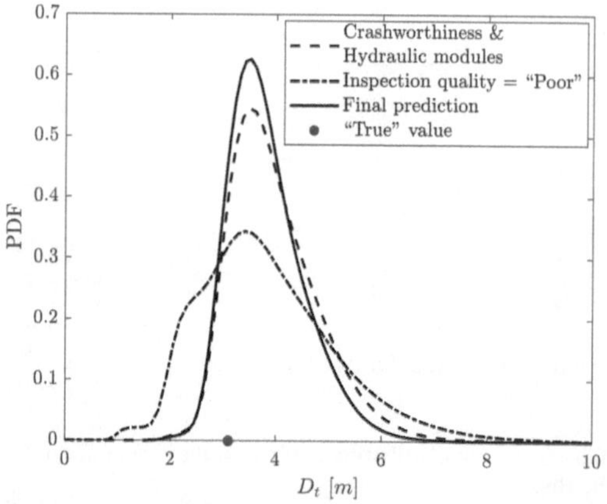

Figure 9. Posterior probability densities of damage vertical extent.

data, water depth data, and inner hull breach observation together.

On the other hand, a poor-quality inspection provides a higher degree of uncertainty in the damage estimate than the combination of crashworthiness and hydraulic modules (see Figure 9). Similar observations are made for vertical damage extent (see Figure 11). These findings indicate that a low-quality underwater inspection can be replaced by onboard measurements and available physics-based models. This is important for reducing costs and mitigating the risks associated with underwater inspections in challenging conditions, such as harsh environments or a moving hull.

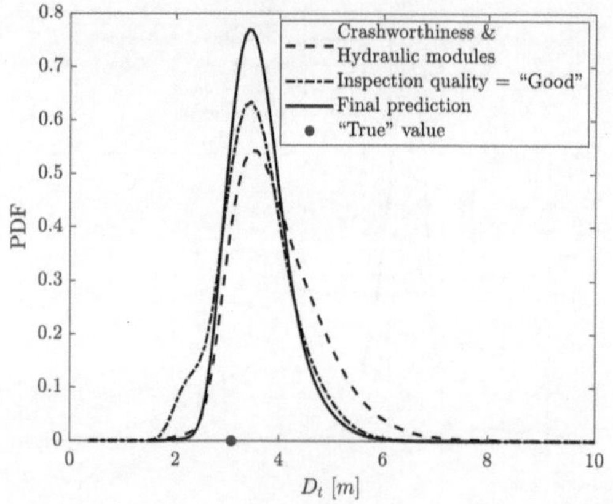

Figure 8. Posterior probability densities of damage width extent.

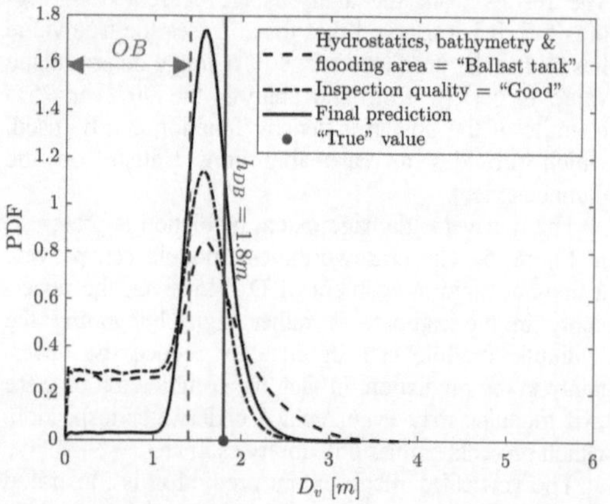

Figure 10. Posterior probability densities of transverse location of damage center.

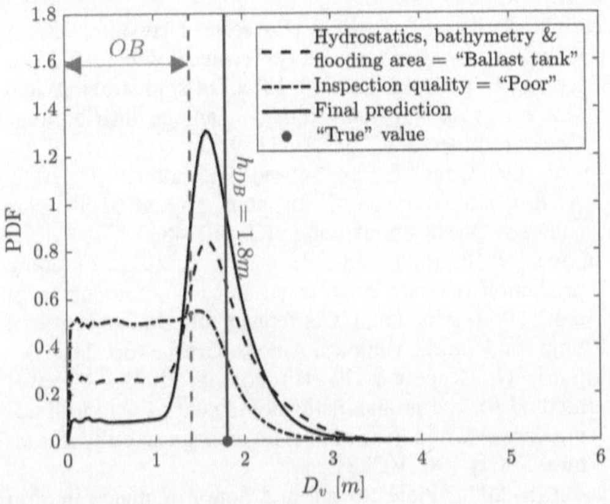

Figure 11. Posterior probability densities of transverse location of damage center.

5 CONCLUDING REMARKS

A Bayesian network for rapid assessment of grounding damage is proposed. The presented BN model offers several advantages over existing approaches, the most important of which are: (i) the introduction of a probabilistic framework for damage extent estimation, (ii) the combination of existing physics-based models (e.g., energy-based crashworthiness models, hydraulic models) and information sources from the incident site in an efficient and intuitive way, and (iii) the provision of near-real-time results.

The BN model is verified through a hypothetical grounding scenario on a Bulk Carrier. We found that the combination of the crashworthiness and hydraulic modules can potentially replace costly underwater inspections; however, further studies on real-world applications are needed for validation. Moreover, the hydrostatics and bathymetric module in combination with the inner hull breach observation can offer very good predictions of vertical damage extent. Finally, the transverse location of the damage can be estimated with very good accuracy if an approximate value of the ground reaction is available. A practical guide for a quick estimation of ground reaction can be found in Dean (2016).

A limitation of the proposed BN model is that it does not account for time-dependent effects, such as progressive flooding. A solution to this problem could be the extension to a dynamic BN model (Murphy 2002). Moreover, the present model does not consider complex hydrodynamic effects as well as the additional damage that may be caused by tides and waves. For example, if a ship runs aground at high tide, it is very likely that further damage may be caused as the tide drops and buoyancy is lost (Nguyen et al. 2011). Finally, an extension of the model to account for grounding on multiple rocks, resulting in multiple damage openings, can be performed.

In terms of model complexity, some parameters are treated as deterministic quantities and integrated into a single node, with limited loss of accuracy. A more refined version of the model could explicitly include these parameters as separate nodes, allowing a better representation of their inherent uncertainties.

In summary, the primary objective of this work is to develop a user-friendly digital assistance tool for rapid damage assessment in real operational conditions; its results can be used as input for hull-girder residual strength analysis using Smith's method. The proposed framework can replace master's judgment and support classification societies, shipowners, and salvage companies in making informed decisions about ship operability after a severe grounding event.

ACKNOWLEDGEMENT

This work is supported by the Alexander Von Humboldt Foundation.

REFERENCES

Animah, I. 2024. Application of Bayesian network in the maritime industry: Comprehensive literature review. Ocean Engineering 302, 117610.

BayesFusion, LLC (2025) GeNIe Modeler. Version 5.0. Available at: https://www.bayesfusion.com/genie.

Cerup-Simonsen, B., Tornqvist, R., Lutzen, M. 2009. A simplified grounding damage prediction method and its application in modern damage stability requirements. Marine Structures 22, 62–83.

Dalheim, Ø.Ø. & Steen, S. 2021. Uncertainty in the real-time estimation of ship speed through water. Ocean Engineering 235, 109423.

Dean, M.S. 2016. Salvage operations. Springer Handbook of Ocean Engineering, 985–1066.

DNV, 2025. Emergency Response Service. https://www.dnv.com/maritime/ship-classification-services/emergency-response-service.

Friis-Hansen, A. 2000. Bayesian networks as a decision support tool in marine applications. Ph.D. thesis. Department of Naval Architecture and Offshore Engineering, Technical University of Denmark.

Georgiadis, D.G., Samuelides, M.S., Straub, D. 2025. Near-real-time ship grounding damage assessment using Bayesian networks. arXiv preprint arXiv:2506.06493.

Hong, L. & Amdahl, J. 2012. Rapid assessment of ship grounding over large contact surfaces. Ships and Offshore Structures 7, 5–19.

IACS 2024. Common Structural Rules for Bulk Carriers and Oil Tankers. International Association of Classification Societies, London, UK.

IMO 2003. Revised interim guidelines for the approval of alternative methods of design and construction of oil tankers – Marine Environment Protection Committee of the Organization by Resolution MEPC 110(49). Technical Report. International Maritime Organization.

Jensen, F.V. & Nielsen, T.D. 2007. Bayesian Networks and Decision Graphs. Springer.

Jiang, D., Wu, B., Cheng, Z., Xue, J., Van Gelder, P. 2021. Towards a probabilistic model for estimation of grounding accidents in fluctuating backwater zone of the Three Gorges Reservoir. Reliability Engineering & System Safety 205, 107239.

Lee, A., VanDerHorn, E., Wang, G. 2012. Residual strength analysis for rapid response damage assessment (RRDA) of steel ship structures, in: International Conference on Offshore Mechanics and Arctic Engineering, American Society of Mechanical Engineers. pp. 491–502.

Li, X., Tang, W. 2019. Structural risk analysis model of damaged membrane LNG carriers after grounding based on Bayesian belief networks. Ocean Engineering 171, 332–344.

Mazaheri, A., Montewka, J., Kujala, P. 2016. Towards an evidence-based probabilistic risk model for ship-grounding accidents. Safety Science 86, 195–210.

Murphy, K.P. 2002. Dynamic Bayesian networks: representation, inference and learning. University of California, Berkeley.

Nguyen, T.H., Amdahl, J., Leira, B.J., Garre, L. 2011. Understanding ship grounding events. Marine Structures 24, 551–569.

Pearl J. & Russell S. 2000. Bayesian Networks. UCLA Cogni-tive Systems Laboratory, Technical Report (R-277).

Pedersen, P. & Zhang, S. 2000. Effect of ship structure and size on grounding and collision damage distributions. Ocean Engineering 27, 1161–1179.

Pineau, J.P., Conti, F., Le Sourne, H., Looten, T., 2022. A fast simulation tool for ship grounding damage analysis. Ocean Engineering 262, 112248.

Ruponen, P., Larmela, M., Pennanen, P. 2012. Flooding prediction onboard a damaged ship. In Proceedings of the 11th International Conference on the Stability of Ships and Ocean Vehicles, Athens, Greece (pp. 23–28).

Taimuri, G., Ruponen, P., Hirdaris, S. 2023. A novel method for the probabilistic assessment of ship grounding damages and their impact on damage stability. Structural Safety 100, 102281.

Zhang, S. 2002. Plate tearing and bottom damage in ship grounding. Marine Structures 15, 101–117.

Trends in Collision and Grounding of Ships and Offshore Structures – Zhu & Guedes Soares (Eds)

Probabilistic prediction of risky ship encounters using Bayesian Logistic Regression

S. Guo, V. Bolbot, L. Lu & O.A. Valdez Banda
Research group on Safe and Efficient Marine and Ship Systems, Marine and Arctic Technology, Department of Energy and Mechanical Engineering, Aalto University, Espoo, Finland
Kotka Maritime Research Centre, Kotka, Finland

ABSTRACT: Ship behavior during encounters has been widely studied to enhance collision risk assessment. However, the complexity of navigation environments makes dynamic modeling and simulation challenging under real-world uncertainties. This study proposes a novel method based on Bayesian logistic regression to probabilistically infer whether a potential encounter scenario will evolve into a risky situation. By learning latent patterns from rich historical encounters, this method offers intuitive and interpretable predictions with uncertainties while avoiding complex contexts modeling. A case study using AIS data from the middle of the Gulf of Finland Reporting (GOFREP) area is introduced to show the effectiveness of the proposed method. Through a comparative validation against the classical logistic regression model, the Bayesian model demonstrates 63.3% improvements in F1 score on uncertain samples. Therefore, it is concluded that the presented method contributes a complementary perspective to support safety management and decision-making for both ship operators and shore-based supervisors.

1 INTRODUCTION

Maritime transportation constitutes a critical component of the global economy, facilitating over 90% of international trade by volume (Fan et al., 2024). The increasing demand for ship transportation has posed significant challenges to navigation safety and efficiency. Ship collisions, which represent the most severe consequences of ship encounters, have been investigated to be one of the main contributors to marine casualties and incidents (EMSA, 2022). In recent years, extensive studies have been conducted to analyze the fundamental causes of maritime incidents and develop solutions for safer maritime operations (Li et al., 2024).

Ship encounter scenario represents a central research subject in maritime safety analysis, encapsulating the dynamic conflict processes arising from changes in ship speed, course, and relative positioning (Guo et al., 2023). Leveraging Automatic Identification System (AIS) data is a commonly adopted approach for studying these scenarios. It helps analyze ship behavior patterns for identifying potential collision risks between encounter candidates, facilitating automated monitoring and intelligent management of maritime traffic, which can effectively mitigate accident occurrence rates (Zhou et al., 2023). Contemporary research on ship encounter scenarios has primarily focused on the following directions: (1) risk assessment and modeling, (2) behavior prediction and decision support, (3) risk intervention and alert management.

Risk assessment and modeling form the foundation for understanding and quantifying collision risks (Cheng et al., 2024). Researchers seek to develop accurate models that predict risky encounters and assess the probability of collisions under various conditions. For example, Liu et al. (2024) introduced a novel collision risk assessment method integrating Velocity Obstacle (VO) with complex network theory for high-risk ship identification in busy waters. On the other hand, behavior prediction and decision support focus on anticipating ship actions and providing operators with insights to avoid dangerous situations. A noteworthy example is the approach developed by Chen et al. (2024), which utilizes AIS data to predict ship behaviors, providing enhanced decision support to help operators avoid dangerous encounters. In contrast, risk intervention and alert management concentrate on real-time detection and mitigation of collision risks. As an illustration, Du et al. (2020) presented a COLREG-compliant collision alert system for stand-on vessels, triggering multi-tiered alerts for proactive collision mitigation

DOI: 10.1201/9781003684404-53

when the give-way ship violates obligations. Collectively, these research areas together contribute to a comprehensive approach to enhancing maritime safety.

Despite significant progress, modeling and simulating ship encounters remain challenging due to the complex interactions among ship dynamics, weather conditions, and human factors (Zhang et al., 2024). Such complexity highlights the shortcomings of existing models in addressing the inherent uncertainties and dynamic factors present in real-world scenarios (Xin et al., 2021). As a result, there is a growing need for sophisticated probabilistic approaches that can better account for these uncertainties and improve the reliability of risky ship encounter predictions.

This study proposes a novel Bayesian logistic regression model to predict the escalation probability of potential ship encounters into risky situations. The presented approach investigates multiple features, including Distance to Closest Point of Approach (DCPA), Time to Closest Point of Approach (TCPA), encounter duration, encounter type, ship speeds, and ship types to ensure accurate and robust prediction. AIS data collected from the Gulf of Finland Reporting (GOFREP) area were used to train and validate the performance of the proposed model. Furthermore, its interpretability in managing uncertainties was evaluated through comparison with the classical logistic regression model.

The remainder of this paper is organized as follows: Section 2 elaborates on the methodology of this study. Section 3 introduces the case study and discusses the results. Section 4 concludes the contributions and limitations of this work, indicating possible directions for future improvements.

2 METHODOLOGY

Bayesian data analysis provides a powerful probabilistic framework for understanding complex maritime encounter scenarios. In contrast to traditional statistical approaches, Bayesian methods enable researchers to incorporate prior knowledge and quantify uncertainties through probabilistic inferences. This approach is particularly valuable in maritime risk assessment, where decisions are always made under uncertainties (Jia et al., 2024). In this paper, we incorporate Bayesian theory into logistic regression to establish a Bayesian logistic regression model for encounter scenario analysis, offering probabilistic prediction of risky ship encounters while accounting for uncertainties. The methodology framework is outlined in Figure 1. Notably, for simplification and to focus on the model design, this study only considers two-ship context.

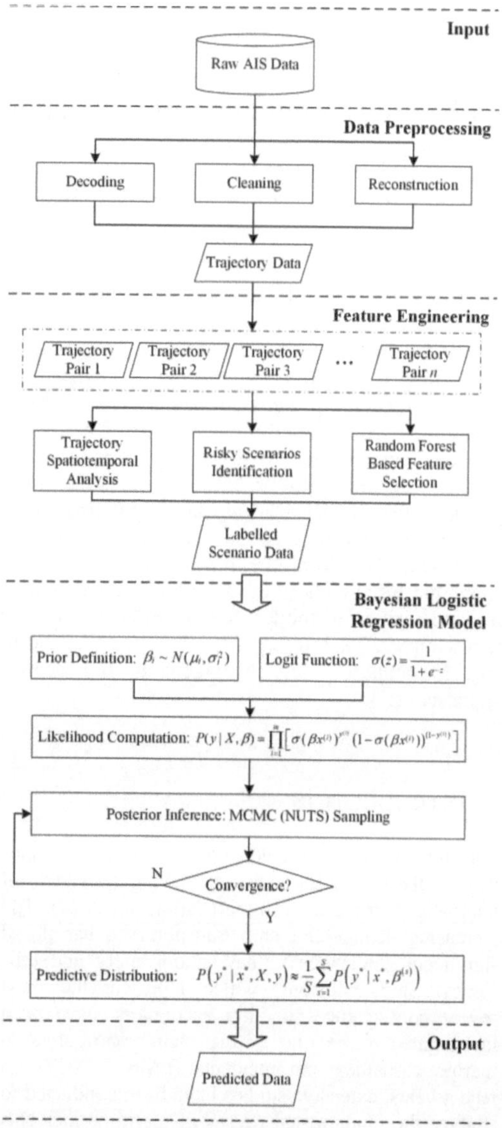

Figure 1. Overview of the methodology.

2.1 *Data preprocessing*

Data preprocessing is critical to transform raw AIS data into high-quality, structured inputs. The decoded AIS messages contain both dynamic and static information (Guo et al., 2024). For the scope of this study, Maritime Mobile Service Identity (MMSI), timestamp, longitude, latitude, Speed Over Ground (SOG), Course Over Ground (COG), ship length, ship width, and ship type are considered to describe ship behavior and characteristics (ITU, 2014). Following the cleaning and reconstruction methods as presented in (Guo et al., 2021a, 2021b), each ship trajectory is denoted by:

$$T_{id} = \{p_1, p_2, \ldots, p_n\} \quad (1)$$

where p_i is the i-th trajectory point denoted by a quintuple:

$$p = (t, lon, lat, sog, cog) \quad (2)$$

with *lon*, *lat*, *sog*, and *cog* representing longitude, latitude, Speed Over Ground (SOG), and Course Over Ground (COG) respectively, describing the behavior of the ship at the sampling time *t*. In addition, *id* is the MMSI of this ship, the corresponding static information is denoted by:

$$SI_{id} = (l_{id}, w_{id}, \tau_{id}) \quad (3)$$

where l_{id}, w_{id}, and τ_{id} represent ship length, width, and type respectively.

2.2 *Feature engineering*

2.2.1 *Potential encounter scenarios extraction*

In the absence of effective risk mitigation measures, ship encounters may ultimately escalate to collisions, which is the most adverse outcome. Thus, in order to precisely use the model to predict risky encounters, we first extract potential encounter scenarios to constitute the input dataset.

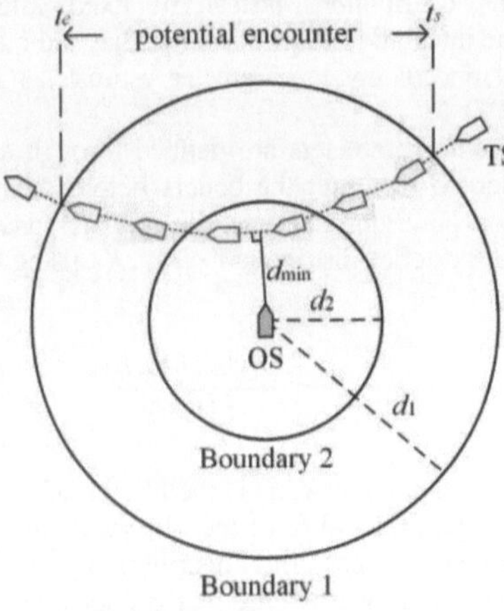

Figure 2. Potential encounter scenario.

As shown in Figure 2, from the perspective of own ship, if there is a target ship enters the d_1 nm zone centered on own ship at time t_s and leaves at t_e, and their minimum distance in this duration is lower than d_2, such a process is identified as a potential encounter scenario, denoted by:

$$E = \{(T_1(t), T_2(t)) | t \in [t_s, t_e], d_t < d_1, d_{\min} < d_2\} \quad (4)$$

In this context, $T_1(t)$ and $T_2(t)$ represent the states of the two ships at time t. d_t computes the distance between two ships at t by transforming the coordinates to the Universal Transverse Mercator (UTM) coordinate system, and $d_{\min}$ is the minimum relative distance of the two ships. With reference to Du et al. (2021), d_1 and d_2 are set to 12 nm and 6 nm as they are the typical radar detection ranges. In addition, during the extraction, linear interpolation is incorporated to ensure temporal alignment of the two trajectories.

2.2.2 *Risky scenarios identification*

A potential encounter scenario is identified as a risky encounter scenario if, at any time, the Distance to Closest Point of Approach (DCPA) is less than a threshold d_{risk} and the Time to Closest Point of Approach (TCPA) is less than the corresponding time t_{risk}. In mathematical terms, for any $t \in [t_s, t_e]$, E is labelled as risky if:

$$\begin{cases} DCPA_E(t) < d_{risk} \\ TCPA_E(t) < t_{risk} \end{cases} \quad (5)$$

Such a definition implies that there is a collision risk worthy of attention. According to (Gang et al., 2016), t_{risk} is defined by

$$t_{risk} = \frac{\sqrt{d_{risk}^2 - DCPA^2}}{v_r} \quad (6)$$

where d_{risk} ranges from 0.6 to 1.1 nm based on relative bearings and v_r is the relative speed. Zhang et al. (2021) further applied the definitions in the Gulf of Finland and validated their reasonableness for identifying risky encounters. On the other hand, as reported by Dinh & Im (2016), navigators typically maintain a 2 nm safety distance from other ships under normal conditions. Rong et al. (2022) mention that pilot's decisions in real scenarios introduce uncertainties due to the multifaceted influences of physical conditions, mental states, and environmental variables. Hence, based on these navigational practices and the goal of predicting unsafe encounters to alert operators in advance, we set d_{risk} to 3 nm for risky encounter scenario determination.

2.2.3 *Feature selection*

To concisely depict ship encounter development while minimizing redundant features, we set two determination boundaries as shown in Figure 2. Boundary 1 is the time when two ships first enter within the 12-

nautical-mile range of each other, and Boundary 2 is when they first enter within the 6-nautical-mile range. At each of these moments, the corresponding DCPA, TCPA, and ship speeds are computed, where encounter types (overtaking, crossing, and head-on) are also determined based on COLREGs (Guo et al., 2023; IMO, 1972). On top of that, the time spent between two boundaries is calculated and used to obtain the rate of change in DCPA and TCPA as well.

The maneuvering capabilities across different ship types may also influence encounter scenarios. Ships are identified into 6 different categories, namely cargo, fishing, tanker, passenger, tug, and others, based on their static information. Consequently, the initial selected feature variables are presented in Table 1.

However, while these features are theoretically justified, their actual contribution to the risky encounter development may vary due to data-specific characteristics, such as noise, redundancy, or limited variability. To address the issue, we employ a Random Forest-based feature importance analysis method to empirically evaluate the relevance of each feature to the target variable (Breiman, 2001). The results of this analysis are discussed in Section 4.

Table 1. Selected features.

Feature	Description
DURATION	Time spent from Boundary 1 to Boundary 2
ENCOUNTER_TYPE_B1	Encounter type at Boundary 1
ENCOUNTER_TYPE_B2	Encounter type at Boundary 2
DCPA_B1	DCPA at Boundary 1
DCPA_B2	DCPA at Boundary 2
TCPA_B1	TCPA at Boundary 1
TCPA_B2	TCPA at Boundary 2
DCPA_RATE	DCPA change rate from Boundary 1 to Boundary 2
TCPA_RATE	TCPA change rate from Boundary 1 to Boundary 2
SHIP1_SPEED_B1	Speed of ship 1 at Boundary 1
SHIP1_SPEED_B2	Speed of ship 1 at Boundary 2
SHIP2_SPEED_B1	Speed of ship 2 at Boundary 1
SHIP2_SPEED_B2	Speed of ship 2 at Boundary 2
SHIP1_LENGTH	Length of ship 1
SHIP1_WIDTH	Width of ship 1
SHIP1_TYPE	Type of ship 1
SHIP2_LENGTH	Length of ship 2
SHIP2_WIDTH	Width of ship 2
SHIP2_TYPE	Type of ship 2

2.3 *Prediction model design*

2.3.1 *Classical logistic regression*

Logistic regression has proven to be an effective method of statistical modeling for predicting binary outcomes (Hilbe, 2009). It provides a principled approach for addressing binary classification problems through sigmoidal transformation of linear predictors. Let X and y denote the dataset consisting of m samples and are expressed as:

$$X = \begin{bmatrix} x_1^{(1)} & x_2^{(1)} & \cdots & x_n^{(1)} \\ x_1^{(2)} & x_2^{(2)} & \cdots & x_n^{(2)} \\ \vdots & \vdots & \ddots & \vdots \\ x_1^{(m)} & x_2^{(m)} & \cdots & x_n^{(m)} \end{bmatrix} \in^{m\times n}, y = \begin{bmatrix} y^{(1)} \\ y^{(2)} \\ \vdots \\ y^{(m)} \end{bmatrix} \in \{0,1\}^m \tag{7}$$

For i-th binary response variable $y^{(i)}$ and input feature vector $x^{(i)}$, the model specifies:

$$P(y^{(i)} = 1|x^{(i)}) = \sigma(\beta x^{(i)}) \tag{8}$$

Here, $\beta = [\beta_1, \beta_2, \ldots, \beta_n]^T \in^n$ represents the model's coefficients and $\sigma(\cdot)$ is the sigmoid function, defined as:

$$\sigma(z) = \frac{1}{1+e^{-z}} \tag{9}$$

While logistic regression provides interpretable odds-ratio metrics through Maximum Likelihood Estimation (MLE), this conventional approach has two key limitations: (1) deterministic parameter estimates disregard estimation uncertainty, and (2) lack of mechanisms to incorporate historical knowledge or domain expertise. These constraints pose significant challenges for safety-critical applications like maritime safety.

2.3.2 *Bayesian logistic regression*

Bayesian logistic regression overcomes these limitations and extends classical logistic regression by treating regression coefficients as random variables with probability distributions instead of fixed values. It allows the integration of prior knowledge and quantification of uncertainty in parameter estimates (Genkin et al., 2007).

The model parameters are defined through a prior distribution to represent the beliefs before seeing the data. Upon observing the data, the prior is updated to obtain the posterior distribution $p(\beta|y, X)$ using Bayes theorem:

$$p(\beta|y, X) = \frac{p(y|X, \beta)p(\beta)}{p(y|X)} \tag{10}$$

In this formula, $p(y|X, \beta)$ is the likelihood function represents the probability of the observed data given the parameters. The prior distribution term $p(\beta)$ equals $p(\beta|X)$, as we assume that β and X are independent. This is because the prior distribution over model parameters is specified before observing any data. $p(y|X)$ functions as the marginal likelihood or evidence to normalize the distribution.

2.3.3 *Model specification and inference*

Building on the Bayesian framework introduced in previous section, we specify Gaussian prior distributions for the regression coefficients $\beta = [\beta_1, \beta_2, \ldots, \beta_p]^T$. For coefficients where no strong prior knowledge is available, we assume $\beta_i \sim N(0, \sigma_i^2)$, reflecting a neutral prior centered at zero. When domain-specific prior is available, we assign $\beta_j \sim N(\mu_j, \sigma_j^2)$, where μ_j is set to a positive or negative value based on expert knowledge, and σ_j^2 controls the strength of the prior belief.

The likelihood function is constructed using the Bernoulli distribution:

$$P(y|X,\beta) = \prod_{i=1}^{m}\left[\sigma\left(\beta x^{(i)}\right)^{y^{(i)}}\left(1-\sigma\left(\beta x^{(i)}\right)\right)^{\left(1-y^{(i)}\right)}\right] \tag{11}$$

This formula links the observed data to the model parameters. Since the posterior distribution $p(\beta|y, X)$ lacks a closed-form solution due to the complex likelihood function and prior structure, Markov Chain Monte Carlo (MCMC) sampling is always used to approximate it. In this study, No-U-Turn Sampler (NUTS), an adaptive variant of Hamiltonian Monte Carlo (HMC) is used to estimate $p(\beta|y, X)$(Hoffman & Gelman, 2014). Subsequently, because the integral required to compute the predictive distribution is intractable due to the complexity of the posterior distribution and likelihood function, predictions can be made by using Monte Carlo estimation:

$$P(y^*|x^*, X, y) \approx \frac{1}{S}\sum_{s=1}^{S} P\left(y^*|x^*, \beta^{(s)}\right) \tag{12}$$

Here, $\beta^{(s)}$ indicates s-th posterior sample of the parameters, S is the total samples, x^* and y^* represent new input data and the corresponding predictions respectively.

3 CASE STUDY

The AIS data used for this study was collected from the middle of the Gulf of Finland Reporting (GOFREP) area. The training dataset covers one-month period of May 2021, whereas the validation dataset for prediction is from May 2022. The experiments were conducted using Python 3.8 and R 4.4.3 on a Windows 10 64-bit computer equipped with an Intel Core i7-12700KF processor and 32 GB RAM.

3.1 *Data preparation*

Figure 3 shows the traffic flow based on the training dataset. Statistically, the training dataset has 2953 trajectories and 513916 points, while the validation dataset contains 2841 trajectories and 504401 points. After feature engineering, the identified numbers of potential encounter scenarios and risky encounter scenarios are listed in Table 2. It is evident that both datasets demonstrate similar numbers of these encounters.

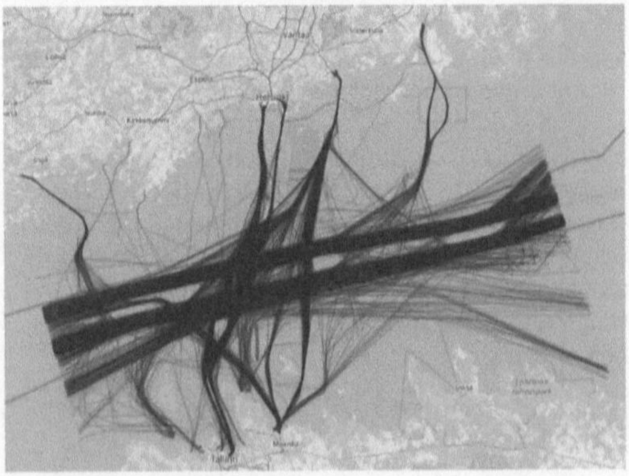

Figure 3. Traffic flow in May 2021.

Table 2. Identified potential and risky encounters.

Dataset	Potential Encounters	Risky Encounters	Risky Ratio
May 2021	5766	2053	35.61%
May 2022	5699	2039	35.78%

Since the initially selected features have both numerical and categorical attributes, we applied Random Forest to identify the most informative ones, ensuring that only the most relevant features are retained to facilitate the training of the Bayesian logistic regression model. Notably, the one-hot encoding technique was applied to the categorical attributes to convert them into numerical form, allowing the Random Forest algorithm to effectively process the data.

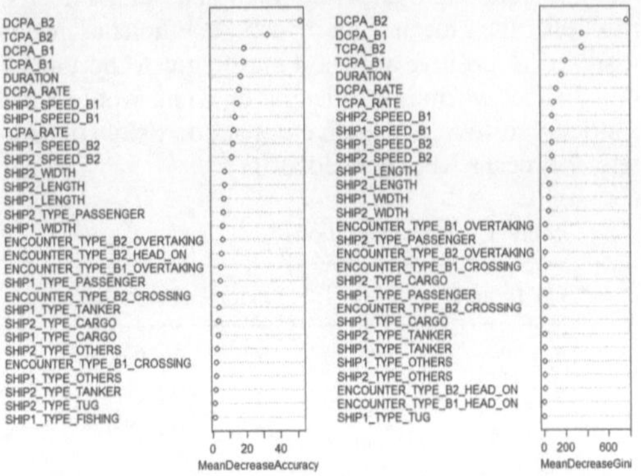

Figure 4. Random Forest-based variable importance analysis.

In Figure 4, mean decrease accuracy and mean decrease Gini are standard feature importance measures used in Random Forest. They reflect a feature's contribution to model accuracy and to reducing node impurity respectively. The results reveal that DCPA_B2 shows significantly higher values for both Mean Decrease Accuracy and mean Decrease Gini compared to other features, followed by TCPA_B2 and DCPA_B1, and then TCPA_B1. This indicates a strong association between these safety indicators and the likelihood of a risky encounter, with DCPA having a greater impact when the relative distance is smaller. Additionally, other numeric variables show considerable contributions, consistently yielding higher metrics scores than categorical variables. Based on these observations, we retained all numeric features while discarding the categorical ones during model training. Finally, the remaining features were standardized to ensure they were on the same scale before being input into the model.

3.2 *Model training*

3.2.1 *Prior definitions*

The specified priors in the experiments are detailed in Table 3. All coefficients were assigned relatively larger standard deviations, indicating that we wanted to impose weak priors, allowing the data to predominantly inform the model parameters. As for the prior means, since smaller DCPA and TCPA imply a more critical situation, their coefficients' means were set to negative values. Additionally, we thought that the state at Boundary 2 affects more to the result compared to the state at Boundary 1, thus the coefficients' absolute values of means at Boundary 2 were slightly higher. Similarly, a short duration between Boundary 1 and Boundary 2 may demonstrate limited reaction time left for the ships. Therefore, the mean for $\beta_{DURATION}$ was set to a negative value, as well as the means for β_{DCPA_RATE} and β_{TCPA_RATE}, indicating that a decrease in both DCPA and TCPA suggests a more urgent situation. Furthermore, since higher speed also leads to reduced maneuverability and less time for reaction, the mean for speed coefficients were assigned to positive values. Finally, given the uncertainty about whether ship length or width would have a strictly positive or negative impact on risky encounters, the means were defined as 0.

Table 3. Prior definitions.

Coefficient	Prior Distribution
β_0	$N \sim (-6, 10^2)$
β_{DCPA_B1}	$N \sim (-0.5, 5^2)$
β_{TCPA_B1}	$N \sim (-0.5, 5^2)$
β_{DCPA_B2}	$N \sim (-1, 5^2)$
β_{TCPA_B2}	$N \sim (-1, 5^2)$
$\beta_{DURATION}$	$N \sim (-1, 5^2)$
β_{DCPA_RATE}	$N \sim (-1, 5^2)$
β_{TCPA_RATE}	$N \sim (-1, 5^2)$
$\beta_{SHIP1\_SPEED\_B1}$	$N \sim (0.5, 5^2)$
$\beta_{SHIP2\_SPEED\_B1}$	$N \sim (0.5, 5^2)$
$\beta_{SHIP1\_SPEED\_B2}$	$N \sim (1, 5^2)$
$\beta_{SHIP2\_SPEED\_B2}$	$N \sim (1, 5^2)$
β_{SHIP1_LENGTH}	$N \sim (0, 5^2)$
β_{SHIP1_WIDTH}	$N \sim (0, 5^2)$
β_{SHIP2_LENGTH}	$N \sim (0, 5^2)$
β_{SHIP2_WIDTH}	$N \sim (0, 5^2)$

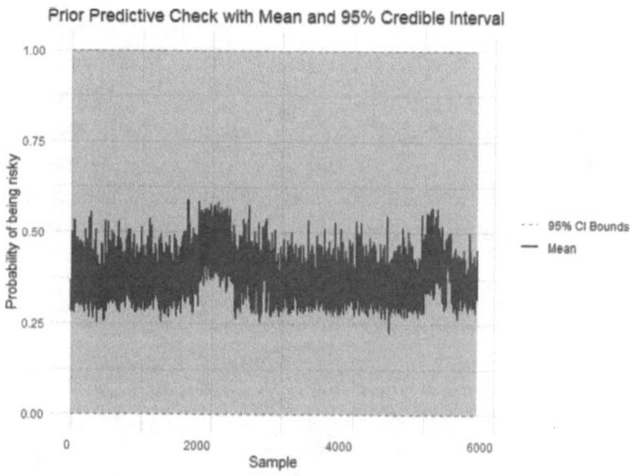

Figure 5. Prior predictive check.

Building on the prior definitions of the model, a prior predictive check was conducted to assess the prior's suitability. As depicted in Figure 5, since the credible interval for the derived samples is centered around 0.35 and spans the entire 0-1 range, we can conclude that the prior aligns with the observed data and is sufficiently flexible to capture latent patterns from the data.

3.2.2 *Posterior analysis*

To evaluate the convergence of the model parameters, we examined the traceplots for all 16 parameters. As shown in Figure 6, the plots indicate that the MCMC sampling has converged, with the chains mixing well and showing no significant trends. This suggests that the model has successfully explored the posterior distribution and the obtained samples are reliable for further analysis.

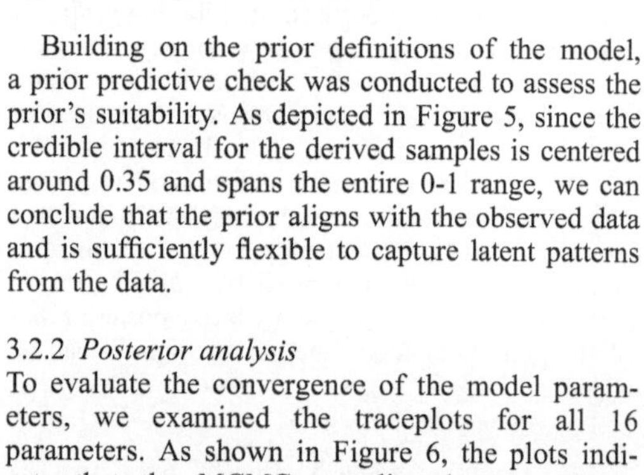

Figure 6. Traceplots for convergence diagnostic.

After confirming convergence, we further evaluated the model's fit to the training data using the Receiver Operating Characteristic (ROC) curve. As illustrated in Figure 7, the ROC curve shows a high area under the curve (AUC) of 0.92, indicating that the model performs well in distinguishing between risky and non-risky encounters within the training data, suggesting a good capture of the patterns.

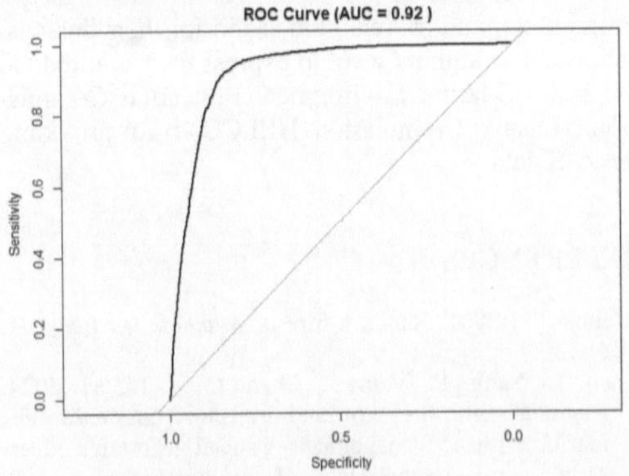

Figure 7. ROC curve on training data.

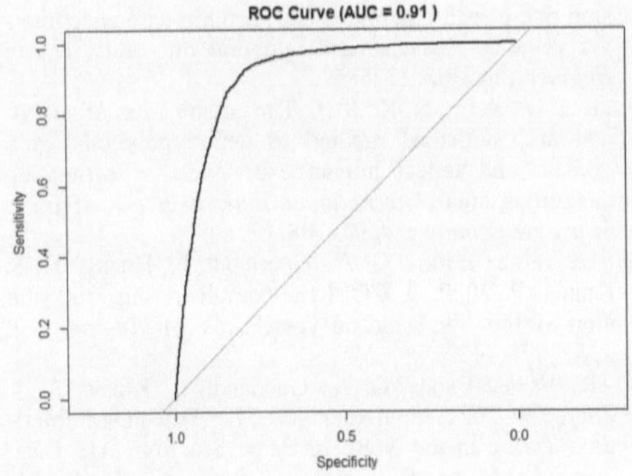

Figure 8. ROC curve on validation data.

3.3 *Model validation*

After training the model, we assessed its generalization ability using the validation dataset. A classical logistic regression model was also trained and predicted on the same data to make a comparison.

First, we used posterior mean predicted probabilities to compute the ROC curve. As depicted in Figure 8, the result indicates that the Bayesian model performs well in correctly classifying both classes.

Then, to fully leverage the advantages of Bayesian prediction, we employed two decision rules based on the posterior predictive distribution:

(1) Rule 1: A sample is classified as positive if the 95th percentile of its posterior predictive distribution exceeds 0.5, and as negative if the 5th percentile falls below 0.5. Samples that do not meet either criterion should be marked as uncertain.

(2) Rule 2: Since Rule 1 can predict uncertain samples, we hypothesize that these samples are risky due to human's continuous efforts to minimize potential risks in real-world navigation.

Consequently, the models' performances were evaluated across all samples, classified samples, and uncertain samples. The evaluation was

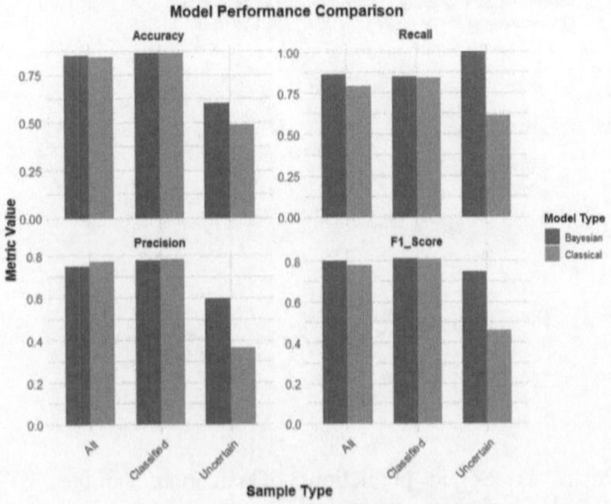

Figure 9. Model performance comparison.

conducted using metrics including accuracy, recall, precision, and F1 score. As shown in Figure 9, the results highlight the advantages of the Bayesian approach, particularly in handling uncertain samples. For uncertain samples, the Bayesian model achieves a recall of 1 due to the employment of Rule 2, yet it still yields a significantly higher precision than the classical model, improving from 0.365 to 0.597. Additionally, the Bayesian model's F1 score of 0.748 for uncertain samples is substantially higher than the classical model's 0.458, demonstrating 63.3% improvements. When considering all samples, the Bayesian model also maintains higher recall and F1 score (0.863 and 0.802) compared to the classical model (0.793 and 0.781). As a supplement, Figure 10 and Figure 11 further illustrate that classical logistic regression tends to make absolute classifications on ambiguous samples, resulting in significant misclassification errors. In contrast, Bayesian logistic regression addresses this limitation by providing predictive distributions, thereby enabling users to come up with more flexible and informed decisions based on the uncertainty estimates.

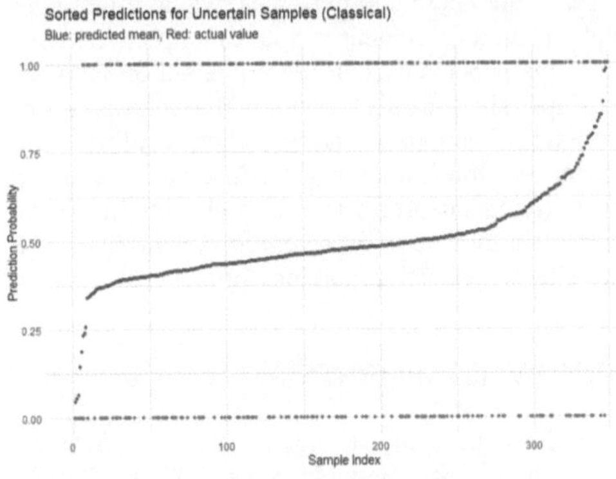

Figure 10. Sorted predictions of uncertain samples by classical logistic regression model.

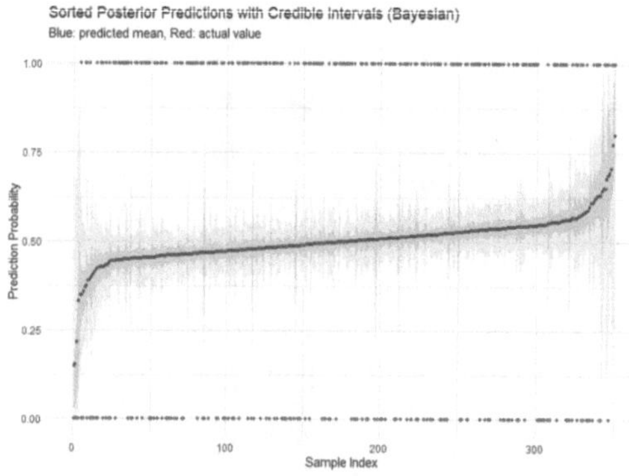

Figure 11. Sorted predictions of uncertain samples by Bayesian logistic regression model.

4 CONCLUSIONS

The inherent complexity of navigational environments poses significant challenges in accurately modeling ship dynamics, particularly in analyzing ship behavior mechanisms for reliable collision risk assessment and effective prevention strategies. To address the challenge, this study presents a novel method for risky ship encounter prediction based on Bayesian logistic regression. The developed model can capture latent patterns from historical data and offer intuitive and interpretable predictive results with uncertainties while avoiding complex context modeling. Experimental results underscore the robustness of Bayesian logistic regression in managing uncertainty and its enhanced performance in risk-sensitive applications like ship navigation, where reliable identification of positive cases is critical. Overall, the Bayesian approach provides a more reliable and effective solution for classification tasks involving uncertainty or imbalanced data.

Despite its achievements, this study is subject to several limitations. First, the datasets for evaluation are restricted to the Gulf of Finland, which may limit the generalizability of the results to other domains or datasets with different characteristics. Second, this study does not account for environmental factors such as water depth and obstacles, which might contribute to the evolvement results of ship encounters. Third, this paper only considers interaction between two ships, significantly reducing the complexity of encounter situations. Finally, when dealing with uncertain samples, adopting a conservative strategy of classifying all predictions as risky might not be optimal. In practice, there could potentially exist more effective decision-making approaches.

ACKNOWLEDGEMENTS

The authors gratefully acknowledge China Scholarship Council (CSC, Grant Number: 202206955019), Safe, ClimAte Resilient Infrastructure (SAFARI) project funded by the European Union's Horizon Europe Programme, and Merenkulun säätiö for their financial support. The authors want to express their gratitude to the Baltic Marine Environment Protection Commission (Helsinki Commission, HELCOM) for providing the AIS data.

REFERENCES

Breiman, L. 2001. Random forests. *Machine learning*, *45*, 5–32.

Chen, P., Yang, F., Mou, J., Chen, L., & Li, M. 2024. Regional ship behavior and trajectory prediction for maritime traffic management: A social generative adversarial network approach. *Ocean Engineering*, *299*, 117186.

Cheng, Z., Chen, P., Mou, J., & Chen, L. 2024. Novel collision risk measurement method for multi-ship encounters via velocity obstacles and temporal proximity. *Ocean Engineering*, *302*, 117585.

Dinh, G. H., & Im, N.-K. 2016. The combination of analytical and statistical method to define polygonal ship domain and reflect human experiences in estimating dangerous area. *International Journal of e-Navigation Maritime Economy*, *4*, 97–108.

Du, L., Valdez Banda, O. A., Goerlandt, F., Huang, Y., & Kujala, P. 2020. A COLREG-compliant ship collision alert system for stand-on vessels. *Ocean Engineering*, *218*, 107866.

Du, L., Valdez Banda, O. A., Goerlandt, F., Kujala, P., & Zhang, W. 2021. Improving Near Miss Detection in Maritime Traffic in the Northern Baltic Sea from AIS Data. *Journal of Marine Science and Engineering*, *9*(2), 180.

EMSA. 2022. *Annual Overview of Marine Casualties and Incidents 2022*.

Fan, L., Zhao, Z., & Yin, J. 2024. Research on dynamic influence mechanism of port state control and ship risk level. *Ocean & Coastal Management*, *250*, 107028.

Gang, L., Wang, Y., Sun, Y., Zhou, L., & Zhang, M. 2016. Estimation of vessel collision risk index based on support vector machine. *Advances in Mechanical Engineering*, *8*(11), 1687814016671250.

Genkin, A., Lewis, D. D., & Madigan, D. 2007. Large-scale Bayesian logistic regression for text categorization. *Technometrics*, *49*(3), 291–304.

Guo, S., Bolbot, V., BahooToroody, A., Valdez Banda, O. A., & Siow, C. L. 2023. Identification of hazardous encounter scenarios using AIS data for collision avoidance system testing. In *Advances in the Collision and Grounding of Ships and Offshore Structures* (pp. 43–50). CRC Press.

Guo, S., Bolbot, V., & Valdez Banda, O. 2024. An adaptive trajectory compression and feature preservation method for maritime traffic analysis. *Ocean Engineering*, *312*, 119189.

Guo, S., Mou, J., Chen, L., & Chen, P. 2021a. An anomaly detection method for AIS trajectory based on kinematic interpolation. *Journal of Marine Science and Engineering*, *9*(6).

Guo, S., Mou, J., Chen, L., & Chen, P. 2021b. Improved kinematic interpolation for AIS trajectory reconstruction. *Ocean Engineering*, *234*.

Hilbe, J. M. 2009. *Logistic regression models*. Chapman and hall/CRC.

Hoffman, M. D., & Gelman, A. 2014. The No-U-Turn sampler: adaptively setting path lengths in Hamiltonian Monte Carlo. *J. Mach. Learn. Res.*, *15*(1), 1593–1623.

Convention on the International Regulations for Preventing Collisions at Sea, 1972.

Recommendation ITU-R M 1371-5. Technical characteristics for an automatic identification system using time division multiple access in the VHF maritime mobile frequency band, 2014.

Jia, C., Ma, J., de Vries, B., & Kouw, W. M. 2024. Bayesian Inference of Collision Avoidance Intent During Ship Encounters. *IEEE Transactions on Automation Science and Engineering*.

Li, H., Çelik, C., Bashir, M., Zou, L., & Yang, Z. 2024. Incorporation of a global perspective into data-driven analysis of maritime collision accident risk. *Reliability Engineering & System Safety*, *249*, 110187.

Liu, Q., Chen, P., Li, M., Chen, L., & Mou, J. 2024. Regional ship collision risk assessment: An integrated approach using velocity obstacle and complex network. *IEEE Transactions on Intelligent Transportation Systems*, 1–15.

Rong, H., Teixeira, A., & Soares, C. G. 2022. Ship collision avoidance behaviour recognition and analysis based on AIS data. *Ocean Engineering*, *245*, 110479.

Xin, X., Liu, K., Yang, Z., Zhang, J., & Wu, X. 2021. A probabilistic risk approach for the collision detection of multi-ships under spatiotemporal movement uncertainty. *Reliability Engineering & System Safety*, *215*, 107772.

Zhang, M., Montewka, J., Manderbacka, T., Kujala, P., & Hirdaris, S. 2021. A big data analytics method for the evaluation of ship-ship collision risk reflecting hydrometeorological conditions. *Reliability Engineering & System Safety*, *213*, 107674.

Zhang, M., Taimuri, G., Zhang, J., Zhang, D, Yan, X., Kujala, P., & Hirdaris, S. 2024. Systems driven intelligent decision support methods for ship collision and grounding prevention: Present status, possible solutions, and challenges. *Reliability Engineering System Safety*, 110489.

Zhou, Y., Daamen, W., Vellinga, T., & Hoogendoorn, S. P. 2023. Ship behavior during encounters in ports and waterways based on AIS data: From theoretical definitions to empirical findings. *Ocean Engineering*, *272*, 113879.

Trends in Collision and Grounding of Ships and Offshore Structures – Zhu & Guedes Soares (Eds)

Improved dynamic window approach for autonomous collision avoidance decision-making algorithm

X. Ma, J. Song & X. Kang
Navigation College, Dalian Maritime University, Dalian, P.R. China

H. Lyu
Navigation College, Dalian Maritime University, Dalian, P.R. China
Key Laboratory of Safety & Security Technology for Autonomous Shipping, Dalian, P.R. China

ABSTRACT: Ship collision avoidance planning is a core issue in the autonomous navigation of Maritime Autonomous Surface Ships (MASS) and must comply with the International Regulations for Preventing Collision at Sea (COLREGs). The Velocity Obstacle (VO) algorithm can determine the velocity vector collision interval and cannot derive an optimal decision. The Dynamic Window Approach (DWA) does not prioritize COLREGs requirements. To address these issues, this paper proposes a ship dynamic autonomous collision avoidance algorithm (VO-IDWA), which first uses the VO algorithm to identify collision risk intervals and then, employs the DWA to simulate and predict the ship's feasible velocity interval, selecting safe collision avoidance decisions. Additionally, the DWA evaluation function incorporates collision risk and COLREGs requirements, comprehensively assessing decision safety and compliance. Simulation results show that the algorithm is feasible and efficient, enabling ships to effectively avoid collision risks according to COLREGs, with good application potential in autonomous ship collision avoidance.

1 INTRODUCTION

With the advancement of intelligent technology, autonomous navigation of Maritime Autonomous Surface Ships (MASS) has become a research hotspot. Key research areas include autonomous ship path planning and collision avoidance technology. Path planning generates a path based on known environmental information and the ship's navigation mission. Collision avoidance technology assesses collision risks with static and dynamic obstacles and executes reasonable maneuvers to reach the planned destination safely.

Numerous scholars have proposed various intelligent collision avoidance algorithms, such as the Artificial Potential Field (Lyu & Yin, 2019), Velocity Obstacle (Zhang et al., 2021), and Dynamic Window Approach (Xu, Yang, Zhou, & Xu, 2024). The VO algorithm, initially proposed by (Fiorini & Shiller, 1998) for robot collision avoidance, converts dynamic collisions into static ones by superimposing position and velocity information, forming a velocity obstacle area that the robot must avoid. (Huangvan Gelder & Wen, 2018) introduced various VO algorithms (LVO, NLVO, PLVO) for marine collision avoidance, noting their dependence on trajectory prediction, which may be imprecise and unable to yield optimal decisions. (HuangChen & van Gelder, 2019) proposed a Generalized Velocity Obstacle (GVO) method considering ship dynamics and distance differences caused by other ships' movements, yet VO algorithms still cannot select the best decision within the velocity interval.

The Dynamic Window Approach (DWA) selects optimal velocity vectors by applying dynamic and kinematic constraints directly to the velocity vector space. Proposed by (D. Fox, 1997) for robot path planning and local collision avoidance, it considers robot dynamics. (Yao et al., 2024) expanded the energy consumption model to the DWA evaluation function and introduced fuzzy logic improvements. (Sun, Wang, Fan, & Mu, 2023) proposed a real-time collision avoidance control algorithm based on dynamic constraints, incorporating rudder angle and danger functions. (Guan & Wang, 2023) combined the A-star algorithm with DWA for COLREGs-compliant collision avoidance. However, DWA is mostly used for static obstacle avoidance with limited research on dynamic obstacles.

(Wang, Meng, Wang, & Wang, 2023) integrated the VO method as a velocity obstacle term in the DWA evaluation function to reduce dynamic obstacle impacts but without considering COLREGs constraints (Lou et al., 2024) used DWA and VO as the first two reward levels for a DDPG algorithm,

DOI: 10.1201/9781003684404-54

enhancing convergence and filtering dangerous behaviors, yet with insufficient real-time performance due to heavy data reliance (Zhou, Wang, Wu, & Yan, 2024) proposed a neural network-based controller with a decision module combining VO and DWA for feasible velocity intervals and actual decisions, but in experiments, the own ship was always the giving-way ship without reinforcing COLREGs constraints.

This paper's main contributions are:

(1) Optimizing velocity decision intervals by integrating VO-derived collision velocity intervals into DWA's decision interval, filtering out dangerous ship behaviors, and enhancing collision avoidance safety.
(2) Improving the DWA evaluation function by adding ship collision and COLREGs evaluation functions to consider rules and other ships' dangers for more reasonable and efficient collision avoidance decisions.

The paper is structured as follows: Chapter 2 introduces VO algorithm principles, ship collision risk calculation, and encounter situations; Chapter 3 presents the improved VO-IDWA algorithm; Chapter 4 validates the algorithm through simulations; Chapter 5 summarizes results, identifies limitations, and suggests future research directions.

2 SHIP COLLISION AVOIDANCE DECISION CONDITIONS

Ship collision risk identification and encounter situation judgment are crucial for decision-making and determining whether to initiate collision avoidance. Historically, collision risk was assessed using DCPA and TCPA between the target ship (TS) and own ship (OS). The VO method, inspired by literature, is used here to identify collision risks by introducing VO and Collision Risk Index (CRI) into DWA's velocity decision interval selection and evaluation function, filtering dangerous behaviors and assessing decision risks. When ships are in danger, COLREGs define their responsibilities and required actions, which are incorporated into the DWA evaluation function to assess compliance.

2.1 *Collision risk identification using the velocity obstacle algorithm (VO)*

Based on collision avoidance experience in open waters with good visibility, the CPA threshold is set at 1 nmile. Collision risk exists when $TCPA > 0$ and $-1nm < CPA < 1nm$. If $TCPA<0$, the TS has passed the closest point and poses no risk.

The VO method intuitively analyzes ship collision risks compared to TCPA and CPA. As shown in Figure 1, the TS's particle P_B is expanded into a circle (expansion radius R_B equals the CPA threshold). The VO algorithm essentially involves the process of the OS's particle P_A evading the TS's P_B expanded circle. The OS's speed is V_0, course is φ_0, the TS's speed is V_1, course is φ_1, BT represents the TS's true bearing, and VR ($VR=V_0-V_1$) respectively denotes the relative course and relative speed of the TS with respect to the OS. Two rays L_{am} and L_{an} are drawn from the OS's P_A tangent to the TS's expanded circle. The area enclosed by these two rays is termed the Relative Collision Region (RCC). If the OS's velocity vector V_0 lies within the sector formed by L_{am} and L_{an}, it indicates a collision risk exists between the two vessels. Since the RCC is defined relative to the velocity space, multiple reference frames may emerge in complex multiple-ship encounter scenarios. Therefore, through the Minkowski sum operation, the RCC is translated along the target ship's velocity vector V_1 to obtain a new Absolute Collision Cone (ACC), $ACC = RCC \oplus V_B$. As shown in the red collision cone area in Figure 1, if the tip of the OS's velocity vector V_0 falls within the ACC, it indicates that a collision risk exists between the two vessels.

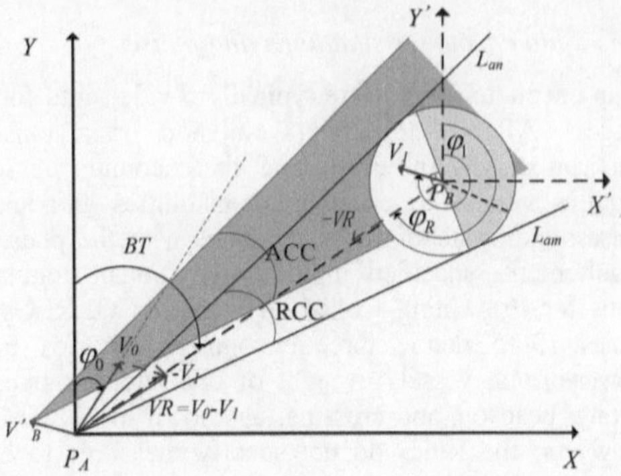

Figure 1. Velocity obstacle algorithm schematic.

2.2 *Ship collision risk index*

The primary challenge in determining ship collision risk lies in the varying factors considered for different vessels' collision risk levels. To address this, experts have proposed the critical concept of Collision Risk Index (CRI), which serves as a key criterion for collision avoidance actions by unmanned vessels. The CRI values range from [0,1] (Hu, Zhang, Tian, Zhang, & Hou, 2020) developed a ship collision risk calculation function based on five traditional factors such as CPA and TCPA. (Korupoju, Kapadia, Vilwathilakam, & Samanta, 2025) incorporated an integrated collision risk calculation function that combines CPA, TCPA, inter-ship distance, relative bearing, and speed ratio.

This paper constructs the ship collision risk using CPA and TCPA values as follows:

$$\begin{gathered} CRI = W|U \\ W = [w_{DCPA}, w_{TCPA}], U = [u_{DCPA}; u_{TCPA}]\circ \end{gathered} \tag{1}$$

The membership function for CPA in the CRI calculation u_{DCPA} is as shown in (2):

$$u_{DCPA} = \begin{cases} 0 & d_2 < |CPA| \\ \frac{1}{2} - \frac{1}{2} \quad \sin\left[\frac{\pi}{d_2 - d_1}\left(|CPA| - \frac{d_1 + d_2}{2}\right)\right] & \\ & d_1 < |CPA| \leq d_2 \\ 1 & d_2 < |CPA| \end{cases} \tag{2}$$

Where d_1 represents the safe passing distance, and d_2 denotes the min allowable distance of closest approach requiring immediate evasive action. The membership function for TCPA in the CRI calculation u_{DCPA} is as shown in (3):

$$u_{TCPA} = \begin{cases} 0 & t_2 < |TCPA| \\ \left(\frac{t_2 - |TCPA|}{t_2 - t_1}\right)^2 & t_1 < |TCPA| \leq t_2 \\ 1 & 0 \leq |TCPA| \leq t_1 \end{cases} \tag{3}$$

$t_1 = d_1/VR, t_2 = d_2/VR.$

2.3 *Ship encounter situations and phases*

Ship encounter phases are typically divided into four stages. After determining collision risk exists between vessels, the third stage, the encounter phase, requires vessels to clarify responsibilities and take corresponding actions in accordance with the phases requirements specified in the International Regulations for Preventing Collisions at Sea (COLREGs). Rules 13-15 define three encounter scenarios for power-driven vessels in sight of one another: overtaking, head-on, and crossing, as shown in Figure 2. However, the Rules do not specify measures to be taken in complex multi-ship encounters where vessels may have conflicting responsibilities and collision avoidance actions toward different targets.

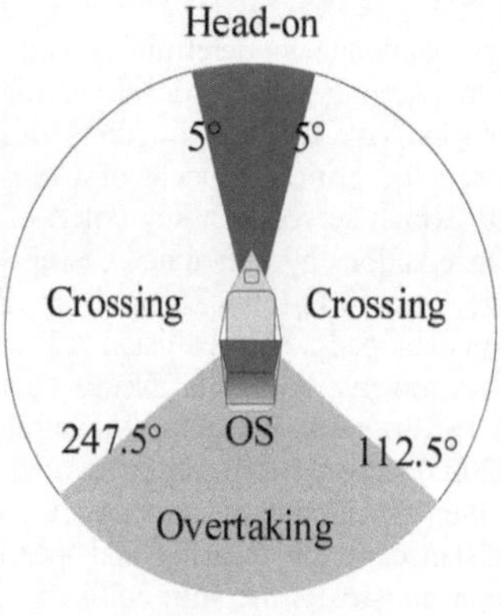

Figure 2. Encounter situations schematic.

(Lyu et al., 2024) exhaustively enumerates various multi-ship encounter scenarios, considering both whether collision risks exist between TSs and whether conflicting collision avoidance actions arise between the OS and TSs in these scenarios.

3 SHIP COLLISION AVOIDANCE DECISION-MAKING BASED ON VO-IDWA

The Dynamic Window Approach (DWA) transforms ship collision avoidance problems into velocity vector interval optimization problems. By modeling the feasible region combining heading and speed, DWA can resolve heading and speed adjustment challenges in collision avoidance decision-making. This paper proposes a collision avoidance algorithm integrating VO and improved DWA, referred to as the VO-IDWA algorithm. As shown in Figure 3, the algorithm processes OS and TS information including position, speed, course, maximum angular acceleration, and maximum acceleration to compute the dynamic velocity sampling interval *Vr*. It then simulates the ship's motion trajectory over a look-ahead time horizon, evaluates these trajectories using the evaluation function $G(v,w)$, selects the optimal velocity command (v,w) with the highest score, and executes this command until reaching the destination.

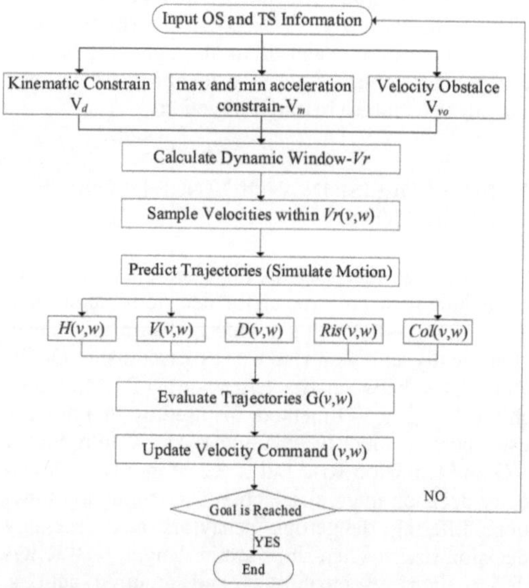

Figure 3. VO-IDWA approach flowchart.

3.1 *Ship velocity vector interval Vr*

The Dynamic Window Approach (DWA) includes the calculation of the ship velocity vector interval *Vr*. to determine the velocity variation range, corresponding to the velocity vector for the next time step. Figure 4 illustrates the velocity vector interval selection diagram. Based on the ship's own settings, the velocity intervalof $V_d \cap V_m$ within the time interval is determined. The VO algorithm calculates V_{vo}, represented by the green velocity interval in Figure 4b. In Figure 4c, the final velocity vector interval of $V_d \cap V_m \cap V_{vo}$, derived as *Vr*; corresponding to the green area enclosed by a black border. In this paper, the initial maximum economic speed of the OS is V_{os}, with a maximum acceleration of $\dot{V}_0$, angular velocity of ω_0, angular acceleration of $\dot{\omega}_0$, and time interval of Δt.

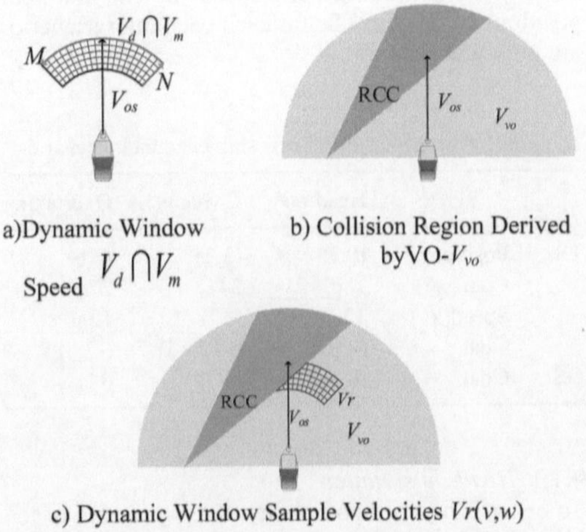

Figure 4. Diagram of the velocity vector interval Vr in the VO-IDWA approach.

The ship itself defines the maximum linear speed and maximum steering angle range as V_d:

$$V_d = \{(v,w)|v \in [0, v_{\max}] \cap w \in [-w_{\max}, w_{\max}]\} \quad (4)$$

Based on its kinematic and dynamic constraints, including max feasible acceleration and angular acceleration, the ship determines the velocity vector variation within a specified time interval as V_m:

$$V_m = \{(v,w)|v \in [V_{os} - \dot{V}_0\Delta t, V_{os} + \dot{V}_0\Delta t] \cap w \in [\omega_0 - \dot{\omega}_0\Delta t, \omega_0 + \dot{\omega}_0\Delta t]\} \quad (5)$$

Considering the kinematics of the TS, including its heading and speed, the VO algorithm is employed to identify collision risks. By analyzing the Relative Collision Cone (RCC) of the OS's velocity vector relative to the TS, and translating this region along the TS's heading at its speed, the Absolute Collision Cone (ACC) of the TS is derived. If the endpoint of the OS's velocity vector V_O does not fall within the ACC, the velocity derived as V_{vo}:

$$V_{vo} = \{(v,w)|U(v,w) \cap ACC = \emptyset\} \quad (6)$$

By integrating the ship's own kinematic and dynamic constraints (maximum acceleration and angular acceleration) with the ACC, as shown in Figure 4c, the feasible velocity vector region Vr is derived to ensure safe navigation:

$$Vr = V_d \cap V_m \cap V_{vo} \quad (7)$$

3.2 *Ship motion trajectory*

During navigation, consideration must be given to the ship's own motion performance and the influence of environmental factors on its heading and speed. In normal navigation of the OS, the aggregated information is denoted as $P_A{}^t = [x, y, \varphi_0, V_1, w]'$. The velocity variation is set $u = [v(i), w(i)]'$, where v represents linear velocity and w angular velocity.

$$P_A{}^{(t+1)} = F|P_A{}^t + B|u \quad (8)$$

$$F = \begin{bmatrix} 1 & 0 & 0 & 0 & 0 \\ 0 & 1 & 0 & 0 & 0 \\ 0 & 0 & 1 & 0 & 0 \\ 0 & 0 & 0 & 0 & 0 \\ 0 & 0 & 0 & 0 & 0 \end{bmatrix},$$

$$B = \begin{bmatrix} \cos(P_A{}^t(3)) \times \Delta t & 0 \\ \sin(P_A{}^t(3)) \times \Delta t & 0 \\ 0 & \Delta t \\ 1 & 0 \\ 0 & 1 \end{bmatrix}.$$

3.3 *The evaluation function G(v,w) of the VO-IDWA*

Based on the feasible velocity vector interval identified in the previous section, the evaluation function of DWA is used to derive safe and efficient ship collision avoidance decisions, executing the highest - scoring set of velocity and heading collision avoidance decisions. The evaluation function of the traditional DWA is formulated from the target bearing, velocity, and obstacle distance sub-functions. In this paper, the evaluation function incorporates a ship collision risk function and a COLREGs compliance sub-function. The collision avoidance behavior derived from the evaluation function is shown:

$$G(v,w) = \alpha(\gamma_1 H(v,w) + \gamma_2 V(v,w) + \gamma_3 D(v,w) + \gamma_4 Ris(v,w) + \gamma_5 Col(v,w)) \quad (9)$$

where $\gamma_1, \gamma_2, \gamma_3, \gamma_4$ and γ_5 are the weighting coefficients for each evaluation factor, andαrepresents the normalization of its five elements.

1) Target Bearing Function $H(v,w)$

The Target Bearing Function calculates the angular difference between the feasible heading interval of the OS and the relative bearing of the target point with respect to the OS, θ.

$$H(v,w) = 180° - \theta \quad (10)$$

2) Speed Function $V(v,w)$

The Speed Function represents the ship's navigational speed. During normal operation, the ship sails at its maximum economic speed. When maneuvering alone is insufficient to avoid collision with other vessels, speed reduction may be combined with course changes. Consequently, the velocity value is determined by its speed function.

$$V(v,w) = vt \tag{11}$$

3) Distance between OS and TS Function *D(v,w)*

In the traditional DWA, the distance function represents the distance between the OS and static obstacles, where the greater the distance, the higher the score. In this paper, the distance between dynamic obstacles is evaluated. Therefore, the evaluation boundary for the distance between ships is set to 2 nmile according to COLREGs. Using the derived feasible velocity vector interval of the ship, a forward simulation is performed over a period of time. At fixed time intervals (set to 30s in this paper), the distance between the OS and TS is calculated, superimposed, and averaged.

$$D(v,w) = \begin{cases} 0 & d \leq 1nmile \\ d - d_{safe} & 1 < 2nmile \\ 1 & d \geq 2nmile \end{cases} \tag{12}$$

4) Risk Evaluation Function *Ris(v,w)*

The ship collision risk, represented by the CRI, quantifies the danger level between ships and takes values between [0, 1]. A higher CRI indicates greater risk. Therefore, the *Ris(v,w)* is designed such that a higher risk corresponds to a lower score in the evaluation function. The relationship between the *Ris(v,w)* and CRI is defined:

$$Ris(v,w) = 1 - CRI \tag{13}$$

5) COLREGs Evaluation Function *Col(v,w)*

The COLREGs impose restrictions on ship collision - avoidance actions. According to different responsibilities, the actions of vessels are constrained. There are two main types of vessels in terms of responsibilities: the Give-way ship and the Stand - on ship. When the OS acts as the Give-way ship, it should take collision-avoidance actions during overtaking and starboard crossing situations. Moreover, the best collision - avoidance action is to pass astern of the TS, with a CPA less than -1 nmile.

$$Col(v,w) = \begin{cases} 1 & CPA < -1nmile \\ 0 & CPA > 1nmile \end{cases} \tag{14}$$

4 EXPERIMENTAL VALIDATION

This paper selects three encounter scenarios specified in COLREGs (Head-on, Crossing, and Overtaking), plus two complex encounter scenarios, to validate the feasibility of the Velocity Obstacle and Improved Dynamic Window Approach (VO-IDWA) algorithm for dynamic autonomous ship collision avoidance.

4.1 *Two-ship encounter scenarios*

In the two-ship encounter experiments, the OS starts at an initial position of (0,−1) in Cartesian coordinates, with an initial heading of 000° and initial speed of 12 kn. The headings, and speeds of TSs and goal positions of OS and TSs for each encounter scenario, are presented in Table 1.

Table 1. TS parameters for two-ship encounter scenarios.

		Head-on	Crossing	Overtaking
TS	Position	(0,15)	(3,3)	(6,6)
	Course(°)	180	344	270
	Speed(kn)	12	9	12
	Goal	(0,0)	(-0.5,15)	(-7,6)
OS	Goal	(0,14)	(0,17)	(0,11)

4.1.1 *Head-on situation*

To better verify the applicability and compliance of the VO-IDWA algorithm, a comparison with the traditional DWA algorithm is conducted using the head-on encounter scenario of two ships. As shown in Figure 5a, the light blue path denotes the collision avoidance trajectory of OS under the traditional DWA algorithm, while the pink path represents the course-keeping and speed-keeping trajectory of TS1 under the same algorithm. In Figure 5b, the blue path illustrates the collision avoidance trajectory of OS under the VO-IDWA algorithm, and the red path shows the collision avoidance trajectory of TS1 under VO-IDWA.

Under the traditional DWA algorithm, OS avoids TS1 to reach the target point with a longer path and a larger steering amplitude, and its collision avoidance behavior does not comply with the Rules. By contrast, both OS and TS1 under the VO-IDWA algorithm adopt collision avoidance actions in line with the Rules, safely arriving at their target points.

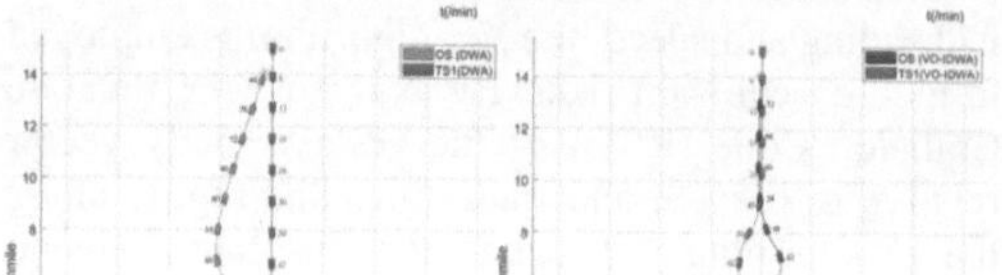

Figure 5. Simulation path diagram of ships in head-on situation.

Figure 6 a, b, and c show the numerical changes in the distance from TS, CPA, and TCPA of OS during navigation under the traditional DWA algorithm and VO-IDWA algorithm, respectively. As indicated by the red dashed line and pink line in Figure 6b, at t=0 min, CPA=0 n mile, indicating a collision risk between the ships. At t=26 min, in the encounter scenario, the two ships turn to starboard while maintaining speed under the VO-IDWA algorithm. At t=35 min, as shown in Figure 7, the course of OS becomes 035°, after which OS turn to right toward the target point. As shown in Figure 6a, at t=42 min, OS reaches the closest approach distance of 1.65 n mile from TS1, under the VO-IDWA algorithm. It returns to the original course at t=60 min and continues navigation, arriving at the target point at t=78 min. Under the traditional DWA algorithm, OS reaches the target point at t=84 min. By comparison, the VO-IDWA algorithm enables OS to avoid TS with a shorter path and less time consumption.

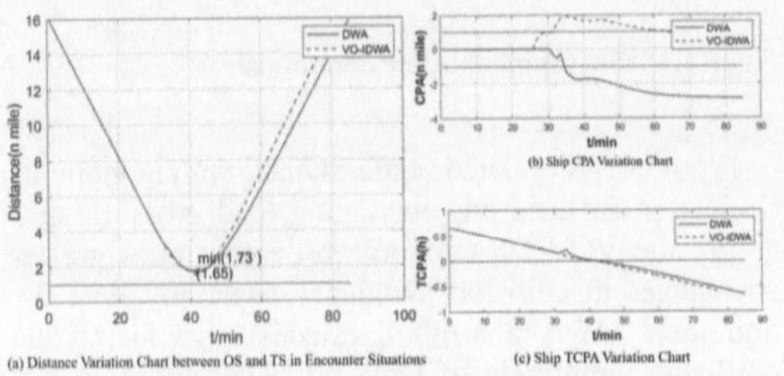

Figure 6. Real-time parameter values in Head-on situation.

Figure 7 shows the detailed parameters of OS during navigation under the traditional DWA algorithm and the VO-IDWA algorithm. Among them, Figure 7a and Figure 7b respectively depict the variations in course, speed, and yaw rate of OS during navigation under the traditional DWA algorithm and the VO-IDWA algorithm.

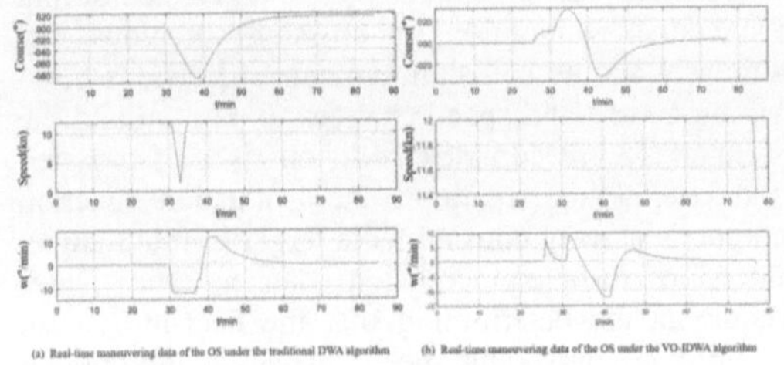

Figure 7. Own ship real-time control data.

4.1.2 *Overtaking situation*

To further verify the applicability and compliance of the VO-IDWA algorithm, a comparison with the traditional DWA algorithm is conducted using the overtaking situation of two ships. For ease of comparing the paths generated by the algorithms, the paths of OS under the two algorithms are plotted in Figure 8a. The light blue path represents OS's collision avoidance trajectory under the traditional DWA algorithm, while the blue path shows OS's trajectory under the VO-IDWA algorithm. In the overtaking situation, TS1 maintains course and speed.

Figure 8b, c, and d illustrate the numerical changes in the distance from TS, CPA and TCPA of OS during navigation under the traditional DWA and VO-IDWA algorithms, respectively. As shown in Figure 8c, at t=0 min, CPA=-0.8 n mile, indicating a collision risk and OS as the give-way vessel. At t=6 min, under the VO-IDWA algorithm, OS turns to starboard in compliance with the Rules. Between t=26 and t=32 min, OS updates collision avoidance commands combining course and speed in real-time under the VO-IDWA evaluation function. At t=32 min, OS passes astern of TS and reaches TS's starboard side, then considers turning toward OS's target point. Thus, from t=32 to t=42 min, OS's course hovers at 020° under the evaluation function's constraint. At t=41 min, OS and TS reach the closest approach distance of 1.75 n mile. OS arrives at the target point and stops moving at t=96 min.

As shown in Figure 8a, OS under the VO-IDWA algorithm passes astern of TS while maintaining a safe approach distance, complying with the Rules. In contrast, OS under the traditional DWA algorithm turns to port to avoid collision, passing ahead of TS with a closest distance of 1.97 n mile.

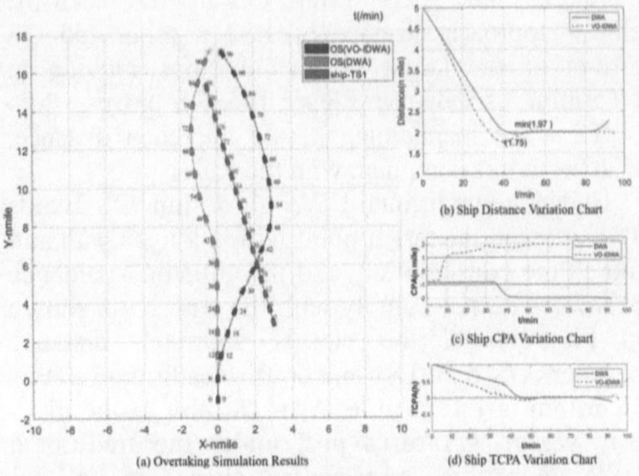

Figure 8. Overtaking situation ship status data plot.

Figure 9 shows the detailed parameters of OS during navigation under the traditional DWA algorithm and the VO-IDWA algorithm. Among them, Figure 9a and Figure 9b respectively depict the variations in course, speed, and yaw rate of OS during navigation under the traditional DWA algorithm and the VO-IDWA algorithm.

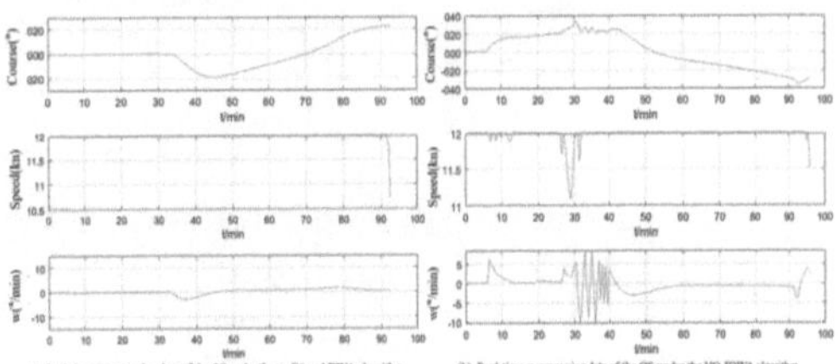

Figure 9. Own ship real-time control data.

4.1.3 *Crossing situation*

To further verify the applicability and compliance of the VO-IDWA algorithm, a comparison with the traditional DWA algorithm is conducted using the crossing situation of two ships. For ease of comparing the paths generated by the algorithms, the paths of OS under the two algorithms are plotted in the same figure. As shown in Figure 10a, the light blue path represents OS's collision avoidance trajectory under the traditional DWA algorithm, while the blue path shows OS's trajectory under the VO-IDWA algorithm. In the crossing situation, TS1 maintains course and speed. Figure 10b, c, and d respectively display the numerical changes in the distance from TS, CPA, and TCPA of OS during navigation under the traditional DWA and VO-IDWA algorithms.

As shown in Figure 10c, at t=0 min, CPA=0.34 n mile, indicating a collision risk between the ships. At t=12 min, as the give-way vessel, OS executes a starboard turn for collision avoidance under the VO-IDWA algorithm. At t=21 min, OS makes a larger starboard turn under the constraint of the VO-IDWA evaluation function. When t=25 min, after OS's course changes to 035°, it turns to port toward the target point. At t=30 min, OS and TS reach the closest approach distance of 1.82 n mile, and OS arrives at the target point and stops moving at t=65 min. As shown in Figure 10a, OS passes astern of TS while maintaining a safe approach distance, with behaviors compliant with the Rules.

Under the traditional DWA algorithm, OS avoids TS1 to reach the target point with a longer path and more time consumption, and its collision avoidance behaviors do not comply with the Rules. As shown in Figure 10b, the closest approach distance between OS and TS1 under the traditional DWA algorithm is 1.17 n mile. At t=18 min, as the give-way vessel, OS turns port under the traditional DWA algorithm, and passes ahead of TS1 at t=48 min. Through comparison, the proposed VO-IDWA algorithm demonstrates good applicability in avoiding dynamic obstacles.

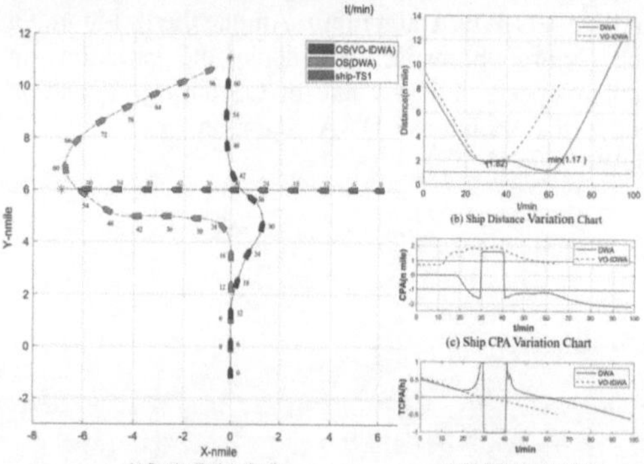

Figure 10. Crossing situation ship status data plot.

Figure 11. shows the detailed parameters of OS during navigation under the traditional DWA algorithm and the VO-IDWA algorithm. Among them, Figure 11a and Figure 11b respectively depict the variations in course, speed, and yaw rate of OS during navigation under the traditional DWA algorithm and the VO-IDWA algorithm.

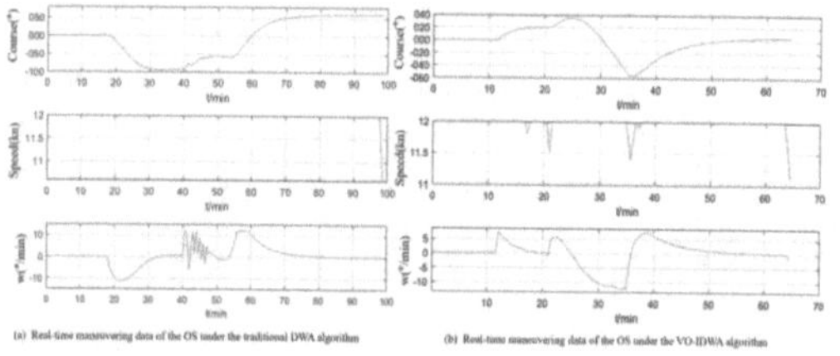

Figure 11. Own ship real-time control data.

From the simulated experiments on autonomous navigation in head-on, overtaking, and crossing situations, the VO-IDWA algorithm demonstrates greater advantages in collision avoidance capability than the traditional DWA algorithm, effectively reducing the oscillation amplitude of yaw rate. The algorithm can not only achieve collision avoidance decisions combining course and speed changes, but also comply with the requirements of the Rules for encounter stages, responsibilities, and collision avoidance behaviors, ensuring safe ship navigation.

Under the VO-IDWA algorithm, decisions vary in different situations: in head-on situations, both OS and TS adopt starboard turn collision avoidance behaviors; in overtaking situations, OS makes a decision to pass around the stern of TS, with coordinated adjustments of course and speed during the course resumption; in crossing situations, OS similarly adopts collision avoidance decisions combining course and speed changes to pass around the stern of TS. In complex collision risks of crossing and overtaking situations, the evaluation function comprehensively considers the target azimuth angle, intership distance, collision danger, and *Rules* requirements, determining that the decision (v, w) combining course and speed changes is the optimal solution, and then executes the combined commands of course and speed changes.

4.2 *Multi-ship encounter scenarios*

To verify the applicability of the algorithm in complex scenarios, multi-ship Simultaneous encounter scenarios were designed for simulation experiments. The position, course, and speed information of the OS and the TS are shown in Table 2.

Figure 12 shows the simulation experiment diagram of ships navigating as per Table 2, successfully avoiding collisions after detecting danger. In Figure 12c, at t=0 min, the DCPA=0 nmile among ships indicated

Table 2. OS and TS parameters for multi-ship encounter scenarios.

	OS	TS1	TS2
Postion	(0,-1)	(3.5,5)	(-3.5,5)
Course(°)	000	240	120
Speed(kn)	12	12	12
Goal	(0,8)	(-4,0)	(4,0)

a dangerous situation between OS, TS1, and TS2. At t=4min, all ships initiated collision avoidance actions under the VO-IDWA algorithm. In Figure 12b, the closest distances at t=20 min were 1.37 nmile between OS and TS1, 1.15nmile between OS and TS2, and 1.23 nmile between TS1 and TS2. The OS reached the goal and ceased movement at t=47 min.

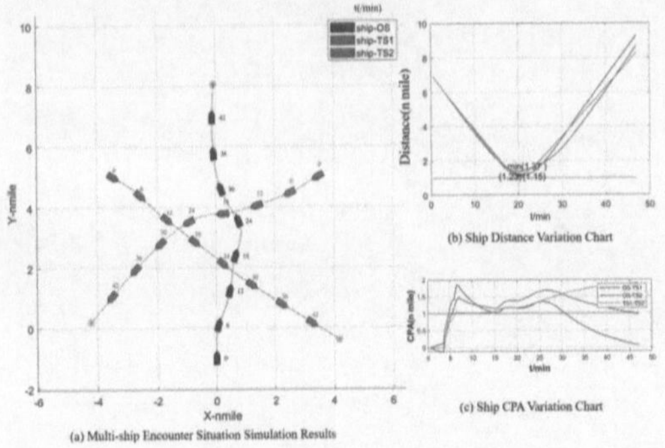

Figure 12. Multi-ship encounter situation ship status data plot.

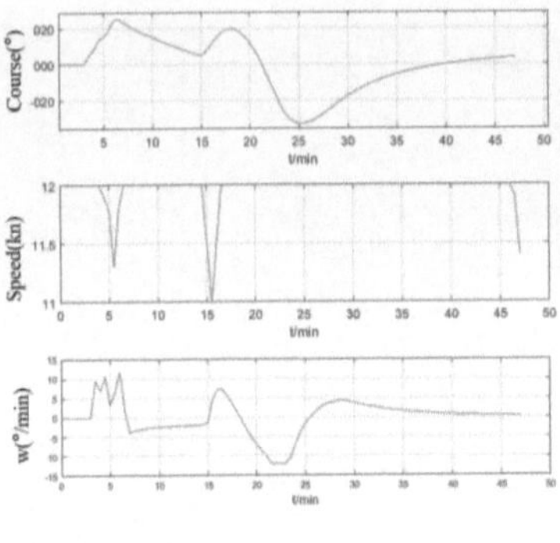

Figure 13. Own ship real-time control data.

Figure 13 shows the detailed parameters of OS during navigation, including OS Course in Figure 13a, Speed in Figure 13b, and Yaw rate in Figure 13c.In Figure 13a, the significant course changes of the OS reflected substantial turning maneuvers to avoid multi-ship collisions; while Figure 13b showed a slight decrease in speed.

5 SUMMARY AND FUTURE

Ship collision avoidance technology continues to face new challenges and opportunities. This paper integrates a ship motion model to address the autonomous collision avoidance problem for Maritime Autonomous Surface Ships (MASS), proposing a dynamic autonomous collision avoidance algorithm (VO-IDWA algorithm) that combines the Velocity Obstacle (VO) method with an Improved Dynamic Window Approach. Simulation experiments verified the feasibility of the improved algorithm. Although preliminary discussions were conducted on multi-ship encounter scenarios, the mutual interactions between ships during such encounters were not thoroughly analyzed. Future research could further explore collision avoidance strategies in multi-ship encounter scenarios, particularly how to coordinate collision avoidance behaviors among multiple vessels in complex environments to ensure overall navigation safety and efficiency.

ACKNOWLEDGEMENTS

National Natural Science Foundation of China (No.52071049).

Dalian City Science and Technology Plan (Key) Project (No.2024JB11PT007).

REFERENCES

Fiorini, P. & Shiller, Z. (1998). Motion planning in dynamic environments using velocity obstacles. *International Journal of Robotics Research*, *17*(7), 760–772.

Fox, D., Burgard, W. & Thrun, S. (1997). The Dynamic Window Approach to Collision Avoidance. *IEEE Robotics & Automation Magazine.vol. 4, no. 1, pp. 23-33.*

Guan, W. & Wang, K. (2023). Autonomous Collision Avoidance of Unmanned Surface Vehicles Based on Improved A-Star and Dynamic Window Approach Algorithms. *IEEE Intelligent Transportation Systems Magazine*, *15*(3), 36–50.

Hu, Y., Zhang, A., Tian, W., Zhang, J. & Hou, Z. (2020). Multi-Ship Collision Avoidance Decision-Making Based on Collision Risk Index. *Journal of Marine Science and Engineering*, *8*(9), 640.

Huang, Y., Chen, L. & van Gelder, P.H.A.J. (2019). Generalized velocity obstacle algorithm for preventing ship collisions at sea. *Ocean Engineering*, *173*, 142–156.

Huang, Y., van Gelder, P. H. A. J. & Wen, Y. (2018). Velocity obstacle algorithms for collision prevention at sea. *Ocean Engineering*, *151*, 308–321.

Korupoju, A. K., Kapadia, V., Vilwathilakam, A.S. & Samanta, A. (2025). Ship Collision Risk Evaluation using AIS and weather data through fuzzy logic and deep learning. *Ocean Engineering*, *318*, 120116.

Lou, M., Yang, X., Hu, J., Zhu, Z., Shen, H., Xiang, Z. & Zhang, B. (2024). A Balanced Collision Avoidance Algorithm for USVs in Complex Environment: A Deep Reinforcement Learning Approach. *IEEE Transactions on Intelligent Transportation Systems*, *25*(12), 21404–21415.

Lyu, H., Ma, X., Tan, G., Yin, Y., Sun, X., Zhang, L., Kang, X. & Song, J. (2024). Identification of Complex Multi-Vessel Encounter Scenarios and Collision Avoidance Decision Modeling for MASSs. *Journal of Marine Science and Engineering*, *12*(8), 1289.

Lyu, H. & Yin, Y. (2019). COLREGS-Constrained Real-time Path Planning for Autonomous Ships Using Modified Artificial Potential Fields. *Journal of Navigation*, *72*(3), 588–608.

Sun, X., Wang, G., Fan, Y. & Mu, D. (2023). Collision avoidance control for unmanned surface vehicle with COLREGs compliance. *Ocean Engineering*, *267*, 113263.

Wang, D., Meng, F., Wang, H. & Wang, W. (2023, 2023-1-1). *An Improved Path-planning Method for Multi-agent by Velocity Obstacle Dynamic Window Approach.*

Xu, D., Yang, J., Zhou, X. & Xu, H. (2024). Hybrid path planning method for USV using bidirectional A and improved DWA considering the manoeuvrability and COLREGs. *Ocean engineering*, *298*, 117210.

Yao, M., Deng, H., Feng, X., Li, P., Li, Y. & Liu, H. (2024). Improved dynamic windows approach based on energy consumption management and fuzzy logic control for local path planning of mobile robots. *Computers & Industrial Engineering*, *187*, 109767.

Zhang, W., Yan, C., Lyu, H., Wang, P., Xue, Z., Li, Z. & Xiao, B., (2021). COLREGS-based Path Planning for Ships at Sea Using Velocity Obstacles. *IEEE Access*, *9*, 32613–32626.

Zhou, W., Wang, Y., Wu, Z. & Yan, H. (2024). Collision Avoidance and Path Point Tracking Control for Underactuated Unmanned Surface Vehicles With Unknown Model Nonlinearity. *IEEE Transactions on Vehicular Technology*, 1–16.

Trends in Collision and Grounding of Ships and Offshore Structures – Zhu & Guedes Soares (Eds)
© 2026 The Author(s), ISBN 978-1-041-16361-9

An autonomous docking path planning algorithm for tugboats based on precise task orientation

G.F. Tan, H.G. Lyu, Y. Li, S.Y. Ji & Y. Yin
School of Navigation, Dalian Maritime University, Dalian, P.R. China

Y.J. Zhao
China Ship Scientific Research Center, with the Taihu Laboratory of Deepsea Technological Science, Wuxi, P.R. China

ABSTRACT: Ports are essential for global trade, and the development of intelligent tugboats is crucial for efficient port operations. However, existing research on autonomous tugboats often overlooks the complex dynamics of docking maneuvers. This paper proposes a novel task-oriented autonomous path planning algorithm for tugboats, based on the Artificial Potential Field (APF) method. The algorithm decomposes the docking target into position, orientation, and velocity, enabling precise and efficient docking with target vessels. Through multi-scenario simulations, the algorithm demonstrates high accuracy and robustness, even in dynamic environments. This study addresses the research gap in intelligent tugboat docking and provides significant technological support for the development of intelligent ports.

1 INTRODUCTION

As a cornerstone of global trade, maritime shipping has significantly propelled regional economic development and facilitated cultural and technological exchanges, primarily due to its low cost, high efficiency, extensive coverage, and strategic importance (Lyu et al., 2023; Lyu, Liu, et al., 2024; Lyu, Ma, et al., 2024; Lyu & Yin, 2019). However, with the rapid expansion of the global economy, the shipping industry is confronted with increasingly stringent challenges in service quality (Shao, Yin, Lyu, & Guedes Soares, 2024; Shao, Yin, Lyu, Guedes Soares, et al., 2024; Sun et al., 2023; Zhang & Zhang, 2015).Ports, as critical terminals for cargo transshipment, are an essential core component of maritime transportation (Ning et al., 2023; Sun et al., 2024; Zheng et al., 2023). Amid the continuous increase in port throughput, tugboats have emerged as vital auxiliary tools for port operations, facing growing responsibilities and pressures. Large vessels, characterized by their poor maneuverability, are unable to dock at ports under their own power and typically require the assistance of tugboats (OS). The OS must provide escorting and auxiliary services in accordance with the requirements of the target ship (TS) to control the posture of the TS and mitigate the risk of damage to personnel, the vessel itself, other ships, and port facilities during the docking process.

In recent years, substantial research has been devoted to the development of autonomous navigation systems for tugboats, with investigations covering various aspects, including escorting and assisting berthing. Escorting involves the tugboat maintaining a predefined position relative to the target ship until the latter safely reaches its berth. This research direction has predominantly been addressed as a path-tracking problem for the escorting position. For instance, an innovative dynamic berthing system integrating artificial potential field-based virtual guidance with a two-dimensional event-triggered iterative learning mechanism was proposed by Zhang et al. (Zhang et al., 2024). This system ensures smooth path planning while significantly reducing the burden on actuators through optimized control input frequency, thereby minimizing wear and tear. A model-following control method based on deep deterministic policy gradient and soft actor-critic algorithms was developed by Li et al., enabling autonomous surface vessels to track the three-degree-of-freedom state of a given target ship (Li et al., 2024). The effectiveness of this method was demonstrated through experiments using scaled tugboat models. Furthermore, an event-triggered control method for tugboat escorting operations was introduced by Liu et al., combining model-free adaptive control with active disturbance rejection control (Liu, Xu, Li, Dong, & Hu, 2025). This approach optimizes control frequency while ensuring system

DOI: 10.1201/9781003684404-55

stability. Building on this foundation, an intelligent escorting control system that integrates path planning and adaptive motion control modules was further developed by Liu et al. (Liu, Xu, Li, Dong, Liu, et al., 2025). The system's performance was validated through full-scale simulator simulations and real-sea trials, demonstrating capabilities comparable to those of experienced tugboat operators. It significantly reduces operator workload and is considered a key technology for future autonomous vessels.

Despite these studies providing important theoretical foundations for tugboat path tracking, most have focused on the development of path-tracking algorithms that treat the predefined escort position as the guidance target, without delving into specific guidance methods. As a result, in scenarios where the target ship's state is highly dynamic, instability in guidance can easily occur. Moreover, these studies have not addressed the complex dynamic process of tugboat active berthing, making it difficult to meet the practical needs of a tugboat approaching a target ship from a distance.

In the field of assisted berthing, tugboats assist target ships in executing decisions such as turning and speed adjustment by performing operations like maintaining course, turning, maneuvering, and speed variation, thereby achieving safer and more efficient berthing. Research in this area typically treats the tugboat and target ship as a coupled system, realizing berthing objectives through coordinated control methods. For example, a multi-objective coordinated control strategy applicable to congested waters was proposed by Zhu et al. (Du et al., 2022). By coordinating multiple tugboats, this strategy enables precise control of trajectory, heading, and speed while also addressing collision avoidance requirements, ensuring efficient and safe ship berthing. Kristiansen et al. explored the control strategies for large ship berthing assisted by tugs from the perspective of dynamics modeling and simulation, providing new methods for improving autonomous berthing control algorithms (Kristiansen et al., 2024). Sano et al. constructed a novel mathematical model framework to accurately describe the coordinated operations between tugs and ships, significantly enhancing the model's sophistication and realism (Sano, 2023). Lee et al. proposed a dynamics model for barge transportation systems and designed a highly robust control system to ensure efficient barge operations (Lee et al., 2022). Bui et al. developed a mathematical model describing the interaction between non-self-propelled vessels and tugs, and solved the thrust allocation problem using a redistributed pseudo-inverse algorithm, reducing the energy requirements of tugs and improving controllability (Bui et al., 2011).

Despite significant progress in the coordinated control of tugboats and target ships, most existing studies have focused on scenarios where tugboats are already in close proximity to the target ships, with limited research on path planning and dynamic adjustment for active docking from a distance. Moreover, the complexity of achieving precise docking in dynamic environments has not been fully addressed, such as how to simultaneously adjust path, speed, and orientation under dynamic conditions to achieve accurate docking. These limitations restrict the application of tugboat autonomous navigation systems in complex port operations, especially in autonomous docking tasks.

Current research achievements have provided a solid theoretical foundation for the intelligent development of tugboats and demonstrated significant value in practical applications. Future intelligent tugboat operations are likely to cover multiple stages, including escorting, docking, and assisted operations. Specifically, tugboats first follow the target ship at an escorting position, then opportunistically perform docking maneuvers with the target ship, and after securing the lines, enter the assisted operation phase until the target ship is safely berthed. While research on escorting and assisted berthing has been extensive, studies on docking maneuvers are relatively scarce. This paper focuses on the path planning algorithm for tugboat docking with the target ship, aiming to achieve precise docking in dynamic environments and thereby ensure efficient integration and coordination of intelligent tugboat operations. Through this innovative research, this paper not only overcomes the limitations of existing studies but also pioneers a new research direction in the field of intelligent tugboat autonomous docking, filling the current research gap and providing important theoretical support and technical assurance for the efficient autonomous operations of intelligent tugboats.

To address the aforementioned requirements, a task-oriented autonomous path planning algorithm for tugboats is proposed in this paper, which aims to meet the task requirements for OS to precisely dock with TS for mooring operations. The path of the OS to the target position is planned accurately, and the heading and speed of the OS are ensured to match the preset target values upon docking. This process constitutes a multi-objective coupled planning problem.

The primary contributions of this paper are summarized as follows:

(1) A novel docking path planning scheme for tugboats, oriented by precise tasks, is designed to achieve the coupled planning of docking position, orientation, and speed.

(2) The effectiveness and robustness of the proposed algorithm are validated through various approaches, including simulation experiments and physical model tests.

The remainder of this paper is organized as follows: The docking path planning method is presented in Chapter 2, which describes the coupled planning of heading, speed, and position under both static and dynamic conditions. The effectiveness of the proposed algorithm is demonstrated through various simulation scenarios in Chapter 3. The work presented in this paper is summarized, and suggestions for future research directions are provided in Chapter 4.

2 DOCKING PATH PLANNING

2.1 *Docking standard line*

To meet the docking requirements of tugs, the position ($\boldsymbol{p}_{ap}$), heading (θ_{ap}), and speed (u_{ap}) targets need to simultaneously satisfy the preset requirements. However, it is difficult to perform coupled planning of these three targets without considering sway. To address this issue, an innovative docking path planning algorithm was proposed. As shown in Figure 1, the concept of a Docking Standard Line (LS) was introduced based on the task requirements, and an orientation method was designed according to LS and the ship's motion characteristics. LS is composed of $\boldsymbol{v}_{ts}$ and $-\boldsymbol{v}_{ap}$, where $\boldsymbol{v}_{ts}$ is the velocity vector of the TS, and $\boldsymbol{v}_{ts}$ is the docking velocity vector, as shown in Eqs. (1) and (2). If the OS is precisely located on the LS, by adjusting its heading and speed to match the target values, OS can remain on the LS for an extended period until it docks with TS.

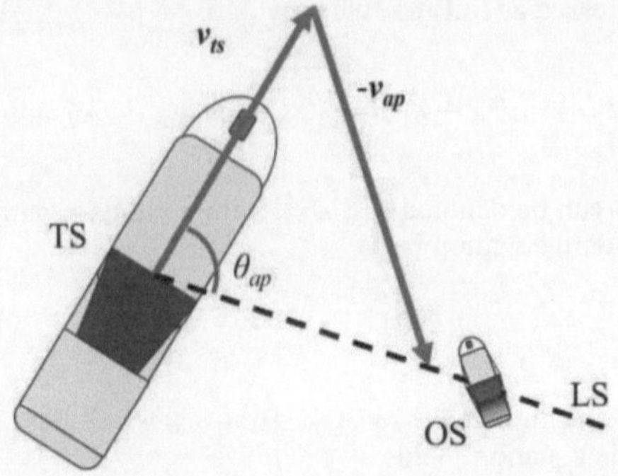

Figure 1. Design of LS.

$$\boldsymbol{v}_{ts} = u_{ts} \cdot (\sin(\varphi_{ts}), \cos(\varphi_{ts})) \tag{1}$$

$$\boldsymbol{v}_{ap} = u_{ap} \cdot (\sin(\varphi_{ts} \pm \theta_{ap}), \cos(\varphi_{ts} \pm \theta_{ap})) \tag{2}$$

where $\boldsymbol{v}_{ts}$ is the speed of TS, φ_{ts} is the heading of TS, u_{ap} is the target docking speed, and θ_{ap} is the target docking angle. If OS docks from the starboard side, θ_{ap} is subtracted; otherwise it is added.

2.2 *Static docking planning*

While controlling the position of OS on the LS and adjusting its navigation state can achieve precise docking, how to realize this objective becomes another challenge. Typically, tug docking operations are conducted at low speeds. If the position of OS is controlled on the LS too early, it will approach TS with the speed of u_{ap}, reducing operational efficiency. The ideal scenario is that OS reaches the LS and contacts TS simultaneously, with its heading and speed matching the target values. To address this, a guidance method for tug docking is designed, as shown in Figure 2. Based on the characteristics of ship motion, ships cannot perform large-angle turns in a short time. The heading adjustment required from the initial state to the docking state should be gradually realized throughout the entire voyage. Assuming no sway, the heading of the ship is aligned with its track. Therefore, assuming TS is stationary, the track line of OS at the end of docking should be tangent to the LS. At each moment, the algorithm calculates the desired heading of OS ($\boldsymbol{v}_{co}$) based on the direction vector of the LS ($\boldsymbol{v}_{ls}$) and the direction vector of OS relative to the beam angle of TS ($\boldsymbol{v}_{tos}$), forming an isosceles triangle. The desired path is tangent to the two sides, with the specific calculation steps as follows.

Let $\boldsymbol{v}_{ls}$ be denoted as (x_{ls}, y_{ls}) and $\boldsymbol{v}_{tos}$ be denoted as (x_{tos}, y_{tos}). The dot product of $\boldsymbol{v}_{ls}$ and $\boldsymbol{v}_{tos}$ is:

$$\boldsymbol{v}_{ls} \cdot \boldsymbol{v}_{tos} = x_{ls}x_{tos} + y_{ls}y_{tos} \tag{3}$$

The magnitude of $\boldsymbol{v}_{tos}$ is given by:

$$\|\boldsymbol{v}_{tos}\| = \sqrt{{x_{tos}}^2 + {y_{tos}}^2} \tag{4}$$

The magnitude of $\boldsymbol{v}_{ls}$ is given by:

$$\|\boldsymbol{v}_{ls}\| = \sqrt{{x_{ls}}^2 + {y_{ls}}^2} \tag{5}$$

The angle θ_1 between two vectors $\boldsymbol{v}_{ls}$ and $\boldsymbol{v}_{tos}$ can be found using the dot product formula:

$$\cos(\theta_1) = \frac{\boldsymbol{v}_{ls} \cdot \boldsymbol{v}_{tos}}{\|\boldsymbol{v}_{ls}\|\|\boldsymbol{v}_{tos}\|} \tag{6}$$

$$\theta_1 = \arccos\left(\frac{x_{ls}x_{tos} + y_{ls}y_{tos}}{\sqrt{{x_{tos}}^2 + {y_{tos}}^2}\sqrt{{x_{ls}}^2 + {y_{ls}}^2}}\right) \tag{7}$$

The direction vector of TS relative to OS ($\boldsymbol{v}_{ot}$) is rotated clockwise by an angle θ_1 to obtain $\boldsymbol{v}_{co}$. Let $\boldsymbol{v}_{ot}$ be denoted as (x_{ot}, y_{ot}). Then, $\boldsymbol{v}_{co}$ can be solved using the following formula.

The rotation matrix for a clockwise rotation by θ_1 is given by:

$$R(\theta_1) = \begin{pmatrix} \cos(\theta_1) & \sin(\theta_1) \\ -\sin(\theta_1) & \cos(\theta_1) \end{pmatrix} \tag{8}$$

When the rotation matrix is applied to $\boldsymbol{v}_{ot}$, the result is $\boldsymbol{v}_{co}$:

$$v_{co} = \begin{pmatrix} \cos(\theta_1) & \sin(\theta_1) \\ -\sin(\theta_1) & \cos(\theta_1) \end{pmatrix} \cdot \begin{pmatrix} x_{ot} \\ y_{ot} \end{pmatrix} \quad (9)$$

It should be noted that the clockwise rotation is only for the current example, and the actual rotation direction is related to the relative positions of the two ships. Different rotation angles θ_r can be set for different relative positions to replace θ_1 in the rotation matrix.

$$\theta_r = \begin{cases} \theta_1 & \text{starboard of TS, behind LS} \\ -\theta_1 & \text{starboard of TS, before LS} \\ -\theta_1 & \text{port of TS, behind LS} \\ \theta_1 & \text{port of TS, before LS} \end{cases} \quad (10)$$

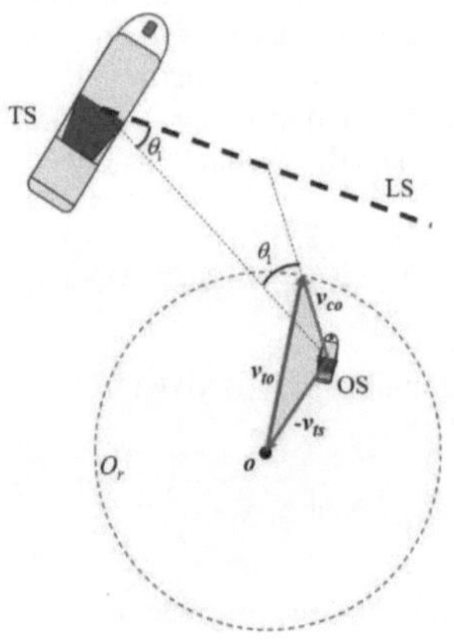

Figure 2. Docking path planning method.

2.3 *Dynamic docking planning*

Assuming that TS is stationary, the aforementioned design scheme has been elaborated in detail. However, if the TS has a nonzero speed, further analysis and adjustment of the existing scheme are required to ensure its effectiveness and applicability. To this end, the relative motion heading is defined as $\boldsymbol{v_{co}}$, as shown in Eq. (11). The algorithm needs to accurately calculate the direction to which the OS should turn based on the current speed state of the OS, so that $\boldsymbol{v_{co}}$ can meet the docking guidance requirements. In the previous design, only the direction of $\boldsymbol{v_{co}}$ was constrained, and its magnitude was not considered. Instead, it was determined by the vector summation result. As shown in Figure 2, the algorithm defines a relative motion vector circle (O_r). The algorithm translates the position of the OS ($\boldsymbol{p_{os}}$) along the vector $-\boldsymbol{v_{ts}}$ to obtain the center of the circle ($\boldsymbol{o}$), with the radius of r being the speed of the OS (u_{os}). If the starting point of $\boldsymbol{v_{co}}$ is set as $\boldsymbol{p_{os}}$, then every point on O_r could potentially be the endpoint of $\boldsymbol{v_{co}}$. We only need to determine the endpoint of $\boldsymbol{v_{co}}$ through the preset vector direction. In simple terms, the magnitude of $\boldsymbol{v_{co}}$ can be determined by the direction ray of $\boldsymbol{v_{co}}$ and O_r. Once the magnitude and direction of $\boldsymbol{v_{to}}$ are determined, the desired heading can be calculated, which will serve as the basis for our vessel to execute the docking maneuver. The detailed formula derivation is shown in Eqs. (12)-(22).

$$\boldsymbol{v_{co}} = \boldsymbol{v_{to}} - \boldsymbol{v_{ts}} \quad (11)$$

In the equation, $\boldsymbol{v_{co}}$ represents the relative motion velocity vector, and $\boldsymbol{v_{to}}$ represents the desired velocity vector of OS.

$$\boldsymbol{o} = \boldsymbol{p_{os}} - \boldsymbol{v_{ts}} \quad (12)$$

In the equation, $\boldsymbol{o}$ represents the center of the circle, $\boldsymbol{v_{ts}}$ represents the velocity vector of TS. $\boldsymbol{p_{os}}$ represents the position of OS, and the vector starting point can be denoted as (x_{os}, y_{os}). The direction vector of the vector can be obtained through. $\boldsymbol{v_{co}}$., denoted as $\boldsymbol{d} = (a, b)$, as shown in Eq. (13).

$$\boldsymbol{d} = \frac{\boldsymbol{v_{co}}}{\|\boldsymbol{v_{co}}\|} \quad (13)$$

The parametric equation of the ray can be expressed as follows, where $n \geq 0$:

$$\boldsymbol{P}(n) = \boldsymbol{P_{os}} + n \cdot \boldsymbol{d} = (x_{os} + an, y_{os} + bn) \quad (14)$$

$\boldsymbol{o}$ can be denoted as (h, k), with a radius r, and its standard equation is:

$$(x - h)^2 + (y - k)^2 = r^2 \quad (15)$$

Substituting $\boldsymbol{P}(n) = (x_{os} + an, y_{os} + bn)$ into the circle equation yields:

$$(x_{os} + an - h)^2 + (y_{os} + bn - k)^2 = r^2 \quad (16)$$

Expanding and rearranging into a quadratic equation in terms of n:

$$\begin{aligned} &(a^2 + b^2)n^2 + 2(a(x_{os} - h) + b(y_{os} - k))n \\ &+ ((x_{os} - h)^2 + (y_{os} - k)^2 - r^2) = 0 \end{aligned} \quad (17)$$

Simplifying Eq. (17) as follows:

$$A = a^2 + b^2 \quad (18)$$

$$B = 2(a(x_{os} - h) + b(y_{os} - k)) \quad (19)$$

$$C = (x_{os} - h)^2 + (y_{os} - k)^2 - r^2 \quad (20)$$

Eq. (17) can be transformed into:

$$An^2 + Bn + C = 0 \quad (21)$$

Using the quadratic formula to find the roots:

$$n = \frac{-B \pm \sqrt{B^2 - 4AC}}{2A} \quad (22)$$

The solution of Eq. (22) yields the value of n corresponding to the intersection point between the line and the circle. If n is positive, it indicates the intersection point of the ray with the circle.

Before solving for n, we need to use the discriminant formula: $D = B^2 - 4AC$ to determine whether there are intersection points. If $D > 0$, there are two real solutions, meaning the line intersects the circle at two points; if $D = 0$, there is one real solution, meaning the line is tangent to the circle; if $D < 0$, there are no real solutions, meaning the line does not intersect the circle. After obtaining n, we can discard the negative solution. If there are two solutions, we take the larger value, $\boldsymbol{P}(n) = (x_{os} + an, y_{os} + bn)$ corresponds to the intersection point of the ray with the circle. At this point, we can obtain $\boldsymbol{v_{to}}$ and the corresponding docking guidance force ($\boldsymbol{F_m}$), as shown in Eqs. (23)-(25). $\boldsymbol{v_{to}}$ will serve as the output of the task planning layer and reflect the task guidance in the subsequent heading

$$\boldsymbol{v_{co}} = \boldsymbol{P}(t) - \boldsymbol{p_{os}} \quad (23)$$

$$\boldsymbol{v_{to}} = \boldsymbol{v_{co}} + \boldsymbol{v_{ts}} \quad (24)$$

$$\boldsymbol{F_m} = k_m \cdot \boldsymbol{v_{to}} \quad (25)$$

Where, k_m represents the position guidance gain coefficient.

2.4 *Ship motion planning*

In this study, the ship motion model was not considered in the algorithm design. Only constraints on steering and speed variation were established. In Section 2.3 of this paper, the desired heading of OS is obtained by using the direction of $\boldsymbol{F_m}$ as the reference. OS adjusts its heading towards the desired heading at each moment, and a turning constraint is set such that the turning amplitude of OS at each moment does not exceed a certain limit, as shown in Eqs. (26) and (27). In terms of speed planning, a gradient planning method is designed. This method first calculates the distance between the two ships. If the distance is less than a preset threshold, a gradient planning is conducted between the maximum speed and the target speed to determine the desired speed. After obtaining the desired speed, OS makes speed adjustment decisions with the desired speed as the target, but the speed change is also subject to certain constraints, as shown in Eqs. (28)-(30). After updating the heading and speed, the new position of the ship at the next moment can be calculated based on this, as shown in Eq. (31).

$$\Delta\theta = \frac{\theta_{(\boldsymbol{v_{os}},\boldsymbol{F_m})}}{\left|\theta_{(\boldsymbol{v_{os}},\boldsymbol{F_m})}\right|} \times \min(\left|\theta_{(\boldsymbol{v_{os}},\boldsymbol{F_m})}\right|, \theta_c) \quad (26)$$

$$\varphi_{os}^{t+1} = \varphi_{os}^{t} + \Delta\theta \quad (27)$$

Where, $\theta_{(\boldsymbol{v_{os}},\boldsymbol{F_m})}$ represents the relative beam angle of $\boldsymbol{F_m}$ with respect to OS, with positive values for the starboard side and negative values for the port side. θ_c is the turnin0067 constraint angle, $\Delta\theta$ is the turning angle, φ_{os}^{t} is the heading of OS at time t,φ_{os}^{t+1} is the heading of OS at time t+1.

$$u_t = (u_m - u_{ap}) \times \frac{\min(d_{(os,ts)}, d_c)}{d_c} + u_{ap} \quad (28)$$

$$\Delta u = \min(u_c, \left\|u_t - u_{os}^{t}\right\|) \times \frac{(u_t - u_{os}^{t})}{\left\|u_t - u_{os}^{t}\right\|} \quad (29)$$

$$u_{os}^{t+1} = u_{os}^{t} + \Delta u \quad (30)$$

Where, u_m is the maximum speed, $d_{(os,ts)}$ is the distance between the two ships, d_c is the distance threshold, u_t is the desired speed, u_{os}^{t} is the speed of OS at time t, u_c is the speed change threshold, Δu is the speed change value, and u_{os}^{t+1} is the speed of OS at time t+1.

3 SIMULATION EXPERIMENTS

3.1 *Ship parameters and experimental design*

When conducting docking experiments, ships have a volume and cannot be treated as point masses. As shown in Figure 3, the length of OS is set to 38 m, and the width is 10 m in this paper. The bow is a semicircular arc with a diameter of 10 m, with the center denoted as $\boldsymbol{p_h}$. $\boldsymbol{P_{os}}$ is the coordinates of OS, located at the center of the ship's body, and $\boldsymbol{p_h}$ can be obtained using Eq. (31). The contact point of the OS is calculated based on the relative motion state between the two ships. The angle between the headings of the two ships is denoted as θ_s, and θ_c is the supplementary angle of θ_s. Using θ_c, the contact point can be determined, as shown in Equation (32). Based on the current motion state of the two ships, the future contact point of the OS is $\boldsymbol{P_t}$. At this time, θ_s is not the preset docking angle. By replacing θ_s with the preset value, the calculated point $\boldsymbol{P_t}$ becomes the target contact point of the OS. When performing docking path planning algorithms, $\boldsymbol{P_t}$ should be used as the coordinate instead of $\boldsymbol{p_{os}}$.

$$\boldsymbol{p_h} = \boldsymbol{p_{os}} + 16.5 \cdot (\sin(\varphi_{os}), \cos(\varphi_{os})) \quad (31)$$

$$
\begin{aligned}
p_t &= p_h + 5 \cdot (\sin(\varphi_{os} - \theta_c), \cos(\varphi_{os} - \theta_c)) \\
&= p_h + 5 \cdot (\sin(\varphi_{os} + \theta_s - \pi/2), \cos(\varphi_{os} + \theta_s - \pi/2))
\end{aligned}
\tag{32}
$$

Where, φ_{os} is the heading of the OS. As shown in Figure 3, since the OS intends to dock on the starboard side of the TS, φ_{os} needs to be subtracted by θ_c. Conversely, if docking on the port side, θ_c should be added.

As shown in Figure 8, eight actual docking positions (numbered P1-8) are set around the TS. In the simulation experiments, the experiment is deemed to be completed when the distance between the target contact point of the OS and the TS's body is less than 0.5 m. In the single-ship docking experiment, the TS is set with a length of 289.6 m and a width of 32.2 m, and the OS is planned to dock at P5 with an angular difference of 10° and an absolute speed of 2 m/s.

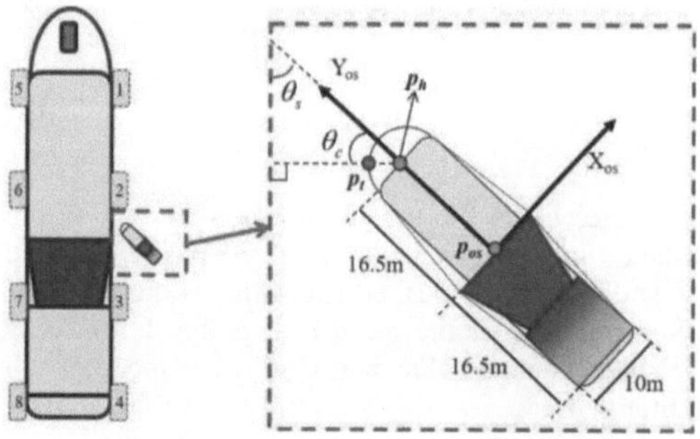

Figure 3. Calculation of docking point for OS.

3.2 *Stern-follow docking approach*

In this experiment, the OS is positioned behind the port side of the TS and plans to dock at point No. 5 with preset values, while the TS is expected to maintain its heading and speed. The initial speed of the TS is set at 1.5 m/s. To verify the robustness of the proposed planning algorithm in a windy and wavy environment, the TS randomly changes its heading by 0.5 degrees and adjusts its speed by 0.01 m/s every second. Figure 4 illustrates the simulation experiment, showing the trajectories and historical positions of both ships. It can be observed that the TS exhibits slight heading changes throughout the maneuver, which is consistent with real-world docking operations. Figure 4 also includes a close-up view at the moment of docking, demonstrating satisfactory docking performance.

Figure 5 records the heading and speed data of both ships during the experiment. From the heading change plot, it is evident that the TS experiences minor heading variations throughout the experiment. Guided by the planning algorithm, the OS continuously adjusts its heading to maintain a small heading difference with the TS at the moment of docking. At docking, the OS has a heading of 86.85°, while the TS has a heading of 77°, resulting in a heading difference of 9.85°, which is close to the preset value. In the speed plot, the two curves converge at the end because the TS's speed is 1.6 m/s, and the preset docking speed is 2 m/s, with the OS achieving an actual docking speed of 2.003 m/s. The data trends in heading and speed demonstrate the high accuracy of the proposed algorithm in single-ship docking experiments.

During the experiment, the docking position, angle, and speed errors were recorded every second and plotted in the docking error curves shown in Figure 6. The Z-axis represents the distance between the docking points of the two vessels. The X-axis indicates the difference between the relative heading angle of the two vessels and the preset docking angle. The Y-axis represents the discrepancy between the actual speed of our vessel and the preset absolute docking speed. As the position error gradually decreases, the angle and speed errors also diminish. Ultimately, the error curve converges toward zero, indicating that the proposed algorithm successfully achieves integrated planning of the three metrics. The final docking errors are as follows: the docking angle error is 0.18°, the docking speed error is 0.003 m/s, and the docking position error is 0.963 m. These results demonstrate the high overall accuracy of the proposed algorithm in achieving coupled planning of docking position, angle, and speed.

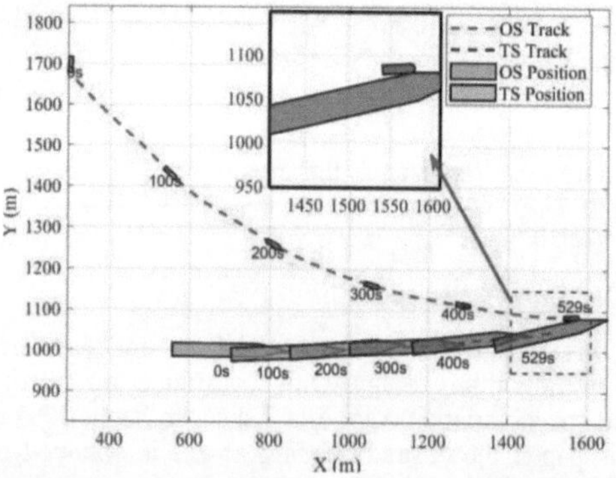

Figure 4. Simulation of stern-follow docking approach.

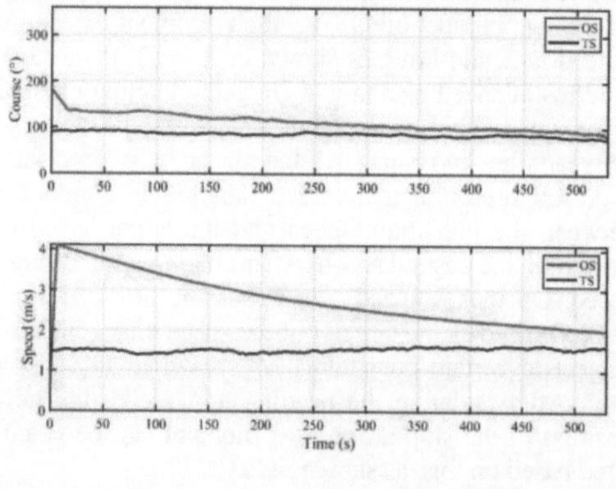

Figure 5. Data charts for 2 ships of stern-follow docking approach.

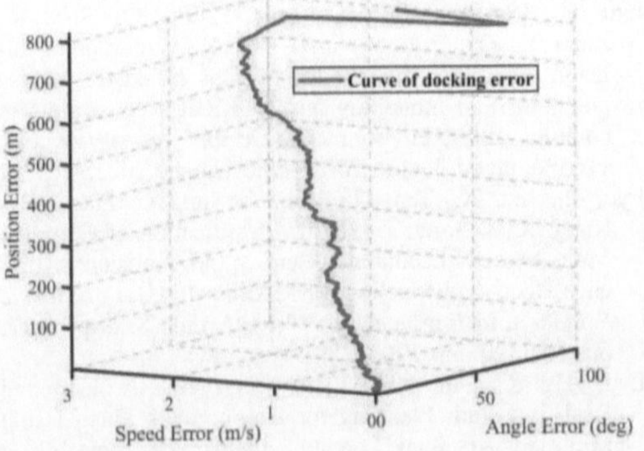

Figure 6. Docking error of stern-follow docking approach.

3.3 *Bow-encounter docking approach*

In this experiment, the OS is positioned before the port side of the TS. Figure 7 illustrates the simulation experiment, showing the trajectories and historical positions of both ships. It can be observed that the TS exhibits slight heading changes throughout the maneuver, which is consistent with real-world docking operations. Figure 7 also includes a close-up view at the moment of docking, demonstrating satisfactory docking performance.

Figure 8 records the heading and speed data of both ships during the experiment. From the heading change plot, it is evident that the TS experiences minor heading variations throughout the experiment. Guided by the planning algorithm, the OS continuously adjusts its heading to maintain a small heading difference with the TS at the moment of docking. At docking, the OS has a heading of 96°, while the TS has a heading of 88.5°, resulting in a heading difference of 7.5°, which is close to the preset value. In the speed plot, the two curves converge at the end because the TS's speed is 2.05 m/s, and the preset docking speed is 2 m/s, with the OS achieving an actual docking speed of 2.004 m/s. The data trends in heading and speed demonstrate the high accuracy of the proposed algorithm in single-ship docking experiments.

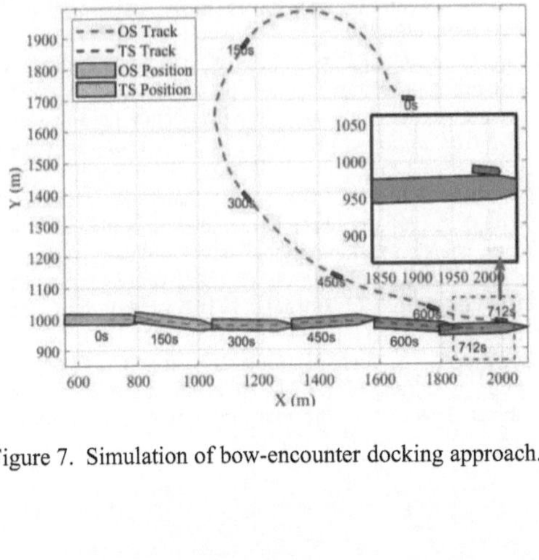

Figure 7. Simulation of bow-encounter docking approach.

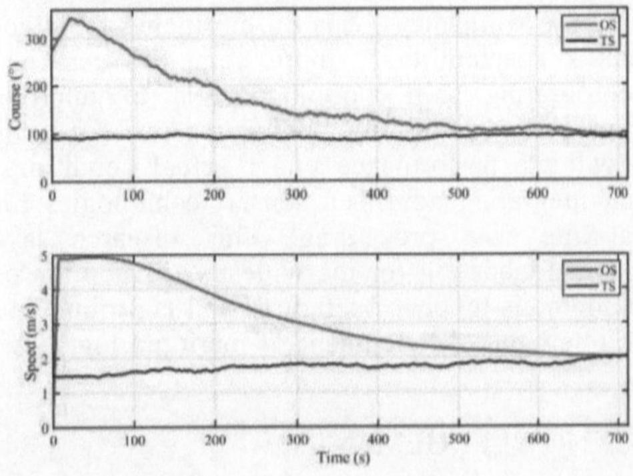

Figure 8. Data charts for 2 ships of bow-encounter docking approach.

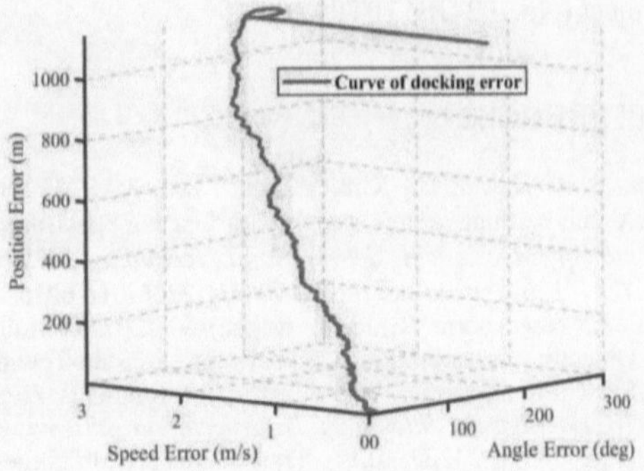

Figure 9. Docking error of bow-encounter docking approach.

The final docking errors are presented in Figure 9: the docking angle error is 2.5°, the docking speed error is 0.004 m/s, and the docking position error is 0.313 m. These results demonstrate the high overall accuracy of the proposed algorithm in achieving coupled planning of docking position, angle, and speed.

4 CONCLUSIONS

This research proposes a novel task-oriented autonomous path planning algorithm for tugboats, addressing the challenge of precise docking with target vessels in dynamic environments. By integrating the APF method and the Docking LS concept, the algorithm achieves multi-objective coupled planning of position, orientation, and velocity, ensuring accurate and efficient docking operations.

Simulation experiments, including stern-approach and bow-approach docking scenarios, demonstrate the algorithm's high accuracy and robustness. The results show minimal docking errors, validating its effectiveness even under varying environmental disturbances and target ship behaviors. This study fills the research gap in intelligent tugboat docking, providing a new direction for autonomous docking algorithms and supporting the development of intelligent ports.

Future work may focus on enhancing the algorithm's adaptability to more complex scenarios, incorporating ship motion models to improve accuracy and reliability, conducting real-sea trials to validate performance under actual conditions, and integrating advanced sensor technologies for real-time data processing. This research lays a solid foundation for the widespread adoption of autonomous tugboat operations and contributes to the intelligent development of maritime logistics.

ACKNOWLEDGEMENTS

National Natural Science Foundation of China (No.52071049).

Dalian City Science and Technology Plan (Key) Project (No.2024JB11PT007)

REFERENCES

Bui, V. P., Kawai, H., Kim, Y. B. & Lee, K. S. (2011). A ship berthing system design with four tug boats. *Journal of Mechanical Science and Technology*, *25*(5), 1257–1264. https://doi.org/10.1007/s12206-011-0215-4.

Du, Z., Negenborn, R. R. & Reppa, V. (2022). Multi-Objective Cooperative Control for a Ship-Towing System in Congested Water Traffic Environments. *IEEE Transactions on Intelligent Transportation Systems*, *23* (12), 24318–24329. IEEE Transactions on Intelligent Transportation Systems. https://doi.org/10.1109/TITS.2022.3208328.

Kristiansen, R., Ørke, H. L. & Gravdahl, J. T. (2024). Modelling and simulation of interaction forces in tugboat-assisted docking of large marine vessels. *Simulation Modelling Practice and Theory*, *130*, 102866. https://doi.org/10.1016/j.simpat.2023.102866.

Lee, D.-H., Huynh, T., Kim, Y.-B. & Park, J.-S. (2022). Motion Control System Design for Barge-Type Surface Ships Using Tugboats. *Journal of Marine Science and Engineering*, *10*(10), Article 10. https://doi.org/10.3390/jmse10101413.

Li, S., Xu, Z., Liu, J., Zhou, K. & Hu, X. (2024). Design and implementation of a model-following controller for autonomous surface ships based on actor–critic reinforcement learning. *Ocean Engineering*, *312*, 119157.https://doi.org/10.1016/j.oceaneng.2024.119157.

Liu, J., Xu, C., Li, S., Dong, Z. & Hu, X. (2025). Event-triggered ADRC-MFAC for intelligent tugs in escort operations with a thrust re-allocation strategy. *Control Engineering Practice*, *156*, 106237.https://doi.org/10.1016/j.conengprac.2024.106237.

Liu, J., Xu, C., Li, S., Dong, Z., Liu, J., Zhao, Y. & Sun, B. (2025). Towards future autonomous tugs: Design and implementation of an intelligent escort control system validated by sea trials. *Advanced Engineering Informatics*, *65*, 103116. https://doi.org/10.1016/j.aei.2025.103116.

Lyu, H., Hao, Z., Li, J., Li G., Sun, X., Zhang, G., Yin, Y., Zhao, Y. & Zhang, L. (2023). Ship Autonomous Collision-Avoidance Strategies—A Comprehensive Review. *Journal of Marine Science and Engineering*, *11*(4), Article 4. https://doi.org/10.3390/jmse11040830.

Lyu, H., Liu, W., Guo, S., Tan, G., Fu, C., Sun, X., Zhao, Y., Zhang, L. & Yin, Y. (2024). Autonomous collision avoidance method for MASSs based on precise potential field modelling and COLREGs constraints in complex sailing environments. *Ocean Engineering*, *292*, 116530.https://doi.org/10.1016/j.oceaneng.2023.116530.

Lyu, H., Ma, X., Tan, G., Yin, Y., Sun, X., Zhang, L., Kang, X. & Song, J. (2024). Identification of Complex Multi-Vessel Encounter Scenarios and Collision Avoidance Decision Modeling for MASSs. *Journal of Marine Science and Engineering*, *12*(8), Article 8. https://doi.org/10.3390/jmse12081289.

Lyu, H. & Yin, Y. (2019). COLREGS-Constrained Real-time Path Planning for Autonomous Ships Using Modified Artificial Potential Fields. *The Journal of Navigation*, *72*(3), 588–608. https://doi.org/10.1017/S0373463318000796.

Ning, J., Ma, Y., Liu, L. & Li, T. (2023). Fuzzy course tracking control of an unmanned surface vehicle with event-triggered mechanism and input quantization. *Ocean Engineering*, *288*, 115704. https://doi.org/10.1016/j.oceaneng.2023.115704.

Sano, M. (2023). Mathematical model and simulation of cooperative manoeuvres among a ship and tugboats. *Brodogradnja*, *74*, 127–148. https://doi.org/10.21278/brod74207.

Shao, Z., Yin, Y., Lyu, H. & Guedes Soares, C. (2024). A robust method for multi object tracking in autonomous ship navigation systems. *Ocean Engineering*, *311*, 118560. https://doi.org/10.1016/j.oceaneng.2024.118560.

Shao, Z., Yin, Y., Lyu, H., Guedes Soares, C., Cheng, T., Jing, Q. & Yang, Z. (2024). An efficient model for small object detection in the maritime environment. *Applied Ocean Research*, *152*, 104194. https://doi.org/10.1016/j.apor.2024.104194.

Sun, S., Lyu, H. & Dong, C. (2023). AIS aided marine radar target tracking in a detection occluded environment. *Ocean Engineering*, *288*, 116133. https://doi.org/10.1016/j.oceaneng.2023.116133.

Sun, S., Lyu, H., Gao, Z. & Yang, X. (2024). Grid Map Assisted Radar Target Tracking in a Detection Occluded Maritime Environment. *IEEE Transactions on Instrumentation and Measurement*, *73*, 1–11. IEEE Transactions on Instrumentation and Measurement. https://doi.org/10.1109/TIM.2024.3381495.

Zhang, G., Shang, X., Li, J. & Huang, J. (2024). A Novel Dynamic Berthing Scheme for An USV: DPFS Guidance and Two-dimensional Event Triggering ILC. *IEEE Transactions on Intelligent Vehicles*, 1–12. IEEE Transactions on Intelligent Vehicles. https://doi.org/10.1109/TIV.2024.3402229.

Zhang, G. & Zhang, X. (2015). A novel DVS guidance principle and robust adaptive path-following control for underactuated ships using low frequency gain-learning. *ISA Transactions*, *56*, 75–85. https://doi.org/10.1016/j.isatra.2014.12.002.

Zheng, K., Zhang, X., Wang, C., Zhang, M. & Cui, H. (2023). A partially observable multi-ship collision avoidance decision-making model based on deep reinforcement learning. *Ocean & Coastal Management*, *242*, 106689. https://doi.org/10.1016/j.ocecoaman.2023.106689.

Shao, Z., Yin, Y., Lyu, H., Guedes Soares, C., Cheng, T., Jing, Q. & Yang, Z. (2024). An efficient model for small object detection in the maritime environment. *Applied*

Ocean Research, *152*, 104194. https://doi.org/10.1016/j.apor.2024.104194.

Sun, S., Lyu, H. & Dong, C. (2023). AIS aided marine radar target tracking in a detection occluded environment. *Ocean Engineering*, *288*, 116133. https://doi.org/10.1016/j.oceaneng.2023.116133.

Sun, S., Lyu, H., Gao, Z. & Yang, X. (2024). Grid Map Assisted Radar Target Tracking in a Detection Occluded Maritime Environment. *IEEE Transactions on Instrumentation and Measurement*, *73*, 1–11. IEEE Transactions on Instrumentation and Measurement. https://doi.org/10.1109/TIM.2024.3381495.

Zhang, G., Shang, X., Li, J. & Huang, J. (2024). A Novel Dynamic Berthing Scheme for An USV: DPFS Guidance and Two-dimensional Event Triggering ILC. *IEEE Transactions on Intelligent Vehicles*, 1–12. IEEE Transactions on Intelligent Vehicles. https://doi.org/10.1109/TIV.2024.3402229.

Zhang, G. & Zhang, X. (2015). A novel DVS guidance principle and robust adaptive path-following control for underactuated ships using low frequency gain-learning. *ISA Transactions*, *56*, 75–85. https://doi.org/10.1016/j.isatra.2014.12.002.

Zheng, K., Zhang, X., Wang, C., Zhang, M. & Cui, H. (2023). A partially observable multi-ship collision avoidance decision-making model based on deep reinforcement learning. *Ocean & Coastal Management*, *242*, 106689. https://doi.org/10.1016/j.ocecoaman.2023.106689.

Zhang, G., Shang, X., Li, J. & Huang, J. (2024). A Novel Dynamic Berthing Scheme for An USV: DPFS Guidance and Two-dimensional Event Triggering ILC. *IEEE Transactions on Intelligent Vehicles*, 1–12. IEEE Transactions on Intelligent Vehicles. https://doi.org/10.1109/TIV.2024.3402229

Zhang, G. & Zhang, X. (2015). A novel DVS guidance principle and robust adaptive path-following control for underactuated ships using low frequency gain-learning. *ISA Transactions*, *56*, 75–85. https://doi.org/10.1016/j.isatra.2014.12.002.

Zheng, K., Zhang, X., Wang, C., Zhang, M. & Cui, H. (2023). A partially observable multi-ship collision avoidance decision-making model based on deep reinforcement learning. *Ocean & Coastal Management*, *242*, 106689. https://doi.org/10.1016/j.ocecoaman.2023.106689.

Trends in Collision and Grounding of Ships and Offshore Structures – Zhu & Guedes Soares (Eds)

An intelligent decision support method for assisted ship maneuvering and route planning using pre-trained deep learning model

M.Y. Zhang
Aalto University, Marine and Arctic Technology, School of Engineering, Espoo, Finland

W.H. Shen
School of Naval Architecture, Ocean and Energy Power Engineering, Wuhan University of Technology, Wuhan, P.R. China

N. Tsoulakos
Laskaridis Shipping Co. Ltd., Athens, Greece

H.D. Wang
MOE Key Laboratory of Marine Intelligent Equipment and System, Shanghai Jiao Tong University, Shanghai, P.R. China

P. Kujala
Tallin University of Technology, Estonian Maritime Academy, Estonia

S. Hirdaris
Global Ship Systems Centre, American Bureau of Shipping, Greece

ABSTRACT: This study presents an intelligent decision support method for ship-assisted maneuvering and route planning using a pre-trained Long Short-Term Memory (LSTM) deep learning model. The model is trained on full-scale data from a Kamsarmax bulk carrier, collected over 69 days at sea and covering 350 parameters sampled at 60-second intervals. It captures the dynamic relationship between control inputs—rudder angle and propeller speed—and ship motion responses including surge, sway, and yaw. The framework enables real-time prediction of maneuvering trajectories, allowing for evaluation and verification of different control strategies under realistic operating conditions. By identifying safe and effective commands, the system supports navigation decisions that reduce risks (grounding or collision). The proposed approach enhances situational awareness, operational safety in advance, and offers a practical solution for data-driven ship maneuvering in intelligent maritime systems. The findings demonstrate that subject to further validation the model holds significant promise for decision support.

1 INTRODUCTION

Intelligent decision support systems form the foundation of ship intelligent navigation and maritime digitalization. Fundamental to the development of such systems is the development of ship maneuvering prediction models that are rapid, accurate, and reliable. These models could be used for the evaluation and verification of maneuvering commands as illustrated in Figure 1. Their ability to provide timely and precise predictions is essential to ensuring navigational safety and operational efficiency, particularly in complex operational conditions (Zhang et al., 2025).

Over the past decades, significant research efforts have been devoted to the development of ship maneuvering prediction models. Broadly, these can be categorized into mathematical and data-driven models, the latter using machine learning. Mathematical maneuvering models typically describe ship behavior as a rigid-body system with six degrees of freedom (6-DOF). They are founded on Newtonian mechanics and maneuvering theory, with widely adopted frameworks including the MMG (Manoeuvring Modelling Group) standard method, the Abkowitz model, and their extensions (Taimuri et al., 2020; Shen et al., 2023). For example, Zhang et al. (2023) conducted a comparative analysis of the conventional MMG model and an automatic berthing model. Their study proposes a modeling approach based on hull hydrodynamics and propulsion systems and displays the importance of maneuvering models as a basis for control algorithm design in terms of mathematical modeling and optimization. Parametric identification methods such as nu-Support Vector Machine (nu-SVM) have been applied to estimate hydrodynamic derivatives within the 3-DOF Abkowitz framework by Wang et al. (2019). The high computational cost, lack of real-time capability, and poor exploitation of historical data inherent in traditional

DOI: 10.1201/9781003684404-56

mathematical models hinder the ability of these methods to facilitate rapid iterative testing (Zhang et al., 2025).

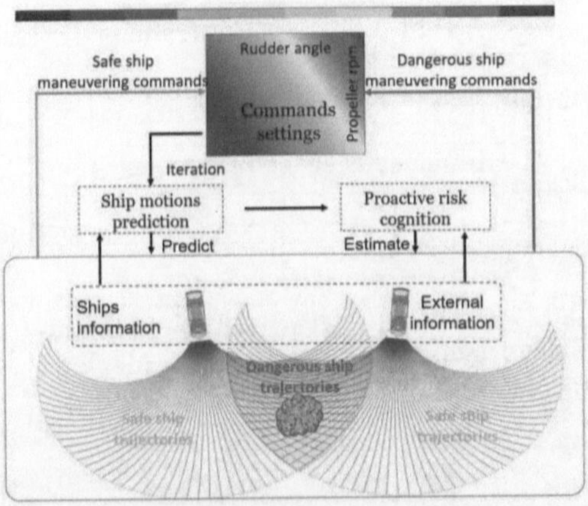

Figure 1. Testing processes of ship manoeuvring commands for safe assisted navigation decisions (Zhang et al., 2025).

Deep learning models excel in capturing the complex, nonlinear, and high-dimensional relationships underlying ship behavior. Artificial Neural Networks (ANN) (Silva et al., 2022), Gaussian Process Regression (GPR) (Ouyang et al., 2021; Ramirez et al., 2018), Long Short-Term Memory (LSTM) networks (Sun et al., 2022; Zhang et al., 2024), and Transformer-based architectures (Zhang et al., 2022) have shown considerable promise. These models capture ship motion features directly from data without relying on predefined physical formulations. A good example is given by Zhang et al. (2022) who introduced a Transformer-based deep learning framework trained on extensive historical datasets combined with 6-DOF dynamics. Notwithstanding this, many of these efforts primarily focused on predicting and verifying ship motions, often without delving into the modeling and optimization of ship navigation decision-making processes. Notably the application of ship maneuvering datasets of real ship operations in training these models remains limited (Shen et al., 2025).

When using historical ship maneuvering data, deep learning models eliminate the need for computationally intensive real-time simulations while ensuring rapid and accurate ship motion predictions. This makes them particularly well-suited for integration into intelligent decision support systems, where real-time evaluation of maneuvering strategies is crucial for the prevention of critical events such as collisions and groundings (Zhang et al., 2025; Shen et al., 2025).

This paper proposes an intelligent decision support method for ship-assisted maneuvering, using a pre-trained deep learning model. The proposed framework integrates the predictive strengths of deep neural networks and theory of ship dynamics to offer reliable, real-time maneuvering guidance. By incorporating realistic operational data during model training and validation, the approach aims to enhance accuracy, robustness, and adaptability across various operations scenarios. Ultimately, this research contributes to the advancement of intelligent, data-driven navigation systems, supporting broader objectives of safe, efficient, and autonomous ship operations.

2 FRAMEWORK OF THE SHIP-ASSISTED MANOEUVRING METHOD USING A DEEP LEARNING MODEL

This study presents a data-driven framework to enhance ship maneuvering safety at sea using deep learning, as shown in Figure 2. At first, sensor data from real operations—covering rudder angles, propeller speeds, and ship motions (surge, sway, yaw) are collected. Then, an LSTM network is trained using time-series data of ship maneuvering data to model the relationship between control inputs and ship dynamics. Upon the prediction of ship trajectories, obstacle avoidance for informed decision-making is possible. The method proposed comprises of three steps namely:

Step I: Data collection onboard. Sensor data from real ship operations, including maneuvering commands (rudder angle, propeller speed) and ship motions (surge, sway, yaw rate), are collected to capture the ship dynamic behavior under various conditions.

Step II: Deep learning model training and validation. A LSTM network is trained using time-sequenced maneuvering commands and ship motion data, enabling accurate prediction of ship dynamics based on historical states and control inputs.

Step III: Ship-assisted maneuvering commands determination. The trained model predicts ship trajectories under different control commands, allowing for real-time evaluation of maneuvering safety and obstacle avoidance, supporting optimal and safe navigation decisions.

2.1 *Data collection onboard*

To support the development of a data-driven framework for the determination of maneuvering commands, a comprehensive dataset was systematically collected from data made available for a Kamsarmax 64,000 DWT bulk carrier operated by Laskaridis Shipping Co. Ltd. The vessel of (Length 199.9 meters, and Beam 32.0 meters), was instrumented with an extensive sensor suite designed to continuously monitor key operational parameters during routine sea voyages (see Figure 3).

The data acquisition process was configured to capture 350 distinct parameters at 60-second intervals. These parameters encompass a wide range of information, including maneuvering commands, engine performance metrics, ship motion responses, loading conditions, and external environmental influences (Zhang et al., 2024).

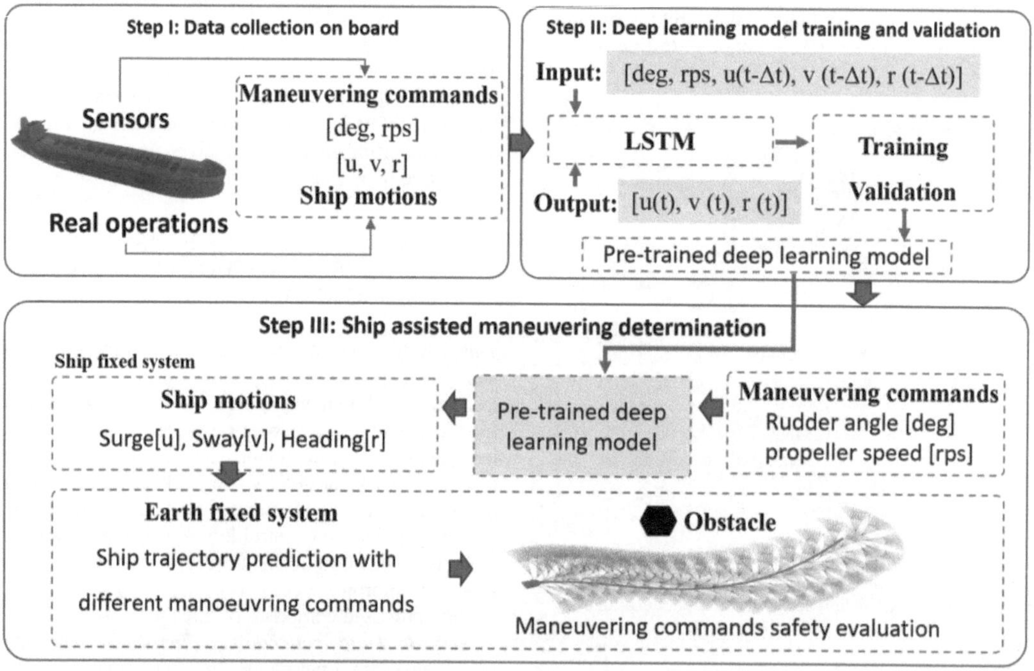

Figure 2. The framework for the safety evaluation of the ship maneuvering commands using deep learning model.

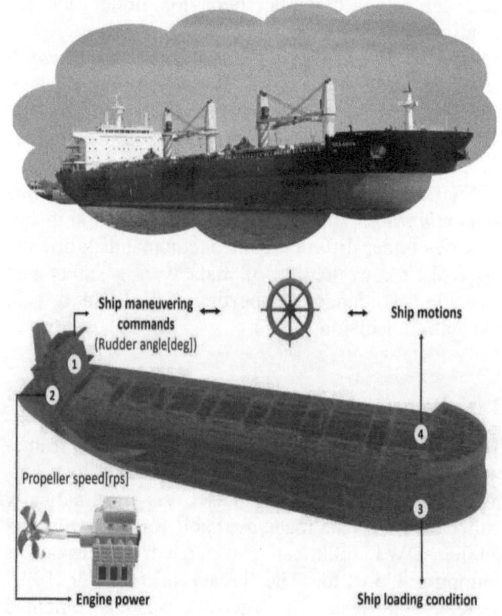

Figure 3. The ship and ship maneuvering data collection at sea.

Within the context of Step I of the framework (Figure 2), particular focus was placed on two critical categories of measurements essential for ship dynamic modeling. First, maneuvering commands, specifically rudder angle (measured in degrees) and propeller speed (measured in revolutions per minute, rpm), were recorded to document the ship control inputs. Second, the ship motion responses were continuously monitored. Those captured surge velocity (u), sway velocity (v), and yaw rate (r) to reflect the ship dynamic behavior at sea.

The collected dataset provides an empirical basis for analyzing the interaction between control actions and ship responses under various operating conditions. This data foundation is crucial for the subsequent training and validation of the deep learning model, thus ensuring its capability to accurately replicate and predict real ship maneuvering dynamics.

2.2 *Deep learning model for the prediction of ship motions*

Following the data collection process described in Section 2.1, a deep learning approach was employed to model the relationship between maneuvering commands and ship motion responses. Specifically, a LSTM based neural network was selected due to its superior capability in handling time-series data and capturing temporal dependencies inherent in ship dynamics. The LSTM architecture is particularly suited for modelling sequential time-series data due to its ability to effectively retain long-term dependencies and mitigate issues such as vanishing gradients, which commonly affect traditional recurrent neural networks (RNNs).

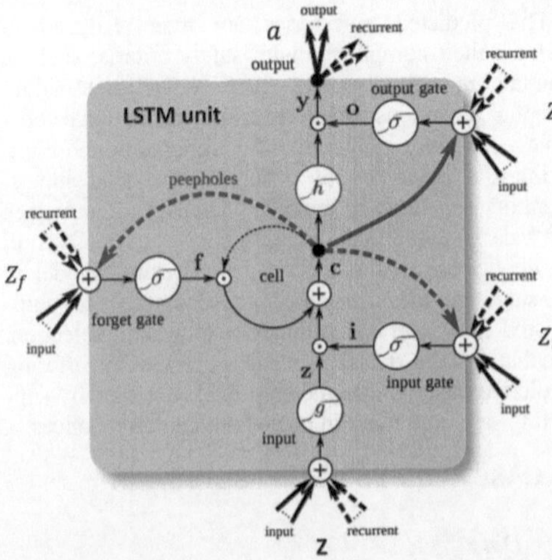

Figure 4. The internal structure of the LSTM unit (Greff et al., 2016).

Each LSTM unit consists of four interconnected components namely an input gate, a forget gate, an output gate, and a memory cell. These gates collaboratively regulate the flow of information, enabling the network to decide which information should be retained, updated, or discarded at each time step. The internal structure of the LSTM unit is illustrated in Figure 4. Specifically, the input at time step t, denoted as x_t, and the hidden state from the previous step, h_{t-1}, are combined and processed to generate control signals for each gate, as Eq. (1-4). The activation functions used are typically sigmoid $f(x) = 1/(1 + e^{\wedge}(-x))$ and hyperbolic tangent functions, which regulate gate outputs within the range [0,1]. The updated memory cell state c_t and the hidden state h_t are computed as Eq. (5-6).

$$Z_f = w_f[h_{t-1}, x_t] + b_f \quad (1)$$

$$Z_i = w_i[h_{t-1}, x_t] + b_i \quad (2)$$

$$Z_o = w_o[h_{t-1}, x_t] + b_o \quad (3)$$

$$Z = w_x[h_{t-1}, x_t] + b_x \quad (4)$$

$$c' = g(Z)f(Z_i) + c_{(t-1)}f(Z_f) \quad (5)$$

$$h_t = h(c')f(Z_o) \quad (6)$$

Here, $g(x)$ and $h(x)$ denote the respective activation functions, and c' represents the candidate memory cell state. The combination of these mechanisms allows the LSTM network to preserve relevant historical information over long time periods, making it highly effective in modeling ship motion responses governed by both immediate and prior control actions.

The dataset described in Section 2.1 was partitioned into training and validation sets. The LSTM model was trained to minimize the mean squared error (MSE) and mean absolute error (MAE) between the predicted and actual ship motion states. The input to the LSTM model comprises sequential data points, including the maneuvering commands—rudder angle (degrees) and propeller speed (rpm)—as well as historical ship motion states, represented by surge velocity u(t-Δt), sway velocity v (t-Δt), and yaw rate r(t-Δt). The model outputs the predicted ship motion states at the current time step *t*, specifically the surge velocity u(t), sway velocity v(t), and yaw rate r(t), as shown in Figure 5. By incorporating data from preceding time steps, the model effectively learns the time-dependent influence of control inputs on ship dynamics in real operational conditions.

Upon completion of the training phase, the model was validated against unseen operational data to evaluate its predictive accuracy. The results demonstrated that the trained LSTM network effectively captures the nonlinear and dynamic characteristics of ship maneuvering, providing reliable predictions of surge, sway, and yaw motions in response to varying control commands, see more in Section 3.2. This trained deep learning model forms the core component of the proposed framework, enabling real-time trajectory prediction and serving as the basis for maneuvering safety assessment in subsequent stages.

2.3 *Ship maneuvering commands testing and selection*

Building upon the predictive capabilities of the pre-trained deep learning model established in Section 2.2, the trained LSTM model was applied to systematically test and evaluate various ship maneuvering command scenarios. The objective of this stage has been to assess the feasibility and safety of different control strategies under real-time operating conditions for ship assisted maneuvering.

In this process, multiple combinations of maneuvering commands corresponding to specific rudder angles and propeller speeds were input to the pre-trained LSTM model. For each command sequence, the model predicted the corresponding ship motion responses, including surge velocity, sway velocity, and yaw rate (rate of turn), over the forecast horizon. At each discrete time step $t_k = t_0 + k \times \Delta t$. The model receives as input as Eq. (7). The LSTM network outputs the predicted ship motion states at time t_k, as shown in Eq. (8).

$$x_k = [\delta(t_K),\ n(t_K),\ u(t_{K-1}),\ v(t_{K-1}),\ r(t_{K-1})] \quad (7)$$

$$\hat{y}_K = [u(t_k),\ v(t_k),\ r(t_k)] \quad (8)$$

where, k = 0, 1, …, N. $\delta(t_k)$ is the rudder angle, n (t_k) is the propeller speed, and $u(t_{k-1})$, $v(t_{k-1})$, and r (t_{k-1}) represent the surge velocity, sway velocity, and yaw rate from the previous time step.

To translate the predicted velocities from the ship-fixed coordinate system to the earth-fixed coordinate system, a coordinate transformation is applied based on the ship's heading angle $\psi(t_k)$. The earth-fixed velocities are computed as Eq. (9-11). Starting from the initial position (x_0, y_0 and heading ψ_0, the ship trajectory is iteratively updated at each time step using Eq. (12-14).

$$\dot{x}(t_K) = u(t_K)\cos\psi(t_K) - v(t_K)\sin\psi(t_K) \quad (9)$$

$$\dot{y}(t_K) = u(t_K)\sin\psi(t_K) - v(t_K)\cos\psi(t_K) \quad (10)$$

$$\dot{\psi}(t_K) = r(t_K) \quad (11)$$

$$\dot{\psi}x(t_{K+1}) = x(t_K) + \dot{x}(t_K)\Delta t \quad (12)$$

$$y(t_{K+1}) = y(t_K) + \dot{y}(t_K)\Delta t \quad (13)$$

$$\psi(t_{K+1}) = \psi(t_K) + r(t_K)\Delta t \quad (14)$$

After N prediction steps, the final position (x_N, y_N) at $t = t_0 + N\Delta t$ is given by Eq. (15-16).

$$x\_N = x_0 + \sum_{K=0}^{n-1} [u(t_K)\cos\psi(t_K) - v(t_K)\sin\psi(t_K)] \cdot \Delta t \quad (15)$$

$$y\_N = y_0 + \sum_{K=0}^{n-1} [u(t_K)\sin\psi(t_K) - v(t_K)\cos\psi(t_K)] \cdot \Delta t \quad (16)$$

By applying this process over the prediction horizon $N\Delta t$, these motion states are subsequently transformed from the ship-fixed coordinate system to the earth-fixed coordinate system, enabling the reconstruction of the ship anticipated trajectory in geographical space, as shown in Figure 6.

The predicted trajectories are then analyzed to evaluate their compliance with safety criteria, such as maintaining safe distances from obstacles, avoiding potential collisions, and ensuring navigational efficiency, see more in Figure 1. By simulating and comparing the outcomes of different command inputs, operators can identify optimal maneuvering strategies that balance safety and operational objectives. Overall, the application of the trained deep learning model in this stage provides a reliable and efficient tool for ship-assisted maneuvering command testing and selection. It enhances the decision-making process by offering predictive insights into the ship behavior, thereby supporting safer and more informed navigation practices.

3 CASE STUDIES AND RESULTS

3.1 *Data*

To validate the intelligent decision support method for ship assisted maneuvering using pre-trained deep learning model, this study utilized a large-scale dataset collected from real sea operations of a Kamsarmax bulk carrier operated by Laskaridis Shipping Co. Ltd. As illustrated in Figure 7, the dataset spans ship voyages between March 11, 2023, and May 19, 2024. In the dataset, each data point along the recorded trajectory includes 350 parameters, capturing a comprehensive profile of ship operation. These parameters encompass maneuvering commands, engine performance indicators, ship motion states, loading conditions, and external environmental factors.

Data were recorded at fixed 60-second intervals to ensure consistent temporal resolution. For the purpose of ship maneuvering prediction model development, this study focused on extracting two primary categories of data: ship maneuvering commands (rudder angle and propeller speed) and ship motion responses (surge, sway, and yaw), as shown in Figure 8. These variables serve as the input-output

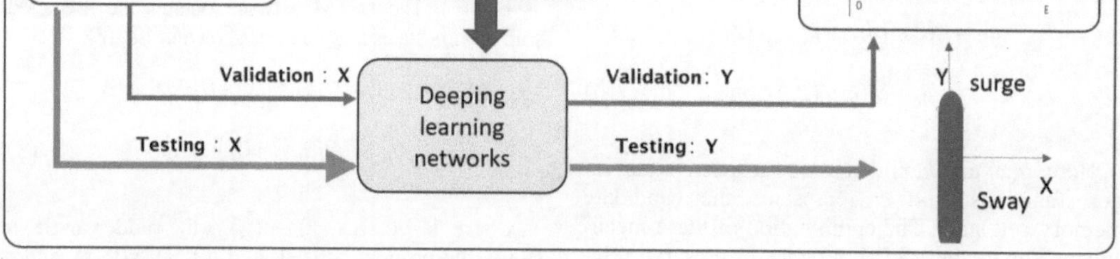

Figure 5. The development of ship maneuvering model using deep learning for the prediction of ship motions.

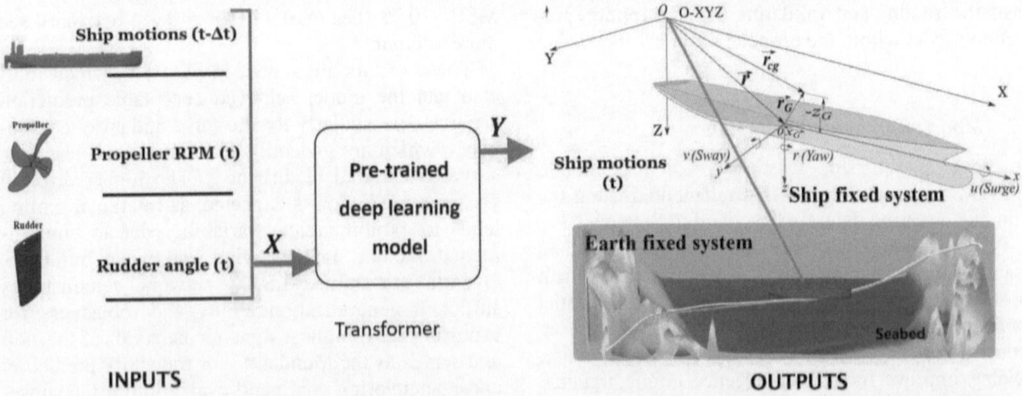

Figure 6. Maneuvering commands safety evaluation using pre-trained deep learning model.3. Case studies and results.

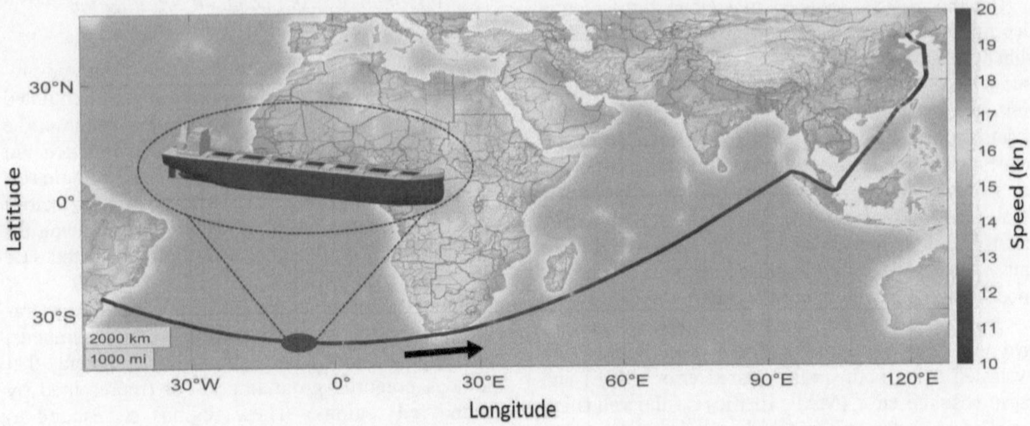

Figure 7. Ship real operations at sea of the bulk carrier from 03.2023 to 05.2024.

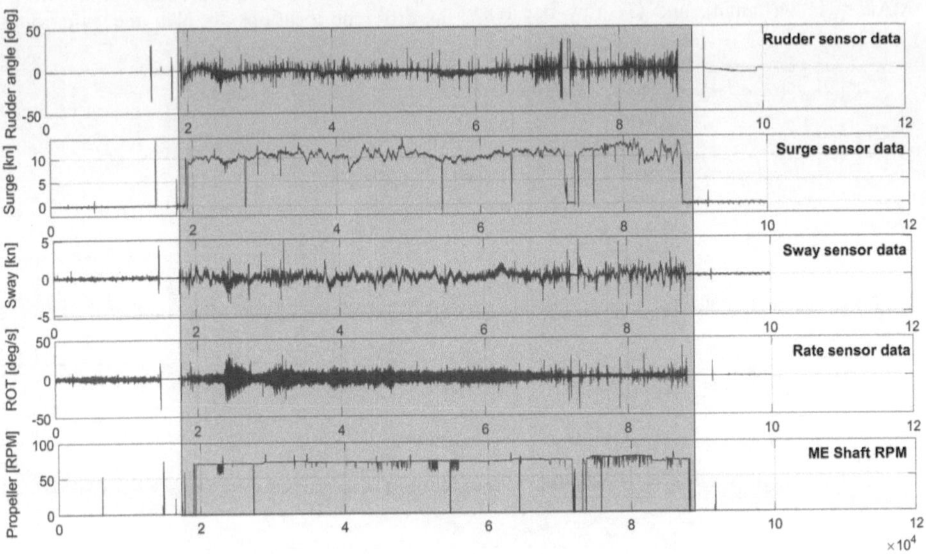

Figure 8. Ship maneuvering commands and the corresponding responses.

basis for training and validating the ship motion prediction model within the proposed framework.

3.2 *Model training and validation*

To develop the ship motion prediction model, a LSTM neural network was implemented using the ship maneuvering datasets described in Section 3.1.

Prior to training, the time-series data was smoothed using a Savitzky–Golay filter to suppress measurement noise while preserving signal characteristics essential for ship motion dynamics. This pre-processing step reduces short-term fluctuations caused by sensor noise, helping improve model convergence during training. All input and output variables were normalized to a standard scale to ensure numerical stability during LSTM training.

The model was designed to capture the temporal dependencies between control inputs and the resulting ship dynamic responses in surge (u), sway (v), and yaw rate (r) motions. The input to the model consisted of time-series sequences with a fixed length of five consecutive time steps. Each input vector included rudder angle, propeller speed, and the corresponding historical motion states (surge velocity $u(t-\Delta t)$, sway velocity $v(t-\Delta t)$, and yaw rate $r(t-\Delta t)$). The network architecture comprised four hidden layers, each with 100 LSTM units. A batch size of 64 was used for training, and the model was optimized using the Adam optimizer with a learning rate of 0.001. The training process was carried out over 50 epochs. Model performance was evaluated using both mean squared error (MSE) and mean absolute error (MAE) metrics on the validation set. The results are summarized as follows: (1) Surge velocity (u): MSE = 1813.21$(m/min)^2$, MAE = 33.32 (m/min), (2) Sway velocity (v): MSE = 212.48$(m/min)^2$, MAE = 11.24(m/min), and (3) Yaw rate (r): MSE = 0.79 $(deg/min)^2$, MAE = 0.48 (deg/min), see more in Figure 9.

These results are shown in Figure 9, which indicate that the model achieved acceptable prediction accuracy, particularly for the surge and sway components, which are generally more stable and less sensitive to external disturbances. The higher error in yaw rate prediction is expected, as rotational motion tends to exhibit greater variability due to environmental factors such as wind and wave influence. Overall, the trained LSTM network demonstrates sufficient generalization ability and robustness for capturing the nonlinear dynamic behavior of the ship and serves as the foundation for trajectory prediction and maneuvering command evaluation in the subsequent framework stages.

3.3 *Ship maneuvering commands testing for grounding avoidance*

To demonstrate the applicability of the proposed prediction model, a grounding scenario was simulated along a predefined sailing route. In this test case, the propeller rotational speed was maintained at a constant value of 71 rps. For each rudder angle setting, the model predicted the ship motion trajectory over 100 discrete time steps (100 mins), allowing for the evaluation of potential deviations from the safe navigation corridor.

Figure 10: presents the predicted ship motion trajectories under various maneuvering commands, superimposed on a navigation environment that includes potential grounding zones (represented by shaded red regions). These regions correspond to hypothetical shallow water areas where grounding risks are high. In this scenario, a series of rudder angle and propeller speed combinations were tested at different locations to plan the safe route. The

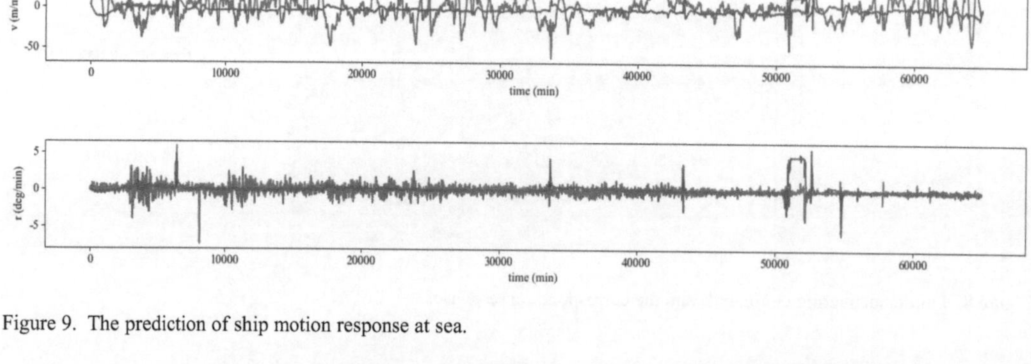

Figure 9. The prediction of ship motion response at sea.

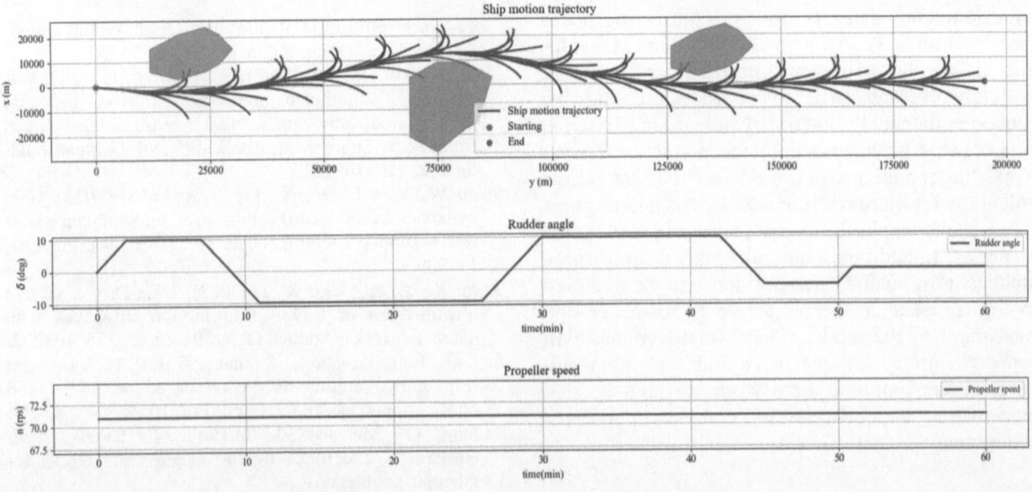

Figure 10. Case of ship manoeuvring commands testing for grounding avoidance (100 minutes per iteration).

resulting trajectories were predicted using the trained LSTM model described in Section 3.2. Unsafe trajectories—i.e., those that intersect with the grounding zones—are highlighted in red. These trajectories illustrate the effect of suboptimal control inputs that may lead to unintended deviations toward hazardous areas. Based on the trajectory predictions, unsafe control commands can be identified and excluded. The system subsequently suggests alternative control actions that result in safe trajectories avoiding shallow waters. This safety-aware adjustment process enables the selection of assisted maneuvering commands that maintain navigational safety.

The lower two subplots in Figure 10 illustrate the applied control sequences over time. The middle panel presents the rudder angle adjustments in the time domain, while the bottom panel displays the corresponding propeller speed inputs. These control actions were selected based on predicted trajectories to ensure safe avoidance of grounding zones while maintaining adherence to the predefined sailing route.

This example highlights the capability of the proposed framework to simulate and evaluate maneuvering outcomes under realistic risk conditions. By leveraging data-driven ship motion prediction, the framework enables proactive decision support for maneuvering, thereby significantly enhancing operational safety in complex navigational environments.

In this grounding avoidance scenario in Figure 10, the propeller speed was fixed to isolate the effect of rudder angle on maneuverability and trajectory prediction. The purpose was not to induce specific maneuvers like zigzag patterns, but rather to systematically test a variety of rudder angle commands under a controlled speed condition. The resulting zigzag-like paths are coincidental and reflect the dynamic responses of the ship to different steering inputs predicted by the model. Among the tested control sequences, those resulting in predicted trajectories that avoid grounding zones (highlighted in green) were selected as safe maneuvering commands. These safe command sequences can be integrated into an intelligent navigation support system to aid real-time decision-making and ensure grounding avoidance.

4 CONCLUSIONS

This study proposed a data-driven framework for ship-assisted maneuvering decision-making based on a pre-trained deep learning model. By utilizing full scale real operational data from a Kamsarmax bulk carrier, an LSTM network was developed and validated to predict ship motion responses. The model achieved satisfactory accuracy in forecasting surge, sway, and yaw behaviors, forming the basis for trajectory simulation and safety evaluation.

A key scientific contribution of this research lies in the development of a maneuvering command method based on a pre-trained deep learning model, that enables real-time assessment of ship control commands. This represents a significant step forward from traditional practice, where maneuvering decisions are made solely based on crew experience without systematic prior testing or trajectory prediction.

However, the current framework does not explicitly incorporate hydro-meteorological factors such as wind, waves, and ocean currents, which can significantly influence the accuracy of ship motion predictions. This limitation arises primarily from the challenge of constructing realistic grounding scenarios using fully coupled environmental datasets for validation purposes (see Section 3.3). As a result, the current evaluation is conducted under simplified conditions.

Future research will focus on addressing the above-mentioned limitation by integrating external environmental dynamics into the prediction framework. Incorporating wind, wave, and current effects will allow for more realistic simulations and improve the model's applicability to real-world navigation scenarios. In addition, more advanced deep learning architectures, such as transformer, will be explored to further enhance the predictive performance and robustness of the model. These improvements will support a more comprehensive and accurate validation of the proposed framework for ship-assisted maneuvering under diverse and complex operational conditions.

ACKNOWLEDGEMENTS

All authors acknowledge the support from their employers. Dr. Mingyang Zhang and Mr. Nikolaos Tsoulakos acknowledge that research funding received from the Horizons Europe project" RETROFIT solutions to achieve 55% GHG reduction by 2030 (RETROFIT55) – Project No.: 101096068". Dr. Mingyang Zhang has received funding from the Merenkulun Säätiö, which is gratefully acknowledged. The views set out in this paper are those of the authors and do not necessarily reflect the views of their sponsors.

REFERENCES

Greff, K., Srivastava, R. K., Koutník, J., Steunebrink, B. R. & Schmidhuber, J. (2016). LSTM: A search space odyssey. IEEE transactions on neural networks and learning systems, 28(10), 2222–2232.

Ouyang, Z. L. & Zou, Z. J. (2021). Nonparametric modeling of ship maneuvering motion based on Gaussian process regression optimized by genetic algorithm. Ocean Engineering, 238, 109699.

Ramirez, W. A., Leong, Z. Q., Nguyen, H.,& Jayasinghe, S. G. (2018). Non-parametric dynamic system identification of ships using multi-output Gaussian Processes. Ocean Engineering, 166, 26–36.

Shen, W., Hu, X., Liu, J., Li, S. & Wang, H. (2025). A pre-trained multi-step prediction informer for ship motion prediction with a mechanism-data dual-driven framework. Engineering Applications of Artificial Intelligence, 139, 109523.

Shen, W., Yao, J., Hu, X., Liu, J. & Li, S. (2023). Ship dynamics model identification based on Semblance least square support vector machine. Ocean Engineering, 287, 115908.

Silva, K. M. & Maki, K. J. (2022). Data-Driven system identification of 6-DoF ship motion in waves with neural networks. Applied Ocean Research, 125, 103222.

Sun, Q., Tang, Z., Gao, J. & Zhang, G. (2022). Short-term ship motion attitude prediction based on LSTM and GPR. Applied Ocean Research, 118, 102927.

Taimuri, G., Matusiak, J., Mikkola, T., Kujala, P. & Hirdaris, S. (2020). A 6-DoF maneuvering model for the rapid estimation

of hydrodynamic actions in deep and shallow waters. Ocean Engineering, 218, 108103.

Wang, Z., Zou, Z. & Guedes Soares, C. (2019). Identification of ship manoeuvring motion based on nu-support vector machine. Ocean Engineering, 183, 270–281.

Zhang, M., Liu, C., Kujala, P. & Hirdaris, S. (2024). Comparison and evaluation of learning capabilities of deep learning methods for predicting ship motions. In International Marine Design Conference.

Zhang, M., Taimuri, G., Zhang, J. & Hirdaris S. (2022). A deep learning method for the prediction of 6-DOF ship motion in real conditions. Proceedings of the Institution of Mechanical Engineers, Part M: Journal of Engineering for the Maritime Environment, Doi 14750902231157852.

Zhang, M., Taimuri, G., Zhang, J., Zhang, D, Yan, X., Kujala, P. & Hirdaris, S. (2025). Systems driven intelligent decision support methods for ship collision and grounding prevention: Present status, possible solutions, and challenges. Reliability Engineering & System Safety, 253, 110489.

Zhang, M., Wang, H., Tsoulakos, N., Kujala, P. & Hirdaris, S. (2024). An AI model for the prediction of ship motions and fuel consumption of a Kamsarmax bulk carrier.

Zhang, S., Wu, Q., Liu, J., He, Y. & Li, S. (2023). State-of-the-art review and future perspectives on maneuvering modeling for automatic ship berthing. Journal of Marine Science and Engineering, 11(9), 1824.

Author index